INTERPRETATION OF MASS SPECTRA OF ORGANIC COMPOUNDS

INTERPRETATION OF MASS SPECTRA OF ORGANIC COMPOUNDS

MYNARD C. HAMMING
Continental Oil Company
Research and Development Department
Ponca City, Oklahoma

NORMAN G. FOSTER
Department of Chemistry
Texas Woman's University
Denton, Texas

1972

ACADEMIC PRESS New York and London

ACADEMIC PRESS, INC.
111 Fifth Avenue, New York, New York 10003

United Kingdom Edition published by
ACADEMIC PRESS, INC. (LONDON) LTD.
24/28 Oval Road, London NW1

LIBRARY OF CONGRESS CATALOG CARD NUMBER: 72-87228

PRINTED IN THE UNITED STATES OF AMERICA

CONTENTS

Chapter 3 The Mass Spectrum

Chapter 4 The Sample—Its Character and Handling

Chapter 5 Fragmentation Reactions—Key to Interpretation of Mass Spectra

Chapter 6 Proceeding with an Interpretation

Chapter 7 The Rectangular Array and Interpretation Maps

Chapter 8 Computerizing Mass Spectral Data

Chapter 9 Correlations Applied to Interpretations

Appendix I Structure-Correlation References for Some Organic and Related Compounds

PREFACE

The interpretation of mass spectra is inherently a "numbers game." Hence it should be possible, in theory, to computerize the entire process of interpretation. Some workers have attempted this, and some have even carried the process further to computerize the operations of the mass spectrometer from start to finish, i.e., introduce sample, push a button, and accept printout data and interpretation, all for a specific quantitative analytical situation. Mass spectrometry deals with ions, both molecular and fragment, and these are characterized by separation according to their atomic mass unit or mass number. Thus it is far easier to digest and adjust to these numbers than to infrared or nuclear magnetic resonance data which, indeed, deal with "numbers" but of much less direct meaning to an organic chemist. Mass spectrometry has grown into a rather extensive field from its beginnings early in this century, but it is still not "standardized" in the same sense that is true for many other instrument fields. It continues to grow and expand because of its numerous and apparently endless applications. Its use is essential to both basic and applied research, and it is part of the sophisticated research "team" developed today.

Because of the many facets involved in the field, i.e., physics, ion-optics, electronics, engineering, high-vacuum techniques, chemistry, people with varied backgrounds become interested in or are assigned to work in mass spectrometry. To bring all these people to a common ground of knowledge is an impossibility. Instead, a blending of a group or team of persons with backgrounds from the above-mentioned areas will usually be achieved. In the process, no one individual is actually a mass spectroscopist, but each becomes a specialist in a given area. For example, there are literature searchers, computer liaison personnel, electronics maintenance personnel, and theoreticians. It is interesting to note that as mass spectrometry is applied to a new field, the new crop of graduate students decides that the field of mass spectrometry was invented for them and their specifically oriented uses. This volume was developed with all these people in mind. Data from the instrument may often be quickly produced, but then problems of interpretation of the data arise.

This treatise was conceived to be an aid in developing facility in the interpretation of organic mass spectra. It was written to assist personnel from the technician to the intermediate managerial level. Although it may appear too elementary to our more experienced colleagues, it should prove helpful to those entering the field regardless of their technical skills and levels of achievement simply because a broad background is necessary for the working mass spectroscopist. Many of the items covered will have been neglected or of little use in the previous technical experience of the novitiate in mass spectrometry.

The prime aim of this work is to introduce the reader to the subject, survey basic instrumentation, introduce simple procedures used, and acquaint one with the existing terminology of mass spectrometry. All of this material is presented in preparation for the utilization of mass spectral data. Since fragmentation of the molecular and daughter ions is the key to interpretation of organic mass spectra, it is covered in some detail. Following this, an introduction to the array format is given which is essential to rapid interpretation because it provides a visual interpretive peak approach which is simultaneously compatible to the interpreter and computer. Believing it to be essential that man control the "idiot" computer, he must have the know-how of all facets of interpretation in order to utilize the computer to the fullest. The balance of the book is concerned with the process of instruction so as to meet the requirements of eventually turning many of the tedious arithmetic calculations and testing procedures over to the computer. Example interpretations, calculations, data-processing procedures, and computer programs are included, even though it is recognized that the many types of computers available prohibit standard usage of the programs. The methodology and logic involved are universal, and hence the examples can stimulate the user to apply these ideas in his own computer system.

The type of stimulation found in the various reviews and advances in mass spectrometry should not be expected in this treatise. Books of the above type will find their places in the developing mass spectrometry laboratory, just as many of the other laboratory service and aid books, such as listings of metastable ion processes and isotopic contributions, become a part of the "team" operation. In a sense, then, this treatise will find its principal use as something a bit more informative than a short treatise to hand to the newcomer to the laboratory who wants to "get aboard—join the team."

Portions of the book were developed from lecture and laboratory materials used in presenting a two-week course in mass spectrometry at the Texas Woman's University. The suggestions of numerous students involved in these courses have been incorporated in portions of the book, particularly in Chapters 2–4. We are indebted to all these students for their comments and assistance. Quite naturally, the content of a two-week course is considerably less than is found in this book, but the material contained herein could be covered in an introductory one-semester course in mass spectrometry. It is hoped that the volume will find multiple-purpose usage in both the university and industrial laboratory.

MYNARD C. HAMMING
NORMAN G. FOSTER

ACKNOWLEDGMENTS

In writing a book, authors have different reasons for their efforts. The inspiration or driving force may come from one or more sources. Among those I (N. G. F.) have recognized are of the contributions of one man, the late Dr. Elton B. Tucker (formerly of the American Oil Company, research department), to the field of mass spectrometry and to my personal life. I was encouraged by Dr. Tucker to return to graduate school to obtain a Ph.D. degree after nearly five years with the company (1946–1951). In addition, the knowledge (acquired for the most part at a later time) of his encouragement of Mr. Harold Wiley (CEC) during the mass spectrometer installation at Standard Oil Company (Indiana), now American Oil Company, and his suggestion to form an ASTM-E type committee on mass spectrometry, plus his efforts in that direction, all indicate the depth of his interest in mass spectrometry and ASTM. His success as a research executive was well known, but his behind-the-scenes efforts are probably unknown to many mass spectrometrists.

During the early years of instrument development methods, it should be noted that men like Dr. Tucker, then head of the Analytical Division, not only utilized all available combinations of techniques for separation and identification of organic species in petroleum and related materials but also provided constant encouragement to the researchers involved in this work. Thus, although the petroleum companies became known for rather large budgets for instrument analysis, this persistence was necessary before other research units would begin to consider expenditures of this magnitude for their analytical interests. This debt is also owed by modern research teams to some of the European workers, as will be described in the text.

I (N. G. F.) am indebted to Seymour Meyerson of American Oil Company for encouragement in the form of discussions, particularly of the history of the early days, for suggestions as to some areas that should be covered in detail, and for review of a chapter of the book. About five years of contact with "Sy" during my employment at Standard Oil Co. (Indiana) did a great deal to develop my interest in mass spectroscopy while working in the hydrocarbon Analysis Group as a separations specialist.

The driving force for the author (M. C. H.) was a desire to provide a book to help those involved in industrial research mass spectrometry laboratories, who are often expected to achieve through their own efforts. The brief personal remarks made to me several years ago by such persons as R. I. Reed, Klaus Biemann, and John Beynon provided a lasting inspiration. A special thanks goes to Fred McLafferty who first guided and inspired me in mass spectrometry and has retained his interest in my activities. My indebtedness goes to many other individuals, many of whom are cited in the references. Appreciation is due Continental Oil Company, Ponca City, Oklahoma for giving approval to publish this book. Individuals cited are Gerald Perkins, Jr., Harrell T. Ford, and David B. Burrows of the Research and Development Department. Too often not cited are those many capable people at Continental Oil Company who in so many ways gave help. Several spectra are from samples generously supplied by E. J. Eisenbraun and his students at Oklahoma State University. My wife, Beverly Jane Hamming, developed an understanding of my idiosyncrasies and did much to encourage me. My daughters, Laura Lee and Heidi Ann, often helped me work the long hours by just being well behaved.

Associated with the production of every book is a rather large number of people. In the case of a college professor, this includes the university itself, his colleagues, his graduate students, supporting personnel at the university, and quite often the members of his family. The Texas Woman's University generously reduced my (N. G. F.) teaching load by three hours for several semesters to assist in the preparation of the manuscript. I am grateful also to Dr. George H. Stewart, Chairman of the Department of Chemistry, and to Dr. Carlton T. Wendel, also of the Chemistry Department, for taking a portion of my teaching load during the fall semester of 1970 to permit completion of the work. I am indebted to a number of my colleagues for helpful discussions and suggestions concerning portions of the text. In particular, I am indebted to Dr. James A. Hardcastle for the organization of the items of biological interest in Appendix I, Dr. Robert S. Davidow for suggestions on the section on the ionizing process, Dr. Lyman R. Caswell for permission to mention parts of his research, and to Dr. James E. Johnson for the advice on nomenclature and from time-to-time on mechanisms from the physical-organic chemist's viewpoint. Considerable help in literature searching and evaluation was given to me by Mrs. Diana Wong-Kiu Shiu Suvannunt and Miss Julie Pei-Min Liao during their graduate programs. I am indebted to Dr. George Vose and Mr. Ted Booker of the electron microscope laboratory in the Texas Woman's University Research Foundation for assistance with and suggestions for the presentation of Figure 3-1. In addition, Dr. Robert A. Fuerst, Professor of Biology, has been of great help and assistance in advising me upon a number of editorial-type problems. For their assistance in literature searching, I am grateful to Mrs. Christine F. Lorenz, Mr. James W. Lacy, and Mrs. Susan F. Lacy. To my daughters, Karen, Claire, and Francene, I am indebted for assistance in proofing, and to Karen, for some assistance in the laboratory concerning data for some of the figures. Last, to my wife, Alice Stover Foster, I am in debt, not only for tolerating the usual neglect of family that accompanies the work of an author but also for an assist in the form of a critical grammatical review of over two-thirds of the manuscript.

We both are indebted to the assistance of many of the manufacturers of equipment for their kind permission to reproduce some of the figures and tabular material. In particular, we are indebted to Mr. Ed E. Escher, Mr. Urrie McCleary, and Mr. Charles Johannsen, all of the CEC Division of Bell and Howell, and to Mr. Gerard Kearns of Picker Nuclear. We are grateful to Mr. Hunt Payne of Bell and Howell Co. Film Division for technical assistance concerning spectral papers.

Last, we are indebted to the staff of Academic Press for advice and rapid replies to questions.

CHAPTER 1

INTRODUCTION

I. Scope and Definition

In light of the rapid expansion in mass spectrometry during the last five years, it is readily apparent that the field is too large to be covered adequately in a single treatise. Therefore, it is not surprising to note the appearance of several books dealing with the various specialized areas of mass spectrometry. (A general bibliography appears following the text.) This treatise will specialize in the interpretation of organic mass spectra and will introduce the reader to key references in many of the other major subfields. With the growth in the use of mass spectrometry in chemistry, it is only natural that a large number of chemists working in the fields of organic chemistry, biochemistry, and inorganic chemistry have recently found themselves concerned with the whys and wherefores of mass spectrometry. The analytical and physical disciplines have been strongly represented for a number of years. This volume is intended to aid the organic chemist and biochemist in several ways: We shall (a) summarize the accepted definitions, concepts, procedures, and working language of those already in the field; (b) outline the basic concepts of ionization, fragmentation, and rearrangement of ions as found in mass spectra; (c) present a summary of the instrumentation, sample handling techniques, and interpretive procedures currently employed; (d) by means of the array format of data presentation show how interpretation of complex spectra can be made easier; and (e) suggest a natural extension into computer interpretation. These aims will be considered mainly for mass spectrometer systems of low to medium resolution.

Mass spectrometers and mass spectrographs are instruments that can analyze substances for their constituent atoms, atomic groups, or molecules on the basis of a mass-to-charge ratio (m/e) separation of ions formed from the substance by electron impact or other means. These two types of instruments, while quite distinct, are similar in certain respects. Both have sample introduction systems, an ion source, a means of separating the ions formed, and a detector system. Instruments, ion sources, and detectors are discussed in Chapter 2, and sample introduction systems are covered in Chapter 4.

The term "mass spectrograph" was introduced by Aston in 1919 and refers to instruments that produce a "mass spectrum" on a photographic plate. Just six years later, "mass spectrometer" was used by Smythe and Mattauch (Kiser, 1965a, p. 1). Today the term "mass spectrometer" is applied to those instruments which produce a mass spectrum by bringing a focused beam of ions to a fixed collector. The mass-to-charge ratio ion in focus can be varied by electrical or magnetic field changes. A scan of the mass spectrum is then produced by varying either of the above fields with time. The ion current is detected at the collector, amplified electronically, and recorded.

The term "mass spectrometry" is presently used in a loose sense to include the use of both types of instruments and the studies made with them. Typical studies include isotope abundances, precise mass determinations, a great variety of analytical chemical applications involving both qualitative and quantitative analyses, and appearance potential studies. With such a list, it is not surprising to find applications to research in all of the physical and biological sciences.

The objective of this treatise is to furnish the reader with sufficient insight into many types of instruments, their uses, and usefulness for a particular problem. With this information, he may better understand the details supplied in specific references. With the outlined procedures for interpretation of and guide to computer programs as an aid, the newcomer to the field should be able to progress rapidly to the point of confidence and competence in his interpretation of the mass spectra of organic compounds.

II. Historical Developments

The discovery of positively charged electrical "entities" was made before the turn of the century (Goldstein, 1886). Wien (1898) showed that a beam of positive ions could be deflected using electrical and magnetic fields. Thomson (1913) demonstrated that naturally occurring neon consists of two different atomic weight species (isotopes) with weights of 20 and 22 (g/mole). The term "isotope" was introduced by Soddy (1913). Kiser (1965a) referred to Thomson as the Father of Mass Spectrometry and cited his discovery concerning neon as one of the two most significant mass spectrographic contributions to science, the second being the contribution of Aston. This second achievement was the discovery that the various isotopes of the elements do not have integral masses; they are not simple multiples of a fundamental unit. Aston's fairly elaborate instrument of 1919 (Aston, 1942) found favor with investigators of isotopes for precise mass measurements. Dempster (1918) produced a somewhat less elaborate instrument that was used for the measurement of relative abundances of isotopes. Dempster's instrument could not be used for precise mass measurements; it was better suited for measuring the relative abundance of the ionic species present and for studying electron impact processes in gases. It should be emphasized that these early instruments did not have the benefit of modern high-vacuum technology and thus were operated at higher pressures than are used today. Despite this limitation, Thomson was able to detect, by photographic record, effects which he suggested were due to the dissociation of ions in flight. These dissociated ions were observed as metastable peaks from the metastable transitions of the ions (Chapter 3). Also observed by Thomson were multiple-charged, negative, and fragment ions.

With the electronic developments of the 1920's and 1930's, more sophisticated equipment could be developed. The parallel improvement in vacuum and electronics technology led eventually to an increasing interest in the field of mass spectroscopy. The original interest in commercial mass spectrometry was generated by a group of California engineers headed by Washburn and with majority backing from Herbert Hoover, Jr. (Washburn, 1970). The petroleum companies were interested in some form of instrumentation which could determine the amount of hydrocarbon material present in geological core samples, etc. While the mass spectrometer as designed could do at least a portion of this work and was being developed upon this premise, it was noted that the lighter hydrocarbon substances were much easier to deal with during analysis. In 1942 wartime pressures for aviation gasoline made mass spectrometry more essential to the petroleum industry. The first commercial instrument, built by Consolidated Electrodynamics Corp. (CEC), was delivered to the Atlantic Refining Corp. This instrument was a 180° Dempster geometry unit (CEC Model 21-101), which was used for the analysis of light hydrocarbon gases, usually up to the C_6 range. Within a few years, a substantial number of these instruments were in the laboratories of the American petroleum processing companies. By 1948 experimentation was under way with higher molecular weight hydrocarbons. Using an instrument nominally designed for a resolution of 1–200, O'Neal and Wier (1951), at the Shell Oil Co., attached an oven to the inlet system and produced mass spectra of volatile high molecular weight species (mass 900). By this time, interest in the field had grown to the point at which a reevaluation of all instrumental designs was undertaken. Instruments of a great many varieties are available today, each with its special claims for uses in certain areas.

Improved instrumentation advanced interest in the studies of the processes of ionization, fragmentation, and rearrangement. For ions formed by electron impact, it was found before 1930 that transitions between one electronic state and another follow what has today become known as the Franck–Condon principle (Franck, 1926; Condon, 1928). The term "ionization efficiency curve" was introduced to define the plot of ion current of a given m/e as a function of the energy of the ionizing electron beam. Bleakney (1929, 1932) called attention to the "foot" of the ionization curve and suggested that it is due in part to two or more close-lying ionization potentials and in part to the lack of a monoenergetic electron beam. As proof of these contributions to the foot, he pointed out that helium, with only one low-lying ionization potential, has a smaller foot. Hipple and Stevenson (1943) made the first acceptable direct determinations of the ionization potentials of free radicals, predicted by earlier workers to be among the many uses for the mass spectrometer. Eltenton (1947) was the first to successfully study free radicals with a mass spectrometer.

The discipline of isotope separation was the area in which laymen were introduced to the word "mass spectrometer." In 1922 Aston had anticipated the use of the mass spectrograph for the separation of isotopes (Aston, 1942). After a number of attempts by various workers, two groups succeeded in making isotope separations about the same time. The group of Smythe *et al.* (1934) obtained 1 mg of ^{39}K in a 7-hour run. Oliphant *et al.* (1934) separated and collected as much as 10^{-8} g of a pure lithium isotope. Nier and co-workers (1940) isolated ^{235}U and ^{238}U. From this separated material, it was established that when bombarded with neutrons, the ^{235}U isotope was responsible for the process now called fission. Dunning had postulated this as early as 1939 (Hewlett and Anderson, 1962). The Berkeley 100-in. cyclotron was used by Alvarez and Cornog (1939) to find ^{3}He present in natural helium. Meanwhile, Lawrence headed a project to convert the 37-in. Berkeley cyclotron to a large mass spectrograph. With this instrument, tenths of milligram quantities of ^{235}U were eventually obtained (for a detailed discussion, see Kiser, 1965a).

Thomson had anticipated the application of mass spectrometry even in the area of chemical analysis (that of most interest from this treatise's viewpoint). In his book, he not only suggested using mass spectrometry for determining atomic and molecular weights but claimed that information thus obtained was superior to emission spectrography. He recognized the advantages of the small sample size requirement and even showed the possible use of the mass spectrometer for the identification of the components of air. The first report of actual chemical application appears to have been the work of Conrad (1930), who studied organic compounds. Mass spectrometry has grown into a very large field since these early beginnings.

Although this treatise is aimed primarily at organic structures, organic chemists and biochemists should be aware of the fact that mass spectrometry was used in about half the analyses made of the moon rocks and reported in marathon fashion in the January 1970 issue of *Science*. Typical inorganic applications are the high-temperature studies of the stabilities of tungsten and molybdenum oxyfluorides as reported by Zmbov *et al.* (1969) and the mass spectrometric investigation of the high-temperature reaction of hydrogen with boron carbide given by Steck *et al.* (1969). Of a more theoretical nature are the studies of the mass spectrum and molecular energetics of krypton difluoride by Sessa and McGee (1969) and the report of Holzhauer and McGee (1969) covering the development and use of a variety of reactor designs for studying reactions at cryogenic temperatures and permitting continuous mass spectrometric analysis with no associated sample warm-up.

For the reader who is interested in a more complete history of mass spectrometry, the book of Kiser (1965b) is recommended since it contains reproductions of early data, pictures, and much detailed information. It also contains a section on the instruments which were important historically in the development of mass spectrometry but which are not the prime concern of this treatise.

III. "New Frontier" Developments

During the evolution of mass spectrometry as a specialty field, a number of important events took place, including the formation of commercial instrument manufacturing companies and their "user clinics," the establishment of discussion groups, national and international meetings, societies, and the appearance of entire journals devoted to the speciality. This has all occurred in the past thirty years. Hence, to parallel the history of the mass spectrometer and its applications, one should consider the human elements and personalities involved with the dissemination of information and ideas in mass spectrometry.

Although the installation of the first Consolidated Electrodynamics Corp.

instrument at the Atlantic Refining Corp. went smoothly, the second CEC instrument proved to be quite a problem. According to Meyerson (1969) this unit was installed at the Whiting Technical Service Laboratories of the Standard Oil Co. (Indiana), now the American Oil Co. Harold F. Wiley, formerly vice president and general manager of the CEC division of Bell and Howell Co., was engineer in charge of the final tune-up. Despite the difficulties the installation was successfully made, and the commercial instrument became a success. Mass spectrometry took a turn into new fields.

From 1945 to 1951 CEC group meetings, sponsored and conducted by CEC personnel, were held yearly. The various customer "users" attended and discussed problems and applications. It was apparent that this function had to be taken over by a suitable national technical society since manufacturers other than CEC as well as university-based researchers were interested. Dr. E. B. Tucker of American Oil Co. suggested an American Society for Testing and Materials (ASTM) type E committee to CEC through Wiley (1969). According to Friedel (1969) both Dr. E. B. Tucker and William S. Young (Atlantic Refining Co.) were instrumental in selling the idea to the ASTM Board. R. G. Painter was secretary of ASTM at that time and has been called "a far-seeing individual who managed to convince the ASTM Board to go along with this group on a trial basis." Further, according to Friedel (1969), there were two organizational meetings held at ASTM headquarters in Philadelphia.

The first annual meeting of the ASTM committee E-14 was held at Pittsburgh (1953), cosponsored by E-14 and the Pittsburgh Conference on Analytical Chemistry. It was decided at the 1953 meeting that ASTM E-14 should meet separately because of the size of the Pittsburgh conference. In addition it was decided to return from time to time to the Pittsburgh meetings, but this was never carried out because of the growth in attendance at both meetings. A list of the subsequent meetings in the United States was provided by O'Neal (1969) and is shown in Table 1-1. A possible point of interest is the fact that a Friday afternoon session on gas–liquid chromatography (GLC) [sans connection to mass spectrometry (MS)] at the 1961 Chicago meeting saw the official formation of the ASTM E-19 Committee on Gas–Liquid Chromatography. For several years a number of related papers had been given during the E-14 meetings. Today there are many laboratories operating MS–GLC combinations.

It was apparent from the growth in attendance and increase in papers presented at the E-14 meetings that the character of this organization was indeed changing. In addition, the appearance among the attendees of many more academic people with the support of instrumentation provided by the National Science Foundation broadened the scope of the meeting to a truly national society. Because of the information explosion, two international journals, *Organic Mass Spectrometry* and *Journal of Mass Spectrometry and*

TABLE 1-1

ANNUAL MEETINGS IN THE UNITED STATES ON MASS SPECTROMETRY

Date	No.	Place	Chairman
1953	1	Pittsburgh	William S. Young
1954	2	New Orleans	William S. Young
1955	3	San Francisco	M. J. O'Neal
1956	4	Cincinnati	M. J. O'Neal
1957	5	New York	W. Priestly
1958	6	New Orleans	W. Priestly
1959	7	Los Angeles	R. A. Friedel
1960	8	Atlantic City	R. A. Friedel
1961	9	Chicago	Vernon H. Dibeler
1962	10	New Orleans	Vernon H. Dibeler
1963	11	San Francisco	Russell E. Fox
1964	12	Montreal	Russell E. Fox
1965	13	St. Louis	N. D. Coggeshall
1966	14	Dallas	N. D. Coggeshall
1967	15	Denver	H. M. Rosenstock
1968	16	Pittsburgh	H. M. Rosenstock
1969[a]	17	Dallas	J. L. Franklin
1970	18	San Francisco	J. L. Franklin
1971	19	Atlanta	R. E. Honig

[a] The American Society for Mass Spectrometry (ASMS) was organized. Meetings after 1969 are joint ASMS–ASTM E-14 meetings.

Ion Physics, were first published in 1968. The mailing list of E-14 had grown to over 1500 names of those in attendance for one or more of the last three meetings.

The actual origin of the American Society for Mass Spectrometry (ASMS) stems from a conversation that Henry Rosenstock (National Bureau of Standards), Ralph Harless (Carbide & Carbon), and Joe Franklin (Rice University) had while they were in Denver at the E-14 executive committee meeting in October 1966, planning for the Denver conference. Franklin (1969) said:

> At that time, we were discussing the activities of Committee E-14 and we all voiced the opinion, which I am sure had been forming separately in our minds for a long time, that E-14 really was not a committee but a society and we discussed the desirability of making a move toward establishing on a formal basis what in fact Committee E-14 had already become. Henry Rosenstock ... carried the ball after we had agreed that we should take some definite steps in the direction of forming the Society. He talked to Tom Marshall and some members of ASTM's Council on several occasions during the succeeding year. I think Henry managed to convince Tom Marshall but there was considerable doubt on the part of the Board of ASTM.

Matters seemed to drag a bit at this time, but the executive committee (of E-14) managed to get Harold Bogart, the president elect of ASTM, and Tom Marshall to come to the 1968 Pittsburgh meeting. Rosenstock and Franklin spent a good part of a day with them and demonstrated quite successfully to Bogart that the functions of the group were far broader than those of any committee. At the annual banquet, Bogart made a few remarks, from which it was evident that he was convinced, and after that it was rather easy sailing.

The concurrence of the ASTM Board was obtained, and a lawyer was retained to help with articles of incorporation and the preparation of a legal constitution and bylaws. A discussion session at Pittsburgh in 1968 gave the attendees an opportunity to express opinions and interest in the society. A similar meeting was held during the Dallas Annual Meeting of 1969, and further questions and problems were discussed. Franklin (1969) said, "Ultimately, the group that was interested voted by an overwhelming majority to establish the Society and the rest then became merely a matter of mechanics."

As secretary of E-14, McCrea did most of the detail work on the establishment of the society. Articles of incorporation had to be filed and approved by the courts. This was officially done on 4 August 1969 by the state of Pennsylvania, in the city of Pittsburgh, thus creating the American Society for Mass Spectrometry.

The ASTM Committee E-14 continues to exist, and the chairman of E-14 (Ralph Harless) is ex officio vice-president of the society. It is expected that close relations with E-14 and the ASMS will continue to exist and that the two groups will meet jointly. Committee E-14 will be principally responsible for such matters as standardization of methods, much as was the original intent of the ASTM Board during the early 1950's.

Developments of a similar nature were under way in the United Kingdom and in Europe after World War II. For much of what follows, the authors are indebted to the cooperation of John Beynon (1969), Handel Powell (1969), Professor Knewstubb (1969), R. M. Elliott (1969), and Professor Allan Maccoll. Just as occurred in North America, the development of hydrocarbon analyses required spectroscopic techniques and their development. Most of this was done in the petroleum company laboratories, but two academic groups also had fully developed laboratories studying hydrocarbons. These were Dr. G. B. B. M. Sutherland's laboratories at Cambridge and Dr. H. W. Thompson's at Oxford. According to Bradford (1951) the Hydrocarbon Research Group of the Institute of Petroleum was started as a cooperative group during World War II. Considerable ultraviolet–visible–infrared spectrophotometric work was well under way, and widespread applications were achieved by the end of the war. The individual petroleum company laboratories had been able to equip themselves by 1944, and after the war this work was continued. The Hydrocarbon Research Group also began supporting fundamental research at United Kingdom universities and continues to do so to this day. Areas supported include synthesis of hydrocarbons, hydrocarbon reactions, and spectroscopy of hydrocarbons. The last eventually included mass spectrometry when commercial instruments began to appear in the United Kingdom. Bradford (1951) has written a complete history of the group from its inception to the date of his paper. He has pointed out that the organization is somewhat unique in that it is a voluntary organization with the original sponsoring bodies of the Institute of Petroleum providing support. In 1951 the sponsors included Anglo-Iranian Oil Co. Ltd., Esso Development Co. Ltd., Imperial Chemical Industries Ltd., Manchester Oil Refineries Ltd. & Petrochemicals Ltd., Ministry of Supply, Monsanto Chemicals Ltd., Shell Petroleum Co. Ltd., and Trinidad Leaseholds Ltd. The funds of the group are derived entirely from the contributions of the sponsoring bodies, and their representatives have full control over the disposition of the funds. The group is, however, closely linked with the Institute of Petroleum (which provides its administrative services), and the general research policy is a matter for mutual agreement between the group and the institute.

According to Bradford (1951) one factor that entered into the slower rate of expansion of mass spectrometry in the United Kingdom, and all of western Europe for that matter, was brought about by World War II. Although it was the urgency of wartime necessities that forced the label of routine laboratory instrument upon the mass spectrometer, the decision had been made early in the war for major development work to take place in the "zone of the interior," which put the bench work into the United States and Canada. Therefore, most of the knowledge, instrumentation details, etc., were available to the United States and Canada, and hence a commercial instrument was developed and put to use in the dollar area considerably ahead of the pound area. Plans for commercial production of the mass spectrometer were made immediately after the conclusion of World War II and activated as soon as priorities would permit. By 1949 deliveries of the Metropolitan-Vickers Ltd. instrument were being made to commercial users and universities in the United Kingdom. Bradford also pointed out that government projects conducted at a number of the petroleum companies made some use of mass spectrometry in connection with isotope studies. Of course, when this service was no longer needed, the instruments were used on hydrocarbons for technical service and analytical routine jobs. This was the philosophy applied to the use of other equipment as well.

Bradford (1951) wrote that it was decided by the Hydrocarbon Research Group in 1948 that the application of the mass spectrometer to hydrocarbon work should be aided and encouraged as much as possible. It was agreed that immediate needs could best be served by providing a *meeting ground* for all those interested in the subject. Although hydrocarbon analysis was recognized to be only one of the many aspects of mass spectrometry, it was thought to be to the advantage of all to provide for the development of the subject in its early stages on the broadest possible basis.

A Mass Spectrometry Panel was formed with J. Blears as chairman; it included representatives of the sponsoring companies, the manufacturers of the mass spectrometer, the universities, and various government laboratories. According to Bradford (1951) this panel very quickly proved its value by affording the opportunity for exchange of information among the designers and users. This resulted in a rapid growth of the understanding of instrument requirements and difficulties. A symposium was organized and held in April 1950 at Manchester. The details of other conferences organized by the Mass Spectrometry Panel are shown in Table 1-2 with the information provided by Powell (1969).

TABLE 1-2

INTERNATIONAL CONFERENCES ON MASS SPECTROMETRY

Date	Place	Organizers	Publisher of proceedings
1950	Manchester	HRG[a]	Institute of Petroleum
1953	London	HRG	Institute of Petroleum
1958	London	HRG and ASTM E-14	Pergamon
1961	Oxford	HRG and ASTM E-14	Pergamon
1964	Paris	HRG, ASTM E-14, and GAMS[b]	Institute of Petroleum
1967	Berlin	HRG, ASTM E-14, GAMS and AG Massen DGP[c]	Institute of Petroleum
1970	Brussels	HRG, ASTM E-14, and GAMS	

[a] Hydrocarbon Research Group.
[b] Groupement pour d'Avancement des Methodes Spectrographique.
[c] Allgemein Gesellschaft Massen Deutsche Gesellschaftphysisch.

Bradford (1951) pointed out that the Mass Spectrometry Panel fulfilled the need for a discussion group for the exchange of experience among workers dealing with similar problems. Although the general progress of mass spectrometry was considered to be proceeding at a satisfactory rate, much needed to be done in improving the application of the technique to hydrocarbon analysis and to the study of other hydrocarbon problems. The Hydrocarbon Research Group recognized that the development of analytical techniques had to be done in oil company laboratories and planned cooperative testing. In view of the recognized considerable empirical element, the panel decided to sponsor fundamental research into the principles of mass spectrometry at selected universities. The group began with support of several projects: "Ionization of Hydrocarbons–Ion Sources" at Birmingham University; "Studies of Free Radicals" at King's College, London; and "Studies of Hydrocarbon Cracking and Oxidation" at Liverpool University. According to Powell (1969) this support has continued and, in fact, has been expanded. Birmingham investigated ion–molecule reactions from 1950 to 1969; Glasgow had a program on mass spectra of complex molecules from 1959 to 1969; King's College studied field ionization from 1959 to 1969; and Liverpool changed to the study of mass spectra of certain classes of compounds during 1966 to 1969, while Aston had a project on negative ions from 1966 to 1969. This research was all sponsored by the Mass Spectrometry Panel. In the Western Hemisphere there is really no equivalent to this type of sponsorship because in the United States most granting agencies restrict their grants to a maximum of three years and prefer a title change at the end of that time. Mass spectroscopists may find support from the Petroleum Research Fund, administered by the American Chemical Society, but their projects must compete with all other forms of fundamental research.

For those who are interested in a more recent report on the activities of the Hydrocarbon Research Group, see the Informal Symposium Report of Powell (1965) in which he describes some changes in membership of the sponsoring companies but little change in the operations and philosophy of the group. Three papers follow the report with one from each of the specialty panels, "Spectroscopic," "Hydrocarbon Chemistry," and "Mass Spectrometry" (Dr. Ivor Reed).

Also in the United Kingdom, one finds the Mass Spectrometry Discussion Group, which held its inaugural meeting in January 1961 with R. M. S. Hall as the secretary. The secretaryship passed to Dr. J. Cuthbert in May 1963 and to Professor P. F. Knewstubb in March 1967. According to information supplied by Knewstubb (1969):

> The Group holds one-day meetings at approximately quarterly intervals unless a clash with some larger meeting would arise. The venues are provided by members as and when they feel able to act as hosts and, generally, provide speakers on the work done in their own establishment. The mailing list, which is circulated with information on the meetings, now numbers about 80, but numbers at meetings rarely exceed 50, so that the atmosphere is still very informal. The Group has run successfully so far without formal constitution or any financial arrangements. The cost of catering at each meeting is covered by those attending while the postage, typing, etc., are provided by this Department (Department of Physical Chemistry, University of Cambridge). Membership is achieved by recommendation or request, provided only that it arises through an active interest in mass spectrometry other than routine analysis.

In July 1962 the group began a volunteer effort to collect titles of relevant papers in mass spectrometry. Some searching back in time from that date was accomplished, and the work was continued until the *Mass Spectrometry Bulletin* appeared in November 1966. This activity quite parallels an American group.

A task group of the ASTM Committee E-14 made a literature survey of high molecular weight mass spectrometry for the years 1951–1959. The members of this committee were C. J. Robinson, chairman, G. L. Cook, and W. C. Ferguson, with contributions from V. H. Dibeler, A. Hood, H. E. Lumpkin, F. W. McLafferty, and F. C. Stehling. The survey was continued on a voluntary basis until 1965. At this point, the task was too great to be carried on by individuals and their companies even with the aid of computers as applied by E. M. Emery and M. C. Hamming, who shared the greatest role in the last years of these surveys. The appearance of the *Mass Spectrometry Bulletin* filled the gap.

A third group in the United Kingdom was formed in 1964 according to Elliott (1969), secretary of the Mass Spectrometry Group, "with the object of organizing meetings which would be open not only to practicing mass spectroscopists but to anyone with a casual or potential interest in the subject." The general pattern of these meetings is a two- or three-day conference about every two years. The years that international or European meetings are held are avoided because there is no point in having a conflict with a purely British meeting. Meetings have been held in 1965, 1966, 1968, and 1969, with the next meeting scheduled for 1971. The meetings are usually held in September and usually on university premises. The attendance is usually about 150–200 people.

The program for the September 1968 meetings, according to Maccoll (1968), held at University College, London, included the following sessions: "Ionization and Dissociation Processes" (15 papers), "Organic Chemistry Applications" (17 papers), "Ion–Molecule Reactions" (6 papers), "Instrumentation" (7 papers), and "Inorganic Applications" (3 papers). There was also a discussion on appearance potentials and several brief communications. Although this is a more formal group, it is obvious from the attendance and the number and types of sessions that such meetings are as popular in the United Kingdom as they are in the United States.

IV. Selection of Instrumentation

Frequently, instrumentation is selected in accordance with the specific type of data one needs to obtain. Since this affects the method of interpretation of the data, it is essential to consider instrumentation and its function before examining the methods of interpretation.

In the past, various types of data were sought, not all of which were readily available from a particular instrument. Some workers desired precise mass measurement, while others were interested in the relative abundances of isotopes. Still others were interested in ionization processes. As the versatility of instrumentation increased, the uses of mass spectrometry increased. Today the organic chemist uses the mass spectrometer as a "queen" on the chessboard of analytical identification and structure determination. Actually, there are three "queens," the other two being the infrared and the nuclear magnetic resonance spectrometers. Just as there is no clear-cut line of distinction between organic and inorganic molecules, there is no single instrument best adapted for use in organic mass spectrometry. Therefore, whenever possible, this text will treat instrumentation and the interpretation of spectra quite generally. It is our ultimate aim, however, to bring the reader to an awareness of all of the approaches that may be applied to molecules of reasonable volatility and relatively low molecular weight (under 2000). The concepts developed can be extended to many other areas of mass spectrometry, as the reader sees fit. While mass spectrometers in themselves are quite versatile, it should become apparent to the reader that certain instruments are best adapted to specific types of studies. Vacuum spark sources are not primarily intended for use with organic substances, nor are isotope ratio instruments intended for organic identifications. Similarly, while a given instrument may be excellent for analytical purposes because of its reproducibility, it will not be particularly adapted to the study of ionization processes or even to the study of instrument optics. Since our aim is to deal primarily with the mass spectra of organic compounds, we shall confine our interests to spectra from only a few varieties of mass spectrometers. Similarly, our interests in interpretation will exclude most metallic compounds, their oxides, nitrides, halides, sulfides, etc. A little detail will be presented as to how ionization processes are studied, but the emphasis will be upon fragmentations and how they affect interpretation.

V. A Summary of Interpretation Objectives and Methods

With the accumulation of sufficient background knowledge and skills in interpretation on the part of the mass spectroscopist, mass spectrometry can provide (a) the assignment of specific molecular structures to many unknown substances, (b) the detection of unexpected or trace components in mixtures, and (c) a means of predicting probable reaction mechanisms. It still remains a way of obtaining quantitative analyses rapidly and accurately with a limited number of standards. The objectives and the approach to interpretation are the topics of Chapter 6; however, let us briefly consider these with respect to the interests of the organic chemist.

Initially, the organic chemist may be interested only in molecular weight determinations in his samples since these will partially define the products of a reaction. After isolating synthetic products, he is usually interested in determining functional groups and, if possible, structure. Quite often the inspection of the mass spectrum of even a 95% purity material will reveal the above information. After further purification, the combination of mass spectrometry with ultraviolet and/or infrared spectrophotometry and nuclear magnetic resonance data will usually produce sufficient analytical data for identification purposes. However, there are situations in which the unequivocal establishment of an unknown structure is not possible by these means alone. Practical examples will be provided in subsequent chapters.

Persons working with natural products usually deal with complex mixtures or those containing a number of related compounds. As will be shown in later chapters, these mixtures can be "characterized" to the extent that very useful information may be obtained.

One of the earliest analysis performed with the mass spectrometer was of a mixture of light hydrocarbon gases. Quantitative information may be obtained if n equations (independent linear) in n unknowns can be set up in the usual spectroscopic fashion where calibrations are used to obtain the contribution to a given band or peak by each of the known substances present in the mixture. Trace contaminants or impurities in otherwise similar substances can be determined to varying degrees of precision, depending upon the nature of the problem. The isotopic abundances can be estimated for some elements and, in fact, can be useful in identifications, as will be pointed out in subsequent chapters. High-resolution applications and quantitative analyses will be discussed but in a much briefer fashion than low-resolution mass spectrometry. Interpretation of high-resolution mass spectra will have many future facets, but even today, the types of methods being used are numerous and difficult to classify. Conversely, the quantitative analyses methods are rather routine once the components expected to be in the mixtures are known and suitable calibrations are made.

Sometimes the great wealth of information in a mass spectrum is hidden by the arrangement of the data from the spectrum. For this reason, the rectangular array format presentation of mass spectra is featured. The rectangular array consists of rows formed by listing 14 consecutive peaks (mass/charge ratio, or m/e) which differ by one mass unit. This arrangement produces columns having elements differing by 14 mass units which correspond to one methylene or CH_2 group. A cycling by 14 mass units is continued until the highest or last mass peak of a given spectrum is reached. This type of data presentation is finding wider usage in the catalogs of the American Petroleum Institute Research Project 44 and the Thermodynamics Research Center Data Project in the section of selected mass spectral data—matrix form.

Since we believe that the next few years will bring about the natural development of computer-aided interpretation of mass spectra, with the interpreter seated at the computer keyboard if necessary, we also place great importance upon the array format as a natural path into the use of computerized assistance to interpretation. Later chapters in the book feature this aspect with both illustrative and practical examples.

REFERENCES

Alvarez, L. W., and Cornog, R. (1939). *Phys. Rev.* **56**, 379, 613.
Aston, F. W. (1919). *Phil. Mag.* **38**, 707.
Aston, F. W. (1942). "Mass Spectra and Isotopes," 2nd Ed., p. 257. Arnold, London.
Beynon, J. H. (1969). Private communication.
Bleakney, W. (1929). *Phys. Rev.* **34**, 157.
Bleakney, W. (1932). *Phys. Rev.* **40**, 496.
Bradford, B. W. (1951). *World Petrol. Congr. Proc. 3rd 1951* **6**, 240.
Condon, E. U. (1928). *Phys. Rev.* **32**, 858.
Conrad, R. (1930). *Phys. Z.* **31**, 888.
Dempster, A. J. (1918). *Phys. Rev.* **11**, 316.
Elliott, R. M. (1969). Private communication.
Eltenton, G. C. (1947). *J. Chem. Phys.* **15**, 455.
Franck, J. (1926). *Trans. Faraday Soc.* **21**, 536.
Franklin, J. L. (1969). Private communication.
Friedel, R. A. (1969). Private communication.
Goldstein, E. (1886). *Berlin. Ber.* **39**, 691.
Hewlett, R. G., and Anderson, O. E., Jr. (1962). "The New World, 1939–1946," pp. 14, 668. Penn. State Univ. Press, University Park, Pennsylvania.
Hipple, J. A., and Stevenson, D. P. (1943). *Phys. Rev.* **63**, 121.
Holzhauer, J. K., and McGee, H. A., Jr. (1969). *Anal. Chem.* **41**, 24A.
Kiser, R. W. (1965a). "Introduction to Mass Spectrometry and Its Applications," pp. 1, 12, 85. Prentice-Hall, Englewood Cliffs, New Jersey.
Kiser, R. W. (1965b). "Introduction to Mass Spectrometry and Its Applications," 356 pp. Prentice-Hall, Englewood Cliffs, New Jersey.
Knewstubb, P. F. (1969). Private communication.
Maccoll, A. (1968). *Org. Mass Spectrom.* **1**, 918.
Meyerson, S. (1969). Private communication.
Nier, A. O., Booth, E. T., Dunning, J. R., and Grosse, A. V. (1940). *Phys. Rev.* **57**, 646.
Oliphant, M. L., Shire, E. S., and Crowther, B. M. (1934). *Proc. Roy. Soc. London* **146**, 922.
O'Neal, M. J., Jr. (1969). Private communication.
O'Neal, M. J., Jr., and Wier, T. P. (1951). *Anal. Chem.* **23**, 830.
Powell, H. (1965). *J. Inst. Petrol. London* **51**, 325.
Powell, H. (1969). Private communication.
Sessa, P. A., and McGee, H. A., Jr. (1969). *J. Phys. Chem.* **73**, 2078.
Smythe, W. R., Rumbaugh, L. H., and West, S. S. (1934). *Phys. Rev.* **45**, 724.
Soddy, F. (1913). *Annu. Rep. Progr. Chem.* **10**, 262.
Steck, S. J., Pressley, G. A., Jr., and Stafford, F. E. (1969). *J. Phys. Chem.* **73**, 1000.

Thomson, J. J. (1913). "Rays of Positive Electricity and Their Application to Chemical Analyses." Longmans, Green, London.
Washburn, H. W. (1970). Private communication.
Wien, W. (1898). *Ann. Phys.* (*Leipzig*), **65**, 440.
Wiley, H. F. (1969). Private communication.
Zmbov, K. F., Uy, O. M., and Margrave, J. L. (1969). *J. Phys. Chem.* **73**, 3008.

CHAPTER 2

A SUMMARY OF INSTRUMENTATION

I. Introduction

Mass spectrometers are commercially available to meet the demands of numerous disciplines in science for studying atoms and molecules and their behavior. A basic understanding of the instrumentation is essential for successfully interpreting mass spectral data. It is not the purpose of this chapter to cover the topic of instrumentation in detail or to supply information on the selection of suitable instruments for specific needs. Only those features contributing in important ways to modern commercial instruments are discussed in the following sections. Other texts (many of which are listed in the general bibliography at the end of this volume) cover the topic of instrumentation in much greater detail.

The common feature of all mass spectrometers is the production of ions, the separation of ions by masses, and the collection and recording of the intensities of the ions which are separated. Major differences among these instruments involve the methods of production, separation, and collection of ions. These three operations comprise the basic instrument and are the subject of this chapter. Vital to the system are a suitable vacuum, a means of recording the spectrum, and a sample-introduction system. The latter two subjects are discussed in Chapters 3 and 4.

II. Ion Sources

A. Production of Ions

The ions observed in mass spectrometry are usually produced by (a) electron impact (or bombardment, the term preferred by some workers) of vapor-phase atoms or molecules, (b) the radiation of susceptible substances with ultraviolet light of sufficiently short wavelengths (photoionization source), (c) thermal evaporation of an element from a heated metal surface (surface emission source), (d) the excitation of atoms or molecules to form ions by an electrical spark in a vacuum source between two electrodes of a metal or semiconductor coated with the material to be examined (vacuum spark source), (e) the release of ions from substances near a "point" surface by applying a strong electric field (the field ionization source), or (f) charge exchange with previously ionized gaseous species which are allowed to collide with the substance under examination. Chemical ionization mass spectrometry, introduced by Munson and Field (1966), may be regarded as an outgrowth of this area since according to Field (1968), "the ionization of a substance under investigation is effected by reactions between the molecules of the substance and a set of ions which serve as ionizing reactants." In addition to these general types of ion sources, a few sources available for special applications shall also be discussed.

B. The Electron Impact Source

An example of an electron impact source is shown in Fig. 2-1. Electrons are produced from a heated wire filament a and travel through a slit or grid b across the evacuated space toward an anode or trap T. Vapor atoms or molecules (M^0) are introduced perpendicular to the electron beam through an entrance slit c. Vapor pressures of 10^{-6} torr or less are employed for most applications. The particles suffer a collision or near collision with the electrons at the point x. Ions are formed if the energy of the electron beam is sufficiently great according to the general equation,

$$M^0 + e^- \rightarrow M^+ + 2e^- \qquad (2\text{-}1)$$

A small number, usually only a fraction of a percent, of negative ions are also formed. These ions are not brought out of the source to any appreciable extent

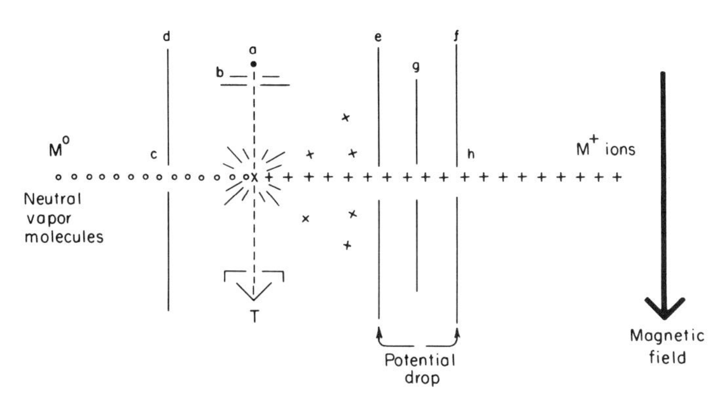

Fig. 2-1. Schematic of an electron impact ion source.

but are discharged by collisions with the walls or positive ions. Any that do come out are discharged since they suffer deflection in the direction opposite to the positive ions in either a magnetic or electrostatic analyzer section. A negative potential with respect to ground is applied across accelerator plates e and f and a relative positive potential of small magnitude (a few volts per centimeter) applied to plate d. Plate d is referred to as a repeller; its function is to start the ions in a path toward the accelerating region and to permit the discharge of negative ions upon collision with the repeller. Those ions entering the accelerating region with the proper trajectory do not encounter the accelerator plates, or collimating slit g, but are accelerated by the potential applied to plates e and f. They pass through exit slit h and enter the analyzer tube on their way to discharge either upon the analyzer wall or upon a collector if the ion is in focus upon the collector.

The electrons emitted from the filament are attracted to slit b by maintaining b at a potential positive with respect to a. Slit b is small, permitting only a finely collimated beam of electrons to enter the ionizing region x. The energy of the impacting electrons is controlled by the potential drop between a and b. This is the ionizing voltage referred to in spectral data and is most commonly used between 0 and 100 eV. If the ionizing beam energy is sufficiently great, ionization and subsequent dissociation or fragmentation may also occur. The positive ions formed are repelled or drawn out of the ion chamber by accelerating voltages of from less than 1000 to over 3000 V. During the acceleration, the positive ions acquire a large kinetic energy in addition to the small thermal (or kT) energy of about 0.03 eV that they possessed at room temperature. The ions may also possess any excess kinetic energy with which they are formed. Usually, this excess kinetic energy is small, but it may be 1 eV or more in certain cases. The electron emission current of the filament a is determined by the current flowing between a and b, while the anode current is determined by the number of electrons reaching the anode T. Only a small fraction of the electrons in the beam encounters molecules; the remaining electrons, therefore, do not enter into the impact equation. The number of ions formed at a given ionizing voltage is directly proportional to the electron current (number of electrons flowing) to which the vapor molecules are exposed. This is one way in which the intensity of the collected ions (and hence the spectrum) can be varied in the ion source. Another method is to increase the amount of sample and vapor within limits of rather low magnitude. Pressures higher than 10^{-5} torr lead to new effects primarily due to increased collisions of particles. This is discussed in detail in Chapter 3. Source design studies continue today, and literature references are sometimes found under the headings "electron optics" or "ion optics."

Since electrons from the filament are one of the "reagents" in the impact equation, one would expect to have to add more reagent from time to time. Commercial instruments all have ion sources with more or less readily replaceable filaments. Many instruments today employ the rhenium filament as suggested by Robinson and Sharkey (1958) for examining organic substances. A very complete treatment of the subject of filaments is to be found in the excellent text of Beynon (1960b).

The net result of ions produced from the impact source is a positive-ion beam exiting at h and producing a current in the range of 10^{-10}–10^{-15} A. The beam consists of ions of different masses and charges but of almost the same kinetic energy. It is well to remember that frequently ions of many different masses are produced and, hence, the ion current collected for a given mass will be only a small fraction of the value given above.

The electron impact ion source is very reliable, gives a small energy spread and a steady ionizing beam during the relatively long scanning time, and is the most frequently employed ion source in modern mass spectrometers.

C. The Photoionization Source

Many ionization processes require energies of 10 eV or more which correspond to photons having wavelengths of about 1240 Å. Thus, photoionization sources must employ electromagnetic radiation in the ultraviolet range. Terenin and Popov (1932a,b) used such a source to produce ions in a mass spectrometer. Watanabe *et al.* (1962) studied photoionization of many substances. Lossing and Tanaka (1956) combined a krypton discharge tube for photoionization with a mass spectrometer system. These workers showed that the ratio of ions collected to the number of quanta entering the ionizing chamber was about 1 in 10^6, which is approximately the same ratio as obtained from 50-eV electrons (Beynon, 1960c). Hurzeler *et al.* (1957, 1958) were the first to describe completely, and to employ, ultraviolet radiation that was continuously variable in energy but essentially monochromatic. They used a monochromator design similar to that of Seya (1952) and Namioka (1954). Ion currents of about 10^{-15} A were detected. It was soon apparent that fine structure in the ionization efficiency curves could be obtained. It was also noted that the spectra consisted mainly of parent-molecule ions, with only slight secondary (weak) spectra accompanying these peaks.

Steiner *et al.* (1961) studied the photoionization of alkanes, while Elder *et al.* (1962), with the same group of co-workers, studied the photoionization of alkyl free radicals. With the use of photons, the absolute energy of the ionizing particles is known and is unaffected by stray electric fields and contact potentials present in most impact sources. Thus, the determination of ionization potentials by photoionization can usually be made to within ± 0.02 eV. Recognizing the analytical possibilities of dealing with simpler spectra, consisting essentially of only parent-molecule ions, Vilesov and Akopyan (1962) applied the technique to mixtures of complex organic compounds and reported on some of the advantages of the photoionization source. Beynon *et al.* (1967) have described an accessory photoionization source for a typical commercial instrument. There are difficulties, and to paraphrase Kiser (1965b), the use of "simpler" mass spectra does not solve all problems; in fact, some analytical problems might become insolvable, e.g., mixtures of isomers.

The laser has also been used as an ionization source. It is an extremely monochromatic source of light. Honig and Woolston (1963) and Honig (1963, 1964) have reported on the use of ruby lasers as sources for the mass spectrometer and have examined inorganic substances such as conductors, semiconductors, and insulators. Beginnings have been made amid difficulties. Sharkey *et al.* (1964) have also experienced difficulties in working with coal. About 20% of the material in the region of laser impact was ejected as solids, while the remaining 80% was vaporized. Under the conditions used, they found methyl and ethyl hydrocarbon fragments which provided little information about the structure of coal (see also characterization techniques in Chapter 6).

On the other hand, characteristic fragmentations have been observed in the m/e 200 range for organic materials such as neoprene, isoprene, amino acids, dyes, coffee beans, etc., by Winter and Azarraga (1969). Their spectra were obtained with a combination laser–electron impact source and a modified Bendix TOF (time-of-flight) mass spectrometer. Samples are inserted directly into the source via a vacuum lock. Viewing and focusing optics are identical, thus permitting the operator to select the "target area" of the sample or samples and observe the results after pulsing the laser. In their application, the laser acts primarily as the vaporizer and the electron beam as the ionizer of the vaporized fragments. Some modifications need to be made to the above TOF system before spectra of laser-induced positive ions can be obtained.

Honig and Woolston (1963) report ion currents up to 500 mA from a similar system.

The selection of such parameters as laser power output, wavelength, and spectral recording techniques offers the investigator analytical "selectivity," insight into preferred fragmentation routes, free radical and chemical reaction studies, etc. It is probably too early to elucidate the problem of laser–matter interaction. It has been shown by Folmer and Azarraga (1969) that laser pyrolysis of polymers and other materials typically pyrolyzed for chromatographic analysis gives simpler chromatograms with much better characterization than does pyrolysis by the usual techniques. Retention times of peaks corresponding to C_{20} fragments have been frequently observed and are probably limited primarily by the columns themselves and the inlet systems. It is almost a certainty that such hardware combinations with conventional mass spectrometers will become a powerful analytical tool.

D. The Surface Emission Source

While little application of this technique to *organic substances* has been found, future workers may well develop modifications that will be applicable. The surface emission ion source is of primary use in analyzing elements which have low ionization potentials. About 30 of the metallic elements lend themselves to analysis by this technique. The surface emission ion source is shown schematically in Fig. 2-2. The substance to be analyzed is applied to the filament tip or ribbon and placed into the ion source unit. Heating the filament in the vacuum of the mass spectrometer system allows some of the material on the filament to vaporize. A portion of the vapor may be ions. Any ions produced are accelerated and collimated into an ion beam by means of slits, as shown in the illustration. The filament must be a very high melting substance. Tungsten, tantalum, and rhenium are used. Note that high voltage is applied to the first slit. One difficulty with this ionization source has been solved by placing a grid between the first slit and the filament. Hess *et al.* (1951) observed

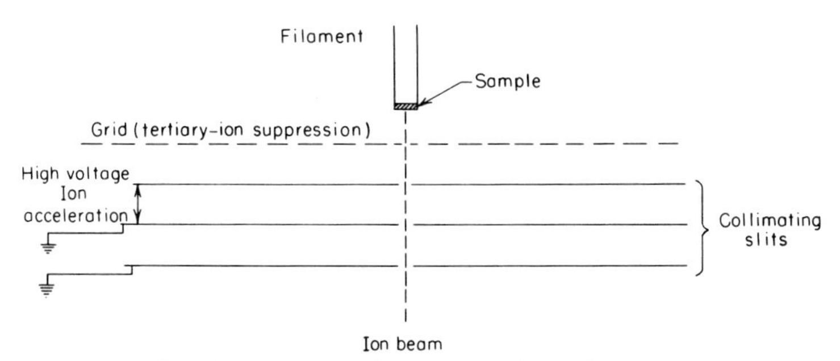

Fig. 2-2. Schematic of a surface emission ion source.

unexpected peaks in the mass spectrum. It was found that these were formed by electron bombardment of the filament with electrons that arise from ions produced from the sample which in turn cause secondary electron emission at the first collimating slit. These secondary electrons are accelerated back toward the filament, striking the sample and causing the formation of "tertiary" ions. The grid prevents the "tertiary" ions from entering the mass spectrometer. Muschlitz *et al.* (1961) described an ion source providing a moderately high intensity beam of negative ions. Crouch (1963) has introduced a method for the thermal ionization of elements with a high ionization potential.

E. The Vacuum Spark Source

Dempster (1935b, 1936) made some studies of the use of spark sources for ions. The vacuum vibrator of Fabry and Perot (1900a,b) gave a variable intensity spark and also a positional shift due to electrode wear. An arc discharge ion source, maintained by electrons passing from the filament to the anode through the vapor being studied, has the disadvantage of requiring high gas pressures (10^{-2} torr). Collimation of the electron beam of the discharge by a magnetic field permits reduction of the pressure to about 10^{-4} torr. Since this source produces intense ion beams, it is useful in the quantity separation of isotopes.

An intermittent hot spark (between metallic electrodes of short gap) is unsuitable for use in the far ultraviolet range because of the high gas pressure necessary to maintain the spark (Millikan *et al.*, 1921). Increasing the frequency of discharge and decreasing the current density produces a spark that gives a good ion yield of the substance comprising the electrodes. A voltage of about 50 kV is required; obviously, under these conditions, the electrodes would melt. This is avoided by the use of pulsed electronic oscillators. The filament of the usual ion source is replaced by a primary electrode of the material to be analyzed or of a tube containing the material to be analyzed. Monoisotopic metals such as gold or aluminum are usually employed for these tubes. The spark discharge carries some of the primary electrode into the vapor phase. A portion of the vapor forms ions and these are then accelerated and collimated into a beam.

The first construction and use of a spark source instrument was reported by Gorman *et al.* (1951) for the analyses of steel samples. A double-focusing mass analyzer is necessary to resolve the spark source ion beam, which has a large energy spread, according to Woolston and Honig (1964). Because of considerable fluctuations in the spark intensity, photographic plates must be employed so as to simultaneously record all ion masses and integrate their ion intensities for the exposure time.

At the energies employed a solid is found to break up almost entirely into its elements. A plethora of multiply charged ions is observed, sometimes making interpretation easy, other times making it difficult. Studies of inorganic solids and organic materials have been reported by Inghram (1953) and Hodgson *et al.* (1963). Organic materials placed in hollow electrodes or "painted" upon a solid electrode have been examined. Fragmentation is usually extreme, producing many fragment ions in the lower mass range. However, doubly charged metalloporphyrin parent ions and dipyrrolemethine fragments have been observed by Foster (1965) on plates obtained from a spark source instrument (see Fig. 3-1).

Interesting applications continue to appear in the literature. The analysis of human hair has been described by Yurachek *et al.* (1969), and the analysis of trace-metal components continues to be a big application. Alvarez *et al.* (1969) described the simultaneous determination of trace elements in platinum by isotope dilution and spark source mass spectrometry.

Small samples may be employed at low pressures (10^{-5} torr), and still a high sensitivity may be obtained. Over half of the elements have detection limits below 3 ppb, according to Brown *et al.* (1963). Sensitivities of most of the elements are of the same order of magnitude, based upon the results of Chakravarty *et al.* (1963). This leads to obvious analytical advantages. Chastagner (1969) has described analysis of microsamples by means of *single-exposure* spark source techniques. Brown and Wolstenholme (1964) have also compared the speed of analyses of geological samples by spark source mass spectrometry with other analytical methods.

The gaseous discharge ion source is often confused with the above types of spark sources. Although this source is of historical interest, there is little to note here other than that gaseous discharges present ion beams with large energy spreads (because of numerous collisions), leading to difficulties in resolution in the mass analyzer. Knewstubb and Tickner (1962a,b) have reviewed work with different gaseous discharge ion sources.

F. The Field Ionization Source

Intense fields can be developed in the vicinity of a metallic point of small radius of curvature when a potential difference of about 5 kV is applied between the point and another electrode spaced 2–5 mm from the point. This produces electric field gradients of 10^7–10^8 V/cm. Müller (1953) first observed that such a field produced the desorption of ions from the alkali metals. If a gas such as hydrogen were present at pressures of 10^{-3} torr, protons could be observed.

Inghram and Gomer (1954, 1955a,b) first used such a source to produce ions and a mass spectrum. The spectra produced were much simpler than electron impact spectra. Parent-molecule ions dominated the spectra almost exclusively. Beynon (1960d) claimed that this ion source is useful in that it can give information on the adsorption process, as demonstrated by Kirschner (1954) and others. It does permit the study of molecules on surfaces and has the added advantage that it can be operated at low temperatures. In this sense it is similar to photoionization sources. The simpler spectra produced may be easier to interpret, but once again the organic chemist is likely to find that isomers cannot be distinguished. This problem can sometimes be alleviated if a combined field ionization and electron impact source is used. Such sources have been fabricated for low-resolution mass spectrometers by Beckey *et al.* (1966) and Brunnée *et al.* (1967). Sources for high-resolution instruments have been constructed by Brunnée (1967), Chait *et al.* (1969), Schulze *et al.* (1969), and Curtis (1970). These ion sources may be quickly changed from one mode of operation to the other, thereby allowing both types of spectra to be obtained on the same sample charge.

Developments in field ionization were discussed at the Mass Spectrometry Conference held in Berlin (Kendrick, 1968) by Block (1968), Beckey (1968), Beckey *et al.* (1968), Wanless (1968), Robertson and Williams (1968), and Schuddemage and Hummel (1968). Earlier reviews are also available by Good and Müller (1956), Müller (1960), and Beckey (1959, 1962, 1963). Weiss and Hutchison (1968) reported on the field ionization mass spectra of some hydrocarbons and monochloro-substituted derivatives, while Mead (1968) extended consideration of this type of spectrometry to waxes. Chait *et al.* (1968) determined elemental composition by the technique, while Beckey and Hey (1968) examined the monoterpenes by the combination of electron impact and field ionization mass spectrometry. Beckey (1969) extended consideration to structure determination and quantitative analysis by field ion mass spectrometry.

G. Chemical Ionization Source

Chemical ionization mass spectrometry was first described by Munson and Field (1966). The developments and discoveries were a natural outgrowth of studies involving charge exchange and ion–molecule reactions (see Chapter 3). Ionization of the substance to be investigated is accomplished by reaction with ions instead of electrons. A set of ions is produced which does not react further with the reaction gas. If a small amount of another material is present, the reaction ions react with this second material to produce a spectrum of ions characteristic of the second material. If methane is used, the ions produced initially are CH_4^+, CH_3^+, CH_2^+, CH^+, C^+, H_2^+ and H^+. Secondary reactions occur to produce new ions:

$$CH_4^+ + CH_4^0 \rightarrow CH_3^0 + CH_5^+ \quad (2\text{-}2)$$

Obviously, the reaction is truly chemical in that the transfer of massive entities is involved. The pressure at the source for a chemical ionization system is about 1–2 torrs, very high by usual mass spectrometric standards. Under these conditions, the reactions are of the ion–molecule type, and of course, products are formed which have suffered collision after formation. This is unlike the low-pressure mass spectrometer systems.

The mass spectrometer employed by Munson and Field (1966) and described by Field (1968) was a conventional magnetic deflection instrument with a 12-in. radius of curvature. It was operable at high pressures because it was equipped with high-capacity differential pumping and utilized an ion source which was relatively gas tight. These workers claimed that at a source pressure of 1 torr of methane, the pressure in the region immediately outside the ionization chamber was about 2×10^{-3} torr, while that in the analyzer was about 2×10^{-5} torr. A mass resolution of about 1600 was achieved, which was comparable to the resolution obtained from the same instrument when used in a low-pressure service configuration. Other than the pumping capacity, chemical ionization measurements did not pose special service and maintenance problems. No doubt, special sources will be forthcoming in the future.

While the technique is new, it holds much promise both analytically and for assistance in structure determination. Its most promising results appear to be in the added understanding of gaseous ionic organic chemistry, as described in Chapter 3, Section I.

H. Miscellaneous Sources

Lozier (1930a,b) constructed a tube for the study of positive and negative ions. Although the apparatus was not a mass spectrometer, it is closely related and has been useful in research on ions. Tate and Lozier (1932) described a more elaborate and refined apparatus. Use of the tube permits the determination of the minimum energy required to produce ions of a given kinetic energy. Either positive or negative ions can be collected if the polarities of the applied voltages are reversed. The tube has been used to study diatomic molecules such as CO. Kiser (1965a) outlined a complete procedure for identification of the various products formed by dissociation and the energies associated with each.

Many new researchers in the field of mass spectrometry will encounter the phrase "use of a Fox gun" or "Fox electron gun" in discussions with their co-workers. In attempts to obtain a nearly monoenergetic electron beam for ionization potential studies, Fox and Hickam (1951) and Fox *et al.* (1955) constructed a source with a grid placed between the filament and ionization region. By determining the difference in ion abundances produced when two different potentials were applied to the grid, the amount of ions produced from electrons of a very narrow energy spread could be measured.

An ion source for the study of ion–molecule reactions has been described by Cermák and Herman (1961). Although of novel design, the source has a very significant advantage over other sources in that essentially only secondary ions are collected (i.e., ions used in the original ion–molecule reaction are excluded). The resulting spectrum is thus considerably easier to interpret, and interferences are absent.

Tandem mass spectrometers employing two sources have been described by Lindholm (1953) and Fedorenko (1954) and have been used to study ion–molecule reactions. This is another area in which fundamental studies will greatly benefit from further development of instrumentation.

Testerman *et al.* (1965) have studied cold electron sources with emphasis on their use in mass spectrometric applications. The designs and the properties of this source have been thoroughly described, and the suggestion has been made that these sources could find application in calibration work and in space flight instrumentation where, of course, the source temperature might vary over a tremendous range.

A series of articles by Bonazzola and Chiavassa (1964), Grodzins *et al.* (1965), and Stewart (1966) describes the development of still another impact species source, the neutron mass spectrometer. The neutron time-of-flight spectrometer seems to be entering the firm stage of development, and it may well contribute important new information to the field of mass spectrometry.

The Leger-type ion source (Leger, 1955) was developed for a special application of mass spectrometry—the study of free radicals. A supply of free radicals, essentially free from collisions, must flow through the ionization source and be aided in flow by a high-capacity, high-speed vacuum pump. This is shown schematically in Fig. 2-3. The quartz nipple has a fine opening

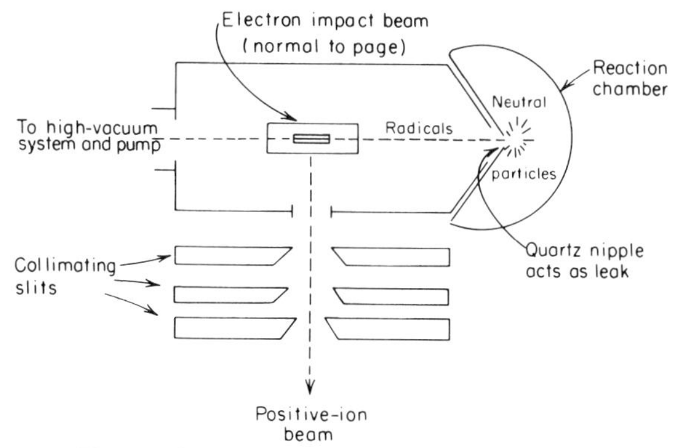

Fig. 2-3. Schematic of a Leger-type ion source.

which forms a "leak" into the ion source chamber. Phillips and Schiff (1962a,b,c) and Foner and Hudson (1962) used a microwave discharge to produce atoms and also used sampling methods different from those shown in Fig. 2-3. Direct electron impact studies of free radicals are possible. Lossing (1954a,b, 1956, 1957) has reported the ionization potentials of a considerable number of free radicals.

III. Ion Beam Separation Methods—A Means of Categorizing Instruments

Because the means of focusing the ion beam in a mass spectrometer is a useful criterion for distinguishing the various capabilities of instruments, it is used here to provide the reader with a guide to the background knowledge of instrumentation. Focusing the ion beam improves the separation between adjacent masses in register and at the same time increases the intensity of the ion beam. Therefore, the measurement of the position of the ion beam is made more precise, and the "sensitivity" of the instrument is increased. Among the different types of focusing are velocity focusing, direction focusing, double focusing, and "time" focusing.

A. The Parabola Mass Spectrograph (Thomson's Instrument)

The parabola mass spectrograph (Thomson, 1911) does not provide a focused ion beam because the beam is passed through parallel electric and magnetic fields. After deflection, the beam impinges on a photographic plate to make a parabolic curve for each ion species of varying m/e, the vertex of the parabola lying at the undeflected position of the ion beam. The position of an ion on the parabola is a function of its momentum. Positive ions of a given m/e are recorded on one-half of the parabola, and negative ions of the same m/e are recorded on the mirror image portion of the "locus." Thus, Thomson could easily detect both positive and negative ions. This type of instrument provides a means of studying the steps in dissociation of molecular ions, and it has proven to be useful in studying ion source characteristics (see Beynon, 1960a, for additional details).

B. Velocity Focusing (Aston's Instrument)

Velocity focusing is achieved when an ion beam is focused such that the ions travel with a distribution of velocities but in the same initial direction; that is, they are directionally collimated. Aston's instruments (Aston, 1919, 1927, 1937) utilized first an electrostatic field and then a magnetic field to focus

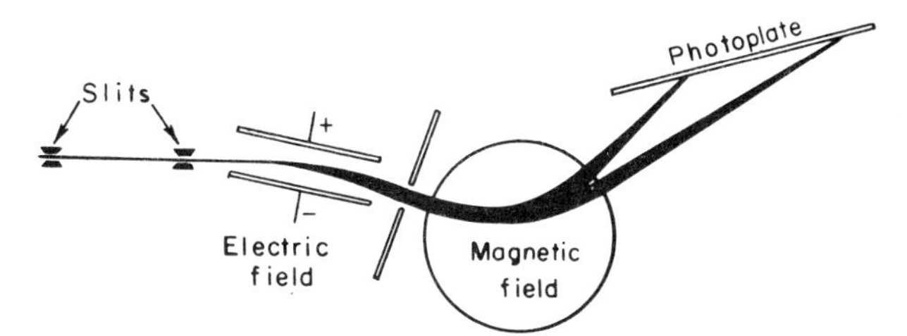

Fig. 2-4. Schematic of Aston's mass spectrometer.

the ion beam. By means of this system and the arrangement shown in Fig. 2-4, he was able to achieve velocity focusing of the ion beam.

No provision was made for direction focusing. Resolution was better than that obtained with the parabola spectrograph. A photographic plate could be used to record the mass spectrum for a time such as to increase the sensitivity of this type of instrument as compared to the early electrometer-type detectors.

C. Direction Focusing (Dempster's Instrument)

Direction focusing is achieved when an ion beam is focused such that the ions move with the same velocity for a range of different initial directions. This method was discovered by Classen (1907, 1908) and employed by Dempster (1918) for his first instrument. Dempster's instrument is shown in Fig. 2-5 in schematic form.

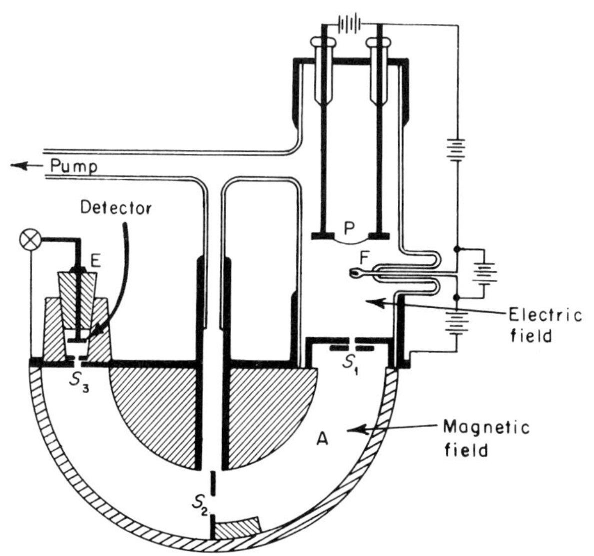

Fig. 2-5. Schematic of Dempster's mass spectrometer.

Ions were produced in two different ways: (1) by emitting electrons from a hot filament F and causing them to impact samples on a platinum strip P and (2) by heating samples applied directly to the platinum strip. A potential drop between P and the adjustable slit S_1 provided the accelerating potential V for the ions. Ions of nearly the same energies were produced. After passing through the adjustable slit S_1, a magnetic field of about 3000 G perpendicular to the plane of the figure bent the ion in a circular path of radius r. Slit S_2 was used in the analyzer section A to reduce the number of reflected ions and/or electrons reaching the exit or collector slit S_3, which was also adjustable. A quadrant electrometer E was used to measure the ion current.

In comparing the Aston and Dempster instruments, it should be noted that Aston used his deflecting fields as "prisms," while Dempster used the magnetic field as a "lens." Since there is no velocity focusing in the Dempster instrument, one must provide an essentially monoenergetic beam. Dempster deduced that the geometrical limit of resolution is

$$M/\Delta M = r/(S_1 + S_3) \tag{2-3}$$

A maximum value for the resolution should occur when both S_1 and S_3 approach zero.

Dempster's design employed ions initially accelerated in an electric field eV entering a uniform magnetic field at right angles to its direction of motion and deflected through 180° in a magnetic field of strength H. Since the ions of mass m and charge e enter the electrostatic field with normal gas-flow velocities, they do possess an initial velocity. This is neglected in the simplified treatment here since the final velocity after acceleration is large compared to the initial velocity. The ions gain kinetic energy T by falling through a potential V (in ergs per electrostatic units). Then

$$T = \tfrac{1}{2}mv^2 = eV \tag{2-4}$$

where e is the charge in electrostatic units, m is the mass in grams, and v is the velocity of the ion in centimeters per second. When the accelerated ions enter the magnetic field (perpendicular to the direction of the positive-ion beam), each ion experiences a force at right angles to both its direction of motion as accelerated and the direction of the magnetic field, thereby deflecting or bending the ion beam.

In the magnetic field the ions experience a force F_M in dynes, where

$$F_M = Hev/c \tag{2-5}$$

and where c is the velocity of light in a vacuum ($c = 2.9979 \times 10^{10}$ cm/second), and H is in oersteds. The force of the magnetic field is equal and opposed to the centrifugal force F_C. Hence

$$F_M = F_C \tag{2-6}$$

and since

$$F_C = mv^2/r \tag{2-7}$$

then

$$mv^2/r = Hev/c \tag{2-8}$$

which, upon rearranging terms, becomes

$$v = Her/mc \tag{2-9}$$

Now the value of v from Eq. (2-9) is substituted into Eq. (2-4) to give

$$\tfrac{1}{2}m(Her/mc)^2 = eV \tag{2-10}$$

or

$$H^2e^2r^2/2mc^2 = eV \tag{2-11}$$

or, in a form most suitable for mass spectrometer use

$$m/e = H^2r^2/2Vc^2 \tag{2-12}$$

Evaluating the constants and changing the units of m to atomic mass units (see Example 2-1) one can write

$$m/z = H^2r^2/20{,}740\,V \tag{2-13}$$

where z is the number of electronic charges per ion, H is in gauss (the permeability is assumed equal to one), r is in centimeters, and V is the potential drop in volts. Since most instruments must fix the radius r, Eq. (2-13) can be expressed as

$$m/z = KH^2/V \tag{2-14}$$

where $K = r^2/20{,}740$ and is characteristic for a given instrument. Equation (2-14) is important and useful in that it shows the dependence of the m/z ratio of the ions upon the accelerating potential V and the magnetic field strength H. The equation indicates that increasing the magnetic field H will focus heavier ions on the detector, while increasing the accelerating potential V will focus lower m/z ions on the detector. While it is possible to calculate the field strength between the pole pieces of an electromagnet from equations involving the amperage applied, the number of turns of wire, and the permeability of the pole pieces, it is a general practice of mass spectroscopists to proceed somewhat differently. A known material containing m/e of known values is introduced to the operating ion source of the instrument and the field strength calibrated for a fixed magnetic current (amperes) from observing a known m/e in focus at a given V.

Example 2-1

The m/e 58 (parent) ion of n-butane appears at a potential of 1200 V at an instrumental magnetic current setting of A amperes. Calculate the field strength H associated with this amperage, A. The radius r is 6.00 in. for the instrument in question.

Using Eq. (2-12) and rearranging

$$H^2 = (m/z)(2Vc^2/r^2) \tag{2-15}$$

and since r must be in centimeters and V in ergs per electrostatic units,

$$r = 6.00 \times 2.54 = 15.24 \text{ cm} \quad \text{and} \quad V = 1200/300 = 4.00 \text{ erg/esu}$$

Substituting these values in Eq. (2-15) gives

$$H^2 = \frac{2 \times 4.00 \times (58.07825/6.024) \times 10^{-23} \times 9 \times 10^{20}}{(15.24)^2 \times 4.802 \times 10^{-10}}$$

where atomic weights are based upon carbon = 12.00000, Avogadro's number ($N = 6.024 \times 10^{23}$), and the electronic charge is 4.802×10^{-10} esu. Note that the c^2 (9×10^{20}) is necessary to convert electrostatic units to the second power into electromagnet units to the second power. Upon solving,

$$H^2 = 6.224 \times 10^6 \quad \text{or} \quad H = 2494.7 \text{ G}$$

It is simpler to employ Eq. (2-13) in a rearranged form

$$H^2 = \frac{58.07825 \times 20{,}740 \times 1200}{(15.24)^2} = 6223.5 \times 10^6 \text{ or } H = 2494.7 \text{ G} \tag{2-16}$$

Example 2-2

Using the same mass spectrometer as in Example 1 and keeping the magnetic field constant at a rounded value of 2495 G, (a) what V would be necessary to put the m/z 29 ion ($C_2H_5^+$) in focus? (b) Similarly, what V would be necessary to have an ion at m/z 78 appear in focus ($C_6H_6^+$)?

(a) Using Eq. (2-13) in a rearranged form,

$$V_a = \frac{H^2 r^2}{20{,}740 \times m/z} \tag{2-17}$$

or substituting in values,

$$V_a = \frac{(2495)^2 \times (15.48)^2}{20{,}740 \times 29.0391} = 2400 \text{ V}$$

(b) Using a similar approach to determine V_b,

$$V_b = \frac{(2495)^2 \times (15.48)^2}{20{,}740 \times 78.0468} = 885.5 \text{ V}$$

We now see that we could have used

$$V_{i(H)} = V_{58(H)} \times \frac{58.078}{(m/z)_i} \tag{2-18}$$

so that

$$V_{78(H)} = \frac{1200 \times 58.078}{78.0468} = 893.0 \text{ V}$$

Note the effect of rounding off the field to 2495 Gauss.

It can be seen that successively lighter ions will be in focus as the potential is increased based upon the calculations given above. For magnetic scanning, an equation similar to Eq. (2-18) can be used because from Eq. (2-14) one can note that the mass-to-charge ratio is proportional to the square of the magnetic field and, hence,

$$(m/z)_{H_1} = a(H_1)^2 \tag{2-19}$$

and

$$(m/z)_{H_2} = a(H_2)^2 \tag{2-20}$$

where a is a proportionality constant. The ratio of these two equations can be written

$$\frac{(m/z)_{H_1}}{(m/z)_{H_2}} = \frac{(H_1)^2}{(H_2)^2} = \left(\frac{H_1}{H_2}\right)^2 \tag{2-21}$$

This equation is useful for constant V and changing H.

Example 2-3

For a magnetic scanning mass spectrometer, the naphthalene parent ion at m/e 128.0626 was observed at $H = 2315$ G. What magnetic field would be necessary to focus the parent ion of ethylbenzene which would occur at m/e 106.0782?

Equation (2-21) is employed:

$$(H_1/2315)^2 = 106.0782/128.0626 = 0.82833$$

from which

$$H_1/2315 = 0.9101 \qquad \text{or} \qquad H_1 = 0.9101 \times 2315 = 2106.9 \text{ G}$$

Very often the nominal mass (integral) values are used in these equations because the error in field strength would be less than 1 G and not significant in comparison to the reproducibility of operating conditions in most mass spectrometers.

D. The Sector Instrument

1. General Description

It was shown by Barber (1933) and by Stephens and Hughes (1934) that the 180° geometry is simply a special case of the "sector" magnetic field. The focusing action of any wedge-shaped magnetic field would be similar. Thus, by utilizing a wedge-shaped magnetic field with a smaller sector angle, the area of the magnetic field (for a given radius of curvature) can be reduced (see Fig. 2-6). This, in turn, permits a smaller magnet, which consumes less power. A second advantage is that the ion source and collector need not be in the magnetic field, and hence, only the smaller cross-sectional area of the ion beam must be in the field gap. The ease of accessibility of the source and

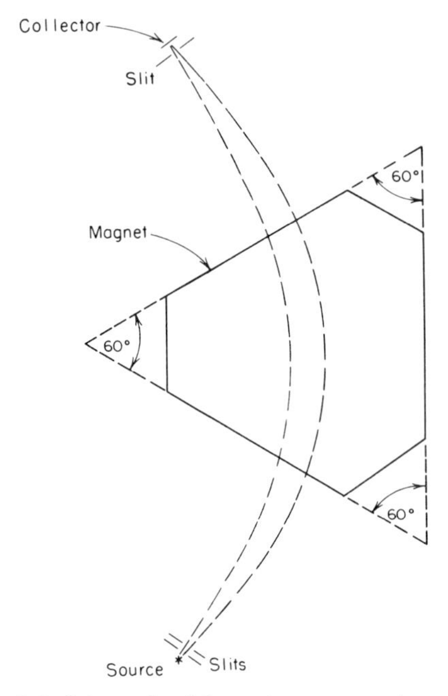

Fig. 2-6. Schematic of the sector mass spectrometer.

collector permits numerous design modifications, such as using the electron multiplier for a collector. For additional details, the reader is referred to Beynon's book, Chapter I (1960a). Nier (1940) reported the completion of a 60° sector instrument with a compact ion source possessing a highly stable electron beam. Since then, many additional sector instruments have been constructed.

Among the commercially available 60° sector instruments of the single-focusing type are units made by Nuclide Analysis Associates, the Atlas-MAT CH-4, Hitachi (Tokyo), and Mitsubishi (Tokyo). The 90° sector instruments are employed in a number of single-focusing commercial instruments, some of which can be expanded to double-focusing instruments in the field. These are employed in the Perkin–Elmer–Hitachi RMU series and in the CEC 490 series, among others, for the magnetic sectors.

2. *The Isotope Ratio Instrument—A Special Application*

The isotope ratio mass spectrometer employs the Nier sector type of magnetic deflection. The conventional system is modified to produce a recording of the intensities of the light and heavy isotopic m/e's simultaneously. This helps to eliminate the effect of fluctuations, however slight, in the intensity of the electron beam. In addition, both m/e's of ions should be accelerated by the same potential in order to keep the ratio of the initial energy of the ion to the accelerating energy essentially constant. This will practically eliminate any "mass discrimination" effects. Since the two ion beams follow different paths, collectors may be placed at the focal point of each ion species. A pair of matched amplifiers, one for each collector, is employed. The ratio of their outputs is measured by a null bridge, thus producing a measurement of the exact ratio of the abundances of ions from each of the isotopes. The accuracy and precision for isotope ratio measurements are discussed by Nielsen (1968).

E. Double-Focusing Theory

The Dempster equation predicts that for a constant magnetic field H a spread in the magnitude of V will produce a spread in r for any given value of m/e. A statistical distribution of initial kinetic energies is to be expected for the gas to be ionized. A Gaussian (or more complex) distribution of the electron beam is also to be expected. Thus, the ionized particles will initially possess thermal energy and will gain kinetic energy during acceleration in the electric field V. The resulting ion beam will occupy a wider area when refocused, and this, in turn, will limit the resolution of mass spectrometers utilizing only direction focusing. Elimination of the energy spread in the ion beam before it enters the magnetic field should result in a considerable increase in resolving power. Although Bartky and Dempster (1933) had suggested such instruments, only a limited portion of the mass spectrum was obtainable (Bondy and Popper, 1933).

Dempster (1935a) published details of a double-focusing instrument having an electrostatic analyzer of 90° and a magnetic analyzer of 180° employed in a tandem arrangement. Bainbridge and Jordan (1936) combined a 127° 17′ direction-focusing electrostatic analyzer with a 60° direction-focusing magnetic analyzer. It is noteworthy that this was the first use of the sector magnetic field, and it happened to be in a double-focusing instrument, the design of which is considered to be of simplest form. This instrument gave twice the dispersion of the Dempster instrument and required only one-third as much field area. The mass scale is essentially linear, whereas in the Dempster instrument the mass scale varies as the square root of the *m/e*.

F. Mattauch–Herzog Geometry

Herzog (1934) and Mattauch and Herzog (1934) published a complete theory for achieving double focusing, and all instrument designs since then can be regarded as special cases of the equations reported. Ions are passed through a radial electrostatic field and are focused according to velocity. A slit placed between the electrostatic analyzer and the magnetic field serves to limit the kinetic energy of the ions to a certain value. Direction focusing is then accomplished in the magnetic field. Resolution claims for such instruments are now reaching values in excess of 50,000. Perhaps the real value of obtaining a resolution in excess of 50,000 lies in the ability to resolve doublets at relatively near masses.

The design of Mattauch and Herzog (Mattauch, 1936a,b) is one of the earliest for double-focusing instruments and has been widely employed in commercial instruments. The CEC Model 21-110 is a popular commercial instrument of this design (Robinson *et al.*, 1961). Associated Electrical Industries Ltd. (AEI) of England market the MS-7, a radiofrequency (rf) spark source instrument well suited to metals analyses, also of the Mattauch–Herzog geometry. A number of other commercial instruments are also available from various manufacturers. Some double-focusing instruments with slightly modified electrostatic fields are now appearing in the medium price range and possess the advantage of being modular. A single-focusing magnetic sector instrument is first purchased; then the electrostatic sector can be added, followed by direct GC interfacing.

The Mattauch–Herzog design is shown schematically in Fig. 2-7. Ions enter through the object slit, pass through the electrostatic sector and the exit slit, and enter the magnetic sector. Here they are focused on the photographic plate at the focal plane. The plate is contained in the magnetic field and a part of the vacuum system. Note that M_1 is the image of an ion of lighter mass, while M_2 is the image of one of higher mass. The spectrum consists of lines having their spacing related to the *m/e* and their exposure blackness related to ion intensity. A microdensitometer is necessary for measurement of both values. Conventional recording, that is, peak to peak, is usually done by utilizing an electron multiplier placed at a fixed position in the focal plane and magnetically scanning the various *m/e* ratios by the detector. The recording of the complete mass spectrum of an organic compound of modest molecular weight would prove to be quite tedious. Usually, the user is interested only in high resolution over a limited mass range or possibly only in a few peaks of a given spectrum. One advantage of the photographic plate is that the entire

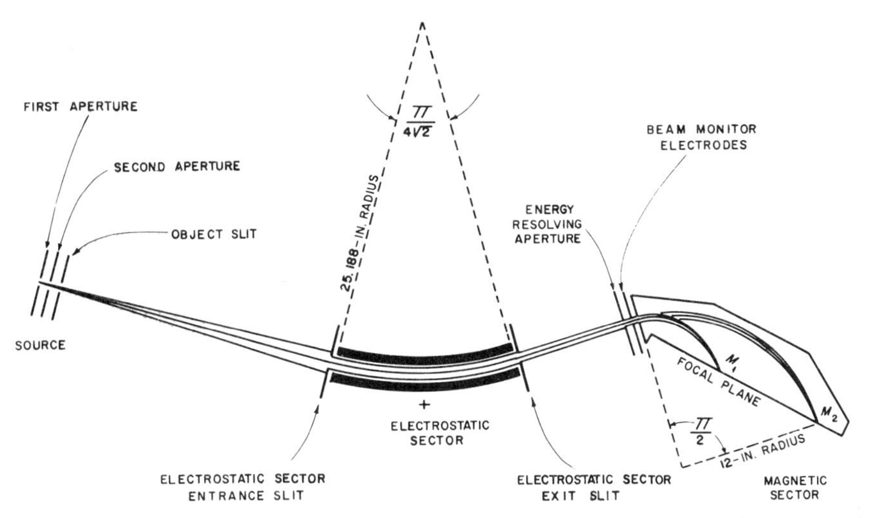

Fig. 2-7. Schematic of Mattauch–Herzog double-focusing geometry. (Courtesy of Consolidated Electrodynamics Corp.)

spectrum is recorded simultaneously, thus averaging out any small instrumental deviations that could, for example, minimize or exaggerate a peak or two during an electrical surge. It is also useful for recording the spectrum in a short period of time. The commercial instrument covers a low to high m/e range of a ratio of 1:30 for one magnetic field setting.

G. Nier–Johnson Geometry

Johnson and Nier (1953) have also designed a double-focusing instrument, as shown in Fig. 2-8. A 90° electrostatic analyzer is followed by a 90° magnetic sector. A collector is placed at the focal point of the magnetic analyzer, and ion beams are brought to the collector by scanning the magnetic field. A prototype of this instrument was used by Beynon (1959) in his early work that showed the value of high-resolution mass spectrometry in examining organic molecules. Elliott *et al.* (1961) have described one commercially available instrument, the AEI-MS-900 series.

H. Trochoidal Geometry Instruments

The cycloidal mass spectrometer was the outgrowth of early attempts to design double-focusing instruments (Hipple and Bleakney, 1936). Bleakney and Hipple (1938) used crossed homogeneous magnetic and electrostatic fields and introduced a beam of positive ions. The ion paths form a prolate trochoid.

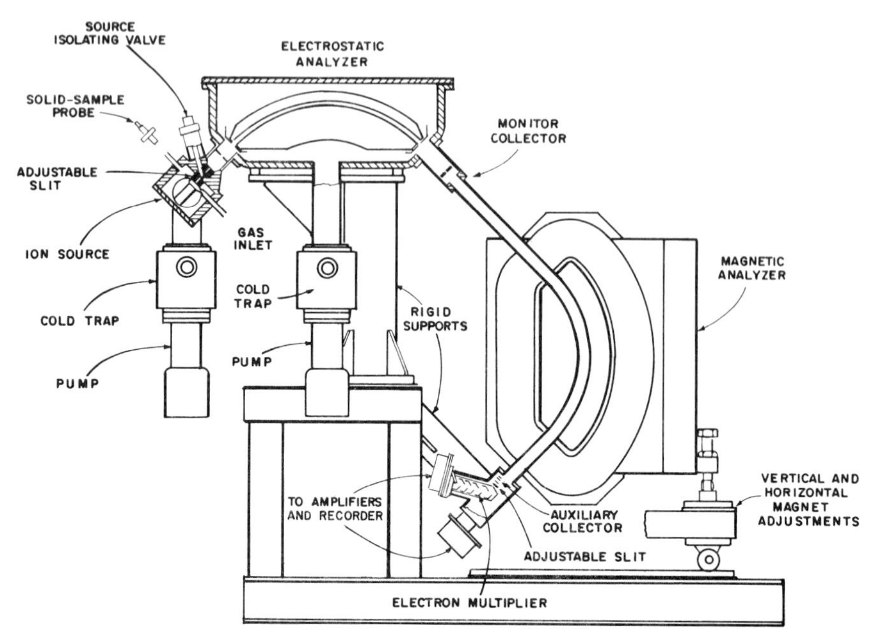

Fig. 2-8. Schematic of Nier–Johnson geometry. (Courtesy of Associated Electrical Industries, Ltd.)

In theory, one obtains a perfect focus and a spatial mass dispersion. Neither the initial values of velocity nor direction of motion of the ions affect the focus. These researchers constructed two instruments, both varieties of the cycloidal type shown in Fig. 2-9. Advantages of such instruments are that only two precise slits are required and a linear mass scale is achieved. This would make it highly desirable for a mass spectrograph. Disadvantages include the large areas of uniform magnetic and electric fields required and the fact that the focal point of ions of one m/e can be traversed by ions of differing m/e on the way to their focal points. A commercial instrument has been described by Robinson and Hall (1956), and recently a second instrument has become available from Varian Associates.

I. Linear Time-of-Flight Mass Spectrometer

The concept of time-of-flight mass spectrometry, as will be described, had been experimented with by Smythe and Mattauch (1932). Stephens (1946) and Wolff and Stephens (1953) reawakened interest in the technique when they successfully applied it in 1953. Katzenstein and Friedland (1955) raised the performance of their instrument to a resolution of about 100. Wiley (1956) and Wiley and McLaren (1955) described an instrument which has achieved

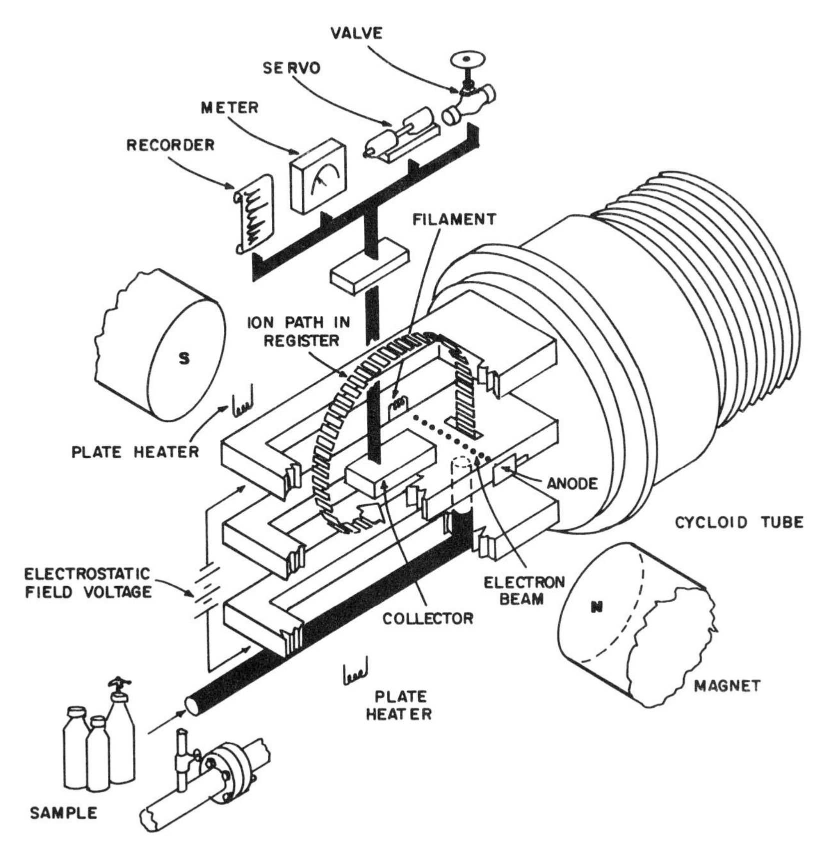

Fig. 2-9. Schematic of the cycloidal mass spectrometer. (Courtesy of Consolidated Electrodynamics Corp.)

success and popularity on a commercial basis. A later improvement in resolution was achieved by Harrington (1960) and by Harrington and Gohlke (1962). The time-of-flight mass spectrometer employs no magnetic fields (see Fig. 2-10). Ions are produced in a conventional manner, such as by electron impact, but a control grid is used to pulse the electron beam when a positive potential is applied. The pulses are very short; the beam is energized for 0.25 μsecond every 100 μseconds. The ions thus produced are accelerated by an electric field pulsed at the same frequency but lagging slightly behind the ionization pulse. In this manner, all of the ions formed receive essentially the same kinetic energy and attain a velocity which depends upon their mass-to-charge ratio. The ions of different m/e separate as they travel down a field-free evacuated drift tube approximately 1 m in length. An electron multiplier is used as a detector at the end of the flight path. The output of the multiplier

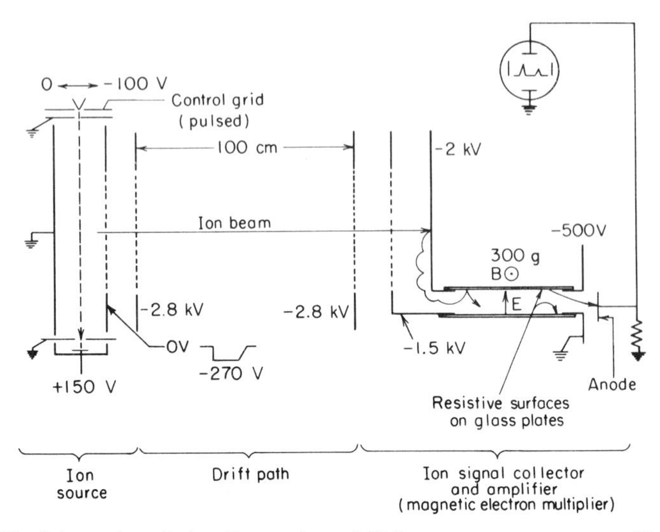

Fig. 2-10. Schematic of the linear time-of-flight mass spectrometer. (From D. B Harrington *in* "Encyclopedia of Spectroscopy" (C. F. Clark, ed.), Reinhold Book Corp., a subsidiary of Chapman-Reinhold, Inc., New York, 1960.)

is used to produce a signal on either a conventional recorder or an oscilloscope screen. Since all of the ions produced in a single pulse are collected during the time of the pulsing, very rapid production of the mass spectrum is possible.

The exact determination of mass is obtained by accurate electronic measurement of the ion flight time. It is directly proportional to the square root of the mass-to-charge ratio; hence, the ratio of the times of flight, t_1 and t_2 of two ions of the same charge and masses m_1 and m_2, from source to collector, is given by

$$t_1/t_2 = \sqrt{m_1{}^+/m_2{}^+} \tag{2-22}$$

For a given instrument, the transit time t (in microseconds) for ions of a given m/e through the length of the drift tube L (in centimeters) under a potential of V (in ergs per electrostatic units) is described by

$$t = L\left(\frac{m}{e}\cdot\frac{1}{2V}\right)^{1/2} \tag{2-23}$$

or in rearranged form

$$m/e = 2Vt^2/L^2 \tag{2-24}$$

Example 2-4

A time-of-flight mass spectrometer has a flight length of 95 cm and employs an accelerating potential of 2950 V. How long will it take a $CO_2{}^+$ ion to travel the distance from the ion source to the detector?

Using Eq. (2-23),

$$t = 95 \times \left(\frac{43.9898}{4.802 \times 10^{-10} \times 6.024 \times 10^{23}}\right)^{1/2} \left(\frac{300}{2 \times 2950}\right)^{1/2}$$

$$= 95 \times 3.899 \times 10^{-7} \times 2.254 \times 10^{-1}$$

$$= 834.893 \times 10^{-8} \text{ second}$$

$$= 8.35\ \mu\text{seconds}$$

Example 2-5

On the same instrument as given in Example 2-4, what would be the difference in the arrival time of the ^{13}C-containing ion of CO_2^+?

$$t_{44} = 8.349\ \mu\text{seconds}$$

$$t_{45} = 8.349 \times (45/44)^{1/2} \text{ (using approximate masses)}$$

$$= 8.349 \times (1.02272)^{1/2}$$

$$= 8.349 \times 1.0112 \qquad \text{or} \qquad t_{45} = 8.442\ \mu\text{seconds}$$

Hence the difference in time Δt,

$$\Delta t = t_{45} - t_{44} = 8.442 - 8.349 = 0.093\ \mu\text{seconds}$$

This is about 0.1 μsecond. This small difference in arrival time illustrates the need for very fast electronics and precise control of those electronics concerned with the pulsing of the ion beam.

Theoretically, the mass range of this type of instrument extends from mass 1 to 5000 for the 1-meter drift tube because only ions having m/e's greater than this would be overtaken by the low-mass ions of the next pulsed cycle. In practice, the resolution obtained is much lower, and it is difficult to determine the approximate mass of particles in the range of 2000–3000 m/e.

Some advantages of this type of instrument are as follows: Precise narrow slits are not required; relatively few ions are lost between the source and the collector, thus making more effective use of the ion beam possible; the pulsing of the beam and ions produces extremely rapid scanning; the vacuum system has few bends and, in fact, is quite open; and the accessibility of the ion source and general geometrical layout permit rugged construction. It does lack resolving power, long-term reproducibility of spectra, and ease of mass determination when compared to the magnetic instruments.

This instrument is an especially valuable tool for studying phenomena of short duration, such as fast kinetic changes, flames, explosions, and shock waves, and for recording spectra of gas chromatographic peaks. Thirion (1966) has prepared a review of this subject.

J. Bennett Tube Mass Spectrometer

Actually, the radiofrequency (rf) mass spectrometer (Bennett, 1950, 1953) is a unit with focusing action in time, and the scanning of various m/e's is accomplished by varying the frequency of the rf field. The unit can be made quite compact since no magnetic field is used, and the electronic requirements are modest. The resolving power is also modest. An improved type of instrument has been described by Ryan and Green (1965).

K. Cyclotron Resonance Mass Spectrometers

A charged particle in a magnetic field describes a circular orbit at an angular velocity given by $\omega = eH/m$. For a circular orbit this implies that the time for the ion to complete an orbit will depend upon its mass according to the equation $t = 2\pi m/He$. Note that this is in accordance with the cyclotron resonance frequency, which is $\omega_c/2\pi$. Ions introduced into such a system could be accelerated to very high energies. This method was first used to produce particles with sufficient energy to cause nuclear disintegrations.

A high-frequency (rf) electric field applied perpendicularly to the direction of the magnetic field with a period equal to the period of revolution of the ion causes the acceleration and a spiraling outwards of the ion until it strikes a target electrode. The above implies resonance of the rf field (f) with the oscillating frequency of the ions of mass m in the magnetic field H of strength B. Equating these,

$$f = eB/2\pi m = \omega_c/2\pi \qquad (2\text{-}25)$$

where e is the electronic charge and ω_c is the critical angular velocity. Note that the use of $H = \mu B$ is made where μ is the magnetic permeability. Normally, in a high vacuum this assumption of $\mu = 1$ is quite correct, but one should consider this to be true strictly for very low ion beam densities.

The net effect of this approach is that masses should be resolved by observation of their times of flight for complete revolutions in a magnetic field. The time relationship shows that a linear mass scale will be obtained in this experiment.

The first application of this principle to mass spectral analysis was accomplished by Hipple *et al.* (1949) and Sommer *et al.* (1951) in the omegatron (Fig. 2-11). The ions of a volatile substance are formed by electron impact within the core of the resonator zone, which is located in a homogeneous magnetic field perpendicular to the page in the right-hand portion of the

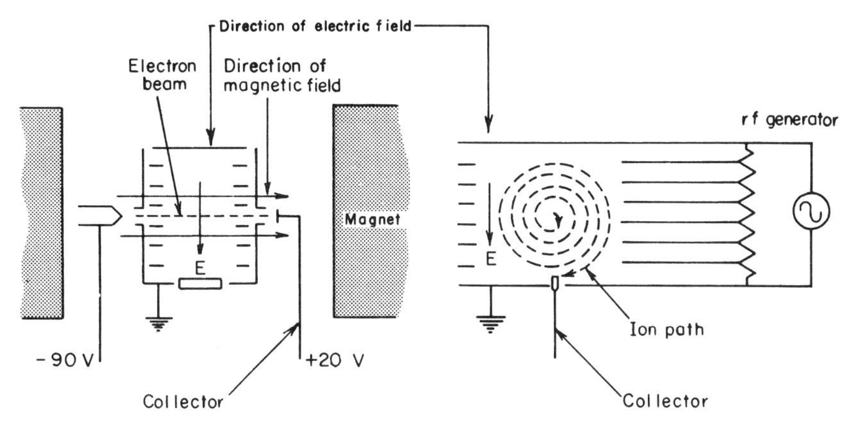

Fig. 2-11. Schematic of the omegatron. (After Hipple *et al.*, 1949.)

drawing. The ions of respective masses are spiraled outwards, in turn, by scanning the frequency of the ac electric field required to produce resonance. The collector electrode is designated by *E*. Many of these instruments are in use, primarily as residual gas analyzers. Mass resolution is usually limited to CO_2 at m/e 44, although about 80 is theoretically possible.

A new instrument, the Syrotron, has been developed by Varian Associates (Lewellyn, 1965), who claim a resolution of 1/700 at mass 140. This instrument will bear watching for special applications, particularly ion–molecule studies. Goode *et al.* (1970) have presented a review of this relatively new form of mass spectrometry and pointed to its wide range of applications.

L. Quadrupole and Monopole Mass Spectrometers

The quadrupole mass filter was developed and described by Paul and Steinwedel (1953). A quadrupole field is capable of separating ions on the basis of their m/e but is unlike most of the previously described instruments in that a magnetic field is not employed. An rf electric field is used for mass analysis. A cross-sectional view of the quadrupole electrodes is shown in Fig. 2-12. Opposing pairs of electrodes are electrically connected and carry a steady dc voltage. The dc voltages are equal but opposite in sign. Superimposed upon this dc voltage is an rf voltage. If ions are introduced into the field parallel to the electrodes (in the Z direction) and preferably at the point of intersection of the X–Y reference lines, they drift in the direction of injection with a constant velocity; but they oscillate in perpendicular directions, the amplitude increasing exponentially with time. Eventually the ion strikes a surface, is discharged, and lost. Geometric and electrical parameters can be chosen so that for a small range of masses the amplitude may be made constant, thus allowing the ions

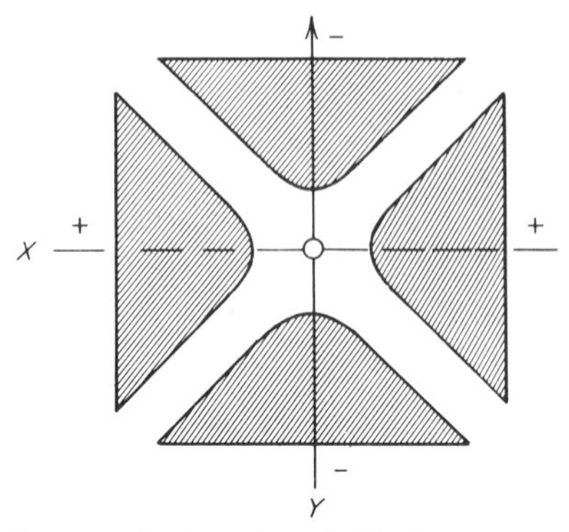

Fig. 2-12. Schematic of the cross section of the field of a quadrupole mass spectrometer.

of that particular mass (or range of masses) to leave the quadrupole field. This theory has been summarized by Kiser (1965d). The ions can be collected by conventional means. A mass spectrum of the ions originally injected can be obtained by regular variation of either the amplitude or rf voltage, which will change the value of the mass of ion transmitted through the field and collected. Originally used as residual gas analyzers, the instruments now find general analytical applications (von Zahn *et al.*, 1962; Brubaker, 1964). Brunnée *et al.* (1964) extended the applications into many areas of vacuum technology. Smith and Cromey (1968) described an inexpensive and bakeable quadrupole mass spectrometer.

A number of the instrument manufacturers now offer quadrupole instruments with various features and options. The Finnegan Corp. offers one in the form of their series 3000—Gas Chromatograph Peak Identifier. The mass spectra can be taken in the "quadrupole mode" or the "program mode," which they claim provides spectra closely matching the existing compilations of mass spectral data. Thus this unit joins those of the older manufacturers such as CEC, Perkin–Elmer, Varian–Atlas MAT, and AEI with entrants in the low price field of mass instruments.

Von Zahn (1963) devised a "monopole" mass spectrometer that produces a quadrupole field from one cylinder and one right-angled electrode. General Electric manufactures a monopole mass spectrometer which, they claim, has unit resolution up to mass 300, is operable to 250°C, has a bakeable tube, and provides a linear mass display. An electron multiplier detector is used, and several readout options are available. The instrument is said to have performance equal to that of the quadrupole for all practical purposes.

M. The Double-Beam Mass Spectrometer

While not distinguishable on the basis of its source or geometry, the double-beam mass spectrometer offers a number of advantages to the researcher. It is of particular value when in use with a GC–MS combination where it can simultaneously record on the same chart the unknown eluting GC peak and a proposed reference compound thought to be the same substance. Marketed by Picker Nuclear in the United States, it is manufactured by AEI Scientific Apparatus, Ltd., Manchester, England. The MS 30 has two ion beams produced by two independent ion sources into which identical or different samples can be admitted. A schematic of the instrument is shown in Fig. 2-13. A common mass analyzer is used, permitting the spectra from the two samples to appear on identical mass scales, run and recorded under simultaneous conditions. The versatility of the arrangement is such that two identical samples can be run under different conditions such as high and low resolution or at high and low ionization voltages, in addition to the comparison technique described above. This type of operation allows chemical mass marking with great utility. The manufacturer claims that since the two spectra are recorded on the same chart the assignment of the mass numbers is easy and permits

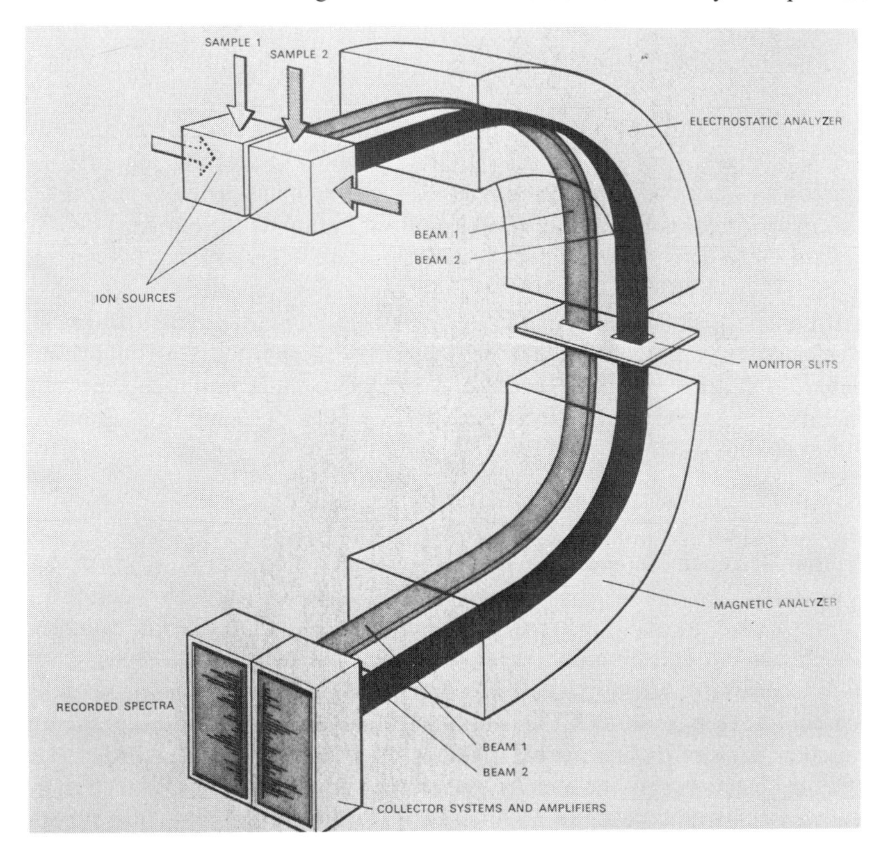

Fig. 2-13. Schematic diagram of first double-beam mass spectrometer. (Courtesy of AEI Scientific Apparatus Ltd. and Picker-Nuclear.)

an accuracy of a few parts per million so that elemental composition can be determined. The unit is modular so that portions of the apparatus can be obtained at different times. The instrument can be directly coupled to an on-line data acquisition and analysis system. It features many additional developments in simple controls and rapid switching of resolution, and it produces linear mass scales on the recording charts. This instrument should prove to be quite useful for many applications beyond those suggested here, and it will allow the chemist to put most of his efforts into utilization of the instrument, its data and interpretation rather than into operating details.

N. Miscellaneous Instruments

A number of special types of instrumentation are beyond the scope of this book, such as the ion beam microprobe, the cascade analyzer, a spectrometer for both positive and negative ions recorded simultaneously, combination isotope separator–mass spectrometers, etc. The reader who is interested in these instruments is referred to White's book (1968). The ion beam microprobe may eventually find application in organic chemistry, although at present it is primarily used on inorganic materials and on solids, in thin-film analysis, surface analyses of metallurgical samples, geology, semiconductor analysis, etc. The extension to organic coatings may well be under investigation at the time of writing this book since samples of pure graphite have been shown to contain CH_4^+ ions from sputtering of the surface. This technique discussed by the GCA Corp. will bear watching (Product Data Sheet, No. 113, 1969).

The instruments discussed here are only a few of many; hence, the reader is referred to some of the other texts listed in the appendix for additional material. A rather complete survey of available commercial instruments and their main characteristics has been recently compiled by Beynon and Fontaine (1967a,b, 1968).

IV. Ion Detection Methods

To complete the basic instrumentation necessary in the mass spectrometer, a means of ion detection and recording of ion intensities for a given set of m/e's is necessary. Although recording depends upon the ion detection method, it should be recognized that the two must be compatible and still suitable for the other parts of the instrument. The recorder response must be faster than the ion detector, and the ion detector must be able to present a "fresh" detection surface to the ion beam for each m/e; otherwise, the resolving power, sensitivity, and accuracy of intensity measurement and the speed of analysis of the resulting data may suffer. Commercial instrumentation has made much progress in these areas, and the details of the various systems are too numerous to consider here. Therefore, only a general picture of the basic methods will be discussed.

A. Fluorescent and Scintillating Screens

A screen which fluoresced when ions bombarded it was the first instrument for detecting positive-ion beams. These screens are still used in conjunction with photomultipliers (Mima and Kanetmatsu, 1965) and, after amplification with oscilloscopic display, for either detection or fine tuning of a specific region of the spectrum. Persistent scintillation screens find use with time-of-flight instruments where photographs of sections of spectra projected on such screens provide a permanent record. The lack of a permanent record with screens reduces their general usefulness.

B. Photoplates as Detectors

Photographic plates were used by Thomson after the discovery that ions could be detected by this means (Koenigsberger and Kutchewski, 1910). Many of the instruments used today still employ photographic plates despite the widely recognized problem of obtaining uniformly high sensitivity. In addition to the ions themselves, other factors that contribute to the image include stray light, secondary electrons produced in the process of stopping the ion beam, the breaking of the silver halide bonds in the impact process, and chemical effects due to the accumulation of neutralized ion beam material in the matrix of the emulsion. Variations in sensitivity of individual peaks may be attributed to this problem, but the use of photoplates does present some advantages. Among these are (a) integration of the entire spectrum from the viewpoint of total ion intensity and (b) the simultaneous recording of the spectrum over the whole range of m/e's covered. This is most important when a source, such as the high-voltage spark, has an output that changes randomly over short periods of time. For the high-resolution instrument it provides a means of physical measurement of differences in atomic mass units by means of the microcomparator. Photoplates also provide a means of long exposure, which can be used to detect very low concentrations of ions. Note that exposure times can be very long or very short since no recorder response time is involved. The use of the plate involves a series of exposures, and this in itself allows far easier intercomparison of spectra. Beynon (1960e) has discussed photoplates in detail. He claims that ions with intensities differing by a factor of 10^8 can be compared, and for ions of mass 200 and energy of 10,000 eV, an ion current of the order of 5×10^{-15} A can be detected.

Photoplates have a number of disadvantages. Chief among these is the fact that the plate must be in the high-vacuum system and must be protected from light. Leaks in the high-vacuum system can become quite difficult and time-consuming to correct despite the present design of rather good vacuum locks, etc. In addition, the plates must be degassed before any spectra are taken. Plates commonly in use will hold several individual spectra; hence, the loss of a plate means the loss of all of the spectra which have been recorded. The penetrating effects of different ions of different mass and energy may differ considerably and therefore the peak intensities may differ. The widths of the image produced vary along the plate, depending upon the type of instrument. The Mattauch geometry causes the image width to be proportional to the square root of the m/e. Corrections can be applied but can become extremely large in some cases. Geometrical discrimination can arise due to different trajectories followed by ions of different m/e's. It is often necessary to make several exposures because intensity measurements must be repeated to detect and eliminate errors from sensitivity changes. The sensitivity increases with the energy of the bombarding ions and is greatest for the lightest ions at a fixed accelerating potential. Surface charges can build up on the plate and may differ in local zones depending upon the intensities of the ions in focus on the plate at a given spot. Finally—the old nemesis of those working with emission spectroscopy and other users of photoplates—the grain size in the emulsion of the plate limits the resolving power.

C. The Faraday Cup As a Detector

It was noted long ago that a sensitive galvanometer was not adequate to detect ion currents below 10^{-9} A, which in itself is often the magnitude of the total ion current to be observed for a given sample. Hence, other electrical or electronic means had to be employed. Thomson was the first (Kiser, 1965c) to use a Faraday cup behind a parabolic slit as a detector for the intensities of positive-ion beams. Note that the photoplate or screen would indicate both positive and negative ions simultaneously but that here one would be restricted to either positive or negative ions on a given detector. A Wilson tilted electroscope was the current measuring device. The Faraday cup or tube is simply a metallic cylinder oriented so that the ion beam traverses the central axis before impingement occurs. The long cylinder minimizes the escape of any reflected ions and secondary electrons produced by ion impact. Shapes other than cylindrical are encountered, but they have the same purpose. An electron suppressor slit is often placed at the entrance of the cup to return to the collector any secondary electrons formed.

A dual collector system which operates independently of fluctuations in ion intensity has been reported by Nier *et al.* (1947).

1. The Electrometer Tube

Since very small ion currents are found at the detector, a means of amplification was sought. Bleakney (1932) replaced the Wilson tilted electroscope with an electrometer tube to give amplification sufficient to drive a galvanometer. This method finds wide application today not only to galvanometers but to various recording devices. A large resistance is placed between the Faraday cup and ground. The voltage drop so developed is impressed upon the grid of the first electrometer tube in the dc preamplifier. This approach demands wide-range, linear, and high-stability dc amplifiers. Alternatives to the electrometer have been developed in the form of the vibrating reed electrometer and some ac circuitry.

2. The Vibrating Reed Electrometer

A basic vibrating reed electrometer developed by Palevsky *et al.* (1947) has proven to be satisfactory for mass spectrometer use. The dc signal from the Faraday cup is imposed upon a vibrating reed condenser. The resulting ac signal is easily amplified and may be read by a meter or used to drive a recorder system. The unit is dependable and rugged, has a very small drift with time, and also has a dc noise level ten times less than the dc electrometer–amplifier systems described above.

Detection sensitivities of 10^{-17}–10^{-18} A have been reported by Inghram *et al.* (1947). Despite the necessary increases in complexity of electronics required, the gain in sensitivity and the stability of the unit proves to be of great value in this application.

D. The Electron Multiplier Detector

The electron multiplier detector collects energetic ions on the metal surface of an electrode (called the cathode in this apparatus). Secondary electrons are emitted and accelerated to another electrode (referred to as a dynode) where additional electrons are emitted. These in turn are accelerated to another dynode (a second stage), and the process is repeated. After about 10–12 stages, all of the electrons are collected on an anode which has a grid protecting it from losing any further secondary electrons. A large electron current can be obtained for each positive ion collected on the cathode, resulting in about a 10^5 or 10^6 multiplication of the signal. Cohen (1943) appears to have been the first to employ such an electron multiplier for ion detection in the mass spectrometer. The use of the electron multiplier requires a large negative potential on the cathode with the anode near or at ground potential. For positive ions, this adds to the energy of the ion beam and causes greater electron emission from the cathode.

Response is rapid, and high sensitivity can be obtained. The use for ion current and abundance measurements is not as satisfactory as may be desired. The difficulty is with amplification in a vacuum which may not be reproducible and where the presence of gases from the collected ions changes with time and sample. It is probable that surface reactions of ions with the cathode also alter day-to-day sensitivities. Foster and Liao (1968) have observed such behavior during studies of organic sulfur compounds.

Magnetic fields also exert a serious effect upon amplification. Thus, modern instruments which allow for magnetic scanning might cause serious problems with electron multiplier detection methods. In some instruments an auxiliary constant magnetic field is applied to produce a more stable operation. In others a Wien filter is used to produce a linear collimated ion beam which encounters the cathode outside of the magnetic field. In still other instruments, a shielding from the magnetic field is accomplished. For the time-of-flight instrument, the Wiley (1956) magnetic electron multiplier was developed. A planar ion cathode was selected by Wiley and McLaren (1955) because it eliminates variations in ion transit time that were experienced with the curved cathode employed in the conventional electrostatic electron multipliers described above. The Wiley multiplier allows the "gating" out of selected ion signals to a separate anode section, permitting the remaining ions to pass on to the anode of the oscilloscope and to produce the spectrum of the material minus the given m/e(s) gated out of the signal. This detector can be used in monitoring directly introduced gas–liquid chromatographic fractions containing the carrier gas. For example, the helium peak can be gated out for fractions where helium is the carrier gas. This prevents not only a very large number of ions from reaching the spectrum detector, which would cause a build-up of charge, but also a momentary overloading of the detector, which must receive other ion signals in a matter of microseconds or less.

REFERENCES

Alvarez, R., Paulsen, P. J., and Kelleher, D. E. (1969). *Anal. Chem.* **41**, 955.
Aston, F. W. (1919). *Phil. Mag.* **38**, 707.
Aston, F. W. (1927). *Proc. Roy. Soc. London* **A115**, 487.
Aston, F. W. (1937). *Proc. Roy. Soc. London* **A163**, 391.
Bainbridge, K. T., and Jordan, E. B. (1936). *Phys. Rev.* **50**, 282.
Barber, N. F. (1933). *Proc. Leeds Phil. Lit. Soc., Sci. Sect.* **2**, 427.
Bartky, W., and Dempster, A. J. (1933). *Phys. Rev.* **33**, 1019.
Beckey, H. D. (1959). *Z. Anal. Chem.* **170**, 359.
Beckey, H. D. (1962). *Z. Naturforsch. A* **17**, 1103.
Beckey, H. D. (1963). *In* "Advances in Mass Spectrometry" (R. M. Elliott, ed.), Vol. II, pp. 1–24. Pergamon, Oxford.
Beckey, H. D. (1968). *Int. J. Mass Spectrom. Ion Phys.* **1**, 93.
Beckey, H. D. (1969). *Angew. Chem.* **8**, 623.

Beckey, H. D., and Hey, H. (1968). *Org. Mass Spectrom.* **1**, 47.
Beckey, H. D., Knoppel, H., Metzinger, G., and Schulze, P. (1966). *In* "Advances in Mass Spectrometry" (W. L. Mead, ed.), Vol. III, pp. 35–67. Institute of Petroleum, London.
Beckey, H. D., Heising, H., Hey, H., and Metzinger, H. G. (1968). *In* "Advances in Mass Spectrometry" (E. Kendrick, ed.), Vol. IV, pp. 817–831. Institute of Petroleum, London.
Bennett, W. H. (1950). *J. Appl. Phys.* **21**, 143.
Bennett, W. H. (1953). *Nat. Bur. Stand. U. S. Circ. No.* 522, 111.
Beynon, J. H. (1959). *In* "Advances in Mass Spectrometry" (J. D. Waldron, ed.), pp. 328–354. Pergamon, Oxford.
Beynon, J. H. (1960a). "Mass Spectrometry and Its Applications to Organic Chemistry," p. 5. Elsevier, Amsterdam.
Beynon, J. H. (1960b). "Mass Spectrometry and Its Applications to Organic Chemistry," pp. 108–111. Elsevier, Amsterdam.
Beynon, J. H. (1960c). "Mass Spectrometry and Its Applications to Organic Chemistry," p. 118. Elsevier, Amsterdam.
Beynon, J. H. (1960d). "Mass Spectrometry and Its Applications to Organic Chemistry," p. 121. Elsevier, Amsterdam.
Beynon, J. H. (1960e). "Mass Spectrometry and Its Applications to Organic Chemistry," pp. 196–197. Elsevier, Amsterdam.
Beynon, J. H., and Fontaine, A. E. (1967a). *Instrum. Rev.* **14**, 470.
Beynon, J. H., and Fontaine, A. E. (1967b). *Instrum. Rev.* **14**, 501.
Beynon, J. H., and Fontaine, A. E. (1968). *Instrum. Rev.* **15**, 34.
Beynon, J. H., Fontaine, A. E., Turner, D. W., Williams, A. E. (1967). *J. Sci. Instrum.* **44**, 283.
Bleakney, W. (1932). *Phys. Rev.* **40**, 496.
Bleakney, W., and Hipple, J. A. (1938). *Phys. Rev.* **53**, 521.
Block, J. (1968). *In* "Advances in Mass Spectrometry" (E. Kendrick, ed.), Vol. IV, pp. 791–816. Institute of Petroleum, London.
Bonazzola, G. C., and Chiavassa, E. (1964). *Nucl. Instrum. Methods* **27**, 41.
Bondy, H., and Popper, K. (1933). *Ann. Phys.* (*Leipzig*) **17**, 425.
Brown, R., and Wolstenholme, W. A. (1964). *Nature* **201**, 598.
Brown, R., Craig, R. D., and Elliott, R. M. (1963). *In* "Advances in Mass Spectrometry" (R. M. Elliott, ed.), Vol. II, p. 141. Pergamon, Oxford.
Brubaker, W. M. (1964). *Rev. Sci. Instrum.* **35**, 1007.
Brunnée, C. (1967). *Z. Naturforsch.* **22**, 121.
Brunnée, C., Delgmann, L., and Kronenberger, K. (1964). *Vak.-Tech.* **13**, 35.
Brunnée, C., Kappus, G., and Maurer, K. H. (1967). *Z. Anal. Chem.* **232**, 17.
Cermák, V., and Herman, Z. (1961). *Nucleonics* **19**, 106.
Chait, E. M., Shannon, T. W., Amy, T. W., and McLafferty, F. W. (1968). *Anal. Chem.* **40**, 835.
Chait, E. M., Shannon, T. W., Perry, W. O., van Lear, G. E., and McLafferty, F. W. (1969). *Int. J. Mass Spectrom. Ion Phys.* **2**, 141.
Chakravarty, B., Venkatasubramanian, V. S., and Duckworth, H. E. (1963). *In* "Advances in Mass Spectrometry" (R. M. Elliott, ed.), Vol. II, p. 128. Pergamon, Oxford.
Chastagner, P. (1969). *Anal. Chem.* **41**, 796.
Clark, C. F., ed. (1960). "Encyclopedia of Spectroscopy." Van Nostrand-Reinhold, Princeton, New Jersey.
Classen, J. (1907). *Jahrb. Hamburg, Wiss. Anst., Beih.*
Classen, J. (1908). *Phys. Z.* **9**, 762.
Cohen, A. (1943). *Phys. Rev.* **63**, 219.

Crouch, E. A. C. (1963). *In* "Advances in Mass Spectrometry" (R. M. Elliott, ed.), Vol. II, p. 157. Pergamon, Oxford.
Curtis, H. M. (1970). Unpublished results. Continental Oil Co. Ponca City, Oklahoma.
Dempster, A. J. (1918). *Phys. Rev.* **11**, 316.
Dempster, A. J. (1935a). *Proc. Amer. Phil. Soc.* **75**, 755.
Dempster, A. J. (1935b). *Nature* **135**, 542.
Dempster, A. J. (1936). *Rev. Sci. Instrum.* **7**, 46.
Elder, F. A., Giese, C. F., Steiner, B., and Inghram, M. G. (1962). *J. Chem. Phys.* **36**, 3292.
Elliott, R. M., Craig, R. D., and Errock, G. A. (1961). *In* "Instruments and Measurements" (H. Von Koch and G. Ljungberg, eds.), p. 271. Academic Press, New York.
Fabry, C., and Perot, A. (1900a). *Acad. Sci.* **130**, 406.
Fabry, C., and Perot, A. (1900b). *J. Phys.* **9**, 381.
Fedorenko, N. V. (1954). *Zh. Tekh. Fiz.* **24**, 769.
Field, F. H. (1968). *Accounts Chem. Res.* **1**, 42.
Folmer, O. F., Jr., and Azarraga, L. V. (1969). *In* "Advances in Chromatography" (A. Zlatkis, ed.), p. 216. Preston Technical Abstracts, Evanston, Illinois.
Foner, S. N., and Hudson, R. L. (1962). *J. Chem. Phys.* **37**, 1662.
Foster, N. G. (1965). Unpublished observations. U. S. Bureau of Mines, Petroleum Research Center, Bartlesville, Oklahoma.
Foster, N. G., and Liao, J. P. (1968). Unpublished observations. Texas Woman's Univ., Denton, Texas.
Fox, R. E., and Hickam, W. M. (1951). *Phys. Rev.* **84**, 859.
Fox, R. E., Hickam, W. M., Grove, D. J., and Kjeldaas, T., Jr. (1955). *Rev. Sci. Instrum.* **26**, 1101.
Good, R. H., Jr., and Müller, E. W. (1956). "Handbuch der Physik," 2nd Ed., Vol. 21, p. 176. Springer-Verlag, Berlin.
Goode, G. C., O'Malley, R. M., Ferrer-Correia, A. J., and Jennings, K. R. (1970). *Nature* **227**, 1093.
Gorman, J. G., Jones, E. J., and Hipple, J. A. (1951). *Anal. Chem.* **23**, 438.
Grodzins, L., Rose, P. H., and Van Degraff, R. J. (1965). *Nucl. Instrum. Methods* **36**, 202.
Harrington, D. B. (1960). *In* "Encyclopedia of Spectroscopy" (C. F. Clark, ed.), pp. 628–647. Van Nostrand-Reinhold, Princeton, New Jersey.
Harrington, D. B., and Gohlke, R. S. (1962). Paper No. 42, Tenth Annual Conference on Mass Spectrometry, ASTM Committee E-14, New Orleans, Louisiana.
Herzog, R. F. K. (1934). *Z. Phys.* **89**, 447.
Hess, D. C., Whetherill, G., and Inghram, M. G. (1951). *Rev. Sci. Instrum.* **22**, 838.
Hipple, J. A., and Bleakney, W. (1936). *Phys. Rev.* **49**, 884.
Hipple, J. A., Sommer, H., and Thomas, H. A. (1949). *Phys. Rev.* **76**, 1877.
Hodgson, F. N., Desjardins, M., and Baun, W. L. (1963). *J. Phys. Chem.* **67**, 1250.
Honig, R. E. (1963). *Appl. Phys. Lett.* **3**, 8.
Honig, R. E. (1964). Paper No. 38, Twelfth Annual Conference on Mass Spectrometry, ASTM Committee E-14, Montreal, Canada.
Honig, R. E., and Woolston, J. R. (1963). *Appl. Phys. Lett.* **2**, 138.
Hurzeler, H. M., Inghram, M. G., and Morrison, J. D. (1957). *J. Chem. Phys.* **27**, 313.
Hurzeler, H. M., Inghram, M. G., and Morrison, J. D. (1958). *J. Chem. Phys.* **28**, 76.
Inghram, M. G. (1953). *J. Phys. Chem.* **57**, 809.
Inghram, M. G., and Gomer, R. (1954). *J. Chem. Phys.* **22**, 1279.
Inghram, M. G., and Gomer, R. (1955a). *J. Amer. Chem. Soc.* **77**, 500.
Inghram, M. G., and Gomer, R. (1955b). *Z. Naturforsch. A* **10**, 863.
Inghram, M. G., Hayden, R. J., and Hess, D. C. (1947). *Phys. Rev.* **72**, 349.

Johnson, E. G., and Nier, A. O. (1953). *Phys. Rev.* **91**, 10.
Katzenstein, H. S., and Friedland, S. S. (1955). *Rev. Sci. Instrum.* **26**, 616.
Kendrick, E., ed. (1968). "Advances in Mass Spectrometry," Vol. IV, p. 971. Institute of Petroleum, London.
Kirschner, F. (1954). *Naturwissenschaften* **41**, 136.
Kiser, R. W. (1965a). "Introduction to Mass Spectrometry and Its Applications," pp. 26–30. Prentice-Hall, Englewood Cliffs, New Jersey.
Kiser, R. W. (1965b). "Introduction to Mass Spectrometry and Its Applications," p. 37. Prentice-Hall, Englewood Cliffs, New Jersey.
Kiser, R. W. (1965c). "Introduction to Mass Spectrometry and Its Applications," p. 59. Prentice-Hall, Englewood Cliffs, New Jersey.
Kiser, R. W. (1965d). "Introduction to Mass Spectrometry and Its Applications," pp. 86–89. Prentice-Hall, Englewood Cliffs, New Jersey.
Knewstubb, P. F., and Tickner, A. W. (1962a). *J. Chem. Phys.* **36**, 674.
Knewstubb, P. F., and Tickner, A. W. (1962b). *J. Chem. Phys.* **36**, 684.
Koenigsberger, J., and Kutchewski, H. (1910). *Phys. Z.* **11**, 666.
Leger, E. G. (1955). *Can. J. Phys.* **33**, 74.
Lewellyn, P. (1965). *Varian Associates Technical Information Bulletin*, fall issue; see also *Chem. Eng. News*, 55 (1965).
Lindholm, E. (1953). *Proc. Phys. Soc. London, Sect. A* **66**, 1068.
Lossing, F. P. (1954a). *J. Chem. Phys.* **27**, 621.
Lossing, F. P. (1954b). *J. Chem. Phys.* **27**, 1489.
Lossing, F. P. (1956). *Can. J. Chem.* **34**, 701.
Lossing, F. P. (1957). *Can. J. Chem.* **35**, 305.
Lossing, F. P., and Tanaka, I. (1956). *J. Chem. Phys.* **25**, 1031.
Lozier, W. W. (1930a). *Phys. Rev.* **36**, 1285.
Lozier, W. W. (1930b). *Phys. Rev.* **36**, 1417.
Mattauch, J. (1936a). *Phys. Rev.* **50**, 617.
Mattauch, J. (1936b). *Phys. Rev.* **50**, 1089.
Mattauch, J., and Herzog, R. F. K. (1934). *Z. Phys.* **89**, 786.
Mead, W. L. (1968). *Anal. Chem.* **40**, 743.
Millikan, R. A., Bowen, I. S., and Sawyer, R. A. (1921). *Astrophys. J.* **53**, 150.
Mima, H., and Kanetmatsu, F. (1965). *Mem. Fac. Eng. Osaka City Univ.* **7**, 179.
Müller, E. W. (1953). *Ergeb. Exakt Naturwiss.* **27**, 290.
Müller, E. W. (1960). *In* "Advances in Electronics and Electron Physics" (L. Marton, ed.), Vol. 13, p. 83. Academic Press, New York.
Munson, M. S. B., and Field, F. H. (1966). *J. Amer. Chem. Soc.* **88**, 2621.
Muschlitz, E. E., Randolph, H. D., and Ratti, J. N. (1961). *Rev. Sci. Instrum.* **33**, 445.
Namioka, T. (1954). *Sci. Light (Tokyo)* **3**, 15.
Nielsen, H. (1968). *In* "Advances in Mass Spectrometry" (E. Kendrick, ed.), Vol. IV, pp. 267–274. Institute of Petroleum, London.
Nier, A. O. (1940). *Rev. Sci. Instrum.* **11**, 212.
Nier, A. O., Ney, E. P., and Inghram, M. G. (1947). *Rev. Sci. Instrum.* **18**, 294.
Palevsky, H., Swank, R. K., and Grenchik, R. (1947). *Rev. Sci. Instrum.* **18**, 298.
Paul, W., and Steinwedel, H. (1953). *Z. Naturforsch. A* **8**, 448.
Phillips, L. F., and Schiff, H. I. (1962a). *J. Chem. Phys.* **36**, 1509.
Phillips, L. F., and Schiff, H. I. (1962b). *J. Chem. Phys.* **36**, 3282.
Phillips, L. F., and Schiff, H. I. (1962c). *J. Chem. Phys.* **37**, 1233.
Robertson, A. J. B., and Williams, P. (1968). *In* "Advances in Mass Spectrometry" (E. Kendrick, ed.), Vol. IV, pp. 847–856. Institute of Petroleum, London.

Robinson, C. F., and Hall, L. G. (1956). *Rev. Sci. Instrum.* **27**, 504.
Robinson, C. F., and Sharkey, A. G. (1958). *Rev. Sci. Instrum.* **29**, 250.
Robinson, C. F., Perkins, G. D., and Bell, N. W. (1961). *In* "Instruments and Measurements" (H. Von Hock and G. Ljungberg, eds.), p. 261. Academic Press, New York.
Ryan, K. R., and Green, J. H. (1965). *J. Sci. Instrum.* **42**, 455.
Schuddemage, H. D. R., and Hummel, D. O. (1968). *In* "Advances in Mass Spectrometry" (E. Kendrick, ed.), Vol. IV, pp. 857–866. Institute of Petroleum, London.
Schulze, P., Simoneit, B. R., and Burlingame, A. L. (1969). *Int. J. Mass Spectrom. Ion Phys.* **2**, 183.
Seya, M. (1952). *Sci. Light (Tokyo)* **2**, 98.
Sharkey, A. G., Schultz, J. L., and Friedel, R. A. (1964). Paper No. 63, Twelfth Annual Conference on Mass Spectrometry, ASTM Committee E-14, Montreal, Canada.
Smith, D. and Cromey, P. R. (1968). *J. Sci. Instrum.* Series 2, 523.
Smythe, W. R., and Mattauch, J. (1932). *Phys. Rev.* **40**, 429.
Sommer, H., Thomas, H. A., and Hipple, J. A. (1951). *Phys. Rev.* **82**, 697.
Steiner, B., Giese, C. F., and Inghram, M. G. (1961). *J. Chem. Phys.* **34**, 189.
Stephens, W. E. (1946). *Phys. Rev.* **69**, 691.
Stephens, W. E., and Hughes, A. L. (1934). *Phys. Rev.* **45**, 123.
Stewart, D. T. (1966). *Nucl. Instrum. Methods* **45**, 341.
Tate, J. T., and Lozier, W. W. (1932). *Phys. Rev.* **39**, 254.
Terenin, A., and Popov, B. (1932a). *Z. Phys.* **75**, 338.
Terenin, A., and Popov, B. (1932b). *Phys. Z. Sowjetunion* **2**, 299.
Testerman, M. K., Raible, R. W., Gilliland, B. E., Williams, J. R., and Grimes, G. B. (1965). *J. Appl. Phys.* **36**, 2939.
Thirion, B. (1966). *Method. Phys. Anal. Ja.–Mr.*, 62.
Thomson, J. J. (1911). *Phil. Mag.* **21**, 225.
Vilesov, F. I., and Akopyan, M. E. (1962). *Prib. Tekh. Eksp.* **7**, 145.
von Zahn, U. (1963). *Rev. Sci. Instrum.* **34**, 1.
von Zahn, U., Gebauer, S., and Paul, W. (1962). Paper No. 41, Tenth Annual Conference on Mass Spectrometry, ASTM Committee E-14, New Orleans, Louisiana.
Wanless, G. G. (1968). *In* "Advances in Mass Spectrometry" (E. Kendrick, ed.), Vol. IV, pp. 833–845. Institute of Petroleum, London.
Watanabe, K., Nakayama, T., and Mottl, J. (1962). *J. Quant. Spectrosc. Radiat. Transfer* **2**, 369.
Weiss, M. J., and Hutchison, D. A. (1968). *J. Chem. Phys.* **48**, 4386.
White, F. H. (1968). "Mass Spectrometry in Science and Technology," 352 pp. Wiley, New York.
Wiley, W. C. (1956). *Science* **124**, 817.
Wiley, W. C., and McLaren, I. H. (1955). *Rev. Sci. Instrum.* **26**, 1150.
Winter, D. L., and Azarraga, L. V. (1969). Unpublished work. Continental Oil Co., Ponca City, Oklahoma.
Wolff, M. M., and Stephens, W. E. (1953). *Rev. Sci. Instrum.* **24**, 616.
Woolston, J. R., and Honig, R. E. (1964). *Rev. Sci. Instrum.* **35**, 69.
Yurachek, J. P., Clemena, G. G., and Harrison, W. W. (1969). *Anal. Chem.* **41**, 1666.

CHAPTER 3

THE MASS SPECTRUM

I. Definitions

A. The Mass Spectrum

The mass spectrum of a substance consists of a record of experimentally determined ion intensities for a set of ion masses. In practice, often only a partial mass spectrum is presented from a given experiment. The data obtained are essentially an *xy* plot.

B. The m/e or m/z

The m/e, m/z, or ion masses occur at some distance, x, from a zero or scan starting point on some recording device as discussed in Chapter 2. The distance depends upon either (a) time, (b) a time-dependent variation of magnetic field strength, or (c) a time-dependent variation of electrostatic field strength. Using a microcomparator on photoplates may cause the record to be in microns.

With a mechanical chart type of recorder, the recorder paper speed can also enter the distance term. The mass scale may be regarded as the x variable.

For many years the use of m/e has dominated the literature, even though as Kiser (1965a) and this treatise have pointed out the m/z term is the more correct one. Therefore, it will be well for the reader to accustom himself to seeing the "mass" listed in the literature in both forms. For the old-timers, it is a difficult task to change over to m/z, and the authors feel that this double usage may continue for a long time because of the ingrained use of m/e. Therefore, even in this work, both forms will be used, more or less interchangeably.

C. The Ion Intensity or Ion Current

The ion intensities or, more correctly, the ion current intensities are commonly referred to as peak intensities or just intensities. This y variable of the spectrum may often appear as arbitrary chart units, such as divisions, millimeters, inches, digitizer printout units, etc., or as amperes, microamperes, etc. The arbitrary nature of the y variable is unimportant, as will be brought out in a later section of this chapter.

D. The "Background" Spectrum

The "background" is the term used in actual operation for the mass spectrum scan obtained as a "blank" before the sample mass spectrum is obtained. Often a given instrument will show very small "background" peaks that arise from trace quantities of materials not completely pumped out of the instrument from previously run samples. Low-volatility materials in the source or introduction system are the likely cause. If one is examining many samples of the same type, paraffin hydrocarbons, for example, a group of low-intensity peaks may be observed as background from ions at m/e 27, 29, 41, 43, 55, 57, 69, 71, etc., which are typical of the major ions found in hydrocarbons. One also often sees peaks arising from traces of air components present either from a slight leak or from air introduced with the samples. Water, carbon dioxide, and even argon are often noted, as are "artifacts" from various sources (see Chapter 4). If ultimate accuracy is desired, the peaks observed in such a "background" spectrum should be subtracted from the sample peaks.

II. Recording the Mass Spectrum

The actual recording of the mass spectrum, which one can regard as raw experimental data, is accomplished by using either display-type records or mechanical records. There are advantages and disadvantages in using both types of records.

A. Display-Type Records

The major display types of records are photographic, oscilloscopic, oscillographic, and pen and ink recorders.

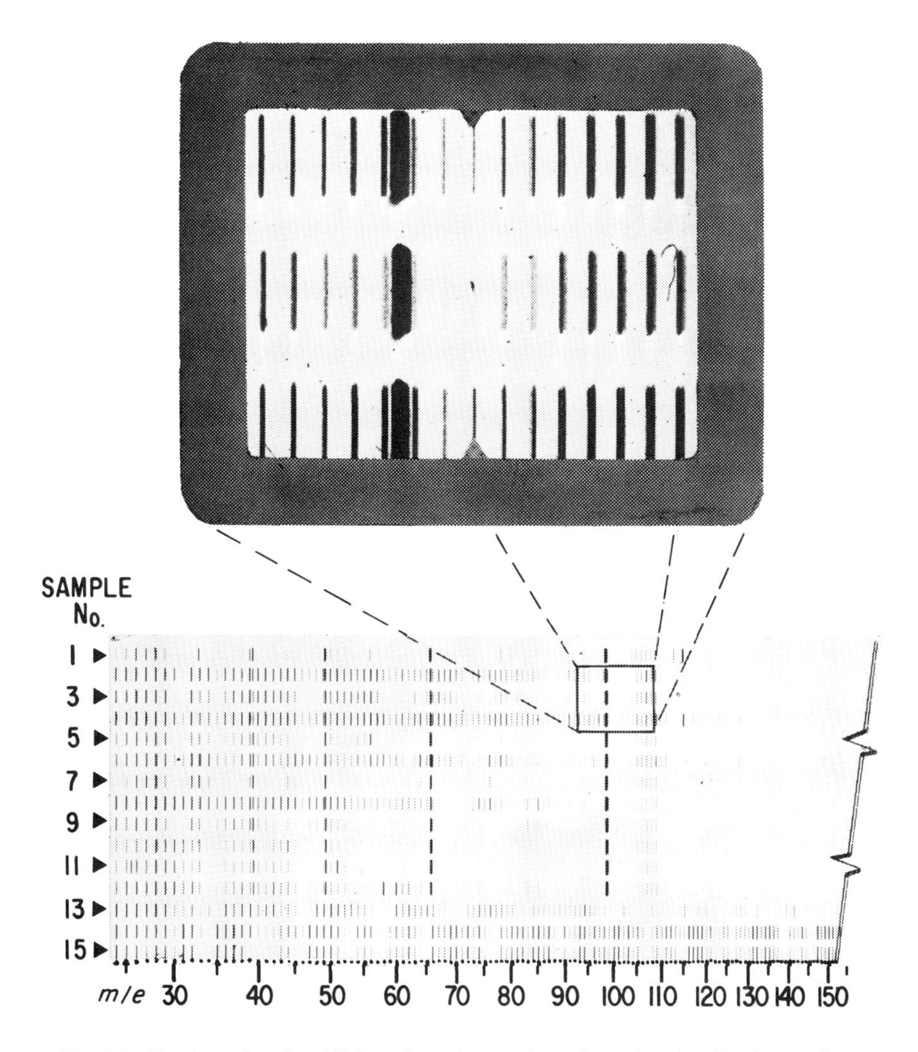

Fig. 3-1. Portion of a (2 × 12)-in. photoplate; enlarged portion is what is seen by the microcomparator. (Courtesy of U.S. Bureau of Mines.)

1. Photoplates

An example of a portion of a photoplate is shown in Fig. 3-1. It should be noted that the x variable would have to be measured in distance in some units from some arbitrary reference point. True, the user can calculate which ion is in focus, but practice has shown that it is far more practical to include a recording on the same photoplate of substances containing ions of exact known mass. A number of materials are available for such "internal" mass standards. In positions 13–15 on the spectral plate are typical mass markers used in spark source mass spectrometry. They are Convalex-10, Fluorolube LG, and Kel-F 200 wax, respectively. It is assumed that successive scans on the photoplate will reproduce a given m/e at exactly the same distance x from the start of the scan point. Thus the use of photoplates usually requires a precision distance measurement by means of a microcomparator.

The y variable or peak intensity is a measurement of optical density for the photoplates. The usual procedures apply as developed by emission spectroscopists a generation or more ago. Kennicott (1965) has listed in detail some of the problems the mass spectroscopist encounters in trying to obtain plates sensitive to low-energy ions where a low background and overall high sensitivity of emulsion are essential to give the highest possible signal-to-noise ratio. He has also pointed out the need for a well-controlled characteristic curve to permit quantitative interpretation of the results. Schuy and Franzen (1962) have made an assessment of photographic spectra evaluation methods for spark source mass analysis, while Ahearn and Malm (1966) have considered background reduction in photoplates for mass spectrometric work.

Because of the optical density measurement, absolute blackening curves of the photographic materials must be obtained for mass spectrometric use. Rudloff (1961) summarized the theory and aspects of this, while Wagner (1964) and Wagner and Mai (1964) have discussed the processes associated with plates exposed to ions of different masses having energies of 250–2500 eV and 2.5–20 keV. Reproducibility in developer conditions is important. Commercially available developers are made by a number of suppliers, but McCrea (1965) has described a simple unit that can be assembled from standard laboratory baths. McCrea (1971) has also updated consideration of the problem of ion-sensitive plate detectors in mass spectrometry by making comparisons of other detectors with the Q-2 plate and Mattauch–Herzog geometry instruments used for organic structure determinations.

An inspection of Fig. 3-1 shows that the intensities of some ions are quite strong; others are too weak to be observed in the lower reproduction but can be seen in the enlargement, typical of what can be seen in the optical microcomparator. Also, imperfections in the processing show up as specks in the emulsion (e.g., pieces of dust incorporated in the emulsion) all of which are integrated by the scanning photometer. The large, intensely dark band just

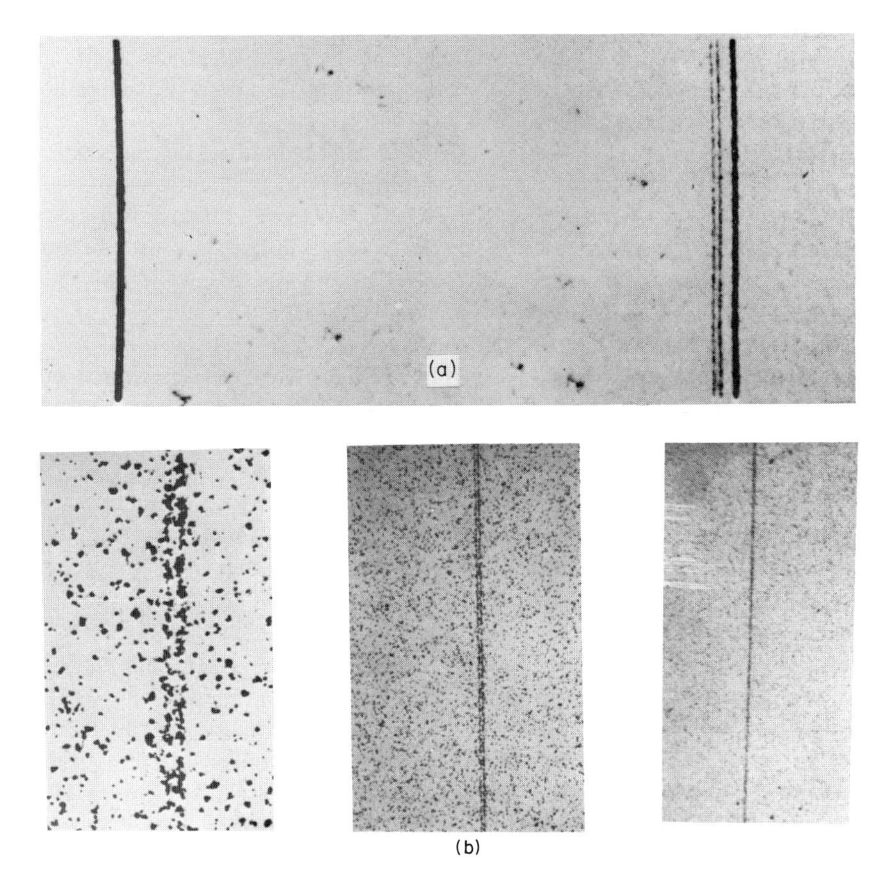

Fig. 3-2. Enlargements of lines from several photoplates. (a) A photoplate from a medium-resolution instrument. Lines from left to right are $C_2H_4^+$, N_2^+, and CO^+, all nominally mass 28; the single line at the right is $C_2H_3^+$ at mass 27. (b) A photoplate doublet on a high-resolution instrument at enlargement values, from left to right, of ×52, ×130, and ×520. The doublet is dimethyl-*o*-nitroanaline ($m/e = 166.07422$) and fluorene ($m/e = 166.07825$). [(a) and (b) courtesy of Consolidated Electrodynamics Corp.]

to the left of the indicator points is the doubly charged gold ion at m/e 98.495 (Au^{2+}), which was used as the electrode to hold the sample. Also apparent on the enlargement are doublets of a number of lines. At 103 on the upper spectrum is an obvious doublet about 0.1 mass unit apart. The more intense doublets further to the right obviously overlap. Overall fogging of the plate is not apparent to the eye and shows only as an increased background level when the recording photometer is run. The gross differences in ion intensity can obviously be "eyeballed," but most individuals do not possess an innate ability to include the necessary logarithm function in their "eyeballing." It is also apparent in this figure how a given ion intensity is spread out through the emulsion along both the x and vertical distances as the ion intensity increases. The entire 12-in. plate runs out to about a mass of 800, but the portion shown here is most instructive. The multiply charged gold ions are very intense: Au^{3+} occurs at 65.65, Au^{4+} is at 49.25, Au^{5+} is apparent between 39 and 40, Au^{6+} can be seen at 31.88, and Au^{7+} appears at 28.14. Using the microcomparator, one can see lines for Au^{11+}, which is reasonable in light of the d electrons present in the gold atom. In this case the gold electrode provides its own internal mass marker but actually is only convenient in the lower mass range.

Sample 11 is also of interest, showing a peak of 25.5 due to the V^{2+} ion arising from the sample of vanadyl porphyrin. The V^+ ion at m/e 51 is obvious, while the vanadyl group apparently is lost independently since an ion can be seen at m/e 67. Sample 12 is a nickel porphyrin and shows the presence of the nickel at m/e 58–64 corresponding to its isotopic composition.

In Fig. 3-2a are three lines from $C_2H_4^+$, N_2^+, and CO^+, all nominally mass 28, at the left of a plate from a somewhat higher-resolution instrument. The line at the right is $C_2H_3^+$ at mass 27. The background problem is quite apparent here. In Fig. 3-2b are spectra from a high-resolution instrument at three different values of enlargement. The doublet is a nominal 40,000 $M/\Delta M$ from dimethyl-*o*-nitroanaline ($m/e = 166.07422$) and fluorene ($m/e = 166.07825$). To the uninitiated, these show the problem of graininess in the emulsion, background, etc., as well as the difficulties with optical density measurements.

The practice with photoplates is to use a microcomparator with a recording densitometer facility. Those in use with emission spectrographic work have been improved by the needs of mass spectrometry. Some of the high-resolution instruments now available have optional recording by electronic means in addition to the photoplate. Conversely, the popularity of time-of-flight instruments for studying fast reactions has led to a report by Moulton and Michael (1965) of a method for recording well-defined and complete mass spectra as a function of time. The impact of mass spectrometry has also produced a commercially available automatic mass spectrum data system (Consolidated Electrodynamics Corp., CEC-21-087), which records mass spectral data of m/e and ion intensity from photoplates.

2. The Oscilloscope

An oscilloscopic trace normally produces no permanent record, which is one of its major drawbacks. However, screens with high persistency are available. A photograph made from an oscilloscope is shown in Fig. 3-3. If the oscilloscope horizontal sweep is set to include all possible m/e's on the x axis, the crowding of peaks becomes the main problem; i.e., the apparent resolution is lost in trying to get the entire record on the scope plate. Similarly,

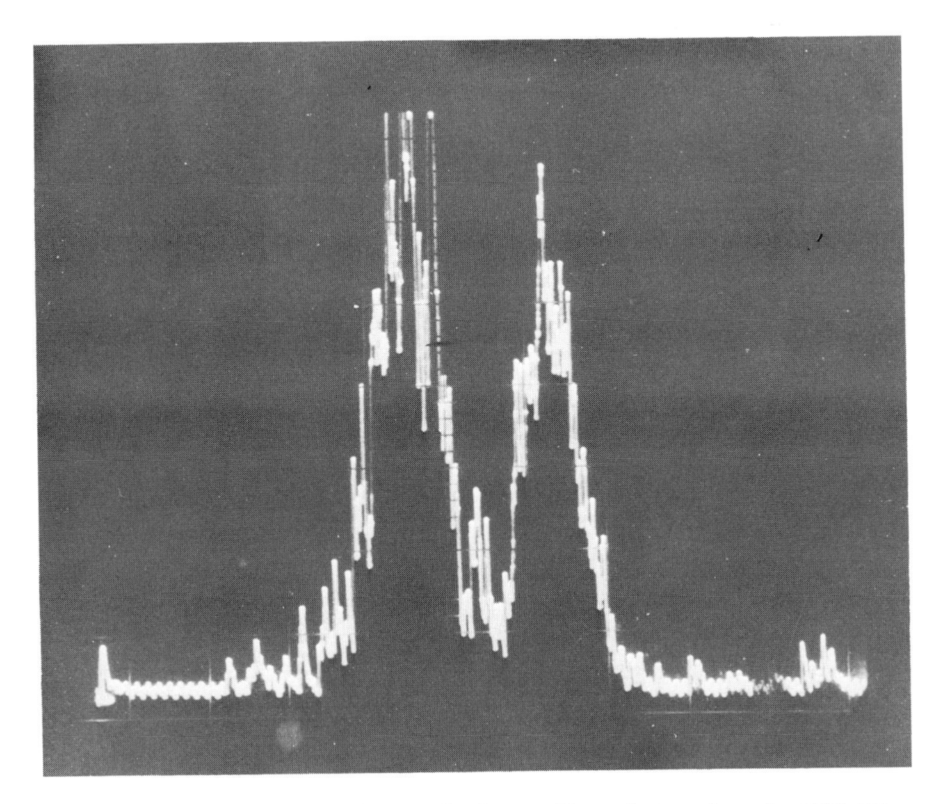

Fig. 3-3. A typical Polaroid photograph of an oscilloscopic trace from the CEC-110. The doublet is C_6H_6 and C_6H_4D ($M/\Delta M = 50{,}500$). (Courtesy of Consolidated Electrodynamics Corp.)

if the *y*-axis controls are set to show the most intense peaks, then the small peaks are lost in the trace of background noise level. Most mass spectroscopists utilize the oscilloscope as an auxiliary recorder for specific regions of the mass spectrum or, perhaps more importantly, as a display device during fine tuning of the instrument to bring out the best possible conditions for a given range of the spectrum. The oscilloscope also finds application in fast-reaction studies where other types of recorders are not usable and for circuitry "trouble-shooting" where it permits a three-dimensional display by use of a *z* axis which can provide considerable insight in some problems.

3. *Oscillographs*

Recording galvanometer traces on a moving photographic paper roll is a popular means of recording the spectrum. An example is shown in Fig. 3-4.

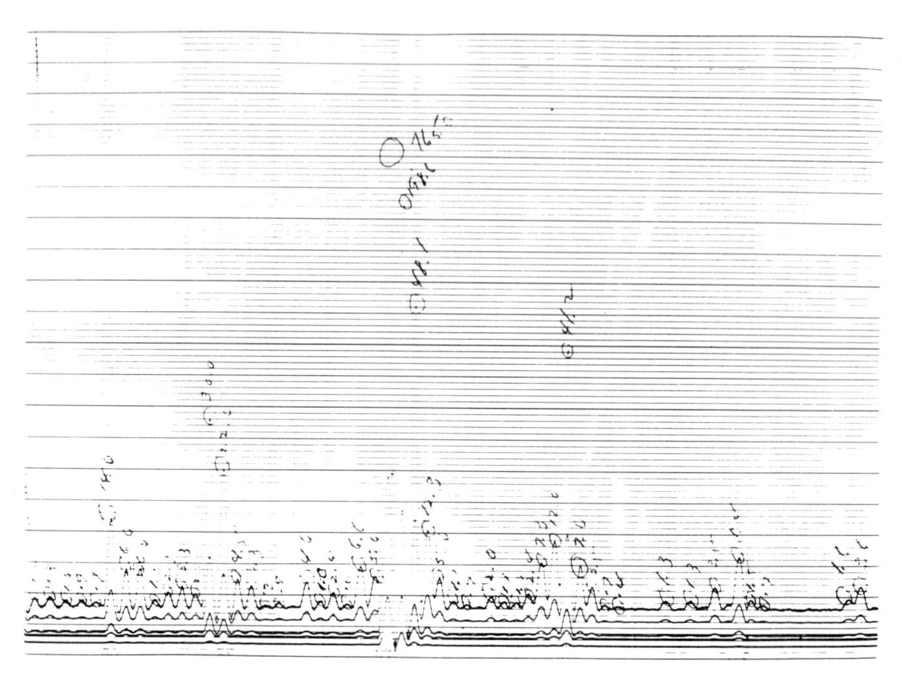

Fig. 3-4. A typical photographic paper record made on a CEC-21-103. The material is 2-*n*-hexylthiophene. Peak intensities are picked by measuring each individual peak and multiplying by the appropriate galvanometer attenuation.

Various instruments have from one to five galvanometer tracings on the chart with attenuation factors varying from 1 to 1000. As a permanent record the photographic method on paper is very satisfactory. (The spectrum can be "eyeballed," but only after development.) Likewise, if properly developed, fixed, washed, and dried, the record can be stored for a long period of time. Such a spectrum can be folded for filing, but admittedly this takes up considerable space in today's data-storage space-conscious world. This type of spectrum has the advantage of permitting the interpreter to make notes on the original, thus automatically filing the notes with the spectrum for future reference. Portions of such spectra can always be traced for use in publications or slide presentation of data. Specific instruments are often equipped with devices that put a fiduciary mark on the spectral chart to indicate the m/e. This is usually done relative to scanning voltages or in some cases on a standard timed basis. Just as in the case of the photoplates, added mass markers can be used; however, in this case the material is charged (usually through a liquid inlet) while the sample is still in the source or inlet and the marker allowed to enter the source region followed by a rescan of the sample plus marker. Even for "probe" inlets and gas–liquid chromatographic effluent direct introduction systems, a suitable mass marker may usually be stored in the inlet system for use as needed. Perfluorokerosenes with relatively noninterfering peaks are usually preferred because of their high volatility and hence low contribution to the background. The Convalex-10 (a mixture of polyphenylethers) leaves much to be desired as do numerous other compounds. Marking kits are commercially available (Varian Associates) and contain sufficiently volatile compounds to give, with considerable accuracy, mass marking out to *m/e* 793. Just as in the case of the photoplates, mass marking by chemical means is accomplished as an alternative to precision distance measurements. Quite often peaks in the vicinity of known ion species can be identified by measurements of distance and the knowledge that certain interfering peaks cannot be present in the material being examined.

A more recent development in oscillography has been light-sensitive paper which does not require development but which may be optionally developed to preserve a permanent recording of the spectrum. The spectra look similar to the ones from the heavier paper used in the regular photographic recording referred to above, but the paper is considerably thinner in texture. It has the advantage of allowing the mass spectroscopist to "eyeball" a portion of the spectrum as it is being recorded. After exposure to ambient light sources other than daylight a period of 15–20 seconds is all that is normally required to produce a legible tracing. This type of record has the disadvantage that the entire paper continues to develop during exposure to light. The records may also be permanentized, if desired, by spray coatings (different materials are used on different film manufacturers' products) and preserved. After a few years, even when stored in a dark file cabinet, the undeveloped or unsprayed records fade badly. Data obtained from these types of records are shown in Figs. 3-8 to 3-12 and others.

4. *Pen and Ink Recorders*

Pen and ink recordings are no longer commonly used on quality instruments because the response time is too slow for the rapid scan systems now in vogue. An example of such a recording is shown in Fig. 3-5. The record is from a CEC-620 instrument and despite the modest resolution gives a usable spectrum. Wijnbeigen *et al.* (1964) reported on the use of a logarithmic pen recording of mass spectral peaks with a high-speed pen recorder wherein a logarithmically calibrated rule permits results to be immediately available.

B. Mechanical Records

Mechanical records may also be obtained either separately or simultaneously with photographic records, depending upon the individual mass spectrometry installation.

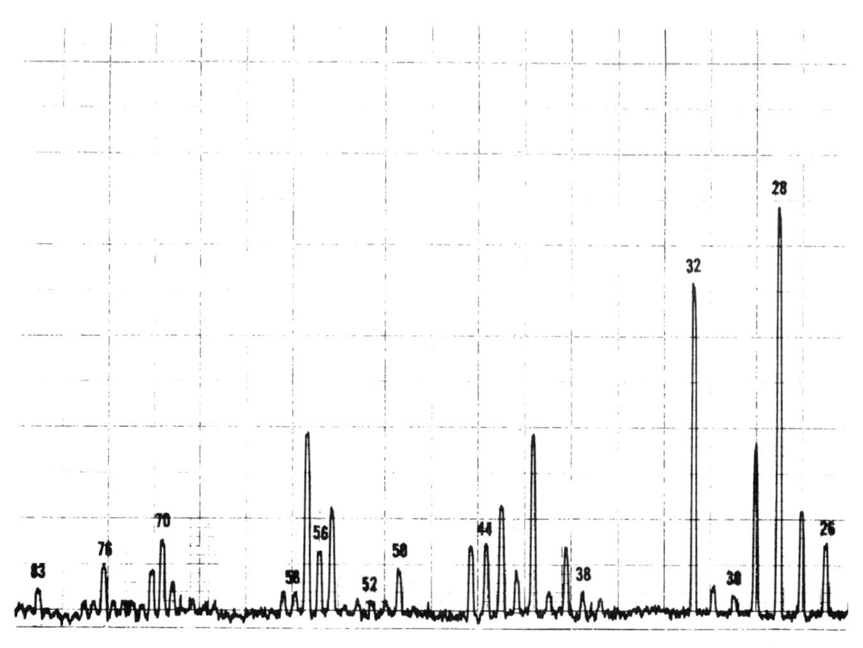

Fig. 3-5. A typical pen-recorded paper strip chart. (Courtesy of Consolidated Electrodynamics Corp.)

1. Digitizers

For several years, a number of sources (Infotronics Corp. and Nonlinear Systems) have made available digitizer units which convert the ion intensity signals and the m/e in focus through suitable electronic circuitry to produce a digital output that can be printed on paper tape, punched on computer tape or cards, and/or put on magnetic tape. This may be done simultaneously or individually, depending upon the financial opulence of the mass spectrometry installation. Thomason (1963) described a solid-state unit that permitted high stability and rapid response.

2. Printed Tape

An example of the paper printout from a CEC-34-201 digitizer is shown in Fig. 3-6. In the example given, the first six lines of information are coded entries for the institutional record of the spectrum and operating conditions. The first line indicates that the date run was 11-08-(6)1 and the day run number was 002. The second line contains a project charge number followed by the time run, 0935, based on a 2400 day. Line three is for sample identification; the 4 with no preceding entries indicates simply *n*-butane, according to the local laboratory code. The remaining three lines indicate the accelerating

```
1 1 0 8 1 0 0 2
  5 0 8 0 9 3 5
              4

  3 3 0 0 0 9 5
  1 2 2 2 0 7 0
  3 5 8 0 0 2 5

  2 7 0 3 6 3 8
  2 7 2 1 0 0 2
  2 8 1 3 4 9 8
  2 8 3 1 0 0 4
  2 9 1 3 5 5 0
  2 9 3 1 0 0 3
  3 0 1 1 1 2 8
  3 2 1 1 1 9 5
  3 6 1 1 0 3 0
  3 7 1 1 2 6 9
  3 8 2 1 4 4 8
  3 9 2 2 8 2 1
  4 0 2 1 3 5 7
  4 1 2 3 3 9 8
  4 2 2 2 4 9 8
  4 3 2 4 4 0 1
  4 3 6 1 0 0 8
  4 4 3 1 4 0 5
  4 8 2 1 0 1 8
  4 9 3 1 0 9 8
  5 0 2 1 2 7 4
  5 1 2 1 2 0 4
  5 2 3 1 0 4 6
  5 3 3 1 1 2 8
  5 4 3 1 0 3 1
  5 5 2 1 1 4 2
  5 6 3 1 1 1 3
  5 7 4 1 3 2 1
  5 8 4 2 4 6 9
  5 9 3 1 0 5 9

          9 4 6

  2 6 9 3 6 3 0
  2 7 3 1 0 0 5
  2 8 1 3 5 9 8
  2 8 3 1 0 0 6
  2 9 2 3 5 4 4
  2 9 4 1 0 0 0
  3 0 0 1 1 2 5
  3 2 0 1 4 7 7
  3 6 2 1 0 2 8
  3 7 2 1 2 6 8
  3 8 2 1 4 4 1
  3 9 2 2 8 1 0
  4 0 3 1 3 6 1
  4 1 2 3 3 9 5
  4 2 2 2 4 9 1
  4 3 2 4 3 9 7
  4 3 6 1 0 0 4
  4 4 3 1 4 0 4
  4 8 1 1 0 1 6
  4 9 3 1 0 9 7
  5 0 1 1 2 7 4
  5 1 3 1 1 9 8
  5 2 2 1 0 4 5
  5 3 3 1 1 2 8
  5 4 3 1 0 3 1
  5 5 3 1 1 4 0
  5 6 3 1 1 1 1
  5 7 3 1 3 1 8
  5 8 3 2 4 6 7
  5 9 3 1 0 6 5
```

Fig. 3-6. A typical paper tape printout from a mass digitizer unit.

voltage used, the ionizing current, the scan rate, inlet system used, its temperature, the type of sample charged (solid, liquid, gas), the ionizing voltage, the micromanometer reading, if used, and/or the amount of sample charged. This information is quite flexible and can be designed to fit the needs of each individual laboratory. The keeping of records on sample conditions is absolutely essential to good laboratory operations and is simplified immensely by this method. Researchers who request mass spectrometric analyses may forget that operational details are needed to obtain reproducible recordings across moderately long periods of time. The information given above was sufficient for the instrument being used, but other instruments may require the recording of additional variables.

The actual spectrum starts after a spaced gap and utilizes the first four digits to give mass numbers, i.e., up to 999.9 to the nearest tenth mass unit. The central, underlined number indicates an attenuation factor to be applied to the three-digit entries appearing on the right side of the strip. These digits are arbitrary ion intensities, uncorrected for background contributions, and require multiplication by the attenuation factor. The numbers 1, 2, 3, 4, and 5 may appear in the attenuation factor column. Associated with these are the respective factors 1, 3, 10, 30, and 100 which give the correct ion intensity when the three digits are multiplied by the associated factor. The mass spectroscopist is directly presented with a listing of m/e versus ion intensities that can be examined to give some idea of the most important peaks. Note that the two spectra of n-butane are reasonably reproducible, but an examination of the m/e 32 (and 28) ions indicates that a substantial increase has occurred. The inlet system must have an appreciable "leak" since air is entering the system at a rather rapid rate. Note that the repeated spectrum from the same initial charge was started exactly 11 minutes after the sample was first introduced into the inlet (9:46–9:35). Normally, with such a light gas, the manufacturer calls for a 1-minute time period between introduction of the sample to the source and the commencement of a scan. This is to allow sufficient time for the sample to build up pressure to an essentially constant flow rate. Depending upon the instrument used with the digitizer, peaks as close as 0.2 m/e could be accurately recorded.

A review of the expected accuracy, reproducibility, etc., for this unit has been given by Grubb and Vander Haar (1962). This particular printout unit could be replaced by one made by several other manufacturers and still accomplish about the same type of performance. Such units are usually purchased independent of the digitizer, but it is wise to consider all aspects of compatibility when obtaining such equipment. While the use of the digitizer is no panacea, it can markedly cut down the time of interpretation if photographic or other oscillographic types of records are in use. The "picking of peaks" by a "peak picker" can be an expensive and time-consuming process. It takes a graduate student about 2 hours to measure peak heights by hand using a millimeter rule and covering a mass range of 27–250. Industrial laboratories tend to spend more funds for the automated equipment and thus reduce this time markedly.

3. *Punch Tape*

As an alternate or for use in conjunction with this digitizer, one can purchase a punching unit which will simultaneously produce computer punch tape carrying the same information that appears on the printed tape. Many tape forms are available, but compatibility with a local computer is a necessity.

4. *Punch Card*

One may also purchase a card punch that will put these same data onto computer punch cards, making them suitable for feeding directly to the computer. The first job of the computer in any event is to store the correct ion intensity value in a location associated with the m/e value of that peak. Hence, the problem of multiplying by the attenuation factors is usually cleared up even before data storage occurs. The initial and final cards of a typical set of cards carrying such data are shown in Fig. 3-7.

One of the most commonly used punches, an IBM 523 summary punch, picks up the data from contact closures in the base of a spectrum printer, such as the Clary Model 1900. The closure configuration is transferred in

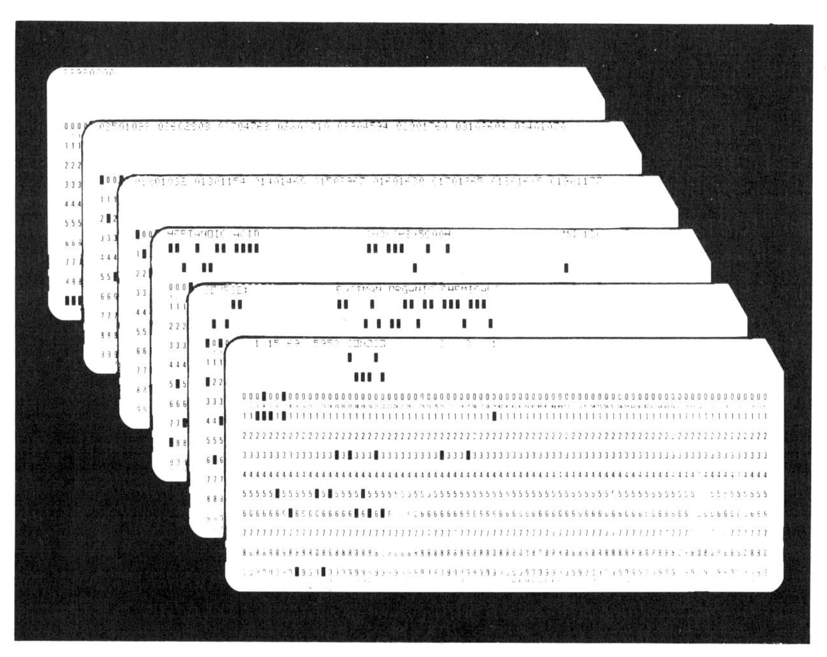

Fig. 3-7. Typical punch cards from a spectrum. The first card denotes the selection of options. The next two cards describe the sample. Following cards (only two of several cards are shown) carry m/e and intensity data for eight peaks. The last card is a terminator.

parallel to a single card row. In the normal operating mode, spectra of several hundred peaks thus generate several hundred cards. An IBM 026 serializing punch was used by Winter (1968). It has the advantages of low-cost leasing, availability, IBM-performed maintenance, and speed compatibility (approximately two data words per second). It also serves as a general punching terminal for other mass spectrometric uses during "nondedicated" time. Few modifications are required of the printer and punch. The auxiliary program drum assembly, with appropriate program card, scans the printer contact matrix, column by column, through a cheap and simple interface. Up to eight data words are stored per card.

5. *Magnetic Tape*

A number of mass spectrometric installations record mass spectra directly from the collector on magnetic tape. For some computers this is an advantage, but the mass spectroscopist has no way of checking the tape for errors without calling for a computer printout of the raw data. Editing can then be done, and the processing of the mass spectrum is completed by the computer. The ability to edit the record at this point saves considerable time later, particularly when an extra peak, such as the argon peak at m/e 40, shows up with a typical hydrocarbon fragment. Once again, the reader is cautioned that compatibility of speeds must exist between the tape recording facility, the computer, and the scanning of the mass spectrometer. A suitable interfacing must be arranged. Since the computer can take over the job of searching and is faster than the human, libraries of mass spectra stored on magnetic tape are coming into vogue. These are commercially available at a substantial cost. A good example of magnetic tape recording of analytical data is given by Issenberg *et al.* (1965).

III. Instrumental Variables Affecting the Mass Spectrum

On the assumption that our readers are not experienced with the production of mass spectra, the reminder is given that the production and collection of ions are dependent upon pressure; source temperature; the sample; ionizing electron energies; many instrumental characteristics such as slits, ion optics, ion current, and accelerating voltages; and the types of ions to be observed. Hence, one must become familiar with some of the above variables before considering the presentation and interpretation of mass spectral data.

The use of mass spectrometric instrumental control circuits requires knowledge of some basic electronic techniques. The manufacturers of most commercial instruments provide adequate instruction manuals, as well as instruction with the field installation of the spectrometer; some companies even offer clinics and training programs in electronics and other techniques. In lieu of this the reader may find of interest and importance the chapter of Frost (1963) covering electronic techniques for mass spectrometry.

A. Ionizing Voltage

The ionizing voltage or the energy of the electrons doing the impacting and causing the ionization is an important variable. If just sufficient energy is supplied when the electron impacts the molecule, the first process to occur should be the formation of the molecule ion. If excess energy is imparted, the molecule ion formed may suffer a decomposition or fragmentation by cleavage of a weak bond in the molecule ion. Usually a daughter ion and a neutral species are the products of the fragmentation or dissociation that has occurred. Thus, as the ionizing voltage is increased (assuming all other variables are constant), the mass spectrum shows ions other than the molecule ion. Also to be noted is the fact that the ion intensities are in general decreased as the ionizing voltage is decreased. This is shown in Fig. 3-8 where a portion

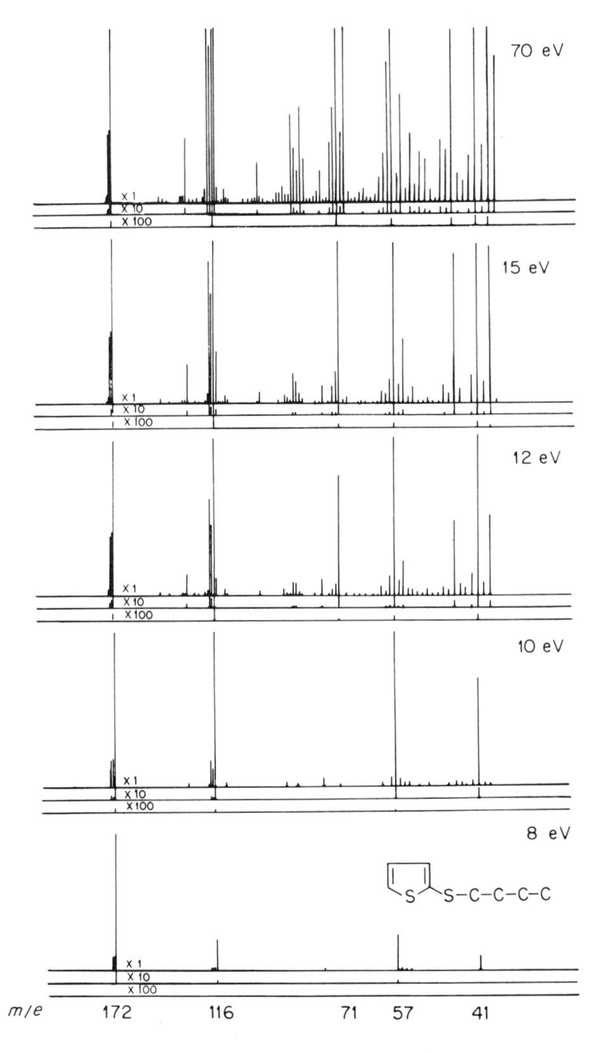

Fig. 3-8. Variation of mass spectrum with variation of ionization voltage from 70 to 8 eV. A CEC-21-104 medium-resolution instrument, electrostatically scanned at rate 7, initial $V = 3550$ V; 10 μA ionizing current, 4.0-mil slits, 7.3 magnet current (4100 G), source temperature 250°C; source pressure 8×10^{-8} torr; paper rate 1/4 in./second; electron multiplier detector. Compound: 1-(2-thienyl)-1-thiapentane, MW = 172.

of the scan is presented at ionizing voltages of 70, 15, 12, 10, and 8 eV, as measured on the meter. The molecule ion is marked 172, the molecular weight (MW) of 1-(2-thienyl)-1-thiapentane. Some prominent fragment ions are also labeled and all indicate the manner in which ion intensity decreases with decreased ionization voltage. Disappearance indicates that one is below the appearance (or ionization) potential for that ion (see Chapter 5 for a detailed discussion).

The scanning rate and chart rates can be altered to present a more detailed spectrum, as will be seen later. This gives rise to definite differences in peak shape, as will be obvious, compared to some of the other spectra presented.

B. Ionizing Current

The ionizing current or the electron current of the impacting electron beam is a measure of the number of electrons in the ionizing beam. If the density of electrons in the beam is increased, one would expect the number of ionizing collisions or near collisions to increase. That this is the case can be seen for the molecule ion peak intensities plotted in Fig. 3-9 where the ionizing current is varied from 0 to 18 μA. The spectrum in Fig. 3-8 employed 10 μA. The spectra were obtained with all other variables constant and using 70-V ionizing electrons.

C. The Accelerating Voltage—Electrostatic Scanning

The ions formed in the source are first subjected to an accelerating voltage before they enter the constant magnetic field of the analyzer tube. It was shown in Chapter 2 how to calculate the ion in focus for a given accelerating potential, etc. If one varies that voltage, a scan is made in which successively higher m/e's appear in focus as the voltage is lowered from the starting value. The scans in Fig. 3-8 were made with initial accelerating voltage at 3550 V. If a lower accelerating voltage had been used, the spectrum scan would have started at a higher m/e. If 4000 V had been used, the starting m/e of the scan would have been about mass 30–31. If the lower accelerating voltage had been used, the lower m/e peaks would have disappeared from the spectrum since the accelerating voltage would not have been high enough to put them in focus and thereby detect them. It should be readily apparent that the mass spectroscopist frequently is interested only in a very small portion of the spectrum and will, perhaps, repeat that region several times to observe details. When this is required the entire range of m/e's need not be scanned, and hence the accelerating voltage may be set to start the scan at any desired point.

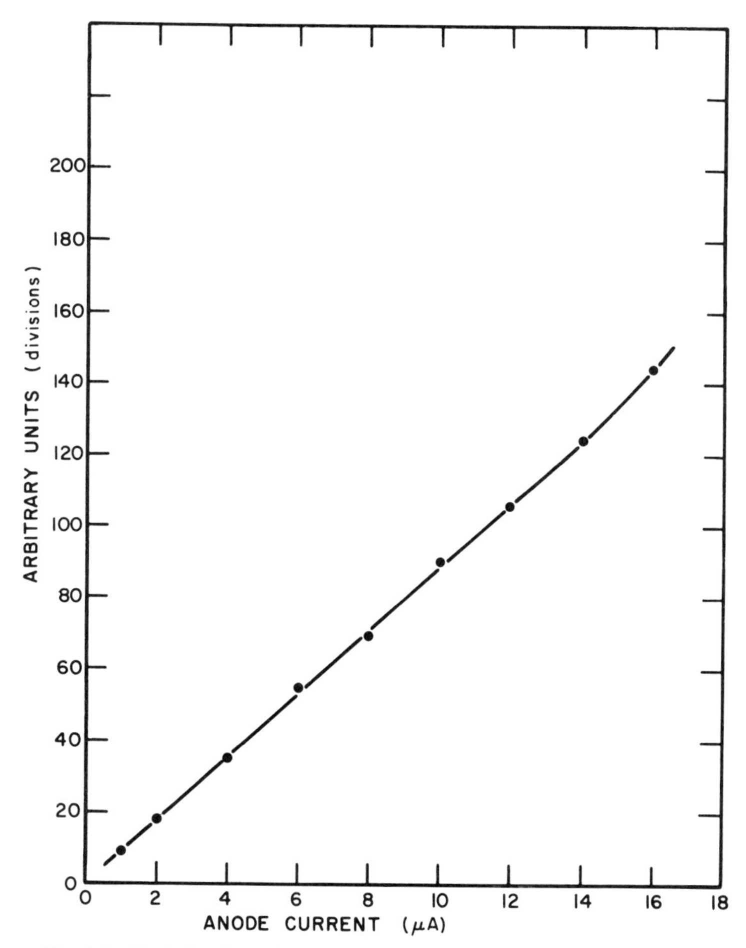

Fig. 3-9. Variation in molecular ion peak intensity with ionizing current.

D. The Magnetic Field Strength

In the above example a constant magnetic field was used. However, if one sets the constant magnetic field to another value, one finds an alteration in the range of *m/e* scanned even if all scans were started at the same accelerating voltage. This is to be expected on the basis of the equations presented in Chapter 2, but it has been restated here to emphasize the point.

In the scan shown in Fig. 3-8 the magnetic field used allows one to begin at about *m/e* 36 (with 3600-V accelerating voltage) and continue to about 360. This change in *m/e* through a 10-fold range is characteristic of the performance of the CEC-104 instrument. If one wishes to scan out to about mass 1000, the beginning of the scan is made so that about *m/e* 90 appears as the lowest peak. All calibrations for the *m/e* in focus can be done by calculation (Eq. 2-12 or 2-13) or by the use of mass markers of the electronic or chemical type. A careful inspection of the peak heights shows that despite the equivalence of sample size the ion intensities in regions where the *m/e*'s overlap for scans of differing magnetic field strength are not equivalent (see also Fig. 3-10). This is due to the fact that mass discrimination is occurring in the instrument. A discussion of this problem was given by Coggeshall (1944) and reviewed by Washburn and Berry (1946) and again by Berry (1950). Attempts have been made to standardize spectral presentation through the American Society for Testing and Materials (ASTM) Committee E-14 on Mass Spectrometry. Such attempts depend upon a general agreement between individuals working in a wide variety of laboratories with ever-advancing techniques. Thus, putting into practice the recommended procedure, even by the individuals who formulated the recommendation, is difficult. The effects of discrimination have been arbitrarily eliminated for most analytical purposes by using *m/e* 90 as the "switch-over" point. This is not satisfactory for the researcher doing theoretical studies in which a range of *m/e* coverage is selected that will include most of the ions formed upon electron impact.

E. Magnetic Scanning

Many of the more modern instruments are also equipped to scan magnetically; that is, by continuously varying the magnetic field while holding the accelerating potential constant. Figure 3-10 shows the results of scanning magnetically from 16 to 3 A while holding the accelerating potential constant at 3500, 3000, 2500, and 2000 V, respectively. The nonlinearity with respect to *m/z* is apparent. Comparing Fig. 3-10 with Figs. 3-8, 3-11, and 3-14 indicates the variation in spectral quality that should be expected. Note that the 172 ion was not seen in the 3500-V scan. All of the instrumental variables were the same as those for Fig. 3-8, except, of course, that magnetic scanning from 16 to 3 A at a rate of 6 on the CEC-104 was conducted while the accelerating potential was held constant at the indicated values. The ionization voltage was 70 eV. It will become apparent later that the recorded spectrum can be improved by using a faster paper rate and a slower scanning rate. A quality equal to that of electrostatic scanning (see also Fig. 3-14) can be achieved. At this point, one may well raise the question: Why use magnetic scanning for samples at all? The reason for its use is the source beam discrimination mentioned in the previous section. Schaeffer (1950) showed that magnetic scanning gave data which showed no systematic shift in abundances of krypton

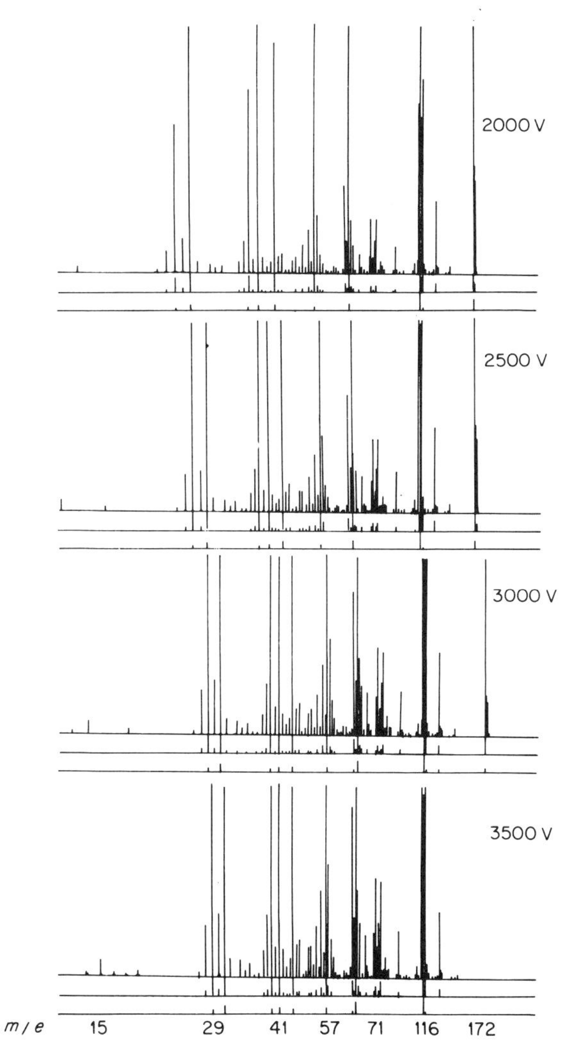

Fig. 3-10. Magnetic scanning of spectra at various constant accelerating potentials (V). Conditions identical with those in Fig. 3-8, except magnetic scanning at rate 6 used for magnet current of 16–3 A, or 8300–1800 G. Ionization voltage = 70 eV.

isotopes whereas when electrostatic scanning was employed, such shifts in data, due to the discriminations, or "voltage effects," did appear. A brief discussion is to be found in Beynon (1960d). The effects can also be seen in the spectrum given in Fig. 3-10.

F. Variation of Slit Opening to Detector

Just as in any optical-type spectrophotometer, opening or closing the slits to the ion beam detector causes a corresponding increase or decrease in signal intensity. Similar, too, is the gain or loss of resolution. In Fig. 3-11, the peak at m/z 40 is shown for the sample containing a little argon (from air) as originally shown in Fig. 3-8. The respective slitwidths are 20, 10, 5, 2, 1, and $\frac{1}{2}$ mil. It is immediately apparent that peak shape, as well as peak height, changes with slitwidths. The peak height gets smaller as the slit closes, but the peak shape shows considerably improved resolution as the slit narrows. In fact at 5 mils,

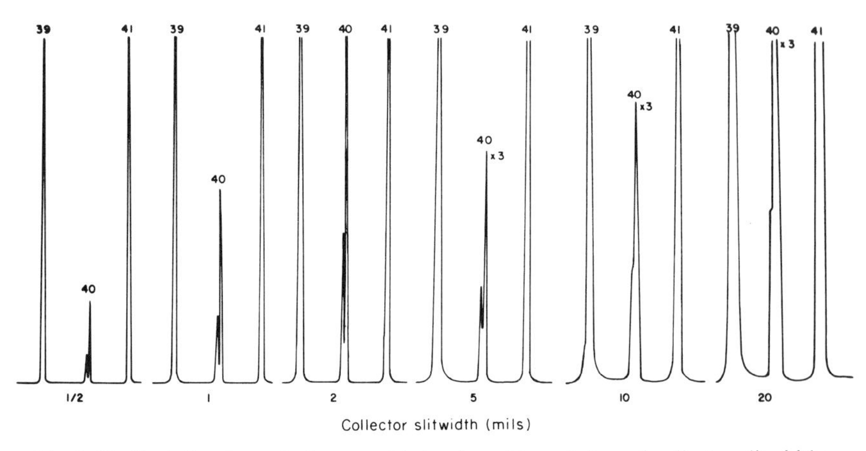

Fig. 3-11. Variation in peak shape and intensity with variation of collector slitwidths.

the presence of two peaks is clearly shown, while at 10 mils, the separation is suggested but certainly not achieved.

G. Definition of Resolution

The above discussion, while introducing the reader to the fact that resolution can be changed, also points out the need for a definition of resolution. In a practical sense, resolution is the ability of a mass spectrometer to discriminate between two adjacent peaks. Many "numerical" definitions of resolution are in the literature and are too numerous to list completely. A number possess merit and, although the present discussion shall be limited, the consideration of additional details in Biemann (1962), Kiser (1965b), and Beynon (1960a), will prove to be of value when formal reporting of resolution is desired. From Chapter 2 and a consideration of the Dempster and Aston instruments, the reader should have noted that for a single-focusing instrument the resolution increases with an increase in radius and with an increase in the accelerating potential. Since the radius is usually fixed by the manufacturer, this becomes a constant for any given commercial instrument. It was probably also recognized that the widths of the source exit slit and the collector slit determined the extent of overlap of the recorded peaks. The difficulty with Dempster's resolution definition in Eq. (2-3) is that $S_1 + S_3$ has to approach zero,

$$M/\Delta M = r/(S_1 + S_3)$$

Note that the denominator amounts to the sum of the ion beam width plus the collector slit width (in centimeters). The equation also assumes that the recorded ion intensity falls to zero, if only momentarily, between the peaks involved. Actually, various aberrations increase the peak width and gas scattering (Beynon, 1960b) produces "tails" on either side of a peak. In effect, the peak intensities fall off to zero, asymptotically on either side of the peak. A major item in the category of aberrations is the possibility of local space-charge effects, either in the source, analyzer tube, or at the detector, which may vary from instrument to instrument and may even vary with time, cleanliness of source, etc., in a given instrument. A second important item would be the variation in the kinetic energy possessed by ions of the same m/e value. Note that doubly charged ions of $2m/2$ would fall at the same nominal m/e value as the singly charged ion of mass m. For quantitative analysis, it is necessary that the maximum of an ion peak height be an accurate measure of the intensity of the ion beam producing the peak. There must be no appreciable contribution by an adjacent peak to the height of the peak to be measured. For qualitative analysis, it is necessary only that the peaks be clearly distinguishable.

For the purpose of this treatise, the 10% valley definition of resolving power has been selected as satisfactory. Several parameters of the instrument must be known or stated to define and measure resolution. When the ion beam reaches the collector, it has a finite width (W), which is due to the width of the source exit slit (S_1 in the Dempster equation), plus any aberrations of the ion optical system. Each ion beam reaching the collector has been dispersed by the magnetic field and is separated by some finite distance from an adjacent beam. The distance is called the dispersion, which is measured from the center point of one ion beam to the center of the next beam. Instrument geometry and the ion mass determine the dispersion according to the equation.

$$d = Kr(\Delta M/M) \tag{3-1}$$

where d is the dispersion, K is a constant and a function of the analyzer design, r is the radius of the ion path, ΔM is the mass difference between the ions in the two beams in question, and M is the nominal mass of the two beams.

Usually this is taken as the higher mass value, not the average. If the dispersion for the given ion pair is greater than the beam width (W), there will be a space or "gap" between the ion beams. The two beams will touch or overlap if the dispersion is equal to or is less than the beam width.

To completely resolve or separate the two adjacent peaks on the recording device, the dispersion must be equal to or greater than the sum of the beam width (W) and the collector slitwidth (C) (the same as S_3 in the Dempster equation). The situation is illustrated in Fig. 3-12, the (a) portion showing

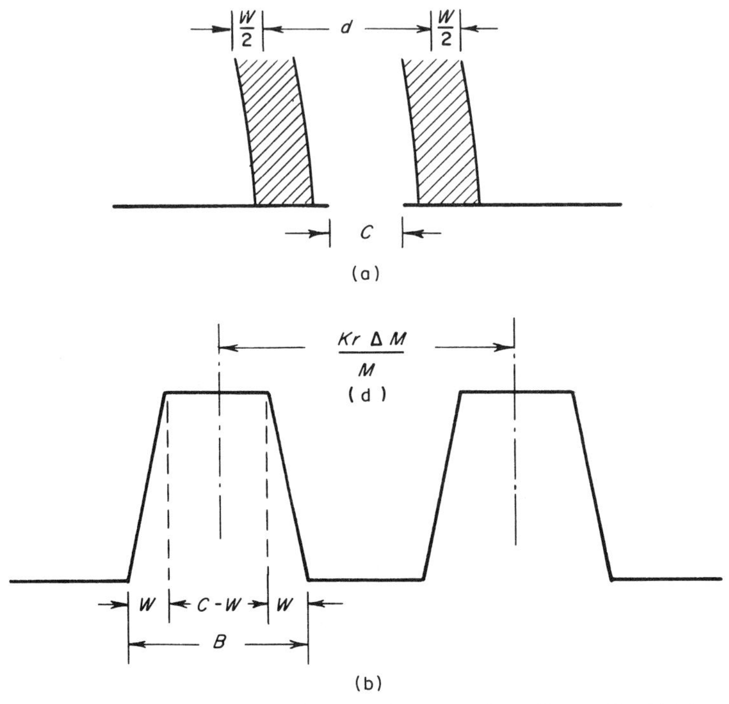

Fig. 3-12. (a) Physical separation of two ion beams at the collector slit. (b) The recorder "view" of the beam separation.

the beam separation at the collector slit and (b) showing the beam separation as viewed at the recorder. If the resolution is complete, this means that the first ion beam has moved across the collector slit and is recorded with no overlap. Just after the first ion beam is off the collector, the second ion beam moves on to the collector. For complete resolution there would be a zero valley and $W + C = d$. The figures assume sharp beams with no tailing. Since the resolution,

$$R = M/\Delta M = Kr/(W + C) \tag{3-2}$$

it would appear that the problem is completely resolved. However, both W and C are dimensions inside the analyzer and may not be known to someone having a mass spectrum and desiring to estimate the resolving power of the instrument used. Despite this, the recorded peak can be labeled as shown in Fig. 3-12b because all the dimensions are *proportional* to their respective dimensions in the analyzer.

To pursue this to a practical result, the recorded spectrum distances can be measured and designated as follows: Peak base is B (at the respective galvanometer baseline) and the dispersion is d, as used before (but remember that it is proportional to the d in Eq. (3-1), not equal to it). Then using the same terms in a proportional sense, the following relationship exists:

$$\frac{B}{W+C} = \frac{d}{Kr\Delta M/M} \tag{3-3}$$

or, on rearranging,

$$W + C = BKr\Delta M/dM \tag{3-4}$$

and, upon substitution into Eq. (3-2),

$$R = dM/B\Delta M \tag{3-5}$$

where R is resolution, d is the measured distance on the record between two adjacent peaks, B is the measured base of one of the peaks, M is the nominal mass of the peak whose base is measured, and ΔM is the mass difference between the two peaks.

The case of complete, minimum separation between the peaks occurs for $d = B$. Under these conditions a 0% valley (baseline separation) exists between these two peaks. Then the resolution (0% valley definition) is simply $M/\Delta M$.

Example 3-1

For N_2 and CO molecular ions, what resolution would be required to just separate these ions?

Using the exact mass value for the nitrogen atom of 14.003074, it can be seen that

the nitrogen molecule mass	= 28.006148
Similarly for C = 12.000000 and O = 15.994915, CO	= 27.994915
From which the ΔM	= 0.011233

Hence, substitution into Eq. (3-2) leads to a required resolution of

$$R = \frac{M}{\Delta M} = \frac{28.006148}{0.011233} = 2493.2 \approx 2500$$

Example 3-2

For the N_2–CO doublet occurring at mass 200, what resolution would be required?

From Example 3-1 we have the difference in mass and, therefore,

$$R = \frac{M}{\Delta M} = \frac{200.0}{0.011233} = 17{,}804.68 \approx 17{,}800$$

One should note that this hypothetical case would indicate that the balance of the molecule atomic content is the same, i.e., alkyl groups, etc., but such that the mass spectroscopist has the choice between the molecule which contains a CO grouping in some fashion or else two nitrogen atoms. The structure need not be specified to decide upon the resolution required.

If d is greater than B on the record (as in the illustration), then the ratio d/B times M expresses a higher mass at which two peaks having original mass difference of ΔM would have minimum complete separation between them. With a linear rule and knowledge of the masses involved, anyone inspecting the mass spectrum can closely estimate the resolution achieved. The logic of the above case can be reversed and the corollary statement made that two peaks at mass M_1 and M_2 would have a 0% valley separation if the mass difference were $B\Delta M/d$.

Example 3-3

In a mass spectrum two peaks of equal height are observed at m/z 84.00 and 85.00. Their centers are separated by 20 mm on the recording. The baseline width of each of the peaks is 5.0 mm. What resolution is theoretically being achieved?

By observation, $d = 20.0$ mm, $B = 5.0$ mm, $\Delta M = 1.00$ at $M = 85.00$. Hence, by Eq. (3-5),

$$R = \frac{dM}{B\Delta M} = \frac{20 \times 85.00}{5 \times 1.00} = 4.0 \times 85 = 340$$

It was stated earlier in this section that the 10% valley definition of resolution would be considered acceptable. In light of the above discussion, one may well ask what the difference is between 0, 2, 5, 10%, etc., valley definitions of resolution and how the measurements are taken from the record to be interpreted for these various definitions. Figure 3-13 illustrates the two most commonly occurring definitions of resolution today. The valley definition in Fig. 3-13a is discussed first. Two peaks of equal height partially overlap. Since the superimposed portions of the peak are additive, the peak height where the peaks overlap would be a for each peak. The observed height of the

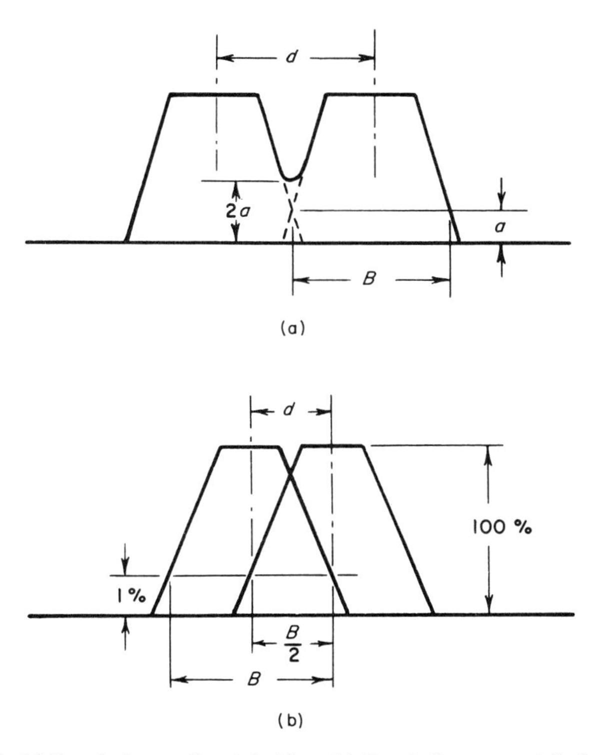

Fig. 3-13. (a) Resolution: valley definition. (b) Resolution: cross talk definition.

composite curve would be $2a$. If the value of $2a$ is 10% of the peak height, then the resolution is stated to be a specified value with a 10% valley. Similarly, for the value of $2a$ equal to 1, 2, 5%, etc., other definitions of resolution can be written. In the preceding paragraph, it was shown how to obtain the resolution with a 0% valley definition. The person desiring to use the 10% valley definition wishes to know at what value of M he can expect to find the same resolution value as obtained with the 0% valley definition. In more general terms, for any valley definition: What value of d and B taken from the record will lead to a $2a$ valley definition of resolution? In Fig. 3-13 a new baseline is drawn through the point of intersection of the two peaks. Disregarding the portions of the peaks below this new baseline, one would then have a pair of peaks which are just minimally completely resolved. By definition $d = B$ for this 0% valley resolution. In reality, the peaks have a $2a$% valley. Hence, if the base of a peak is measured at a height, a, above the baseline and the measured value of B is used in Eq. (3-5), the resulting value of resolution will have a $2a$ valley definition. For example, B measured at 1% of peak height will lead to a 2% valley definition; if measured at 5% of peak height, a 10% valley definition; etc. Note that no contribution to the top of the adjacent peak occurs with a 10% valley.

The "cross talk" definition of resolution will be encountered by the mass spectroscopist. Resolution specifying 1% cross talk is most commonly seen; that is, two peaks of equal height overlap to the extent that each contributes 1% of its height to the top of the other. Considering Fig. 3-13b, one can note that if the overlap of the peaks were exactly 50% (equivalent to 0% cross talk), d would be equal to $B/2$ when B is measured at the baseline. Hence, when the overlap is such that 1% cross talk occurs, d will equal $B/2$ when B is measured at 1% of the peak height. To calculate resolution with a cross talk resolution of a of peak height, the following equation is used:

$$R = 2dM/B\Delta M \tag{3-6}$$

Note that all parameters are defined as in the previous section covering resolution, and idealized peak shapes are used.

Table 3-1 provides a number of pairs of substances, both gases and liquids, which may be used in determining the resolution at a given m/e for a given instrument. Proper proportions of the substances should be used to give equal intensity peaks for the doublet.

The reader is cautioned against "glib" usage of resolution terms for mass spectrometry and should be judicious when specifying the resolution of an instrument employed in a given spectrum. It should be obvious that any attempt to classify instruments as to resolving power will not meet with universal appeal, nor should this be expected. For purposes of this text, the following arbitrary ranges are classified:

Low-Resolution Instruments. Instruments giving resolutions of 200/1 or less. These include older commercial instruments, most isotope ratio instruments, leak detectors, process-type instruments, and, particularly, instruments used in association with metabolic studies, gas analysis, etc.

Medium-Resolution Instruments. Instruments giving resolution of from 500/1 to about 5000/1. Included would be many of the modern single-focusing commercial instruments in general analytical use for organic materials, structure identification, reaction studies (i.e., the TOF), etc.

High-Resolution Instruments. Instruments giving resolution from about 10,000/1 to 100,000/1 or above. These include most of the double-focusing instruments of the Mattauch–Herzog or Nier–Johnson designs and some other special instruments.

Hopefully, the passage of ten years' time may see further subdivisions of these classifications. Perhaps there will be agreement on standard definitions and possibly another change in power of ten toward even higher resolution.

TABLE 3-1

SOME COMMON SUBSTANCES WITH DOUBLETS USEFUL FOR DETERMINING RESOLUTION

Substances	Nominal m/e	Doublet ions	Exact mass of ions	ΔM	Resolution $(M/\Delta M)$	Pressure ratio
Carbon tetrachloride	84	$C^{37}Cl^{35}Cl$	83.93475	0.15914	528	1/6
Cyclohexane		C_6H_{12}	84.09389			
Propane	44	C_3H_8	44.06259	0.07277	604	5/1
Carbon dioxide		CO_2	43.98982			
Ethane	28	C_2H_6	28.03129	0.03638	769	1/1
Carbon monoxide		CO	27.99491			
Bromobenzene	156	$C_6H_5{}^{79}Br$	155.95751	0.13638	1,144	1/1
Dimethyl-naphthalene		$C_{12}H_{12}$	156.09389			
Neon	20	Ne^+	19.99244	0.01125	1,778	1/1.4
Argon		Ar^{2+}	19.98119			
Nitrogen	28	N_2	28.00614	0.011123	2,493	1/1
Carbon monoxide		CO	27.99491			
Styrene	103	C_8H_7	103.054772	0.012576	8,195	3/1
Benzonitrile		C_7H_5N	103.042196			
Dimethyl-quinoline	156[a]	$C_{11}H_{10}N$	156.081320	0.012576	12,400	1/2[a]
Dimethyl-naphthalene		$C_{12}H_{12}$	156.093896			
1-Octene	71	C_5H_{11}	71.086071	0.004470	15,900	None
		$^{13}C^{12}C_4H_{10}$	71.081601			
Toluene	92	C_7H_8	92.062597	0.004470	20,922	1/10
Xylene		$^{13}C^{12}C_6H_7$	92.058127			
Dimethyl-naphthalene	142[a]	$^{13}C^{12}C_{10}H_9$	142.073777	0.004470	31,800	1,000/1[a]
Methyl-naphthalene		$C_{11}H_{10}$	142.078247			

[a] A 200°C sample introduction temperature required.

It should be noted that the instrument used will indirectly but in a very marked way affect the mass spectrum obtained. The resolution of the instrument has practical limits; this, in turn, is expressed in the form of the mass spectrum observed for a given substance. For example, with a modest-resolution instrument, say 500/1, the molecule 6-dodecanone ($C_{12}H_{24}O$) containing 37 individual atoms produces at least 187 peaks (from *m/e* 12 to 300) in its mass spectrum. These peaks are considered as nominal integer masses, because the instrument used does not have the resolving power to see the nominal mass 43 ion peaks broken up into the possible fragments of $C_3H_7^+$, $C_2H_3O^+$, and $^{13}C_1C_2H_6^+$, all of which are probably present, and might double the number of peaks observed at high resolution. Nevertheless, it is generally accepted that organic molecules produce more than five times as many ion species as the number of atoms making up the molecule.

H. Scanning Speeds

In general, the speed at which a scan is made affects the ease of processing the mass spectral data and, for paper recording, produces a problem in the ease of reading the peaks. For mechanically digitized or punched data, the response to a peak change must be faster than the rate at which peaks are produced. For oscillographic recordings, the combination of GLC with mass spectrometry may call for an extremely rapid scanning rate of the spectrum, from mass 12 through 200 perhaps, in as little as 10 seconds. To compensate for this, many instruments have fast paper travel speeds which, in turn, spread the distances between peaks out to more readable values. Most instruments that provide for both electrostatic and magnetic scanning also provide some variation of time to cover the range of V or H to be covered during the scan. Combined with the variations available for chart or paper speeds, a multiplicity of changes occurs in the spectrum produced. A portion of the spectrum that appears in Fig. 3-8 is shown in Fig. 3-14. The spectrum in Fig. 3-8 was obtained at a paper rate of $\frac{1}{4}$ in./second and an electrostatic scan rate indicated as 7. In Fig. 3-14d the same charged material was scanned with all other conditions identical except that a paper rate of 1 in./second was combined with an electrostatic scan rate of 7. The distance between peaks is identical; thus one can conclude that the electrostatic scan rate 7 is 4 times that of 9. Hence, compensation for the faster scan rate with an increased paper rate may be achieved. However, the sacrifice in resolution (and peak height) is obvious.

In the upper portion of the figure, a part of the spectrum of the same compound run at electrostatic scan rate 5 shows another fourfold increase in the rate at which the peaks are scanned. The $\frac{1}{4}$ in./second paper rate now shows peaks from *m/e* 51 to about 86 covered in 1 in. of paper. The compression is rather extreme, and much detail is lost. The main fragments are still identifiable and capable of numeration with a reasonable certainty. Increasing the paper rate increases the certainty of numeration but shows the reduction in

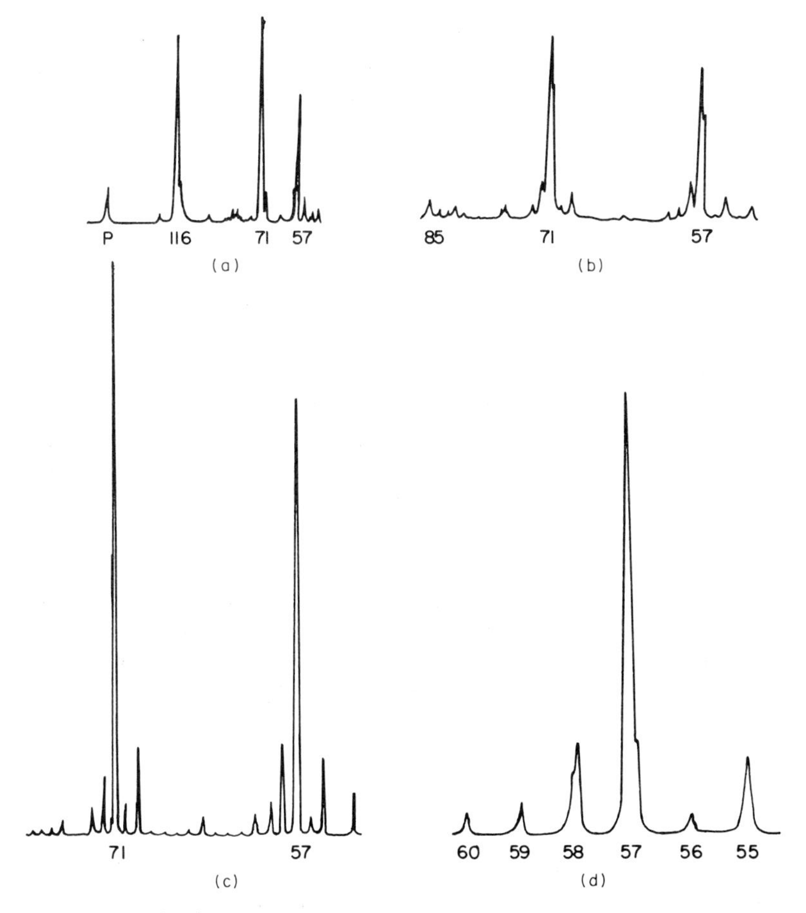

Fig. 3-14. Variation in spectrum due to electrostatic scanning speeds and paper speeds. Scan rate and paper rate, respectively, are (a) 5, $\frac{1}{4}$ in./second; (b) 5, 1 in./second; (c) 7, $\frac{1}{4}$ in./second; (d) 7, 1 in./second.

resolution which is suffered because of the rapid scanning. The expected further reduction in peak height is also experienced. With magnetic scan rate changes there is a similar problem.

It is apparent from these data that, at the will of the operator, a given instrument can produce spectra of varying quality usable for various purposes.

I. The Repellers

The repellers used in various instruments have an effect upon the spectrum in the form of observed ion intensities. Increasing the repeller potentials increases the observed peak heights, all other factors being held constant. This type of change is often useful in detecting metastable peaks of low intensity and is also helpful in low-ionization work. In some instruments the repeller voltage may be regarded as adding to the ionizing voltage applied to the electrons.

J. Focus Adjustments

Every instrument should be tuned (usually according to the manufacturer's instructions) to produce high-intensity peaks. Depending again upon the design of the individual instrument, a slight change in focusing potentials leads to drastic changes in the observed peak heights. It is beyond the scope of this work to do anything more than issue a warning to the new mass spectroscopist that he must consider these latter two variables as they affect the spectrum.

For the mass spectrometrist who becomes interested in the physics and instrumental side of the field, a consideration of ion optics is of great importance. Elliott (1963) has discussed ion source optics and the types of "ion lenses" commonly employed in the source. A very complete treatment of the subject of ion optics is to be found in the chapter by Kerwin (1963). The motions of ions in uniform magnetic fields, fringing fields, and inhomogeneous fields, as well as optimum geometrical conditions for obtaining the best possible focused ion beam at the collector, are discussed.

K. The Detector Type

The type of detector employed markedly affects the observed spectrum, as does the amplification of the signal. For a 1-μliter sample, the peak heights obtained when one is using the electrometer detector are quite modest in comparison to those obtained with the electron multiplier detector. In fact, for this size charge, the largest peaks on the electron multiplier run have to be "off record"; that is, the lowest attenuated galvanometer hits the top of the recording scale. The reason is obvious. The electron multiplier detector at its highest gain may give sensitivity changes of 500–2000 times that of the electrometer detection mode, at a modest gain. The peak width changes correspondingly, and hence the spectroscopist has to select optimum conditions for each individual case.

IV. Sample Variables Affecting the Mass Spectrum

While some indication of sample effects (because of resolution or the lack thereof) has been discussed previously, a number of other important problems arise from the character of the sample.

A. Chemical Composition

Depending upon the substance to be examined, a considerable spread can be expected in the number of *m/e*'s to be observed. For a simple example, consider nitrogen. The nitrogen molecule is composed of two nitrogen atoms, and nitrogen has two naturally occurring isotopes, ^{14}N and ^{15}N. The abundance of ^{15}N is only 0.367%, and that of the ^{14}N isotope is 99.633%. The mass spectrum shows peaks at nominal masses 29, 28, 15, and 14, with the 29 and 15 peaks being quite small in view of the abundances given. This is all the spectrum that should be expected to be observed if metastable transitions are ignored. For argon, matters are even simpler, as they are for all the rare gases. The molecule ion at nominal mass 40 is observed and possibly the trace isotopes at *m/e* 38 and 36. If the ionizing voltage is somewhat above the second ionization potential, about 22 eV, the doubly charged ion at mass 20 can be seen. Since argon is present to about 1.09% in air, the 40 peak is commonly observed if a substantial vacuum system leak exists. The *m/e* 20 peak is observed at times in the normal air spectrum, provided sufficient sample has been entered. As a much more complex system, consider again the 6-dodecanone. This compound can produce fragments ranging in *m/e* from 1 (for hydrogen) to about *m/e* 188 due to the molecule ion *m/e* 184 and the naturally occurring isotopes of ^{13}C, deuterium, and oxygen isotopes at 17 and 18 mass units. Note that most organic compounds cannot produce ions from *m/e* 20 to 23 because no combination of atoms can give such an *m/e*. Familiarity with compounds and ions develops expertise in examining records for the mass spectroscopist. Gross mixtures produce many ion peaks and thus make the problem considerably more complex than when observing a pure substance.

The chemical composition can also affect the spectrum in that fragment-ion stabilities from the parent substance may be of many different orders of magnitude. For example, neopentane produces essentially no *m/e* 72 peak. It can be observed only at low ionizing voltages and then when one takes great patience and care to utilize the instrument to the fullest degree of sensitivity. The reason is rather obvious; the loss of an electron from the molecule weakens a methyl bond which, upon cleavage, produces the relatively stable *m/e* 57 ion, a *t*-butyl carbonium ion. Thus chain branching, the presence of a number of "weak" bonds in a molecule, and even the presence of certain functional groups lead to ease of fragmentation of the molecule ion or even of the first daughter ions. Needless to say, this problem also affects the sensitivities of various substances, because the parent ion may decompose more easily in some cases than in others. Actually, the stability of similar ions in isomers may differ to such an extent as to produce a considerably different pattern. The various isomeric hexanes are an example of this and are discussed in detail in Chapter 5.

B. The Amount of Sample

Most scientists would immediately recognize that reducing the amount of sample available for ionization would reduce the number of ions produced and would therefore give a spectrum with less intense peaks. Because of the several modes of sample introduction (see Chapter 4), two problems are presented to the mass spectroscopist. These are analytical reproducibility of the spectrum and a method of interrelating spectra for quantitative purposes.

The spectrum actually observed and the peak sensitivities depend upon the number of moles of substance being ionized. Thus it is most convenient to compare the mass spectrum of two substances when equimolar amounts have been used. If gases are used, this calls for equal pressures of each gas, assuming ideal gas law behavior or that Dalton's law of partial pressures holds for the mixture. The temperature of the inlet system becomes a problem here. At 150°C one would not be concerned about methane, but certainly one would be concerned with anthracene. Because of the latter's melting point, one would be concerned with the validity of pressure measurement and the fact that much of the sample introduced would not reach the ionizing region. Thus volatility and reliable gas pressure may affect the spectrum and, in fact, could hinder the interpretation.

If liquid samples are charged, say 1 μliter in a suitable syringe, the actual amount of material (in moles) reaching the ionization chamber would not be equal unless by mere chance the liquid molar volumes (and the volatility) of the two substances were the same. On most instruments 1 μliter of benzene will give a very good spectrum with large peaks. However, 1 μliter of dodecylbenzene run in an identical instrument will probably give a useful spectrum; but the peak sizes will be markedly reduced. On a per mole basis, the number of peaks and the sum of their intensities should go up as the molecular weight of the compound increases (see also Section VI, C of this chapter).

C. The Sample Temperature—Volatility

From the above discussion, one can readily see that temperature of the gas particles at the time of ionization should influence the spectrum. The main effect is one of volatility, but small differences because of the thermal energy may also be observed. Most of the excess energy acquired in the ionization process is due to the ion acceleration, but kT energy of the vapor particle would be different at different inlet temperatures. Some mixture spectra can be affected by temperature; i.e., consider the fate of anthracene dissolved in benzene. One will be well below its pure melting point, the other very volatile. Accurate analysis would present a challenge.

D. Wall Reactions

Reactions of the sample with the walls of the inlet system have been observed in the mass spectra of some substances. Hydrogen exchange reactions with the walls of the inlet system can be readily detected when deuterated molecules containing deuterium in labile positions are examined. As an example of this, consider the data shown in Fig. 3-15 (Higgins and Foster, 1962, 1963). The deuterated ethylene glycol species was made by exchange with heavy water and was to be used in an attempt to synthesize other deuterated species. The data were taken on a CEC-102 modified mass spectrometer using a liquid charging system. Successive mass spectra were run as a matter of course when examining isotopically labeled species. The ratios of the 32/31 peak were used to estimate the percentage of deuteration, but the various scans show a decrease of deuterium content with time. This is taken as proof of the reactivity of the labile deuterium positions with the wall. The samples were allowed to remain in the heated inlet system, and scans were started at the times indicated. Note that the analyst could estimate the maximum amount of deuteration by extrapolation to time zero—the time the sample was charged into the heated inlet system. The samples were run in numerical order and hence one might well ask if the "wall" of the inlet system was becoming saturated with deuterium. Normal pump-out time was employed. Some unlabeled materials were run after sample III and then followed by more labeled glycols. The unlabeled materials did not show deuteration, but no labile position was available. The labeled glycols run after this showed similar behavior to the first three, so the conclusion was reached that these materials adsorbed on the wall were gradually desorbed during the pump-out period. Nevertheless, the analyst must be aware that the possibility exists of this type of exchange, or erroneous results will be obtained.

E. Sample Decomposition in the Inlet or Source

Some materials may well be thermally unstable and produce some products of ordinary thermal decomposition, which in turn may become ionized and thus become a part of the observed mass spectrum. If one suspects such an

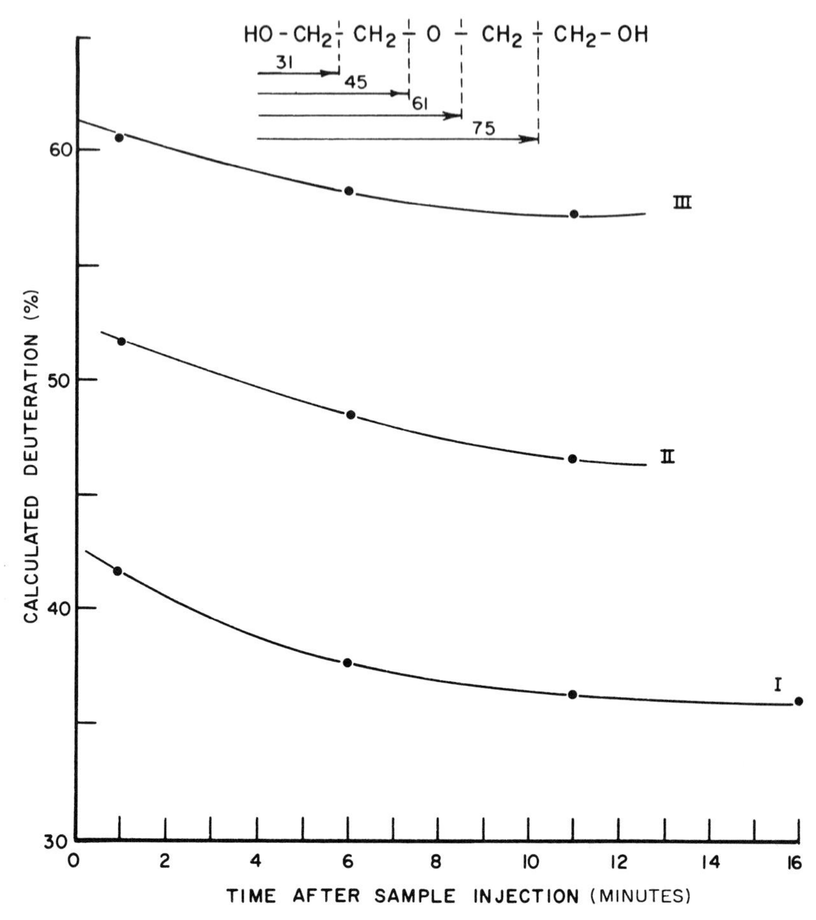

Fig. 3-15. Deuterium–hydrogen exchange at the wall of an all-glass inlet system. Variation in apparent percent deuteration in three partially deuterated samples (at the hydroxyl group) of diethylene glycol with time after sample injection. The percentage is based upon the *m/e* 32/31 ratio.

event on a prior chemical basis, a check can be made by carefully comparing the spectrum obtained at a number of different inlet system temperatures. Sometimes such a temperature dependence can be useful to the analyst, but more often it leads to the production of "peaks" of an unexpected nature, which may possibly render the interpretation of the mass spectrum difficult if not actually impossible. Meyerson and McCollum (1959) have given such an example for the alkylphenyl sulfones and sulfoxides; about 14% of the ions formed came from bimolecular processes which are pressure dependent.

Foster *et al.* (1968, 1969) observed an *m/e* 64 peak corresponding to the

S_2^+ ion in the mass spectra of thienylthiaalkanes and benzo[*b*]thienylthia-alkanes. Since the two sulfur atoms in these molecules are separated by a carbon atom, a rearrangement would be necessary for formation of the 64 ion comprised of S_2^+. Thermal decomposition in the heated inlet system was a distinct possibility. To rule out this possibility, several compounds were kept in the inlet system for a considerable length of time and parent-ion intensity compared to the ion at *m/e* 64. No pattern change was observed which would indicate thermal decomposition. Therefore, the rearrangement was taking place in some fashion during the ionization process.

Ukai *et al.* (1967) observed the growth of the M + 2 peak with time in the examination of 1,2-quinones. It may well be that the necessary hydrogen was obtained from a wall reaction as described previously, but these workers also saw ions characteristic of the original material, less some hydrogen atoms. Since these also increased in intensity in time, it appears that a reductive reaction was taking place in the mass spectrometer. One should not be too surprised to encounter reactions such as this or possibly an auto-Cannizarro reaction under the proper circumstances and with the right precursors.

V. Types of Ions Observed

To assist in illustrating the types of ions observed in a mass spectrum, Fig. 3-16 is presented of a scan made under conditions to permit careful scrutiny of the actual ions one may encounter. For preparation of reference spectra or research-grade spectra for mechanism studies it is advisable to use a slow enough scan rate to obtain the type or resolution shown. Likewise it is best to have a paper rate which permits some seeing of detail, etc. The material used was 1-(2-thienyl)-1-thiapentane, having a molecular weight of 172. This ion can be noted quite easily. The very small peak at *m/e* 40 (on the low-mass side) is an argon peak due to a trace of air that was introduced with the sample. (Almost all liquid samples contain some dissolved air unless they have been vacuum transferred and stored under a vacuum.) This is often of use in beginning the numbering of such a sample without the benefit of an added mass marking substance. It also indicates, to some extent, the resolution obtained. The ions at *m/e* 64 and 66 are quite obviously doublets and arise from S_2^+ and isotopic ions, as described in the previous section for these types of molecules.

A. Molecular Ions

When a substance is exposed to impact by electrons, the following general mass spectrometer reaction equations can be written

$$M + e^- \rightarrow M^+ + 2e^- \tag{3-7}$$

Hence, the neutral species loses a net single electron to produce a single charged molecular ion. If singly charged and from a stable compound, the ion must contain an odd number of electrons. If from a neutral radical, an even-electron ion will result. Considerable detailed discussion of these ions will be presented

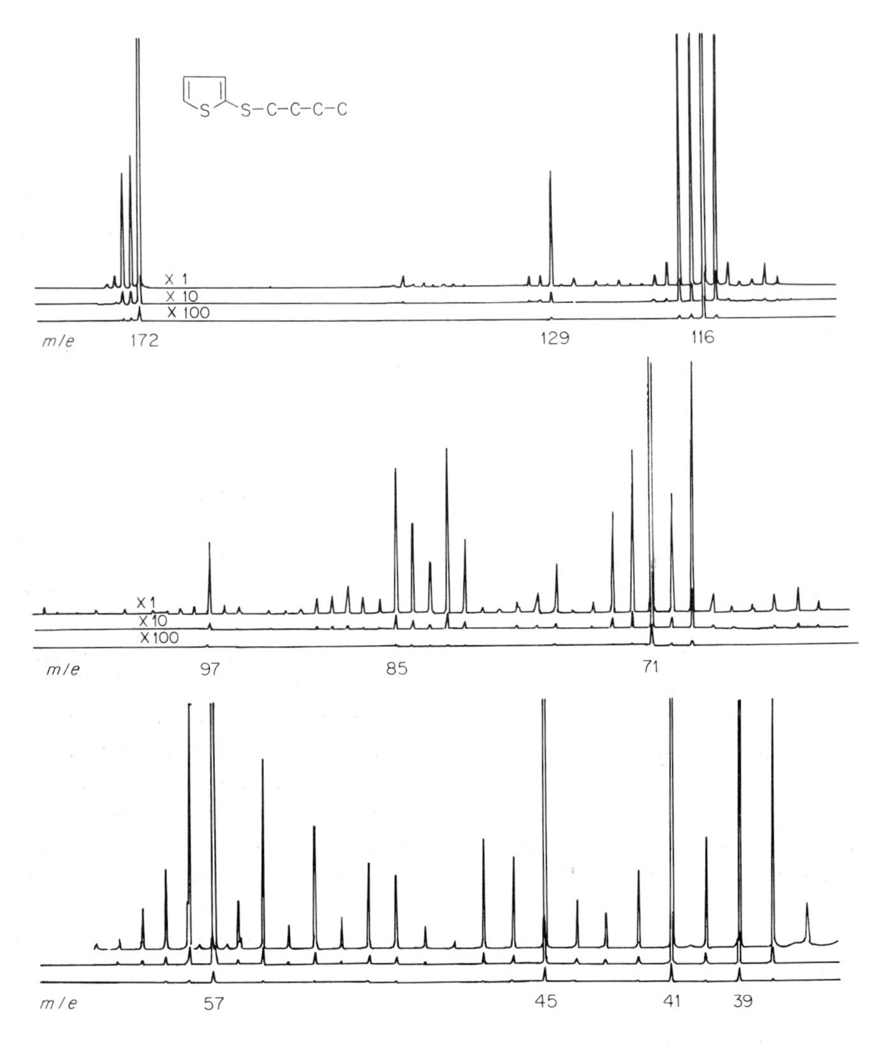

Fig. 3-16. Typical oscillographic recording of a 70 eV spectrum at medium resolution. The compound is 1-(2-thienyl)-1-thiapentane, MW = 172. Conditions the same as Fig. 3-8 except that electrostatic scan rate 9 employed to yield a record capable of producing a spectrum suitable for reference work and illustration of a number of the types of ions observed.

in later sections and in Chapter 5. The molecular ion in Fig. 3-16 occurs at *m/e* 172.

B. Isotopic Ions

Many of the elements possess naturally occurring isotopes such as ^{2}H, ^{13}C, ^{15}N, ^{17}O, ^{18}O, ^{34}S, etc. Some of these occur to an appreciable extent, and their presence in a molecule leads to a different *m/e* than the ordinary mono-isotopic peak at the nominal mass. For example, CO_2 produces a molecular ion at *m/e* 44 but also produces an ion at *m/e* 45 because of the isotope $^{13}C^{16}O_2$. Since the natural abundance of ^{13}C is 1.107% and that of ^{12}C is 98.893%, one would expect to find the 44 peak about 100 times as intense as the 45 peak. This is what is observed. Hydrogen sulfide would be expected to show a peak at *m/e* 34, but it also shows a small peak at *m/e* 35 because of the presence of the ^{33}S isotope. A moderate peak at *m/e* 36 arising from the ^{34}S atom is also present. Sulfur-34 occurs to a little more than 4% in natural abundance. For the naturally occurring isotopes, their exact masses, and their abundances see Appendix IV. The presence of the two sulfur atoms in this molecule is quite obvious in connection with the parent ion-peak of Fig. 3-16 where the parent ion +2 is about 8.8% of the intensity of the molecular ion.

C. Fragment Ions

Under the low pressures involved, some of the molecular ions will be brought through the analyzer tube to the detector. Some of the molecular ions containing energy in excess of the amount required for ionization will "fragment," a unimolecular process. In actuality, this is the dissociation of the ion moiety into two products, one of which carries the charge while the other becomes a neutral particle. The fragment or "daughter" ion may then be accelerated and brought through the analyzer tube to be detected. Its intensity may be greater or less than the molecular ion intensity, depending upon the ionization voltage and other factors. Also, the daughter ion itself may possess sufficient energy for a further dissociation to produce a fragment ion of still smaller *m/e* and another neutral particle. A substance is said to have a "fragmentation pattern" because this process of fragmentation is reproducible for a given substance in a given mass spectrometer under duplicate conditions of operation and a like amount of sample. In earlier literature one frequently encounters the term "cracking pattern" as a synonym for fragmentation pattern.

All of the ions in the spectrum in Fig. 3-16 with the exception of the molecular ion and its isotopes are "fragment ions." The two sulfur atoms must still be with the ion moiety occurring at *m/e* 116 on the basis of the M + 2 peak. Similarly, the *m/e* 129 peak also reflects the presence of two sulfur atoms in the ion. Ions at *m/e* 76–80 still reflect the presence of two sulfurs and probably comprise CS_2^+, CS_2H^+, $CS_2H_2^+$, and their isotopes. The other ion of the doublet is probably a hydrocarbon variety. Ions at *m/e* 88–93 and 56–60 obviously contain sulfur (always on the low-mass side due to the sulfur exact mass being 31.972073 and thus being below the corresponding hydrocarbon ion).

D. Rearrangement Ions

These ions arise by the transfer to another bonding location of one or more atoms in an ion moiety before subsequent cleavage occurs. In many cases, the process is concerted; that is, the rearrangement takes place simultaneously with the cleavage or fragmentation. Usually, it is obvious that the ions produced could not have been produced from the original ion entering the reaction without such a rearrangement occurring. Carbon skeletal changes are much rarer than hydrogen transfers, suggesting that the atomic weight and size may have a good deal to do with the processes. As an example, even at low voltages long-chain monoalkylbenzenes (1-phenylalkanes) produce the ion at *m/e* 92 (corresponding to the parent *m/e* of toluene).

$$C_6H_5\text{—}CH_2\text{—}CH_2\text{—}CH_2\text{—}(CH_2)_n\text{—}CH_3 \longrightarrow [C_6H_5\text{—}CH_3]^+ + CH_2\text{=}CH\text{—}R\cdot$$

Thus cleavage at the weakest carbon–carbon bond in the molecule, beta to the ring, will occur concurrent with a hydrogen atom shift from the alkyl-chain portion to form an even *m/e* ion at 92. Intense ions at *m/e* 91 also appear, but the existence of an appreciable amount of the rearrangement ion is shown by the excess intensity of the 92 ion above the amount expected from isotopic contributions from the 91 ion. Other rearrangements involve methyl groups and other hydrocarbon groups, as will be shown in considerable detail in Chapter 5.

The ion at *m/e* 116 is the ion of largest intensity in Fig. 3-16, and a consideration of its MW leads to the conclusion that it is a rearrangement ion on the basis of what is given above. The thiophene ring contains C_4H_3S– for a total of 83 amu, the second sulfur atom adds 32 to give a mass of 115. To get the 116 ion, a hydrogen atom would have to shift to the ionic species from somewhere on the alkyl chain as the molecular ion fragmented to produce the *m/e* 116 species. Note that such an ion can arise only from the molecular ion of higher MW or some other fragment ion of higher or equal MW, but possessing a different structural configuration. As we shall see later on, this ion is quite stable.

E. Multiply Charged Ions

The reaction

$$M + ne^- \rightarrow M^{n+} + (n+1)e^- \tag{3-8}$$

can occur in the electron impact source. Doubly and triply charged ions are thus observed. In most organic molecules, these ions occur to a relatively small extent, primarily because relatively low ionization energies are employed and also because many organic moieties can find energetically favorable paths for further fragmentation before very much energy is absorbed by the ion. As one might guess, those organic species capable of allowing charge separation with various resonance forms tend to show some double and even triply charged ions, as shown by Meyerson and Vander Haar (1962). These ions produce a peak at nonintegral m/e values, most noticeably if m is odd and e is 2.

As one can anticipate, the spectrum in Fig. 3-16 shows a number of doubly charged ions. The ion at m/e 40.5 probably arises from an ion of mass 81, but which is doubly charged ($81/2 = 40.5$). The presence of doubly charged ions at m/e 57.5 and 56.5 is obvious. These must arise from 115^{2+} and 113^{2+} species, respectively. Note that if the 116^{2+} ion exists it is covered by the singly charged ion at m/e 58 and hence is not detectable unless one has extremely high resolution. A reexamination of the upper 70 eV spectrum in Fig. 3-8 at this point is warranted. The doubly charged ions at 56.5 and 57.5 can be discerned with difficulty. When the ionizing voltage was at 15 eV, there was insufficient energy to form the doubly charged ions. Most doubly charged ions disappear below voltages of about 22.5 eV. While examining Figs. 3-8 and 3-16, note the region between 40.5 and 41 where a small metastable peak is expressing itself. Note that at 15 eV, this disturbance, which may well suggest simply noise at first, does not disappear. Metastable peaks are discussed in the next section and require special attention.

As suggested at the beginning of this section, ease of charge separation may have a good deal to do with stabilizing doubly charged ions in organic ion species. However, charge separation alone is not responsible since Newton (1964) claims observations of triply charged CS_2 under ordinary operating conditions and using nominal 70 V electrons. The peak occurs at m/e $25\frac{1}{3}$. Newton concludes that stable states of triply charged ions of small molecules can exist provided that there is a sufficient number of nonbonding or delocalized electrons in the molecule. Dewar and Rona (1965) have reported doubly charged ions in the mass spectra of organoboron derivatives.

For inorganic mass spectrometry, the situation is rather different. Multiply charged ions of most metallic species are easy to observe and occur rather commonly. In the extreme high-energy sources of the spark source instrument, ions from the gold electrode can be seen showing the loss of eleven electrons (see Fig. 3-1). Trace metals in the electrode even show species out to four electrons lost. In the older mass spectrometers employing mercury anywhere in the system, it was common practice to use the singly and doubly charged mercury isotopic ions as an internal marker. The ions could be seen; why not use them? Winters and Kiser (1966) reported doubly charged transition-metal carbonyl ions. They observed an interesting fact. Within experimental error, the second ionization potential of the $M(CO)_x$ species corresponds to the second ionization potential of the respective metal atom. The element with the lowest second ionization potential among those examined was molybdenum at 15.717 eV. Obviously, these are quite in contrast to the organic molecules.

While Meyerson and Vander Haar (1962) indicated that fragmentation of doubly charged ions into another doubly charged ion and a neutral fragment occurred, the possibility of doubly charged ions decomposing into two singly charged fragments should be considered. McCulloh *et al.* (1965) developed a new technique for observing the decomposition of multiply charged polyatomic ions. From their work they observed numerous decompositions that led to the formation of positive-ion pairs. Significant fractions of fragment ions with particularly high kinetic energy are formed by these processes.

A minor type of ion that may also be observed is one formed from polymeric molecules. Leckenby *et al.* (1964) and Robbins and Leckenby (1965) have observed double molecule ions in some of the rare gases, carbon dioxide, nitrogen, oxygen, and nitrous oxide. In condensing steam they observed the preferential formation of even-numbered polymers such as $(H_2O)_2$ and $(H_2O)_4$. Kebarle and his co-workers have also been active in this research field. The latest contributions are those of Searles *et al.* (1969), wherein gas-phase hydration studies of the positive alkali ions and the negative halide ions were made, and of Good *et al.* (1969), wherein the mechanism of the thermal ion–molecule reaction of oxygen with water vapor was studied. Thus, clustered, polyatomic, or double molecule ions will certainly present a challenging new area for studies of ions by means of mass spectrometry.

F. Metastable Ions

The examination of mass spectra of organic substances usually discloses small, diffuse peaks which frequently occur at nonintegral masses (see Fig. 3-16 in the region between m/e 36.5 and 41.0). These peaks are useful because they often define fragmentation processes between the ions appearing in the spectrum. For fragment ions to be formed in the ion source of a mass spectrometer, the decomposition reaction(s) must proceed with a rate of about 10^6/second or faster. If the rate of the reaction is in the neighborhood of

10^5/second, the decomposition will occur while the ion is in transit, according to Hipple and Condon (1945). Momigny (1961) and Coggeshall (1962) extended consideration to wider ranges of ion source time residence. Decompositions occurring in the accelerating region, before the ion in transit has entered the analyzer tube, produce the "metastable peaks." To be correct, it should be recognized that these ions are the products of metastable transitions.

By equating the radius of curvature of the metastable and ordinary ions, Hipple *et al.* (1946) derived an equation to express the apparent mass m^* at which a peak would appear because of ions from such a metastable transition as $m_i^+ \rightarrow m_f^+ + m^0$. It is

$$m^* = \frac{m_f^2}{m_i}\left(\frac{V_1}{V}\right) + m_f\left(1 - \frac{V_1}{V}\right) \tag{3-9}$$

where m_i is the initial mass of the ion decomposing, m_f is the final mass of the decomposed ion, V is the full potential of the accelerating region, and V_1 is the potential of the accelerating region at the point where the decomposition occurs. Beynon (1960c) pointed out that conditions are much more complex than indicated by the above equations, and hence many of the ions are not recorded. The directional-focusing properties of the sector magnetic fields cause ions that decompose close to the entrance slit of the analyzer section to be favored in recording. Thus, the value of V_1 approaches V and results in

$$m^* = m_f^2/m_i \tag{3-10}$$

Beynon claimed that careful mass measurements on heptadecane indicate that the maximum intensity of an observed metastable peak from the transition $(C_4H_9)^+ \rightarrow (C_3H_5)^+ + CH_4{}^0$ which occurs at mass 29.5 occurs 0.06 amu higher than given by Eq. (3-3). This corresponds to a value for V_1/V in Eq. (3-8) of 0.995, or almost unity. Hipple (1947) also has shown that the equations apply to all sector instruments but that the efficiency of ion collection is greater by a factor of 3 for the 60° sector instruments than for the 180° instruments. In the area of qualitative identification, this is one advantage of a 60° sector instrument over others.

Newton (1966) has discussed the shape and position on the mass scale of metastable ion peaks in Dempster-type instruments. Vestal (1964) has discussed isotope effects of metastable transitions in mass spectra, while Muccini *et al.* (1964) have shown that the pressure dependence of the diffuse peak at m/e 39.1 in n-alkanes from C_3 to C_{10} proves that the metastable transition is not a collision-induced dissociation (see also Section H). Rhodes *et al.* (1966) have described a method for deriving metastable ion transitions in hydrocarbon mass spectra, which may assist the mass spectroscopist in his early stages of studying new and unknown types of compounds. Aplin *et al.*

(1965) have described a logarithmic response circuit modification for assisting in the recording of metastable ion peaks.

Metastable peaks, in addition to being recognized by their more diffuse recording of the peak, usually at nonintegral mass values, may be verified by making several instrumental changes when recording the spectrum. (a) The collector slits may be increased in size to allow more of the "diffuse" peak to be recorded. Thus, with an increase in slitwidths, the metastable ion peak will increase in size relative to the ordinary peaks. Conversely, if the slits are closed to approach the highest possible resolution of an instrument, the metastable ion peaks will disappear. Some instruments used primarily for quantitative analyses may be equipped with metastable peak suppressors to prevent interference with the accurate measurement of peak heights by the metastable ion peaks. (b) Increasing the repeller voltages of the ion source will tend to increase the relative peak heights of the metastable ion peaks compared to normal ions. This is shown for the data of Hipple *et al.* (1946) in Fig. 3-17. Metastable peaks vary linearly with pressure, just as normal peaks do; they also vary with the ionizing current.

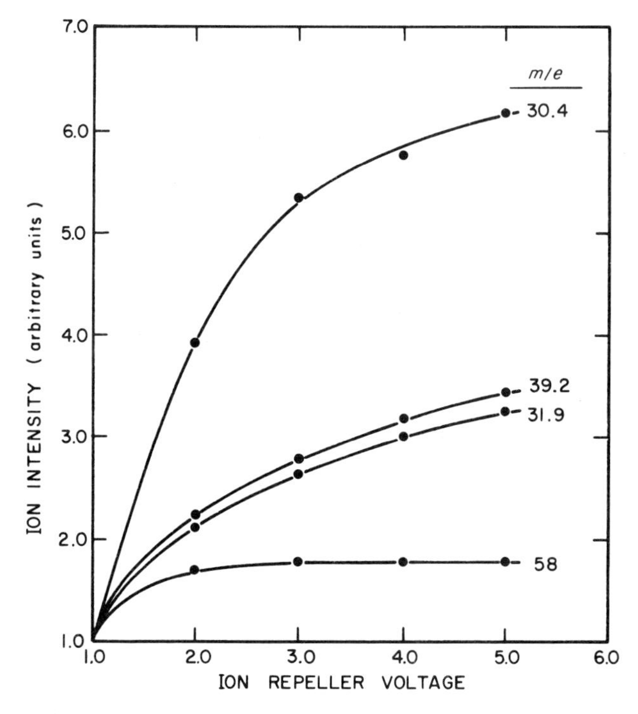

Fig. 3-17. Effect of variation of repeller voltage upon observed ion current. (After Hipple *et al.*, 1946. Courtesy of *The Physical Review*.)

Metastable peaks are never of large intensity when compared to the parent or other intense ions in the mass spectrum. Thus, this leads to the general statement that the presence of a metastable peak proves that a given process is taking place but, conversely, the *absence of the metastable peak cannot be accepted as proof that the process is not taking place.*

Double-focusing mass spectrometers can also produce metastable peaks but to some extent the techniques to be used depend upon the geometry of the analyzers. Futrell *et al.* (1965) discussed the problem in connection with Mattauch–Herzog geometry and reported the presence of a metastable peak for the transition $C_3H_8^+ \rightarrow C_3H_7^+ + H$ in propane. This transition was predicted by the quasi-equilibrium theory but not observed in sector instruments. Sasaki *et al.* (1968) also described detection of metastable peaks in the Mattauch–Herzog spectrometer. Beynon *et al.* (1964) have discussed the techniques used in Nier–Roberts geometry instruments for distinguishing between metastable ions formed during transit in the electrostatic analyzer and those that form before or after this section. Jennings (1965) has extended this technique to obtain a pure metastable mass spectrum from which the normal ions are excluded. Jennings (1966) has also demonstrated that consecutive metastable transitions in a double-focusing mass spectrometer can be detected.

Beynon and Fontaine (1967) reexamined the shapes of metastable peaks and considered decompositions occurring throughout the whole flight path. They calculated the effects on or discriminations against product ions and discussed theoretically the effects of finite beam width and height and the angular spreading and kinetic energy of separation of fragments during the decomposition. They also provided practical examples.

Beynon *et al.* (1965) have discussed the dissociation of metastable ions in mass spectrometers with release of internal energy. The presence of "flat-top" peaks, corresponding to metastable ions in the mass spectra of organic compounds, is often noted. The excess energy involved can be estimated from the width of these peaks and a knowledge of the various operating parameters of the mass spectrometer. Measurements of the retarding potentials necessary to suppress a given metastable peak can be used to confirm the estimate made from the peak widths mentioned above. In a subsequent paper, Beynon *et al.* (1968b) discussed the variation of the energy released in metastable decompositions as a function of the lifetime of the decomposing ion. They claimed that if a metastable ion species always decomposed with release of the same amount of kinetic energy, then the sides of the metastable peak would be as steep as those of a normal mass peak. Since metastable peaks are "diffuse" it is probable that there is a release of a range or distribution of kinetic energies in all such decompositions.

Metastable peaks have also been observed for decompositions of double charged ions (Mohler *et al.*, 1954) and proven to arise from fragment ions formed with kinetic energy in excess of thermal energy, sometimes by 2–3 eV (Olmstead *et al.*, 1964). The Lawrence Radiation Laboratory group (Olmstead *et al.*, 1964) pointed out that in the mass spectra of many organic substances CH_3^+ ions are formed with excess kinetic energy and give rise to a satellite peak on the high-mass side of the ordinary *m/e* 15 peak. The ions may be formed by the dissociation of a highly excited state singly charged ion or more likely from the dissociation of a double charged ion into the CH_3^+ ion and a further ion. McLafferty and Bursey (1967) have reported a case of a specific rearrangement of doubly charged ions from the molecule ion of benzophenone as proven by a metastable peak. The abundance of the $(M - CO)^{2+}$ ion is over 10%. These authors point out that doubly charged ions may involve either *no* or *two* unpaired electrons. Aromatic systems would be expected to favor the loss of electron pairs while for charge-separated ions, the homolytic cleavage of the diradical ion would lead to two odd-electron ions formed with the aid of the kinetic energy of coulombic repulsion. Speculation about the structure of the doubly charged ion formed gives a number of highly conjugated systems with charge separation as the most favored type. The mechanism of how this occurs is an open question.

Beynon and Fontaine (1966) have applied the earlier work of Beynon *et al.* (1965) on release of internal energy and its relation to metastable peak shape to doubly charged ions. They have claimed that structural information about stable fragments and the possible intercharge distance in these ions can be inferred. An excellent example (including illustration and methodology) of a metastable peak arising from doubly charged benzene ions is given by Higgins and Jennings (1965). Their work will allow the interested reader to pursue this area further. Higgins and Jennings (1966) have also examined other aromatic hydrocarbons and have shown that consideration of these types of peaks in higher molecular weight aromatic species should prove fruitful for structure elucidation.

Shug (1964) has shown that the statistical theory of mass spectra [see Chapter 5 and Rosenstock *et al.* (1952, 1955)] gives a reasonable interpretation of the metastable peaks occurring in mass spectra. He employed metastable lifetime measurements to obtain useful information in some of the cases he considered. An approximate method for computing the ratios of parent-ion to metastable peak intensities by means of the statistical theory was presented. Conceding that this is an extreme point of view from that taken by Hipple and his co-workers (1945–1947) in which the parent ions were assumed to dissociate at a single rate, Shug considers that the agreement with statistical theory, which predicts essentially a continuous distribution of rate constants for each dissociation, provides further support for that theory. McLafferty has also considered metastable ion characteristics in further detail in a series of three communications. The most significant of these theoretically (McLafferty and

Pike, 1967) is one that considers the variation of metastable ion abundances in mass spectra of compounds with vibrational degrees of freedom. They established that the abundance of a particular metastable ion peak relative to the abundance of ions of the same elemental composition as the precursor is inversely proportional to the number of vibrational degrees of freedom in the original molecule. This would support the general observation that metastable peaks of larger intensity are often observed when the process involves the breakdown of aromatic species having fewer C–H vibrational possibilities.

Both Shannon *et al.* (1967) and Beynon (1968) have shown the importance and methodology of using metastable ions to aid in the interpretation of mass spectra. The reader will be well advised to watch for developments of additional ways in which the metastable ions appearing in mass spectra can be utilized to assist his interpretation.

G. Negative Ions

The presence of negative ions in the mass spectrometer has been known since the time of the pioneering work of Thomson. However, far less work has been done in this area. There are two prime reasons for this: (a) The production efficiency of positive ions is from three to many times greater than that of negative ions in the 50–100 eV range employed in most of today's instruments; (b) negative ions are often formed with excess kinetic energy which in turn reduces the collection efficiency. The improvements in instrument design to permit reversal of electric and magnetic fields, more precise control of slits, repellers, etc., with their resulting increase in sensitivity has led to more progress in recent years. Negative ions are of importance not only because of their possible contribution to the thermodynamics of ion formation, but also because they are a major factor in such processes as ionosphere physics, the formation of intermediates in polymerization, and from the analytical viewpoint they permit accurate isotopic abundance determinations for some elements and substances, i.e., halogens, etc. Reese *et al.* (1956) demonstrated that a modern instrument could readily be used to study negative ions and emphasized (Reese *et al.*, 1958) the reasons why fewer such studies had been made. Measurements were difficult to carry out, and the peak heights of the various ions were usually far smaller than in the case of the positive ions. Khvostenko (1964) carried out investigations of thiophenes and their homologs and showed that under identical conditions (within the scope of parameters that could be kept identical, i.e., source temperature, pressure, etc.) the negative ions produced were only about 4% of the amount of positive ions. Thus, even if not detected, one would suspect that perhaps a similar small amount of negative ions were being formed during ordinary positive-ion investigation of these same species. The negative ions would, of course, contribute to the overall energetic balance of the system. Dibeler *et al.* (1957) reported an exception to the general rule for the spectrum of perchloryl fluoride, in which the most abundant ion, positive or negative, is ClO_3^-. Since then, other cases have been reported. The interested reader is referred to the excellent chapter, "Negative Ion Mass Spectra," by Melton (1963a) for many additional details in this area.

Negative ions are known to be formed in the gas phase by four different mechanisms. Three of these are the consequence of electron impact, and, using the classification system of Melton (1963b), these reactions are the following:

1. Dissociative resonance capture.

$$AB + e^- \rightarrow A + B^- \tag{3-11}$$

2. Resonance capture.

$$AB + e^- \rightarrow AB^- \tag{3-12}$$

3. Ion-pair production.

$$AB + e^- \rightarrow A^+ + B^- + e^- \tag{3-13}$$

As far as is known processes (1) and (2) are operative only for the production of negative ions by electron impact. Process (3) can be induced by any ionizing radiation, e.g., photons, alpha particles, electrons, etc. For the detailed thermodynamics behind these reactions the reader is referred to Melton (1963b). A fourth mechanism is also known for the production of gas-phase negative ions. In view of a controversial nomenclature situation in which others have used "charge exchange," "charge transfer," or "double electron attachment" Melton (1963b) has adopted the term "charge permutation."

4. Charge permutation.

$$[A'B^+] + AB \rightarrow [A'B^-] + (AB) \tag{3-14}$$

The brackets indicate an energy in excess of 500 eV per ion, and parentheses indicate an unknown species and charge. This mechanism is of much more importance in ions produced by radiation rather than by direct electron impact. The interested reader is therefore referred to additional sources for further information. The energetics of negative-ion formation are discussed in great detail in Massey (1950), Massey and Burhop (1952), and Loeb (1955).

Although the reversal of fields in the mass spectrometer permits negative ions to be observed, their intensities are low. As a consequence, the measurement of negative-ion appearance potentials is considerably more difficult than that of positive-ion appearance potentials. The retarding potential difference method of Fox *et al.* (1951) has given results on standard reference materials to within 0.1 eV of the spectroscopic values for positive ions. Comparable accuracy can be expected when this method is used for negative ions. Processes (1) and (2) above are not like the ordinary processes for the formation of positive ions wherein the electrons emitted can carry away any excess energy as kinetic energy. In the positive-ion case, then, the process can occur at any energy above the threshold. By contrast, in negative-ion formation there are no electrons to carry away the excess energy, so both (1) and (2) above are resonance processes. For this reason the shape of the ionization curve for negative-ion production is very different from that of positive ions. Using the retarding potential difference method, Hickam and Fox (1956) measured the resonance capture peak SF_6^- in the spectrum of sulfur hexafluoride. A pulse-operated electron beam of monoenergetic electrons of known energy was employed. Their results indicate that the capture process occurs at less than 0.1 eV and over an energy range not greater than 0.5 eV. It should be noted that this compound has the highest known dielectric strength for a gas. Hickam and Berg (1959) pointed out that this property might be associated with the ease with which electrons could be picked up before sufficient energy was acquired to initiate breakdown. Since no dissociation is involved in this case, it is a typical type (2) reaction.

If dissociation occurs with the capture a different set of energetics should be expected. The shape of the ionization efficiency curve is still narrow (and sharp, as it was for the SF_6). Kiser (1965c) gave as a thermodynamic example the case for formation of a negative oxygen ion from carbon monoxide,

$$CO + e^- \rightarrow C + O^-$$

which utilizes the data of Lozier (1934). Lozier's 9.6 eV value and the SF_6 value of 0.08 eV discussed above are frequently used standards. Kiser (1965d) tabulated appearance potentials associated with some electron capture processes and with some ion-pair production processes. For the comparison with the possible case in the example of carbon monoxide,

$$CO + e^- \rightarrow C^+ + O^- \tag{3-15}$$

The O^- value of 20.9 eV is taken from the work of Hagstrum and Tate (1939, 1940).

Dillard and Franklin (1968) have reported the onset values for the negative ions from carbonyl sulfide and carbon disulfide as determined on a time-of-flight mass spectrometer. In order to study ion–molecule reactions of negative ions they have also increased the source pressure to several microns (see also next section on ion–molecule reactions developed essentially from the positive-ion viewpoint).

In addition to the measurement of appearance potentials other desirable instrument capabilities as listed by Melton (1963a) include high detection sensitivity; a corrosion-resistant ionizing electron filament; precise control of the ionizing radiation, whether electrons, photons, etc.; a high-temperature ion source; and high-speed differential pumping. Melton also described the reasons for each need and outlined the various solutions that are available to the mass spectroscopist.

A number of compounds have been investigated for both positive and negative ions (as in the above-mentioned work of Khvostenko, 1964). Perhaps the most explicit was that of the methane molecule performed by Melton and Rudolph (1959) and Melton and Rosenstock (1957), the former report concerning the negative-ion spectra and the latter the positive-ion spectra. Melton (1963c) made the graphical comparison of results, which showed that for the positive-ion spectra the relative abundance of ions was CH_4^+, 100%; CH_3^+, 80%; CH_2^+, 10%; CH^+, ≈3%; and C^+, less than 1%. Conversely, the negative ion spectra showed C^-, 100%, CH^-, ≈85%; CH_2^-, ≈25%; and CH_3^-, ≈2%; the CH_4^- ion was not detected. Thus the abundances reflected the stabilities of the various ion structures as was expected. Another important point is that for the 90 eV electrons used, the ratio of CH_4^+ to C^- was about 10,000 to 1.

A number of compounds have been investigated for general features of the negative-ion spectra which could aid in determination of molecular structures and the relation to chemical properties. Melton and Rudolph (1959) also investigated the alcohol series, while Ropp and Melton (1958) considered some aliphatic acids. Mann *et al.* (1940) reported that at low source pressures the negative-ion mass spectrum of ammonia consists of H^- and NH_2^- ions. Undoubtedly NH_3^- would be produced by resonance capture (process 2) at higher pressures. Melton (1963d) also presented a summary table of negative ions observed in halogenated compounds and references associated with each. Quite obviously the halogens would be the most useful group of compounds in negative-ion spectroscopy. Gohlke and Thompson (1968) have shown that the marker (perfluorokerosene) used by many mass spectroscopists is also a useful internal reference for negative-ion mass spectrometry.

Melton (1969) pointed out that negative-ion spectra can often be of great assistance in structure determination because rearrangement processes do not occur as they do in positive-ion spectra. Hydrogen movements are not seen, except in a negative-ion–molecule reaction. Sometimes it can be shown that ions pass through a negative-ion intermediate as part of the fragmentation process. These must be of the charge permutation type (process 4), further examples of which are given later. Isotopic ratios of some substances such as the halogens and other species containing more electronegative atoms lend themselves to negative-ion mass spectrometry. Taylor and Grimsrud (1969) have described the advantages for chlorine isotope ratios found by examination of methyl chloride samples. They have also given a procedure for the conversion of inorganic chloride samples to gaseous methyl chloride in a manner that will insure isotopic identity of the methyl chloride from the chloride precursor. Von Ardenne *et al.* (1961) have reported the negative-ion spectra of a number of condensed aromatic hydrocarbons in which the ions were generated by electron capture. The method offers even more intense molecular ions than in positive spectra, and also by varying the ionizing current offers promise for the determination of the number of hydrogen and carbon atoms present in a hydrocarbon molecule without the interference of the usual fragments, etc.

Melton (1961) has reported on the transient species observed during the catalytic reaction,

$$D_2 + CO_2 \xrightarrow{Pt} D_2O + CO \tag{3-16}$$

Of six transient species found, four were in the negative-ion spectrum. They were D^-, DCO_2^-, CO_3^-, and DCO_3^-. It should be apparent that mass spectroscopists studying reaction catalysis should employ both positive- and negative-ion spectroscopy. Research involving radiation of any kind can also produce ions of both the positive and negative type; therefore, the technique is essential to any thorough study of such processes in which radiation effects are being considered.

Among other interesting negative ions researched is the Cl_3^- ion reported by Melton *et al.* (1958) in the spectrum of chlorine gas. It is unfortunate that it was impossible to determine whether the ion resulted from the neutral Cl_3 atom, the stability of which had been predicted, or if the ion was the result of a collision:

$$Cl_2 + Cl^- \rightarrow Cl_3^-$$

Negative ions can arise from the collision of a neutral molecule and a positive ion, or conversely, positive ions arise from the collision of a negative ion and a neutral particle. Actually, all of these processes are similar to the charge permutation (process 4), except that the energy of the particles and the nature of the resulting structures are not accurately known, and they all need not be of such a high energy content. Henglein (1954) reported the formation of negative C_2H^- and C_2^- ions from $C_2H_2^+$, and Melton (1958) and Ropp and Melton (1958) demonstrated that negative O^-, $(OH)^-$, and $(HCOO)^-$ ions can be formed from collisions of krypton atoms with positively charged formic acid ions. Melton also showed that neutral helium atoms colliding with the negative $(HCOO)^-$ ion from formic acid produced the series of positive ions $(HCOO)^+$, $(COO)^+$, $(HCO)^+$, and $(CO)^+$. One can readily picture why Melton prefers to call the process *charge permutation*. It does work both ways.

The charge permutation process also takes place with other small and large particles. Dukel'skii *et al.* (1956a,b) described collision between neutral gases that transformed He^+ into He^-, while Fogel and Krupnik (1956) and Fogel *et al.* (1956a,b; 1957a,b) performed the transition of H^+ to H^-. Collisions between negative halogen ions and neutral molecules were considered by Dukel'skii (1955) and Dukel'skii and Fedorenko (1955), who also found that the transfer of electrons to an acceptor such as Ag from donors such as Sb^-, Sb_2^-, Sb_3^- occurred if the electron affinity of the acceptor was not less than that of the donor. Negative ions containing Si, Ge, Sn, and Pb have been reported by Dukel'skii and Sokolov (1957).

The field of negative-ion mass spectrometry should continue to expand and prove to be an important technique for many purposes. To the reader who has found some stimulation it is suggested that considerable additional literature material is available. The important point for the organic-oriented mass spectroscopist to remember is that these events can happen, although admittedly they are usually of small importance as source pressures are low. From the applications viewpoint, the suggestions and postulations of Melton should provide a start into consideration of the technique for organic molecules containing hetero atoms.

H. Ion–Molecule Reactions

Peaks from intermolecular reactions between ionized species and un-ionized molecules in the ion source are observed in the mass spectra of some substances. Usually, a substantial percentage of the gaseous molecules entering the electron impact source are not ionized, and these neutral molecules may well undergo collisions with ions already formed in the source. As a consequence, a single atom or even a group of atoms may be abstracted from the neutral molecule by the colliding ion resulting in an ion of a larger mass than the original molecular ion. Since this is a bimolecular process, it should be second order and show a linear dependence of the ion intensity with the square of the sample pressure. This provides a convenient means for the mass spectroscopist to check on the origin of such ions. An examination of spectra obtained at several different sample pressures is usually conclusive. Fortunately, at the low pressures employed for most analytical work, the number of such collisions is small and the amount of ions formed by this process is also small. Boudart (1961) estimates only 1 in 10,000 at 10^{-6} mm pressure. Several exceptions are notable: The abstraction of hydrogen atoms gives rise to ions at $M + 1$ in the spectra of some ethers, esters, amines, amino esters, and nitriles. In all of these cases, the molecular ion has a low stability resulting in cleavage to fragment ions. While the M + 1 peak is useful for finding the molecular weight in such cases, it can be misleading and interferes with the determination of the isotopic profile of the molecular ion. It frequently can present a problem in interpreting the spectrum of an unknown material.

Ion–molecule reactions occur by a number of processes, and the mass spectroscopist must at least be aware of the various possibilities that may present themselves in the course of interpreting organic mass spectra. Following the system of Melton (1963e), the following mechanisms are given:

1. Simple charge transfer wherein an electron is transferred.

$$A'B^+ + AB \rightarrow A'B + AB^+ \tag{3-17}$$

2. Charge-transfer-induced dissociation; B is a free radical.

$$A'B^+ + AB \rightarrow A'B + A^+ + B \tag{3-18}$$

3. Radical or atom abstraction.

$$A'B^+ + AB \rightarrow A'B_2{}^+ + A \tag{3-19}$$

or

$$A'B^+ + AB \rightarrow A'^+ + AB_2 \tag{3-20}$$

This process will obviously interfere with isotope abundance measurements,

4. Ionic abstraction.

$$A'B^+ + AB \rightarrow A' + AB_2{}^+ \tag{3-21}$$

or

$$A'B^+ + AB \rightarrow A'B_2 + A^+ \tag{3-22}$$

Melton (1963f) claimed that the formation of the $CH_5{}^+$ ion can occur by either process 3 or 4.

$$CH_4 + CH_4{}^+ \rightarrow CH_5{}^+ + CH_3 \tag{3-23}$$

Abramson and Futrell (1967) have examined various ion–molecule reactions of normally encountered fragment ions from cyclohexane and other hydrocarbons and used both conventional high-pressure mass spectrometers as well as a tandem mass spectrometer in their investigation. A general conclusion from their work is that all the fragment ions investigated reacted in a hydride transfer reaction of the type

$$R^+ + C_6H_{12} \rightarrow RH + C_6H_{12}{}^+ \tag{3-24}$$

They also observed that the $C_3H_6{}^+$ and $C_4H_8{}^+$ ions abstracted $H_2{}^-$ to produce the $C_6H_{10}{}^+$ ion. These reactions are typical of the variety that can occur in the two types of mechanisms given in 4 above.

5. Attachment.

$$A'B^+ + AB \rightarrow A'BAB^+ \tag{3-25}$$

In this process, the excess kinetic energy cannot be dissipated but the aggregate can subsequently lose energy in collision with a neutral molecule.

6. Collision-induced dissociation.

$$A'B^+ + AB \rightarrow A'^+ + B + AB \tag{3-26}$$

or

$$A'B^{2+} + AB \rightarrow A'^+ + B + AB^+ \tag{3-27}$$

Both of these processes require high kinetic energy, but were noted in early instruments.

7. Charge permutation.

$$A'B^+ + AB \rightarrow A'^- + \text{other products} \quad (3\text{-}28)$$

or

$$A'B^+ + AB \rightarrow A'B^- + AB^{2+} \quad (3\text{-}29)$$

These processes require ions of high kinetic energy.

8. Ionization.

$$A'B^+ + AB \rightarrow A'B^+ + AB^+ + e^- \quad (3\text{-}30)$$

This is just a special case of ionization, but it does require a high-energy reactant species, such as a negative ion or atom (500 eV), to induce the reaction.

9. Excitation. This is in reality simply a case of exchange of energy between the reactant ion and the neutral molecule which results in the degradation of the energy of the ion and an enhancement of the energy of the neutral molecule. If sufficient energy is imparted, the neutral molecule may dissociate. Only quantum restrictions apply to the energy transferred and, hence, electronically or vibrationally excited states (also rotational and translational) may be formed. The excited neutral molecule dissociates if the vibrational energy exceeds that of the critical dissociation limit. The reaction may be represented by

$$A'B^+ + AB \rightarrow A'B^+ + AB^* \quad (3\text{-}31)$$

Because of the increased use of ionization potential data from mass spectrometric measurements and the comparison with photoionization processes, the organic mass spectroscopist must be aware of the behavior of ion–molecule interactions in the instrument as well as the possible spurious ions that can cause difficulty in interpretations. A brief discussion is therefore included dealing with the effects on these aspects of mass spectrometry.

An extensive bibliography of ion–molecule reactions was prepared by Harllee *et al.* (1966), covering the period of January 1900 to March 1966. Langevin's (1905) study concerning such reactions and their contribution to kinetic theory was the earliest entry in this compilation. The second entry was Dempster's (1916), "The Ionization and Dissociation of Hydrogen Molecules and the Formation of H_3"; Smythe (1924) and Thomson (1924) were the next entries; and only Thomson's title "Recombination of Gaseous Ions, etc." hinted at the application to mass spectrometry by the appearance of the word "ions." Another paper by Smythe (1925) followed and it was not until the sixth entry, a study by Hogness and Harkness (1928), that a title included the name "mass-spectrograph." It is interesting to note than an effort is still being made to incorporate the words "mass spectrometer" or synonyms into titles of publications. This does much to assist in proper abstracting.

Progress was steady, albeit slow, with 20 more entries appearing through the ensuing years to 1948. These were studies fundamental to the field of chemical physics. Mitchell *et al.* (1948) wrote a report called "Secondary Processes of Ion Production in the Mass Spectrometer" while Franklin's (1952) study, "Carbonium Ion Reactions—A Study of Their Rates and Mechanisms," seems to have opened a new era for mass spectrometry. In the same year, Tal'rose and Lyubimova (1952) submitted a paper called "Secondary Processes in the Ion Source of the Mass Spectrograph," and Lindholm (1953) reported "Ionization and Fragmentation of Molecules by Bombardment with Atomic Ions," thus insuring that research with ion–molecule reactions was underway.

By 1960 over 100 additional papers had appeared, and the rate of appearance in the literature is still increasing. While there are far too many to introduce to the readers of this treatise, several papers are of note or interest. Meisels *et al.* (1956) discussed ion–molecule reactions in radiation chemistry while Stevenson and Schissler (1955) discussed the rate of gaseous reactions of the type that produce a charge exchange. A most interesting report is that of Lampe and Field (1957), who showed that the protonated methane ion can be formed in the mass spectrometer at pressures somewhat above those normally encountered in mass spectrometry. Franklin *et al.* (1963) were the first to use a title containing the words "Chemi-Ionization and Ion-Molecule Reactions in Gases," thus starting a new subfield which is of considerable importance and merits separate attention. Field and Lampe (1958) made a study of hydride ion transfer reactions, while Hunt *et al.* (1964) reported spurious fragments from charge transfer and dissociation reactions in a TOF instrument. Ferguson *et al.* (1965) also described studies in a linear, pulsed TOF instrument in which the products of ionic collision processes were observed. As an example of the contributions to organic mass spectra the reader may wish to examine the work of Derwish *et al.* (1964), the "Mass-Spectrometric Study of Ion–Molecule Reactions in Propane." The experimental techniques, results, and implications about secondary ions for such a "simple" system are very enlightening.

Melton (1963e) and Stevenson (1963) have written chapters of advanced treatises that contain considerable information for a background in depth on ion–molecule reactions. Both chapters contain rate data and examples of theoretical applications of these to molecular energetics. Knewstubb (1969) has added a text, "Mass Spectrometer and Ion–Molecule Reactions," which gives considerable detail about methods of generating ions and the applications and theories of ion–molecule reactions.

I. Chemical Ionization

The term "chemical ionization mass spectrometry" was first described by Munson and Field (1966a), as reported in Chapter 2 wherein the necessary instrument–source modifications for this technique were described. Their developments and discoveries seem to be a natural outgrowth of studies involving charge exchange and ion–molecule reactions as carried on by this group for some time previous.

Field (1968) has produced an excellent review paper summarizing the results of investigations on at least six types of compounds and their behavior upon chemical ionization. Field (1968) has pointed out that only beginnings have been made in this area, and conditions used have been such as to hopefully lead to analytically useful (and reproducible) results on organic compounds. It should be recognized rather immediately that chemical ionization processes are not governed by Franck–Condon considerations, as compared to both electron impact and photon impact ionization, since massive entities, protons and larger ions or molecules, are involved. Chemical ionization processes should approximate adiabatic processes, which are slow equilibrium adjustments of electronic states and atomic positions. Additionally, the chemical ionization processes differ by the fact that the product ions contain an even number of electrons. This is in direct contrast to the formation of odd-electron ions in electron impact, photon impact, and field ionization mass spectrometry. As a result of these facts, the mass spectra obtained from the chemical ionization of materials are different from the spectra produced by the other ionization processes. Most chemical ionization mass spectra are relatively simple, and for many compounds significant amounts of ions are found at m/e values near the molecular weight of the compound.

The technique employed is to introduce a reaction gas into the ionization chamber of a mass spectrometer at a pressure of about 1 torr. The reaction gas is ionized by electrons to produce the reaction ions. If a small amount of another gaseous material is present in the mixture, the reaction ions react with the second material to produce a spectrum of ions characteristic of the second material. Most of the work of Munson and Field utilized methane, which produces primary ions by the reaction with electrons of

$$CH_4 + e^- \rightarrow CH_4^+, CH_3^+, CH_2^+, CH^+, C^+, H_2^+, \text{and } H^+ + 2e^- \qquad (3\text{-}32)$$

The ionization of the second material or additive will be negligible if the ratio of methane to the second material is about 1000 : 1 or so. If methane is regarded as the primary material, then product (secondary) ions are formed at virtually every collision by reactions that were well established by Field and Munson (1965):

$$CH_4^+ + CH_4 \rightarrow CH_5^+ + CH_3 \qquad (3\text{-}33)$$

$$CH_3^+ + CH_4 \rightarrow C_2H_5^+ + H_2 \qquad (3\text{-}34)$$

$$CH_2^+ + CH_4 \nearrow C_2H_4^+ + H_2 \qquad (3\text{-}35)$$

$$CH_2^+ + CH_4 \searrow C_2H_3^+ + H_2 + H \qquad (3\text{-}36)$$

$$CH^+ + CH_4 \rightarrow C_2H_2^+ + H_2 + H \tag{3-37}$$

$$C_2H_5^+ + CH_4 \rightarrow C_3H_7^+ + H_2 \tag{3-38}$$

$$C_2H_3^+ + CH_4 \rightarrow C_3H_5^+ + H_2 \tag{3-39}$$

$$C_2H_2^+ + CH_4 \rightarrow \text{Polymer} \tag{3-40}$$

While small relative concentrations of ions such as $C_3H_7^+$, $C_2H_2^+$, $C_3H_3^+$, $C_3H_4^+$, and $C_4H_9^+$ are formed, these ions do not make any appreciable contribution to the total ionization observed. These authors also demonstrated that for pressures above 0.5 torr, the relative intensities of the CH_5^+ and the $C_2H_5^+$ ions produced from the methane are substantially independent of pressure. If the additive concentration is kept low as described above, the reactions of the primary ions of methane with the additive is negligible by comparison, and the rapid reaction for the methane ions should be predominant.

Since the ions at *m/e* 16 and 15 from methane comprise about 90% of the total ionization from electron impact [see API-44 Mass Spectral Data Serial No. 1; Zwolinski (1968)] the major secondary ions should be (and are) the CH_5^+ and $C_2H_5^+$ ions, which also are about 90% of the total ionization under chemical ionization conditions. The $C_3H_5^+$ ion is about half of the remaining ion current. Hence, if methane is the species which is being used to accomplish the chemical ionization, it is actually the two ions CH_5^+ and $C_2H_5^+$ reacting with the additive which produce the major part of the chemical ionization mass spectrum of the additives.

Earlier investigations of Munson and Field (1965a,b) and Munson *et al.* (1964) have led to some generalizations. First, the CH_5^+ and $C_2H_5^+$ react mostly by donating a proton or abstracting a hydride ion. The reactions are

$$CH_5^+ + BH \rightarrow BH_2^+ + CH_4 \tag{3-41}$$

$$C_2H_5^+ + BH \rightarrow BH_2^+ + C_2H_4 \tag{3-42}$$

$$C_2H_5^+ + BH \rightarrow B^+ + C_2H_6 \tag{3-43}$$

where B represents any organic group (or possibly inorganic group). The important distinction between these reactions and conventional electron impact reactions is that for methane the energy available in these collision-induced ionizations is of the magnitude of 0.20 eV. Hence the even-electron ions produced (see Chapter 5) will not be energy rich, and its lifetime will be correspondingly increased.

These same authors developed the discussion in the following terms: First, methane is a weak Bronsted base, the CH_5^+ ion is a strong Bronsted acid, and proton transfer reactions to bases stronger than CH_4 will occur. Second, if the reactions are sufficiently exothermic, then the BH_2^+ ions produced above will dissociate,

$$BH_2^+ \rightarrow B^+ + H_2 \quad (3\text{-}44)$$

$$BH_2^+ \rightarrow A_i^+ + B_i \quad (3\text{-}45)$$

where A_i^+ is one possible fragment ion. Munson and Field (1965a) reasoned that since C_2H_4 is a stronger Bronsted base than CH_4, then $C_2H_5^+$ is a weaker Bronsted acid than CH_5^+. Proton transfer may still occur from the $C_2H_5^+$ moiety to stronger bases, Munson and Field (1965b) having observed this for water and ammonia. The competing reaction (3-44) is a hydride transfer reaction in which the $C_2H_5^+$ acts as a Lewis acid. These were observed by Field and Munson (1967).

For both types of reactions of the $C_2H_5^+$ ion, product ions may be formed with sufficient energy to decompose. Hence, one would expect to observe such a series of fragment ions characteristic of the added material. These characteristic mass spectra are observed and are the essence of the major differences between electron impact and chemical ionization spectra.

In a subsequent series of papers, Munson and Field examined the chemical ionization of a number of compounds. Esters were considered in the first of these (Munson and Field, 1966b) and the spectra explained in terms of the main attack at the carbonyl group by the CH_5^+, $C_2H_5^+$, and $C_3H_5^+$ ions. Proton transfer reactions occurred to give $(M + 1)^+$ ions. Some of these in turn dissociated to give $RCO_2H_2^+$, RCO^+, and alkyl ions, R'^+, from the alcohol chain. For some compounds, the protonated alcohols were produced. Collision-stabilized addition of the $C_2H_5^+$ and $C_3H_5^+$ occurred with some of the esters as well as displacement reactions in which these ions were added to the molecule and an olefin moiety was expelled. These spectra are typical of those compounds which possess a dipole in the molecule. Moran and Hamill (1963) have shown that the presence of a permanent dipole in a molecule markedly increases the cross section for an ion–molecule reaction. This certainly must be the case for this group of compounds.

The paraffins were the first compounds for which chemical ionization mass spectra were determined, but the paper appeared in an "Advances in Chemistry Series" (Field *et al.*, 1966) at about the same time as the ester work. To illustrate the difference between chemical ionization and electron impact, Field (1968) utilized the spectrum of *n*-octadecane. This is presented in Fig. 3-18, where both types of spectra are given. The reader should note that the graphs shown were computer plotted from the original data. For the chemical ionization spectrum, the $(M - 1)^+$ ion is the dominant ion; while in the electron impact spectrum, the ion is insignificant. The remaining ions are alkyl ions of the +1 series occurring at each carbon number and in approximately equal in-

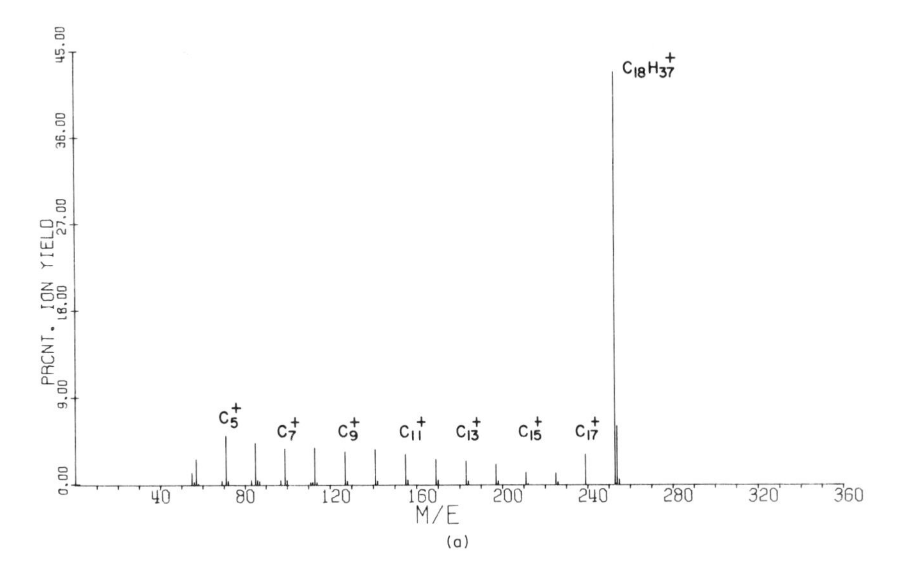

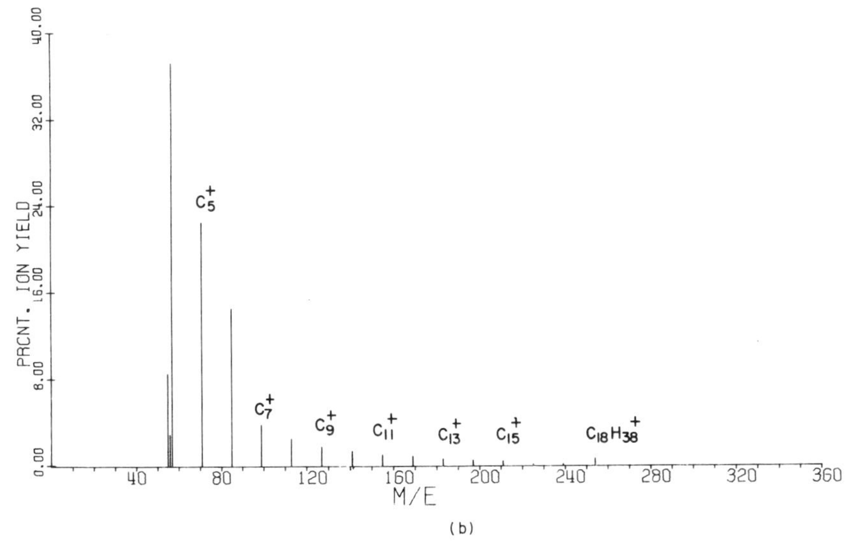

Fig. 3-18. Comparison of typical chemical ionization (a) with electron impact ionization (b). Spectrum is of *n*-octadecane. Graphs are computer plotted. Reactant is CH_4; pressure, P_{CH_4}, is 1.0 torr. [Reprinted from F. H. Field, *Accounts Chem. Res.* **1**, 45 (1968). Copyright (1968) by the American Chemical Society. Reprinted by permission of the copyright owner.]

tensities. Field concluded that the "chemical ionization processes in paraffins are quite different from electron impact processes." The electron impact spectrum of the *n*-octadecane is quite typical, possessing more intense ion peaks corresponding to *m/e* 43, 57, 71, etc., and decreasing in intensity very markedly with increasing carbon number. Field also concluded that "the $(M-1)^+$ ion is as dominant in $n\text{-}C_{44}H_{90}$ as it is in smaller paraffins, but by contrast, the M^+ intensities produced by electron impact exhibit a monotonic decrease as the size of the normal paraffin increases." The work of Field and his co-workers also contains considerable thermodynamic data and justification for their conclusions as to the course of chemical ionization mechanisms in the case of the compounds studied. They show for the *n*-octadecane, as an example, that the formation of the $(M-1)^+$ ion from the normal paraffin occurs by H^- abstraction from the molecule, either as a direct hydride ion abstraction with $C_2H_5{}^+$ or as a dissociative proton transfer from $CH_5{}^+$. In either case, the reactions are written to produce a secondary octadecyl ion and with the heat of reaction being exothermic. The formation of the primary octadecyl ion would be slightly endothermic and, thus, relatively improbable as a reaction product.

Field and Munson (1967) have reported the chemical ionization mass spectra of 28 compounds of the cycloparaffins class. The $(M-1)^+$ ions are the most intense ions in the spectra of all but one of these compounds. These authors have developed correlations of the intensities of the $(M-1)^+$ ions with the structure of the molecules for the alkylcyclopentanes and alkylcyclohexanes. The mass spectra give information about the presence of methyl groups in the structures of the cycloalkane molecules in the form of $(M-15)^+$ ions. This is not characteristic of electron impact spectra and is another of the number of differences observed. Cyclic olefin ions are also formed which give information about the ring present.

Munson and Field (1967) have reported on the chemical ionization spectra of 21 alkylbenzenes and two alkylnaphthalenes. The addition reactions of H^+ and $C_2H_5{}^+$ are observed, in contrast to the paraffins. Hydride abstraction does occur for the aliphatic hydrogen atoms but not for the aromatic hydrogens. Alkyl ion and olefin displacement reactions are noted for some of the branched alkylbenzenes. For the tertiary amyl compound the $C_5H_{11}{}^+$ ion is about 80% of the additive ionization. The $(M-1)^+$ peak intensities for methyl and ethyl substitution are shown to be related to the relative number of ring and side chain carbon atoms in the molecules. If larger side chains are present, particularly branched ones, these are stripped from the ring as alkyl ions, thus reducing the $(M+H)^+$ ion peak intensities. Hence, the correlation developed has a limited usefulness. The M^+ ions are usually observed, but the intensities are extremely small; $(M-H)^+$ ions were observed for all of the compounds reported except for benzene itself. Munson and Field (1967) showed that the hydrogen atom is removed from the side chain for all the alkylbenzenes, and that this was to be expected upon the basis of the thermodynamics, i.e., abstraction from the alpha position of the side chain is exothermic. It is suggested here that the reader refer to the article for additional details on this and their proposed mechanisms for the dealkylation reactions that occur. This article is the first indication of the additional benefit of chemical ionization, that of aiding in delineation of fragmentation and other reaction mechanisms going on in mass spectrometry.

In article VI of the series, Field (1967) reported detailed chemical ionization studies of the C_7H_8 isomers, toluene, cycloheptatriene, and norbornadiene. Toluene-α-d_3 and cycloheptatriene-7-*d* were also examined. Thus, in this article the hint given in the earlier article is emphasized in that the ionization processes are postulated, which in these cases involve quite conventional carbonium ion chemistry. A disproportionately high amount of D^- abstraction occurs when compared to the amount of deuterium and hydrogen in the molecules. Thus, it is concluded that in the hydride abstraction process by chemical ionization, the hydrogens do not become equivalent. The abstraction occurs in the case of the toluene from the methyl groups and from the methylene group in cycloheptatriene. Protonation of the toluene or cycloheptratriene followed by the loss of H_2 does not appear to occur. From the isotopic distributions observed in the formation of benzenium and propyl ions from cycloheptatriene-7-*d*, the tentative conclusion is drawn that randomization of hydrogens occurs in the processes forming these ions. The suggestion is made that the addition of $C_2H_5^+$ (or, less likely, H^+) to the cycloheptatriene occurs by rearrangement leading to the formation of the product ion.

It was noted in this article that in the chemical ionization formation of $(M - 1)^+$ ions, isotopic mixing does not occur. This is in contrast to the electron impact ionization behavior, and seems to be the consequence of the fact that $(M - 1)^+$ ion formation in chemical ionization occurs by hydride abstraction by a Lewis acid, a process which does not appear to involve mixing and exchange. No analog for this process exists in electron impact ionization phenomena. Field (1967) also concluded, "It would appear that the ions involved in these transformations are sufficiently activated and have lifetimes prior to ultimate decomposition long enough to permit migration of hydrogen atoms around the molecule. In this respect, these chemical ionization processes conform on the one hand to the mobility observed in electron-impact ionization and dissociation of C_7H_8 isomers, and on the other hand, they conform to the very well known lability of carbonium ions in condensed phase organic chemistry."

Field (1968) pointed out that chemical ionization mass spectrometry is still very new and that the determination of the extent of its utility will require much further work. Most of the work of Field and his co-workers has been done with methane; however, probing results from other reactants give indications that very different spectra, as predicted, are produced. The reactants used were propane, isobutane, water, methanol, rare gases, and mixtures of rare gases in benzene. The major ions in propane (s-$C_3H_7^+$) and isobutane (t-$C_4H_9^+$) react primarily by hydride ion abstraction. The t-butyl ion is a milder reagent than the secondary propyl, and both are milder than the ethyl ion found in methane. The degree of fragmentation of the additive materials is less when these reactants are used than with methane as a reactant.

With water and methanol, protonated entities are formed almost exclusively. These can be monomers, dimers, trimers, etc., depending upon the identity and pressure of the reactant substance. Field reported that by using water as a reactant, selective ionization of an equimolar mixture of 1-decanol and n-decane resulted in ions from the decanol appearing. The ions were $(M + 1)^+$ and $(M + H_3O)^+$. Since no ions appeared from the decane, this ability to effect differential ionization of compound types should be of great importance in practical studies.

The rare gases act as reactants by abstracting an electron from the material to be ionized, producing an odd-electron ion. The spectra are thus somewhat similar to spectra from electron impact, but with the degree of fragmentation occurring dependent upon the identity of the rare gas used as a reactant. As one might expect, if this is similar to a charge exchange situation, xenon produces the least fragmentation and helium the most fragmentation.

Chemical ionization mass spectrometry has now expanded into at least a second generation of workers and into a new area of interest. Fales *et al.* (1969) have reported on the extension of the technique to a number of high molecular weight compounds of biological interest. As expected, somewhat simpler fragmentations are encountered, thus making the method an improvement over the electron impact examination of the same compounds. In some steroidal ketones the variation in the abundance of the peak corresponding to loss of water suggests that the location of the carbonyl group in such steroid nuclei might be determined by using chemical ionization spectra. In a subsequent paper, Fales *et al.* (1970) reported the chemical ionization (CI) of alkaloids and indicated that both spectra (electron impact, EI, and CI) are complementary in that they supply structural data by their differences in fragmentation. The quasimolecular ion $(M + 1)^+$ is invariably more relatively abundant in the CI mode than the EI mode. Ziffer *et al.* (1970) have used the CI technique to study the structure of the photodimers of cyclic α,β-unsaturated ketones, and Milne *et al.* (1970) have applied the technique to studies of amino acids. The reader will be well advised to keep abreast of the many new developments which will be forthcoming in this expanding area of mass spectrometry.

VI. Presentation of Mass Spectral Data for Interpretation

A. Spectral Conventions

Complete or reference mass spectra are usually presented for study, interpretation, or catalog use by converting the *xy* data of m/z versus raw peak height or ion intensity to a tabularized column of m/z and percent relative intensity and/or percent total ion intensity. Mention must be made here of partial mass spectra, often obtained over limited regions for various purposes, such as location of a metastable peak, verification of atomic composition of selected m/e by means of a high-resolution instrument, or perhaps even making a photorecord of isotope ratios in a limited region or just the variations of several peaks during a chemical reaction monitored by a time-of-flight instrument. Obviously, these are only partial mass spectra and not reference or complete mass spectra. Still, they are valuable working spectra.

1. *Percent Relative Intensity*

To obtain relative intensity from the raw peak height of the "*i*th" ion, the ion of largest intensity in the spectrum is selected. The intensity of this ion peak is called the base peak intensity, *B*. All of the raw peak heights, p_i, are multiplied by 100 and divided by *B* to give the percent relative intensity, %RI (relative to the largest ion):

$$(p_i \times 100)/B = \%\mathrm{RI} \tag{3-46}$$

Tabulation by increasing order of m/z is the standard method.

2. *Percent Total Ion Intensity*

Often it is of great advantage to the interpreter or correlator of spectra to utilize spectra presented in the total ion intensity or total ion current form. To obtain the percent total ion intensity ($\%\Sigma_i$), the p_i are summed to give the total ion intensity. Then the various peaks, p_i, are multiplied by 100 and divided by the sum of the p_i to give the percent total ion intensity:

$$(p_i \times 100)/\sum p_i = \%\Sigma_i \tag{3-47}$$

It is customary to indicate the m/e range covered in the summation by subscripts such as 12–200, 37–350, etc., or Σ_{37-350}.

3. *Sensitivity*

This term is used in numerous different ways to express a connection between the amount of material in the sample and the resulting ion intensity. Usually, it is given for the ion of largest intensity or the base peak. Often, the molecule ion is used, but any other suitable ion may be used for a given problem. While it should be recognized that the ion intensity is ultimately measured in amperes (to some negative power of 10) for all of the electrical detection systems or in terms of photoplate intensity of a "line" the presentation may be in almost any arbitrary units which are relatable to the above quantities. In addition, a number of instrumental variables must be specified, such as the ionizing voltage, the ionizing current (microamperes, usually), the sample inlet temperature, etc. Most important, however, is some form of quantity of sample measurement, such as micrograms, microliters (or lambda), or microns of gas pressure, etc. Thus, sensitivities are often expressed as microamperes of ion current for a selected peak per micron of sample, as chart divisions per micron of gas, or as divisions per microliter per microampere (of ionizing current). The type of sample often dictates the system to be used (refer again to Sections IV,B and C). The sensitivities of various substances vary considerably. Molecular ion intensities of a number of compounds are discussed in Chapter 5.

B. Presentation as a Bar Graph

From the calculated spectral data, a bar graph of the most important ions may be made and shown to illustrate some features of the spectra. An example is given in Fig. 3-19. Note that the bar graph may have the left ordinate labeled in %RI, while the right ordinate may be in $\%\sum_i$. The abscissa is given in m/z units. The bar graph has the advantage of permitting the exclusion of a number of peaks not of interest to the discussion, thus emphasizing the largest peaks in the spectrum. It allows fragmentation pathways to be directly suggested and usually contains far less data than the tabularized forms of mass spectra. Disadvantages outweigh advantages. Among the disadvantages are the inability to show features at low ion intensities, including most doubly charged ions and metastable ions, and the relative inability to discern the isotopic profiles of prominent ions. The bar graph does not usually appear in mass spectral data catalogs. The example shown in Fig. 3-19 is far superior

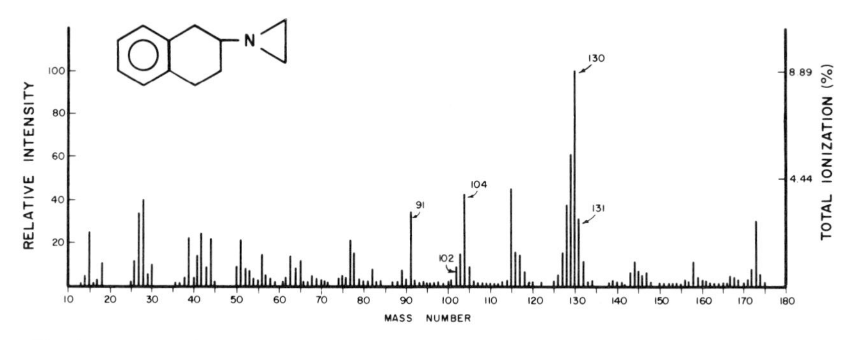

Fig. 3-19. Example of a bar graph presentation of a mass spectrum of 1-(1,2,3,4-tetrahydro-2-naphthyl)aziridine. (Courtesy of Continental Oil Co.)

to many seen in the literature. Bar graph presentation using a computer plotter is shown in Fig. 3-18.

C. Tabulation by Percent Relative Intensity

This method of presentation has the advantage of allowing all observed data to be on a single page, clearly visible to the interpreter, and it still shows doubly charged ions and metastable peaks. It permits the consideration of isotopic profiles and, at a glance, allows the mass spectroscopist to inspect complete spectral data, including sensitivity data, instrument variables, etc. In other words, it is a complete or reference mass spectrum. (It can be considered a reference spectrum only if the sample purity is very high and independent means of structure verification such as infrared spectra, nuclear magnetic resonance spectra, etc., were employed. All instrumental variables should be specified in addition to these requirements of purity.) An example is given in Fig. 3-20.

D. Tabulation by Total Ion Intensity

Presentation by this method has all the advantages of the percent relative intensity method plus the advantages given below. A glance tells what percentage of the ionized material remains as the molecular ion, and a summation of discernible parts of fragmentation pathways indicates the approximate proportion of ions formed by that pathway. Summation of selected fragments from these spectra also assists in interpretation. An example of this type of presentation is shown in Fig. 3-21. Total ionization data can also be used to check sensitivity data, for a specific compound or set of compounds, when the mass spectra have been determined in several laboratories.

Total ionization is defined as the sum of all of the ion intensities for all of the ions in the mass spectrum multiplied by the sensitivity of the base peak (ion of largest intensity). The sensitivity is usually expressed in units of current per unit of pressure with the specific problem determining the individual dimensions to be used. Total ionization of hydrocarbons was reported by Mohler *et al.* (1950), who described the values for a number of compounds containing only C and H. While there is no absolute relationship between total ionization and structure, Ötvos and Stevenson (1956) showed that cross sections of molecules for ionization by electrons could be related to Bethe's (1930) theoretical development of the ionization cross section. Here the cross section, Q_{nl}^{i}, of an atomic electron with quantum numbers, (n, l), is approximately proportional to the mean square radius of the electron shell (n, l). Ötvos and Stevenson also suggested that their results show that cross sections for molecules are a constitutive property to a good approximation. This

Ratio of Mass to Charge	Relative Intensity At magnetic field of 4800 gauss, For ionizing voltages of 70 volts	gauss / volts
57	4.72	
58	1.29	
59	1.63	
60	.19	
61	.29	
62	.22	
63	.79	
64	.40	
65	2.78	
66	2.40	
67	2.31	
68	.41	
69	1.78	
70	.95	
71	3.08	
72	.34	
73	.55	
74	.21	
75	.33	
76	.31	
77	5.66	
78	2.00	
79	2.90	
80	.41	
81	.78	
82	.43	
83	.70	
84	2.74	
85	1.26	
86	.24	
87	.43	
88	.14	
89	.51	

Ratio of Mass to Charge	Relative Intensity At magnetic field of 4800 gauss, For ionizing voltages of 70 volts	gauss / volts
90	.37	
91	3.81	
92	.71	
93	.81	
94	.23	
95	1.16	
96	.78	
97	23.33	
98	2.40	
99	2.00	
100	.31	
101	.29	
102	.24	
103	.79	
104	.60	
105	.92	
106	.22	
107	.36	
108	.64	
109	2.14	
110	15.12	
111	16.53	
112	3.18	
113	1.29	
114	.34	
115	.64	
116	.33	
117	.47	
118	.14	
119	.43	
120	.16	
121	.75	
122	.96	

Ratio of Mass to Charge	Relative Intensity At magnetic field of 4800 gauss, For ionizing voltages of 70 volts	gauss / volts
123	16.74	
124	2.88	
125	2.69	
126	.47	
127	.31	
128	.24	
129	.24	
130	.13	
131	.18	
133	.25	
134	.53	
135	1.17	
136	.53	
137	3.74	
138	.43	
139	.90	
140	.22	
141	.25	
142	1.19	
143	.50	
144	.46	
145	.37	
146	.23	
147	.41	
148	.20	
149	.54	
151	1.28	
152	.40	
153	.71	
154	.26	

Ratio of Mass to Charge	Relative Intensity At magnetic field of 4800 gauss, For ionizing voltages of 70 volts	gauss / volts
156	1.17	
157	.21	
161	.15	
162	.16	
163	.26	
165	.42	
166	.26	
167	1.14	
168	.39	
169	.86	
170	.17	
171	.24	
172	.09	
173	.11	
174	.10	
175	.12	
176	.16	
177	.23	
178	.31	
179	.81	
181	100.00	
182	14.02	
183	5.88	
184	.78	
185	.25	
193	.57	
195	90.00	
196	12.26	

COMPOUND

Name: 2-*n*-Hexyl-5-*n*-heptylthiophene

Molecular Weight	Molecular Formula	Approximate Boiling Point	Approximate Freezing Point	Approximate Density
266.49	$C_{17}H_{30}S$	°C	°C	g/ml at °C

Semi-structural Formula:

C–C–C–C–C–C–C(thiophene ring, S)C–C–C–C–C–C

Source: Professor R. W. Higgins (Welch Foundation Grant) Texas Woman's University Denton, Texas

Purity: by Mass Spectrometer 99.3 mole percent

Total Ionization for Compound: div/micron/microampere; div/lambda/microampere

Sensitivity for Base Peak: 74.97 div/micron; div/lambda/microampere

MASS SPECTROMETER

Maker and Model: Consolidated Model 21-102 (Modified)

Ionizing Current	Ion Chamber Temperature	Vapor Temperature	Collector Slit Width
9.5 microamperes	250 °C	135 °C	mils

Sample Pressure and Basis of Measurement: Micromanometer

LABORATORY: U.S. Bureau of Mines Bartlesville Petroleum Research Center Bartlesville, Oklahoma

STANDARDS

n-BUTANE	*n*-HEXADECANE
Magnetic Field: 4800 gauss	Magnetic Field: gauss
Total Ionization: div/micron/microampere; div/lambda/microampere	Total Ionization: div/micron/microampere; div/lambda/microampere
Sensitivity at mass/charge 43: 35.0 div/micron; div/lambda/microampere; $I/\Sigma I$	Sensitivity at mass/charge 57: div/micron/microampere; div/lambda/microampere; $I/\Sigma I$

mass/charge	Relative Intensity	mass/charge	Relative Intensity	mass/charge	Relative Intensity
15		57		141	
27		71		155	
29		85		169	
43		99		183	
58		113		197	
		127		226	

Additional Information:

Date of Measurement: February 5, 1960

Fig. 3-20. A typical mass spectrum of 2-*n*-hexyl-5-*n*-heptylthiophene, expressed in relative intensities. (From the Americal Petroleum Institute, Research Project 44, Serial No. 2074, Texas A & M University, 31 October 1966; contributed by the U.S. Bureau of Mines, Bartlesville, Oklahoma.)

Ratio of Mass to Charge	Relative Intensity — At magnetic field of / For ionizing voltages of		Ratio of Mass to Charge	Relative Intensity — At magnetic field of / For ionizing voltages of		Ratio of Mass to Charge	Relative Intensity — At magnetic field of / For ionizing voltages of		Ratio of Mass to Charge	Relative Intensity — At magnetic field of / For ionizing voltages of	
	4800 gauss, 70 volts	gauss, volts		oersteds, volts	oersteds, volts		oersteds, volts	oersteds, volts		oersteds, volts	oersteds, volts
197	4.84										
198	.59										
199	.12										
207	.35										
208	.25										
209	1.82										
210	.40										
211	.22										
221	.47										
222	.14										
223	1.26										
224	.33										
225	.18										
227	.08										
228	.08										
230	.16										
233	.23										
237	.82										
238	.28										
250	.23										
251	.70										
252	.09										
254	.06										
255	.04										
258	.04										
259	.04										
260	.08										
261	.03										
262	.09										
264	.42										
265	.68										
266 p	32.87										
267	4.88										
268	1.94										
269	.25										
280	.22	Impurity									

Ratio of Mass to Charge	Relative Intensity At magnetic field of 2994 oersteds, for ionizing voltages of 70 volts	Relative Intensity At magnetic field of 1785 oersteds, for ionizing voltages of 70 volts	Ratio of Mass to Charge	Relative Intensity At magnetic field of 2994 oersteds, for ionizing voltages of 70 volts	Relative Intensity At magnetic field of 3672 oersteds, for ionizing voltages of 70 volts	Ratio of Mass to Charge	Relative Intensity At magnetic field of 2994 oersteds, for ionizing voltages of 70 volts	Relative Intensity At magnetic field of 3672 oersteds, for ionizing voltages of 70 volts	Ratio of Mass to Charge	Relative Intensity At magnetic field of 2994 oersteds, for ionizing voltages of 70 volts	Relative Intensity At magnetic field of 3672 oersteds, for ionizing voltages of 70 volts
12		12	55	39	37	80	54	63	111	21	26
13		8	56	24	25	80.5	25				
14		42	57	131	174	81	155	205	113	148	195
15		260	57.5	23		82	311	411	114	187	241
			58	261	346	83	86	109	115	1773	2272
26		129	59	127	147	84	78	97	116	264	336
27		698	59.5	11		84.5	16		117	96	120
28		104	60	45	49	85	131	159	118	17	21
29		382	60.5	80		85.5	74		119	61	88
30		12	61	156	207	86	229	298	120	112	146
			62	567	739	86.5	21		121	551	752
32		25	63	1475	1835	87	273	368	122	58	83
33		14	64	198	247	88	171	215	123	29	44
34		24	65	285	341	89	770	990			
			66	75	82	90	230	288	126	74	96
37		50	66.5	16		91	291	349	127	159	207
38		165	67	175	199	92	33	33	128	469	608
39	1967	1002	67.5	28		93	295	400	129	124	167
40	117	58	68	24	32	94	91	109	130	17	19
41	552	312	69	2106	2560	95	151	197	131		17
42	54	25	70	244	304	96	32	40	132		84
43	51	36	71	398	483	97	88	108	133		101
44	33	20	72	76	87	98	72	94	134		1293
45	3012	1942	72.5	81		99	57	69	135		258
46	74	49	73	169	201	100	29	28	136		75
47	157	99	73.5	280		101	229	287	137		18
			74	474	622	102	885	1147			
49	38		74.5	24		103	1526	1861	139		91
50	631		75	546	695	104	134	169	140		29
51	1143		76	303	380	105	40	51	141		167
52	203		77	1582	1873	106	64	90	142		129
53	142		78	178	222	107	32	46	143		37
54	28		78.5	20		108	248	331			
			79	253	286	109	56	74	145		712
			79.5	51		110	36	49	146		553

COMPOUND

2-*n*-Propylbenzo[*b*]thiophene (2-*n*-Propyl-1-thiaindene)

Molecular Weight	Molecular Formula	Approximate Boiling Point °C
176.27	$C_{11}H_{12}S$	

Semi-structural Formula

(benzo[b]thiophene ring with S) —$CH_2CH_2CH_3$

Source: D. S. Rao, PRF Fellow, Bureau of Mines, Laramie, Wyoming, Through API RP 48

Purity: mole percent

MASS SPECTROMETER

Maker and Model: Consolidated No. 21-103C

Ionizing Current: 19 microamperes

Basis of Pressure Measurement: Internal Standard - n-Hexadecane

LABORATORY: Laramie Petroleum Research Center, U. S. Bureau of Mines, Laramie, Wyoming

Relative Intensity

mass/charge		mass/charge	
n-Butane (At		oersteds)	
15		43	
27		58	
29			
n-Hexadecane (At	*	oersteds)	
57	10000	141	244
71	5466	155	202
85	3376	169	155
99	666	185	100
113	450	197	53
127	325	226	214

Sensitivity

	n-Butane	*n*-Hexadecane
mass/charge	43	57
div/micron		
div/mm³		
$I/\Sigma I$		17835

ΣI (See note on reverse side)

Temperatures

Inlet Vapor	250 °C
Ionization Chamber	250 °C

Additional Information

Relative intensities shown in body of table are based on Tot. I = $\sum$ m/e 12-38 at 1785 oersteds + $\sum$ m/e 39-100 at 2994 oersteds + $\sum$ m/e 100-204 at 3672 oersteds.

* Relative intensities at m/e 57, 71, 85, 99, at field of 2994 oersteds, all others at field of 3672 oersteds.

Date of measurement: December 29, 1959

Fig. 3-21. A typical mass spectrum of 2-*n*-propylbenzo[*b*]thiophene (2-*n*-propyl-1-thiaindene), expressed in percent total ion intensity. (From the American Petroleum Institute, Research Project 44, Serial No. 1850, Agricultural and Mechanical College of Texas, 30 April 1962; contributed by the U.S. Bureau of Mines, Laramie, Wyoming.)

Ratio of Mass to Charge	Relative Intensity		Ratio of Mass to Charge	Relative Intensity		Ratio of Mass to Charge	Relative Intensity		Ratio of Mass to Charge	Relative Intensity	
	At magnetic field of			At magnetic field of			At magnetic field of			At magnetic field of	
	oersteds	3672 oersteds		oersteds	oersteds		oersteds	oersteds		oersteds	oersteds
	For ionizing voltages of			For ionizing voltages of			For ionizing voltages of			For ionizing voltages of	
	volts	70 volts		volts	volts		volts	volts		volts	volts
147		39532									
148		5522									
149		2036									
150		240									
151		22									
157		12									
158		173									
159		161									
160		334									
161		477									
162		74									
163		28									
169		16									
170		23									
171		174									
172		62									
173		143									
174		93									
175		303									
176 P		11020									
177		1395									
178		543									
179		61									
190		10									
204		15									

NOTE: $\sum I$ as used in sensitivity section equals $\dfrac{\sum I \text{ sample/unit weight}}{\sum I \text{ n-Hexadecane/unit weight}}$

where n-Hexadecane was the internal standard

$\sum I = 0.766$

approximation allowed the development of "type analysis" or characterization of complex mixtures such as gasolines, kerosenes, and other high molecular weight materials by Hood (1958), Crable and Coggeshall (1958), and Clerc *et al.* (1955) (see also the example in Chapter 6). The additivity concept is satisfactory for many other types of organic molecules, and Lampe *et al.* (1957) have amply demonstrated this for the hydrocarbons.

For the hydrocarbons, the *principle of additivity* can be given as

$$\sum_i I = K(n_H Q_i^H + n_C Q_i^C) \tag{3-48}$$

where $\sum_i I$ is the total ionization, Q_i^H and Q_i^C are the ionization cross sections for hydrogen and carbon atoms, n_H and n_C are the number of hydrogen atoms and carbon atoms, respectively, in a hydrocarbon molecule, and K is a proportionality constant. Using Ötvos and Stevenson's relative total ionization cross sections, where $Q_i^H = 1.00$, the above equation becomes

$$\sum_i I_i = K(n_H + 4.16 n_C) \tag{3-49}$$

As an example of the application of this equation, consider the data in Fig. 3-22. Using the data of Foster *et al.* (1964, 1965) for 37 alkylthiophene mass spectra determined on the same instrument and normalized to equal *n*-butane sensitivities, the total ionization was calculated. The ordinate axis is in arbitrary divisions and thus is involved with K, the proportionality constant above. The original data were plotted in terms of number of carbon atoms in the side chains, since this represented a constant increase of CH_2 per carbon atom.

The 37 alkylthiophene mass spectra were obtained at an inlet temperature of 135°C, while the entries for the 2-methylthiophene and thiophene are from the data of Kinney and Cook (1952) obtained on a room-temperature inlet system (API-44 No. 500, 914). The slight temperature effect should be present because of increased molecular vibrations, rotations, and translations with temperature. The additivity of atomic ionization cross sections to give molecular cross sections neglects temperature effects.

Kiser (1965d) warned that comparison of such data on different instruments may be difficult. Although it was originally expected that such measurements would be independent of the instrument, observation and experience have shown otherwise. Thus, the data in Fig. 3-23 illustrate some of the problems of utilizing such data.

A second point of note is the variation within isomers containing the same number of atoms. The isoalkyl derivatives seem to fall below the general line of data, the neopentyl compound first showing a marked deviation while the 2-(4-methylpentyl)— and 2-2-(ethylhexyl)— compounds continue on the low side.

Another effect that is noticeable is the changing of the magnetic field for the higher molecular weight compounds. Those compounds with 10 carbon

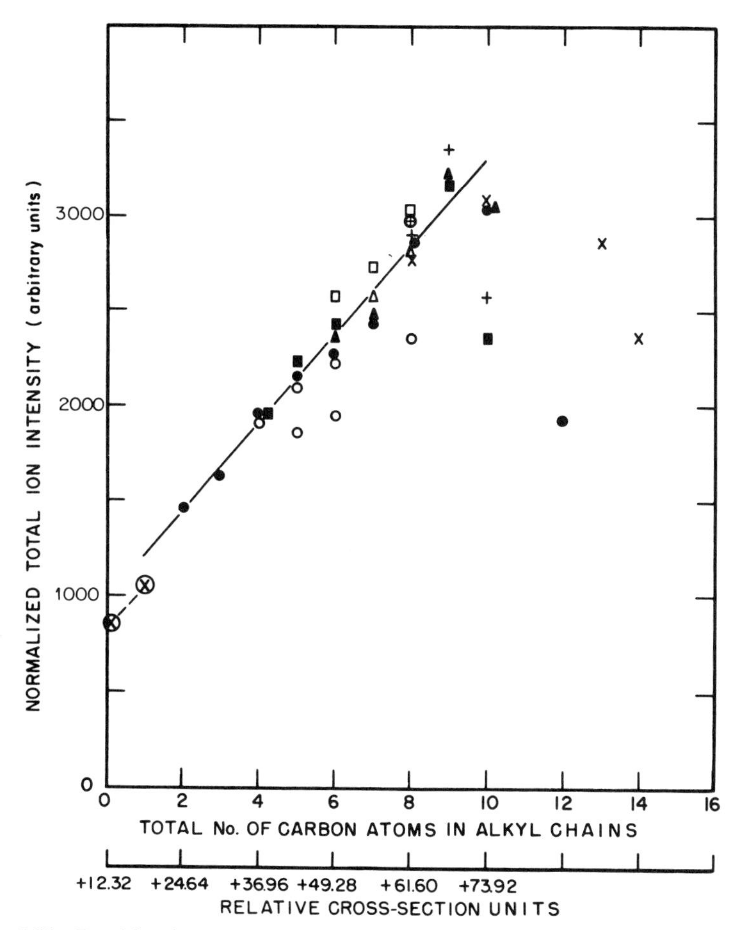

Fig. 3-22. Total ion intensity data on the class, alkylthiophenes. Key: ●, 2-*n*-alkylthiophenes; ○, 2-isoalkylthiophenes; ■, 2-ethyl-5-*n*-alkylthiophenes; □, 2-ethyl-5-isoalkylthiophenes; ▲, 2-*n*-propyl-5-*n*-alkylthiophenes; △, 2-*n*-propyl-5-isoalkylthiophenes; ×, 2,5-di-*n*-alkylthiophenes; ⊕, 2-*n*-alkyl-5-isoalkylthiophenes; +, 2,5-diisoalkylthiophenes; ⊗, API-44 Serial Nos. 500 and 914.

atoms or more as substituents were run at higher magnetic fields. The first reason for a decrease is that the lower m/e's were not scanned, but these could add, at most, only about 20% to the total ionization. The second reason for the decrease is the discrimination against the higher m/e peaks in this instrument as described by Berry (1950). Although the magnetic field was raised to help compensate for the discrimination, it is obvious that it was insufficient. A third reason, and probably the biggest contributor to the deviations, is that the very high molecular weights of these compounds caused the micromanometer in the inlet to indicate too high a pressure for the amount of material actually reaching the ion source. This could, of course, be due to volatility, wherein the dodecyl-substituted compound would possibly have a somewhat lower vapor pressure than the two disubstituted molecules having a higher molecular weight.

Since the mass spectra were run for reference purposes only, the data reflect the variations that one might expect to find for a set of spectra that was *not* run to prove anything about total ion intensities. The calculations were made after the publications appeared. The spectra of additional thiophenes, if corrected to the same relative intensity for *n*-butane, should show total ionization values along or near the general data line. Because of the very obvious problems with volatility, etc., for the compounds above nine carbon atoms on the side chain, a least-squares line was made for only the compounds having eight carbon atoms or less as substituents. This line falls slightly under the "general" line given but has the same intercept as the "general" line. From the treatment that follows, it is apparent that the sensitivity of the intercept value to the slope of the line is very critical. Hence, taking any subclass of compounds, highly branched, di-substituted, etc., will lead to a different intercept value. Since the ionization cross section for the sulfur atom can be estimated from the intercept, a consideration of these data is of interest.

Extrapolation of the line to intercept the x axis results in a value of between -4.4 and -4.5 units. The x axis represents methylene (CH_2) groups. To convert to the Ötvos and Stevenson relative ionization cross-section units, one must recall that 6.16 cross-section units would equal the two hydrogen atoms plus the carbon atom of the methylene group. Using the value of 4.5 methylene units $\times$ 6.16 cross-section units per methylene unit results in a total of 27.72 cross-section units assignable to the thiophene ring. Note that by the method used here the third terminal methyl hydrogen atom was not accounted for in the plot but was left with the central thiophene moiety. Thus, one must account for the total four hydrogens of thiophene. The carbon and hydrogen atoms in the ring account for $4 \times (1.00 + 4.16) = 20.64$ cross-section units. On this basis, the sulfur atom in the ring must account for the 7.08 or about 7.1 cross-section units. The value of 7.1 is too low when compared to the 12.8 units for S in CS_2 and H_2S, as given by Ötvos and Stevenson (1956). To indicate the magnitude of change in cross-section value from the resulting change in intercept, consider the two points from the room-temperature data for thiophene and methylthiophene. A -4.0 intercept results in 24.64 cross-section units for the thiophene ring and yields a value for sulfur, by the same procedure as given above, of only 4.00, which is less than the carbon atom. Including all of the high molecular weight data would obviously lead to a huge cross-section value for the sulfur atom. In any case, the nature of the sulfur atom in the thiophene ring should not be expected to correspond to the sulfur atom environments in the two molecules Ötvos and Stevenson observed. In fact, if ring expansion occurs as postulated by Foster *et al.* (1965) and Foster and Higgins (1968), then the sulfur atom would be the 10-electron type involving *d* orbitals!

It seems that Ötvos and Stevenson's claim of additivity of atomic ionization cross sections is about as valid as the claim of additivity of atomic refractions. Agafonov and Agafonov (1964a,b) discussed this relationship and presented equations correlating these quantities with cross-section data. In the cases examined, the cross section for ionization on electron impact and total ionization can be determined with average errors of approximately 8 and 13%, respectively. Also given is the conclusion that for many compounds total ionization can be determined by calculation from atomic refractions with an error of 4.7–7.7%. Careful inspection of the refractive index–density data supplied by API-44 (Zwolinski, 1968, tables a) for the various hydrocarbon isomers leads to the conclusion that although they have the same atom content, isomeric structures show a variation in molar refraction values. Nevertheless, this is the type of general relationship that may bear fruit at a later time. In as convincing a manner, from plots of ionization cross-section versus polarizability, Lampe *et al.* (1957) showed a nearly linear relationship when 75-eV electrons were used for determining the Q_i. Perhaps further consideration of the interactions of the Maxwell relationship—D (dielectric constant) $= n_\infty^2$ (refractive index at infinite wavelength)—might lead to use of the mass spectrometer for microscopic dielectric constant determinations.

When using this extremely simple approach to obtain the value for the ionization cross section of the sulfur atom in alkylthiophenes, one should not expect great accuracy. To be more confident of the validity of the values he obtains, the spectroscopist should employ total ion intensity data from all of the available positional isomers possible. It should be noted here that only thirty-seven 2- and 2,5-alkylthiophenes were available and, even then, not all of the possible isomers were available.

The mass spectroscopist should note that the use of total ionization data will at least supply a crosscheck as to the approximate validity of the sensitivity data.

E. The Rectangular Array

In the last few years, an increasing number of spectra have appeared in reference catalogs in the format of the "rectangular array" (Hamming and Grigsby, 1966). An example is shown in Fig. 3-23. In this case, the data for m/e are presented as percent total ion intensity, but the m/e's are organized in columns and rows. The m/e's increase 1 mass unit in going from left to right in the row. The row contains 14 entries before a second row is started. Thus,

RELATIVE INTENSITIES

The ratio of $\frac{\text{mass}}{\text{charge}}$ is indicated by the superscript in the upper left hand corner of each block.

Magnetic Field: 2398 gauss to m/e 90; 4138 gauss from m/e 91

Electron Energy: 70 volts

									12	13: 19	14: 79	15: 1017	16: 33
17	18	19	20	21	22	23	24	25	26: 409	27: **2595**	28: 762	29: 628	30: 22
31	32	33	34	35	36	37: 86	38: 337	39: **2543**	40: 324	41: 884	42: 64	43: 58	44: 32
45	46	47	48	49: 28	50: 612	51: **1680**	52: 491	53: 808	54: 69	55: 85	56: 29	57: 91	58: 168
59	60	61: 74	62: 345	63: 1171	64: 826	65: 1119	66: 133	67: 133	68: 23	69: 80	70: 242	71: 614	72: 156
73: 93	74: 272	75: 365	76: 339	77: 1484	78: 584	79: 535	80: 154	81: 23	82	83	84	85: 33	86: 98
87: 173	88: 69	89: 374	90: 85	91: 1498	92: 151	93: 58	94	95	96: 29	97	98: 53	99: 39	100: 23
101: 116	102: 400	103: 534	104: 175	105: 1243	106: 127	107: 23	108	109: 23	110: 23	111: 18	112	113: 75	114: 115
115: **3145**	116: 754	117: 877	118: 192	119: 278	120: 50	121	122	123	124	125: 50	126: 157	127: 1171	128: **3712**
129: **3478**	130: **3390**	131: 752	132: 105	133: 49	134	135	136	137	138	139: 181	140: 93	141: 851	142: 447
143: 926	144: 803	145: **34373**	146: **4139**	147: 246	148	149	150	151: 29	152: 66	153: 56	154: 23	155: 45	156: 28
157: 58	158: 126	159: 633	160: p **6687**	161: 858	162: 57	163	164	165	166	167	168	169	170
171	172	173	174	175	176	177	178	179	180	181	182	183	184
185	186	187	188	189	190	191	192	193	194	195	196	197	198
199	200	201	202	203	204	205	206	207	208	209	210	211	212
213	214	215	216	217	218	219	220	221	222	223	224	225	226
227	228	229	230	231	232	233	234	235	236	237	238	239	240

IONS WITH FRACTIONAL RATIO OF MASS TO CHARGE

mass/charge	Relative Intensity	mass/charge	Relative Intensity	mass/charge	Relative Intensity	mass/charge	Relative Intensity
	2398 gauss		2398 gauss				
51.5	23	76.5	55				
57.5	269	77.5	50				
58.5	34	78.5	46				
63.5	140	79.5	44				
64.5	254	80.5	20				
65.5	35						
69.5	132						
70.5	628						
71.5	514						
72.5	561						
75.5	34						

Total Ionization for Compound: div/micron/microampere; div/lambda/microampere

n-BUTANE

Magnetic Field: 2398 gauss

Sensitivity at $\frac{\text{mass}}{\text{charge}}$ 43: 10.30 div/micron/microampere; div/lambda/microampere

mass/charge	Relative Intensity
15	8.57
29	33.92
43	100.00
58	11.17

n-HEXADECANE

Magnetic Field: 2398 / 4138 gauss

Total Ionization: div/micron/microampere; div/lambda/microampere

mass/charge	Relative Intensity	mass/charge	Relative Intensity
57	100.00	141	2.18
71	56.74	155	1.95
85	35.17	169	1.61
99	4.93	183	1.08
113	3.30	197	0.62
127	2.64	226	3.53

COMPOUND

Name: 1,5,7-Trimethyl-[2,3-dihydroindene] (1,5,7-Trimethylindan)

Molecular Weight	Molecular Formula	Approximate Boiling Point	Approximate Freezing Point	Approximate Density
160.12	$C_{12}H_{16}$	68°C at 2 mm	°C	g/ml at °C

Semi-structural Formula:

Source: R. C. Bansal and E. J. Eisenbraun, API Research Project 58A, Department of Chemistry, Oklahoma State University, Stillwater, Oklahoma

Purity: Sample purified by GC

LABORATORY: Continental Oil Company, Ponca City, Oklahoma

MASS SPECTROMETER

Maker and Model: Consolidated Model No. 21-103C

Ionizing Current	Ion Chamber Temperature	Vapor Temperature	Collector Slit Width
10.0 microamperes	230 °C	320 °C	30 mils to m/e 90; 7 mils from m/e 91

Sample Pressure and Basis of Measurement:

Additional Information:
Sensitivity for n-C_4, I/ΣI 43: 0.298
Sensitivity for n-C_{16}, I/ΣI 57: 0.175
Ion beam: Focused at high sensitivity
Recording system: Consolidated Mascot
Scan rate and magnet current adjusted for repetition rate of one peak per second
Mass spectral data edited by: Mynard C. Hamming

Note: The intensities set in bold face type in the matrix are the ten most intense peaks in this spectrum.

Date of Measurement: September 24, 1964

Fig. 3-23. A typical mass spectrum of 1,5,7-trimethyl-[2,3-dihydroindene] (1,5,7-trimethylindan), expressed in "matrix format." (From the American Petroleum Institute, Research Project 44, Serial No. 197-m, Texas A & M University, 31 October 1968; contributed by the Continental Oil Company, Ponca City, Oklahoma.)

columns of related m/e's can be considered in which the ion above and below a selected m/e ion will be either 14 mass units less or 14 mass units more. For organic mass spectrometry in which homologous series of compounds are to be examined, the data are organized in a manner whereby the natural differences of the CH_2 group can be readily visualized. A complete discussion of the array and its use is the subject of Chapter 7.

VII. Catalogs of Mass Spectral Data

A number of catalogs of mass spectral data, organized in several different ways, are now available. Considerable use of some of these catalogs is made throughout this book. The beginning mass spectroscopist will find it to his advantage to accumulate considerable mass spectral reference data. The many approaches to studying and cataloging these types of data should be considered, and it should also be recognized that a complete listing of formats is an impossibility. These collections are most useful for the identification of unknown substances when known samples are unavailable and the empirical means of comparison of spectra must be considered.

A. Data of the API-44–TRC Catalogs

A considerable number of years ago the decision was made by the American Petroleum Institute (API) to support a long-term project for the collection and dissemination of information on thermodynamics properties of hydrocarbons and related materials including spectroscopic data. The project was originally set up in 1942 at the National Bureau of Standards, probably because of wartime conditions and the fact that an outstanding group of thermodynamicists had gravitated to those laboratories. In 1949 the project and a considerable number of the personnel under the direction of Dr. Frederick Rossini moved to Carnegie Technological Institute at Pittsburgh, Pennsylvania. The Manufacturing Chemists Association Project was begun in 1955 at Carnegie. When Dr. Rossini became graduate dean of sciences at Notre Dame University in 1961 the two projects were relocated at Texas A & M University at College Station, Texas. Dr. Bruno J. Zwolinski became the director of these projects and also is the director of the Thermodynamics Research Center Data Project supported by the Texas A & M Research Foundation.

In the intervening years this collection has become recognized as one of high-quality mass spectra. These mass spectral data are presented on looseleaf sheets in either the standard tabular form with relative intensities or in the convenient matrix format (rectangular array) with intensities expressed in percent total ion yield. In the matrix format, the ten principal intensities are shown in boldface type. Additional information is usually provided on certain physical properties in addition to the molecular weight, the compound classification, and the structural formula. Close attention is given to the operating conditions of the instrument employed and to the source and purity of the sample. Acknowledgment is given to the investigator as well as the contributing laboratory on each mass spectral data sheet.

After transfer of the project to Texas A & M, the Thermodynamics Research Center Data Project was formed to carry on the function of the former Manufacturing Chemists Association Research Project. These projects also produced a large collection of mass spectra. The total mass spectra listed as of 31 October 1968 was 2531.

An API-44-TRC comprehensive index has been prepared (Zwolinski, 1968) which greatly aids in the effective use of mass spectral data published since 1947 in the catalogs of API Research Project 44, the Thermodynamics Research Center (TRC). and the Manufacturing Chemists Association (MCA). (Several other categories of data, both thermodynamic and spectroscopic, are included in the same index.) This index has two principal parts, compound name index and formula index. The reference entries for both the name index and formula are identical, which means that either the compound name or the formula can be used to locate indicators for finding spectra. Both name and formula are shown with reference entries, which is a convenience when making a survey of spectra for certain compounds within a given class. For example, assume that for a possible structural indentification of a $C_8H_{18}O$ alcohol it is desired to locate all mass spectra of such alcohols published by API-44 or TRC. The formula index makes possible the elimination of such compounds as ethers, etc., having the same formula as $C_8H_{18}O$, by the listing of compound names. In this example, 24 indicators are provided for finding reference standards. In some cases, more than one spectrum has been published on the same compound; this becomes readily apparent. Any one compound can quickly be found in the compound name index.

B. Indexes of the ASTM (American Society for Testing and Materials)

An index is published by the ASTM (Special Technical Publication No. 356) which lists the six strongest peaks, but the spectra are indexed by molecular weights and the four strongest peaks. The number of spectra listed was 2531 as of 31 October 1968.

An earlier enterprise of CEC involved blank punch cards, which contained a listing of all of the mass spectral peaks above *m/e* 12 and 1% relative intensity. In looking for an unknown, just as in the ASTM infrared card deck case, the user pushed a long needle-type tool through the proper *m/e* value of a high-intensity peak possessed by his unknown compound. All compounds not having an intense peak at that m/e would hang onto the needle, while those punched out because of the presence of an intense peak would fall out of the deck when raised above the desk level. Often this procedure was helpful if one had advance knowledge of the molecular composition of the unknown since decks were accumulated for molecules containing oxygen, nitrogen, sulfur, etc., and the process was not as tedious as might first appear. Only a few hundred cards were available on a prepunched reference spectra basis. Blank cards could be purchased and made up in the user's laboratory. One could also "punch the deck" for MW, bp, elemental content, and other properties if the deck was set up in that fashion and all physical property data were available. In a sense, this was the computer sorting system of the 1950's.

C. ASTM E-14 "Uncertified Mass Spectra"

During the formation of the ASTM E-14 committee, it became apparent that a cooperative program of sharing spectra on compounds available to only a few laboratories would be a big help to mass spectroscopists the world around. As a consequence a subcommittee (no. IV) set up a program whereby spectra were circulated to those requesting the spectra on an "uncertified" basis. This means that the purity of the compound was not guaranteed, that only a partial mass spectrum was available, or that the operating conditions were not necessarily carefully controlled or even set near what the "ordinary" laboratory might use. The committee gathered over 2000 spectra for release in this category. Some of the spectra were later published by API-44 or MCA after the quality, purity, and authenticity of the compounds involved were ascertained.

A large collection of spectra was released by the Dow Chemical Co. in Midland, Michigan (Gohlke, 1963). These spectra were distributed in cooperation with the ASTM E-14 committee and gave a boost to the subcommittee's program.

D. The Mass Spectrometry Data Center

The E-14 program has waned in recent years because of the appearance of two new journals in mass spectrometry and the work of the Mass Spectrometry Data Center under the direction of Dr. R. G. Ridley. Today this center collects and distributes mass spectral data on a worldwide basis in three different forms: (a) printed sheets, (b) 16-mm microfilm, and (c) magnetic tape. Literature sources of mass spectral information and titles are noted in the *Mass Spectrometry Bulletin*, published by the Mass Spectrometry Data Center. Individual authors are invited to contribute tabular low-resolution mass spectra. Where possible, input on cards, magnetic tape, or punched tape is requested to permit computer collation, etc. The center has devised a "preferred" format for machine-readable data, and all the input information is punched or converted into this format. This allows the complete spectrum, details of the compound's properties and purity, and the instrument conditions to be stored and reported with the mass spectrum.

An "Eight Peak Index of Mass Spectra" is made available by the Mass Spectrometry Data Center (see Appendix III). This is a compilation of essential mass spectral data of over 17,000 organic compounds. The arrangement of the data gives a chemist an ideal, direct compound-identification aid with spectra indexed by (a) molecular weight subordered on formula, (b) molecular weight subordered on fragment-ion *m/e* values, and (c) fragment-ion *m/e* values.

E. Miscellaneous Indexes and Catalogs

With the appearance in 1968 of the new journal *Organic Mass Spectrometry*, in addition to the subject index, a chemical compound index also appeared. The index is ordered by molecular weight but includes the molecular formula, the compound name, the index number, and the page number in that volume on which the spectra appears. This will be of great assistance to the organic mass spectroscopist since he will be able to check the literature up to the current year for spectra. Beyond this, of course, he must rely on the journals and other sources of information.

Tatematsu and Tsuchiya (1968, 1969) have devised an index in a series of volumes by which structures are visually depicted. The visual presentation provides a convenient means of finding similar compounds which have been discussed in the literature of the year covered. Each structure has a literature reference cited with it. By this arrangement, similar structures from different references for the year are grouped together. Thus this index goes considerably beyond the simple listing of spectra and provides a means of comparing spectra, structures, and discussions of fragmentation to be found in the literature. A compound type of classification index makes it possible to quickly locate the visual presentation of the various structures of a given class of compounds. For example, the fragmentation of 21 different unsaturated aliphatic esters was discussed in three different publications in the 1966 literature. While no tabulation of peak intensities is presented, this scheme of indexing has great practical value. It becomes a "dictionary" of the compounds discussed in the literature which are not, for the most part, included in other published indexes of spectral data. The compounds covered in the volume are also grouped by molecular weights, thus making a second partial indexing possible.

A type of subject index listing names of compounds (and other descriptive terms) has been provided by McLafferty and Pinzelik (1967). A current series of subject indexes of literature has been supplied by Capellen *et al.* (1967). By the use of such indexes, compound names or types of compounds can be found which have been examined for their mass spectra, and their fragmentations discussed, provided only that the names appear in the title of the article (or have been added). For example, McLafferty and Pinzelik (1967) show sulfones in their key-word index, which with other words in the title, such as diarylsulfones, give a further subdivision within the sulfones. Often this type of subdivision is very useful because the mass spectra of the alkylsulfones, alkylarylsulfones and diarylsulfones happen to give rather different mass spectra due to rearrangement processes that take place.

Another new collection of mass spectra appears in three volumes entitled "The Atlas of Mass Spectral Data" (Stenhagen *et al.*, 1969). This source contains approximately 6000 organic compounds arranged according to molecular weight. Spectra not included in the atlas are, and will appear in a publication called the *Archives of Mass Spectral Data* (Stenhagen *et al.*, 1970). This is a collection of reliable mass spectra which have been carefully checked prior to publication by an experienced mass spectrometrist. The editors are E. Stenhagen and S. Abrahamsson of the University of Gothenburg and F. W. McLafferty of Cornell University.

Cornu and Massot (1966, 1967) have indexed over 6000 mass spectra by reference numbers, by molecular weight, and by molecular and fragment-ion values. These authors have excluded what was, in their judgment, poor data from the original spectra released by various organizations.

Structural significance of peaks and some common elemental compositions are quickly found in an American Chemical Society publication authored by McLafferty (1963).

These compilations, while very useful in identification and interpretation work, should be used, if possible, with the original spectral data so that additional spectral details are available. Proof of identification is much more convincing when intensities from 50 or 60 peaks are shown to be in close agreement than when only the six or ten indexed peaks are used. In addition, it should be remembered that the total of all these compilations cannot be all inclusive, nor can the indexes be completely free of errors.

Any laboratory doing extensive identification work will find it an advantage to secure as many mass spectra as possible. The first advantage is in their being available for comparison. Their auxiliary use, in conjunction with one or two members of a given series of compounds run on the same instrument under conditions identical to those used on an unknown, will permit correlation with the reference data to a much higher degree of accuracy. The second advantage is that additional identification schemes based on data which are not available from published compilations can be developed to meet a particular need. With the use of computers, tabulation cost can be reduced and can often become part of a larger computer routine. Such a technique has been developed at Continental Oil Co. by which standard spectra are tabulated in various ways by a computer program (Hamming *et al.*, 1967) using several subroutines. One selection is of the six peaks of highest mass numbers which have an intensity of greater than 0.5% of the observed total ion yield. Such an index has often been found to be more efficacious for identifications than the 10 strongest peak indexes.

REFERENCES

Abramson, P., and Futrell, J. H. (1967). *J. Phys. Chem.* **71**, 3791.

Agafonov, I. L., and Agafonov, A. L. (1964a). *Zh. Fiz. Khim.* **38**, 187.

Agafonov, I. L., and Agafonov, A. L. (1964b). *Russ. J. Phys. Chem.* **38**, 26.

Ahearn, A. J., and Malm, D. L. (1966). *Appl. Spectrosc.* **20**, 411.

Alpin, R. T., Budzikiewicz, H., Horn, H. S., and Lederberg, J. (1965). *Anal. Chem.* **37**, 776.

Berry, C. E. (1950). *Phys. Rev.* **78**, 597.

Bethe, H. (1930). *Ann. Phys.* (*Leipzig*) **5**, 352.

Beynon, J. H. (1960a). "Mass Spectrometry and Its Applications to Organic Chemistry," pp. 38–42. Elsevier, Amsterdam.

Beynon, J. H. (1960b). "Mass Spectrometry and Its Applications to Organic Chemistry," p. 52. Elsevier, Amsterdam.

Beynon, J. H. (1960c). "Mass Spectrometry and Its Applications to Organic Chemistry," p. 252. Elsevier, Amsterdam.

Beynon, J. H. (1960d). "Mass Spectrometry and Its Applications to Organic Chemistry," pp. 64, 65. Elsevier, Amsterdam.

Beynon, J. H. (1968). *In* "Advances in Mass Spectrometry" (E. Kendrick, ed.), Vol. IV, pp. 123–138. Institute of Petroleum, London.

Beynon, J. H., and Fontaine, A. E. (1966). *Chem. Commun.* 717.

Beynon, J. H., and Fontaine, A. E. (1967). *Z. Naturforsch. A* **22**, 334.

Beynon, J. H., Saunders, R. A., and Williams, A. E. (1964). *Nature* **204**, 67.

Beynon, J. H., Saunders, R. A., and Williams, A. E. (1965). *Z. Naturforsch. A* **20**, 180.

Beynon, J. H., Saunders, R. A., and Williams, A. E. (1968a). "The Mass Spectra of Organic Molecules," 510 pp. Elsevier, New York.

Beynon, J. H., Hopkinson, J. A., and Lester, G. R. (1968b). *Int. J. Mass Spectrom. Ion Phys.* **1**, 343.

Biemann, K. (1962). "Mass Spectrometry: Organic Chemical Applications," p. 13. McGraw-Hill, New York.

Boudart, M. (1961). Paper No. 2, Fourth Symposium on Hydrocarbon Chemistry, Houston, Texas.

Capellen, J., Svec, H. J., Jordan, J. R., and Watkins, W. J. (1967). "Bibliography of Mass Spectroscopy Literature, Compiled by Computer Method," IS-1829, 336 pp. Division of Technical Information, U. S. Atomic Energy Commission, Oak Ridge, Tennessee.

Clerc, R. J., Hood, A., and O'Neal, M. J., Jr. (1955). *Anal. Chem.* **27**, 868.

Coggeshall, N. D. (1944). *J. Chem. Phys.* **12**, 19.

Coggeshall, N. D. (1962). *J. Chem. Phys.* **37**, 2167.
Cornu, A., and Massot, R. (1966). "Compilation of Mass Spectral Data." Heyden and Son, London.
Cornu, A., and Massot, R. (1967). "First Supplement to Compilation of Mass Spectral Data." Heyden and Son, London.
Crable, G. F., and Coggeshall, N. D. (1958). *Anal. Chem.* **30**, 310.
Dempster, A. J. (1916). *Phil. Mag.* **31**, 438.
Derwish, G. A. W., Galli, A., Giardini-Guidoni, A., and Volpi, G. G. (1964). *J. Chem. Phys.* **41**, 2998.
Dewar, M. J. S., and Rona, P. (1965). *J. Amer. Chem. Soc.* **87**, 5510.
Dibeler, V. H., Reese, R. M., and Mann, D. E. (1957). *J. Chem. Phys.* **27**, 176.
Dillard, J. G., and Franklin, J. L. (1968). *J. Chem. Phys.* **48**, 2349.
Dukel'skii, V. M. (1955). *Dokl. Akad. Nauk SSSR* **105**, 955.
Dukel'skii, V. M., and Fedorenko, N. V. (1955). *Zh. Eksp. Teor. Fiz.* **29**, 473.
Dukel'skii, V. M., and Sokolov, V. M. (1957). *Zh. Eksp. Teor. Fiz.* **32**, 394.
Dukel'skii, V. M., Afrosimov, V. V., and Fedorenko, N. V. (1956a). *Zh. Eksp. Teor. Fiz.* **30**, 792.
Dukel'skii, V. M., Afrosimov, V. V., and Fedorenko, N. V. (1956b). *Sov. Phys. JETP* **3**, 764.
Elliott, R. M. (1963). *In* "Mass Spectrometry" (C. A. McDowell, ed.), pp. 97–100. McGraw-Hill, New York.
Fales, H. M., Milne, G. W. A., and Vestal, M. L. (1969). *J. Amer. Chem. Soc.* **91**, 3682.
Fales, H. M., Lloyd, H. A., and Milne, G. W. A. (1970). *J. Amer. Chem. Soc.* **92**, 1590.
Ferguson, R. E., McCulloh, K. E., and Rosenstock, H. M. (1965). *J. Chem. Phys.* **42**, 100.
Field, F. H. (1967). *J. Amer. Chem. Soc.* **89**, 5328.
Field, F. H. (1968). *Accounts Chem. Res.* **1**, 42.
Field, F. H., and Lampe, F. W. (1958). *J. Amer. Chem. Soc.* **80**, 5587.
Field, F. H., and Munson, M. S. B. (1965). *J. Amer. Chem. Soc.* **87**, 3289.
Field, F. H., and Munson, M. S. B. (1967). *J. Amer. Chem. Soc.* **89**, 4272.
Field, F. H., Munson, M. S. B., and Becker, D. A. (1966). "Advances in Chemistry Series," No. 58, p. 167. American Chemical Society, Washington, D.C.
Fogel, Ya. M., and Krupnik, L. I. (1956). *Sov. Phys. JETP* **21**, 252.
Fogel, Ya. M., Mitin, R. V., and Koval, A. G. (1956a). *Zh. Eksp. Teor. Fiz.* **31**, 397.
Fogel, Ya. M., Krupnik, L. I., and Ankudinov, V. A. (1956b). *Zh. Eksp. Teor. Fiz.* **31**, 569.
Fogel, Ya. M., Krupnik, L. I., and Ankudinov, V. A. (1957a). *Sov. Phys. Tech. Phys.* **1**. 1181.
Fogel, Ya. M., Ankudinov, V. A., and Slabospitskii, R. E. (1957b). *Zh. Eksp. Teor. Fiz,* **32**, 453.
Foster, N. G., and Higgins, R. W. (1968). *Org. Mass Spectrom.* **1**, 191.
Foster, N. G., Hirsch, D. E., Kendall, R. F., and Eccleston, B. H. (1964). *U.S. Bur. Mines Rep. Invest.* No. 6433.
Foster, N. G., Hirsch, D. E., Kendall, R. F., and Eccleston, B. H. (1965). *U.S. Bur. Mines Rep. Invest.* No. 6671.
Foster, N. G., Shiu, D. W.-K., and Higgins, R. W. (1968). Paper No. 71, Sixteenth Annual Conference on Mass Spectrometry, ASTM Committee E-14, Pittsburgh, Pennsylvania.
Foster, N. G., Liao, J., and Higgins, R. W. (1969). Paper No. 162, Seventeenth Annual Conference on Mass Spectrometry, ASTM Committee E-14, Dallas, Texas.
Fox, R. E., Hickam, W. M., Kjeldaas, T., Jr., and Grove, D. J. (1951). *Phys. Rev.* **84**, 859.
Franklin, J. L. (1952). *Trans. Faraday Soc.* **48**, 443.
Franklin, J. L., Munson, M. S. B., and Field, F. H. (1963). *Progr. Astronaut. Aeronaut.* **12**, 67.

Frost, D. C. (1963). *In* "Mass Spectrometry" (C. A. McDowell, ed.), pp. 179–200, McGraw-Hill, New York.
Futrell, J. H., Ryan, K. R., and Sieck, L. W. (1965). *J. Chem. Phys.* **43**, 1932.
Gohlke, R. S. (1963). "Uncertified Mass Spectral Data." Dow Chemical Company, Midland, Michigan.
Gohike, R. S., and Thompson, L. H. (1968). *Anal. Chem.* **40**, 1004.
Good, A., Durgen, D. A., and Kebarle, P. (1969). Paper No. 150, Seventeenth Annual Conference on Mass Spectrometry, ASTM Committee E-14, Dallas, Texas.
Grubb, H. M., and Vander Haar, R. W. (1962). Paper No. 55, Tenth Annual Conference on Mass Spectrometry, ASTM Committee E-14, New Orleans, Louisiana.
Hagstrum, H. D., and Tate, J. T. (1939). *Phys. Rev.* **55**, 1136.
Hagstrum, H. D., and Tate, J. T. (1940). *Phys. Rev.* **59**, 354.
Hamming, M. C., and Grigsby, R. D. (1966). Twelfth Tetrasectional of the Oklahoma Section of the American Chemical Society, Stillwater, Oklahoma.
Hamming, M. C., Wright, W. M., Gartside, H. G., Ford, H. T., and Haley, J. (1967). Paper No. 16, Fifteenth Annual Conference on Mass Spectrometry, ASTM Committee E-14, Denver, Colorado.
Harllee, F. N., Rosenstock, H. M., and Herron, J. T. (1966). *Nat. Bur. Stand. U.S. Tech. Note* No. 291.
Henglein, A. (1954). "Applied Mass Spectrometry," p. 158. Institute of Petroleum, London.
Hickam, W. M., and Berg, D. (1959). *In* "Advances in Mass Spectrometry" (J. D. Waldron, ed.), pp. 458–472, Pergamon, Oxford.
Hickam, W. M., and Fox, R. E. (1956). *J. Chem. Phys.* **25**, 642.
Higgins, R. W., and Foster, N. G. (1962). Unpublished results.
Higgins, R. W., and Foster, N. G. (1963). Unpublished results.
Higgins, W., and Jennings, K. R. (1965). *Chem. Commun.* **6**, 99.
Higgins, W., and Jennings, K. R. (1966). *Trans. Faraday Soc.* **62**, 97.
Hipple, J. A. (1947). *Phys. Rev.* **71**, 594.
Hipple, J. A., and Condon, E. U. (1945). *Phys. Rev.* **68**, 54.
Hipple, J. A., Fox, R. E., and Condon, E. U. (1946). *Phys. Rev.* **69**, 347.
Hogness, T. R., and Harkness, R. W. (1928). *Phys. Rev.* **32**, 784.
Hood, A. (1958). *Anal. Chem.* **30**, 1218.
Hunt, W. W., Jr., Huffman, R. E., and McGee, K. E. (1964). *Rev. Sci. Instrum.* **35**, 82.
Issenberg, P., Bazinet, M. L., and Merritt, C., Jr. (1965). *Anal. Chem.* **37**, 1074.
Jennings, K. R. (1965). *J. Chem. Phys.* **43**, 4176.
Jennings, K. R. (1966). *Chem. Commun.* 283.
Kennicott, P. R. (1965). *Anal. Chem.* **37**, 313.
Kerwin, L. (1963). *In* "Mass Spectrometry" (C. A. McDowell, ed.), pp. 104–178. McGraw-Hill, New York.
Khvostenko, V. I. (1964). *Khim. Seraorg. Soedin. Soderzh. Neftyakh Nefteprod. Akad. Nauk SSSR Bashkir. Filial* **6**, 240.
Kinney, I. W., Jr., and Cook, G. L. (1952). *Anal. Chem.* **24**, 1391.
Kiser, R. W. (1965a). "Introduction to Mass Spectrometry and Its Applications," p. 46. Prentice-Hall, Englewood Cliffs, New Jersey.
Kiser, R. W. (1965b). "Introduction to Mass Spectrometry and Its Applications," pp. 55–57. Prentice-Hall, Englewood Cliffs, New Jersey.
Kiser, R. W. (1965c). "Introduction to Mass Spectrometry and Its Applications," p. 193. Prentice-Hall, Englewood Cliffs, New Jersey.
Kiser, R. W. (1965d). "Introduction to Mass Spectrometry and Its Applications," p. 194. Prentice-Hall, Englewood Cliffs, New Jersey.

Knewstubb, P. F. (1969). "Mass Spectrometry and Ion–Molecule Reactions," p. 136. Cambridge Univ. Press, London.
Lampe, F. W., and Field, F. H. (1957). *J. Amer. Chem. Soc.* **79**, 4244.
Lampe, F. W., Franklin, J. L., and Field, F. H. (1957). *J. Amer. Chem. Soc.* **79**, 6129.
Langevin, M. P. (1905). *Ann. Chim. Phys.* **5**, 245.
Leckenby, R. E., Robbins, E. J., and Trevalion, P. A. (1964). *Proc. Roy. Soc.* **280**, 409.
Lindholm, E. (1953). *Proc. Phys. Soc. Ser. A* **66**, 1068.
Loeb, L. B. (1955). "Basic Processes of Gaseous Electronics." Univ. of California Press, Berkeley.
Lozier, W. W. (1934). *Phys. Rev.* **46**, 268.
McCrea, J. M. (1965). *Appl. Spectrosc.* **19**, 61.
McCrea, J. M. (1971). *Appl. Spectrosc.* **25**, 246.
McCulloh, K. E., Sharp, T. E., and Rosenstock, H. M. (1965). *J. Chem. Phys.* **42**, 3501.
McLafferty, F. W. (1963). "Mass Spectral Correlations," *Advan. Chem. Ser.* **40**, 117 pp. Am. Chem Soc., Washington, D.C.
McLafferty, F. W., and Bursey, M. M. (1967). *Chem. Commun.* 533.
McLafferty, F. W., and Pike, W. T. (1967). *J. Amer. Chem. Soc.* **89**, 5951.
McLafferty, F. W., and Pinzelik, J. (1967). "Index and Bibliography of Mass Spectrometry." Wiley (Interscience), New York.
Mann, M. M., Hustrulid, A., and Tate, J. T. (1940). *Phys. Rev.* **58**, 340.
Massey, H. S. W. (1950). "Negative Ions." Cambridge Univ. Press, London.
Massey, H. S. W., and Burhop, E. H. S. (1952). "Electronic and Ionic Impact Phenomena." Oxford Univ. Press, London.
Meisels, G. G., Hamill, W. H., and Williams, R. R., Jr. (1956). *J. Chem. Phys.* **25**, 790 .
Melton, C. E. (1958). *J. Chem. Phys.* **28**, 359.
Melton, C. E. (1961). *J. Chem. Phys.* **35**, 1751.
Melton, C. E. (1963a). *In* "Mass Spectrometry of Organic Ions" (F. W. McLafferty, ed.), pp. 163–205. Academic Press, New York.
Melton, C. E. (1963b). *In* "Mass Spectrometry of Organic Ions" (F. W. McLafferty, ed.), p. 165. Academic Press, New York.
Melton, C. E. (1963c). *In* "Mass Spectrometry of Organic Ions" (F. W. McLafferty, ed.), p. 172. Academic Press, New York.
Melton, C. E. (1963d). *In* "Mass Spectrometry of Organic Ions" (F. W. McLafferty, ed.), p. 179. Academic Press, New York.
Melton, C. E. (1963e). *In* "Mass Spectrometry of Organic Ions" (F. W. McLafferty, ed.), p. 67. Academic Press, New York.
Melton, C. E. (1963f). *In* "Mass Spectrometry of Organic Ions" (F. W. McLafferty, ed.), pp. 65–115. Academic Press, New York.
Melton, C. E. (1969). Paper No. 63, Seventeenth Annual Conference on Mass Spectrometry, ASTM Committee E-14, Dallas, Texas.
Melton, C. E., and Rosenstock, H. M. (1957). *J. Chem. Phys.* **26**, 568.
Melton, C. E., and Rudolph, P. S. (1959). *J. Chem. Phys.* **31**, 1485.
Melton, C. E., Ropp, G. A., and Rudolph, P. S. (1958). *J. Chem. Phys.* **29**, 968.
Meyerson, S., and McCollum, J. D. (1959). Division of Physical Chemistry of the American Chemical Society, 136th Meeting, Atlantic City, New Jersey, and private communication.
Meyerson, S., and Vander Haar, R. W. (1962). *J. Chem. Phys.* **37**, 2458.
Milne, G. W. A., Axenrod, T., and Fales, H. M. (1970). *J. Amer. Chem. Soc.* **92**, 5170.
Mitchell, J. J., Perkins, R. H., and Coleman, F. (1948). *J. Chem. Phys.* **16**, 835.
Mohler, F. L., Williamson, L. L., and Dean, H. (1950). *J. Res. Nat. Bur. Stand.* **45**, 235.
Mohler, F. L., Dibeler, V. H., and Reese, R. M. (1954). *J. Chem. Phys.* **22**, 394.

Momigny, J. (1961). *Bull. Soc. Chim. Belg.* **70**, 291.
Moran, T. F., and Hamill, W. H. (1963). *J. Chem. Phys.* **39**, 1413.
Moulton, D. McL., and Michael, J. V. (1965). *Rev. Sci. Instrum.* **36**, 226.
Muccini, G. A., Hamill, W. H., and Barker, R. (1964). *J. Phys. Chem.* **68**, 261.
Munson, M. S. B., and Field, F. H. (1965a). *J. Amer. Chem. Soc.* **87**, 3294.
Munson, M. S. B., and Field, F. H. (1965b). *J. Amer. Chem. Soc.* **87**, 4242.
Munson, M. S. B., and Field, F. H. (1966a). *J. Amer. Chem. Soc.* **88**, 2621.
Munson, M. S. B., and Field, F. H. (1966b). *J. Amer. Chem. Soc.* **88**, 4337.
Munson, M. S. B., and Field, F. H. (1967). *J. Amer. Chem. Soc.* **89**, 1047.
Munson, M. S. B., Franklin, J. L., Field, F. H. (1964). *J. Phys. Chem.* **68**, 3098.
Newton, A. S. (1964). *J. Chem. Phys.* **40**, 607.
Newton, A. S. (1966). *J. Chem. Phys.* **44**, 4015.
Olmstead, J., III, Street, K., Jr., and Newton, A. S. (1964). *J. Chem. Phys.* **40**, 2114.
Ötvos, J. W., and Stevenson, D. P. (1956). *J. Amer. Chem. Soc.* **78**, 546.
Reese, R. M., Dibeler, V. H., and Mohler, F. L. (1956). *J. Res. Nat. Bur. Stand.* **57**, 367.
Reese, R. M., Dibeler, V. H., and Franklin, J. L. (1958). *J. Chem. Phys.* **29**, 880.
Rhodes, R. E., Barber, M., and Anderson, R. L. (1966). *Anal. Chem.* **38**, 48.
Robbins, E. J., and Leckenby, R. E. (1965). *Nature* **206**, 1253.
Ropp, G. A., and Melton, C. E. (1958). *J. Amer. Chem. Soc.* **80**, 3509.
Rosenstock, H. M., Wallenstein, M. B., Wahrhaftig, A. L., and Eyring, H. (1952). *Proc. Nat. Acad. Sci. U.S.* **38**, 667.
Rosenstock, H. M., Wahrhaftig, A. L., and Eyring, H. (1955). *J. Chem. Phys.* **23**, 2200.
Rudloff, W. (1961). *Z. Naturforsch. A* **16**, 1263.
Sasaki, S., Watanabe, E., Itagaki, Y., Aoyama, T., Yamauchi, E. (1968). *Anal. Chem.* **40**, 1000.
Schaeffer, O. A. (1950). *J. Chem. Phys.* **18**, 1681.
Schuy, K. D., and Franzen, J. (1962). *Fresenius' Z. Anal. Chem.* **225**, 260.
Searles, S. K., Dzidic, I., Arshadi, M., Yamdagni, R., and Kebarle, P. (1969). Paper No. 149, Seventeenth Annual Conference on Mass Spectrometry, ASTM Committee E-14, Dallas, Texas.
Shannon, T. W., Mead, T. E., Warner, C. G., and McLafferty, F. W. (1967). *Anal. Chem.* **39**, 1748.
Shug, J. C. (1964). *J. Chem. Phys.* **40**, 1286.
Smythe, H. D. (1924). *J. Franklin Inst.* **198**, 795.
Smythe, H. D. (1925). *Phys. Rev.* **25**, 452.
Stenhagen, E., Abrahamsson, S., and McLafferty, F. W. (1969). "Atlas of Mass Spectral Data," Vols. I–III, 2266 pp. Wiley (Interscience), New York.
Stenhagen, E., Abrahamsson, S., and McLafferty, F. W. (1970). "The Archives of Mass Spectral Data," Vol. I, 784 pp. Wiley (Interscience), New York.
Stevenson, D. P. (1963). *In* "Mass Spectrometry" (C. A. McDowell, ed.), p. 589. McGraw-Hill, New York.
Stevenson, D. P., and Schissler, D. O. (1955). *J. Chem. Phys.* **23**, 1353.
Tal'rose, V. L., and Lyubimova, A. K. (1952). *Dokl. Akad. Nauk SSSR* **86**, 909.
Tatematsu, A., and Tsuchiya, T. (1968). "Structure Indexed Literature of Organic Mass Spectra—1966," 275 pp. Academic Press of Japan, Japan.
Tatematsu, A., and Tsuchiya, T. (1969). "Structure Indexed Literature of Organic Mass Spectra—1967," 496 pp. Academic Press of Japan, Japan.
Taylor, J. W., and Grimsrud, E. P. (1969). *Anal. Chem.* **41**, 805.
Thomason, E. M. (1963). *Anal. Chem.* **35**, 2155.
Thomson, J. J. (1924). *Phil. Mag.* **47**, 337.

Ukai, S., Hirose, K., Tatematsu, A., and Goto, T. (1967). *Tetrahedron Lett.* **49**, 4999.
Vestal, M. L. (1964). *J. Chem. Phys.* **41**, 3997.
Von Ardenne, M., Steinfelder, K., and Tummler, R. (1961). *Angew. Chem.* **73**, 136.
Wagner, H. (1964). *Ann. Phys. (Leipzig)* **13**, 189.
Wagner, H., and Mai, H. (1964). *Z. Naturforsch. A* **19**, 1624.
Washburn, H. W., and Berry, C. E. (1946). *Phys. Rev.* **70**, 559.
Wijnbeigen, J. J., Tromp, F. M., and Van de Vijver, W. J. (1964). *J. Sci. Instrum.* **41**, 574.
Winter, D. L. (1968). Unpublished work. Continental Oil Co., Ponca City, Oklahoma.
Winters, R. E., and Kiser, R. W. (1966). *J. Phys. Chem.* **70**, 1680.
Ziffer, H., Fales, H. M., Milne, G. W. A., and Field, F. H. (1970). *J. Amer. Chem. Soc.* **92**, 1597.
Zwolinski, B. J. (1968). "Comprehensive Index of API 44-TRC Selected Data on Thermodynamics and Spectroscopy," 507 pp. Thermodynamics Research Center, Department of Chemistry, Texas A&M University, College Station, Texas.

CHAPTER 4

THE SAMPLE—ITS CHARACTER AND HANDLING

I. Applicability of Mass Spectrometry to a Wide Range of Sample Types

Very few other instrumental methods of analysis take so little sample to produce so much detailed information; Biemann (1962a) claimed 1 μmole in a volume of 3 liters at 150°C. Frequently, no further sample preparation is required for examination in the mass spectrometer. When further handling is necessary, a number of standard microanalytical procedures, both physical and chemical, are well known and easily applied. Gases, liquids, and solids may be examined by the use of ordinary sampling devices. No special cells are required as in infrared, ultraviolet, or Raman spectrophotometry, nor are sample holders necessary as in nuclear magnetic resonance (NMR). Preparation of the sample for the actual mass analysis is usually not very time-consuming or complex. However, it should be borne in mind by the mass spectroscopist that his "customers" may have spent many man-hours obtaining an extremely important and hard-won sample. In this case, the sample preparation time becomes enormous.

The spectroscopist has an obligation to his customers in the form of proper programming of samples for introduction to the mass spectrometer. It should be noted that the relatively high cost of mass spectrometry operations necessitates multiple research use of the instrument. The expense of an instrument reserved for a single project can seldom be justified. Hence, the spectroscopist inherits the unhappy lot of frequently having to inform his customer just when a specific sample can be run. As a general rule, an instrument has an adequate vacuum pumping system so that most volatile samples can be examined and then pumped out in a reasonable amount of time. Despite this, experience has shown that it is desirable to run very volatile materials ahead of high molecular weight, low volatility materials so as to avoid contamination of the source with the latter substances. These might create a considerably higher background that could interfere with analysis of the highly volatile materials. For example, paraffin waxes produce ion peaks at m/e 43, 57, 71, among others, in the background. These could interfere with hydrocarbon analysis and particularly with quantitative analyses. In addition, some substances require several runs to "condition" the source to that type of sample. Halogens, oxygenated materials, sulfur, and nitrogen compounds are prominent among the many that may exhibit this problem. Continuous running of such substances seems to promote the buildup of surface deposits which in time interfere with the proper operation of the source. Some laboratories routinely receive a potpourri of organic materials. Judicious assignment of the order in which samples are run is essential to prolonged satisfactory operations.

Ultimately, sample handling in mass spectrometry depends upon three primary problems: (a) the type of inlet system, (b) the mechanism of introducing representative samples of the various types to the given inlet system, and (c) the requirement of either physical or chemical (or both) separation techniques. These are discussed below in detail.

As yet this book has not dealt with the mechanics of assembling an operable high-vacuum system. In facing the problems of (b), particularly as posed above, one must recognize that the attainment of high-speed, high-efficiency vacuum systems depends upon seals and valves in addition to pump mechanics, diffusion pumps, cold traps, ion-getter pumps, etc. These details are beyond the scope of this book, but the reader is referred to Beynon (1960) for a rather complete background of the types of seals and valves that have been used in mass spectrometry. One should remember that the desired end results may not be achieved if there is any general failure in the vacuum system or if the sample cannot be introduced to the system and brought to the ionizing region in a representative manner.

Usually a laboratory develops its own characteristic accessories for the introduction of samples to the mass spectrometer, including perhaps several types of inlet systems and sample-handling techniques. No all-inclusive general method of sample handling is possible, nor should it be recommended. Rather, the mass spectroscopist should anticipate the types of sample-handling problems to be encountered and then utilize whatever modifications of the available systems are indicated for such problems. A selection of the more commonly used methods is included in the material that follows.

II. Types of Inlet Systems

A. Glass Inlet Systems

Because of the early development of ion sources in glass tube enclosures, it was considered relatively easy to attach a glass inlet system, the designs of which were simply extensions of known vacuum techniques. The various functional parts of the inlet system are shown schematically in Fig. 4-1. In essence, the system provides a means of introducing the sample into a pre-evacuated chamber which can be isolated from the ion source and which possesses an expansion volume or reservoir large enough to hold a considerable supply of the sample. The samples to be discussed first are considered as volatilized liquid or solid samples or gases at ambient temperature. A pressure

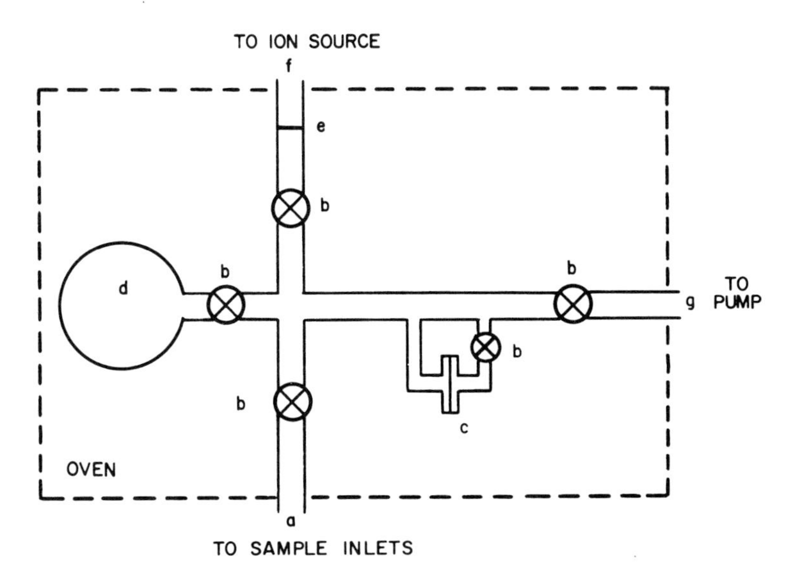

Fig. 4-1. Schematic components of a multiple-purpose inlet system. a, Large-diameter heated port to gas, liquid, and solid inlets; b, suitable isolation valves; c, micromanometer or other pressure measuring device; d, sample reservoir expansion volume; e, capillary or gold leak; f, heated tube to ion source; and g, large-diameter tubing to diffusion and mechanical vacuum pumps.

of about 10^{-5} torr is needed in the reservoir to insure that fresh sample molecules will be available in the ion source, which provides ions during the scanning time of the spectrum. A very small decrease in reservoir pressure with time should occur so that the processes in the source involved with electrostatically (or magnetically) scanning the spectrum can be regarded as "dynamic, steady-state" processes. To accomplish this, a capillary or metal-foil leak is inserted into the line, as indicated in Fig. 4-1 at e, causing about 1% of the sample in the reservoir to enter the source per unit time. Metal-foil leaks are known to plug with various organic materials such as polymers, peroxides, etc., depending upon the types of sample passed through the leak. Capillary leaks usually do not exhibit this problem but are known to plug, probably due to a buildup of surface deposits of organic materials. Kiser (1965) gave a detailed theoretical treatment of leaks and molecular flow, as applicable to mass spectrometry. A method for constructing a glass capillary leak developed by McKelvey (1962) resulted in a unit with few plugging problems. The glass did not react with samples being analyzed.

The arrangement of the manometer, as shown in Fig. 4-1, is optional. In some inlets the pressure is measured at the reservoir; in others it is measured outside the reservoir. Suitable mass spectra are obtained on most instruments if a pressure of about 10^{-5}–10^{-7} torr is maintained in the source. With high-sensitivity detectors and efficient collectors, this can be reduced by at least a factor of 10. In some inlets, no pressure measurement is made in the micron range; instead a pressure is read on a mercury manometer of a small gas sample in a calibrated volume. The gas in the calibrated volume at known pressure is released into the expansion volume, and a calculated gas pressure in the calibrated reservoir is obtained assuming ideal gas law behavior. Although there are other known means of measuring pressures in the micron range, the diaphragm manometer or micromanometer has become the most widely used (Alpert *et al.*, 1951; Baxter, 1953; Becker and Stehl, 1952; Cook and Danby, 1953; Dibeler and Cordero, 1951; Dighton, 1953; Opstelton and Warmoltz, 1954; Pressey, 1953; Williams and Everson, 1958). These references are given primarily to indicate the number of modifications and different designs that usually arise once a problem is researched. Other developments may involve the efforts of numerous technical persons until a generally acceptable solution to the problem is found.

The expansion volume can be eliminated in the case of the time-of-flight instrument or made quite small for instruments which use a rapid scan system. Suitable valving allows a cutoff of the expansion volume from the high-vacuum system and from the injection port(s), as proposed by Biemann (1962a). This permits a very small sample to be examined, albeit not with the degree of precision obtained from a standard-sized sample. This valving is also used in conjunction with direct GLC inlets and will be discussed in a later section.

It was soon found that existing vacuum techniques, even with the use of large-bore vacuum stopcocks, produce problems in mass spectrometry. The source is contaminated by stopcock greases, mercury from manometers and covered glass frits in the inlet, and by samples themselves. This is exhibited by deposits on the glass which form conductive coatings. Additional objections to glass inlets (and sources) are the ease of breakage and difficulties with cleaning and handling. The early inlets were operated at ambient temperature, and hence only materials with sufficient volatility at 25°C (gases, for all practical purposes) were investigated. The pressures placed on laboratory work during World War II due to both the contamination problem (and risk of precious manpower time to clean up) and the unavailability of experimentation time were a major restriction to the general expansion of mass spectrometric techniques. Another factor was the complexity of spectra when mixtures of the C_5 and higher hydrocarbons were examined. One of the authors (NGF) well remembers assisting in working out the fractionation steps (high-temperature distillation) necessary for analysis of hydrocarbon materials containing benzene and other compounds in the C_6–C_7 paraffin boiling range (Meyerson, 1953).

B. Metal Inlet Systems

All-metal inlet systems were contrived with the advent of precision needle valves that were bakeable and had seats and tips made from stainless steel, monel, or nickel (and other materials). The all-metal inlet systems were more stable than the all-glass systems because of the inherent problems with glass strains and breakage. Manipulation and repairs by technicians were possible, and hence these systems grew in popularity. For high-temperature work, these inlets could be operated at about 200°C if suitable O-ring materials, lead gaskets, and stainless steel valves were used. It should be noted that in most mass spectrometers a graded glass-to-metal seal is employed somewhere in order to isolate the source electrically from the inlet system. The development of the metal inlet was an important contribution to evolving the high-temperature inlet systems in current use.

C. The Gallium Valve—Heated Inlet Systems

A major breakthrough in mass spectrometry was provided by the work of O'Neal and Wier (1951) and O'Neal (1954). This was the development of a completely unitized high-temperature inlet system for organic substances in which high-boiling hydrocarbons, such as lubricating oils and waxes (O'Neal, 1953, 1954; O'Neal *et al.*, 1955) and, with a later modification, asphalts (Clerc and O'Neal, 1961), were examined with some success. The O'Neal system featured gallium-covered sinters in place of the mercury used in room-temperature systems. The inlet system is shown schematically in Fig. 4-2. Gallium is a liquid between 30° and 1983°C at atmospheric pressure, a greater temperature range than for any other metal. Thus gallium was an improvement over mercury used in the cold inlet systems. Mercury has a vapor pressure of 0.1 mm at 82°C and 1 mm at 126.4°C and would be a health hazard as well as a source contaminant if used in a heated inlet system. Gallium resembles aluminum chemically in that it forms an oxide coating upon heating in air. The coating floats on the surface, wets glass, and complicates the problem of contamination, since the coating plus additional gallium always adheres to a micropipette when it is dipped through the surface of the gallium to the

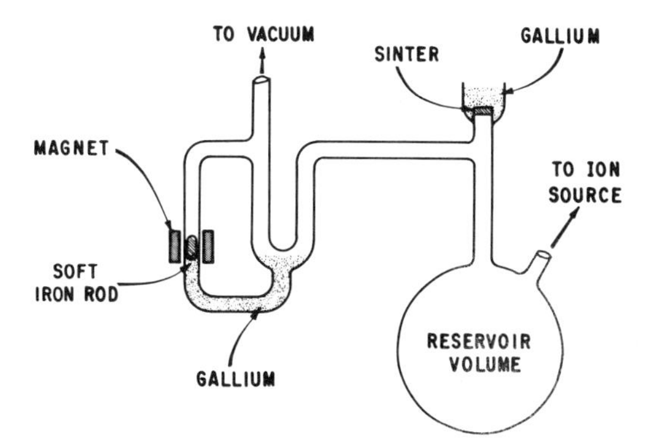

Fig. 4-2. Heated inlet system. (After design of O'Neal and Wier, 1951. Courtesy of *Analytical Chemistry.*)

glass frit. Gallium is constantly lost in this fashion. For the above reasons, gallium sinters are more likely to become plugged than are mercury sinters. It is customary to include several sinters in parallel on an inlet system so that work is not immediately halted with the plugging of a sinter in the course of a series of spectral measurements. A depth of about 1 cm of gallium is needed to insure a leak-free seal, and the level must be maintained by regular addition of the metal.

Beynon and Nicholson (1956) designed a gallium valve that did not involve a sinter but utilized a magnetic lift. Thus it was usable in a heated inlet system. An improvement upon this idea was reported by Lumpkin and Taylor (1961): The gallium is contained in a cup and can be lifted by a magnet external to the heating system, but the entire valve can be opened to the atmosphere and a solid sample placed in the glass sample cup, which in turn is placed upon the pedestal as shown in Fig. 4-3. The unit is reattached to the heated inlet system and the air reevacuated. This is followed by closing off the pump valve

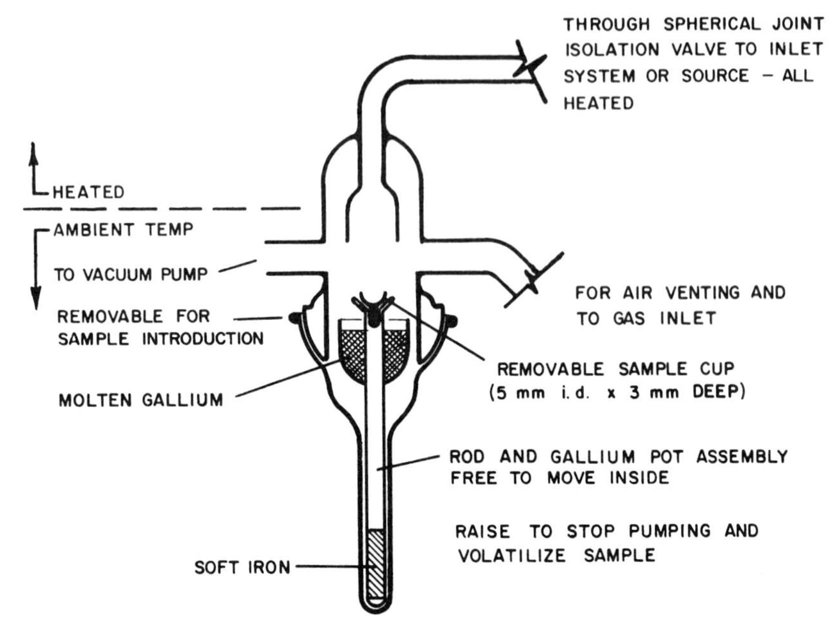

Fig. 4-3. Mass spectrometer heated inlet system valve for solids. (After Lumpkin and Taylor, 1961. Courtesy of *Analytical Chemistry*.)

and opening a valve to the heated inlet expansion chamber, while at the same time the solid sample is gradually heated to develop sufficient sample vapor pressure. Some difficulties occur because samples with low vapor pressure take considerable time to vaporize in this fashion. Conversely, a sample may volatilize so rapidly as to be partially or completely lost during the evacuation of air and the time required for lifting.

The development of the heated inlet system was a rather slow and laborious process involving many trials and some errors. Because of the rather large expense involved, the addition of heated inlet systems to existing mass spectrometers was slow to develop. An informal survey of sample inlet systems operating above 100°C was reported by Caldecourt (1958). This was made under the auspices of the American Society for Testing and Materials (ASTM) Committee E-14 on Mass Spectrometry. Twenty-eight replies were received from 57 inquiries (a typical response) mailed to selected individuals in companies, governmental agencies, or other institutions on the Committee E-14 mailing list. Only 19 groups reported having heated inlet systems. In addition, the National Bureau of Standards had a furnace for high-mass work with polymers. Although other laboratories were in the process of developing heated inlet systems, the figure of 19 is some indication of the progress made in the

7 years that passed from the time of the O'Neal and Wier developments. In the ensuing years, the use of heated inlet systems has become almost routine with the basic mass spectrometric installation.

The 1958 survey also reported that metal parts might react with samples or cause catalytic cracking. For petroleum compounds, this was observed when the sample system was used in the 300°–400°C range. No appreciable cracking appeared to occur in stainless steel systems operated below 250°C. The consensus of opinion was that the metal parts of the ion source could react with samples and could give as much or more trouble than the metal parts in the sample system.

Ion sources and inlet systems are generally not operated above 300°C because of the spectral pattern changes and the possibility of thermal decomposition influencing the observed spectrum. In heated inlet systems the practice has been to admit the sample into the system so that it is in contact with a heated reservoir of a temperature slightly higher or equal to the oven temperature surrounding the expansion volume and valves. Cold spots in any location should be avoided because unintended fractionation of the materials in the sample (if a mixture) might occur. The precipitation of high molecular weight materials of low volatility has been reported by many workers. The reduced sensitivity may make the analysis difficult, if not impossible. The volatility effect is quite obvious in connection with the total ionization data presented in Chapter 3, Fig. 3-22. In line with this reasoning, the sample transfer tube from the inlet oven to the ion source should be maintained at a temperature equal to or higher than the oven. The ion source temperature should ordinarily be the highest temperature of all of these; hence, the operator must select suitable conditions for the problem at hand. Since the source temperature influences the fragmentation pattern, this temperature must be selected first. Note that for various designs of instruments, the source temperature can be an extremely sensitive and important variable, whereas for other instruments it is not as important.

Caldecourt's report (1958) also indicated that gallium was the cause of some maintenance problems on both the valves and the fritted disk. The disks required replacement every 4–8 weeks because of fouling. The glass inlet systems, because of their low reactivity and catalytic activity, were considered preferable to the metal systems.

A number of designs of heated inlet systems employing valves of spherically ground glass joints operated by magnetic lifts and internal to the oven have appeared in the literature and on commercial instruments. Typical among these types of systems is one described by Peterson (1962a,b) which may be used as an all-glass heated inlet system to above 300°C. This is shown schematically in Fig. 4-4. Solid samples can be introduced into this system provided the sample is in a glass capillary. The remains of the capillary stay in the

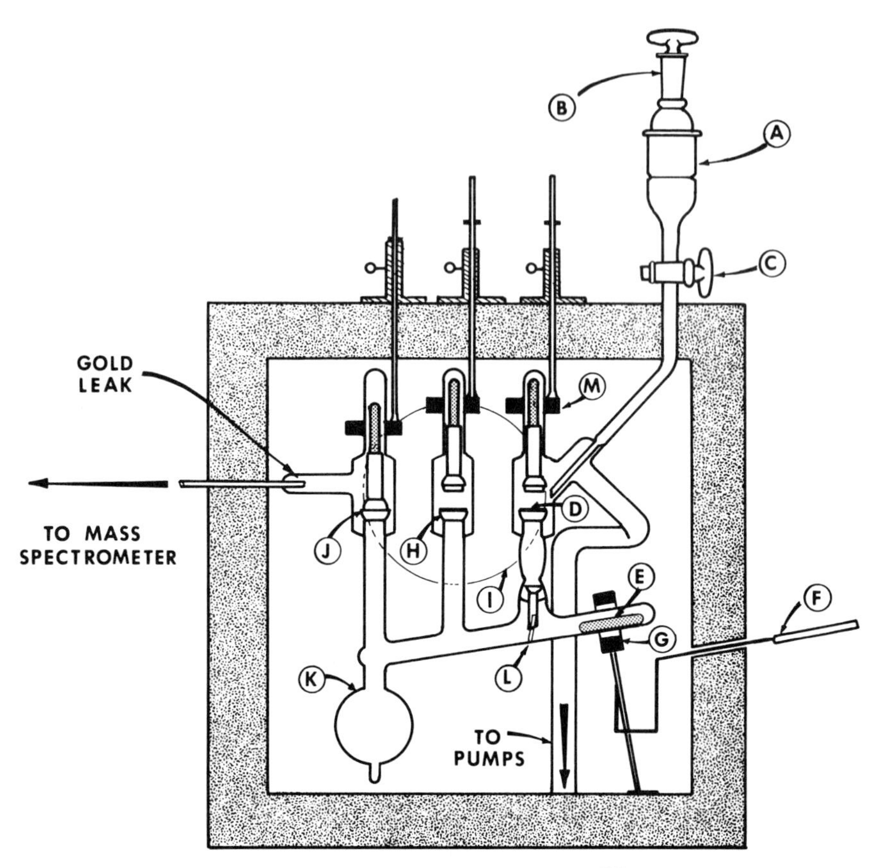

Fig. 4-4. Typical heated inlet system (Peterson, 1962a). A, ₸ joint to introduce sealed sample capillary; B, rod to align capillary for sequential introduction; C, stopcock to permit capillary to enter oven; D, ball and socket glass valve to isolate vacuum pump; E, glass-encapsulated iron rod to break sample capillary at position L; F, rod to move magnet, G; H, ground glass spherical joint valve to expansion volume, I; J, valve to isolate gold leak to ion source; K, collection point for empty capillary tubes; L, capillary position when broken by iron rod, E; M, Magnet to lift glass valve. (Courtesy of *Analytical Chemistry*.)

system after breakage and are present until the system is shut down to clean and change the reservoir. The glass remains can have sample adsorbed on the walls and can cause background or even contamination problems. By placing the sample in a capillary, pump out of the air admitted with the capillary is possible. After the stopcock to the pumping system is closed, the capillary is broken to admit the solid (or liquid) sample into the expansion volume.

Meyerson *et al.* (1963) developed an inlet system which utilizes indium capillary tubes. These can be filled, crimped to seal, and weighed before introduction into the heated inlet system. In addition to the reproducibility of the sample charged, they also permit preliminary evacuation, but only for the period of time before the indium tubule melts. Thus the operator has to be alert and follow a specified time routine very rigorously. The indium also presents difficulties with halogen compounds, primarily the chlorides. A sample of chloroform gives peaks from indium chloride, an obvious reaction product. Other materials also cause problems, but sometimes the products serve as spectrum markers.

Caldecourt (1955) reported a method using Teflon buckets and seals in an alternate arrangement. The bucket is left in the reservoir after each addition of sample and the outer vacuum seal is replaced each time with a Teflon seal. Just as in the above two cases, the Teflon material accumulates in the reservoir and has to be removed during a shutdown period. The operating temperature is limited to 200°C with the Teflon.

The problem of thermolabile organic substances has drawn the attention of Brunnée (1966), who has reported an inlet system setting limits to the decomposition that can occur during the vaporization process in the ion source. In general, the mass spectroscopist must be constantly aware of the problems that can be introduced in the mass spectrum due to such thermal instability of the sample while in the inlet system.

In all the cases described above, the sample is external to the source, having been introduced to the source through an inlet system. An obvious development was the internal sample system; that is, a system which permits direct introduction of the sample into the source. Less sample is required compared to the reservoir, in which only 1–2% of the vapor sample is used in scanning. The cold inlet system requires anywhere from about 10^{-5} torr to 1 standard ml of gas sample. In the heated inlet system this is reduced to 100 μg as a typical sample or about 1 μg as a minimum. The direct probe inlet reduces the sample size to a range of about 0.1–20 μg, while the GLC and direct evaporation introduction systems allow the minimum sample to be reduced to about 0.01 μg.

D. Direct Probe Inlets

Early developments of direct inlets include (a) the vacuum lock of Stevens (1953), which was especially useful with surface ionization sources; (b) a somewhat modified version of this system used by Echo and Morgan (1957); and (c) another design of the sliding vacuum lock used by Roberts and Walsh (1955), which permitted the sample container to be placed very close to the ionization chamber. In all of these units a double O-ring seal prevented leakage of gas from atmospheric pressure into the first pumping stage. Isolation from the pumps was then accomplished by further sliding of the sample into its final position, with the attendant closure of the vacuum port. With suitable machine tolerances and care in handling, the O-ring seals and other seals permitted the vacuum to be maintained at 10^{-6} torr during the final introduction into the source.

Most of the commercial instruments today are optionally equipped with a direct probe inlet accessory. In general, these inlets utilize double or triple O-ring seals, a preevacuation system, a vacuum-tight valve that can be opened to the source when the introduction chamber has reached the approximate source pressure, and a heating system that permits volatilization, pyrolysis, or a thermal wire heating of the sample very close to the source region. In some cases glass capillary tubules are simply inserted into the heater, usually of a ceramic type with a close fit to insure good heat transfer. The heating units are controlled externally and thermocouples are included so that a temperature-programmed pyrolysis or volatilization can be used in examination of the sample. The details of construction of these inlets are quite varied, depending upon the geometry and construction of the individual sources, which, of course, vary from instrument to instrument, depending upon the manufacturer.

E. The Knudsen Cell Outlet

The principle involved in the Knudsen cell outlet is the production of a molecular beam of a relatively involatile substance by evaporation from a furnace. The solid sample is placed in a crucible (usually tantalum with a graphite liner) having a molecular leak into the ion source, arranged to give an intersection of the molecular beam and the electron beam. The details of this apparatus were described by Chupka and Inghram (1955). Heating the material in the crucible to a sufficient temperature produces evaporation and a molecular beam. The beam is usually defined by a series of slits before reaching the electron beam. Both resistance and induction heating may be employed with the sample crucible. The temperature of the material in the crucible can be obtained by sighting with an optical pyrometer. Using this type of internal sample system, one can determine vapor pressures, heats of vaporization, bond dissociation energies, and the various types of species present in the vapor. Samples containing a mixture of substances with differing vapor pressures are difficult to handle because of the variation in sensitivity, which is, after all, a measure of the vapor pressure of the species. The design of Chupka and Inghram (1955) was used to determine directly the much disputed value of the latent heat of vaporization of graphite, but this type of source has found additional important applications. Berkowitz and Marquert (1963) used the technique to examine different forms of sulfur. Fragments consisting of S_2–S_8 were observed; it was noted that rhombic sulfur produced more S_8 ions, while "Engels" sulfur showed an excess of S_6 ions. This was taken by these workers to reflect the actual structure of the solid form. Berkowitz and Chupka (1964) discussed the necessity of obtaining a true thermal equilibrium in the cell in order to obtain meaningful results. The technique continues to find considerable application in theoretical studies, particularly in the inorganic areas of chemistry.

For organic materials the Knudsen cell finds limited use, but de Mayo and Reed (1956) have obtained mass spectra of the higher terpenoids and the steroids and have determined accurate molecular weights of the compounds involved. The use of the Knudsen cell has been fruitful in degradation studies of polystyrene (Bradt *et al.*, 1953), a series of other polymers and crepe rubber (Bradt and Mohler, 1955a), and for a molecular weight distribution in fluorinated polyphenyls (Bradt and Mohler, 1955b). In the latter case, the molecules are extremely stable and can be evaporated at high temperatures out of a furnace which is directly attached to the ionization source. The caution is made, however, that even in this case the entire sample is probably not "seen" or examined but only the more volatile species in the mixture can be observed. The unevaporated material remains in the cell, giving a low average molecular weight for the entire sample. This is a typical problem encountered when molecular weight determinations of mixtures are done by mass spectrometry. All is not accounted for unless the entire sample is completely volatile.

The method can be made quantitative for the volatile species only if the ion currents observed for the various species are integrated over the time that it takes for the whole sample to evaporate. Sensitivities for the various components can be determined if pure samples of the materials thought to be present are examined in exactly the same fashion as the mixture. Bradt and Mohler (1958) reported an analysis or "characterization" of a mixture of tri-, tetra-, penta-, hexa-, and octaphenyls by this means.

III. Techniques of Sample Introduction

All of the usual precautions that the analytical chemist considers in sample handling apply to samples for the mass spectrometer, i.e., volatility, chemical reactivity, personal hazards in handling, representative nature, etc. When working with gas samples from reactions, reagent tanks, etc., the submitter usually regulates the pressure to well under 2 atm if a glass bulb contains the sample. Commercial containers and their shipment come under regulations of governmental agencies the world around. While these regulations differ, usually the containers must allow 20% vapor space (or "outage") at temperatures up to 140°F. (These temperatures are encountered in the Iranian oil refineries and in a number of other places.) In addition, some "dry" gas samples may be under very high pressures. The handling of such containers and sampling devices to obtain representative samples and their removal to containers suitable for laboratory work is described by Askevold *et al.* (1950). Any researcher contemplating working with such gas samples, for reference, known sample blends, etc., would be well advised to become as thoroughly familiar as possible with the above procedures. All too frequently good analyses of gas samples are impossible because of the poor sample-handling techniques employed before the analysis.

Mass spectrometry laboratories handling gas samples should have a vacuum transfer system complete with diffusion pump, Toepler pump, McLoed gage, and a manifold for attaching various sampling devices. These may be either gas bulbs or small capillary tubes for sealing off samples of liquids. The gas sample bulbs should have vacuum stopcocks and freezeout tips to permit pumping off of excess air or nitrogen from the pure condensed gases, such as methane, ethane, etc., all of which can be solidified with liquid nitrogen. The sample transferring system of this type is of value in handling volatile reference liquids which can be divided into a half dozen or more small capillary tubes filled with the identical reference sample. These can be used as needed, assure duplication of calibration materials, and considerably reduce the cost of such calibration materials.

A. Introduction of Gas Samples

There are several methods of introducing gases to the mass spectrometer, depending upon the inlet systems and the research problem. A schematic of a room-temperature gas introduction system is shown in Fig. 4-5.

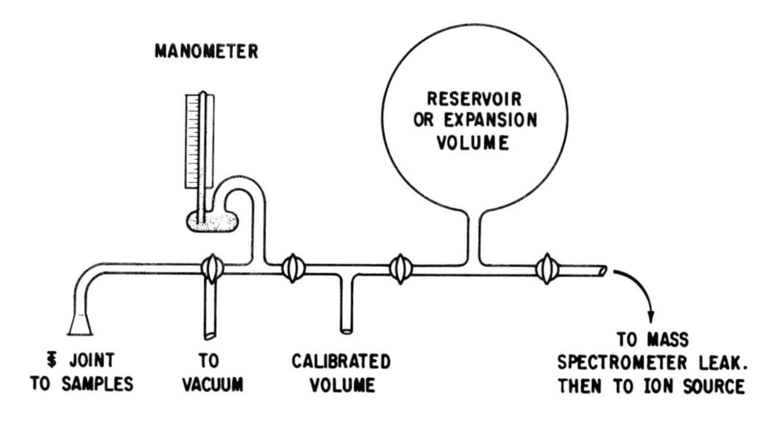

Fig. 4-5. Simple room-temperature gas inlet system.

1. Direct Introduction from a Gas Syringe

A number of syringe manufacturers have available gas-tight syringes complete with valves so that a gas sample can be obtained from a system through a rubber or plastic septum. After the syringe is filled to the appropriate volume, the valve on the attached needle is closed and the sample tagged and brought to the mass spectrometer for analysis. If appropriate, the entire volume of the syringe can be admitted to the mass spectrometer through a septum after preevacuation of the volume between the needle tip and the valve. If only an aliquot sample is desired, introduction can be made via a calibrated volume region, filling the volume to a predetermined pressure as read on a manometer. The predetermined pressure depends upon the materials being analyzed and their sensitivities. The qualified mass spectroscopist appraises the situation in advance. If a quantitative analysis is required (as discussed in Chapter 6), complete calibrations for all components are usually necessary. If only a qualitative analysis is desired, the amount of gas sample depends upon the interest in trace components and the level of accuracy. This method of introduction is quite simple.

2. Introduction from a Gas Sample Bulb

A suitable gas sample bulb with vacuum stopcock and standard taper or spherical joint adaptable to the gas inlet of the sample port is also a popular means of introducing a gas sample. In this case, the attached sample bulb is preevacuated up to the confining stopcock and, with the pump out of the system, a small amount of the gas is admitted to the calibrated volume of the gas inlet. Usually a predetermined pressure as read on the manometer is the criterion for the amount of gas to be used. If the operator overshoots this pressure by a substantial amount, the mass spectrometer may have a sample much too large to handle without encountering source problems. In such an event it is best to pump out the sample and then readmit the proper pressure into the calibration volume. After the calibration volume is isolated, with its known pressure content, the valve can be opened, admitting the sample to the expansion volume of the inlet. This, in turn, will give a pressure of about 10^{-5}–10^{-7} torr in the source. If a micromanometer or other pressure reading device is included with the inlet system, this pressure is the sample pressure to be recorded for analytical purposes. This should be the most accurately readable pressure of the system and is most reproducible in making a comparison to calibration pressures and other sample pressure measurements. This is true regardless of the type of leak between the reservoir volume and the source. Apiezon greases or other hydrocarbon types such as "Kel-F" grease of inert fluorocarbons may be used on these stopcocks. The silicone type of grease is not encouraged (see Section V). The type selected is usually dictated by the type of gas sample involved.

3. Introduction from a Constant-Volume Gas Pipette

With mercury-covered frits and a room-temperature inlet system, another method of introducing a calibrated amount of gas sample is available to the mass spectroscopist. A mercury-sealed gas storage system is shown in Fig. 4-6 in which the mercury-covered frit acts as a seal. As long as the amount

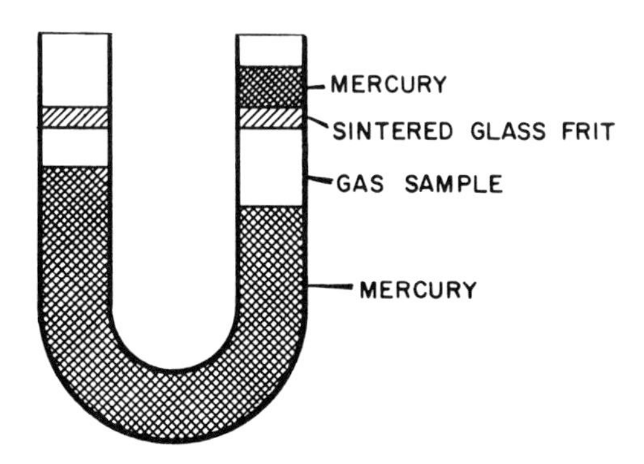

Fig. 4-6. Mercury-sealed gas storage system. (Courtesy of Consolidated Electrodynamics Corp.)

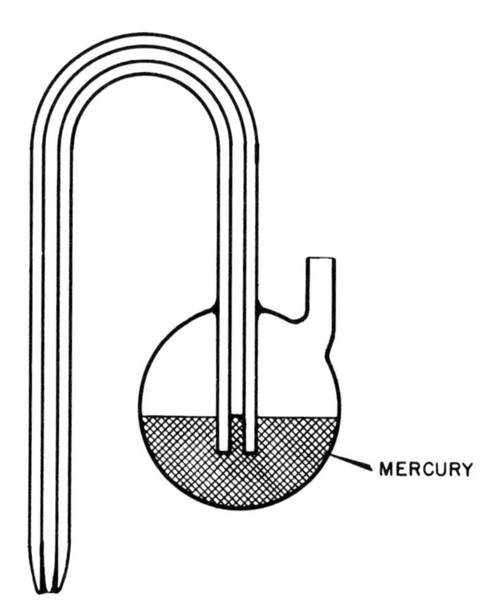

Fig. 4-7. Mercury-sealed constant-volume gas pipette. (Courtesy of Consolidated Electrodynamics Corp.)

of mercury above the frit is maintained, there is no gas leakage. These tubes may be stored for moderately long periods of time at ambient temperatures, thus providing a source of a reference gas. A constant-volume gas pipette is used with this apparatus and is depicted in Fig. 4-7. Mercury is pushed into the capillary to fill it entirely. Then the pipette is filled by contacting the tip of the pipette with the upper surface of the frit of a storage bottle such as discussed above. The release of pressure on the other side of the mercury pipette allows gas to flow into the pipette. The pipette is now submerged below the surface of the mercury-covered frit inlet to the system shown in Fig. 4-5. The gas sample is drawn into the mass spectrometer inlet. As an alternative to the above system, Heymann and Keur (1965) described an all-metal gas pipette which is claimed to be reproducible to about 1%. The volume of the pipette remained unchanged to within 2–5% after opening and closing the pipette 243 times.

4. Miscellaneous Methods of Introduction

Many special adaptations of mass spectrometers to analytical problems exist. Along with these are many interesting means of gas sample introduction. Perhaps the most intriguing is the study of the upper atmosphere by means of rocket-mounted mass spectrometers. Nier and his co-workers have done so much in this field that their work is known even to the lay public. In these mass spectrometers, an explosive port simply opens the entire mass source to the "outside," which in this case is the upper atmosphere, passing by the electron beam as a dynamic sample. Gaseous radicals are usually studied by generation in the source itself, while "leak detectors" simply have the vessel to be tested attached directly to the source. The variations of these types of introduction are too numerous to cover in this treatise, but Silverman and Oyama (1968) have reported an automatic apparatus for sampling and preparing gases for mass spectral analysis that may stimulate the development of new gadgetry in the near future. They described an automatic apparatus which samples the gas phase above a microbial methane-metabolizing system every two hours; separates and measures the individual gas components, hydrogen, oxygen, nitrogen, carbon dioxide, and methane, by dual-column gas chromatography; and combusts methane quantitatively to carbon dioxide. It collects the metabolic CO_2 and the CO_2 derived from methane combustion in separate containers for a later mass spectral analysis of carbon isotope ratios. This is quite elaborate compared to a simple gas bulb.

B. Introduction of Liquid Samples

One of the most important distinctions between gas and liquid (or solid) samples is that in the case of the gas sample a constant-volume introduction means a constant-mole introduction at the same temperature and pressure. Assuming ideal gas law behavior, pressures of gases are proportional to moles, thus making calculations and comparisons much easier. The liquid volume introduced may well be constant within a percent, but the number of moles introduced depends upon the molar volume of the substance, that is, the milliliters occupied by 1 g molecular weight of the compound (usually at 25°C). Thus the introduction of a constant volume of liquid sample for a series of compounds means a decreasing number of moles introduced as the molecular weight increases. To offset this problem, relative sensitivities are used, as will be shown in Chapter 6.

1. Direct Introduction from a Liquid Syringe

The wide variety of liquid microliter-size syringes provides many different methods of introducing samples. The sample is drawn into the needle, injected through a septum into the inlet system, and the syringe is withdrawn. The septum may be in a cool region, but most instruments today utilize a septum heated to the highest temperature consistent with the material in the septum and with the remaining temperatures of the system. The syringe systems are quite flexible and permit introduction of constant volumes, or they may be calibrated to "deliver" a fixed amount. Needle tips are also optional, some having the port at the side so as not to plug the needle with a "coring" from the septum. Despite the best of care, needles are known to plug for various reasons. For quantitative work, several of the same type should be available. A needle is assumed to be dirty if picked up in the laboratory and should be cleaned with a suitable solvent or series of solvents before being used. The number of punctures per septum before failure due to leakage is a matter of needle size, temperature, type of samples, and, of course, the septum material. Usually an isolation valve is employed after the septum and before the reservoir volume so that air leaking in through the septum is not a major problem. Any air leakage contributes to the total pressure of the system and leads to erroneous sensitivity results. Scanning the background will suffice to show the presence of m/e 28 and 32 peaks due to air and indicate the necessity of changing the septum.

One of the principal sources of error when using syringes for liquid samples is the possible evaporation of highly volatile materials in a mixture of considerable volatility range. Rapid handling and experience with the technique usually permits the operator to develop considerable skill. Another source of error is the technique of inserting the needle itself into the septum. If necessary, one should back off the plunger so that a little air is in the needle. Then, when the needle first contacts the hot septum, evaporation of the liquid will not force liquid out of the needle. A rapid insertion of the needle, followed by movement of the plunger, will prevent large losses of the sample. Some samples not only soften the septum but dissolve in the substance, leading to a large background. If this is suspected, the best course of action is to replace the septum rather frequently.

Precision micropipettes or burettes equipped with needles may also be used for introduction, but the simplicity of the syringe with needle results in much greater usage.

2. *Introduction of a Liquid from a Breakoff Sample Bulb*

This type of introduction system, although used far less than other methods, may be employed when a stored reference material is in a small capillary and the entire contents are to be used. Stored isotopically labeled materials lend themselves to this type of handling because of the desirability of keeping a sealed sample for relatively long periods of time. Two simple accessories are shown in Fig. 4-8. In the (a) portion, the stopcock should be large enough to permit passage of at least a portion of the glass capillary that is to be broken off. Evacuation of the undesired air can be made before the tip of the sample capillary is broken. The volume of liquid material should be small enough to be completely vaporized into the volume available at the desired temperature. If the type of inlet shown in (b) is used, glass or Teflon inserts can be used to bring the breakoff tip of the sample tube to the proper position. The inserts may be cleaned and reused, allowing the spectroscopist considerable flexibility. If only an aliquot portion of the sample is to be used, the material in the attachment can be recovered by using a sample-handling system as described earlier.

3. *Introduction from a Constant-Volume Pipette*

When mercury sinter inlets were in use, the constant-volume pipettes of the types shown in Fig. 4-9 were used (some may still be used). These pipettes proved to be expensive, and they were found to be plugged rather frequently by the mercury oxide in the sinter. The pipette was filled with the liquid to be analyzed by means of a filling pipette. Then the constant volume pipette was dipped below the surface of the mercury in the sinter and touched to the sinter. The sample immediately flowed into the evacuated system. The surface of the mercury had to be above the upper tip of the pipette to insure that only the sample entered the mass spectrometer. The pipette filled with mercury at this time. The pipettes were cleaned with nitric and/or hydrochloric acid, rinsed, and dried before reuse.

An improved introduction system, the mercury-sealed orifice method, has been described by Meyerson (1953), Yarborough (1953), and Davis and McCrea (1957). In this method the pipette is flushed with a considerable amount of mercury to insure complete introduction of the liquid sample and the removal of any surface-held liquid. The inlet is shown in Fig. 4-10. The vertical tube is about 1 m long, with the bottom reservoir being open to atmosphere,

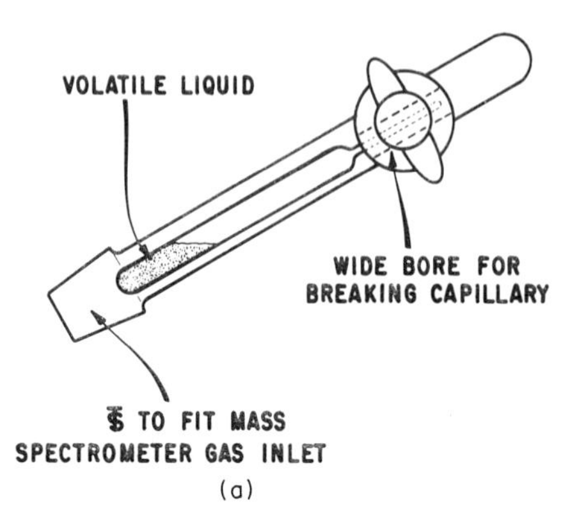

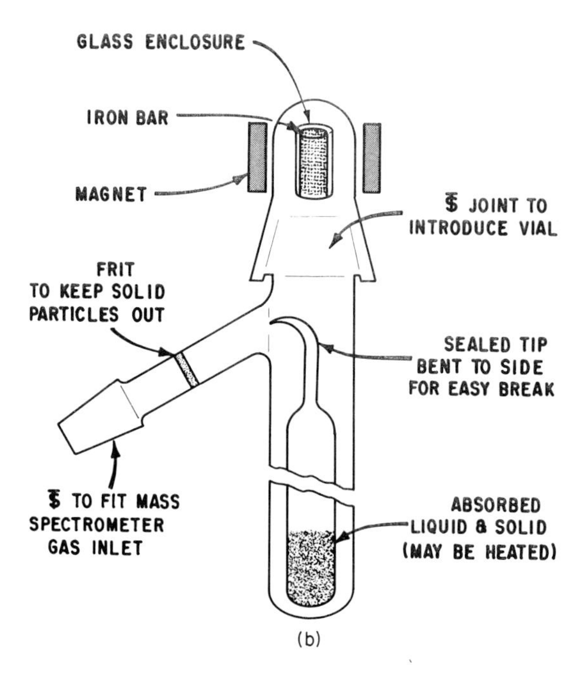

Fig. 4-8. Two devices for introduction of a sample from a sealed sample tube. (a) Volatile liquids in capillary tubes, (b) liquid or solid in larger tube.

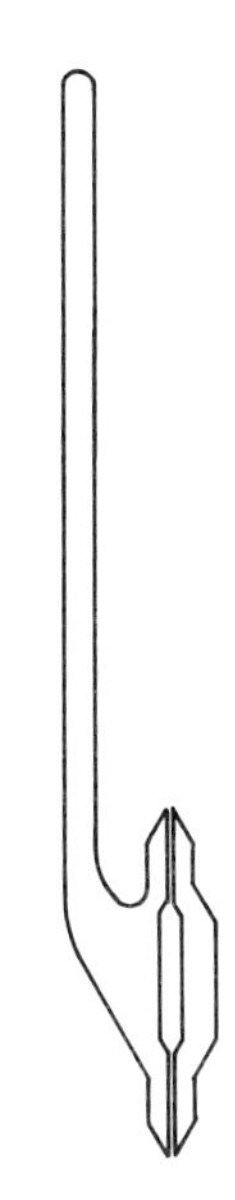

Fig. 4-9. Constant-volume micropipette.

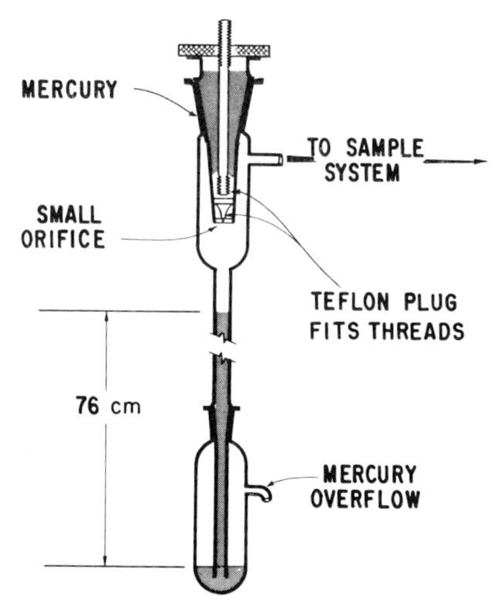

Fig. 4-10. Mercury-sealed orifice for sample introduction. [From Meyerson (1953). Courtesy of *Analytical Chemistry*.]

thus making the mercury in the long arm a part of a barometer. An isolation valve is between the tube to the sample system and the reservoir. This permits the frequent removal of the unit to clean the orifice and to change mercury. Mercury can be added at the top from time to time and, since this mercury comes through the orifice, contaminants do not accumulate except in the lower mercury reservoir. The mercury can be cleaned periodically. An overflow tube is provided in the event the operator forgets about the lower level of mercury. With such a large head of mercury at the top and a good contact at the orifice, insured by a ground tip on the pipette, a substantial amount of mercury may be passed through the pipette without affecting the mass spectrometer operations. Note that peaks from mercury (including doubly charged ions) are always observed in spectra from such an inlet. These are of some use as mass markers. A glass or Teflon plug is used to keep the orifice plugged until the operator wishes to charge a sample. Since the system volume is relatively small, the evacuation period is short. The system is only operable at room temperature.

C. Introduction of Solids

1. *Introduction in a Solvent*

Because of the precision and ease of using the liquid systems described above, solid samples were introduced dissolved in liquid solvents. Despite its appeal, this presents problems. The flashing-in of the sample (particularly in a room-temperature inlet system) naturally produces the cooling experienced with the expansion of any gas. Freezing-out of the solid may occur, and it can then simply lie on top of mercury (from the orifice inlet) until it warms up a bit. Since it is a solid to begin with, it may not melt even though its relative vapor pressure in the system may be substantial. Therefore, only a small portion of the sample may get to the source, but a large portion of the solvent will get to the source. This dilution may make the peaks due to the solid extremely small. Another problem is the character of the sample itself. If it is a mixture, some types of compounds present may be far more soluble in the solvent and hence show up in greater concentration than was present in the sample. The converse effect is also obvious. The biggest objection, when all of the above can be effectively accepted, is that the solvent itself has peaks and hence should be of relatively low molecular weight, producing no peaks to interfere with the sample. Therefore, except for special cases in which a solvent is used to wash high molecular weight materials out of a tube or system, this method is seldom used. Even in liquid–liquid chromatography fractions, eluents are usually evaporated off so as to remove them from the original material.

2. *Introduction as a Solid*

Almost without exception, solid samples are weighed before introduction. Even for a qualitative "look–see," the spectroscopist does not want too much material in the ion source or inlet. Samples may be introduced in capillaries, glass cups, metal tubules, or, in some instances, pressed into micropellets. All of these techniques require great care, and the experienced microanalyst is at an advantage. The amount of sample to be used varies greatly, usually depending upon the volatility of the substance. The analyst can only guess in many cases as to the material which may be present, and hence more errors in amount of sample are made in this technique than in almost any other. After all, if the liquid sample introduced, say 2 μliters, is too small for the analyst's purpose, it can be either added to or pumped out and a new and larger amount of sample introduced with relative ease. This is not true for the case of solid samples where, at least in some instances, all of the sample available from careful extraction by a laborious series involving a natural product has been given to the mass spectrometrist. The fine white powder turns out to be a total of 10 μg. Before deciding how much of this to introduce, the person requesting the analysis should obviously be consulted. An error on the large side usually produces a dirty source or, as in the case of a peroxide, ozonide, or other relatively unstable substance, it can produce a minor disaster requiring at least the cleaning of the entire source. The mass spectroscopist must make a decision and perhaps pay the consequences. In some laboratories the consequences are considered worth the risk, and nearly any solid sample is loaded without weighing. Sensitivity cannot be measured in such cases.

Solids in ampuls such as shown in Fig. 4-8 can be run if the entire system can be heated up to the source temperature. Again, too much material can be a problem, while too little may make one wish for larger peaks, especially for those interesting details of interpretation.

D. Direct Probe Inlets

The direct probe inlet method allows a much smaller amount of sample to give an interpretable mass spectrum than when a sample is loaded in an evacuated reservoir. From a single introduction of a sample consisting of a mixture, several different spectra can usually be recorded at various temperature intervals. This temperature variation technique "fractionates" a sample, which aids significantly in the identification of components in a mixture.

The main problem, in addition to all of the precautions as to sample size mentioned in the preceding section, is cleanliness. Every effort should be made to keep the probe and all of its parts which come into contact with the source or source vacuum immaculately clean. Gloves are suggested for handling the probe and for any sample containers that must be used. Hopefully, tongs or forceps can be used; but sometimes the hand must guide, in which case gloves are essential if one wishes to avoid seeing "fatty acid peaks" or other artifacts introduced by sample handling. A second problem is that, unknown to the spectroscopist, the sample has been recrystallized from "X" and contains inclusions of the solvent with the crystals. Once into the section where preliminary pumping takes place, a small explosion of the crystals may take place, blowing the sample all over the section. When put into the source section, very little of the sample may remain, and hence erroneous conclusions may be drawn. This may occur to a lesser extent, but obviously the sensitivity observed is incorrect. These same problems may well plague workers with Knudsen cells or other vacuum lock systems where preliminary pumping occurs.

E. Direct GC Inlets

Although gas–liquid chromatography (GC or GLC) has been used as a means of separation much as other physical means have been used (see Section IV,A), the combination of GC with the mass spectrometer as a detector is now recognized as a powerful analytical tool. The simplification of a very small sample of a complex mixture by fractionation into one or, at most, several components by means of GC has enlarged the horizons of natural products research and enhanced the possibilities of identifying many trace-quantity substances in such materials. The first combined use was reported by Holmes and Morrell (1957) in which a consolidated Electrodynamics Corp. (CEC) Model 21-103 was used to detect light hydrocarbons from a GC column. Dietz and Klaas (1958) reported a 2-year experience with analysis of hydrocarbons through C_7 and showed how a ratio of paraffins to naphthenes could be obtained through the gasoline range. Gohlke (1959, 1962) used the Bendix TOF instrument with a gas chromatograph and employed photographic and oscillographic detection methods. Lindemann and Annis (1960) reported using the CEC-21-103B instrument with a capillary column for the separation of the known hydrocarbons up to C_{11}. Lindeman and Le Tourneau (1960) described the identification of C_8 and C_9 bicyclic paraffins in petroleum using the combination of GC–MS. Levy *et al.* (1961) used a high-temperature GC with the mass spectrometer to make identification of paraffin waxes. Levy *et al.* (1963) used the GC–TOF combination for studies of reactions. Simmons and Kelly (1961) reported general methods for the quantitative analysis of complex samples using the combination. Widmer and Gaumann (1962) examined phenylcyclohexane by this technique, while Henneberg (1961) reported using the combination while monitoring a single mass peak. Ebert (1961) and Miller (1963) used a manifold for intermediate trapping and holding of samples between the GC and the MS. McFadden *et al.* (1963) used a capillary column and mass spectrometer for the identification of trace flavors and odoriferous materials. Brunnée *et al.* (1962) connected an Atlas CH4 instrument to a capillary column for continuous mass spectral scanning of the effluent, while Dorsey *et al.* (1963) used the cycloidal CEC Model 21-620 for detection of a mixture of 16 isomeric C_9 hydrocarbons from a capillary column.

All of these workers experienced an assorted set of problems, chief of which was the presence of the large excess of helium carrier gas used in GC. Reduction of the amount present seemed to offer the best possibility, although Bendix attempted to use an "electrical gating-out" of the helium peak before it reached the detector. Vacuum system pumping capacities were also increased.

Problems with mass spectrometer detection sensitivities led to the use of the vibrating reed and electron multiplier. Recording of data was also a problem, but Klaver and Le Tourneau (1963) and Klaver and Teeter (1963) described an improved digitizer fast enough to keep up with the fastest scan rates then available on the CEC instruments.

Two papers appeared almost simultaneously in 1964 to report a marked improvement in the technique. One method by Ryhage and another by Watson and Biemann gave a means of removing helium selectively and thus increasing the sample-to-helium ratio in the effluent by large factors. Ryhage (1964) used two molecular separators coupled in series between the column and the gas inlet line of the mass spectrometer. The separators were built on the principles of Becker (1961). From these Ryhage claimed a sample-to-helium ratio of 100/1. The Atlas CH4 was used with a range of m/e 12–500 scanned and recorded in 1–2 seconds, with the necessary modification for such fast scanning given by Ryhage (1962). Twenty-seven components were identified from a 200-μg sample of methylated fatty acids from butter fat. The separation of a mixture of C_{19}–C_{30} (paraffin) hydrocarbons was also described. Ryhage *et al.* (1965) reported using a capillary column with the same mass spectrometer described above in which interpretable mass spectra were obtained with samples as small as 20 ng.

The Watson and Biemann (1964) apparatus employed an all-glass, valveless, pressure reduction system to continuously feed the effluent from a GC column to the CEC Model 21-110 Mattauch–Herzog mass spectrometer. About 20% of the column effluent enters the pressure reduction system. The pressure drop is accomplished by using a fritted glass tube as a stream splitter, the outside of the tube being maintained at less than 1 mm of mercury pressure. The faster effusion of the helium results in a 10-fold or greater enrichment of the compound of interest which enters the mass spectrometer. The system can be heated to 300°C if desired. The entire mass spectrum is recorded on the photographic plate, which is placed in the focal plane of a CEC-110 spectrometer.

Simultaneous recording and signal integration on the photographic plate eliminates the problem of intensity variations during the emergence time of the GC peak. Up to 30 spectra can be exposed on one photographic plate during the GC run. Thus, even for complex mixtures, the operator can obtain considerable information. Watson and Biemann (1964) claimed that the use of a high-resolution mass spectrometer has two additional advantages. First, mass identification is very reliable and can be done so accurately as to permit the determination of the elemental composition of the fraction and of the fragments formed by electron impact. Second, a calibration compound can also be introduced from the regular inlet system so as to provide a mass marker. With high resolution the mass marker peaks do not interfere with the peaks from the GC effluent, i.e., a perfluorokerosene would not interfere with a hydrocarbon because the peaks would be resolved. Hogg (1969) has described a variable leak inlet system for reference compounds used in high-resolution mass measurement. Watson and Biemann (1965) reported applications of their system with minor modifications wherein methyl esters of fatty acids in concentrations of 2–4 μg were identified. Several alkaloid mixtures were also examined, thus establishing the technique as quite applicable to these high molecular weight species and complex structures.

Teranishi *et al.* (1964) calculated column efficiencies for capillary GC columns in which one end of the column is at the mass spectrometer inlet pressure. They found no significant loss in efficiency but did observe evidence for stationary liquid phase bleeding. This places some limitations on the combination analysis, and an upper temperature for stability of a given stationary phase must be determined. McFadden and Day (1964) discussed in detail the question: How fast can you scan a mass spectrum? One is reminded again that the limitations in the amplifier and recording systems are the important factors (see Chapters 2 and 3).

Banner *et al.* (1964) reported on an AEI-MS-12 capillary column combination. An all-glass continuous inlet system was used with a capillary lead from the GC column. Source pressures were in the vicinity of 5 by 10^{-4} torr. The unit is rapid scanning, low in cost, and has a maximum resolution of about 2500.

Improvements and modifications of interfaces for the GC–MS system continue to appear in the literature. Grayson and Wolf (1967) described the use of a fritted glass tube and a porous Teflon capillary tube for the combination. Hawes *et al.* (1969) described a separator composed of all glass but containing a silicone rubber membrane. It may be heated and, if necessary, temperature programmed in the same oven as the column. No auxiliary pumping is required. A commercial device constructed of stainless steel and utilizing a membrane separator is available (Cangal Inc.). Some advantages and limitations of membrane separators have been discussed by Black *et al.* (1969). The unit can operate at optimum temperatures between 200° and 250°C and, since the inlet end of the separator is at atmospheric pressure, the operation of the GC is not affected. This permits simultaneous flame ionization detection and mass spectrometer monitoring. It can also be used with capillary or packed columns. The CEC jet separator is a single-stage unit, similar to the Ryhage type of separator described above. It operates in about the same fashion and is commercially available as a unit. Markey (1970) has described an improved glass frit interface for combined GC–MS use.

Studier and Hayatsu (1968) have used the TOF instrument with open-tube chromatography for the analyses of complex mixtures of hydrocarbons. Since so many current applications of the technique appear in the literature, it is impossible to give more than a sample of the sort of materials being studied by this technique. The reader will be well advised to examine the annual reviews and other literature sources in the field wherein his interests lie. For the geochemists, Modzeleski *et al.* (1968) have used the GC–MS for identifying normal and branched-chain alkanes in sedimentary rocks. Also, Eglinton *et al.* (1968a) reported on the hydroxy acids and fatty acids in a 5000-year-old sediment from the English Lake District. In another article of the series, Eglinton *et al.* (1968b) described studies of long-chain hydroxy acids. Tornabene *et al.* (1967) have identified fatty acids and aliphatic hydrocarbons in *Sarcina lutea* by GC–MS. Vetter *et al.* (1967) have examined phytylubichinon, vitamin K_1, and vitamin K_2, while Heyns *et al.* (1967) described separation and identifications of permethylated furanoside, pyranoside, and septanosides of D(+)-galactose.

As might be expected, many combinations of instruments with GC are being considered. Scott *et al.* (1966) reported using a method that permits solute bands to be developed from a gas chromatograph with no loss of resolution. The unit is fully automated, uses a high-efficiency glass column of 30,000 theoretical plates, and is connected to a conventional mass spectrometer and infrared spectrophotometer. Thus the eluates can produce simultaneous infrared mass spectra with the entire unit, including scale expansions run automatically. Presumably, however, *someone* introduces the sample and preselects the variable conditions. Nevertheless, this is progress, and several commercial units are approaching this degree of automation. Hites and Biemann (1968) have added a dimension by combining the GC–MS system with a computer system on-line for the analyses of very complex mixtures.

Vollmin *et al.* (1966) have extended pyrolysis studies to structure elucidation by thermal fragmentation and the combination technique. Organic molecules were coated upon the surface of a ferromagnetic conductor. The conductor was warmed up to the Curie temperature of the ferromagnetic material in about 20–30 mseconds by high-frequency induction heating. Helium carrier gas was used to bring the fragments into the gas chromatograph, and the eluents from this were fed directly to a Hitachi (Perkin–Elmer) RMU-6D rapid scanning mass spectrometer. In less than 3 seconds, a mass spectrum of each pyrolysis fragment was obtained. The earlier warning about the short-lived nature of these fragments still holds. It is very likely that 3 seconds are many lifetimes for the majority of the thermal fragments formed. The combination is still popular, as testified by a new instrument described by Delaney (1969).

IV. Separation Techniques

A. Preliminary GC Fractionation

It is now history that the advent of gas–liquid chromatography brought drastic changes in the analytical and control procedures for many substances, mixtures, process streams, and products in the drug and chemical industries. Within a relatively short span of time, this new (or rediscovered) technique had expanded into biochemistry and numerous areas of the biological and the microbiological fields. The petroleum companies were involved with in-depth consideration of GC in the early 1950's. Many companies had replaced the low-temperature fractional distillation columns (LTD's) used for control of process streams with GC analysis. Research was done on solid supports, liquid substrates, liquid-phase materials, carrier gases, and column design factors such as length, diameter, flow rates, operational temperature, and even detector designs. As an example, consider detector designs in which a modified thermal conductivity cell was used and described by Schmauch and Dinerstein (1960). Nerheim (1963) described a gas-density balance detector, while McWilliam and Dewar (1958) developed a useful flame ionization detector. Lovelock (1961) reported several additional types of ionization cross section and argon detectors. The latter required argon as a carrier gas. Rouayheb *et al.* (1962) reported application of the Lovelock diode detector to capillary columns to produce quantitative hydrocarbon analyses. Lovelock and Lipsky (1960) described the electron capture detector, which measured the loss of signal due to recombination phenomena instead of measuring an enhanced electrical current. The above list is just a sample; numerous other workers have reported improvements in detector modification and new systems, the complete discussion of which is beyond the scope of this treatise. A synopsis of the important developments and details of methods appears in the texts of Phillips (1956), Desty (1957), Pecsok (1959), Keulemans (1959), Scott (1960), Noebels *et al.* (1961), Dal Nogare and Juvet (1962), Brenner *et al.* (1962), Purnell (1962), and Szymanski (1963).

For an introduction to the subject of gas–liquid chromatography, the reader is referred to the excellent chapter in the text of Willard *et al.* (1965). "Adsorption and Chromatography" by Cassidy (1951), Volume V of the Weissberger (1945) series, "Techniques of Organic Chemistry," will give considerable background to those who are working with the mass spectrometer but who have not had experience with these techniques in separation. A new addition to this series, Volume XIII, on gas chromatography alone, was authored by Schupp (1968). See also Schupp and Lewis (1968). By the mid-1950's the use of GC was replacing the mass spectrometer for control purposes, thus freeing mass spectrometrists for further research developments. The prime movement of the time appeared to be toward the heated inlet system so as to extend the range of m/e's that could be examined.

As applications of GC expanded, it was soon found that retention times (or volumes) were not unique when the number of isomeric substances increased. Thus, in a sense, one was confronted with the same problem as experienced in fractional distillation; namely, that the various isomers overlapped and occurred in varying concentrations in the same fraction, no matter how efficiently the column was operated or how many "theoretical plates" it could be made to exhibit. Since the gas chromatograph utilized such small amounts of materials, a new form of an old challenge appeared in the organic analysis field—namely, identification of fractions collected from the gas chromatograph by extension of existing spectrometric means. Quite naturally, the small sample size of these fractions was of primary concern. It is difficult to try to classify the work as semimicro, micro-, or what have you, but the problem is just the same—identification of complex organic materials.

Mass spectrometry and infrared spectrophotometry offered the most promise at that time, and hence trapping systems were devised. Thompson *et al.* (1960a) used a stainless steel baffle needle trap that could be attached to the Luer-lok fitted exit tube of a GC column. Materials could be trapped in this needle quite effectively, particularly with the aid of a spiral baffle which tended to efficiently condense samples that insisted on coming off the columns in particulate forms as "smokes." Dry ice was most often used to chill the needle while trapping the fraction. When the effluent to be trapped had passed the detector and trap, the needle was simply disconnected from the GC column and a rubber stopper or Teflon plug inserted to seal one end of the trap. The other end could be sealed by simple insertion into an old septum. Some models of these needles were later equipped with a machined thread and stainless steel cap. If the latter were employed, the insertion needle had to be reattached before introduction to the mass spectrometer. Introduction was accomplished through a septum in the mass spectrometer inlet system. The needle trap was kept cooled and sealed until preevacuation of the carrier gas and possible air present had been accomplished. At this point, the mass spectrometer vacuum pump system valve was closed and the contents of the baffle needle trap warmed up causing the sample to flow into the inlet system. On occasion heat guns were employed with high-boiling materials but only after determining that the materials were not thermally labile.

Thompson *et al.* (1956, 1959) reported using such a trapping device for the identification of some sulfur compounds in Wasson crude oil (see also Fig. 4-11). In conjunction with the latter article it was found that fractional precipitation was occurring in the needle trap due to differences in freezing point between the toluene and methylthiophene isomers that occurred in the same retention-time trap. Addition of a drop or two of solvent washed the trapped material into an infrared microcell. This showed a spectrum of essentially toluene, but addition of another drop of solvent followed by direct introduction into the microcell (Perkin–Elmer design) showed an infrared spectrum of the 2- and 3-methylthiophene isomers. This problem was experienced often enough to indicate that the analyst should be extremely careful with the trapped materials, particularly if they were as hard to obtain as the materials reported in the above work.

Earlier work by this group had been done with capillary tubes, spiral glass capillary tubes, glass wool in capillaries (not very reproducible in their makeup), and simple U-traps. Another group at the Bartlesville Petroleum Research Center, U.S. Bureau of Mines, had also attempted to utilize glass U-traps with T-stopcocks that permitted simultaneous bypassing of the trap and isolation of the trap contents. Thereby the upset to column effluent flow was kept to a minimum. Difficulties were experienced with wall contamination and grease contamination of the stopcocks of the trap. Extremely tedious procedures had to be employed to clean the traps. Despite the problems, Hurn and Davis (1958), Hurn *et al.* (1958, 1959), Hughes and Hurn (1960) and Hughes *et al.* (1961) have reported identifications of hydrocarbon and oxygenated materials in automobile exhaust gases. Except in conjunction with prep-scale GC units, the general use of such glass traps has been abandoned.

The techniques with the needle-type trap continued to be improved. Its use in subsequent identifications was described by Coleman *et al.* (1961) in which a total of three isothermal distillations (starting with a 55-gallon drum of Wasson crude oil), pentane deasphaltening, a molecular distillation, a silica gel adsorption, a "brush still" fractionation, and an alumina gel adsorption were employed. A 200°–225°C boiling fraction was obtained which represented 0.6% by weight of the original crude oil. From this fraction benzo[*b*]thiophene and its 2-methyl and 3-methyl isomers were identified by mass spectrometry and ultraviolet (UV) and infrared spectrophotometry. Subsequently, confirmation by microdesulfurization techniques (described later in this chapter) was achieved. The minimum concentration of benzo[*b*]thiophene found was 0.00002% of the original crude oil. Nine GC traps were taken to adequately insure the validity of the analyses given in this article, which represented a good cross-section view of the "art" of fraction collection for further spectroscopic examination from GC at that time.

It should be noted that, as capillary GC columns developed, the direct GC–MS combination was about the only hope of identifying such small quantities of materials. Infrared was more or less ignored, but as anyone familiar with organic identification knows, there are times when infrared is essential. Hence, improvements in trapping techniques continued. The same baffle needle described above has been used by Kendall (1967) in a method for the rapid transfer of GC fractions into infrared cavity cells. The needle is attached directly to the cavity cell by means of a Teflon adaptor, and then the sample is centrifuged directly into the cell. Solvents are not required if the trapped quantity is greater than 0.6 μliter and the boiling point of the material exceeds 100°C. Solvents (2.5 μliters) can be added if less material is available or to assist if the material to be collected is rather volatile. This method is much faster than other methods of collection such as:

1. Trapping directly into chilled powdered potassium bromide, which is then pelleted.
2. Trapping in capillary tubes containing solvent.
3. Trapping in thermoelectrically cooled cells, such as are used in multiple-reflection attachments.
4. Passing through a special fraction collector in an ultramicrocavity cell which requires beam-condensing equipment not necessarily available to all laboratories.

Moisture condensation on the infrared cell can be kept to a minimum by this technique. A final advantage is that the baffle needle can be closed with a threaded stainless steel cap and a Teflon plug at the Luer-lok end which permits storage until the analyses can be made. Obviously, these advantages also hold for the mass spectroscopist. A number of such systems are described in chromatographic literature; the example given in detail here illustrates some of the problems that can be encountered in practice.

Other types of trapping systems have been reported for collection of the sample from the gas chromatograph and introduction to the mass spectrometer. These systems involve collection of the sample on a solid surface and exposure of the surface to the mass spectrometer inlet system while cooled so as to permit an evacuation of carrier gas, air, etc. A release of the sample is accomplished by heating in most cases. An uncooled capillary trap for high-boiling components was reported by Dal Nogare and Juvet (1962). A simple capillary cooled with ice or dry ice in a paper cup was described by Hoffmann and Silveira (1964). Moshonas and Hunter (1964) described a simple device to facilitate the transfer of submicroliter liquid samples directly from small infrared plates or cavity cells into the mass spectrometer. The design allows for recondensation of the remaining sample following mass spectral analysis and thus appeals to those with samples that are difficult to obtain in large quantities. Amy *et al.* (1965) employed a capillary tube containing the same packing as the column. Molecular sieves have been investigated by Cartwright and Heywood (1966). Damico *et al.* (1967) used activated charcoal to adsorb the sample, and Bierl *et al.* (1968) devised a short column containing the column packing which was cooled and subsequently heated in a stream of carrier gas. Scanlan *et al.* (1968) have collected fractions from a packed GC column in 0.03-in. i.d. by 6-in. length stainless steel capillary tubing and introduced the samples without splitting into a modified inlet to a capillary column which employed the mass spectrometer as a detector. It has been applied to identifications in the flavor chemistry area. An article by Burson and Kenner (1969) describes a capillary tubing trap that can be employed for gases as low boiling as methane. A small constant stream of liquid nitrogen keeps the gases in the trap liquified while the helium is being bypassed. The sealed capillary tubes may be stored. These authors claimed that 70–80% of the component to be trapped is retained by using this bypass system. Elimination of the bypass results in higher air peaks but also in higher trapping efficiency.

It should be apparent that employing GC as a means of preliminary separation requires the development of a high order of skill in dealing with small sample sizes and their collection and transfer in a quantitative fashion into the mass spectrometer.

One problem noted early in GC work continues to be bothersome to those desiring separations of high molecular weight materials. This was the fact that as column temperatures were increased, the column liquid phase began to "bleed" out of the column (Teranishi *et al.*, 1964). At high temperatures this problem became appreciable, but solid substrates which were useful for separating out more polar materials from less polar materials became popular. As programmed-temperature chromatography was introduced, the problem reappeared and was reemphasized. For details of the technique see the Harris and Habgood (1966) treatise, and in particular note their solution of a dual-column technique. Although their technique solved the problem of "bleeding" in GC work alone, it is not applicable to the GC–MS combination if high sensitivity is desired. Levy *et al.* (1969) claimed that they could not satisfactorily use the dual-column technique when the minimum detectable quantity was 10^{-10}–10^{-12} g/second. In temperature-programmed operation of the combination, the continually changing rate of bleeding produces a continual increase in intensity of ions from the liquid or even solid substrate. This, in turn, changes the background spectral pattern and not only complicates the overall operation but limits the system sensitivity. To alleviate the problem, a column-bleed absorption technique was employed by Levy *et al.* (1969). An 8-in. × $\frac{1}{8}$-in. o.d. column packed with 15% wt/wt Carbowax 20M-TPC (polyethylene glycolterphthalic acid ester) on Anakrom ABS, 60–80 mesh, conditioned at 245°C for 2 hours was connected between the outlet of the analytical column and the stream splitter to serve as a bleed-absorbing column. For a test mixture of C_{10}–C_{12} normal paraffins, bleeding occurred at about 105°C on a temperature program run with the column starting from 50°C and going to 197°C at an average rate of 4°C/minute with the GC column alone. With the bleed absorption column in, sudden bleeding was not noticed until a temperature of 193°C was reached. One can predict that many variations of this technique will doubtless be reported in the literature in the next few years.

In studies of liquid-phase materials, McKinney *et al.* (1968) used the MS probe inlet to study the stability of 15 different materials commonly employed in gas–liquid chromatography.

B. Other Physical Fractionation Methods

The mass spectroscopist receives samples that have been obtained from a great variety of physical separation methods and must bear in mind the possible effects the method might have on the analysis. Since distillation had been used for many years in simplifying mixtures for analyses by spectroscopic methods, it was quite natural that this would be used for fractions to be examined by mass spectrometry. An example of this type of application was given by Meyerson (1953) whereby the fractional distillation simplified the mixture to the point where it was possible to obtain a quantitative analysis for all the components present. In such an event, the distillation requirements were rather high, perhaps some 30 theoretical plates. Examination of fractions from a molecular distillation, on the other hand, would be difficult to conduct quantitatively because of the extreme overlapping usually present in a one-plate distillation. There may well be cases in which the material to be examined contains a trace of low-boiling solvent in a system of high molecular weight material. In this case, the molecular distillation would produce a first distillate that could probably be readily identified by mass spectrometry.

Extraction is another process that can be employed in many forms. Liquid–liquid, countercurrent, Soxhlet, and adductive types are all well known. Each may introduce materials that will show up in the mass spectrometric analysis of the fractions so obtained. Similarly, liquid–liquid chromatography, paper chromatography, thin-layer chromatography, and the combination of electrophoresis with "curtain" chromatography (sometimes referred to as electrochromatography) all may be used before fractions are given to the mass spectroscopist. All of these methods exhibit some (or all) of the basic problems that cause difficulties in examination of the fractions. The problems are usually (a) overlap of similar substances due to incomplete separation and (b) contamination of the fractions with substances from the extractants, eluents, solvents, or even the solid absorbents.

Biemann (1962b) has reported examination of samples from thin-layer chromatography (TLC) wherein the material thought to contain a "spot" of interest is scraped from the plate and examined. The probe introduction system allows one to introduce the entire solid sample and volatilize off (perhaps selectively, thus adding an additional dimension of separation) the material absorbed. For details on the techniques of thin-layer chromatography, Volume XII of the Weissberger (1945) series by Kirchner (1967) is recommended. Seifert and Teeter (1969) have used preparative thin-layer chromatography and high-resolution mass spectrometry for the analysis and identification of carboxylic acids from a fraction of high interfacial activity isolated from a California crude oil. Similarly, Struck *et al.* (1968), in examining the constituents of the cotton bud, reported the identification of the major high molecular weight hydrocarbons from both the bud and the flower. Thin-layer chromatography was employed on petroleum ether extracts, followed with examination by gas chromatographic and mass spectral techniques. An interesting result of this is that the even numbered hydrocarbons (C_{24}–C_{32}), although present, constitute a relatively minor portion of the total hydrocarbon fraction. The C_{23}–C_{31} odd-numbered hydrocarbons dominate the composition of the material. Hence, being petroleum oriented, these authors presumed that this natural product material (if it gets into a sediment on its way to a marl and thence to the production of petroleum) has suffered some form of oxidation or reduction, or possibly bacterial action has removed the odd carbon atom resulting in even-carbon-number domination of the distribution of the fatty acids observed in many crude oil extracts. Sometimes results such as observed above can lead to such far-reaching conclusions, or at least to new ideas, which thus help develop new avenues of research.

Paper chromatographic separations permit an actual "cutting out" of the sample-containing portion of the paper to be extracted by a suitable solvent. The sample may then be concentrated, placed into the capillary tube, and the remaining solvent evaporated. The direct probe inlet allows volatilization into the ion source. Artifacts from the paper may be introduced on occasion, but a blank extraction run on the paper leads to confidence in the mass spectrum obtained. Carbonization of the paper occurs if the sample is not extracted from the paper, thus leading to possible additional problems from contamination.

Probe introductions should be done cautiously, since a thermally sensitive compound or even an explosive compound might be in the fraction. One of the authors (NGF) experienced a problem with a known ozonide, in the direct probe, when inadvertent rapid heating led to an explosion while the instrument was scanning. The electrostatic scanning potential was about

2800 V when the decomposition occurred but fluctuated over 1200 volts. The explosion probably produced a plethora of ions, radicals, nascent oxygen atoms, etc., and also gross fluctuations in source conditions. Actually, no useful data could be obtained, but the excruciating, prolonged cleanup of a thoroughly contaminated source and instrument was experienced. (The source proper was exchanged.)

Fractions from electrochromatography usually present problems only of solvent contamination. Some attempts have been made to utilize this method of separation in nonaqueous solvent systems, one such adaptation being reported by Johansen *et al.* (1962). A number of other techniques of separation are available but are beyond the scope of this treatise. The Weissberger (1945) series, "Techniques of Organic Chemistry," will suggest many additional possibilities. For additional details on techniques used in connection with the gross mixtures encountered in petroleum, refer to the texts by Brooks *et al.* (1954) and Gruse and Stevens (1960). Snyder and Buell (1962) have summarized the separation of oxygen, nitrogen, and sulfur compounds by means of solid absorbents and ion-exchange resins.

C. Chemical Fractionation—Macrosamples

In this section the idea of chemical separation is applied to those situations in which a sufficient amount of sample exists to consider ordinary separation methods, similar to the physical separations of distillation, extraction, etc. Although the methods described may be applied on a microchemical basis, the majority of research people will utilize these techniques on larger quantities of materials and/or as part of preliminary separation schemes. Several examples of adduct formation are used rather extensively in petroleum separations. Urea adducts are formed to establish concentrates rich in paraffins separated from those rich in isoparaffins so that further separation means such as GC or TLC will lead to simpler mixtures for examination in the infrared or mass spectrometer. Similarly trinitrobenzene adducts are used to separate and concentrate polynuclear aromatics from other materials, particularly the monoalkyl aromatics and partially hydrogenated molecules such as the tetrahydronaphthalenes. An example is to be found in Coleman *et al.* (1961). In practically all cases, this type of technique is confined to treatment of the material at some intermediate stage in the separation process. One could add many other examples to this section, but they can be found in the Weissberger series.

Sometimes the mass spectroscopist finds evidence for ion peaks not originally expected in the sample under investigation. Usually this is due to some unsuspected reaction that has occurred between materials present in the sample. So many materials could prove to be incompatible in this respect that it is impossible to list them here. One unusual adduct has been described which may stimulate one to consider the samples from the chemical interaction viewpoint: Dienes form adducts with thiols, according to Oswald *et al.* (1962); if this occurs one might well expect some strange peaks in addition to those arising from both pure compounds.

To deal with the problem of regeneration of materials from their adducts or chemical derivatives, Bazinet and Merritt (1962) developed an attachment for the CEC instrument whereby the sample could be introduced and regenerated *in situ* in the inlet system. Mason *et al.* (1965) reported using this idea in conjunction with the identification of carbonyl compounds regenerated from their 2,4-dinitrophenylhydrazones. This was an extension of the procedure of Ralls (1964) which was intended primarily for GC fractionation but which may well find use in the GC–MS combination.

D. Chemical Pretreatment—Microsamples

In addition to the adduct formation type of separation referred to in the previous section, chemical pretreatment of the fraction or even a pure material is often done. The pretreatment may be done in order to:

1. Convert the material to a derivative of higher vapor pressure
2. Convert the material to a substance having a prominent and therefore easily identified molecular weight ion
3. Remove or displace the peaks from interfering substances or contaminants
4. Form substances which have reference spectra for comparison
5. Degrade the substance to smaller molecules that can be examined by mass spectrometry
6. Obtain additional information useful in the interpretation of the mass spectrum of the untreated substance.

It should be recognized that many samples may lend themselves to the direct probe inlet type of introduction and thus avoid all (or at least some) of the reasons for treatment given above.

Biemann (1962c) gave a number of criteria for the usefulness of chemical pretreatment reactions. These criteria are expanded and included here.

1. The reactions should be simple one-step reactions requiring no isolation and purification of an intermediate.
2. The reactions should be as complete as possible so as to insure a high yield without interfering side or competing reactions. This assures no confusion of the issue.

3. The procedure should be applicable to small samples. Although Biemann (1962c) claimed 1 mg, it would seem that today's requirements will scale this amount down considerably depending upon the circumstances.

4. The introduction of a chemical group should not give rise to additional cleavages and resulting new fragmentation peaks, unless this would aid in the interpretation of the resulting spectra.

5. The group(s) introduced should not increase the molecular weight beyond a range convenient to the instrument being used. This could be important for a polyfunctional molecule. Consider the case of a tribasic acid which is converted to a perfluoroester. The molecular weight would go up 330 mass units for a perfluoroethyl alcohol reactant.

6. Because of the small amounts of materials required, expensive reagents and their reactions can be utilized to considerable advantage in this area of microchemistry, whereas their use would be prohibitive in classical organic syntheses.

7. The safety of laboratory personnel should be considered, particularly when explosive or thermolabile compounds might result from reactions with other classes or components in the mixture not known to be or expected to be present. It is generally recognized that almost any organic mixture accumulates peroxides upon standing. Many substances, such as olefins, unsaturated oxygenated materials, and others too numerous to list, show phenomenal growth of peroxide number with short periods of time. Care must be exercised at all times despite the very small amounts of materials.

1. Conversion to a More Volatile Derivative

Organic carboxy acids, amides, sulfonic acids, and multihydroxy compounds are among the numerous types of compounds encountered that have a relative volatility considerably lower than that expected due to the molecular weight alone. Compounds of the zwitterion type usually possess extremely low volatility. Acetic acid with a molecular weight of just 60 has a much higher boiling point than *n*-butane with a molecular weight of 58. The differences in vapor pressure are well known, even to the novice organic chemist. It was with the series of fatty acids, which have such low vapor pressures, that organic chemists first proceeded to prepare the methyl esters. These in turn had sufficient vapor pressure that separations by fractional distillation could be accomplished. Hence, the mass spectroscopist seeks the same assistance for his samples.

a. $RCOOH$, RSO_3H, $RCONH_2$. Conversion of compounds with free carboxyl groups may be accomplished in a number of ways. Some are as follows:

$$\mathrm{R{-}COOH + CH_2N_2 \longrightarrow R{-}COOCH_3 + N_2} \tag{4-1}$$

$$\mathrm{R{-}COOH + R'{-}OH \xrightarrow{\text{Dry HCl}} R{-}COOR' + H_2O} \tag{4-2}$$

$$R{-}COOH + \xrightarrow{SOCl_2} R{-}COCl \xrightarrow{R{-}OH} R{-}COOR' \qquad (4\text{-}3)$$

The reader will doubtless think of a number of additional approaches to this problem and is encouraged to do so. In a later section similar chemical conversions are shown that result in a marked increase in molecular weight as well as change in volatility. These three reactions illustrate a general problem with chemical pretreatment. The diazomethane reaction is appealing because it meets the criterion of simplicity. The reactant is added in ether; the reaction occurs, with bubbles of nitrogen observed if one is fortunate; then the solvent is evaporated. The desired ester should be all that remains. However, the diazomethane may undergo side reactions with other functional groups in the molecule, or with other substances in the mixture. Similarly, thionyl chloride can lead to reactions with dicarboxylic acids to form anhydrides or other similar dehydration reactions. It is safest to go by the acid-catalyzed esterification route and accept the difficulty of removing the water formed and the possible problems from excess HCl. Thus one can see that, although the criteria presented are general, chemical considerations may override the approach to be used.

Teeter (1967) reported the use of α,α,ω-trihydroperfluoroalcohols to make volatile derivatives of carboxylic acids. The compounds are stable and the spectra are similar to the methyl esters, except that the ions containing the fluoroalcohol and carbonyl portions of the molecule are removed to higher-mass regions. This aids in the identification of the substance or impurities.

Sulfonic acids can be esterified after conversion to the acid chloride:

$$R{-}SO_3H \xrightarrow{PCl_5} RSO_2Cl \xrightarrow{CH_3OH} RSO_2CH_3 \qquad (4\text{-}4)$$

For salt forms, conventional methods such as extraction are used after conversion of the aqueous salt solution to the acid.

Amides are particularly nonvolatile. They can be reduced to amines with $LiAlH_4$:

$$R{-}CONH_2 \xrightarrow{LiAlH_4} R{-}CH_2NH_2 \qquad (4\text{-}5)$$

Biemann *et al.* (1959) claimed that even polyamides can be made volatile by reduction to polyamines with this reagent. Hydrolysis of the amide to acid, followed by esterification as described above for the acids, can be done. Biemann (1962d) cautioned against the use of this method when the amide group happens to connect two parts of the molecule as it does in the peptides.

b. Zwitterionic Molecules. The acidic or basic group must be removed in the case of zwitterionic molecules. Esterification of the acid group with alcohol and dry HCl is usually preferred because the diazomethane will react with the amino groups present. Biemann and McCloskey (1962) described spectra from both the acid and base moiety of the salt where both are sufficiently volatile. The hydrochlorides or picrates of amines were cited as examples in which dissociation without decomposition upon heating occurs, leading to useful spectra.

c. Hydroxides. The presence of hydroxide groups may present problems, but for alcoholic or phenolic types several useful approaches are available. These groups may be esterified, acetylated, or etherified. An example is

$$R{-}OH + (CH_3CO)_2 \xrightarrow[\text{pyridine}]{H_2SO_4 \text{ or}} R{-}OCOCH_3 + CH_3COOH \tag{4-6}$$

These groups can also be methylated, provided side reactions do not interfere:

$$R{-}OH \text{ (phenols only)} + CH_2N_2 \longrightarrow R{-}OCH_3 + N_2 \tag{4-7}$$

$$R{-}OH \text{ (phenols only)} + (CH_3)_2SO_4 \xrightarrow{-OH} R{-}OCH_3 + CH_3SO_4H \tag{4-8}$$

Here one finds problems when using diazomethane, a reagent otherwise preferred because of the procedural simplicity. It can react with carbonyl compounds to give higher homologs; it can methylate amino groups, even when protonated; it can esterify carboxyl groups as shown above in Eq. (4-7); and it can add to double and triple bonds forming pyrazolines or pyrazoles. As a consequence, the use of this reagent is more practical after one has first obtained all the available information about the sample and has examined a spectrum from the untreated material. Then the comparison with a spectrum, after treatment, frequently supplies considerable additional help in interpretation of the original spectrum. If dimethyl sulfate is used for the methylation of phenols [Eq. (4-8)] quaternization or methylation of additionally present amino groups may occur. Biemann (1961) cited a means of preventing this when sarpagine, a phenolic alkaloid, was converted into a more volatile derivative. Trimethylanilinium hydroxide was used:

$$R{-}OH \text{(phenols only)} + C_6H_5\overset{+}{N}(CH_3)_3 \xrightarrow{OH^-} R{-}OCH_3 + C_6H_5N(CH_3)_2 \tag{4-9}$$

Conversion of an alcohol to a tosylate or halide is a method of eliminating the effect of the hydroxyl group on the compound's polarity. Reduction with lithium aluminum hydride after the reaction leads to the introduction of a hydrogen atom in place of the hydroxyl group. The reactions are as follows:

$$R{-}OH \xrightarrow[\text{pyridine}]{p\text{-}CH_3C_6H_4SO_2Cl} ROSO_2{-}C_6H_4CH_3\text{-}p \xrightarrow{LiAlH_4} RH + HSO_3{-}C_6H_4CH_3\text{-}p \tag{4-10}$$

$$\text{R—OH} \xrightarrow{\text{SOCl}_2} \text{R—Cl} \xrightarrow{\text{LiAlH}_4} \text{R—H} \tag{4-11}$$

Lithium aluminum deuteride has been employed by Biemann (1962d), Biemann and Friedmann-Spiteller (1961a,b), and Biemann and Vetter (1960) to indicate the position from which the hydroxide group was removed. The tosylates may also be reduced with Raney nickel, leading to their removal, according to Biemann (1962e). Biemann (1962e) also pointed out that Raney nickel with D_2 does not lead to positional labeling because of exchange with hydrogen atoms in other parts of the molecule.

In most cases a tertiary hydroxyl group cannot be removed by the methods shown. Steric hindrance may restrict acylation, while ease of elimination of water may prevent any chemical treatment that leads to satisfactory results.

A very useful treatment with hexamethyldisilazane was developed by Sharkey *et al.* (1957) to assist in this type of problem. The general reaction is

$$2\text{R—OH} + [(\text{CH}_3)_3\text{Si}]_2\text{NH} \rightarrow 2\text{R—OSi(CH}_3)_3 + \text{NH}_3 \tag{4-12}$$

The problem was typical of what a mass spectroscopist might encounter. Oxygenated materials from coal processing were being examined for identification and classification of types of compounds obtained from the processing. Carbonyl compounds, acids, as well as hydroxy compounds were present. The hydroxy compounds are well known for their small molecular ion intensities. These would easily be lost in the fragment peaks and their isotopic contributions from other substances present in the mixture. Positive identification required knowledge of the molecular ions and thus the molecular weight of the type of compound being dealt with. To make matters more difficult, gross mixtures containing a spread of four or more carbon numbers were often encountered.

The silyl ethers not only permitted the detection of the molecular ions but gave increased sensitivity since the vapor pressures of the silyl compounds were higher than the corresponding hydroxy compounds. Perhaps the most important benefit for the mass spectroscopist was the fact that the trimethylsilyl (TMS) ethers gave molecular ions at 72 mass units above the molecular ions of the original compounds. Also the characteristic isotopic distribution of the silicon gave an immediately recognizable group of peaks. With the addition of one group, the molecule ions moved to 72 mass units higher than the original. This amounts to a five-carbon-number change in molecular weight. Thus one can readily see how the ion peaks (including fragments) would be raised to give a bimodal distribution (two envelopes for the parent peaks) separating the reacted from the unreacted material in the mass spectrum. Sharkey *et al.* (1958) in another article described the use of silyl ethers with the aid of low ionization voltage techniques to analyze high-boiling tar acids from petroleum and coal in a rapid fashion that did not require the isolation of pure materials.

Sharkey *et al.* (1957) have shown that tertiary alcohols can be silylated, thus opening a path for handling compounds of this type which are difficult, if not impossible, to deal with by the chemical conversions described above. Silylation has been applied to both alcoholic and phenolic hydroxy groups. The silyl ethers may also be subjected to GLC and with their higher vapor pressures have lower retention times, thus aiding in further separation for identification purposes. Teeter (1962) pointed out that the reagent converts free carboxyl and amino groups into silyl derivatives and that these possibilities must be kept in mind when utilizing this method. The silylamines and esters are easy to hydrolyze to the amines and acids due to their extreme sensitivity to water.

The TMS derivatives enjoyed a long period of use by mass spectroscopists and served well to assist in opening up new avenues of research and understanding in many areas. The natural introspection of the scientist leads to reconsideration of what appear to be simple systems. Thus, even silyl derivatives have been examined for rearrangements and fragmentation pathways so as to explain the mass spectra observed for some derivatives. Diekman and Djerassi (1967) have examined some steroid trimethylsilyl ethers by deuterium and substituent labeling. Diekman *et al.* (1967) expanded the study to include amines and sulfides and claimed that the basic fragmentation modes were essentially the same as in the case of the silyl ethers. In addition, triethylsilyl ethers were found to behave quite differently than the trimethyl compounds. Diekman *et al.* (1968) also reported on the trimethylsilyl ethers of aliphatic glycols and related compounds and in a later article (Diekman *et al.*, 1969) the ω-phenoxyalkanoic acids. McCloskey *et al.* (1968) have described the use of deuterium-labeled trimethylsilyl derivatives and have examined variously functionalized long-chain (aliphatic) compounds. They and Draffan and McCloskey (1968) showed that large-ring transition states are involved in the rearrangements described. The use of trimethylsilyl derivatives of the aliphatic hydroxy acids has been shown to be a facile method of double-bond location by Eglinton *et al.* (1968b). The same general method was apparently developed simultaneously and reported by Capella and Zorzut (1968). Petersson (1969) has extended the silylation to determine the mass spectra of alditols and reported excellent correlation with the structures. The problem of rearrangement persists in many cases, however, and Gustafsson *et al.* (1969) in examining additional steroids reached two important conclusions: (1) the TMS group has a high migratory ability, quite similar to hydrogen; (2) as a corollary, the technique of elemental mapping (Biemann *et al.*, 1964) with TMS derivatives is inevitably subject to uncertainties. Since the TMS derivatives have come to play an important part in analytical mass spectrometry, it appears that future developments must be followed carefully, and a continuous revision of the valuable spectra obtained must be made in light of these studies.

As an outgrowth of the silyl derivatives expansion the preparation of a new cyclic silyl dioxy of a *cis*-diol was reported by Kelly (1969). While the main purpose was to modify polar steroids to give them higher volatility for gas chromatographic separations, it is obvious that the success of the effort will be translated into mass spectrometric use.

Silylation reagents that are highly reactive donors are most desirable. Birkofer *et al.* (1963) first described the reagent, bis(trimethylsilyl)acetamide (BSA), while a detailed review by Klebe *et al.* appeared in 1966. They reported at least 11 types of applications (primarily to biological products) in which the use of silylation has produced volatile products from otherwise highly intractable substances such as carbohydrates, peptides, and flavonoid compounds, in addition to those types already discussed in this chapter. The article includes considerable detail on the similarities and differences in behavior of the materials upon treatment and gives sufficient details about reaction conditions to be a definite aid to the analyst preparing these types of compounds. A "Bibliography of Silylation, Synthetic Methods and Analytical Uses" is available from the Pierce Chemical Co., Rockford, Illinois (1967).

The conversion of amines to their trifluoroacetamides followed by chromatography on polar and nonpolar columns has been reported by Pailer and Hubsch (1966). The mass spectra of these types of derivatives have been examined by Saxby (1969), who found strong peaks at $(M - CF_3)^+$ and at CF_3^+. Nevertheless, the increase in volatility will make this technique another useful one for relatively nonvolatile compounds. Saxby (1968) appears to have started a series of articles on the mass spectrometry of volatile derivatives.

Up to this point, the compounds of interest were usually organic in origin, even though they might be nonvolatile or essentially of such low volatility that the main problem was to get some derivative that would vaporize and could thus be identified or analyzed by mass spectrometry. This same technique can be applied to inorganic substances as long as they can be converted to a derivative that is volatile; i.e., the metal carbonyls lend themselves to mass spectrometric examination but in themselves may be difficult to deal with, particularly from the quantitative viewpoint. However, Booker *et al.* (1969) reported the determination of chromium as chromium(III) hexafluoroacetylacetonate. These authors have examined all the other methods of such chromium analysis and have shown that their method is rapid and reproducible and may be applied to nanogram-size samples or to larger samples by aliquoting. They process from 20–30 such samples per day on an MS-9 (AEI) double-focusing mass spectrometer with a modification of removing the gas inlet capillary and replacing it with a flange having a 12/18 inner Ŧ joint. A special inlet system is attached at this point. This technique may well be one to watch, but if it is run routinely on such a caliber instrument it would appear to be a retrogression in the analytical sense of employing such an instrument for routine purposes. However, if the justification at the research source is sufficient, such a use can be made of the instrument. The concept may prove to be very helpful to those in the trace-elements field, particularly biology, where trace-metal content in biological materials is becoming a point of great interest.

2. Conversion to Produce an Identifiable Molecular Ion

Often the mass spectroscopist puts the highest priority on obtaining the molecular weight or weights of materials present in the sample but accepts any additional information available after this is determined. Hence, conversions may be made to produce an identifiable molecular ion whether or not this also includes an increase in volatility (the main desire described in the previous section).

From the preceding sections the mass spectroscopist can readily see that the application of esterification, acetylation, silylation, etc., is not only a means of increasing the volatility but adds to the chances of identifying the parent ion.

Oxidation of alcohols, many of which have very low intensity molecular ions, produces a ketone which has a considerably stronger intensity molecular ion. The equation is as follows:

$$R-\underset{\underset{OH}{|}}{\overset{\overset{H}{|}}{C}}-R' + CrO_3 \xrightarrow[\text{pyridine}]{\text{In}} R-\underset{\underset{O}{\|}}{C}-R' \qquad (4\text{-}13)$$

Since ketones exhibit a rearrangement (Chapter 5) this type of conversion also assists in determination of the substitution, if any, at the carbon atom next to the carbinol group when the alcohol is oxidized to a ketone.

In some cases the preparation of an anilide leads to a strong molecular ion because of the stabilizing influence of the aromatic ring. Butyric acid is an example of a compound which produces such a small molecule ion and a large M + 1 ion that determination of the isotope profile is impossible. Conversion to the anilide alters the situation and permits examination of a strong molecular ion, which in turn leads to more accuracy in the determination of the isotopic profile. The reaction is a two-step reaction:

$$CH_3(CH_2)_2COOH \xrightarrow[(2)\ C_6H_5NH_2]{(1)\ PBr_3} CH_3(CH_2)_2\underset{\underset{O}{\|}}{C}-\underset{\underset{H}{|}}{N}-C_6H_5 \qquad (4\text{-}14)$$

Harless and Crabb (1969) reported the use of ethylene oxide adducts of commercial-grade nonylphenoxypoly(oxyethylene)ethanol (Tergitol nonionic NP surfactants) and their separation by chromatography to produce identifiable mass spectra. Molecular ions up to mass 616 for the 1–9-mole ethylene oxide adducts of nonylphenol produced clear-cut spectra with a small number of fragment ions. Hence, this would appear to be a major advance in obtaining data from the otherwise low-intensity molecular ions from alcohols. In the particular work reported the direct inlet probe was employed with a TOF instrument, but liquid introduction could probably be done with lower molecular weight materials.

Another method of improving the intensity of the molecular ion peaks is the conversion of substances like the dipeptides to *N*-benzyl(oxycarbonyl) dipeptide alkyl esters which give abundant molecular ions, according to Aplin *et al.* (1968). The molecular ions lose a benzyloxy radical and also show the easily recognized cleavages of the peptide bonds and side chains. The corresponding amino acid and dipeptide phenylthioesters have also been investigated. The peak of highest *m/e* in the spectra of these compounds corresponds to the acylium ion formed by the loss of a phenylthio radical. These authors also suggested that the phenylthioesters may find application in sequential analysis by mass spectrometry by "triggering" the stepwise elimination of amino acid residues from the carboxy end. Pfaender (1967) has made similar claims, providing data on the sequential analysis utilizing the *N*-benzyloxylcarbonyltryptophylpeptide ester.

This particular area of mass spectrometry is certain to expand in the future. It appears that many diverse types of treatment will be required for specific types of substances.

3. *Removal of Interferences* (*or Contaminants*) *by Conversion*

There will be many apparent instances from the material discussed previously that fit this category in the reader's mind. There are several noteworthy examples that can be given to assist the mass spectroscopist in initiating creative thinking in this area.

An early problem for petroleum mass spectroscopists concerned the differentiation of the olefins and cycloparaffins that occurred in petroleum stocks after refining, thermal or catalytic cracking, etc. The ordinary monoalkenes have the general formula C_nH_{2n}. So do the cycloparaffins. Hence the molecular ions (and a good many of the fragment ions of these classes of compounds) were similar and contributed to the same *m/e* ion peaks, thus making analysis extremely difficult if not impossible. As a consequence, it was quite natural to call upon simple organic chemistry to treat out the olefins. While it is true that this could be done, the treatments proved to be either complex, or incomplete; the case of the silver mercuric nitrate adducts of olefins or even the addition of halogens are examples of difficulties one could get into. In the first case, other materials were removed when one was working with the liquid hydrocarbons (the method was reasonably good for Orsat analysis and the accuracy that was desired there). In the case of addition of halogens (or even the dry acid, HCl, in the case of the McMillan apparatus of the 1940's) side reactions occurred, and the resulting material after treatment contained many artifacts and contaminants. Thus these methods were impossible to employ in conjunction with mass spectrometry. From the above, it is apparent that the treatment used must not only meet the requirements stated earlier but should permit a *before and after* analysis on representative materials.

Brown (1951) developed a method utilizing as an auxiliary analysis the bromine number of a naphtha fraction. This method then could establish the hydrocarbon types present on the basis of the fragment ions typical of classes of hydrocarbons normally expected to be present in crude oils and therefore in the resulting naphthas. The assumptions made were (a) that the bromine number could be converted to percent olefin present (a reasonable assumption since one has the approximate molecular weight distribution from the mass spectral data) and (b) that the cycloparaffin distribution by carbon numbers was essentially identical to the olefins, with a small correction being made for the C_5 and C_6 carbon-number ranges.

Another popular method of the time was the "fluorescent indicator adsorption method" (FIA), which utilized a sample of about 0.75 ml in a special glass adsorption column packed with activated silica gel. The sample and a drop of fluorescent dye mixture were placed upon the silica gel. A small amount of isopropyl alcohol was added as an eluent. After a time, a separation into several layers or regions of materials was obvious. The liquid front of the clear material containing the paraffins and naphthenes (cycloparaffins) remained clear in the presence of an ultraviolet light source. The olefinic materials followed this section and contained a yellow fluorescing dye which made it easy to discern the location of these compounds. The next portion of the column was clear (water-white) with ordinary light but it was bright blue in the presence of the ultraviolet light source. This material included all of the aromatic compounds, oxygenates, sulfur, and nitrogen-containing compounds in the material being examined. The isopropyl alcohol eluent was water-white in both types of light; hence the end of the wetted sample portion could be determined easily. The FIA analysis was accomplished by measurement of the length occupied by each of the above sections of materials present in a precision bore glass tubing of 1.60–1.65 mm i.d. Thus an approximate volume percent analysis of types could be obtained rapidly and with good repeatability. The entire method is published as ASTM-1319-66T or IP 156/67 and appears in "Manual on Hydrocarbon Analysis" (1968), ASTM Special Technical

Publication No. 332A. Enterprising mass spectroscopists used this technique, followed by attempts to cut out the olefin portion of the glass tube and introduce this separated material into the mass spectrometer. Qualitatively, this was excellent but left much to be desired quantitatively. Similarly, if one were interested only in the paraffin and naphthenes portion of the petroleum stock, controlled sulfonation would remove all but these classes.

A method of conversion of the olefins by the formation of an essentially nonvolatile adduct was developed by Mikkelsen *et al.* (1958). This method involved running the mass spectrum of the sample before and after treatment with benzenesulfenyl chloride. The olefins in a hydrocarbon mixture react quantitatively with benzenesulfenyl chloride to form a high-boiling addition product. The excess reagent is removed by reaction with a drop of mercury. By comparison of the mass spectra *before and after* treatment, olefins, monocycloparaffins, the CODA compounds (cycloolefins, diolefins, and acetylenes), and dicycloparaffin mass peaks could be determined. From this the composition of the sample could be calculated using a modification of the matrix developed for the hydrocarbon type of analysis reported by Brown (1951). The reagent in this development was reported to have a tendency to decompose, evolving hydrogen chloride upon prolonged storage, and therefore was not to be sealed in an ampule or a screw-cap bottle since sufficient pressure might develop to cause an explosion. Despite the precautions, the laboratory developing the method was forced to notify all users (ASTM E-14 cooperating groups) after several "experiences" indicating that the material was too dangerous. The basic method is good and could be redeveloped if some enterprising organic chemist wants to utilize the benzenesulfenyl chloride reagent prepared by an *in situ* reaction.

The fact that the determination of olefin molecular weights and double-bond positions is still of interest is evidenced by a number of recent publications. Teeter (1968) has shown that the addition of thiolacetic acid to olefins produces derivatives which can be helpful in mass spectrometric investigations of olefins. The monoolefins, diolefins, and cyclic olefins all react and maintain their differences in the products. Some correlations between structure and the fragmentation patterns was also reported. Earnshaw *et al.* (1968) utilized an adaptation of the hydrogenation method of Osburn *et al.* (1966) to deuterate the double bonds in linear olefins. Known compounds were used to support their study "which shows that the mass spectrum of any linear monoolefin after deuteration is predictable to the extent that the compounds can be identified as to chain length and double bond position." These authors are attempting to extend the technique out to C_{22} where many isomers are not available as reference compounds. It is their hope that prediction of the spectra can be made and that this will lead to a new method for olefinic analysis. Teeter *et al.* (1966) have reported using the GC–MS combination for the analysis of α-olefins up to C_{20} in materials with and without catalytic hydrogenation. Another technique used to locate a double bond in a long, straight-chain compound is to form the epoxide and then open the epoxide ring with dimethylamine:

$$R'\text{—CH=CH—R} \rightarrow R'\text{—}\underset{\diagdown O \diagup}{\text{CH—CH}}\text{—R} \rightarrow R'\text{—}\underset{OH}{\text{CH}}\text{—}\underset{N(CH_3)_2}{\text{CH}}\text{—R} + R'\text{—}\underset{N(CH_3)_2}{\text{CH}}\text{—}\underset{OH}{\text{CH}}\text{—R}$$

The molecular ions from these types of substances fragment readily to yield the ions $R\text{–}CH{=}N^{+}(CH_3)_2$ and $R'\text{–}CH{=}N^{+}\text{–}(CH_3)_2$, thus directly showing the spectroscopist the values of R and R′ and thus the position of the double bond. In a branched, long-chain alcohol, if any doubt still exists about the position of a side chain (after rearrangements have been considered, as described in Chapter 5), reduction is useful.

Mounts and Dutton (1965) have employed hydrogen as a carrier gas in the chromatographic pathway and attached a "pre" column to the injection port of a GC instrument. The "pre" column contained a catalyst with which they hydrogenated methyl esters of unsaturated fatty acids.

Niehaus and Ryhage (1967, 1968) have developed a method using the GC–MS combination plus chemical treatment techniques for determination of double-bond positions in polyunsaturated fatty acids. They have shown results obtained with the isomeric octadecatrienoic acids, α-linolenic and γ-linolenic acids. In the 1968 article, they reported fatty acids and hydroxy fatty acids containing one to five double bonds oxidized by permanganate or osmium tetroxide to polyhydroxy acids. These in turn were converted to polymethoxymethyl esters with dimethylsulfinyl carbanion and methyl iodide. The GC–MS combination was now used for analysis. Since characteristic fragmentation between methoxyl-substituted carbon atoms occurred, this allowed the determination of the positions of the methoxyl groups. This in turn indicated the positions of the original double bonds in the fatty acids before treatment. Ambiguities in fragment identification were eliminated by determination of the elemental composition. The LKB-9000* single-focusing instrument with a peak-matching accessory was used. An accuracy of 10 ppm in determination of the molecular weight was claimed. Another modification of this instrument has been made which allows the simultaneous recording of the intensities of three different *m/e* peaks as a function of time. It has permitted the determination of the mixture composition of derivatives from these fatty acids which were not resolved by gas chromatography. An accuracy of $\pm 10\,\%$ was claimed. While the same general techniques seem to be employed, the complexities of some natural product mixtures and extracts will put quite

* LKB Instruments, Ltd.

a demand upon the mass spectrometrist and his associates in the identification game.

Richter *et al.* (1969) reported the analysis of acidic fractions of Green River formation oil shale by solvent extraction followed by chemical transformations. The oxo acids content was substantiated by borohydride reduction followed by silylation of the hydroxy derivatives thus formed. La Lau (1960) used reduction of hetero aromatic compounds to resolve a complex mixture of compounds, both homocyclic and heterocyclic, in petroleum distillates. The effort was made on a relatively large scale, using tin and hydrochloric acid, under which conditions the heterocyclic aromatic rings were reduced. This permitted a quantitative estimation of these compounds (predominantly pyridines and quinolines) in the mixture. Fragment peaks from the nitrogen-containing compounds appear due to the loss of an alkyl group. Their interferences with the molecular ion peaks of the ordinary aromatics of the corresponding carbon number are eliminated upon reduction. Note that ethyl pyridine would give an ion at the molecular weight less 15 mass units appearing at m/e 92, which is the molecular ion peak of toluene. La Lau was one of the earliest to separate and identify materials from petroleum showing heteroatomic five-membered fused rings.

Kinney and Cook (1952) suggested reductive desulfurization of hydrocarbons for determination of the alkyl substituents of thiophenes. Hydrocarbons would be the product and could thus indicate the substitution pattern on the molecules. This is indicated below where the respective thiophene isomers, all of the same molecular weight, yield respectively upon hydrogen reduction, heptane, 2-methylhexane, 3-methylhexane, and 2,3-dimethylpentane.

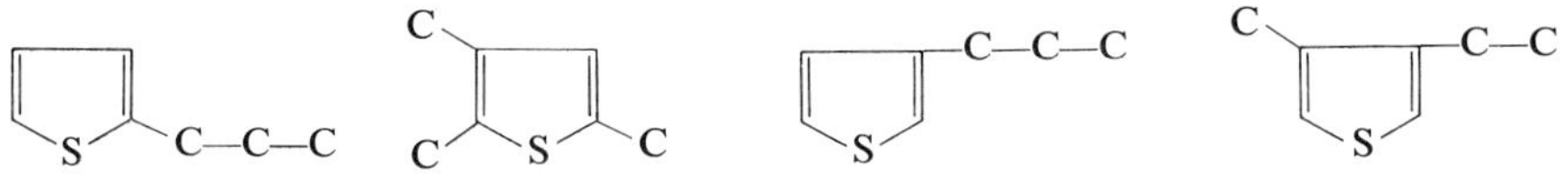

Note that 2-methyl-5-ethylthiophene would also yield *n*-heptane, so the method has limitations. Subsequently, Richardson *et al.* (1961) reported on the infrared spectra and correlations which included the overtone region of some alkylthiophenes. These can be used for determination of some substitutional information.

There are other techniques of organic chemistry which suggest themselves, such as ozonation and the addition of nitrogen tetroxide, all of which ultimately lead to the cleavage of the double bond and the formation of lower molecular weight compounds of the oxygenated family. These in turn may lead to useful information but, as often as not, require further derivatives to be made because of the low molecular ion intensities of the alcohol, low volatility of the acids, etc. (see Section IV,E). Each additional handling procedure, of course, reduces the analytical accuracy and reproducibility, so the mass spectroscopist is well advised to keep the procedures brief and direct if at all possible. If other separational techniques are involved, it is usually the case that the project is in the hands of some other researcher, and the mass spectroscopist need merely advise on the procedures to be followed until the samples reach him for analysis.

In line with the advent of direct methods, the GC–MS and even a computer in-line operation have been described. Perhaps one of the most important developments of auxiliary aid to this team is the technique of microhydrogenation. This was first described by Thompson *et al.* (1960a) in a report pertaining to desulfurization. A second report by Thompson *et al.* (1960b) described the same reductive technique in terms of deoxygenation. A pair of reports describing denitrogenation and dehalogenation appeared in 1962 by Thompson *et al.* (1962a,b). For the interested laboratory worker, an excellent summary with considerable detail appeared in a *Bureau of Mines Report of Investigations* by Thompson *et al.* (1962c). An application to solid samples was described by Thompson *et al.* (1965) while several other summary articles appeared by these same authors in 1967 (see Thompson *et al.*, 1967a,b). The essence of this method is that a trapped GC fraction is transferred to an apparatus especially designed to carry the trapped material (as it warms up and is vaporized) into a catalytic hydrogenation oven. A stream of hydrogen acts as a carrier and passes through the trap so as to flush out all the trapped material into the hydrogenation oven. At the exit end of the oven hydrogenation products are trapped in a cooled needle of the same design. This needle may be used to reintroduce the hydrogenated material into a GC column, into a mass spectrometer, or into an infrared cell of the type discussed earlier (Kendall, 1967).

The complete apparatus is pictured in Fig. 4-11. It consists of an aluminum

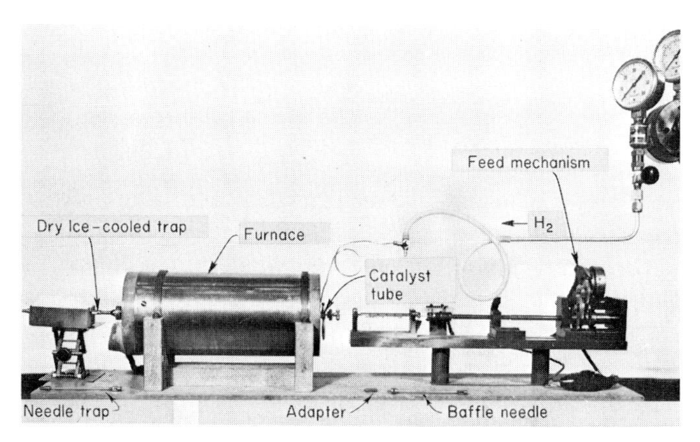

Fig. 4-11. Microhydrogenation apparatus. Note the baffle needle in foreground. [From Thompson *et al.* (1967a). Courtesy of the U.S. Bureau of Mines.]

reaction tube packed with catalyst, a furnace for heating the reaction chamber to the desired temperature, a motor-driven microliter-size syringe for charging quantities as large as 0.01 ml of pure compounds in the liquid state, and the necessary adapters, valves, and connectors for handling very small trapped samples from a GC column. Also in the figure is a special baffle needle "trap" for collecting GC fractions. The reaction tube has an outside diameter of approximately 0.5 in. and has a 0.136-in. bore that is filled with a catalyst usually consisting of 0.5% palladium on activated alumina or 5% platinum on porous glass.

Previously trapped GC samples may be introduced into the reaction tube by removal of the syringe and mechanical drive, which allows positioning of the stainless steel tube trap containing the sample. The traps are equipped with a taper at the entrance end and a threaded adapter at the other end to permit rapid installation of the trap between the catalyst bed in the reaction tube and a three-way valve connected to the regulated hydrogen source. By using this valve, the hydrogen stream can be diverted through the trap rather than directly into the reaction tube as is done during catalyst "sweeping" and with injection of pure liquids. For the GC fractions, the trap is warmed gradually, the hydrogen gas sweeping the contents through the furnace. Details of the various GC columns used with this equipment are described in the above-mentioned publications by Thompson *et al.*

The catalyst and temperature conditions for the apparatus are variables that, in most cases, can remain surprisingly fixed. The ordinary hydrogenation of hydrocarbons, deoxygenation, desulfurization, and dehalogenation may all be accomplished at temperatures of about 200°C in the reaction tube and employing a catalyst of $\frac{1}{8}$-in. diameter alumina pellets coated with 0.5% palladium. Only in the case of denitrogenation is a different catalyst used. For this purpose, 20–28 mesh ground porous Pyrex glass coated with 5% (weight) platinic chloride is used. The glass plus catalyst is placed in the reaction tube, put into the oven, and reduced with hydrogen at 200°C. It is then ready for use.

Thompson *et al.* (1965) devised and described a solids introduction system that was essential for use with solids in the melting point range of 90°–175°C. Originally, solids had been introduced in solution but, once the liquid material was in the entrance to the reaction chamber, the lower-boiling (and inert) solvent flashed off leaving the solid deposit behind. In some cases solidification in the needle or at the relatively cool base of the syringe occurred, thus blocking the passage of material and hindering the overall introduction.

The following examples of substances examined by Thompson and his co-workers give a general idea of the usefulness of the technique. Recalling that GC is used herewith, consider the case of the alkylthiophenes presented earlier by Kinney and Cook (1952). If the materials reduced were examined after a prior GC separation followed by hydrogenation by the Thompson technique, one would expect to find the same hydrocarbons except that in the latter case the GC separation would have probably separated the 2-*n*-propylthiophene from the 2-methyl-5-ethylthiophene, thus giving positive identifications to both compounds. One should remember that often mass spectrometric correlations with structure (such as described in Chapter 5) are either not available or may be too incomplete to be conclusive. In this present case, infrared examination of the material trapped from the GC would have yielded the identity of the substitutions upon the basis of the Richardson *et al.* (1961) work.

As a further example, consider 2-*t*-butyl-5-nonylthiophene. Desulfurization should lead to 2,2-dimethylpentadecane as follows:

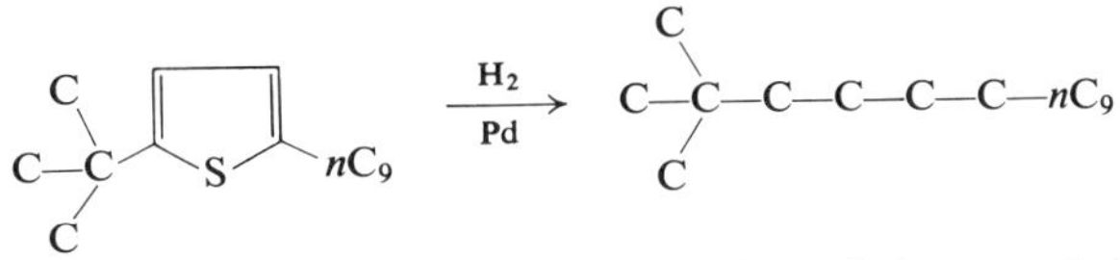

$$(C)_3C\text{-}(C_4H_2S)\text{-}nC_9 \xrightarrow[Pd]{H_2} C\text{—}C(C)(C)\text{—}C\text{—}C\text{—}C\text{—}C\text{—}nC_9$$

The ring is opened and the sulfur atom removed, and the remaining carbon skeletal moiety becomes saturated at every location. In Table 4-1, typical results of the hydrogenation of materials reported by Thompson *et al.* for various compounds are given.

TABLE 4-1[a]

TYPICAL MICROHYDROGENATION REACTIONS[b]

Name	Structural formula	Products
Amyl cyclohexyl sulfide	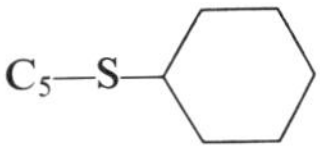 C_5—S—(cyclohexyl)	nC_5 + (cyclohexane) 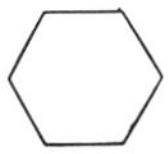
Cyclopentyl cyclohexyl sulfide	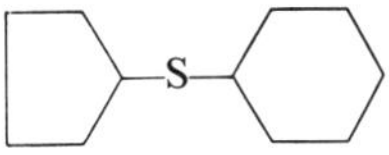 (cyclopentyl)—S—(cyclohexyl)	 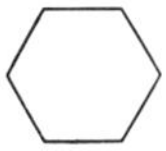 (cyclopentane) + (cyclohexane)
3-Hexanethiol (250° C)	C_2—C(SH)—C_3	nC_6
2-Methylthiacyclopentane	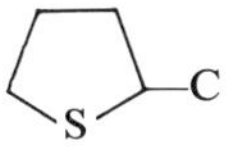 (thiacyclopentane ring)—C	nC_5
3-Methylethiacyclohexane	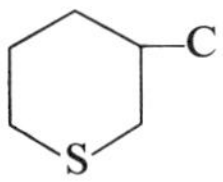 (thiacyclohexane ring)—C	C_3—C(C)—C

[a] From Thompson *et al.* (1967a). Courtesy of the U.S. Bureau of Mines.
[b] Hydrogenations at 200°C ± 10° except where indicated.

TABLE 4-1 *(continued)*

Name	Structural formula	Products
3-Methylbenzo[*b*]thiophene (170° C)	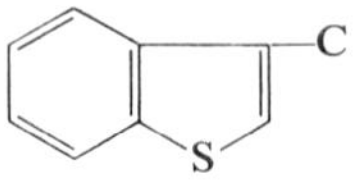 C, S	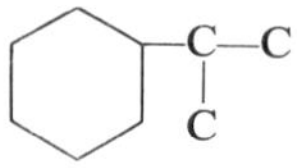 C—C, C
2-(4-Thiadecyl)thiophene	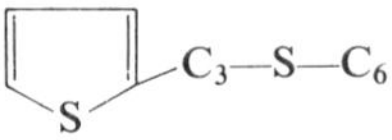 S, C_3—S—C_6	$nC_7 + nC_6$
3-Pentanol (350° C)	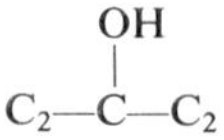 OH; C_2—C—C_2	nC_5
Cyclohexanol	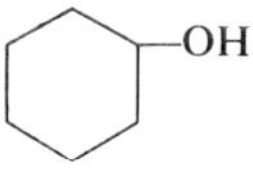 OH	
Phenol	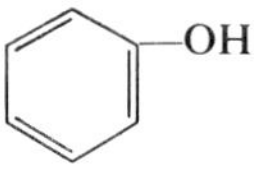 OH	
Benzoic acid	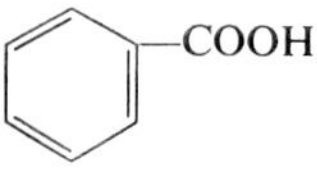 COOH	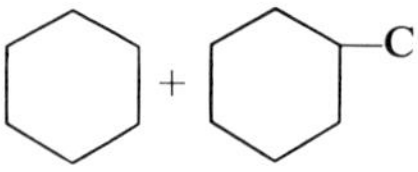 + C
Propyl ether	C_3—O—C_3	C_3
Furfural (245° C)	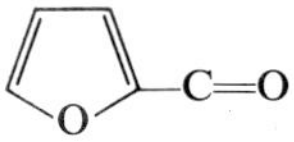 O, C=O	$nC_5 + nC_4$
Isopropyl propionate	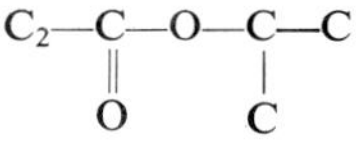 C_2—C(=O)—O—C(C)—C	C_3
2,5-Dimethylpyrole	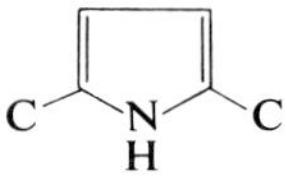 C, N, H, C	nC_6
3-Methylpiperidine	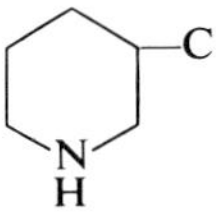 C, N, H	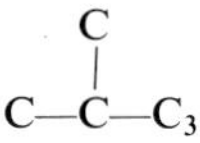 C—C(C)—C_3
Nitrobenzene	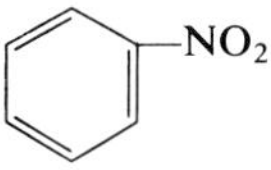 NO_2	
Benzonitrile	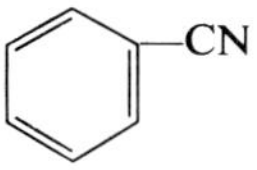 CN	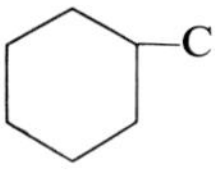 C

TABLE 4-1 *(continued)*

Name	Structural formula	Products
Quinoline	N	—nC_3
Carbazole	N H	
1-Chlorohexane	C_6—Cl	nC_6
Bromocyclohexane	—Br	
Iodobenzene	—I	
α-Bromotoluene	—C—Br	—C
1-Fluoronaphthalene	—F	and
1-Bromonaphthalene	—Br	(*cis* and *trans*)
α-*o*-Dichlorotoluene	—C—Cl, Cl	—C
Benzotrifluoride	—C—F_3	—C
2-Ethylhexyl chloride	C_4—C(—C_2)—C—Cl	C_4—C(—C_2)—C

The various sulfur compounds exhibit about what one would expect, assuming always that the carbon–sulfur bonds are broken and saturation occurs. There were no surprises here, all predicted products being found; however, the temperatures required were somewhat interesting. The case of the 3-methylbenzo[*b*]thiophene may prove of commercial interest eventually because these classes of compounds are present in relatively high concentrations in fuel oils, diesel fuel, and some jet fuels. Removal of the sulfur by a refining processs may eventually have to be done to assist in cutting down pollution problems, but this would not solve the problem of reactions of very reactive non-sulfur-containing intermediates from the unburned fuels. Similarly, the thiols are relatively thermally unstable but here they required a high reaction temperature. For the oxygenates, the corresponding 3-pentanol also required a high temperature; the hydrocarbon product was, as expected, normal pentane. Note that for the benzoic acid and furfural two products are formed. This "anomaly" is present only in the hydrogenolysis of the oxygenated compounds and was never observed with sulfur compounds. In oxygen compounds, in which the oxygen is attached to a primary carbon atom, cleavage occurs at two points: at the oxygen–carbon bond and at the carbon–carbon bond adjacent to the oxygen. For example,

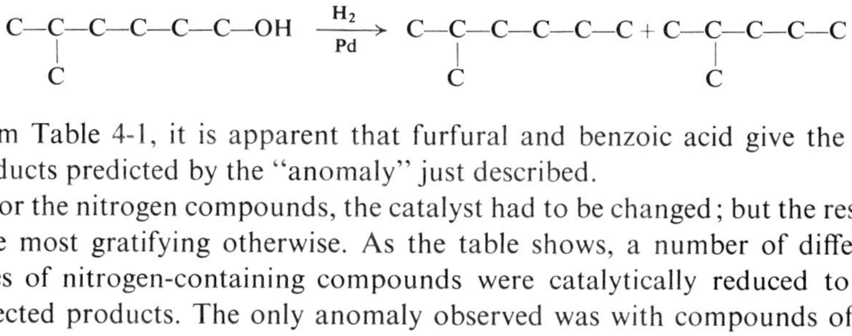

$$\underset{\displaystyle\mathrm{C}}{\mathrm{C{-}\underset{|}{C}{-}C{-}C{-}C{-}C{-}OH}} \xrightarrow[\mathrm{Pd}]{H_2} \mathrm{C{-}\underset{|}{C}{-}C{-}C{-}C{-}C} + \mathrm{C{-}\underset{|}{C}{-}C{-}C{-}C}$$

(with a C substituent on the second carbon of each chain)

From Table 4-1, it is apparent that furfural and benzoic acid give the two products predicted by the "anomaly" just described.

For the nitrogen compounds, the catalyst had to be changed; but the results were most gratifying otherwise. As the table shows, a number of different types of nitrogen-containing compounds were catalytically reduced to the expected products. The only anomaly observed was with compounds of the carbazole type. The substituted quinolines behaved in accordance with the expected removal of the nitrogen atom and saturation of the remaining carbon skeletal moiety. For the carbazoles, however, regardless of whether 1,2,3, or 9 substituted, hydrogenation resulted in the removal of the alkyl group! Infrared spectra were obtained on all the compounds to insure the positional substitution before this anomaly was readily accepted as a fact.

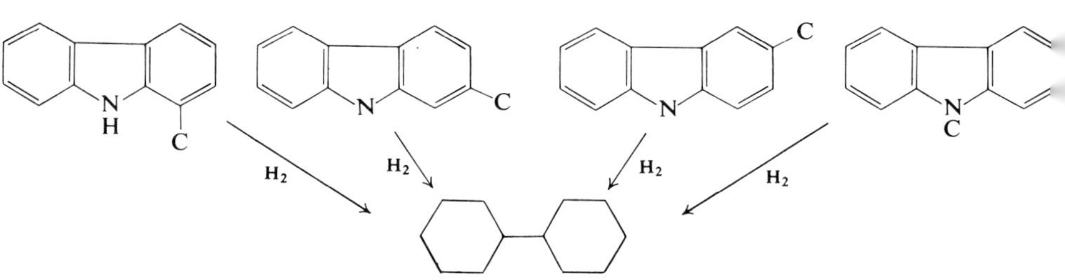

N H C — N C — C N — N C

H_2 H_2 H_2 H_2

The matter was not investigated fully, but Thompson *et al.* (1962c) thought that the alkyl-chain removal might be a function of the activity of the catalyst. This has not been confirmed. It is known that methyl biphenyls are easily hydrogenated to the corresponding methyl bicyclohexyls with no loss of the methyl group. Similarly, the methyldibenzothiophenes produce the analogous methyl bicyclohexyls.

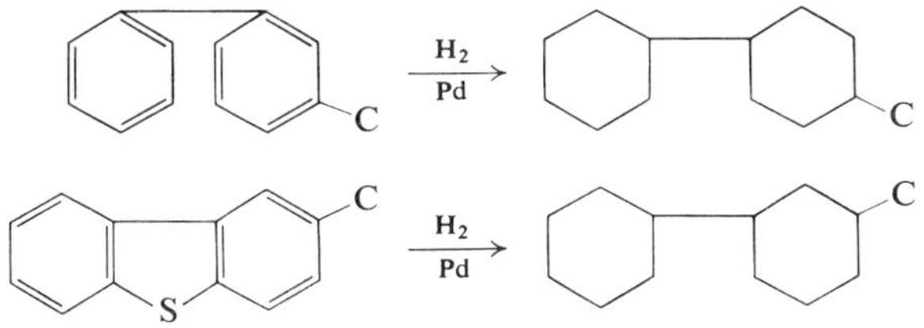

The halogenated compounds listed were among a number investigated and showed that matters were relatively straightforward; in other words, the expected products were obtained. It is to be noted that the last compound was similar to those oxygenated compounds which showed anomalous behavior. It produced only the hydrocarbon from cleavage of the carbon chlorine bond. The other di- and trisubstituted halogen compounds produced the expected products in each case.

For the two-ring compounds, products of the fully saturated and partially saturated variety were observed. These were the substituted tetralins as well as the decalins. Thompson *et al.* (1962c) suggested that perhaps the hydrogen acceptor reaction, "as probably first observed by Wieland (1912)" and discussed by Atkins *et al.* (1941), Blue and Engle (1951), Corson (1955), and Linstead *et al.* (1937), can explain in part the appearance of the tetralins in these products. In support of their suggestion was the desulfurization of 2-methylbenzo[*b*]thiophene in the absence of hydrogen but in the presence of decahydronaphthalene. The products were *n*-propylbenzene and naphthalene. While the mechanism and cause are not clear, the net result must be accepted as a part of microhydrogenation problems.

This phenomenon was investigated further from the temperature viewpoint. Thompson *et al.* (1967a) examined the hydrogenolysis of cyclohexanol at three temperatures, 200°, 275°, and 350°C. They found cyclohexane to be the predominant product (about 99 %) at 200°C; at 275°C the ratio of cyclohexane to benzene was about 2/1; and at 350°C the product was about 99 % benzene and only a small amount of cyclohexane. Thus temperature is no doubt a major factor, but activation energies and entropy factors must enter the picture. Thompson and his co-workers pointed out that this is to be thermodynamically expected, and the user of the technique must recognize that it can occur. Normally, the presence of these products does not interfere with the identification work. In the Thompson *et al.* (1962c) work, perfluoroheptane, C_7F_{16}, was completely resistant to hydrogenation under the experimental conditions of time, temperature, and catalyst.

Thompson *et al.* (1965) reported expected products from the hydrogenation of substances such as tetraphenyltin, triphenylarsine, triphenylstibine, and ferrocene. In Table 4-2 are found some additional miscellaneous compounds and the hydrocarbon products of their hydrogenolysis. The variety of compounds is sufficient to indicate the rather general and predictable behavior under the hydrogenation conditions used.

TABLE 4-2[a]

MISCELLANEOUS COMPOUNDS AND THEIR HYDROCARBON HYDROGENOLYSIS PRODUCTS

Reactant	Product expected and obtained
O N—S—S—N O	C_2
N N N	C—C C + C_2 + C_3
H_2—N— —CH_2—P=O (O—C_2H_5, O—C_2H_5)	—C + C_2
Cl Cl Cl Cl O C O C O	C C
HN N NH N H	C_3
C_4 C_4 Sn S—C_{12} S—C_{12}	C_4, C_{12}

[a] From Thompson *et al.* (1967a). Courtesy of the U.S. Bureau of Mines.

Since the microhydrogenation technique was developed in conjunction with identification and separation studies of the American Petroleum Institute Research Project 48, "Sulfur in Petroleum," we should demonstrate the usefulness of the technique in such identifications. By a combination of distillation, adsorption, and chemical techniques (Coleman *et al.*, 1965), a thiol concentrate boiling between 111° and 150°C was obtained from a Wasson, Texas, crude oil distillate. The concentrate consisted only of thiols, no other types of compounds being present. A distillate subfraction obtained by a concentric tube distillation of the thiol concentrate and referred to as "fraction 10," was examined by GC, the results being presented in Fig. 4-12. Table 4-3 lists some

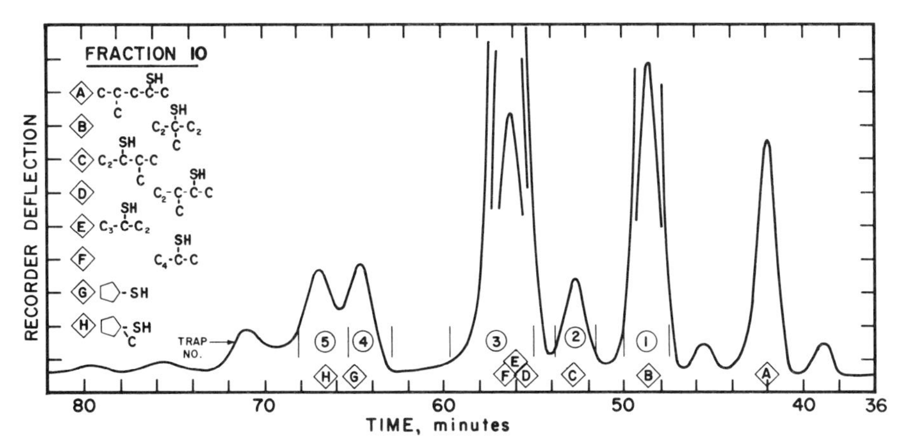

Fig. 4-12. Gas–liquid chromatogram of fraction 10. Column: dimensions, $\frac{1}{4}$ in o.d.. by 40 ft long; substrate, Di-2-ethylhexyltetraphenyl phthalate (20 g/100 g, 30–42 mesh F. B.). Conditions: temperature, 120°C; gas flow rate, 60 ml/minute helium; charge, 2 μliters. [From Thompson *et al.* (1967a). Courtesy of the U.S. Bureau of Mines.]

of the possible thiols present and the retention time for the compound on the same column under the same operating conditions as used for the fraction. The desulfurization product is also listed. In Table 4-4 are listed data for traps 1 and 3 indicating the time interval of the trap and the desulfurization product.

a. Trap No. 1. It is immediately apparent from Table 4-3 that only the first four compounds could have produced the indicated hydrocarbon. The retention time essentially excludes consideration of all compounds except 3-methyl-3-pentanethiol. Mass spectrometry and infrared spectrophotometry confirmed this. It should be remembered that compounds for reference quality mass spectra were sometimes hard to come by and, in addition, for mixtures of these isomers, it was considered essential to have structure confirmation by the combination of all available data.

TABLE 4-3[a]

RETENTION DATA—SOME C_6 THIOLS AND THEIR HYDROGENOLYSIS PRODUCTS

Possible thiol precursors		Desulfurization product expected	Possibly present in trap	
Retention time (minutes)	Formula		No. 1	No. 3
70.1	C—C—C(C)—C—C—SH	C—C—C(C)—C—C		
55.4	C—C—C(C)—C(SH)—C			X
48.6	C—C—C(SH)(C)—C—C		X	
67.6	C—C—C(C—SH)—C—C			
83.7	C—C—C—C—C—C—SH	C—C—C—C—C—C		
56.4	C—C—C—C—C(SH)—C			X
56.0	C—C—C—C(SH)—C—C			X

[a] From Thompson *et al.* (1967a). Courtesy of the U.S. Bureau of Mines.

b. Trap No. 3. Three possible compounds, on the basis of retention time, can cause the appearance of the hydrocarbon products found upon hydrogenation of the materials in trap No. 3. The compound 3-methyl-2-pentanethiol is the only likely compound to cause the 3-methylpentane product. However, either or both 2-hexanethiol or 3-hexanethiol could cause the appearance of normal hexane. An infrared spectrum of additionally trapped material from this retention time gave a spectrum which showed that both the 2-hexanethiol and 3-hexanethiol were present, with the former present in the larger amount.

The consideration of the two cases presented here should convince the

TABLE 4-4[a]

GAS CHROMATOGRAPHY DATA FROM "FRACTION 10"

Trap No.	Time interval of trapping (minutes)	Desulfurization product from trap
1	47.5–50.0	C—C—C—C—C | C
3	55.0–59.7	C—C—C—C—C | C C—C—C—C—C—C

[a] From Thompson *et al.* (1967a). Courtesy of the U.S. Bureau of Mines.

researcher interested in microquantities and identification that this is a technique to be added to the standard kit.

Thompson *et al.* (1967b) also reported the examination of flavone, a material of biological interest. In this case the expected hydrogenation product, 1,3-dicyclohexylpropane, was found:

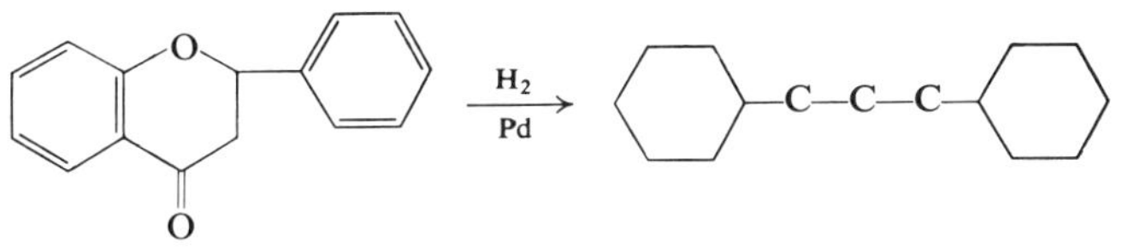

O O $\xrightarrow[Pd]{H_2}$ C—C—C

These workers also hydrogenated biotin (vitamin H) to yield the expected results despite the presence of three different hetero atoms:

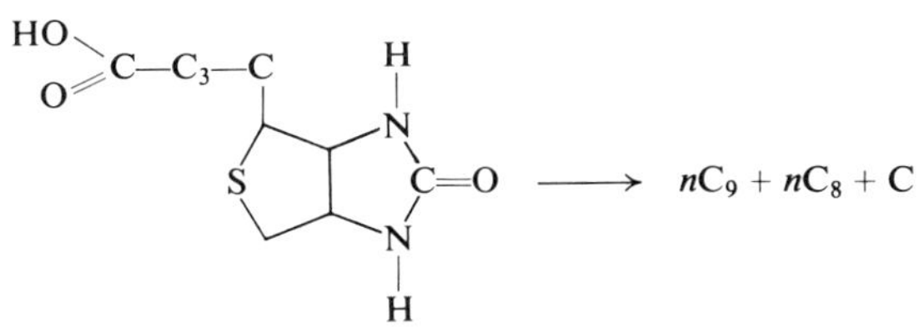

HO O C—C$_3$—C H N S C=O N H $\longrightarrow$ $nC_9 + nC_8 + C$

An added benefit from this technique is that the laboratory utilizing mass spectrometry and GC can have access to small amounts of necessary reference compounds prepared on the spot. Needless to say, when reference spectra in mass spectrometry, infrared, or proton resonance spectrometry are not available, this is an excellent place to start delineation of the unknown structure.

The technique of microhydrogenation has been recognized by workers in other fields, particularly those in flavors and in other natural products. Beroza (1962) began to apply the method of Thompson *et al.* (1962a) to the determination of insect sex attractants and described his approach as one of carbon-skeleton chromatography. It is quite obvious from the preceding discussion that this is quite true since the skeleton is retained in most cases; the problem of the carbazoles, however, raises a warning flag in some instances. Without one of these methyl (or possibly alkyl) groups, perhaps the attractant would not be the real thing as observed in nature. This problem has become relatively minor due to considerable progress by a number of workers. Beroza and Acree (1964) reported that the use of the carbon skeleton, so determined, was an aid in identification and structure studies. The hydrogenation unit was attached directly to the GC column thus avoiding the necessity of trapping small quantities of materials. A flame ionization detector was used so that the products from hydrogenation such as H_2O, H_2S, HCl, HI, HBr, CO, and CO_2 did not interfere with the operation. In the Thompson and co-workers case, these gases were simply vented and not trapped unless the light hydrocarbons were of interest. Beroza (1970) has given a complete review of chemical structure by gas chromatography, some of which will interest mass spectroscopists.

Silverstein *et al.* (1968) used carbon-skeleton chromatography to deduce the structure of brevicomin, an ingredient of the sex attractant of the Western pine beetle. Nonane was the product found upon hydrogenation:

CH_2—CH—CH—CH_2—CH_3
CH_2 O O $\longrightarrow$ nC_9
CH_2—C—CH_3

Synthesis of the proposed structure was done to verify the compound. This group used a combination technique described by Brownlee and Silverstein (1968) to examine the extremely small samples involved. Adhikary and Harkness (1969) have used a catalyst of platinum on siliconized glass beads to allow the hydrogenation of steroids.

Beroza (1970) pointed to the advantages of mass spectrometric detection for GC work, and yet mass spectrometry does not replace some of the more sensitive detectors present in nature. Bayer and Anders (1959) claimed that male moths exhibited a characteristic wing movement and dance when exposed to as little as 10^{-17} g of female sex attractant per milliliter of carrier gas isolated by means of GC. The human nose has been used to detect the active ingredients of flavor and odor essences at levels that do not register on a gas chromatograph, according to Beroza (1970). Crocker and Sjostrum (1949) reported odor thresholds with the nose for a number of common substances. They claimed that vanillin could be detected in concentrations as low as 1.7×10^{-7} ppm. Skatol, or 3-methylindole, came in second at 3.3×10^{-7} ppm; synthetic musk was third at 4.2×10^{-6} ppm; while the public's probable choice, mercaptan ("skunk oil"), came in fourth at 3.3×10^{-5} ppm. One of Thompson's co-workers, Ralph Hopkins, is a very fine GC aromatics detector, being able to call his components by the "nose" up through the C_{10} aromatics. Mass spectrometry has always agreed with his identifications.

In other spectroscopic fields, one should point out that if the figure for concentration of vanillin detectable by the average human nose is correct, then a long-path infrared cell taking in air plus the vanillin as a contaminant at that concentration would have to be about 200 miles long to give an interpretable spectrum (Kagarise and Saunders, 1960)! Despite the detection problem, the field of separation and identification is still an interesting one and full of solvable challenges.

E. Deliberate Degradation

As a technique to accompany mass spectrometry, deliberate degradation is frequently done, but it is usually not given the notice ascribed to the previously described techniques. This is because often more questions about the origin of lower molecular weight fragments are raised than can be answered. This brings to mind one of the layman's impressions about mass spectrometry—that all one has to do is examine the m/e 16 peak for the total oxygen present, or, even better, the concept that the total m/e 28 peak (from C=O) can be used to determine the amount of oxygenated material in a substance. Generally, the questions are grave and unanswerable, but there are a number of instances in which the mass spectroscopist can profitably employ the technique.

Ozonation is an example of one technique that in most cases results in the breakup of a larger molecule into predictable fragments which can be identified. The fragments can usually be reassembled to obtain an idea of the original structure. The value of ozonolysis has been demonstrated by Davison and Dutton (1966) and Nickell and Privett (1966), who used it to locate double bonds in the methyl esters of unsaturated fatty acids. Beroza and Bierl (1966, 1969) recognized the value and described a microozonizer suitable for the quantities of materials obtainable from GC work. They claimed a maximum time of 15 seconds to indicate the presence of excess ozone and, thus, the completion of the reaction. The ozonide is regenerated by triphenylphosphine to form aldehydes or ketones depending upon whether or not the olefin is substituted. Stein and Nicolaides (1962) have also used this reagent. Identification of the products leads to reconstruction of the original moiety. Details are to be found in the article of Beroza (1970).

In complex molecules it is sometimes possible to improve the identification chances by chemical degradation such as the partial hydrolysis of large peptides. At best, this also leads to problems with technique and to questions about what comes from where. It may be worthwhile if one is beginning a program of investigation and only desires information that will guide his choice of approach. However, the presence of these types of samples in the source would not necessarily be welcomed by the mass spectroscopist. Arrangements should be made and the project undertaken jointly by the investigator and the mass spectroscopist. In the case of the hydrolysis of esters, one normally produces molecules with very low molecular ion intensities; hence, the alcohols that result would not be preferred moieties in any investigation. Usually the conversion or removal of functional groups should be done to increase either the sample volatility or the intensity of the molecular ions. Reversal of technique is permitted, particularly if one hydrolyzes off a group of long-chain alcohols in the C_{30}–C_{50} range. This material can be converted to identifiable materials (or fragments) by the various techniques already described.

Deliberate fractionation of a sample should be attempted when solvent impurities will be removed in the process. This will assist the interpretation of the mass spectrum obtained in that peaks in the *m/e* range below that of the solvent will not be masked by a known material of no interest or use in the identification. Likewise, fractionation to enrich a sample in impurities is possible. Bokhofen and Theeuwen (1959) have described a useful device for this purpose. The direct probe inlet often lends itself to this use, and its versatility may surprise the mass spectroscopist at first. It may be difficult at times to differentiate between the process of probe distillation and pyrolysis, but a bit of experience will go a long way.

Pyrolysis of samples into the mass spectrometer source has also become an important aspect of solids examination. The release of relatively volatile portions of the material may permit deductions as to the original composition. Zemany (1952) described a method by which one can examine polymers and other compounds of very high molecular weight. The original structures are related to their pyrolysis products, as suggested by Lane *et al.* (1954). In this work it was found that very rapid high-temperature pyrolysis results in the simplest products. If a linear homopolymer is examined, the monomer will be the main volatile component observed by high-temperature pyrolysis. However, if the pyrolysis is conducted slowly, and at lower temperatures, the products are usually indicative of a much more complex mixture. The researcher is cautioned that under these conditions of pyrolysis, the mass spectrometer will not observe all of the fragments formed, simply because many very reactive free radicals and very low molecular weight compounds could form and react with one another long before they enter the mass spectrometer.

Happ and Maier (1964) have extended the concept of pyrolysis studies by mass spectrometric means to include the nonvolatile portion. In their method the volatiles and nonvolatiles are separated and collected simultaneously and each portion can be examined in a mass spectrometer equipped with a heated inlet system. The demarcation line between volatile and nonvolatile is quite a "gray" one and, in reality, it depends upon the individual case of just "what is what."

Sharkey *et al.* (1963) provided an excellent piece of work for the mass spectroscopist to consider in detail if he is interested in this area. The title is simply "Advances in Coal Spectrometry—Mass Spectrometry," but the techniques and data included exceed one's expectation. Investigations of pyrolysis products and extracts of coal and also whole coal are reported. In synopsis it was shown that (a) Aromatic hydrocarbons obtained by the vacuum pyrolysis of coal below 300°C possibly were sorbed material rather than decomposition products. (b) Thirteen structural types with molecular weights from 78 to 400 were identified in pyridine extracts. These constituted one-quarter or more of the coal, by weight. (c) Material extracted at room temperature with a methanol–benzene solvent and a vacuum pyrolysis condensate obtained at 450°C contained similar components. The extract and the condensate had approximately the same average molecular weights (200–300). (d) The material extracted from coal at room temperature was altered even by mild thermal treatment. (e) Changes in the various alkyl series occurring in thermally treated coal extracts could be followed quantitatively. (f) Similarities in the spectra of coal and graphite were indicated in an investigation using a mass spectrometer equipped with a spark source. Determinations on a coal-ash sample by this technique showed good agreement with data obtained by emission spectroscopy. The wealth of information on techniques and additional references will be very helpful to anyone dealing with similar materials. Holden and Robb (1958) had actually pyrolyzed the coal in a specially designed mass spectrometer ion source and included their results with a discussion of mass spectrometry of substances of low volatility. Reed (1960) has also described the mass spectrometry of some coal samples and associated techniques that should be considered.

Clerc and O'Neal (1961) made a preliminary investigation of the possibility of examining asphalts by means of mass spectrometry. A modification of the ion source was made to include a looped Nichrome wire heater upon which the asphalt sample was "painted." Carbon tetrachloride was used as a solvent, and after evaporation the asphalt coating remained on the heater. The heater was mounted in its ion source position and the instrument was brought to operating conditions. After the normal background of the newly evacuated instrument had reached reasonable levels, the heater voltage was increased until ions appeared. The sample was heated to between 200° and 300°C to produce normal vapor flow conditions. A reproducible mass spectrum was obtained for over a 1-hour period, which indicated that very little thermal decomposition was occurring but that ions from the sample caused the spectrum obtained. During this time no significant change in molecular weight distribution was observed, which indicated that little, if any, fractionation was going on. Clerc and O'Neal (1961) pointed out that if cracking of the material had occurred, the light hydrocarbons would have been observed in significant quantities. A pressure rise in the high-vacuum system should have also been observed. Hence, it was thought that no significant decomposition of the sample occurred and that the spectra obtained represented the material placed upon the heater. The article also gives data on the types of substances which probably occurred in the asphalt. This type of work indicates the necessary steps one must take to check such factors as decomposition when using today's version of the probe inlet, which is used in a similar fashion. Pyrolysis of materials in the direct probe inlet should be an area of much future promise.

A degradation profile comparison technique for mass spectrometric analysis of water-soluble cellulose ethers has been reported by Harless and Anderson (1970). In this method a time-of-flight instrument was used, and a direct probe inlet which accepts quartz tubes 1 mm i.d. by 10 mm length contained the sample. About 1-mg amounts of the cellulose materials were heated progressively until significant intensities of ionic species were detected. This usually occurred at about 150°–170°C, following which about 12 mass spectra were recorded for each sample during a 20-minute period at an ionizing potential of 70 eV. Spectra were recorded intermittently from the onset of degradation to decomposition at about 450°–500°C. The degradation profile was obtained for selected significant ions such as *m/e* 15, 31, 45, 59, 103, and 161. Comparison of commercial products by this means provides additional analytical information not readily obtainable in the past.

V. Misinterpretation Associated with the Sample

Although there are a number of problems associated with the sample and its possible misinterpretation, the most important is not immediately apparent. The operator is associated with the sample! Errors are bound to occur in this connection, and it is well to point these out before proceeding to the nonhuman difficulties. To cite an example, the operator may forget to close the isolation valve between the vacuum pump and the inlet system and still proceed to introduce a sample. Except with the probe inlet, the chances are that the sample is gone. Scanning will not produce any interpretable peaks unless the sample has extremely low volatility. Even so, if it is a mixture, fractionation will have occurred. Similarly, the valve to the source may well be closed (glass ball and socket joints have been known to stick, inadvertently) by accident or omission. Again, no scan is produced, just a "background" spectrum. Other human problems are (a) scanning the wrong range of m/e values for the type of material to be examined because the instrument settings were not adjusted, (b) ionization voltage left at a very low value and not raised to the nominal operating value, (c) slits left too narrow, (d) amplifier gain too low; (e) detector option improperly set up, and (f) possible loss of a highly volatile sample before liquid introduction is properly accomplished. There are many more, as the experienced mass spectroscopist soon discovers. If some of the above-listed items are not personally experienced in the first year of operation, you are not operating, you are supervising! Among the nonhuman factors that are the real causes of problems are a number that are quite common and must be anticipated and accounted for.

A. Background Effects

The background spectrum can be attributed to memory effects in the mass spectrometer. Literally speaking, the sample is *never* completely pumped out; a few molecules are always left clinging to a wall or a portion of the source. To be practical, however, most vacuum systems are so efficient that this becomes a minor problem. Obviously, the problem is aggravated by the continued examination of samples of low volatility. In fact, for instruments with high-temperature inlet systems, it is not uncommon to find spots of condensed high molecular weight material on a so-called cold spot when the source is opened up after some period of operation. These must be eliminated if proper spectra are to be obtained. In addition, this type of deposit obviously contributes to the instrument's background spectrum. As a general rule, it is judicious to run a background between every sample. This may be done without making a recording simply by examining the scan by means of an oscilloscope accessory to monitor the instrument detection system. Alternatively, if the best possible quantitative results are desired, the background spectrum should be recorded just before introduction of the sample under the identical conditions at which the sample is to be run. Then, the individual mass peaks, at the various m/e's observed, can be evaluated and subtracted from the mass spectrum of the sample if this proves necessary. For example, if a laboratory is running many samples associated with one or two types of materials, a background inevitably builds up. In petroleum products the background peaks show up at m/e's corresponding to 29, 43, 57, 71, and 85, with the largest probably being the 43 peak. These are due to the many alkyl fragments that are produced in hydrocarbon spectra. They could also be due to ketones or possibly other oxygenated species which give peaks at these same nominal masses. A bit of experience goes a long way in telling an operator what will be a reasonable background and what might be unusual. It is customary to record at least one background (usually the first run of the day) even if the remainder of the background checks are taken on the oscilloscope.

In time, the background peaks may build up to the point where it is essential to reduce them again to a negligible point. Most instruments have "bakeout" systems incorporated into their design. In this case, the analyzer tubing is heated to about 250°C for a period of time, until the peaks in the background decrease in size, or until the source pressure (if measurable) returns to a low value (depending upon the instrument's individual use and history). It should be remembered that the heating coil creates an inductive field that may interfere with the magnetic field of the instrument, although some instruments can be run with the bakeout heater on at all times. This is an advantage if only high molecular weight materials are being run. The unit can be used for a longer period of time.

When baking out fails to give a satisfactory reduction in background, it is customary to turn to means of cleaning that begin to approach the "last resort stage" before actually shutting down the instrument and performing a jeweler's precision job of cleaning (not to be undertaken without consultation with the instrument manufacturer, who will probably prefer to send in the factory representative). In the installation at the U.S. Bureau of Mines at Laramie, Wyoming, high molecular weight materials were the usual menu for the instrument. G. L. Cook devised the addition of pyrrolidone as a last treatment to remove everything else from the surface of the inlet system, source, analyzer, and detector. This usually helps to keep the instrument working for a time when one is in the middle of a specific program when time is not available for a major shutdown or when instrument conditions must be as reproducible as possible. This "solvent" may not leave conditions exactly as they were, but it is worth a try. Remember, a shutdown will have to occur shortly after this is done. Meyerson (1956) has used water as a solvent for displacement of species thought to be absorbed and then observed the species in the resulting mass spectrum. Normally, water (and pyrrolidone, for that matter) is rather hard on the filament and source, leading to some temporary sensitivity changes, possibly due to removal of polar species absorbed on the filament.

B. Inhomogeneity of the Sample

While this problem may be anticipated in some cases, it usually is not noticeable until interpretation of the spectrum is begun, or possibly even at a later time. Improper sampling may often be the prime cause. If one recalls the problems with light hydrocarbons being dissolved in heavier materials, the effect should be obvious. Butane injection into gasoline is well known for giving quicker cold-weather starting. The reason is that the more volatile component vaporizes at the low temperature and is available for combustion.

In the mass spectrometer, then, one may expect to find enhanced peaks from low-boiling solvents used in the chemist's treatment of the sample that is being examined. Every effort should be made on the part of the mass spectroscopist and the person submitting the sample to consider the homogeneity of the sample. The vaporization or "flashing-in" of a liquid sample is still a one-plate fractionation in a sense and so the volatility effect may become quite serious. It may leave an unexpectedly low volatility component behind in the introducing syringe. Adsorption of polar materials on the walls of the inlet, source, etc., may well lead to some expected component or components not being observed in the routinely scanned sample. That is why, ideally, a bit of eyeballing on a spectral chart, which amounts to editing the spectrum, assures the mass spectroscopist that he has a spectrum worth processing further. If adsorption is suspected, a "chaser" such as the water described above, is called for. In this case, both the first spectrum and the spectrum with "chaser" should be kept and examined in detail. Often, this tells the interpreter a good deal. As indicated in Section IV, deliberate fractionation is a very good technique *sometimes*.

C. Gross Volatility Differences

Under the conditions in which a low-volatility trace component is suspected, the molecule ion peak may be all that is detectable. Possibly the trace component has a weak parent ion and is detectable only on the basis of some fragment peak. If this be the case, the trace-component molecule ion can be called with confidence only if it is quite distinguishable from other peaks coming from the remainder of the sample. In addition, it is easier if it has a higher molecular weight. The same must be said about the fragment peak. Perhaps treatment that concentrates the contaminant will improve the situation. A treatment of such a small amount of material may reduce the total concentration appreciably, or it might even result in a total loss. Newton (1953) described techniques that can be employed for noncondensable impurities in condensable gases.

A reverse of the above situation occurs when one is looking for contaminants in air or water, such as in the current problems of pollution. Quiram *et al.* (1954) worked with air samples containing traces of light hydrocarbons in the range of 3–8 ppm and described the techniques associated with the problem. In the area of water pollution, Melpolder *et al.* (1953) developed a method for determination (qualitative and quantitative) of traces of volatile substances contributing to the taste and odor of water. While there have been many additional articles in this area in the ensuing years, further consideration is beyond the scope of this book.

D. Decomposition of the Sample

The unexpected decomposition of the sample, or a portion of a mixture, leads to results that are difficult to interpret. The sample pressure changes with time if this is a sufficiently rapid process. If it is a violent process, there will be a small explosion. Chances are that no useful information will be obtained from the spectrum. Fortunately, in most cases, the decomposition is rather slow, in fact, slow enough to be insidious but perhaps not negligible. If thermal instability is suspected due to the observation of ions formed by thermal processes, the sample material should be left in the inlet system and scanned a number of times. This should be done for the time the mass spectroscopist is willing to allocate to such a problem, perhaps an hour, a day, or even a week, if this is of interest for the particular sample. If the process is rather rapid, observing low molecular weight fragment and rearrangement ions often tells something about the original material and about the mechanism and products of thermal decomposition. For example, m/e 15–44 must be scanned to insure observation of expulsions of methyl through carbon dioxide (mass 44) groups. If only a portion of the sample is decomposing, a much more difficult situation presents itself. In this case, the amount of decomposing material may be small enough to not show appreciable changes in most of its peaks. Also, the spectra obtained must be very carefully examined for evidence of decomposition, and merely watching pressure (or total source ion intensity, if one has the unit) does not disclose any useful information to the operator while the sample is being scanned at the instrument. It may be surprising that many materials exhibit traces of thermal degradation in heated inlet systems with the oven at 175°C and the source temperature about 250°C. Above these temperatures, organic substances of many types exhibit an appreciable amount of thermal cracking.

E. Artifacts from Contaminants

Materials present in the sample (labeled pure) often shock the neophyte mass spectroscopist. Sometimes the purity claimed versus the purity determined is an inverse relationship but guaranteed to raise the pressure (blood). Discounting mislabeling and the basic problems of purity, one soon discovers that just as at the market, the price includes a surprising amount of solvent, whether water or otherwise. Water is difficult to deal with because of the reduction in sensitivity usually experienced. In addition, one does not like to heat organic materials, particularly if they are to be used as a secondary reference in the mass spectrometer. Usually, in-house fractions or samples have something known about them, and the risk is low under these conditions.

Commercial-grade solvents often contain "stabilizers" and other materials, added or remaining from the chemical process, which may lead to difficulties. An example is the trace of methanol usually found in chloroform which, presents a small but confounding OH absorption band to the beginner in infrared spectrophotometry.

Stopcock greases are a main source of contamination. A number of excellent greases, producing a background of very low intensity, are available (consult the instrument manufacturer for his recommendations). The sample may have been exposed at some stage or another to materials that cause artifacts to appear in the mass spectrum. Silicones are to be avoided if at all possible. Although the triplet of peaks characteristic of the silicon atom is readily detected by the mass spectroscopist, the silicone greases interfere with *more* than interpretation. Once present, the grease has properties that are quite detrimental to the source. The material persists for a long time and gradually flows across the surfaces, giving a thin-layer coating. This leads to alterations of the electric fields and ultimately to poor ion focusing conditions, etc., according to McLafferty (1957). The peaks occur at m/e 133–135, 207–209, 281–283, 355–357, and 429–431. The difference in mass of 74 is reminiscent of the silyl ether derivatives. Hydrocarbon greases present peaks at the usual fragments of m/e 43, 57, 71, etc.

Oils from the diffusion pumps in the system are often a problem. McLafferty and Gohlke (1959) described the major peaks observable in Octoil or dioctyl phthalate. The molecular ion at 390 is minor, but the m/e 149 is the most intense ion. Also appearing in the spectra should be an ion at m/e 167 ($\sim$23% of the 149 ion) and one at 279 ($\sim$10%), in addition to other minor ions. Biemann (1962f) discussed the fragmentation that leads to these ions.

Dibutyl phthalate, sometimes used as a plasticizer, also gives an intense ion at m/e 149. Hence, if an ion of these m/e's shows up and is otherwise unexplainable, it may be from a contaminant. Quayle (1959) has reported the mass spectrum of tributyl phosphate, a plasticizer, which has been eluted from plastic tubing by solvents or even by steam quite frequently. It has a MW of 266, but a weak molecular ion intensity is observed. The compound produces a large ion at m/e 99 and appreciable ions at m/e 155 and 211.

The various vacuum systems may allow materials to backflow into the source region. Admittedly this takes place in very small amounts but sometimes can be a real problem. Ion-getter pumps can lead to a problem in that from time to time it seems that the surface is crowded with absorbed materials. Apparently a stage is reached where changes occur, and a form of regurgitation is noticed, usually when it is time to replace the liner. A bit of forepumping to reduce the pressure sweeps this out, and normal operations can usually be resumed in a few hours. Barrington (1965) also reported on the suppression of excited species from these pumps by means of a Zeolite trap. It was demonstrated that, even after baffles were installed, a signal could be detected on a 20-stage electron multiplier. Since argon was the main residual gas in the system, the effect was attributed to photons radiated by metastable argon atoms which were able to migrate from the pump into the detector (multiplier) section.

Column substrates bleed, as described earlier in this chapter; in addition, eluents in the column may introduce impurities as described above or remove them from the solid material. This is always true of extraction processes in which plastic tubing is employed, such as in the very popular rotary evaporators. One experience of these authors involved the removal of petroleum waxes absorbed upon a commercial silica gel. This material had been prepared batchwise in a drying oven that had an unknown leak from the furnace section into the oven and drying chambers. Since fuel oil was burned in the furnace section, the batch produced some very interesting identifications from some unusual samples until the cause of the problem was discovered.

Unexpected reactions between solvents or materials in the sample produce some very real interpretative problems, and hence editing the spectrum before fully processing through a calculation procedure is always a good thing to do. In addition, one should not be surprised to observe peroxides and other materials such as hydroperoxides from time to time. Exchange with the walls, usually confined to labile hydrogen, may really cause difficulties, particularly with $M + 1$ peak determinations and isotope ratios. Lastly, of course, reactions with the metal or glass walls (and contaminants held there) should be anticipated, depending upon the type of material being examined.

In light of these authors' observations of a number of "unexpected components" in various fractions, the best advice seems to be: Good liaison between the mass spectroscopist and the sample submitter is essential to achieve satisfactory results.

VI. Combination of the Mass Spectrometer with Other Identification Techniques

Even though deeply involved in mass spectrometry, these authors recognize that the mass spectrometer is only one of the three "queens" upon the analytical-identification "chessboard" (the other two being infrared and nuclear magnetic resonance). Therefore, methods should be developed involving any combination that will accomplish the required study or identification. True, if one is studying mass spectrometric fragmentation, ion–molecule reactions, etc., then the "means" of using the mass spectrometer includes the "ends." From the analytical-identification viewpoint, however, one must at this point accept the fact that mass spectrometry has some limitations. When combined with other means of identification and the newest tools of separation the team approach to identification, structure determination, and analysis is unbeatable.

Often times the mass spectroscopist finds that the problem cannot be solved (for his first time) because many substitutional isomers simply fragment the same, and since they give the same molecular ion, identification is therefore not possible. A typical example is the case of *meta*- and *para*-xylenes (1,3- and 1,4-dimethylbenzenes), in which the molecule ion is at m/e 106, and the large ions at m/e's 105, 92, 91, 77, 78, and 79 are all of just about the same intensities. Therefore, identification is not possible by means of mass spectrometry alone. The other "queens" can readily do this job, infrared being classical. This is a simple example, and one can cite far more complex problems.

An interesting example of the use of a combination of techniques was reported by Flanagan *et al.* (1966) for determining the composition of the C_nH_{2n-8} fraction from an alkylation reaction. The question was whether these compounds were substituted indans or tetralins. The combination of spectroscopic techniques was not able to resolve this problem. By mass spectrometry the "retro-Diels–Alder reaction" technique of Biemann (1962g) indicated the reaction product as about 37% tetralin. By NMR the same sample was indicated to be 50% tetralin. The IR parameters led to the conclusion that this fraction contained a negligible indan content. The UV data indicated a 85/15 tetralin–indan ratio. These discrepancies were resolved by subjecting the sample to an aromatic reaction. Dehydrogenative aromatization reactions are known to give 1,4-dialkylnaphthalenes from 1,4-dialkyltetralins (Shishido and Nozaki, 1944). Thus the sample was reacted. The product was separated by GC and the same spectroscopic methods used. This time all these methods led to the conclusion that the structures in question were unequivocally that of tetralin rather than indan. However, in the course of this proof, *cis* and *trans* isomers of 1-ethyl-3-heptylindan were examined. Again the results of the different techniques were not in agreement mainly due to lack of reference standards. Often at this point such problems must wait for the availability of additional spectral reference standards, which may be very difficult to obtain.

Polyfunctional molecules are also often very difficult to identify by using one technique. One must apply every means at his disposal to differentiate compounds in such a case. As an example, consider the following problem. In studies of cyclic imides [article VII in the series by Caswell *et al.* (1968) concerning carboxyphthalimidoacetic acids], it was desired to determine which of the following members (or both) of each pair were present in the products:

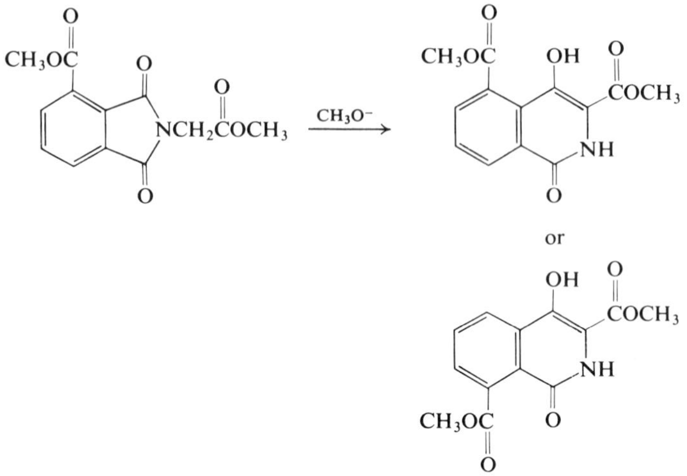

$\xrightarrow{CH_3O^-}$

or

I

$\xrightarrow{CH_3O^-}$

or

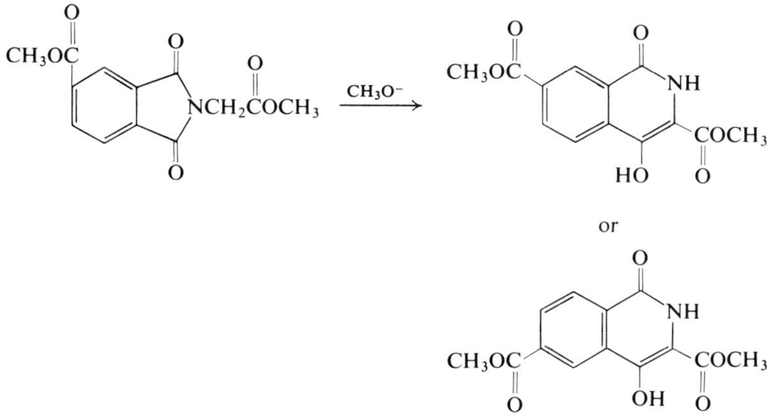

II

These workers pointed out that "although the spectral data are sufficient to identify these Gabriel–Coleman products as derivatives of 3-carbomethoxy-4-hydroxyl-1-2*H*-isoquinolone, the insolubility and instability of these compounds frustrate further structural studies." The products I and II may exist in either or both forms, and the current status of the problem is to work out a scheme, utilizing all the known techniques that can lead to conclusive results. The infrared spectrum has been obtained in solid form in pellets. These types of compounds show broad bands throughout the carbonyl absorption region and hydrogen bonding interactions are evident, but that is about all that one can learn. Most reactions alter the structure so markedly as to raise the problem of what the precursor's structure was. The instability, insolubility, and relative nonvolatility make this an excellent challenge. It may remain a problem for some time, but work continues on it at the Texas Woman's University.

There are problems with handling very complex mixtures that should be considered before one sets out to make an "analysis." If the range of molecular weights of substances present is very great, quantitative results are extremely difficult to achieve, unless one is interested in only one type of compound or even in a single compound. Thus, as is done with other methods, mass spectroscopists frequently employ the term "characterization" rather than analysis. The term really means that recognition of the complex nature of the mixture precludes the analysis being any more than semiquantative at best. Also, not all components are necessarily identified, but select groups are known (or previously identified), and an estimate of the relative amounts of materials or carbon-number distributions of the members of one or more classes may be all that is desired (and obtained). An example of such a "characterization" is given in Chapter 6 and illustrates what can be obtained with patience and a relatively small amount of preliminary assistance from mass spectrometry.

Frequently, the word "characterization" is used with a somewhat different connotation by persons in the biological and biomedical fields. Basically this type of use is one of structural determination and intercomparison of synthetic and natural materials occurring in these fields of research. An excellent example is from the work of a team of biomedical researchers, originating in the Department of Physiology and Lipid Research Institute of the Baylor College of Medicine, Houston, Texas. The work of Burgus *et al.* (1969a,b, 1970a,b) utilizes all types of instrumental and separational techniques for this type of "characterization." The 1970a article title begins with the word "characterization." The paper of Desiderio *et al.* (1970) is entitled "The Primary Structure of the Hypothalamic Hypophysiotropic Thyroid Stimulating Hormone Releasing Factor of Ovine Origin," While this work was presented from the mass spectroscopist's viewpoint, infrared and nuclear magnetic resonance data were also given for the substance isolated with the assistance of thin-layer chromatography. The structure of the substance was firmly established, but it was the mass spectrometric data which proved to be the most convincing with regard to the composition of the various fragment ions observed. Logical reassembly of these fragments led to the identity of the molecule and to its structure.

Needless to say, one of the main points of all structural determinations is to have confidence in the results obtained. After all, in the drug industry the structure must be completely determined for the company to have protection during disclosure and patent applications. Hence, the team of instruments today must produce a set of data which is compatible and consistent with the entire analysis picture as determined by the research group. For example, the infrared spectra of a sample shows an OH band, yet this is not compatible with other spectroscopic data. Since it is present it must be resolvable and explained in some way, i.e., another group causing the peak, an artifact, a reaction-induced band possibly due to differing treatment before obtaining the various spectra, etc. The satisfaction of complete concordance of all the techniques and certainty of the end result is a reward in itself and justifies the many man-hours spent in the identification.

REFERENCES

Adhikary, P. M., and Harkness, R. A. (1969). *Anal. Chem.* **41**, 470.

Alpert, D. C., Matland, J., and McCoubrey, A. O. (1951). *Rev. Sci. Instrum.* **22**, 370.

Amy, J. W., Chait, E. M., Baitinger, W. E., and McLafferty, F. W. (1965). *Anal. Chem.* **37**, 1265.

Aplin, R. T., Jones, J. H., and Liberek, B. (1968). *J. Chem. Soc. C*, 1011.

Askevold, R. J., Tucker, E. B., Stross, M. J., and Webb, G. M. (1950). Handling samples of gaseous and liquid hydrocarbons, *in* "Analytical Methods." Technical Advisory Committee, Petroleum Industry War Council, pp. 93–103. American Petroleum Institute, New York.

Atkins, H., Richards, L. M., and Davis, J. W. (1941). *J. Amer. Chem. Soc.* **63**, 1320.

Banner, A. E., Elliott, R. M., and Kelly, W. (1964). *Gas Chromatogr.* (*London Inst. Petrol.*), p. 180.

Barrington, A. E. (1965). *Rev. Sci. Instrum.* **36**, 549.

Baxter, I. G. (1953). *J. Sci. Instrum.* **30**, 456.

Bayer, E., and Anders, F. (1959). *Naturwissenschaften* **46**, 380.

Bazinet, M. L., and Merritt, C., Jr. (1962). *Anal. Chem.* **34**, 1143.

Becker, E. W. (1961). "Separation of Isotopes," p. 360. Newnes, London.

Becker, E. W., and Stehl, O. (1952). *Z. Angew. Phys.* **4**, 20.

Berkowitz, J., and Chupka, W. A. (1964). *J. Chem. Phys.* **40**, 287.

Berkowitz, J., and Marquert, J. R. (1963). *J. Chem. Phys.* **39**, 275.

Beroza, M. (1962). *Anal. Chem.* **34**, 1801.

Beroza, M. (1970). *Accounts Chem. Res.* **3**, 33.

Beroza, M., and Acree, F., Jr. (1964). *J. Ass. Agr. Chem.* **47**, 1.

Beroza, M., and Bierl, B. A. (1966). *Anal. Chem.* **38**, 1976.

Beroza, M., and Bierl, B. A. (1969). *Mikrochim. Acta*, p. 720.

Beynon, J. H. (1960). "Mass Spectrometry and Its Applications to Organic Chemistry," pp. 132–144. Elsevier, Amsterdam.

Beynon, J. H., and Nicholson, G. R. (1956). *J. Sci. Instrum.* **33**, 376.

Biemann, K. (1961). *J. Amer. Chem. Soc.* **83**, 4801.

Biemann, K. (1962a). "Mass Spectrometry: Organic Chemical Applications," p. 23. McGraw-Hill, New York.

Biemann, K. (1962b). "Mass Spectrometry: Organic Chemical Applications," p. 192. McGraw-Hill, New York.

Biemann, K. (1962c). "Mass Spectrometry: Organic Chemical Applications," pp. 176, 177. McGraw-Hill, New York.

Biemann, K. (1962d). "Mass Spectrometry: Organic Chemical Applications," p. 180. McGraw-Hill, New York.

Biemann, K. (1962e). "Mass Spectrometry: Organic Chemical Applications," p. 179. McGraw-Hill, New York.
Biemann, K. (1962f). "Mass Spectrometry: Organic Chemical Applications," pp. 170–172. McGraw-Hill, New York.
Biemann, K. (1962g). "Mass Spectrometry: Organic Chemical Applications," p. 102. McGraw-Hill, New York.
Biemann, K., and Friedmann-Spiteller, M. (1961a). *Tetrahedron Lett.* **2**, 68.
Biemann, K., and Friedmann-Spiteller, M. (1961b). *J. Amer. Chem. Soc.* **83**, 4805.
Biemann, K., and McCloskey, J. A. (1962). *J. Amer. Chem. Soc.* **84**, 3192.
Biemann, K., and Vetter, W. (1960). *Biochem. Biophys. Res. Commun.* **3**, 578.
Biemann, K., Gapp, F., and Seibl, J. (1959). *J. Amer. Chem. Soc.* **81**, 2274.
Biemann, K., Bommer, P., and Desiderio, D. M. (1964). *Tetrahedron Lett.* p. 1725.
Bierl, B. A., Beroza, M., and Ruth, J. M. (1968). *J. Gas Chromatogr.* **6**, 286.
Birkofer, L., Ritter, A., and Giessler, W. (1963). *Angew. Chem.* **75**, 93.
Black, D. R., Flath, R. A., and Teranishi, R. (1969). *J. Chromatogr. Sci.* **7**, 284.
Blue, R. W., and Engle, C. J. (1951). *Ind. Eng. Chem.* **43**, 494.
Bokhofen, C., and Theeuwen, H. J. (1959). *In* "Advances in Mass Spectrometry" (J. D. Waldron, ed.), p. 222. Pergamon, Oxford.
Booker, J. L., Isenhour, T. L., and Sievers, R. E. (1969). *Anal. Chem.* **41**, 1705.
Bradt, P., and Mohler, F. L. (1955a). *J. Res. Nat. Bur. Stand.* **55**, 323.
Bradt, P., and Mohler, F. L. (1955b). *Anal. Chem.* **27**, 875.
Bradt, P., and Mohler, F. L. (1958). *J. Res. Nat. Bur. Stand.* **60**, 143.
Bradt, P., Dibeler, V. H., and Mohler, F. L. (1953). *J. Res. Nat. Bur. Stand.* **50**, 201.
Brenner, N., Callen, J. E., and Weiss, M. D., eds. (1962). "Gas Chromatography." Academic Press, New York.
Brooks, B. T., Kurtz, S. S., Jr., Boord, C. E., and Schmerling, L. (1954). "The Chemistry of Petroleum Hydrocarbons," Vol. I. Van Nostrand-Reinhold, New York.
Brown, R. A. (1951). *Anal. Chem.* **23**, 430.
Brownlee, R. G., and Silverstein, R. M. (1968). *Anal. Chem.* **40**, 2077.
Brunnée, C. (1966). *Z. Anal. Chem.* **217**, 333.
Brunnée, C., Jenckel, L., and Kronenberger, K. (1962). *Z. Anal. Chem.* **189**, 50.
Burgus, R., Dunn, T. F., Desiderio, D. M., Jr., Vale, W., and Guillemin, R. (1969a). *C. R. Acad. Sci.* **269**, 226.
Burgus, R., Dunn, T. F., Desiderio, D. M., Jr., and Guillemin, R. (1969b). *C. R. Acad. Sci.* **269**, 1870.
Burgus, R., Dunn, T. F., Desiderio, D. M., Jr., Ward, D. N., Vale, W., and Guillemin, R. (1970a). *Nature* **226**, 321.
Burgus, R., Dunn, T. F., Desiderio, D. M., Jr., Ward, D. N., Vale, W., Guillemin, R., Felix, A. M., Gillessen, D., and Studer, R. (1970b). *Endocrinology* **86**, 573.
Burson, K. R., and Kenner, C. T. (1969). *J. Chromatogr. Sci.* **7**, 63.
Caldecourt, V. J. (1955). *Anal. Chem.* **27**, 1670.
Caldecourt, V. J. (1958). Subcommittee III Report, ASTM E-14 Committee on Mass Spectrometry, June, New Orleans, Louisiana.
Capella, P., and Zorzut, C. M. (1968). *Anal. Chem.* **40**, 1458.
Cartwright, M., and Heywood, A. (1966). *Analyst* **91**, 337.
Cassidy, H. G. (1951). "Adsorption and Chromatography," Vol. V of "Techniques of Organic Chemistry" (A. Weissberger, ed.). Wiley (Interscience), New York.
Caswell, L. R., Haggard, R. A., and Yung, D. C. (1968). *J. Heterocycl. Chem.* **5**, 865.
Chupka, W. A., and Inghram, M. G. (1955). *J. Phys. Chem.* **59**, 100.
Clerc, R. J., and O'Neal, M. J., Jr. (1961). *Anal. Chem.* **33**, 380.

Coleman, H. J., Thompson, C. J., Hopkins, R. L., Foster, N. G., Whisman, M. L., and Richardson, D. M. (1961). *J. Chem. Eng. Data* **6**, 464.
Coleman, H. J., Thompson, C. J., Hopkins, R. L., and Rall, H. T. (1965). *J. Chem. Eng. Data* **10**, 80.
Cook, D. B., and Danby, C. J. (1953). *J. Sci. Instrum.* **30**, 238.
Corson, B. B. (1955). *In* "The Chemistry of Petroleum Hydrocarbons" (B. T. Brooks, S. S. Kurtz, C. E. Boord, and L. Schmerling, eds.), Vol. III, p. 310. Van Nostrand-Reinhold, Princeton, New Jersey.
Crocker, E. C., and Sjostrom, L. B. (1949). *Chem. Eng. News* **27**, 1924.
Dal Nogare, S., and Juvet, R. S., Jr. (1962). "Gas–Liquid Chromatography," p. 265. Wiley (Interscience), New York.
Damico, J. N., Wong, N. P., and Sphon, J. A. (1967). *Anal. Chem.* **39**, 1045.
Davis, R. E., and McCrea, J. M. (1957). *Anal. Chem.* **29**, 1114.
Davison, V. L., and Dutton, J. H. (1966). *Anal. Chem.* **38**, 1302.
Delaney, E. B. (1969). *Amer. Lab.* June, p. 19.
de Mayo, P., and Reed, R. I. (1956). *Chem. Ind.* (*London*), 1481.
Desiderio, D. M., Jr., Burgus, R., Dunn, T. F., Vale, W., and Guillemin, R. (1970). Paper No. D6. Eighteenth Annual Conference on Mass Spectrometry, ASTM Committee E-14, San Francisco, California.
Desty, D. H., ed. (1957). "Vapour Phase Chromatography." Academic Press, New York.
Dibeler, V. H., and Cordero, F. (1951). *J. Res. Nat. Bur. Stand.* **46**, 1.
Diekman, J., and Djerassi, C. (1967). *J. Org. Chem.* **32**, 1005.
Diekman, J., Thomson, J. B., and Djerassi, C. (1967). *J. Org. Chem.* **32**, 3904.
Diekman, J., Thomson, J. B., and Djerassi, C. (1968). *J. Org. Chem.* **33**, 2271.
Diekman, J., Thomson, J. B., and Djerassi, C. (1969). *J. Org. Chem.* **34**, 3147.
Dietz, W. A., and Klaas, P. J. (1958). Paper No. 20. Sixth Annual Conference on Mass Spectrometry, ASTM Committee E-14, New Orleans, Louisiana.
Dighton, D. T. R. (1953). *Electronic Eng.* **25**, 66.
Dorsey, J. A., Hunt, R. H., and O'Neal, M. J., Jr. (1963). *Anal. Chem.* **35**, 511.
Draffan, S. H., and McCloskey, J. A. (1968). Paper No. 67. Sixteenth Annual Conference on Mass Spectrometry, ASTM Committee E-14, Pittsburgh, Pennsylvania.
Earnshaw, D. G., Doolittle, F. G., and Decora, A. W. (1968). Paper No. 65. Sixteenth Annual Conference on Mass Spectrometry, ASTM Committee E-14, Pittsburgh, Pennsylvania.
Ebert, A. A. (1961). *Anal. Chem.* **33**, 1865.
Echo, M. W., and Morgan, T. D. (1957). *Anal. Chem.* **29**, 1593.
Eglinton, G., Hunneman, D. H., and Douraghi-Zadeh, K. (1968a). *Tetrahedron* **24**, 5929.
Eglinton, G., Hunneman, D. H., and McCormick, A. (1968b). *Org. Mass Spectrom.* **1**, 593.
Flanagan, P. W., Hamming, M. C., and Evens, F. M. (1967). *J.Amer. Oil Chem. Soc.* **44**, 30.
Gohlke, R. S. (1959). *Anal. Chem.* **31**, 535.
Gohlke, R. S. (1962). *Anal. Chem.* **34**, 1332.
Grayson, M. A., and Wolf, C. J. (1967). *Anal. Chem.* **39**, 1438.
Gruse, W. A., and Stevens, D. R. (1960). "The Chemical Technology of Petroleum," 3rd Ed. McGraw-Hill, New York.
Gustafsson, J. Å., Ryhage, R., Sjövall, J., and Moriarty, R. M. (1969). *J. Amer. Chem. Soc.* **91**, 1234.
Happ, G. P., and Maier, D. P. (1964). *Anal. Chem.* **36**, 1678.
Harless, H. R., and Anderson, R. L. (1970). *Text. Res. J.* **40**, 448.
Harless, H. R., and Crabb, N. T. (1969). *J. Amer. Oil Chem. Soc.* **46**, 238.
Harris, W. E., and Habgood, H. W. (1966). "Programmed Temperature Gas Chromatography," p. 228. Wiley, New York.

Hawes, J. E., Mallaby, R., and Williams, V. P. (1969). *J. Chromatogr. Sci.* **7**, 690.
Henneberg, D. (1961). *Z. Anal. Chem.* **183**, 12.
Heymann, D., and Keur, E. (1965). *J. Sci. Instrum.* **42**, 121.
Heyns, K., Muller, D., Stute, R., and Paulsen, H. (1967). *Chem. Ber.* **100**, 2664.
Hites, R. A., and Biemann, K. (1968). *Anal. Chem.* **40**, 1217.
Hoffmann, R. L., and Silveira, A., Jr. (1964). *J. Gas Chromatogr.* **2**, 107.
Hogg, A. M. (1969). *J. Sci. Instrum. Ser.* 2. **2**, 289.
Holden, H. W., and Robb, J. C. (1958). *Nature* **182**, 340.
Holmes, J. C., and Morrell, F. A. (1957). *Appl. Spectrosc.* **11**, 86.
Hughes, K. J., and Hurn, R. W. (1960). *J. Air Pollut. Contr. Ass.* **10**, 367.
Hughes, K. J., Hurn, R. W., and Edwards, F. G. (1961). *In* "Gas Chromatography" (H. J. Noebels, R. F. Wahl, and N. Brenner, eds.), p. 171. Academic Press, New York.
Hurn, R. W., and Davis, T. C. (1958). *Proc. Amer. Petrol. Inst.* **38** (III), 353.
Hurn, R. W., Hughes, K. J., and Chase, J. O. (1958). Society of Automotive Engineers Annual Meeting, January 13–17, Detroit, Michigan.
Hurn, R. W., Chase, J. O., and Hughes, K. J. (1959). *Ann. N.Y. Acad. Sci.* **72**, 675.
Johansen, R. T., Heemstra, R. J., and Dunning, H. N. (1962). *Proc. Amer. Petrol. Inst.* **42** (VIII), **60**.
Kagarise, R. E., and Saunders, R. A. (1960). "The Analysis of Contaminants in Aviators' Breathing Oxygen," *Naval Res. Lab. Rept.* 5554, October 19.
Kelly, R. W. (1969). *Tetrahedron Lett.* **12**, 967.
Kendall, R. F. (1967). *Appl. Spectrosc.* **21**, 31.
Keulemans, A. I. M. (1959). "Gas Chromatography," 2nd ed., Van Nostrand-Reinhold, Princeton, New Jersey.
Kinney, I. W., Jr., and Cook, G. L. (1952). *Anal. Chem.* **24**, 1391.
Kirchner, J. J. (1967). "Thin-Layer Chromatography." Vol. XII of "Techniques of Organic Chemistry" (A. Weissberger, ed.), 788 pp. Wiley (Interscience), New York.
Kiser, R. W. (1965). "Introduction to Mass Spectrometry and Its Applications," pp. 208, 209. Prentice-Hall, Englewood Cliffs, New Jersey.
Klaver, R. F., and Le Tourneau, R. L. (1963). *Proc. Amer. Petrol. Inst.* **43** (III), 254.
Klaver, R. F., and Teeter, R. M. (1963). Paper No. 32. Eleventh Annual Conference on Mass Spectrometry, ASTM Committee E-14, San Francisco, California.
Klebe, J. F., Finkbeiner, H., and White, D. M. (1966). *J. Amer. Chem. Soc.* **88**, 3390.
La Lau, C. (1960). *Anal. Chim. Acta* **22**, 239.
Lane, G. H., Katzenstein, H. S., and Friedland, S. S. (1954). *Phys. Rev.* **93**, 363.
Levy, E. J., Doyle, R. R., Brown, R. A., and Melpolder, F. W. (1961). *Anal. Chem.* **33**, 698.
Levy, E. J., Miller, E. D., and Beggs, J. W. S. (1963). *Anal. Chem.* **35**, 946.
Levy, R. L., Gesser, H., Herman, T. S., and Hougen, F. W. (1969). *Anal. Chem.* **41**, 1480.
Lindeman, L. P., and Annis, J. L. (1960). *Anal. Chem.* **32**, 1742.
Lindeman, L. P., and Le Tourneau, R. L. (1960). Paper No. 81. Eighth Annual Conference on Mass Spectrometry, ASTM Committee E-14, Atlantic City, New Jersey.
Linstead, R. P., Millidge, A. F., Thomas, S. L. S., and Walpole, A. L. (1937). *J. Chem. Soc.* 1146.
Lovelock, J. E. (1961). *Anal. Chem.* **33**, 162.
Lovelock, J. E., and Lipsky, S. R. (1960). *J. Amer. Chem. Soc.* **82**, 431.
Lumpkin, H. E., and Taylor, G. R. (1961). *Anal. Chem.* **33**, 476.
McCloskey, J. A., Stillwell, R. N., and Lawson, A. M. (1968). *Anal. Chem.* **40**, 233.
McFadden, W. H., and Day, E. A. (1964). *Anal. Chem.* **36**, 2362.
McFadden, W. H., Teranishi, R., Black, D. R., and Day, E. A. (1963). *J. Food Sci.* **28**, 316.
McKelvey, E. E. (1962). Unpublished work. Continental Oil Co., Ponca City, Oklahoma.

McKinney, R. W., Light, J. F., and Jordan, R. L. (1968). *J. Gas Chromatogr.* **6**, 97.
McLafferty, F. W. (1957). *Appl. Spectrosc.* **11**, 148.
McLafferty, F. W., and Gohlke, R. S. (1959). *Anal. Chem.* **31**, 2076.
McWilliam, I. G., and Dewar, R. A. (1958). *In* "Gas Chromatography" (D. H. Desty, ed.), Vol. 2, p. 142. Butterworth, London.
Markey, S. P. (1970). *Anal. Chem.* **42**, 306.
Mason, M. E., Johnson, B., and Hamming, M. C. (1965). *Anal. Chem.* **37**, 761.
Melpolder, F. W., Warfield, C. W., and Headington, C. E. (1953). *Anal. Chem.* **25**, 1453.
Meyerson, S. (1953). *Anal. Chem.* **25**, 338.
Meyerson, S. (1956). *Anal. Chem.* **28**, 317.
Meyerson, S., Grubb, H. M., and Vander Haar, R. W. (1963). *J. Chem. Phys.* **39**, 1445.
Mikkelsen, L., Hopkins, R. L., and Yee, D. Y. (1958). *Anal. Chem.* **30**, 317.
Miller, D. O. (1963). *Anal. Chem.* **35**, 2033.
Modzeleski, V. E., MacLeod, W. D., Jr., and Nagy, B. (1968). *Anal. Chem.* **40**, 987.
Moshonas, M. G., and Hunter, G. L. K. (1964). *Appl. Spectrosc.* **18**, 193.
Mounts, T. L., and Dutton, H. J. (1965). *Anal. Chem.* **37**, 641.
Nerheim, A. G. (1963). *Anal. Chem.* **35**, 1640.
Newton, A. S. (1953). *Anal. Chem.* **25**, 1746.
Nickell, E. C., and Privett, O. S. (1966). *Lipids* **1**, 166.
Niehaus, W. G., Jr., and Ryhage, R. (1967). *Tetrahedron Lett.* **49**, 5021.
Niehaus, W. G., Jr., and Ryhage, R. (1968). *Anal. Chem.* **40**, 1840.
Noebels, H. J., Wahl, R. F., and Brenner, N., eds. (1961). "Gas Chromatography." Academic Press, New York.
O'Neal, M. J., Jr. (1953). *Nat. Bur. Stand.* (*U.S.*) *Circ. No.* 522, 217.
O'Neal, M. J., Jr. (1954). "Applied Mass Spectrometry," p. 27. Institute of Petroleum, London.
O'Neal, M. J., Jr., and Wier, T. P. (1951). *Anal. Chem.* **23**, 830.
O'Neal, M. J., Jr., Hood, A., Clerc, R. J., Andre, M. L., and Hines, C. K. (1955). *In* "Fourth World Petroleum Congress," Section V/C, p. 307. Carlo Colombo Publ., Rome.
Opstelton, J. J., and Warmoltz, N. (1954). *Appl. Sci. Res. B* **4**, 329.
Osburn, J. A., Jardine, F. H., Young, J. F., and Wilkinson, G. (1966). *J. Chem. Soc. A*, 1711.
Oswald, A. A., Griesbaum, K., Thaler, W. A., and Hudson, B. E., Jr. (1962). *J. Amer. Chem. Soc.* **84**, 3897.
Pailer, M., and Hubsch, W. J. (1966). *Monatsch. Chem.* **97**, 1541.
Pecsok, R. L., ed. (1959). "Principles and Practice of Gas Chromatography." Wiley, New York.
Peterson, L. E. (1962a). *Anal. Chem.* **34**, 1850.
Peterson, L. E. (1962b). *Chem. Ind.* (*London*), 264.
Petersson, G. (1969). *Tetrahedron* **25**, 4437.
Pfaender, P. (1967). *Liebigs Ann. Chem.* **707**, 209.
Phillips, C. S. G. (1956). "Gas Chromatography," p. 36. Academic Press, New York.
Pressey, D. C. (1953). *J. Sci. Instrum.* **30**, 20.
Purnell, H. (1962). "Gas Chromatography." Wiley, New York.
Quayle, A. (1959). *In* "Advances in Mass Spectrometry" (J. D. Waldron, ed.), p. 365. Pergamon, Oxford.
Quiram, E. R., Metro, S. J., and Lewis, J. B. (1954). *Anal. Chem.* **26**, 352.
Ralls, J. W. (1964). *Anal. Chem.* **36**, 946.
Reed, R. I. (1960). *Fuel* **39**, 341.
Richardson, D. M., Foster, N. G., Eccleston, B. H., and Ward, C. C. (1961). *U.S. Bur. Mines Rep. Invest.* No. 5816.

Richter, W. J., Simoneit, B. R., Smith, D. H., and Burlingame, A. L. (1969). *Anal. Chem.* **41**, 1392.
Roberts, R. H., and Walsh, J. V. (1955). *Rev. Sci. Instrum.* **26**, 890.
Rouayheb, G. M., Folmer, O. F., and Hamilton, W. C. (1962). *Anal. Chim. Acta* **26**, 378.
Ryhage, R. (1962). *Ark. Kemi* **20**, 185.
Ryhage, R. (1964). *Anal. Chem.* **36**, 759.
Ryhage, R., Wikstrom, S., and Waller, G. R. (1965). *Anal. Chem.* **37**, 435.
Saxby, M. J. (1968). *Chem, Ind.* (*London*), 1316.
Saxby, M. J. (1969). *Org. Mass Spectrom.* **2**, 33.
Scanlan, R. A., Arnold, R. G., and Lindsay, R. C. (1968). *J. Gas Chromatogr.* **6**, 372.
Schmauch, L. J., and Dinerstein, R. A. (1960). *Anal. Chem.* **32**, 343.
Schupp, O. E. (1968). "Gas Chromatography," Vol. XIII of "Techniques of Organic Chemistry" (A. Weissberger, ed.), 437 pp. Wiley (Interscience), New York.
Schupp, O. E., and Lewis, J. S. (1968). "Compilation of Gas Chromatographic Data," 2nd ed., 732 pp. ASTM Committee E-19 on Gas Chromatography.
Scott, R. P. W., ed. (1960). "Gas Chromatography." Butterworth, London.
Scott, R. P. W., Fowlis, I. A., Welti, D., and Wilkins, T. (1966). "Gas Chromatography" (A. B. Littlewood, ed.), Butterworth, London.
Seifert, W. K., and Teeter, R. M. (1969). *Anal. Chem.* **41**, 786.
Sharkey, A. G., Jr., Friedel, R. A., and Langer, S. H. (1957). *Anal. Chem.* **29**, 70.
Sharkey, A. G., Jr., Wood, G., Schultz, J. L., Wender, I., and Friedel, R. A. (1958). Paper No. 22. Sixth Annual Conference on Mass Spectrometry, ASTM Committee E-14, New Orleans, Louisiana.
Sharkey, A. G., Jr., Schultz, J. L., and Friedel, R. A. (1963). *U.S. Bur. Mines Rep. Invest.* No. 6318.
Shishido, K., and Nozaki, H. (1944). *J. Soc. Chem. Ind.* (*Japan*) **47**, 819.
Silverman, M. P., and Oyama, V. I. (1968). *Anal. Chem.* **40**, 1833.
Silverstein, R. M., Brownlee, R. G., Bellas, T. E., Wood, D. L., and Browne, L. E. (1968). *Science* **159**, 889.
Simmons, M. C., and Kelly, T. R. (1961). *In* "Second Gas Chromatography International Symposium" (H. J. Noebels, R. F. Wahl, and N. Brenner, eds.), p. 225. Academic Press, New York.
Snyder, L. R., and Buell, B. E. (1962). *Proc. Amer. Petrol. Inst.* **42**, (VIII), 95.
Stein, R. A., and Nicolaides, N. (1962). *J. Lipid Res.* **3**, 476.
Stevens, C. M. (1953). *Rev. Sci. Instrum.* **24**, 148.
Struck, R. F., Frye, J. L., and Shealy, Y. F. (1968). *J. Agr. Food Chem.* **16**, 1028.
Studier, M. H., and Hayatsu, R. (1968). *Anal. Chem.* **40**, 1011.
Szymanski, H. A., ed. (1963). "Lectures on Gas Chromatography." Plenum, New York.
Teeter, R. M. (1962). *In* "Mass Spectrometry: Organic Chemical Applications" (K. Biemann, ed.), p. 179. McGraw-Hill, New York.
Teeter, R. M. (1967). *Anal. Chem.* **39**, 1742.
Teeter, R. M. (1968). Paper No. 66. Sixteenth Annual Conference on Mass Spectrometry, ASTM Committee E-14, Pittsburgh, Pennsylvania.
Teeter, R. M., Spenser, C. F., Green, J. W., and Smithson, L. H. (1966). *J. Amer. Oil Chem. Soc.* **43**, 82.
Teranishi, R., and Mon, T. R. (1968). *In* "Theory and Applications of Gas Chromatography in Industry and Medicine" (H. S. Kroman, and S. R. Bender, eds.), p. 293. Grune & Stratton, New York.
Teranishi, R., Buttery, R. G., McFadden, W. H., Mon, T. R., and Wasserman, J. (1964). *Anal. Chem.* **36**, 1509.

Thompson, C. J., Coleman, H. J., Mikkelsen, L., Yee, D., Ward, C. C., and Rall, H. T. (1956). *Anal. Chem.* **28**, 1384.
Thompson, C. J., Coleman, H. J., Ward, C. C., and Rall, H. T. (1959). *J. Chem. Eng. Data* **4**, 347.
Thompson, C. J., Coleman, H. J., Ward, C. C., and Rall, H. T. (1960a). *Anal. Chem.* **32**, 424.
Thompson, C. J., Coleman, H. J., Hopkins, R. L., Ward, C. C., and Rall, H. T. (1960b). *Anal. Chem.* **32**, 1762.
Thompson, C. J., Coleman, H. J., Ward, C. C., and Rall, H. T. (1962a). *Anal. Chem.* **34**, 151.
Thompson, C. J., Coleman, H. J., Ward, C. C., and Rall, H. T. (1962b). *Anal. Chem.* **34**, 154.
Thompson, C. J., Coleman, H. J., Hopkins, R. L., and Rall, H. T. (1962c). *U.S. Bur. Mines Rep. Invest.* No. 6096.
Thompson, C. J., Coleman, H. J., Hopkins, R. L., and Rall, H. T. (1965). *Anal. Chem.* **37**, 1042.
Thompson, C. J., Coleman, H. J., Hopkins, R. L., and Rall, H. T. (1967a). *J. Gas Chromatogr.* **5**, 1.
Thompson, C. J., Coleman, H. J., Hopkins, R. L., and Rall, H. T. (1967b). *World Petrol. Congr., Proc. 7th* **9**, 93.
Tornabene, T. G., Gelpi, E., and Oro, J. (1967). *J. Bacteriol.* August, 333.
Vetter, W., Vecchi, M., Gutmann, H., Ruegg, R., Walther, W., and Meyer, P. (1967). *Helv. Chim. Acta* **50**, 1866.
Vollmin, J., Kriemler, P., Omura, I., Seibl, J., and Simon, W. (1966). *Microchem. J.* **11**, 73.
Watson, J. T., and Biemann, K. (1964). *Anal. Chem.* **36**, 1135.
Watson, J. T., and Biemann, K. (1965). *Anal. Chem.* **37**, 844.
Weissberger, A., Series ed. (1945–1969). "Techniques of Organic Chemistry," Vols. I–XIV. Wiley (Interscience), New York.
Widmer, H., and Gaumann, T. (1962). *Helv. Chim. Acta* **45**, 2175.
Wieland, H. (1912). *Berichte (DCG)* **45**, 484.
Willard, H. H., Merritt, L. L., Jr., and Dean, J. A. (1965). "Instrumental Methods of Analysis," 4th ed., Chap. XIX. Van Nostrand-Reinhold, Princeton, New Jersey.
Williams, J. L., and Everson, G. F. (1958). *J. Sci. Instrum.* **35**, 97.
Yarborough, V. A. (1953). *Anal. Chem.* **25**, 1914.
Zemany, P. D. (1952). *Anal. Chem.* **24**, 1709.

CHAPTER 5

FRAGMENTATION REACTIONS—KEY TO INTERPRETATION OF MASS SPECTRA

If only the molecular ion and the associated isotopic ions were formed upon ionization in the source of the mass spectrometer, the organic chemist would find little use for the field of mass spectrometry. Fortunately, the organic molecules are ionized and then fragment to produce a pattern of different ions referred to as the mass spectrum. In early mass spectrometry, the patterns were sometimes called "cracking patterns" because of their resemblance to thermal cracking, the name being used in many circles until the 1950's. As the understanding of mass spectra increased, the term was modified to fragmentation pattern, and the door was gradually opened to the study of fragmentation paths. A general understanding of these "chemical reactions" leads to the point where the mass spectra of many compounds can be reasonably predicted. It is with the gross differences in the mass spectra of the polyatomic molecules of organic substances as compared to those of the diatomic molecules more typical of inorganic substances that this treatise is concerned.

The characteristics of organic mass spectra are partially suggested by the term "cracking pattern" used above, that is, they present a "chemical" appearance. Many fragment ions are formed from parent ions by bond rupture processes (and sometimes rearrangements) which resemble those occurring in reaction mechanisms of thermally excited neutral molecules. Some resemblances to processes observed in photoionization are to be noted, but these will be described in some detail in a later section of this chapter. It is also to be noted that isomers may have radically different mass spectra in some cases and practically identical spectra in other cases. This requires an explanation which involves structure. A particular fragmentation process may be either enhanced or hindered by isotopic substitution. The polyatomic spectra also show far more pronounced changes with source temperature than do diatomic spectra. Usually parent-ion intensities show a pronounced negative temperature coefficient, suggestive again of thermal degradations, whereas the fragmention intensities may show a positive, negative, or even a zero temperature coefficient. The excess energies of fragment ions are usually very small (several tenths of a volt or less). Small ions containing a few atoms are often exceptions to this rule. Methyl and ethyl ions are small ions.

Other important characteristics are observed for the polyatomic mass spectra. One of these is that the ionization efficiency curves (as defined later) for various ions do not all have the same shape, particularly near the threshold. Metastable transitions are more commonly observed in polyatomic spectra, and sometimes several such transitions can be found which delineate a fragmentation "path" or sequence of reaction steps connecting important (or intense) fragment ions in the mass spectrum. It is well to remember that sometimes ions fragment by several possible paths and, of course, that a given ion may be formed from more than one precursor. Multiply charged ions are rare in the mass spectra of most aliphatic compounds, but with aromatic compounds, multiple ionization can become very significant. Usually ease of charge separation is a consideration, but the more aromatic the electronic system is, the more likely one is to observe doubly or even triply charged ions. For identification purposes this is a significant aid. While negative ions are sometimes formed, particularly for halogen-, oxygen-, or sulfur-containing compounds, they are generally considerably smaller in abundance than the positive ions. One must remember, however, that this factor may become important in attempting to consider an energy balance for a system of positive ions. The negative ions and processes leading to nonionic excited states must be accounted for.

Taking a rather naive approach to the "cracking pattern" of ethane and ignoring ion–molecule processes, one can consider the ionization and cleavage of the molecule ion in the mass spectrometer as shown in Table 5-1. To form the parent ion, an electron is lost. This is just the ionization process. It is

TABLE 5-1

Mass Spectrometric Cleavage of Ethane[a]

```
    H  H
    |  |
H—C—C—H
    |  |
    H  H
```

Bonds cleaved (net total)	Composition of ion formed	The *m/e* of ion formed	Intensity of ion formed (% total ion yield)
Electron lost, no bonds cleaved	C_2H_6	30	10.1
One of the C—H bonds	C_2H_5	29	9.0
Two of the C—H bonds	C_2H_4	28	42.6
Three of the C—H bonds	C_2H_3	27	15.4
Four of the C—H bonds	C_2H_2	26	11.1
Five of the C—H bonds	C_2H	25	2.2
Six of the C—H bonds	C_2	24	0.5
The only C—C bond	CH_3	15	2.5
The C—C bond and a C—H bond	CH_2	14	1.8
The C—C bond and two C—H bonds	CH	13	0.8
The C—C bond and three C—H bonds	C	12	0.4
Ejection of two hydrogens; two of the C—H bonds	H_2	2	0.4
One of the C—H bonds	H	1	3.2

[a] Isotopic effects are not shown.

assumed that all of the ions observed are then formed from this parent ion, either by simple cleavage of one bond or by a succession of such fragmentations. Note that only three ions can be formed from the parent ion by a simple cleavage (which is really a dissociation), the $C_2H_5^+$ ion, the CH_3^+ ion, and the H^+ ion at *m/e* 29, 15, and 1, respectively. On the basis of the data presented, it appears that the formation of the larger fragment is favored, with the retention of the charge, by an approximate ratio of 3/1. The breaking of a C–C bond, leading to the formation of the methyl group, appears to be less favored. It should also be observed that all of the other ions can be formed only if two or more bonds are broken! Nevertheless, a casual examination indicates a definite trend toward some statistical control of the process taking place. (That is, it is easier to "strip" the hydrogen atoms off of the ethane in steps of one or possibly two at a time, with the completely stripped ion being least favored.) Once the methyl group is formed, it too shows the reduction in ion intensity with each successive loss of a hydrogen. Since each fragmentation requires energy, it would be statistically expected that the energy distribution of the methyl ion would be of some Boltzmann character with a relatively small number possessing the necessary energy to fragment further by the rupture of a second C–H bond to form the methylene ion. In a naive fashion, too, one could consider the appearance of the m/e 2 peak to be due to an ion–atom reaction of the hydrogens (note that one would have to be a neutral, the other a proton) to lead to the formation of the *hydrogen–molecule ion*. It is necessary to consider the possibility of the ejection of two hydrogens and one electron simultaneously from the molecular ion, raising the question: Which hydrogens are ejected, one from each carbon atom, or two from a single carbon atom? Thus it becomes essential to consider the thermodynamics of ion formation and fragmentations of these ions, which, of course, are dissociations.

At one time during the 1930–1945 era, it was suggested by some persons that the molecule be considered a "glob" (not necessarily from the molecular orbital viewpoint), which upon ionization simply broke apart with any possible event occurring. It has long since been observed that this is not true. The mass spectra of polyatomic molecules follow reasonably predictable chemical reactions, in some respects similar to thermal behavior of such substances. As such, these reactions should be amenable to thermodynamic treatment.

I. The Ionization Process

The mass spectrometer reaction equation given earlier is an oversimplification of the ionization process in the mass spectrometer. The equation is

$$\mathrm{M} + e^{-}(\text{fast}) \rightarrow (\mathrm{M}^{+})^{*} + 2e^{-}(\text{slow}) \tag{5-1}$$

This is indeed the net reaction one is interested in for the examination of positive ions. The equation assumes that the electron has sufficient energy to just cause the reaction to occur. If the energy of the electron is below the ionization threshold, the reaction will not occur with that electron. A vibrationally excited molecule ion $(\mathrm{M}^{+})^{*}$ might result if there is a simple exchange of kinetic energy. It can be shown that an electron with 10 eV of energy has a velocity of 1.88×10^{8} cm/second [from Eq. (2-4)]. This is assumed to be sufficient energy to just permit reaction (5-1) to occur. The electron passes a molecule of about 2 Å in diameter and at the velocity given covers this distance in about

$$t = \frac{2 \times 10^{-8}\ \text{cm}}{1.88 \times 10^{8}\ \text{cm/second}}$$

or 1×10^{-16} second. The atoms in the molecule are vibrating about various bond centers, but the frequencies of these vibrations are of the order of 10^{14}/second. Hence an impacting electron passes the molecule in a very small fraction of the vibrational period. As a consequence, little change in the relative positions of the nuclei occurs during any resulting electronic transition. This is in essence the Franck–Condon principle, and a vertical transition is most probable.

In Fig. 5-1A a potential energy diagram shows the favored transition as

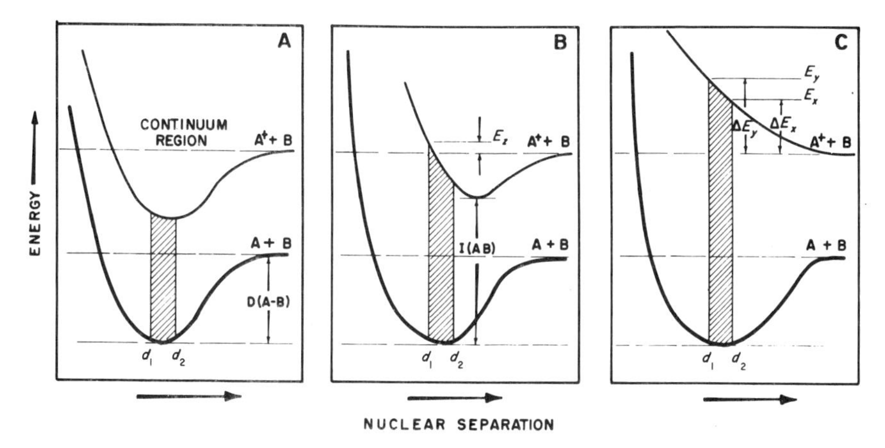

Fig. 5-1. Potential energy curves for dissociation and ionization. [From R. W. Kiser (1965). "Introduction to Mass Spectrometry and Its Applications." Prentice-Hall.]

a "vertical" one. This occurs between d_1 and d_2, the positions of the maximum and minimum internuclear separation due to a given vibrational level. The vertical transition does not necessarily correspond to a $v'' = 0 \rightarrow v' = 0$ transition. Such a 0–0 transition is termed an adiabatic transition and corresponds to the true ionization potential (IP) of the molecule. If the final state is as shown, it lies within the region of discreet vibrational levels of the potential energy curve for the moiety AB^+.

It is also possible for the potential energy curve of a given moiety represented by AB^+ to resemble the curve given in Fig. 5-1B. In this case, a portion of the possible excited states will lie in the continuum. Some of the transitions lead to stable but vibrationally excited AB^+ ions, while the rest lead to fragmentation (dissociation as well as ionization) of the AB^+ ions to form $A^+ + B$ (or $A + B^+$), depending upon the potential energy curve of the upper electronic state at very large internuclear distances. The heterocyclic cleavage leading to the formation of $(A^+ + B^-)$ can come only from the neutral excited species and usually does not contribute significantly to positive-ion mass spectrometry.

A third possibility exists—the final state of a transition may lie totally along a repulsive curve in the continuum of nuclear levels, thus producing only fragment ions since the moiety AB^+ dissociates in all cases. This is shown in Fig. 5-1C.

Note that in case B some fragments are formed with relative excess energies (ΔE) ranging from zero at the continuum energy line to E_z as indicated. In the third example all of the fragments have excess energies ranging from E_x to E_y. The excess energy consists of internal energy and translational (kinetic) energy. To make a more complete examination of the processes, the organic mass spectroscopist must consider the *electronic transition probability*, which is proportional to the square of the vibrational overlap integral* when the variation of electronic perturbational integrals with internuclear separation is small. The reader is referred to Krauss and Dibeler (1963a), Nicholls (1961), Nicholls and Stewart (1962), and the texts by Herzberg (1950) and Melton (1970) for additional information. A number of recent works report on the calculation of these Franck–Condon factors for polyatomic molecules.

For polyatomic molecules, fragmentation paths are often influenced markedly by other factors. Among these are the relative strengths of various bonds (bond dissociation energies), the stabilities of the neutral fragments involved, distribution in various vibrational frequencies, and the degrees of freedom in a molecular ion.

Events other than the ionization given in Eq. (5-1) can occur. For one possibility, the impacting electron may simply join the valence electrons in the "outer" shell to produce a negative molecule ion by electron attachment. This is discussed in Chapter 3.

Not all of the excited species formed need ionize, despite the exchange of energy by the near approach of the electron since the energy is still less than the IP. A certain fraction of the species may have energy in excess of the ionization potential but *need not ionize*. These remain in a superexcited molecular state, according to Krauss and Dibeler (1963b). The detailed discussion of Platzman (1962) pointed out that there is a probability η that ionization will occur. Further, if the difference between the actual energy E and the IP is not very great (between 0 and about 20 eV), η is always appreciably smaller than unity. Platzman also referred to an excited state with E greater than IP as a "superexcited" state.

Considering the case in which the energy E is greater than the IP and is imparted to a molecule, ionization may occur directly (with a probability δ) or, as an alternative, the superexcited state with energy E may be formed with

* The integral over the product of the vibrational eigenfunctions of the two states involved. This amounts to the square of the change of the dipole across the two states, Ψ_A and Ψ_B.

a probability of $(1 - \delta)$ by "direct ionization." Platzman means that the immediate ejection from the molecule of one of the orbital electrons by the action of the disturbing electric field occurs within a time of the order of magnitude of 10^{-16} second, the approximate period of valence electrons. After a brief period of time, the superexcited state may undergo "preionization" (electron emission from the molecule) or indirect ionization. Preionization is also called autoionization. A third possibility is simply the loss of energy to a level below the continuum. Fano (1961) has developed considerable knowledge about the behavior of atoms, wherein a reorganization of the electronic system seems to be all that is involved. In molecules, far less is known about the phenomenon, and Platzman (1962) suggested that it might be necessary "to await a rearrangement of atomic positions, as in the crossing of potential-energy surfaces."

A priori one might expect that $\eta \approx \delta$ or that most of the ionization occurs directly. Platzman (1962) pointed out that values of η (E) obtained from experiments in which the energy E (having 0–20 eV in excess of the IP) is acquired by absorption of a photon (Weissler, 1956), and values from experiments in which it is transferred from an excited atom (Platzman, 1960), are equal, within the respective limits of accuracy. From this he argued, although admitting that it does not prove, conclusively, that in fact $\delta \ll \eta$. Hence most of the ionization appears to be the result of a secondary process. Jesse and Platzman (1962) supported this tentative conclusion by experiments in which it was found that η for some simple hydrocarbon molecules is increased by the substitution of deuterium. They reasoned that if η were equal to δ there would be no isotope effect since δ is independent of isotopic composition. However, the proportion of superexcited states that decompose by preionization is raised, and η is accordingly increased because of the retardation of competitive atomic rearrangements caused in turn by increases in atomic masses and the accompanying changes in bond frequencies. Additional use of this information will be made in later sections in connection with isotopically labeled species and the consideration of fragmentation reactions apparently requiring high energy. For further reading on ionization processes and processes occurring at higher energy levels, the text of Henley and Johnson (1969) should be consulted.

II. Theory of Mass Spectra

From the previous discussion, it is apparent that some relationship should exist between the strengths of the bonds cleaved and the relative abundances of the ions formed. The parent-molecule ion formed in the electron impact process has a certain amount of excitational energy distributed in its electronic and vibrational degrees of freedom. Rosenstock (1952) and Rosenstock *et al.* (1952) devised a theory to explain the electron impact spectrum with its relative intensities of the various ions from large polyatomic molecules. The excited-state parent-molecule ion has a life of sufficient duration to allow several vibrations to occur before decomposition into fragment ions and neutral fragments. In addition, it is assumed that radiationless transitions occur with high probability among the many potential energy surfaces of the parent-molecule ion. This results in a completely random distribution of the excitational energy and hence the parent-molecule ion decomposes only when sufficient energy has been concentrated in the necessary degrees of freedom. Rosenstock and Krauss (1963) presented a chapter entitled the "Quasi-Equilibrium Theory of Mass Spectra," to which the reader is referred for the details necessary to employ the equations developed to calculate a *theoretical* mass spectrum of a substance. Presentation of the equation without full discussion is given here in the hope that the reader will become interested in developing familiarity with this aspect of mass spectrometry.

Rosenstock and Krauss (1963) gave the following general unimolecular expression for "the average reaction rate of an ion or a molecule;

$$k\,(E, E_0) = \frac{1}{h}\left[\sum_{\epsilon=0}^{E-E_0} \frac{\rho^{+}(\epsilon)}{\rho(E)}\right] \tag{5-2}$$

In the equation, E_0 is the activation energy of the reaction, h is Planck's constant, and $\rho(E)$ is the density of states of the reactant molecules. The function $\rho^{+}(\epsilon)$ represents the density of states obtained by including all internal degrees of freedom of the activated complex except the translation along the reaction coordinate."

From the equation, one notes that the $\rho^{+}(\epsilon)$ must be summed over all possible partitions of energy between these degrees of freedom and the one special translational motion, which is the reaction coordinate. It contains a potential energy equal to the activation energy E_a. The kinetic energy of translation along the reaction coordinate can vary from zero to the total internal energy less the activation energy. Rosenstock and Krauss (1963) pointed out that this rate expression is perfectly general but contains two simplifications: (a) The transmission coefficient [see Glasstone *et al.* (1941) and Eyring *et al.* (1944) for definitions] is unity, and (b) quantum mechanical tunneling effects are unimportant. This is discussed in detail in their chapter. They indicated that the rate calculated from Eq. (5-2) represents an average over all the possible rates of passage over the potential barrier. The rate constant is thus a dissociation probability and does not represent a precise lifetime. For the reader expecting a more conventional form of absolute rate theory equation,

Rosenstock and Krauss have emphasized an important difference between the assumed kinetic processes and the processes encountered in ordinary gas kinetics. This difference is the absence of a temperature as encountered in ordinary kinetics where molecular collisions continually energize and de-energize the molecules. With this condition, the equilibrium assumption of absolute reaction rate theory leads to the corollary that the distribution (a Maxwell–Boltzmann type) of the reactants and activated complexes among the respective accessible states may be described if temperature is used as a parameter. In mass spectrometry, at pressures of 10^{-6} torr, collisions are very rare. Even so, particles that suffer collisions do not enter into recorded spectral data. Hence the ionization processes can lead to completely different excitation energy distributions than might be expected in ordinary gas kinetics. Magee (1952) has shown that when Eq. (5-2) is averaged over a Boltzmann distribution of energy which is characterized by a temperature, the resulting absolute rate theory expression containing partition functions is obtained. The work of Rosenstock and Krauss should be consulted for details of how this general rate expression is converted into a specific form which yields a rate constant. Kiser (1965a) also gave details of the conversion to a form used for calculation of rates.

While the application of Eq. (5-2) has met with considerable success, there have been some problems. Chupka (1959), using a study of metastable ions, has concluded that only a fraction of the total number of oscillators seems to be accessible to provide the activation energy for the decompositions. Using a correction of $(N-1)/2$ or $(N-1)/3$ for the term $(N-1)$, where N is the total number of internal degrees of freedom as used in the specific form equations, leads to better agreement between calculated and experimental mass spectra.

Kiser (1965b) showed an example calculation of a mass spectrum of thiacyclobutane as obtained from the specific form of Eq. (5-2) starting from calculated heats of reaction and estimated activation energies. The comparison is good for the six major ions formed, which comprise 84% of the total ion intensities.

Kiser (1965c) also gave an example calculation involving 2,3-dithiabutane and used the geometric mean frequency $\bar{\nu}$ obtained from the frequencies assigned to that molecule by Scott *et al.* (1950). All $3n-6$ frequencies were multiplied together and the $(3n-6)$th root taken. From this he obtained the value of $h\bar{\nu}$ that appears in another specific form of Eq. (5-2) used to calculate the respective k's. Calculations are tedious but can be aided markedly by computer use. Because of increased availability of the computer and the training of today's organic chemists, it is probable that considerable additional testing will be done and support provided for the quasi-equilibrium theory of mass spectra.

III. Energetics of Electron Impact Processes

From the mass spectrometric equation and the foregoing discussions, it should be apparent that the molecular ion, if singly charged, has lost one electron and is an odd-electron ion. Also if the moiety AB^+ fragments to give an ion A^+ and a neutral fragment B, the fragment ion A^+ must have an even number of electrons, while B carries an odd number of electrons, is a free radical, and possibly possesses excess energy. It should be noted that the term "molecular ion" can be misleading. Most persons associate this name with the ion formed by electron impact with the molecular species of a substance. For clarity this ion is referred to as the parent-molecule ion or molecular ion and is designated by $M^{+\cdot}$ to distinguish it from the rearrangement or fragment type of molecule ions, designated $[X]^{+\cdot}$, that appear in mass spectra. For example, the $H_2S^{+\cdot}$ ion appears in the mass spectrum of ethanethiol, probably from the following process:

$$[C_2H_5SH]^{+\cdot} \rightarrow [C_2H_4]^0 + [H_2S]^{+\cdot} \tag{5-3}$$

Odd-electron ion	Even-electron neutral particle	Odd-electron ion

Similarly, the M − 18 (parent-molecule ion less mass 18) ion from alcohols is a rearrangement molecule ion, and the equation is

$$[C_5H_{11}OH]^{+\cdot} \rightarrow [H_2O]^0 + [C_5H_{10}]^{+\cdot} \tag{5-4}$$

Odd-electron ion	Even-electron neutral moiety	Odd-electron ion

This is the same $C_5H_{10}^+$ ion that would appear from either pentene-1 or pentene-2, but not necessarily possessing the same energy distribution in its vibrational levels. As a third example, dialkyldisulfides of the type shown produce a rearrangement molecule ion according to the following equation:

$$[C_2H_5\text{—S—S—}C_2H_5]^{+\cdot} \rightarrow 2[C_2H_4]^0 + [H_2S_2]^{+\cdot} \tag{5-5}$$

Odd-electron ion	Neutral moiety	Odd-electron ion

Molecule ions can also be fragment ions as will be seen later in specific examples of fragmentation processes. The brackets are used to designate ions when the locus of the charge localization is not defined. To avoid problems when large organic ions are involved, the brackets are omitted and the symbol $\urcorner^{+\cdot}$ or $\urcorner^{+}$, as the case may be, is placed at the upper right-hand corner of the formula.

A. The Appearance Potential and the Ionization Potential

The *appearance potential* (AP) of a given ion is the minimum energy required to produce the ion and any accompanying neutral moieties from a given molecule, ion, or radical. This definition covers the possible cases presented in Fig. 5-1 including the possibility that the products, whether neutral or ionic or both, might be in excited states rather than in the ground state. Hence it is possible for experimentally determined values to be somewhat greater than theoretical values. From the above discussion and Fig. 5-1, it is apparent that the term *ionization potential* (IP) is merely a special case of the term appearance potential where ionization occurs without dissociation. In other words,

$$\mathrm{AP}(\mathrm{M}^{+\cdot}) = \mathrm{IP}(\mathrm{M}^{+\cdot}) = \Delta H_f(\mathrm{M}^{+\cdot}) - \Delta H_f(\mathrm{M}) \tag{5-6}$$

where ΔH_f is the heat of formation.

For the general reaction of electron impact one may write

$$\mathrm{M}^{+\cdot} + e^-(\text{fast}) \rightarrow \mathrm{F}^+ + \mathrm{N}_i + 2e^-(\text{slow}) \tag{5-7}$$

where F^+ is a fragment ion and N_i represents the neutral fragment(s). Then the appearance potential (AP) of the fragment ion F^+ is

$$\mathrm{AP}(\mathrm{F}^+) = \Delta H_f(\mathrm{F}^+) + \sum_i [\Delta H_f(N_i) - \Delta H_f(\mathrm{M})] \tag{5-8}$$

As indicated above, the equation holds provided that the AP determined experimentally corresponds to the various species in their ground states, and no excess energy is involved for the process shown in Eq. (5-7). Similar equations are used for other processes such as negative-ion formation.

Experimental determination of AP and IP values usually involves the determination and evaluation of an ionization efficiency curve. A typical partial ionization efficiency curve for a monatomic gas (argon) is shown in Fig. 5-2. The curve is simply a plot of the ion current of a given m/e as a function of the energy of the electron beam. Honig (1948) has shown that the portion of the curve in section a is exponential in character, that in section b is intermediate in character, while that portion in section c is essentially linear. Typical curves usually reach a maximum intensity at 20–50 V above the ion threshold, although this is not shown here. Beyond the maximum, the intensity of the ion current gradually degreases with a further increase in electron energy. This is due to either fragmentation reactions of the ion or formation of multiply charged species due to the available higher energies. Bleakney *et al.* (1937) have given empirically determined formulas for the overall shape of the curve. A similar formula has been used by Lorquet (1960).

A critical review by Krauss and Dibeler (1963b) of the problems associated with appearance potential data of organic molecules should be read by those

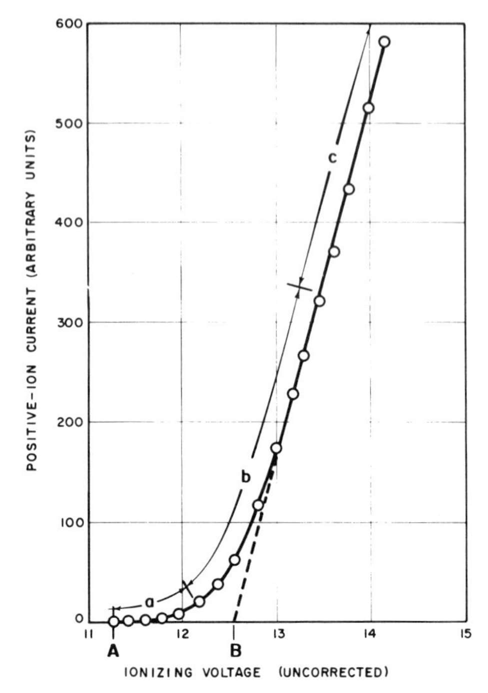

Fig. 5-2. Typical ionization efficiency curve.

interested in utilizing such data since the material is beyond the scope of this treatise.

B. The Electron Energy Spread—Source Effects

Three main features of appearance potential data are discussed to some extent because of a quite general importance to mass spectroscopy. These are (a) the electron energy spread, (b) potential gradients within the ionization chamber and contact potentials, and (c) the techniques used to obtain AP and IP from ion efficiency curves.

The electrons that are emitted from the heated filament in the source have appreciable thermal energies. The resulting electron energy distribution is essentially Maxwellian in nature and according to Honig (1948) can be represented by

$$dN(U) = (4\pi mA/h^3)\, U \exp[-(\theta + u)/kT]\, dU \qquad (5\text{-}9)$$

where $dN(U)$ is the number of electrons with energy between U and $(U + dU)$ emitted from the heated filament per second, U is the thermal energy in ergs of the electrons of mass m (in grams), h is Planck's constant in erg-seconds,

k is Boltzmann's constant, A is the surface area in square centimeters of the filament with a work function (θ), and T is the absolute temperature of the filament. A plot of $dN(U)/dU$ versus U leads to a Boltzmann distribution giving a maximum at U. Kiser (1965d), using $\theta = 4.50$ eV, a filament 1 cm long and 5 mils in diameter, and an assumed temperature of $T = 2500°$K, gave such a plot. From it, the thermal energy spread of the emitted electrons varied from 0.2 to 0.4 eV at the above temperature where $kT = 0.215$ eV. It is obvious why a number of workers have gone to various types of sources to produce narrower limits for the electron energy spread (see Chapter 2). Theoretical studies have been carried out by King and Long (1958), Friedman *et al.* (1957), and Eyring and Wahrhaftig (1961) on the effect of electron energy on mass spectra.

The "foot" (section a) of the ionization efficiency curve shown in Fig. 5-2 is associated with this spread in electron energy. Also responsible for a part of the "foot" is the presence of possible states lying close in energy to the ground state. Then ionization may occur from a state which is thermally populated.

Inherent source and instrumental problems may also cause a change in the electron beam energy distribution. Some gas-phase molecules affect the work function of the filament by chemical reactions. This varies with the types of substances possibly present in only trace quantities in the material to be examined, and thus the effects may be irregular and unpredictable. The repeatability of a given determination may be interfered with. Contact potentials in the ion source and the applied repeller voltage may also affect the electron energy distribution. Cleaning the source may be necessary. Some of these effects can be reduced if the masses of the ions being considered and their AP's (unknown and calibrating ion) are kept as similar and close together as possible. Pulsing the electron beam and the repellers is also a help in this regard. The Fox *et al.* (1951, 1955) *method of retarding potential differences* (as mentioned in Chapter 2) employs a modified ion source which includes these features.

A study of effects of instrument geometry and sources upon ionization and appearance potentials employing azomethane has been reported by Prasil and Forst (1968). Carpenter *et al.* (1967) described the source dependence upon temperature and its effects upon the ionization voltage studies of some labeled aliphatic ethers. Additional details are to be found in Kiser's text (Kiser, 1965e).

C. Experimental Techniques for Obtaining the AP and IP

In all of the methods below a calibrating gas and the gaseous sample to be studied are mixed intimately (usually in the mass spectrometer inlet system) in about equal concentrations and then admitted to the ion source. The ionization voltage and ion current values are then recorded for both ions, the unknown and the "internal" standard reference ion.

The *linear extrapolation method* was introduced by Vought (1947) and, although no longer in use for accurate determinations of an AP, does give an upper limit for ionization and the AP. This is illustrated in Fig. 5-2 where the linear portion of the curve is extrapolated to zero ion current to give the intercept value at B. Comparing this to the linear extrapolation value of the reference material gives the electron volt value of the ionization potential of the desired unknown ion. Because variations in the "foot" of the curve cause considerable variation in accuracy, this method is not favored. In practically all cases the AP or IP determined in this fashion is higher than the true value. The upper limit application is obvious as is the simplicity of interpretation.

The *initial break method* has the obvious drawback, as illustrated in Fig. 5-2, of requiring the precise determination of the intersection of the curve and the energy axis. Sensitivity at the lower ionizing voltages is very low, and with the asymptotic approach of the curve there is no sharp break, thus compounding the difficulties. Waldron and Wood (1952) showed that the value of the ionization potential determined by the initial break method decreases as the sample pressure in the ion source is increased. This should be anticipated from the parallel sensitivity change that occurs as ionization voltage decreases. The method is little used today other than in the *vanishing current method* with a Fox source.

The *extrapolated voltage difference method* of Warren (1950) has been found to be reasonably satisfactory. In plotting the data, the ion current scale (i) of either the sample ion or the calibrating ion is arbitrarily adjusted to make the linear portions of the ionization efficiency curves parallel. This is shown in Fig. 5-3A and B. An expanded plot of the foot of the curve and the beginning of the parallel sections is prepared, as in C. Next, voltage differences (ΔV) at various (small) values of the ion current (i_+) are obtained from the curves. A final plot (D) is made of the ΔV's at various values of i_+ and a linear extrapolation to $i_+ = 0$ is made. The value of ΔV at $i_+ = 0$ is designated as ΔV_0 and is the difference between the AP of the calibrating ion and the unknown ion. The method gives good results, according to Kiser (1965f), if the two ions have an abundance greater than about 3% of the largest ion in the spectrum. The extrapolations are normally made over the range of 0.05–2.0% of the 50–75-eV ionizing current.

The *semilogarithmic plot method* was developed by Lossing *et al.* (1951), who found that plots of the logarithm of the ion current versus electron energy for parent-molecule ions from many substances were parallel in the region of about 1% of the 50-eV current. Values stated to be reproducible to ± 0.01 eV were obtained. The authors of this treatise have used the method in their

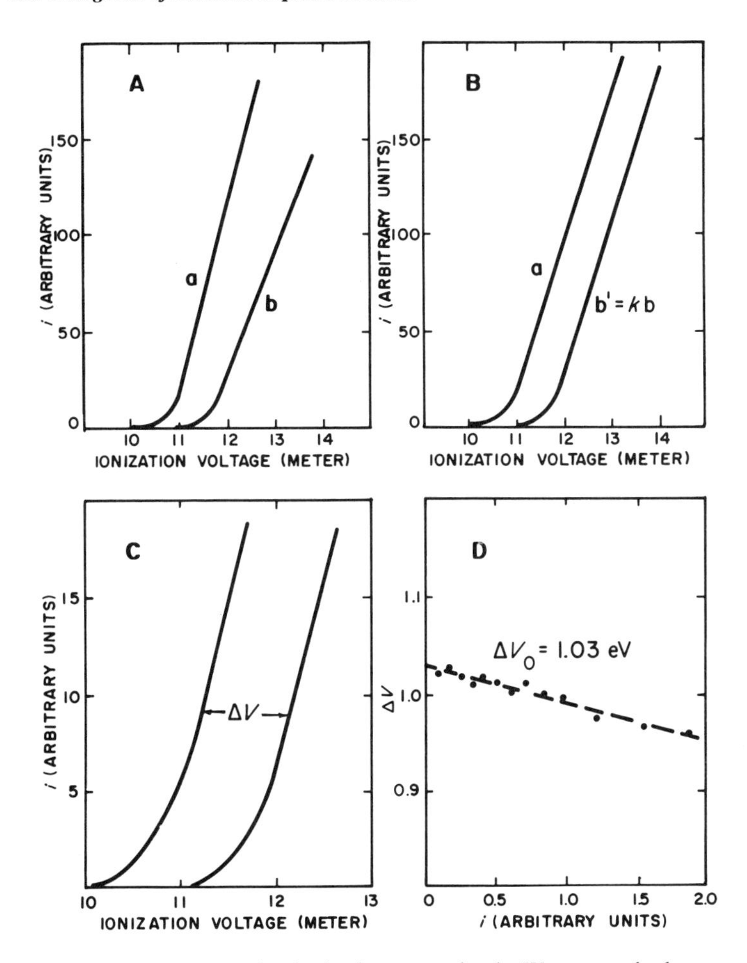

Fig. 5-3. Interpreting ionization curves by the Warren method.

laboratories and found reproducibility to vary from 0.05 to 0.10 eV, depending upon the substance. Similarly, the range of parallelism varies considerably, sometimes from 2–3% down to about 0.05% of the 70-eV ion current.

Morrison (1953) has shown that additional information can be obtained from the ionization efficiency curve if the second derivative is determined. It is proportional to both the energy distribution of the electron beam centered at the threshold energy and a function called the scale factor, σ_0, which is proportional to the transition probability of the ion electronic state at that energy. This technique is coming into wider use as shown by the work of Shapiro and Turk (1969) in the next section.

The *critical slope method* was devised by Honig (1948) and employs a semilogarithmic plot of electron energy versus the percentage of 50-eV ion current. The data plot is thus similar to that of the previous method. The difference is the voltage at which a line of slope $\frac{1}{2}kT$ or $\frac{2}{3}kT$ becomes tangent to the curve. Honig assumed that the probability of ionization was proportional to the square of the *excess electron energy* above the "critical" or ionization potential. His relationship was

$$\ln N_i(V) = \ln [(V_c - V) + 3kT] - [(\theta + V_c - V)/kT] + C \qquad (5\text{-}10)$$

In the equation, $N_i(V)$ is the number of ions produced per second at an energy of V, V_c is the critical (or ionization) potential, k is Boltzmann's constant, T is absolute temperature, θ is the work function of the filament emitting the electrons, and C is a constant. A plot of $\ln N_i(V)$ against E gives a straight line of slope M, given by

$$M = \left(\frac{1}{kT}\right) - \left[\frac{1}{(V_c - V) + 3kT}\right] \qquad (5\text{-}11)$$

which at $V = V_c$ reduces to

$$M = (1/kT) - (1/3kT) \qquad \text{or} \qquad M = 2/3kT \qquad (5\text{-}12)$$

If a first-power dependence of the ionization probability on the electron energy *in excess* of the ionization potential is used one obtains

$$M = 1/2kT \qquad (5\text{-}13)$$

The method gives satisfactory results for determination of ionization potentials and for some appearance potentials. Once more information is obtained about ionization probabilities, this may become a very dependable method. It is better than the linear extrapolation method.

The *energy compensation technique* has been developed and reported by Kiser and Gallegos (1962). The ion currents of the known calibrating gas and the gas to be measured are measured at 50 eV and recorded on separate channels of a dual-channel recorder. The sensitivity of the two amplifiers (one for each ion output) is increased by convenient multiplying factors from 100 to 1000 times. Then the electron energy is decreased until the ion current intensity reads the same for each ion as it did at 50 eV. The difference in the voltages is taken as the difference in the appearance potentials of the calibrating gas and the unknown being studied. Kiser (1965g) claimed that although this technique was used only on a TOF instrument, it should be useful for other types of instruments. Kiser also gave some comparative data by several different methods, which indicates good agreement in general and, to some extent, more consistent data is obtained.

The *ionization potentials of multiply charged ions* can also be determined from ion current versus electron energy plots. Dorman and Morrison (1959a,b, 1960, 1961a,b,c) and Fox (1960a,b, 1961) have studied various monatomic and polyatomic ions. The ionization efficiency curves are formed by plotting the electron energy in electron volts in the usual fashion versus the *n*th root of the ion current in arbitrary units. Assuming one is interested in the Kr^{2+} ion appearance potential, the square root of the ion current observed for the Kr^{2+} is used in the plot and then treated just the same as in any of the foregoing cases, i.e., extrapolation of the linear portion of graph, semilogarithmic plot, etc.

As will be seen in a later section, thermochemical calculations allow the determination of ionization potentials or appearance potentials from heat of formation data provided the necessary values are in the literature. Nicholson (1958) gave a critical review of the various methods of determining the values from the data. He set up criteria for nine different methods. More recently Collin (1965) included a summary of the status of the "art" in a survey of theoretical problems and experimental problems encountered in studies of excitation phenomena and ionization.

D. Methods for Calculating the AP and IP

It is possible that some workers in the field of mass spectrometry have not realized the importance of areas other than analytical, and yet they have had a number of years of experience in the field. Ionization potentials are one of the most important properties of a molecule. From the physical chemist's viewpoint, determination is laborious, and hence a look at possible methods of calculation is considered. The IP gives an indication of electronegativity, bond order, and bond energy (or strength). Purely theoretical calculations of ionization potentials beyond the usual helium calculations made in elementary quantum mechanics are almost impossible. Naturally, then, one turns to approximate methods in the hope of obtaining by semiempirical means the desired ionization potential. Several approaches are briefly presented here to acquaint the newcomer to the field with the possibilities.

The *equivalent orbital method* was developed by Hall (1951, 1953) and Lennard-Jones and Hall (1952a,b). Essentially the procedure involves the removal of an electron from a molecular orbital distributed over the entire molecule. This results in ionization. There are corresponding ionization potentials for each of the various molecular orbitals of the molecule. Further, these ionization potentials are equal to the negative of the energies of the molecular orbitals. The lowest ionization potential corresponds to the orbital of highest energy, etc. The molecular orbitals, in turn, may be expressed in terms of equivalent orbitals by means of a secular equation. The necessary parameters for these equations are determined from experimental data. Hall employed the results of Honig (1948) for the ionization potentials of the alkanes. One can readily see that to employ a calculation method of this type a set of self-consistent ionization potentials is necessary. Such compilations have been accumulated over the past twenty years and have been published. Additional information will be given in a later section of this chapter.

The *group orbital method* is derived in a sense from the preceding method. For branched-chain paraffins and other substituted paraffins the equivalent orbital method becomes extremely complicated. Hence, Lennard-Jones and Hall (1952a,b) showed that for planar molecules which have an orbital antisymmetric in the plane of the molecule, the secular equation can be reduced to where the terms refer to characteristic groups. For a relatively simple molecule like ethylene the parameters are reduced to the ionization potential of C–H, the ionization potential of $CH_2{=}CH_2$, the H–C=C interaction parameter, the H–H interaction parameter on the same carbon atom, and the H–H interaction parameter on the adjacent carbon atom. With only five parameters, the secular equation results in a five-square determinant, which is equal to zero. Again, experimental data is fed back into the equations to determine the parameters indicated. Kiser (1965g) gave some rather simple example calculations of data for both this method and the equivalent orbital method.

Franklin (1954) has extended the group orbital concept to calculate ionization potentials for molecules such as alkylbenzenes, alkenes, aldehydes, amines, alcohols, ethers, carboxylic acids and esters, ketones, alkanes, cycloalkanes, and alkyl halides. This work has been accumulated into a table of group equivalent values by Franklin *et al.* (1969) for the convenience of those working with ionization potential calculations. It should be noted that the IP of free radicals may also be calculated by this method. The method yields calculated values for nonplanar compounds of the type which cannot be treated by the Hall approaches. Kiser (1965h) gave a table of calculated results wherein agreement within 0.2–3 eV of the experimental values is shown. He cautioned, however, that extreme faith should not be placed in the results of the calculations by the group orbital method but pointed out that the method is very useful in attempting to approximate an as yet undetermined ionization potential. There is little doubt that there will be many research papers concerning this approach in the future.

Svec and Junk (1967) pointed out that despite the reports of Kiser (1960) and Field and Franklin (1957) and the review works of Krauss and Dibeler (1963b) and McDowell (1963), only one article (Junk and Svec, 1963) reported on the energetics for the ionization and fragmentation of alkanes having more than one substituent. From their studies they concluded that if a bifunctional or multifunctional molecule is to be investigated, the concept of molecular ionization as opposed to "isolated ionization" leads to problems. The net effect is that "if the concept of 'isolated ionization' is correct then the ionization of substituted alkanes is more a property of the substituent united atom (where NH_2 and COOH are considered as a single atom) combined with the adjacent carbon atoms than of the molecule as a whole." While the molecular orbital type of approach is appealing, the concept of isolated ionization is too. This is particularly true for those working with hetero atoms in the system where the first ionization is usually associated with the nonbonded pair of electrons (if present) on the hetero atom in question. Svec and Junk pointed out that in this respect the charge resulting from ionization may be regarded as truly localized, as is often assumed in the texts of Biemann, McLafferty and Budzikiewicz, Djerassi and Williams, and mentioned in the literature by Sharkey *et al.* (1959), Beynon *et al.* (1959), and Chupka and Berkowitz (1960). Svec and Junk cautioned that it is probably not correct to correlate the intensities of the fragment-ion currents produced by 70-eV electrons with those produced by electrons whose energy is near the ionization threshold.

It appears to the authors of this book that there is merit in both approaches, and it may well be that hetero atoms in a Hückel π-bonded system may be treated by the molecular orbital approach, while those restricted to other systems would be treated by the isolated ionization approach. In any event, it may not always be safe to assume that a nitrogen atom, for example, in an azasilane π-ring system is the site of ionization, etc., even though subsequent events show a bond cleavage to occur at that position. The problem is open to workers of the future.

Kaufman and Koski (1960) have proposed the use of a set of constants called the δ_K values to permit quantitative calculations of the change in the ionization potentials of free radicals and molecules with substitution of the parent molecule of radical. These values are additive within a given series, and the method also involves very simple calculations.

An experimental method to parallel this has appeared in the work of Shapiro and Turk (1969). Relative free-radical stabilities are determined by a hypothetical reaction which involves a transfer of a phenylaminoethyl radical instead of a hydrogen atom as was done by Martin *et al.* (1966). The basis of the method will be discussed in the following section on thermochemistry. Shapiro and Turk (1969) also concluded that the ionization occurs by removal of an electron from the anilino portion of the molecules, in accordance with the ideas of Junk and Svec (1963).

The ω *technique* was developed by Streitwieser and Nair (1959). Molecular orbital calculations of the ionization potentials of organic compounds using this technique have been carried out for a large number of molecules. The technique involves the Hückel molecular orbital method, modified by the introduction of a parameter, ω, to obtain agreement with experiment. In general the value of $\omega = 1.4$ was found to be satisfactory (Streitwieser, 1960).

Basically a charge distribution on each carbon atom can be calculated and then used to calculate a new coulomb integral. Using the new charge distribution that results, the calculation is repeated. Reiteration is continued until the system is self-consistent. The final result is the energy of the positive ion, just as in the earlier molecular orbital methods. Details on parameters, equations, and calculations are to be found in the text of Streitwieser (1961).

E. Thermochemistry of Ionization and Fragmentation Processes

In the previous parts of this section, the significance of the IP, AP, methods of calculation or experimental determination of these quantities, as well as some instrumental factors that affect their determination, have been discussed. It is possible to apply the IP and AP in a thermochemical fashion to assist in elucidation of many problems. Among these are (a) identification of a probable ion structure through the use of the appearance potential, (b) decision between two possible complementary reactions on the basis of heats of formation, (c) determination of bond dissociation energies, (d) determination of proton affinities, and (e) determination of relative free-radical stabilities. All of these applications are extensions of the law of Hess to ions observed in the mass spectrometer.

1. Identification of Ion Structure. Possibly the most classical application of AP data in the field of mass spectrometry is that associated with the structure of the ion appearing at *m/e* 31 from methanol. The ion also appears in most primary alcohols with a large intensity. Cummings and Bleakney (1940) determined the AP of the *m/e* 31 ion to be 11.9 ± 0.1 eV. The reaction to produce this ion is

$$CH_3OH + e^- \rightarrow CH_3O^+ + H + 2e^-, \qquad AP = 11.9 \text{ eV} \tag{5-14}$$

This is the only possible reaction that can be written. One can question the structure of the *m/e* 31 ion; i.e., is it CH_3O^+ or CH_2OH^+? But since the $AP = 11.9$ eV $= \Delta H_r$ (heat of reaction) we can write $AP = 11.9 \text{ eV} \times 23.06$ kcal/mole eV $= 274.5$ kcal/mole and, further,

$$\Delta H_r = 274.5 = \Delta H_f(CH_3O^+) + \Delta H_f(H) - \Delta H_f(CH_3OH) \tag{5-15}$$

This can be rearranged to give

$$\Delta H_f(CH_3O^+) = 274.5 + \Delta H_f(CH_3OH) - \Delta H_f(H) \tag{5-16}$$

Using the known ΔH_f of $(CH_3OH)_g = -48.1$ and $\Delta H_f(H) = +52.1$, then

$$\Delta H_f(CH_3O^+) = 274.5 + (-48.1) - 52.1 \tag{5-17}$$

or

$$\Delta H_f(CH_3O^+) = 174.3 \text{ kcal/mole} \tag{5-18}$$

Thus we have calculated the heat of formation of the ion at *m/e* 31; this contains the atoms CH_3O, but the structure is not known. Examining the data for this ion from a number of alcohols, including those in Franklin *et al.* (1969), an average of about 172 kcal/mole ± 6% is found, with no attempt being made to discard any data. Hence if one assumes that the ion formed at *m/e* 31 for all of these alcohols is the same ion, then there is strong suggestive evidence that the ion has the structure CH_2OH^+. If the structure were CH_3O^+, rearrangements would be required in all of the alcohols greater than methanol in molecular weight. If a cyclic structure is assumed, then one can compare the calculated heat of formation with those for the propylene oxide and other epoxides in the compilation of Franklin *et al.* (1969). Four such compounds are listed, producing an *m/e* 31 ion of composition CH_3O. The resulting heats of formation vary from 207 to 220 kcal/mole, too large to be considered at all similar to the 174 kcal/mole of the alcohol type of ion. Another set of data is available for the ion at *m/e* 31 produced from CH_3ONO with a heat of formation of 235 kcal/mole, an even greater difference than the cyclic structure. Hence, these data provide further evidence that the structure of the *m/e* 31 ion from the alcohols is CH_2OH^+ and that

$$\Delta H_f(CH_2OH^+) \approx 172 \text{ kcal/mole.}$$

It has also been pointed out by McLafferty (1963a) that resonance stabilization of this ion can occur. The forms involved are

$$CH_2{=}\overset{+}{O}H \leftrightarrow {}^{+}CH_2{-}OH$$

2. Choice Between Two Complementary Reactions. When studying fragmentation, one often has this choice, and the rule of Stevenson (1951) must then be utilized. In the reaction below, a molecule may be ionized and dissociated either by the process

$$R_1R_2 \rightarrow [R_1R_2]^+ \rightarrow R_1^+ + R_2 \quad (5\text{-}19)$$

or by the complementary reaction

$$R_1R_2 \rightarrow [R_1R_2^*]^+ \rightarrow R_2^+ + R_1 \quad (5\text{-}20)$$

Stevenson's rule states that only when $IP(R_1)$ is less than $IP(R_2)$ will the fragments be found in their lowest states or without kinetic energy. Only a few exceptions to this rule have been observed, and these are restricted to molecules of high symmetry or those containing very few atoms. McLafferty (1963a) restated that "the positive charge should reside on the fragment of lowest ionization potential after the bond cleavage has occurred." This rule *must* be kept in mind for all competitive fragmentation reactions.

An example of a choice between two complementary reactions occurring upon electron impact is taken from the work of Kiser and Hobrock (1962). For the cyclopropylcyanide molecule the reactions are

$$C_3H_5CN + e^- \rightarrow CN\cdot + C_3H_5^+ + 2e^-, \qquad AP = \Delta H_r = 12.7 \text{ eV} \qquad (5\text{-}21)$$

or

$$C_3H_5CN + e^- \rightarrow CN^+ + C_3H_5\cdot + 2e^-, \qquad AP = \Delta H_r = 19.5 \text{ eV} \qquad (5\text{-}22)$$

The data from Kiser and Hobrock (1962) are shown in the following tabulation:

Ion		AP or IP	
m/e	Formula	eV	kcal
67	$C_3H_5CN^+$	11.2	258
41	$C_3H_5^+$	12.7	293
26	CN^+	19.5	450

From Field and Franklin (1957), $\Delta H_f^+(CN) = 425$ kcal. From Dibeler *et al.* (1961), IP(CN) = 14.55 eV = 335.5 kcal. In the manner of ordinary thermochemical methodology, subtracting Eq. (5-22) from Eq. (5-21),

$$CN^+ + C_3H_5\cdot \rightarrow CN\cdot + C_3H_5^+, \qquad -6.8 \text{ eV} \qquad (5\text{-}23)$$

and for the IP(CN) given above,

$$CN\cdot \rightarrow CN^+, \qquad +14.55 \text{ eV} \qquad (5\text{-}24)$$

Adding Eqs. (5-23) and (5-24),

$$C_3H_5\cdot \rightarrow C_3H_5^+, \qquad 7.8 \text{ eV or } 180 \text{ kcal} \qquad (5\text{-}25)$$

which is the IP($C_3H_5\cdot$).

To obtain $\Delta H_f(CN\cdot)$ from the IP(CN),

$$\Delta H_r = 335.5 \text{ kcal} = \Delta H_f^+(CN) - \Delta H_f(CN\cdot) \qquad (5\text{-}26)$$

or

$$335.5 = 425 - \Delta H_f(CN\cdot)$$

hence

$$\Delta H_f(CN\cdot) = 89.5 \text{ kcal} \qquad (5\text{-}27)$$

No experimental value for the heat of formation of the cyclopropylcyanide was available in the literature; hence Kiser and Hobrock used the Franklin (1949, 1953) group method. An estimated value of 43 kcal/mole was obtained for the $\Delta H_f(C_3H_5CN)$.

From Eq. (5-22) one can find

$$\text{AP} \equiv \Delta H_r = \Delta H_f^+(\text{CN}) + \Delta H_f(\text{C}_3\text{H}_5\cdot) - \Delta H_f(\text{C}_3\text{H}_5\text{CN}) \quad (5\text{-}28)$$

or

$$450 = 425 + \Delta H_f(\text{C}_3\text{H}_5\cdot) - 43$$

from which

$$\Delta H_f(\text{C}_3\text{H}_5\cdot) = 69 \text{ kcal} \quad (5\text{-}29)$$

Treating Eq. (5-21) in the same fashion,

$$\text{AP} \equiv \Delta H_r = \Delta H_f(\text{CN}\cdot) + \Delta H_f^+(\text{C}_3\text{H}_5) - \Delta H_f(\text{C}_3\text{H}_5\text{CN}) \quad (5\text{-}30)$$

or

$$293 = 89.5 + \Delta H_f^+(\text{C}_3\text{H}_5) - 43$$

from which

$$\Delta H_f^+(\text{C}_3\text{H}_5) = 246.5 \text{ kcal} \quad (5\text{-}31)$$

Finally from the IP(C_3H_5CN) or $M \rightarrow M^+$ we find

$$\text{IP} \equiv \Delta H_r = \Delta H_f(\text{M}^+) - \Delta H_f(\text{M}) \quad (5\text{-}32)$$

or

$$258 = \Delta H_f(\text{M}^+) - 43$$

or

$$\Delta H_f^+(\text{C}_3\text{H}_5\text{CN}) = 301 \text{ kcal} \quad (5\text{-}33)$$

Thus, from the appearance potential data for three ions, a known heat of formation, and an estimated heat of formation, the ΔH_f's of four ionic and neutral species have been obtained. It is apparent that a few accurate measurements can go a long way in providing thermochemical information. To simplify these calculations rounded values were used and the $\pm$ values of the original data, as supplied by Kiser and Hobrock, were ignored. Suffice to say that an error as large as 1 eV (about 23 kcal) could not be made, and hence the data should be regarded in this light.

These workers raised the question as to whether the C_3H_5 radical and ion for which these energetics have been calculated are cyclic. Pottie and Lossing (1961) and Stevenson (1958) have called attention to the danger of attempting to deduce the energetic properties of cycloalkyl radicals or ions from AP data for the cycloalkanes. The value of 246.5 kcal/mole shown is significantly greater than the value of 220 kcal/mole reported by Field and Franklin (1957) and McDowell *et al.* (1956). This all suggests that the $C_3H_5^+$ ion observed in this case is cyclic. To assign the 26 kcal/mole difference as excitational energy to the allyl ion seems rather unreasonable to Kiser and Hobrock (1962). They also indicated that the 68 kcal/mole for the cyclopropyl radical is appropriately greater than the 30 kcal/mole heat of formation reported by Swarc (1950). These authors allow that the value of 68 may be too high by as much as 12 kcal (or about 0.5 eV) as a maximum.

From the above example, it is apparent that low-ionization studies of molecules and ions in the mass spectrometer have much to tell us about reaction mechanisms. Those utilizing such data are urged to reconsider the accuracy with which they are currently measuring IP and AP and to use care in calibrations.

3. Determination of Bond Dissociation Energies. A reexamination of Eq. (5-7) can lead to additional information obtained by mass spectrometry:

$$M^+ + e^-(\text{fast}) \rightarrow F^+ + \cdot N_i + 2e^-(\text{slow})$$

Rewriting this as a case in which M is simply diatomic,

$$AB \rightarrow A^+ + B + e^- \tag{5-34}$$

then the energetic statement can be written as

$$AP(A^+) = \Delta H_r = \Delta H_f(A^+) + \Delta H_f(B) - \Delta H_f(AB) \tag{5-35}$$

just as has been used in the past discussions. This may also be written as

$$AP(A^+) = \Delta H_r = D(A\text{—}B) + IP(A) \tag{5-36}$$

where the more familiar expression of D (A–B) is the bond dissociation energy. If the IP(A) is known and the AP(A) is determined experimentally, then D(A–B) may be calculated. Mann *et al.* (1940) determined an AP of the OH^+ ion for reaction (5-37) to be 18.8 eV. Hence if IP(OH) = 13.53 eV,

$$H_2O + e^- \rightarrow OH^+ + H + 2e^- \tag{5-37}$$

or

$$D(HO\text{–}H) + IP(OH) = 18.8 \text{ ev} \tag{5-38}$$

then

$$D(HO\text{–}H) = 18.8 - 13.5 = 5.3 \text{ ev}$$

or

$$D(HO\text{–}H) = 122 \text{ kcal/mole} \tag{5-39}$$

This value compares reasonably well to the value of 117.5 kcal/mole given in the compilation of Cottrell (1958).

4. Determination of Proton Affinities. Lampe *et al.* (1961) have shown that it is possible to determine the upper limits of proton affinities from appearance potential studies of ion–molecule reactions involving hydrogen (or deuterium). In this case the heats of formation of the MH^+ species are calculated, and if the heat of formation of M is known, then the proton affinity (PA) for M in the reaction

$$M + H^+ \rightarrow (MH^+) \tag{5-40}$$

can be calculated by

$$PA(M) = -\Delta H_r = \Delta H_f(M) + \Delta H_f(H) - \Delta H_f(MH^+) \tag{5-41}$$

Kiser (1965g) reported that direct determination of proton affinities in molecules of the type $(C_2H_5)_nX$ which often yield ions of the type $XH_n{}^+$ and XH_{n+1}^+ can be made. These ions arise by processes such as

$$(C_2H_5)_nX + e^- \rightarrow n(C_2H_4) + XH_n{}^+ + 2e^- \tag{5-42}$$

and

$$(C_2H_5)_nX + e^- \rightarrow (n-1)C_2H_4 + C_2H_3 + XH_{n+1}^+ + 2e \tag{5-43}$$

If the AP of (XH_{n+1}^+) for the process given in Eq. (5-43) can be determined and if the heat of formation of $(C_2H_5)_nX$ is known, the heat of formation of XH_{n+1}^+ can be calculated. If XH_n is associated with the M of Eq. (5-43), then XH_{n+1}^+ can be identified with MH^+. Kiser concluded that under these conditions the proton affinity of XH_n can be obtained directly from mass spectrometric data. Haney and Franklin (1969) have reported additional proton affinities for various molecules.

5. *Relative Free-Radical Stabilities.* While the field of free-radical mass spectrometry has been active for a long time, a discussion of this work is beyond the scope of this treatise. The chapter by Harrison (1963) and by Stevenson (1963) plus the early articles of Stevenson and Hipple (1942a,b) started interest in this field. Lossing and his co-workers have published over 40 papers in a series entitled "Free Radicals by Mass Spectrometry" (McAllister and Lossing, 1969). From the discussion that follows and the involvement of heats of formation of free radicals and other thermodynamically useful data as shown above, the mass spectroscopist will realize that he is heavily dependent upon the efforts of other workers in this area.

Martin *et al.* (1966) defined the stabilization energy (SE) of a substituted methyl radical, relative to methyl, as the negative of the heat of reaction of the following hypothetical hydrogen transfer reaction:

$$CH_3X + \cdot CH_2X + CH_4 \tag{5-44}$$

$$SE(\cdot CH_2X) \equiv -\Delta H_r = D(H\text{—}CH_3) - D(H\text{—}CH_2X) \tag{5-45}$$

If the IP of $(\cdot CH_2X)$ radicals is known or can be determined, then the bond dissociation D(H–CH$_2$X) can be determined from AP measurements of the reaction,

$$CH_3X + e^- \rightarrow [CH_2X]^+ + H\cdot + 2e^- \tag{5-46}$$

Then using the same ideas as presented above,

$$AP[CH_2X]^+ = IP(\cdot CH_2X) + D(H{-}CH_2X) \qquad (5\text{-}47)$$

and still following the methodology of thermochemistry,

$$AP[H^+] = IP(H\cdot) + D(H{-}CH_3) \qquad (5\text{-}48)$$

$$AP'[H^+] = IP(H\cdot) + D(H{-}CH_2X) \qquad (5\text{-}49)$$

Subtracting Eq. (5-49) from Eq. (5-48),

$$AP[H^+] - AP'[H^+] = D(H{-}CH_3) - D(H{-}CH_2X) \equiv SE(\cdot CH_2X) \qquad (5\text{-}50)$$

Hence, without knowledge of the values of the IP of the $\cdot CH_2X$ radicals, it is theoretically possible to obtain the value of SE for these radicals by an AP difference method. The electron impact equations used are

$$CH_4 + e^- \rightarrow [H^+] + CH_3\cdot + 2e^- \qquad (5\text{-}51)$$

$$CH_3X + e^- \rightarrow [H^+] + \cdot CH_2X + 2e^- \qquad (5\text{-}52)$$

Shapiro and Turk (1969) pointed out that these reactions are violations of Stevenson's rule. Therefore the reactions should be expected to require energy in excess of the true appearance potential at threshold, thus leading to high values for the measured AP values. A second problem is that the formation of $[H^+]$ from substituted methanes is not a favored process and also may arise from primary or secondary daughter ions.

Shapiro and Turk (1969) devised a system for direct determination of a modified $SE(\cdot CH_2X)$ using the β-substituted *N*-ethylanilines. At low ionization energies these decompose primarily to substituted methyl radicals and a phenylaminomethyl cation (structure assumed):

$$\phi NH{-}CH_2CH_2X + e^- \rightarrow (\phi{-}NH{-}\overset{+}{C}H_2 \leftrightarrow \phi NH{=}\overset{+}{C}H_2) + \cdot CH_2X \qquad (5\text{-}53)$$

Their hypothetical reaction involves a transfer of a phenylaminomethyl radical instead of the customary hydrogen atom. The overall equation is

$$\begin{aligned} AP[\phi NHCH_2]^+ - AP'[\phi NHCH_2]^+ &= D(\phi NHCH_2{-}CH_3) - D(\phi NHCH_2{-}CH_2X) \\ &\equiv SE(\cdot CH_2X) \end{aligned} \qquad (5\text{-}54)$$

These workers pointed out that there are quantitative differences between the method and that of Martin *et al.* (1966) simply because the dissociation equation involved for their system is not necessarily equal to the hydrogen–methyl system. By measuring the necessary AP's and IP's the determination of the bond dissociation

$$AP[\phi NH{-}CH_2]^+ - IP(\phi NHCH_2{-}CH_2X) = D[\phi NHCH_2{-}CH_2X]^+ = \epsilon \qquad (5\text{-}55)$$

(i.e., activation energies, ϵ) is made for the molecular ions of the β-substituted *N*-ethylanilines. Shapiro and Turk (1969) made a comparison of their data with that of Martin *et al.* (1966), which is quite good, but the values reported in their article are higher than those in the literature. Two reasons for this are cited. First, Martin *et al.* (1966) utilized the making and breaking of carbon–hydrogen bonds, whereas the Shapiro and Turk (1969) method involves the making and breaking of carbon–carbon bonds. One major source of error is the problem of excess energy, which they necessarily assumed cancels out as being equal in all the reactions. They did not find metastable evidence for released kinetic energy. A big advantage of this method is that relative stabilities of radicals, difficult or impossible to measure by other means, can be determined. The chemical steps necessary to attach the radical to be studied are relatively simple to carry out. The authors pointed out that small differences such as those between ethyl and isopropyl cannot be distinguished.

Additional methods of qualitatively determining the ease of fragmentation of bonds in molecules have been used. Howe and Williams (1969a) have presented an approach wherein they express the "ease of fragmentation" in terms of the daughter-ion–parent-ion ratio at a low voltage. They also relate the metastable ion peaks to the activation energy. This approach will possibly bear fruit with time and should be kept in mind by the mass spectroscopist.

Franklin (1963) presented an excellent review of this area of mass spectrometry which, while still expanding, has been fitted into the standard physical chemistry of gaseous ionic reactions.

IV. General Principles of Fragmentation

A. Introduction

Although some fragmentation studies were made before the advent of the commercial mass spectrometer, the vast bulk of accumulated knowledge is the result of research conducted during the past twenty-five years. From such experience has come a considerable number of "rules" obtained by observation, intuition, and formal proof of a mechanism. Nevertheless, most of the rules are empirical and must be considered so until the body of knowledge of fragmentation in mass spectrometry begins to prove cohesive and fully predictable. While large gaps still exist in several areas, new workers are busy building a framework of the needed fundamental knowledge in these particular areas.

It was only natural that the early workers compared their observations of fragmentation with known chemical reactions and known stabilities of ions. Delfosse and Bleakney (1939) noted low yields of allene ions ($C_3H_4^+$) from propane and propylene. This was surprising since the ion corresponds to a stable compound. They suggested that ions with an even number of electrons might be more stable than those with an odd number. This is because all the electrons of the even-electron ion are paired, whereas the odd-electron ion must be a radical ion. Thus, the organic nature of mass spectrometry was beginning to enter into a field that had previously been largely accepted as a part of inorganic and physical chemistry. Some of the original work on organic compounds proved a bit discouraging in that rearrangements were observed and ions appeared in the mass spectra that were not expected to appear. Washburn *et al.* (1945) observed the loss of methyl groups from 2,3,3- and 2,2,4-trimethylpentane, but not from 2,3,4-trimethylpentane. They concluded that this indicated ready cleavage of the bonds adjacent to a quaternary carbon atom. They also observed a $C_4H_9^+$ ion ($m/e = 57$) in the spectrum of 2,3,4-trimethylpentane. This ion could not appear by simple bond cleavage but had to arise by a hydrogen rearrangement and multiple bond fragmentation processes. This was hardly in concordance with the prevalent idea about a single bond breaking. They also noted the somewhat disturbing fact that similar compounds could produce quite dissimilar mass spectra. Fortunately, research in mass spectrometry did not slow down. (This was probably due to the large expenditures for equipment by some of the major petroleum companies and the need to justify the instrument to the management.) McLafferty (1963a) pointed out that the proper explanation and elucidation of such phenomena can make them useful for structure determination. Armed with the knowledge that *things might be different in mass spectrometric reactions*, the early workers reconsidered some of the problems. The major differences between these reactions and known chemical reactions were that (a) a high degree of excitation accompanied the ionization process; (b) the ionizing electron removed an electron, thus changing the stability; and (c) these were gaseous ions, and no solvents were present to "soak" up the excess energies.

The problems of comparing the events in the mass spectrometer with thermal ionization, photoionization, radiation chemistry, and photochemistry require restraint and careful consideration. There are many similarities and many differences. Platzman (1962) discussed in considerable detail the implications for radiation chemistry and pointed to the use of low-energy excitation wavelengths like Hg 2537 Å or Hg 1849 Å that are especially likely to be misleading in attempting to correlate radiolysis with photolysis at a single wavelength. With the advent of photochemistry as a popular part of the modern organic chemist's background, it is likely that numerous encounters between the mass spectroscopist and the photochemist over processes will occur. Hopefully, knowledge in bridging the gaps will accumulate rapidly, resulting in a much increased understanding of these processes.

Behavior of organic materials under pyrolysis is well known. Some mass spectrometric reactions resemble those of pyrolysis. Blades and Gilderson (1960) showed that alcohols and acetals lose a molecule of water, just as they do in pyrolysis. Many similar examples are known in mass spectrometry today.

Fragmentation of the phenylalkyl ketones by electron impact shows some similarities to and some differences from their fragmentation by photochemical methods. Meyerson (1964) has shown that what the mass spectroscopist considered to be relatively large differences in fragmentation were not considered significant by the photochemist. Alpha cleavages were observed in both techniques, but the mass spectrometer introduced prominently β cleavage with a rearrangement of a γ hydrogen (a Norrish type II mechanism).

As another example of a reasonable comparison, the Kolbe electrolysis of aliphatic acids is considered to involve initial homolytic cleavage of the alkyl–carbonyl bond, similar to events in the mass spectrometer.

The extensive work of Newton and Strom (1958) showed the similarities between radiolysis products and fragmentation in the mass spectrometer. For the interested reader, a number of recent articles show teams extending research in this general area. Basson and van der Linde (1969) correlated mass spectral and radiolysis data for propanol. This was part of a series of radiation chemistry studies. La Londe and Davis (1969) reported on the photolysis and mass spectra of *trans*-cinnamimide after the compound was separated from the solution in which the photolysis was carried out. Plimmer *et al.* (1969) reported on the photochemical desulfurization of methylthio-*S*-triazines, which they indicated was similar to observed mass spectral fragmentation. Fenselau and Wang (1969) observed dimeric photoproducts in the mass spectral examination of these materials, once again indicating a difference between the two processes. It is well that mass spectroscopists are involved in this area since they can greatly assist the photochemist when these types of differences occur. A further study on aromatic ketones by Matsuura and Kitaura (1969) has appeared in the literature. They correlated mass spectra with photoreactions and indicated that competition between intramolecular hydrogen abstraction and fragmentation occurs. The abstraction leads to formation of benzocyclobutenol, thereby containing a less bulky alkyl side chain. The fragmentation leads to cleavage of the carbonyl–alkyl bond, similar to the results of Meyerson (1964).

It should be noted, and probably already has been suspected, that up to this point this treatise has taken the approach, for the most part, of the physical chemist. This is quite intentional because it is assumed that more organic chemists than physical chemists will apply the material in this volume to their work. Since the physical and thermodynamical aspects are essential to produce a working knowledge of mass spectrometry, it was decided to present this viewpoint first. McLafferty (1963b) has pointed out that there are two approaches to the study of the field, the quantum mechanical–molecular orbital or the intuitive–empirical approach.

Believing that a bit of both types of background is good and that the ultimate mass spectroscopist must be accomplished in the fields of engineering, electronics, physics, chemistry, mathematics, computer sciences, economics, and public relations, it is not too much to ask the organic chemist to read a bit of physical chemistry. Let us also point out that physical chemists are basically purists and will undoubtedly find considerable fault with the oversimplified presentation of some of the material covering their areas. It is hard to change people, whether physical chemists or organic chemists. It should be remembered that the transition from benzene ring systems to the tropylium ion was really quite an adjustment, particularly if one considers the fact that at least two generations of organic chemists had been thoroughly indoctrinated in the six-membered ring.

The organic chemist is more used to talking in terms of bond labilities, inductive effects, and resonance possibilities, whereas the physical chemist will call the first item *relative bond stabilities* and is concerned about the even- or odd-electron character of the ion. He is concerned that the homolytic cleavage of M (without electron impact, but with excess energy) can lead to two charged radical ions (odd electron), as opposed to the relative increase in stability if the process is the fragmentation of an odd-electron molecular ion to produce an even-electron fragment ion and a neutral radical. For the organic chemist, however, the mention of the inductive effect, such as occurs in the formation of the tropylium ion to the extent of 57% of Σ_i from β-chloroethylbenzene, brings out a recognizable organic process. Substituent effects that lead to increased electron densities on a given bond tend to stabilize that bond, while if the opposite occurs, that is, the reduction of the electron densities, a bond should be weakened and cleavage at that bond promoted. Typical examples are the *p*-aminophenyl ketones in which the α carbon bond between the ring and the carbonyl group is strengthened, while the presence of a *p*-nitro group leads to increased cleavage. Hammett's ρ and σ relations enter into mass spectrometry, as shall be shown later in this chapter. Stereoisomers and *cis–trans* compounds also show interestingly different mass spectra. Regardless of the approach used, one ultimately realizes that the main considerations of fragmentation in mass spectrometry deal with (a) the labilities of the bonds in the excited-state ions, (b) the relative stabilities of the potential fragment ions and the neutrals formed by competing processes, (c) the relative probabilities of entering a decomposition path, and (d) the possibility of rearrangement occurring involving concerted cleavages of bonds through a cyclic transition state. For complex molecules many paths exist, and only the major paths can be described and delineated in most cases. Because of these facts it must be restated: *The rules developed are only empirical.*

For purposes of simplifying the discussion, we shall consider only the formation of positive ions and ignore processes that may form negative ions. Gallegos and Kiser (1961) (see also Kiser, 1965d) have used the term "clastogram" to describe the fragmentation changes that occur with increasing ionization voltage. This is accomplished by plotting the abundance of the various ions produced at the given energy of the impacting electrons. An example is presented in Fig. 5-4 for the compound 1-(2-thienyl)-1-thiapentane. It should be noted that most organic substances have ionization potentials between 7 and 15 eV and that one ordinarily finds the molecular ion to be the only ion (plus isotopic ions) appearing at the lowest energy. This example is no exception. As other ions appear, the percent total ion intensity due to the M^+ ion decreases as competitive fragmentation processes occur. Only a few major ions are presented to show the general behavior. Note that the ion at *m*/*e* 116 (a rearrangement ion) becomes the ion of greatest intensity at 70 eV, while the parent-molecular ion is now only about one-half the intensity of the 116 ion. As the electron energy is increased, the molecular ions are formed with excess energy. When sufficient energy is available in some M^+ ions, fragmentation occurs and the relative intensity of the new ion moiety increases. It should be recalled that as the electron impacting energy is increased, the total number of M^+ ions also increases. Hence, both ions actually increase in intensity, even though the percent total ion intensity changes due to the fragmentation of the molecular ion.

In using the clastogram (or other low ionizing voltage data) the mass spectroscopist will become aware of the observation of Honig (1948) that within a homologous series of compounds, the ionization potential decreases with increasing chain length. There is a small effect with chain branching, but this is not easy to detect unless one has an almost monoenergetic electron beam. Note also that doubly charged positive ions usually appear at voltages above 22 eV. Further, when they do appear this represents the removal of two electrons from the same ion. The argon doubly charged ion appears at *m*/*e* 20 and, if present, can be used as a guide for the appearance potential of other ions. Wacks and Dibeler (1959) have discussed the process of doubly charged ion formation in detail for aromatic moieties, while Kaminsky and Chupka (1959) have concluded that for the alkanes a considerably smaller percentage of the ions should be doubly charged. An early report on the alkanes was made by Mohler (1948) while Meyerson and Vander Haar (1962) discussed the doubly charged ions in terms of charge separation.

A *general* case can be set up for a hypothetical molecule JKLM. The molecular ion is formed, followed by fragmentation.

$$\mathrm{JKLM} + e^{+}(\text{fast}) \rightarrow \mathrm{JKLM}^{+\cdot} + 2e^{-}(\text{slow}) \tag{5-56}$$

Fragmentation possibilities*:

$$[\mathrm{JKLM}]^{+\cdot} \xrightarrow{k_1} \mathrm{J}^{+} + \mathrm{KLM}^{\cdot} \tag{5-57}$$

$$\xrightarrow{k_2} \mathrm{JK}^{+} + \mathrm{LM}^{\cdot} \tag{5-58}$$

$$\rightarrow \mathrm{J}^{+} + \mathrm{K}^{\cdot} \tag{5-59}$$

$$\rightarrow \mathrm{J}^{\cdot} + \mathrm{K}^{+} \tag{5-60}$$

$$\xrightarrow{k_3} \mathrm{JKL}^{+} + \mathrm{M}^{\cdot} \text{ etc.} \tag{5-61}$$

$$\xrightarrow{k_4} \mathrm{LM}^{+} + \mathrm{JK}^{\cdot} \tag{5-62}$$

$$\rightarrow \mathrm{L}^{+} + \mathrm{M}^{\cdot} \tag{5-63}$$

$$\rightarrow \mathrm{L}^{\cdot} + \mathrm{M}^{+} \tag{5-64}$$

$$\xrightarrow{k_5} \mathrm{JM}^{\cdot} + \mathrm{KL}^{+}, \text{ etc. (internal rearrangement, cyclic)} \tag{5-65}$$

$$\xrightarrow{k_6} \mathrm{JM}^{+} + \mathrm{KL}^{\cdot}, \text{ etc.} \tag{5-66}$$

$$\xrightarrow{k_7} \mathrm{JL}^{\cdot} + \mathrm{KM}^{+}, \text{ etc. (scrambling of atoms)} \tag{5-67}$$

Even though a very minor event at low source pressures, an ion may collide with a neutral molecule causing ionization.

$$[\mathrm{JKLM}]^{+} + [\mathrm{JKLM}]^{0} \rightarrow [\mathrm{JKLM \cdot JKLM}]^{+} \tag{5-68}$$

which in turn may decompose to

$$[\mathrm{JKLMJ}]^{+} + \mathrm{KLM}\cdot \text{ etc.} \tag{5-69}$$

If an ion–molecule process is thought to be entering the fragmentation study, the sample pressure should be doubled and then redoubled so as to see if the ion in question, and hence the process, is pressure dependent. If so, a second-order process is assured, and the ion–molecule reaction is occurring.

* This scheme is adapted from K. Biemann (1962). "Mass Spectrometry: Organic Chemical Applications." McGraw-Hill, New York. Used with permission of McGraw-Hill Book Company.

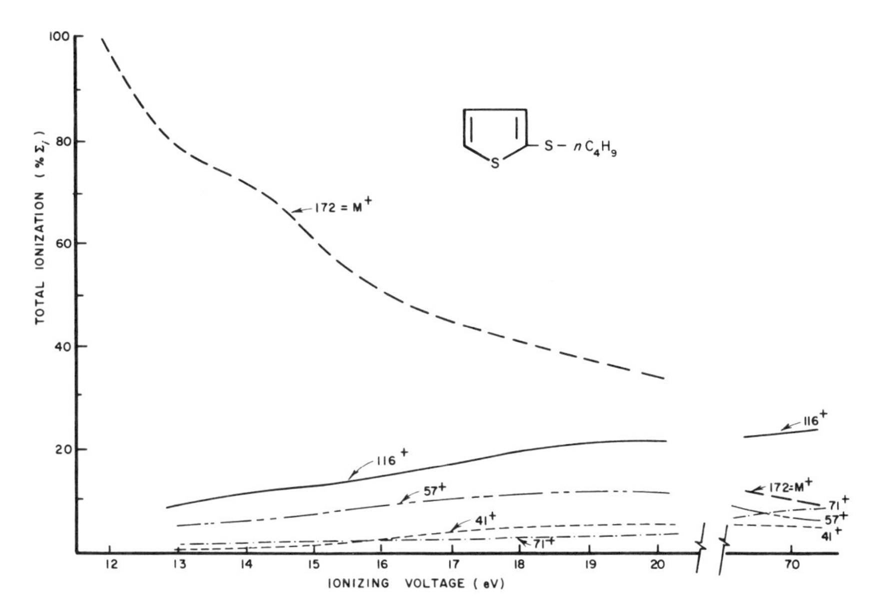

Fig. 5-4. A typical clastogram showing the change in abundance of selected ions with ionizing voltage.

The first four processes shown comprise simple cleavage, the remainder requiring a rearrangement. Field and Franklin (1957) pointed out that the energy of activation for rearrangements in ionic moieties is generally quite low as compared to neutrals, and thus one encounters these processes frequently in mass spectrometry. They also indicated that the various rates depend upon both the energy and entropy of activation for the intermediate state in the process. It should be noted that not only are the fragment ions formed from the molecular ion at rates dependent upon the various k's, but they also decompose to further fragments by additional k's. For the neutrals in the various processes given, the only certain identification is via the observation of a metastable peak for the transition to produce that neutral species. The nonbonding electron pair in heteroatomic systems is a favorable site for ionization. This is also in accordance with the comparative strengths of the molecules as Lewis bases. The molecule ions are more stable and, therefore, of higher intensity if the k's are small. Concurrent with this, the energies of activation for fragmentation processes from that ion are high. The molecular ions are usually more prominent in molecules that are unsaturated or contain groups such as $C{=}O$, NO_2, $C{\equiv}N$, and SO_2 as compared to molecules containing groups such as NH_2, SH, and OH. In some of these cases, when the molecular ions are of very low intensities, chemical conversions, as given in Chapter 4, to produce a derivative with a larger molecular ion intensity are necessary. Yeo and Williams (1968) have discussed molecular ion abundances in relation to their ionization potentials and suggested the selection of a derivative with a very low ionization potential.

In the previous discussions of this chapter we have been concerned with the ionization process, the theory of mass spectra, and the energetics of electron impact processes. It is now appropriate to apply these ideas and develop the so-called "rules" of fragmentation which can lead to interpretation of the mass spectrum. Fragmentation of the molecular ion indeed leads to the eventual mass spectrum observed. Thus all of the fragments can be considered as secondary or tertiary (or further) decompositions arising from this primary ion. The stability, composition, and structure of the molecular ion must be considered when examining the fragmentation reactions that might arise. Determination of the composition of the ion is discussed in Chapter 6. For the reader wishing to review organic reactions and mechanisms, the texts of Liberles (1968) and March (1968) may be helpful.

B. Variation in Molecular Ion Intensities

The chemical composition and structure of the molecule has a great deal to do with the ultimate stability of the resulting molecular ion of the species. Table 5-2 gives data for an assortment of molecular ions, broken down into four categories according to atomic content. Considering first the characteristics of the hydrocarbon ions, a study of the table allows one to estimate the molecular ion (M^+) intensity in terms of the total ion intensity that should be expected for a given hydrocarbon type of compound. Hence, even without a reference spectrum, the mass spectroscopist, with a little practice, can suggest a probable strength for the molecular ion intensity. In a sense, the intensity of the ion M^+ indicates that a sizable energy level change is required for fragmentation to occur. The unsubstituted fused-ring systems provide molecular ions which literally "have no place to go" as far as fragmentation is concerned. The stability of such fused-ring aromatic systems is well known, and the fragmentation of the rings in these systems requires considerable energy in view of the π-aromatic character and the bond strengths. The change caused by "unfusing" the rings is reflected in the considerable change in molecular ion intensity from naphthalene to a biphenyl system. The most noteworthy item in the upper portion of the table is the great reduction in M^+ intensity for the multiring systems which are substituted with long-chain alkyl groups. This arises because the presence of the alkyl chains now provides lower-energy fragmentation routes. Common among these, as will be shown, is cleavage of the bond beta from the ring system in the alkyl chain. It is interesting to note that the hexahydropyrene joins this type of change with the saturation of two of its fused rings. Once the alkyl chain is fairly long, the effect tends to level off. Compare the 1-pentadecylphenanthrene with the 9-dodecylphenanthrene as an illustration of this.

Steric effects are also important. As an example, compare the 7-*n*-butyl-1-*n*-hexylnaphthalene ($M^+ = 16.5\%$) with the 2-*n*-butyl-3-*n*-hexylnaphthalene ($M^+ = 12.2\%$). The alkyl chains are adjacent in the 2-3 compound and isolated in the 7-1 compound. The approximately 25% reduction in the molecular ion intensity can be attributed to the "proximity" problem arising in the 2-3 compound. Steric effects for the *cis*- and *trans*-decalin compounds, while small in terms of the total ionization, are still significant and in line with what one might anticipate. This is even more striking for the 1-ethyl-*trans*-3-*n*-heptylindane where the M^+ intensity is only 2.34%. It should be especially noted for this compound that the low intensity comes about in part because of the long-chain (*n*-heptyl) substitution in addition to the configuration.

Another recognizable generalization is the further reduction in the M^+ ion intensity with the addition of saturation as shown for the tetralin–naphthalenes and the decalin–tetralins.

Changes to be expected with chain branching become evident by comparing the *n*-butylbenzene and the *t*-butylbenzene, the difference between the two M^+ intensities being 1.2%. One may ask about variations among spectra from different instruments, and the answer is that indeed the variations might be great among instruments with different geometries or even differing ion sources. From the qualitative viewpoint, the relationship will still follow these trends. Two values for *n*-butylbenzene are given, both obtained from spectra run on Dempster geometry instruments with all other conditions being equal. This allows some consideration of the variation between instruments of the same design and geometry. Thus, the difference of 1.2% between the *n*-butyl- and *t*-butylbenzenes is considered to be quite real. This difference persists between the 2-*n*-butylthiophene (9.96%) and the 2-isobutylthiophene (8.91%). The *t*-butyl compounds are expected to show the greatest effect since they are the most highly branched.

When considering the paraffins, the chain branching effect becomes more important because the paraffin ion intensity for the molecular ion is usually small for the *n*-alkanes. The M^+ intensity for *n*-decane is only 1.4% of the total ionization. Detecting the molecular ion for a highly branched alkane should become very difficult. The M^+ ion of the 3,3,5-trimethylheptane approaches zero intensity. In the case of neopentane, the novice is really challenged. Even at low ionization voltages, the parent ion may be very difficult to detect. Note that chemical treatment will not help in this case.

Biemann (1962a) utilized the compounds *allo*-ocimene and myrcene to illustrate the problems with the olefins. Each compound contains three π bonds and, as a consequence, at least a moderately intense M^+ ion should be

TABLE 5-2

SOME MOLECULAR ION INTENSITIES[a]

CH	$\%\Sigma_i$	CHO	$\%\Sigma_i$	CHN	$\%\Sigma_i$	CHS	$\%\Sigma_i$
						2-Phenylthiophene (Meyerson and Fields, 1968)	49.8
Naphthalene (Conoco)	45.6					3-Phenylthiophene (*ibid.*)	48.7
						Benzothiophene (API 917)	45.0
Phenanthrene (Conoco)	42.9			Acridine (API 639)	41.4	Dibenzothiophene (API 2095)	42.5
Pyrene (API 959)	40.8			Quinoline (Conoco)	39.6		
		Dibenzofuran (MCA 101)	38.9	Carbazole (MCA 106)	39.0		
				2,4-Dimethylquinoline	35.6		
Biphenyl (API 171-m)	33.5			Indole (API 623)	34.8		
		Phenol (MCA 66)	29.4	Aniline (API 1232)	34.4		
		Benzofuran (MCA 86)	28.0	Pyridine (API 1534)	27.6		
		Acetophenone (MCA 87)	25.7	8-*n*-Propylquinoline	26.2		
Acenaphthene	23.5	Furan (Conoco)	25.0			Benzenethiol (API 1387)	23.8
2-*n*-Dodecylpyrene (API 131-m)	17.5			*N*-Methylpyrrole (API 248-m)	19.3		
9-*n*-Dodecylanthracene (API 11-m)	16.7						
				2-Methylpyrazine (API 249-m)	18.4	2-Methyl-7-ethylbenzo[*b*]-thiophene (API 1951)	15.3
7-*n*-Butyl-1-*n*-hexylnaphthalene (API 206-m)	16.5						
				2,5-Dimethylpyrazine (API 250-m)	16.7	1-Phenyl-1-(2-thienyl)methane (API 2165)	13.2
9-*n*-Dodecylphenanthrene (API 10-m)	16.2						
						2-(1-Thiapentyl)naphthalene (Foster *et al.*, 1970)	12.8
Hexahydropyrene (API 131-m)	16.1						
1-Pentadecylphenanthrene (API 12-m)	15.5					1-(1-Thiapentyl)naphthalene (Foster *et al.*, 1970)	12.4
2-*n*-Butyl-3-*n*-hexylnaphthalene (API 100-m)	12.2					2-*n*-Propylbenzo(*b*)thiophene (API 1850)	11.0

2-*n*-Decylindan (API 178-m)	10.9						
Tetrahydronaphthalene (API 201-m)	10.6			Piperidine (API 2102)	10.2	4,5,6,7-Tetrahydro-1-(2-Benzo-[*b*]thienyl)-1-thiapentane (API 2170)	10.0
6-*n*-Decyltetralin (API 94-m)	10.2						
n-Butylbenzene (Biemann[b])	8.26			1-Methylpyrrolidine (API 174-m)	9.78	1-(2-benzo[*b*]thienyl)-1-(thiapentane (API 2166)	9.50
n-Butylbenzene (API 110-m)	8.22					2-*n*-Butylthiophene (uncertified spectra)	9.96
trans-Decalin (API 234-m)	8.22					2-Isobutylthiophene (uncertified spectra)	8.91
6-*n*-Butyltetralin (API 93-m)	7.23					1-(3-Benzo[*b*]thienyl)-1-thiapentane (API 2167)	8.65
t-Butylbenzene (Biemann)	7.00					1-(2-Thienyl)-1-butanone (Foster and Higgins, 1969)	7.79
cis-Decalin (API 233-m)	6.72					1-(2-Thienyl)-1-thiapentane (API 2124)	7.72
allo-Ocimene (Biemann)	6.40					2-*n*-Hexyl-5-*n*-heptylthiophene (API 2074)	7.30
1,4-Dimethyl-5-*n*-octyldecalin (API 177-m)	3.67					2-(2,2-Dimethylpropyl)-thiophene (API 2047)	6.63
5,8-Dimethyl-1-*n*-octyltetralin (API 134-m)	2.57			*n*-Octylamine (API 246-m)	3.67	Dicyclopentylthiamethane (Dooley *et al.*, 1970)	5.24
1-Ethyl-*trans*-3-*n*-heptylindan (API 200-m)	2.34			2-Methylpiperazine (Biemann)	2.14	1-(2-Thienyl)-1-thiahexane (API 2125)	5.15
n-Decane (Biemann)	1.41	Methylnonanoate (Biemann)	1.10	Diamylamine (Biemann)	1.14	6-Thiaundecane (Biemann)	3.70
Myrcene (Biemann)	1.00	3-Nonanone (Biemann)	0.50	*n*-Decylamine (Biemann)	0.50	2-(Cyclopentylthio)thiophene (API 2125)	3.35
Cyclododecane (Biemann)	0.88	Diamyl ether (Biemann)	0.33			2-Ethyl-5-(2-ethylbutylthio)-thiophene (API 2131)	2.89
Pinane, 2,6,6-trimethylbicyclo-[3.1.1]heptane (MCA 46)	0.45	3-Nonanol (Biemann)	0.05			1-Decanethiol (API 2086)	2.53
3,3,5-Trimethylheptane	0.007	*n*-Decanol (Biemann)	0.002			1-(2-Thienyl)-1-hexanone (Foster and Higgins, 1969)	2.07
2,2-Dimethylpropane	0.000						

[a] Equal intensities are approximately in line.
[b] Biemann (1962b) is the source of all Biemann data in this table.

expected. The location of the double bonds, however, has a good deal to do with the intensity. The conjugated system of the *allo*-ocimene leads to a considerably more intense parent ion (6.40%) as compared to the isolated double bonds in the myrcene with a low-intensity molecular ion (1.00%). The

$$(H_3C)_2C{=}CH{-}CH{=}CH{-}C(CH_3){=}CH{-}CH_3$$

allo-Ocimene

$$(H_3C)_2C{=}CH{-}CH_2 \,\vdots\, CH_2{-}C({=}CH_2){-}CH{=}CH_2$$

Weak bond

Myrcene

myrcene has a weak bond that can fragment easily. It is expected that the general principles given above will be easily applied to molecular structure situations encountered by the mass spectroscopist. There will be some rather interesting exceptions, some of which will be touched upon in later discussions in this chapter.

A few examples of compounds from the oxygenates and the nitrogen group are included to permit cross-comparisons and points of reference for other compounds. The presence of the hetero atom generally increases the M^+ intensities for the oxygen and nitrogen compounds only when included in aromatic or π-bonded systems. For the more aliphatic situations, the converse is true. The 3-nonanone represents a decrease over the alkane, while the diamyl ether represents an even further drop in intensity compared to a normal alkane. The alkanols have notoriously weak molecular ions and, as shown in the previous chapter, usually require chemical treatment to produce a compound of larger molecular ion peak to even ascertain the molecular weight. Generally speaking, the molecular ion should be about 1% to insure detection. If of lower intensity, the uncertainty of identification as a molecular ion peak increases rapidly. The *n*-decylamine shows some improvement over the corresponding alcohol; however, shortening the chain length two carbon atoms proves a more marked increase. In general, the vertical changes among compounds in the oxygen and nitrogen columns can be rationalized in terms of the principles already given.

The extensive number of sulfur compounds is given to show the behavior of the sulfur atom itself. Just as mentioned in Chapter 3 in connection with total ion intensity data, this increase in molecular ion intensities for the sulfur compounds seems to be associated with the increase in ionization cross section. This appears to be the case, particularly if one considers the series of compounds, *n*-decane, *n*-decanol, *n*-decylamine, and 1-decanethiol. The last-named compound exhibits an M^+ percent of 2.53, considerably larger than that of the alcohol, amine, or even the *n*-decane. Further, if we consider the comparison of naphthalene (44.3%) with the benzothiophene (45.0%), then the observed M^+ percents are in fair agreement with the ionization-cross-section ratios. This supports the ionization-cross-section argument. Included in Table 5-3 are data on quinoline, benzofuran, biphenyl, and 2-phenylthiophene which

TABLE 5-3

IONIZATION CROSS SECTIONS COMPARED TO INTENSITIES OF MOLECULAR IONS (%M^+)

Molecule	Ionization cross section	Intensities (%M^+)
2-Phenylthiophene	62.40	49.8
Biphenyl	59.92	33.5
Benzothiophene	51·08	45.0
Naphthalene	49.60	44.3
Quinoline	48.28	39.6
Benzofuran	42.57	25.0

compare the M^+ intensities with the total ionization cross section. It should be noted that these molecules are all either fused-ring systems or joined aromatic ring systems such as the biphenyl. The comparisons are excellent except for the biphenyl where obviously some other effect is taking over. A plot of the last four compounds proves to be almost linear, suggesting that further comparisons could be made. We shall stop here rather than enter into an area that could develop into a series of seminars concerning the "massachor"; it might well be interesting and informative but could lead to little in the way of practical results. Suffice to say that these authors believe that the relationships here are only approximate and reflect the differences in the orbital bonding conditions of the sulfur atom; that is, the ionization cross section for sulfur in decanethiol may be a considerably different value than for the sulfur atom in an aromatic system. Research in this area might well be warranted to improve understanding of mass spectra.

In addition to making the comparison of the sulfur compounds to the hydrocarbons as above, one can observe, in Table 5-2, some further subtle differences between these compounds. The addition of alkyl groups to the benzothiophene nucleus reduces the parent-ion intensity to 11.0% in the case of the 2-*n*-propyl compound, the longest-chain spectrum available in the literature. This reduction is somewhat below the 12.2% found for the 2-*n*-butyl-3-*n*-hexylnaphthalene. In addition, the substitution of a sulfur in the 1 position of the alkyl group to produce the benzothienyl thiaalkanes reduces the M^+ ion intensity to 9.54%. Similarly, the change of the alkylnaphthalene to a 1-thiaalkylnaphthalene brings the two isomers presented down to 12.4% (α-substituted) and 12.8% (β-substituted). Hence, the thia sulfur atom produces a good location for fragmentation to take place. Somewhat surprisingly, the saturation of the benzothienyl thiapentane compound increases the M^+ ion intensity to 10.1%, about a 0.5% rise. This may only be a case of "push–pull" effect on fragmentation, as will be discussed a little later. Another point of comparison possible from this portion of the table is the apparently real but slight position effects in the case of the 2- and 3-substituted positions with respect to the sulfur atom. The 1.1% lower M^+ peak for the 3-phenylthiophene would be expected by the organic chemist on the basis of his ideas about stability of the positions. This is also supported by a difference of about 0.9% in the M^+ ion intensities of the 2-benzothienyl and 3-benzothienyl thiaalkanes reported. This is quite reassuring. As in the case of the benzenes, the substituted thiophenes display the effects of chain branching, culminating in the 2,2-dimethylpropylthiophene (neopentylthiophene) where the M^+ is 6.63% compared to 9.96% for the 2-*n*-butylthiophene. A disubstituted 2-*n*-hexyl-5-*n*-heptylthiophene of considerably higher molecular weight still shows only 7.30% of M^+, considerably lower than the monoalkylthiophenes, neopentyl excepted. An interesting comparison can be made between the 1-(2-thienyl)-1-butanone and the 1-(2-thienyl)-1-thiapentane. The M^+'s are approximately equal, probably within experimental error. Fragmentation processes that closely resemble one another are taking place, as will be demonstrated later, so that the oxygenated atom has not yet lowered the M^+ intensity. By lengthening the alkyl chain just two carbons, the hexanone compound shows an M^+ of only 2.07%, well down into the "oxygenated aliphatic" category. For the more alkyl types of sulfur compounds, the relationships closely parallel those for the hydrocarbons but with somewhat larger M^+ ions for each corresponding compound. Even the disubstituted highly branched 2-ethyl-5-(2-ethylbutylthio)thiophene shows an M^+ of 2.89%, about where one might expect such a structure to occur in the table.

Further study of the table and additions of compounds of interest to the individual worker will prove of value in research work and routine analysis. Accumulation of such data is encouraged. The prime use will be in checking the fit of a synthetic standard (made in your laboratory) or an unknown into the M^+ pattern. The reader should note that considerable work is saved if the presentation of data is done in the total ion intensity manner and in the array format, to be fully explained in Chapter 7.

C. Major Fragment Ions

The mass spectroscopist is tempted to examine the 70-eV mass spectrum and on the basis of rather simple chemistry decide the major processes that occur in fragmentation of the molecular ion to produce the large fragment ions. This can often be misleading, and an example is the appearance of alkane spectra. From the time of the earliest correlation studies of Washburn *et al.* (1945), mass spectroscopists have considered the mass spectra of the alkanes to be the most simple spectra possible or the most complex, depending upon the point of view (Meyerson, 1965). The naiveté of our earlier consideration of the ethane spectra is that the C_2 *n*-alkane, ethane, is hardly representative of the series of *n*-alkanes and in fact displays a number of fragmentations which become rather minor as the molecular weight increases in the series. The first problem with the ethane spectrum is that only 5.9% of the ions formed come from the breaking of the C–C bond. Thus, one might get the idea that *n*-octane behaves the same way and produces essentially only ions clustered around the molecular ion. These ions could arise from the stripping off of successive hydrogen atoms. This is far from the real case, in which the preponderance of fragment ions occurs with three to five carbons in the charged moiety. As a matter of fact, the sum of the ions occurring at *m*/*e* 29, 43, 57, 71, and 85 and the higher homologous peaks of this (+1) series (C_nH_{2n+1}) accounts for from two-thirds to more than three-fourths of the total ionization of the paraffin compounds. While this does indeed suggest to the mass spectroscopist that he can use this fact for class identification purposes, he might well be looking at a series of ions of the form CH_3–CO^+ (or homologs). In Chapter 9 some examples are given in which the series shows more than one maximum, indicating chain branching.

In the work of Magat (1951) and Magat and Viallard (1951), several important features were established. Probably most important was the proof that during ionization and fragmentation the singly charged molecular ion decomposed into a single charged ion and a neutral moiety, as was shown earlier in this chapter. They also found the abundance of the H^+ and H_2^+ to be small, indicating the preference of charge retention with the large fragment whenever a C–H bond was involved in the fragmentation. They also noted that the ratios of the abundances of $CH_2^{+\cdot}$ to CH_3^+ and $C_2H_4^{+\cdot}$ to $C_2H_5^+$ decreased for successive members of the paraffin (and olefin) series with increasing chain length and eventually tended to approach a constant value. This supported the more logical thermodynamic view that only one C–C bond was broken during a given decomposition event rather than the view held by some workers that the ionized moieties seemed to literally fragment into any and all possible combinations.

It should be recalled by the reader that the observed mass spectrum is an averaged recording of the results of a host of fragmentation events, most of which represent only the results of a relatively few successive fragmentations to a given moiety. For example, a sample of *n*-heptane after ionization of the molecular ion can produce cleavages as shown to produce fragments of *m*/*e*

15, 29, 43, 57, 71, 85, and by a C–H cleavage, the *m/e* 99. Cleavage of the methyl group is not energetically favored, and hence the ions of the 29–57

```
    ¦ 99 ¦ 85 ¦ 71 ¦ 57 ¦ 43 ¦ 29 ¦ 15 ¦  1
    ¦    ¦    ¦    ¦    ¦    ¦    ¦    ¦
    ¦ H  ¦ H  ¦ H  ¦ H  ¦ H  ¦ H  ¦ H  ¦
    ¦ |  ¦ |  ¦ |  ¦ |  ¦ |  ¦ |  ¦ |  ¦
  H-+-C--+-C--+-C--+-C--+-C--+-C--+-C--+-H
    ¦ |  ¦ |  ¦ |  ¦ |  ¦ |  ¦ |  ¦ |  ¦
    ¦ H  ¦ H  ¦ H  ¦ H  ¦ H  ¦ H  ¦ H  ¦
    ¦    ¦    ¦    ¦    ¦    ¦    ¦    ¦
  1 ¦ 15 ¦ 29 ¦ 43 ¦ 57 ¦ 71 ¦ 85 ¦ 99 ¦
```

series predominate. With so many other avenues of fragmentation that will remove the excess energy of the excited species, the breaking of the C–H bond to produce a nonoscillator, the neutral H atom, simply is not favored. Viallard and Magat (1949) showed that by the time *n*-octane was reached, only 0.008% of the ions produced were formed by such a C–H cleavage with no breaking of a C–C bond. Obviously then the ethane molecule with over 80% of the ions produced by cleavage of a C–H bond is certainly *atypical* of the hydrocarbons.

Ethane has been presented here to illustrate that with small molecules, *fragmentation events may occur which are not at all representative of the series* but which must be dealt with by the mass spectroscopist at all times. This is why so many workers have received encouragement to investigate a series out to C_6–H_8 alkyl chain lengths so as to make their research more useful. It is the habit of organic chemists to consider that they have examined the "soxhylates" when they have reported the methoxy, acetoxy, hydroxy, formyl, an amine, nitrile or mercaptan derivatives, and perhaps a few halides. They may well report on the mass spectra of these compounds, but without the availability of a long chain or two (with quite a few oscillators to absorb excess energy), they are not able to generalize their results to aid other spectroscopists to the fullest extent possible. A further reason for the longer alkyl chains is the problem of rearrangements, which will be discussed a bit later. Viallard and Magat (1949) also noted the decrease of parent-ion intensity with increase of the source temperature. They also noted the reduction was more pronounced with increasing branch chain length.

Thus, it became clear fairly early in the research on mass spectra and the correlation of spectra with structure that the spectroscopist had to explain the mass spectrum in terms of a number of reactions which led to the various large ions observed but that this did not have to be done in a single step or in an unusual reaction step. Rearrangement ions did require a bit more understanding and research. An attempt is made in the next section to condense the knowledge gained from examination of mass spectra and correlation studies through the years and to summarize the main types of fragmentation reactions that occur.

D. Fragmentation Reactions—Classification Schemes

It is customary for one to learn about a series of reactions in terms of a classification scheme. We know of no better way to accomplish this, and hence we utilize a system similar to that developed by Biemann (1962b) in his fine text. Some additions and new supporting material are provided, and the reader is thus encouraged to refer to the original work for some details not included here. While almost any classification system might prove useful, we have selected Biemann's system because of its ease of development and because to some extent, it parallels, historically, the significant growth of knowledge of mass spectra. It has been used in teaching classes for seven years. Throughout the fragmentation discussions we shall employ the symbols and abbreviations in use in the *Journal of Organic Mass Spectrometry* and developed in an editorial review on nomenclature (Budzikiewicz, 1969). As these various symbols are introduced, they will be explained. A number of these abbreviations have already been introduced in this chapter, i.e., AP and IP. A summary is presented first, followed by discussion and examples of each type of fragmentation.

1. Simple Cleavage (Eight Simple Types of Cleavage)

$$(A_1)\quad \left[\text{—C—C—}\right]^{+\cdot} \longrightarrow \text{—C}^{+} + {}^{\cdot}\text{C} \qquad (5\text{-}70)$$

This is the cleavage of an odd-electron ion to produce an even-electron ion and a neutral radical.

$$(A_2)\quad \left[\text{—C}_{\gamma}\text{—C}_{\beta}\text{—C}_{\alpha}\text{—}\right]^{+} \longrightarrow \text{—C}_{\gamma}^{+} + \text{C=C} \qquad (5\text{-}71)$$

This cleavage, as indicated by the wavy line, of an even-electron ion produces another even-electron ion and a neutral olefin, or other neutral particle. An example of the latter is the ejection of a neutral CO particle from an ion as shown in Eq. (5-72).

$$C_6H_5 \nmid C{\equiv}\overset{+}{O} \longrightarrow C_6H_5^{+} + CO \qquad (5\text{-}72)$$

$$(A_3)\quad \left[C{=}C{-}C \nmid C \right]^{+\cdot} \longrightarrow C{=}C{-}C^{+} + {}^{\cdot}C{-} \qquad (5\text{-}73)$$

The cleavage of a bond beta to the double bond in an allylic cation leads to a resonance-stabilized even-electron ion and a neutral fragment. This also extends to dienes and ynes:

$$C{=}C{-}C \nmid C{-}C{=}C \quad \text{and} \quad C{\equiv}C{-}C \nmid C{-}C$$

$$(A_4)\quad \left[C_6H_5{-}CH_2 \nmid R \right]^{+\cdot} \longrightarrow \left[C_6H_5{-}CH_2 \right]^{+} + R^{\cdot} \longrightarrow \left[C_7H_7^{+} \right] \qquad (5\text{-}74)$$

This is the cleavage of a bond beta to a π-bonded ring system wherein the odd-electron molecular ion produces an even-electron ion and a neutral fragment. The ring plus CH_2 group usually undergo a rearrangement producing a ring expansion. This will be discussed in detail later.

$$(A_5)\quad \left[{-}C{-}X \right]^{+\cdot} \longrightarrow {-}C^{+} + {}^{\cdot}X \qquad (5\text{-}75)$$

This is very similar to case A_1 but deserves to be classified separately because the A_1 fragmentation scheme is so prevalent among alkyl types of molecules and ions where the subsequent events are very different from those involving a hetero atom or group. The X is commonly a halide, OR, SR, or NR_2 ($R = H$ or alkyl, aryl, etc.).

$$(B)\quad \left[R{-}\underset{|X}{C}{-}R \right]^{+\cdot} \longrightarrow R^{\cdot} + {}^{+}\underset{|X}{C}{-}R \longleftrightarrow \underset{\|X^{+}}{C}{-}R \qquad (5\text{-}76)$$

This fragmentation produces an even-electron fragment ion from an odd-electron ion. A neutral radical, usually alkyl, is also produced. As shown by the two-headed arrow, a pair of electrons can be shifted from the hetero atom localizing the charge on that atom.

$$\text{(C)}\quad R'\text{—}\underset{\underset{|O|}{\|}}{C^+} \longleftarrow \left[R'\text{—}\underset{\underset{|O|}{\|}}{C}\text{—}R^+\right]^{+\bullet} \longrightarrow {}^+\underset{\underset{|O|}{\|}}{C}\text{—}R \longleftrightarrow \underset{\underset{^+O|}{|||}}{C}\text{—}R \qquad (5\text{-}77)$$

This is a cleavage of an odd-electron ion to produce an even-electron oxygenated species and a neutral radical. The effect of different R groups may be noted here. Also this cleavage produces the oxygenated species from an even-electron ion if two electrons are shifted, which subsequently undergoes the A_2 type of cleavage to eject the neutral CO. Note that either product ion can have resonance forms involving the oxygen–carbon triple bond.

$$\text{(D)}\quad [\text{cyclohexene}]^{+\bullet} \longrightarrow [\text{butadiene}]^{+\bullet} + \text{C}{=}\text{C} \qquad (5\text{-}78)$$

This is the retro-Diels–Alder (RDA) ejection of a neutral olefinic particle. It arises from an odd-electron ion and produces an odd-electron ion fragment which is considered less stable because of its odd-electron character. Like A_2, this type of cleavage involves the shifting or rearranging of two electrons but differs from A_2 in that two bonds are cleaved. Two new bonds are made, however, keeping the energetics near a balance. This same advantage could be claimed for A_2 in that a new bond is made for the one that is broken. It has been customary for mass spectroscopists to keep reactions involving a two-electron shift classified with cleavages as opposed to rearrangements wherein one or more atoms wind up in a new moiety attached to different atoms than they were in the molecular ion (or precursor ion, if not the molecular ion).

2. Rearrangements

$$(E_1)\quad \left[(C)_n\begin{matrix}\text{C—X}\\ \\ \text{C—H}\end{matrix}\right]^{+\bullet} \longrightarrow \left[(C)_n\begin{matrix}\text{C}\\ |\\ \text{C}\end{matrix}\right]^{+\bullet} + HX \qquad (5\text{-}79)$$

Elimination of a neutral molecule involving a cyclic intermediate and two electron shifts (not a pair, but each singly as indicated by the single-headed arrow or "fishhook") from some molecular ions occurs in a random manner. A most typical example is the elimination of water from an alcohol.

$$(E_2)\quad \left[\text{cyclic B–D(–R)–H–A}\right]^{+\cdot} \longrightarrow \left[\text{B}{=}\text{C–C–A}\right]^{+\cdot} + \begin{matrix}D{-}R\\ |\\ H\end{matrix} \qquad (5\text{-}80)$$

Elimination of a neutral molecule between adjacent groups on a *cis* double bond or an aromatic system is the second reaction of this type observed in mass spectrometry. The odd-electron ion produces another odd-electron ion and a neutral molecule. *ortho*-Disubstituted aromatic systems frequently exhibit such behavior, and it is referred to as an "*ortho*" effect. One of the functional groups must be able to provide a transferable hydrogen while the other must provide the remaining portion of the neutral molecule that is being eliminated.

$$(F)\quad \left[R{-}CH_2{-}CH_2{-}CH(X)\right]^{+} \longrightarrow \begin{matrix}R & & H\\ & C{=}C & \\ H & & H\end{matrix} + {}^{+}CH_2{-}\bar{X} \longleftrightarrow H_2C^{+}{=}X \qquad (5\text{-}81)$$

This is the first rearrangement given for fragment ions. The next two examples can also be regarded as primarily for fragment ions. In reaction (5-81) the ion must reach the resonance form shown, whence the rearrangement process can proceed, probably involving a four-centered transition state. This type of rearrangement produces ions in the same series as the fragment ion, thus confusing the analytical situation to some extent. The reason is that a neutral olefin is ejected, which always has a mass of $(n + 1) \times 14$, and thus the resulting ion will be the $(F - 28)^+$, $(F - 42)^+$, $(F - 56)^+$, etc., all in the same series.

$$(G)\quad R{-}CH_2{-}CH_2{-}\overset{+}{X}{=}CH_2 \longrightarrow \begin{matrix}R & & H\\ & C{=}C & \\ H & & H\end{matrix} + H\overset{+}{X}{=}CH_2 \qquad (5\text{-}82)$$

After simple fission has occurred in the molecular ion of an ether, thiaalkane, or a secondary or tertiary amine, a stable "onium" ion is formed. This can eliminate a neutral olefin moiety, provided only that the alkyl group remaining is two or more carbon atoms long. The net effect is that an even-electron ion produces another smaller-mass even-electron ion, and again if a simple olefin is eliminated, the ion formed is still in the same series as the precursor ion.

$$\text{(H)}\quad \overset{\curvearrowleft H \curvearrowright}{A\quad E^{+\cdot}} \; B-C-D\!=\!E,\ D-G \longrightarrow \underset{B}{\overset{A}{\|}} + H-E^{+\cdot}-D(=C)-G \qquad (5\text{-}83)$$

This is the well-known McLafferty rearrangement process. It is depicted in such general terms because of its widespread occurrence. A multiple bond must be between D and E, and a hydrogen atom (H) must be available on the atom gamma from the D–E multiple bond. The atoms at A–E and the group G can be quite varied. A number of examples will be discussed in detail.

E. Examples of Typical and Nontypical Fragmentations

1. *The A_1 Type of Fragmentation*

The alkanes can be used to illustrate some A_1 type of cleavage behavior as well as some A_2 cleavages in some cases.

In Table 5-4 are presented the partial mass spectra of the six isomeric hexanes, API-44 Serial Nos. 147–151. An examination will show a number of points of interest to the mass spectroscopist. First, the $M^{+\cdot}$ ion of the normal compound is of largest intensity, as one might anticipate, there simply being fewer pathways to cleavage than there are for the other isomers. Conversely, the $M^{+\cdot}$ ion of the most highly branched species, the 2,2-dimethylbutane (or neohexane), has a very weak intensity. Interestingly, the other dimethylbutane shows a molecular ion of the same intensity as that of the 2-methyl- and 3-methylpentanes. One can conclude that the presence of *gem*-dimethyl branching leads to this sort of behavior. As a corollary, then, by an A_1 cleavage the M – 15 peak is the largest for the 2,2-dimethyl compound at *m/e* 71 (the *m/e* 15 is also larger than for the other compounds). Now the 2,3-dimethyl compound has a bit larger *m/e* 71 peak when compared to the others except for the 2-methylpentane, which at first glance is surprising in that it is fairly large. Examination of the skeletal structure shows this to be strictly terminal branching,

```
C—C—C—C—C      as compared to      C—C—C—C
  |                                  |  |
  C                                  C  C
```

but one could reason that the latter molecule has two terminal methyls, one at each end. This is quite true, but the dominant factor in these spectra is the fact that A_1 cleavage is favored to occur at a point along the chain where branching occurs. This leads then to the production of secondary (or tertiary) carbonium ions as compared to the primary ions. The large ions formed at *m/e* 57 for the 3-methylpentane and the 2,2-dimethylbutane are readily understood on the basis of A_1 cleavage with charge retention at the point of branching. In the former case the ion would be the *sec*-butyl ion while in the latter

TABLE 5-4

PARTIAL MASS SPECTRA—ISOMERIC HEXANES[a]

m/e	n-Hexane C—C—C—C—C—C	2-Methylpentane C—C(C)—C—C—C	3-Methylpentane C—C—C(C)—C—C	2,2-Dimethylbutane C—C(C)(C)—C—C	2,3-Dimethylbutane C—C(C)—C(C)—C
15	5.84	4.76	4.48	9.25	5.62
27	45.4	23.4	30.2	32.8	21.8
28	10.7	2.65	5.89	4.71	2.52
29	60.6	16.0	59.6	47.7	8.68
39	19.7	13.4	14.7	21.6	13.2
40	3.21	2.29	1.99	2.74	2.37
41	70.1	29.0	60.4	55.7	28.3
42	40.9	54.0	4.84	4.56	90.1
43	81.2	100.0	28.5	100.0	100.0
44	2.66	3.22	0.92	3.26	3.22
55	6.61	5.00	7.00	12.1	4.77
56	45.3	4.22	78.3	31.6	0.64
57	100.0	10.1	100.0	98.1	0.44
58	4.40	0.43	4.29	4.28	0.03
69	0.34	0.51	0.53	0.64	0.53
70	0.67	6.39	1.49	3.44	0.60
71	4.96	26.7	5.29	71.2	15.4
72	0.28	1.42	0.29	3.89	0.85
84	0.05	0.01			0.01
85	0.44	0.67	0.64	0.15	0.69
86 M	15.5	3.33	3.52	0.06	3.58
87	1.01	0.22	0.22		0.24
Source:	API 147	API 148	API 149	API 150	API 151

[a] Relative intensities.

it would be the *t*-butyl ion. For the *n*-hexane no point of favored cleavage is expected, yet both 57 and 43 are large-intensity ions. The branching rule seems to explain the presence of the *m/e* 43 peaks for the 2-methylpentane and the 2,3-dimethylbutane, but from this point on, the explanations of the various large-intensity ions at *m/e* 56, 42, 41, 29, and 27 will require a bit more than just simple consecutive cleavages. In fact, how can the ion at *m/e* 43 be formed for the neohexane? No simple cleavage of C–C bonds can possibly lead to such a fragment. Rather, a rearrangement of a hydrogen atom must take place, in addition to the cleavage of at least two C–C bonds. Rosenstock *et al.* (1952) have postulated structures for the $C_3H_7{}^+$ ion (*m/e* 43) which, on the basis of the above discussion, would appear to possess unusual stability.

Grubb and Meyerson (1963a) suggested the formation of a cationated cyclopropane ion to explain the formation of the *m/e* 43 or C_3H_7 ion from a *t*-butyl ion formed by the cleavage of any of the methyl groups from neopentane. The ion at *m/e* 43 from the neohexane may well be formed in the same fashion. Another useful point for the mass spectroscopist involves the ions falling in the series $C_nH_{2n-1}^+$. As n gets larger these ions become less intense in the mass spectra of the alkanes. In fact, they are of appreciable size only if n is in the range of 2–5 and decrease rapidly from $n = 6$ and upwards. Once again this should indicate to the reader that the distribution of the excess energy with which ions are formed is subject in large part to the number of atoms (oscillators) present in the ion species. With few atoms, the energy available promotes the excited ion species to the point where "atypical" reactions are possible which do not occur in systems of larger homologs.

Meyerson *et al.* (1963) have pointed to the mass spectra of the alkanes as perhaps the most complex mass spectra because of the difficulty of correlating the known chemical structure with the relative abundances of the ions detected. Fortunately, according to Hood (1963), for practical applications to the petroleum wax fractions, many of the *n*-paraffins exhibit some of the simplest mass spectra of the series. An example of an application to determination of chain branching is given in Chapter 9.

Despite the complexity of the mass spectra of the alkanes, Dean and Whitehead (1961) and Bendoraitis *et al.* (1962) have used mass spectral analysis to prove the presence of quantities of isoprenoid hydrocarbons in petroleum. The interesting point here is the possible precursors of these hydrocarbons. Chlorophyll contains a phytol or pristol group as it occurs in nature. The above authors reasoned that the large amounts of plant life containing chlorophyll could get into petroleum formation processes simply by the loss of these groups from the chlorophyll rather soon after the detritus entered the marl stage. Thus, all that would be required from the geochemical viewpoint would be a reductive *in situ* condition whereupon the corresponding hydrocarbons would, in time, be formed from the ester and/or alcohols. The presence of

2,6,10,14-tetramethylpentadecane (pristane) and 2,6,10,14-tetramethylhexadecane (phytane) in sedimentary rocks has been reported by Oro *et al.* (1965) and Robinson *et al.* (1965), thus indicating an interesting but perhaps ubiquitous situation in which mass spectrometry continues to play an important part. Similarly, the porphin residues have received much study, including prominent identifications by mass spectrometry. In a rather roundabout way then, the original use of commercial instruments, in part stimulated by the interest of oil in rock, has paid off directly and in a continuing fashion, albeit some 20 years after the original impetus was supplied.

The examples above are given to remind the mass spectroscopist that he can always appreciate the literature containing a few reference spectra of the type of compound he is most interested in at the moment. Failing to find these he can appreciate the correlation studies which will allow him to proceed with identifications even though the degree of certainty may be somewhat less than with the compounds for comparison.

One item of interest in Table 5-4 is the approximate reproducibility and accuracy for the various isotopic peaks. This is to be found at *m/e*'s 44, 58, and 72 where little interference (or rearrangement ions) is expected.

From the collective studies of the alkanes have come some general rules which can assist the mass spectroscopist in his consideration of the fragmentation of other molecules containing alkyl substituents. Within a given series of such compounds, the following principles will be expected to hold:

1. The relative height of the M^+ ion peak is greatest for the *n*-alkyl compounds and decreases with increasing chain branching.
2. The loss of a fragment containing a single carbon atom rarely occurs unless the compound contains methyl side chains. *Gem*-dimethyl substitution increases such fragmentation markedly.
3. Fragmentation is favored at highly branched carbon atoms.
4. The C_3 and C_4 alkyl ions are often of large intensity.
5. Ions of odd *m/e* tend to be more abundant than ions of even *m/e*. (Note that the reverse of this is true for molecules with odd numbers of nitrogen atoms.)
6. A secondary fragmentation process involving the loss of one or two hydrogens or a neutral olefin tends to produce ions of odd *m/e*. (Again, the reverse is true for molecules with odd numbers of nitrogen atoms.)
7. A high degree of branching in the compound may produce prominent ions at even *m/e*'s. This is taken to indicate the presence of two separate side chains.

Careful consideration of the mass spectrum can lead to discovery of metastable peaks, which can always assist in determination of the fragmentation paths in the compound in question.

2. *The A_2 Type of Fragmentation*

The mass spectra of the alkanes also show evidences for the A_2 type of cleavage whereby an even-electron ion and a neutral olefin (or acetylene) are produced from an even-electron fragment ion. Beynon *et al.* (1961) have shown that this type of fragmentation does not lead to the accumulation of lower fragments found in most alkane spectra. Metastable ions in the alkane spectra were observed that established the process

$$C_4H_7^+ \xrightarrow{*} C_2H_5^+ + C_2H_2^0 \quad \text{(see selected API spectra, } C_8\text{'s)} \tag{5-84}$$

while even the loss of two hydrogen atoms (process not defined) from ethane to produce an ion at m/e 28 was supported by a metastable peak in the earliest of studies. These types of cleavages produced ions of lower mass in the -1 series from precursors in the $+1$ series. Beynon *et al.* (1968a) discussed the spectra of all C_8 alkanes in considerable detail for the worker interested in the alkanes.

The balance of this section will attempt to illustrate only the most important parts of studies with various molecular species, primarily to serve as examples of a specific type of cleavage or rearrangement. For the working mass spectroscopist details of the behavior of any given class of compounds are of extreme importance. Rather than include a summary of the behavior of all of the classes of compounds, which in reality is an impossibility, we have elected to list the titles of selected spectra–structure correlation papers in Appendix I. It should be noted that to expand correlations too far often leads to generalizations which are at best only partial truths and which in some cases can be very misleading.

Appendix I is intended only as an in-hand guide to literature on specific compounds and classes of compounds. It is not intended as a complete compilation. The text of Silverstein and Bassler (1967) includes a brief summary of some of the more important compounds the organic chemist may encounter. Any listing could not be complete, of course, and it is probable that the only approach to such a problem is through a computerized indexing of the literature. This has been done by Capellen *et al.* (1970a,b) under the auspices of the Atomic Energy Commission to produce compound (by atomic content), author, key-word, and title indexes of the literature, beginning with the year 1965. The reader is urged to utilize these types of references in light of the great abundance of material now appearing in the literature. Svec (1970) gave figures for the total number of references covered by these reports as follows: 1965, 1591 entries; 1966, about 2900 entries; 1967, 8243 entries; 1968, 9879 entries; and the 1969 report, which is still being prepared, will include only about 5000 entries. Thus, it appears that the information "explosion" in mass spectrometry is slowing down. Svec pointed out that the lower figures for 1965 and 1966 reflect, at least in part, the efforts to get the computerized program under way and second the fact that not all the possible journals were covered in 1965, new ones being added in significant numbers during the first two years. In view of the above total of more than 27,500 entries for the past five years, one wonders a bit about how to organize this appendix and how much to include.

Some of the newer journals, *Organic Mass Spectrometry* being the leader, have started to include annual indexes listing the compounds discussed in the journal for that year. These will prove of assistance to the worker looking for a specific compound rather than a class. The index also lists, separately, the various isotopically labeled species discussed in articles in the journal.

The compilation of Appendix I is far from complete, but it contains one or more older references to the class of compounds plus some newer references (appearing since 1964). It should be reemphasized here that too often the newcomer to a field fails to realize that the pioneers did considerable service work in the form of correlation papers and that these works contain foundations upon which modern labeling studies, etc., have been built into complete mechanism studies. From all of these studies have come the generalizations referred to as "rules." The authors are aware of many apparently little "tidbits" appearing in these works which suggest new avenues of research for solving what then appeared to be a minor part of the problem and to which the original workers could not devote additional time. It would be a pleasure to hear papers about some of these problems rather than papers with the greater portion covering work previously reported 10 or more years ago at a mass spectrometry meeting. Hence, as a more positive thought, the plea is made to extend research into these areas to close up the gaps in understanding of the behavior of molecules under electron impact.

3. *The A_3 Type of Cleavage*

This type of cleavage occurs in compounds possessing an allylic double bond, the β scission of which leads to an ion which can undergo resonance stabilization. In the usual process, an odd-electron ion produces an even-electron ion and a neutral fragment. It commonly occurs in olefins and has been extended to the dienes and ynes. Mohler (1948) observed that the M^+ ion was greater in unsaturated compounds (the olefins) than in the corresponding alkanes. This rule was shown to hold true for only the lower members of the series once the mass spectra of the C_6 and higher alkenes could be considered. The reader should recall that these were the years when the inclusion of benzene or a hexane isomer in MS samples was not advised because of the contamination problems in the source! Likewise, the conclusions reached were often based upon a rather limited number of spectra.

In light of the changes in "rules," authors should be encouraged to examine more compounds in a series before reaching conclusions that may have to be modified later or which become a part of the exceptions rather than the "rules." Beynon *et al.* (1968b) stated that Mohler's observation implies that unsaturated parent ions of low mass are more stable than the corresponding saturated ions, and introduction of further unsaturation appears to enhance the stability still further.

Mohler (1948) noted that the alkenes had a greater tendency to produce rearrangement ions than the alkanes. As a consequence, the deduction of structures for the alkenes was not easily, and often not unequivocally, done. An early labeling study by George (1962), utilizing ^{13}C-propene-1, demonstrated that although the 55% isotopic enrichment could be accounted for, the position of the labeled carbon atom had been randomized. Formation of a cyclic intermediate in the source was suggested as a possible occurrence.

It should be noted that the A_3 type of cleavage,

$$\left[\mathrm{C{=}C{-}C{-}\!\!\vdots\!\!{-}C{-}C{-}}\right]^{+\cdot} \longrightarrow \left[\mathrm{C{=}C{-}C}\right]^{+} + {}^{\cdot}\mathrm{C{-}C{-}} \quad (5\text{-}85)$$

leads directly to the series of ions at *m/e* 27, 41, 55, 69, 83, etc., all of the $C_nH_{2n-1}^+$ series. These are characteristic of the olefins and are used, as demonstrated in Chapter 6, in connection with type analysis. Unfortunately, these ions also appear in the spectra of the alkylcycloalkanes. Polyakova and Khmel'nitskii (1961) have reported on correlations with olefinic types of compounds, while Coggeshall (1960, 1963) has compared the behavior of the *n*-alkanes and the 1-alkenes under electron impact. Mohler (1948) noted some of the similarities among the dienes and suggested extensive hydrogen atom migration or what is tantamount to double-bond migrations. Certainly the alkenes and alkadienes are a most difficult group of compounds to analyze by mass spectrometry.

The alkylcycloalkanes have been somewhat neglected after being initially examined by workers at the National Bureau of Standards (NBS) and reported by Mohler (1948) and Mohler *et al.* (1954). An account of the problems of ring fragmentation and events with short-chain substituted compounds, including labeling studies, is given in Budzikiewicz *et al.* (1967a). In all correlation work, it is important to have a rather large number of high-purity (preferably 99.5 + %) isomeric compounds of each series to be investigated. Mass spectra should be obtained on the finest type of available equipment operated by top specialists under the supervision of highest-quality research personnel. Reference-quality spectra require more care and attention than would be given to most analytical procedures since many types of information may be obtained

TABLE 5-5

Partial Mass Spectra of Some Cyclohexanes[a]

m/e	Cyclo-hexane	Methyl-cyclo-hexane	Ethyl-cyclo-hexane	1,1-Di-methyl-cyclo-hexane	*n*-Propyl-cyclo-hexane	Isopropyl-cyclo-hexane	1,1,3-Tri-methyl-cyclo-hexane	*n*-Butyl-cyclo-hexane	Isobutyl-cyclo-hexane	*sec*-Butyl-cyclo-hexane	*t*-Butyl-cyclo-hexane
41	65.1	58.9	49.7	48.3	45.7	56.2	46.2	43.3	53.0	70.4	38.4
42	30.1	36.2	12.5	7.52	9.00	9.29	8.16	8.35	6.93	7.64	3.15
43	14.0	8.87	8.95	14.9	9.22	16.6	16.3	10.8	22.4	11.3	6.62
55	35.0	83.0	72.1	76.6	70.3	85.1	50.9	68.8	87.2	100.0	25.0
56	100.0	30.5	13.0	43.8	11.9	10.2	26.8	12.2	25.0	43.0	100.0
57	4.91	4.92	1.72	12.9	2.85	3.59	10.2	3.58	12.3	15.7	71.6
67	3.26	4.88	11.1	3.94	12.3	27.0	5.03	14.1	15.2	32.3	11.1
68	1.84	9.80	2.70	5.65	3.35	2.60	2.66	3.52	4.56	5.06	0.89
69	22.4	23.0	9.13	49.6	9.15	17.8	78.0	9.67	21.2	40.8	13.3
70	1.22	22.0	3.28	11.1	4.50	1.86	13.6	5.10	4.28	4.71	0.91
82	0.22	14.3	40.8	0.37	51.6	84.9	1.50	58.9	51.4	88.7	14.4
83	4.61	100.0	100.0	2.68	100.0	100.0	27.1	100.0	100.0	99.9	13.1
84	74.9 M	6.57	8.69	4.45	7.63	7.16	3.12	7.87	7.77	6.70	0.89
97		2.31	0.26	100.0	2.24	0.23	0.33	2.88	26.0	0.90	0.21
98		43.8 M	0.02	7.55	0.81	0.11	0.58	0.80	2.38	0.14	0.03
111			0.48	0.12	0.08	1.87	100.0	1.18	0.31	26.0	0.08
112			20.3 M	5.30 M		0.16	8.63	0.57	0.48	2.23	0.02
125					0.18	0.28	0.17	0.02	2.53	0.61	3.42
126					19.5 M	15.7 M	2.42 M		0.24	0.06	0.33
139								0.05	0.11	0.11	0.02
140								14.4 M	14.5 M	14.3 M	0.97 M
Source:	API 216	API 217	API 218	API 219	API 226	API 227	API 228	API 229	API 230	API 231	API 232

[a] Relative intensities.

from these types of spectra. The results of work done under such circumstances can have meaning for quite some time. Some spectra, obtained at the NBS laboratories and under the conditions given above, appear in the catalogs of API-44 spectra for some cyclohexanes. Partial mass spectral data are presented in Table 5-5. If the reason for requesting spectra containing longer alkyl chains than C_3 is not apparent to the reader, a reminder is given that often the longer chains will help consume some of the excess energy by which the molecular ion may be formed and, by redistribution of that energy through the oscillators in the chain, lead to lower-energy fragmentation routes. These routes may not resemble those available to the lowest members of the series which possess relatively few atoms, but such routes may lead to analytically useful information about the molecule in question.

The molecular ions of the cyclohexanes are usually about three times as intense as the corresponding alkane compounds. As expected, the molecular ion intensities of the branched species are smaller. This, in itself, indicates greater stability of the molecular ions. Note the gradual decrease of the M^+ ion intensity for the *n*-alkyl compounds with increasing alkyl chain length. This continues regularly for the compounds until for the nC_{20} or eicosyl derivative (API No. 870) the $M^+ = 3.79\%$ relative intensity, while the *m/e* 83 ion is still the base peak. With a modest alkyl chain present, α cleavage at the ring seems to be reasonable and might be predicted on the basis that the resulting cleaved-ring ion would be a secondary ion as opposed to the primary (and odd-electron) molecular ion:

$$\left[C_6H_{11}\text{—}CH_2\text{—}CH_2\text{—R} \right]^{+\cdot} \longrightarrow C_6H_{11}^{+} + {}^{\cdot}CH_2\text{—R} \qquad (5\text{-}86)$$

By an ordinary A_1 type of cleavage, an *even*-electron ion is formed at *m/e* 83 for the *n*-alkylcyclohexanes plus the neutral radical. The dimethyl and trimethyl compounds show a prominent M − 15 peak. In the case of the cyclohexane itself, the molecular ion is an *odd*-electron ion and, of course, occurs at *m/e* 84. It should be no surprise that fragmentation paths are different for the cyclohexane as distinguished from its longer alkyl chain derivatives. As the alkyl chain gets larger, the parent-ion intensity decreases. This occurs modestly for the *n*-alkylcyclohexanes but is marked for the *t*-butylcyclohexane. The ion at *m/e* 83 remains the largest ion for all of the compounds except the *t*-butyl. Note the 83 peak is reported as 99.9% for the *sec*-butyl compound, but this might well be 101% if run on this same instrument the next day. More will be said about this type of problem later in this section. Hence, if these data were presented as a bar graph, the reader would have little difficulty in deciding that the *m/e* 83 ion was an important ion in this series of compounds.

The tabular presentation, as made here, illustrates some details of the spectra which could not be conveniently read from a bar graph and thereby could not easily be assimilated by the reader.

Perhaps it should also be remembered at this point that the analyst is more interested in the correlations that can be put to use than he is in the mechanism of fragmentation, as fascinating as that subject is to most organic mass spectroscopists. Consequently, the low *m/e* 83 for the *t*-butyl will naturally attract attention. A first thought is that perhaps the 83 ion did not win the charge competition with the *t*-butyl carbonium ion, which can form quite readily by this same type of cleavage and, according to the simplest notions, should be energetically favored over the secondary 83 ion. This must be true in part to account for the presence of the large *m/e* 57 ion. As a crosscheck, however, the *m/e* 57 ion intensity in the other molecules must be considered. Excluding the cyclohexane and the methyl derivative, one notes a gradual increase in intensity of the 57 ion for the *n*-alkyl compounds. (The thoughtful person is now looking for a mass spectrum of *n*-pentyl- and/or n-hexylcyclohexane.) The branched-chain compounds, except for the isopropyl member, show *m/e* 57 to be about three to five times that of the *n*-alkyl compounds. A good correlations worker files this in his personal memory bank or feeds it to the computer as a checkpoint for later use in his identification development. The *t*-butyl compound is then an exception, but the exception is apparently readily explainable. Perhaps labeling studies should be made to establish this as a simple cleavage, but considerable evidence on other *t*-butyl-substituted compounds, some with labeling, would lead one to bypass this particular problem for the present. Of more interest from such a study would be to find out if the largest ion (at *m/e* 56) was formed from the *t*-butyl moiety or by the hydrogen transfer of one of the *t*-butyl hydrogens to the ring moiety with the formation of a neutral cyclohexane particle and an odd-electron ion. It is much more probable that the *m/e* 56 ion is formed from ring fragmentation.

It is interesting to note that for all of the listed compounds except the *t*-butyl (and cyclohexane itself), a metastable peak is observed for the production of the 83 ion by a direct process from the parent ion. Hence, α cleavage of the A_1 variety is definitely operating. Why then be concerned about problems with the cyclohexanes since this appears to be a rather simple situation? Further examination of the spectra, however, quickly leads to many questions. Why should the *m/e* 82 peak be so large for the *n*-alkyl spectra? Is the *m/e* 82 ion formed by the loss of a single hydrogen neutral from the *m/e* 83 peak? This process would be contrary to what would be expected. The API-44 spectral sheets report metastable peaks for the process of the molecular ion producing the *m/e* 82 directly! This process would compete with the formation of the *m/e* 83 peak and would represent the formation of a neutral alkane fragment, the hydrogen atom presumably coming from the cyclohexyl ring, thus leaving a rearrangement *m/e* 82 peak, probably an excited-state cyclohexane ion and/or hexadiene and an odd-electron ion. For the two molecules with branching on the α carbon atom (isopropyl and secondary butyl), the 82 peak is considerably larger. In the spectra given, no metastable ions were presented to show any further fragmentation path for the ion at *m/e* 82. The ions at *m/e* 54, which would correspond to a retro-Diels–Alder cleavage (see later in this section) with ejection of a neutral ethylene particle, were not significantly large. If the amount of ion at 82 for the methylcyclohexane is assumed to represent the amount formed from the 83 by a hydrogen loss, which is about 15%, then the bulk of the 82 for the *n*-alkyl compounds must come about from the path suggested above. As a check, the mass spectroscopist notes that the 82 in the case of the 1,1-dimethyl- and 1,1,3-trimethylcyclohexanes is also of this low order of intensity. The *t*-butyl compound, despite its smaller 83 ion, shows a comparable intensity at 82, thus suggesting that this process is indeed competitive with the α cleavage and the loss of the *t*-butyl ion to form *m/e* 57.

Type A_2 cleavage exemplifies an event that occurs with some prominence in methylcyclohexane. The ejection of a neutral ethylene molecule results in the formation of an ion at *m/e* 70. Presumably the molecular ion at 98 is an odd-electron ion, the 70 ion remains an odd-electron ion, and therefore such a process ordinarily would not be favored. If one regards the intensity of the 84 ion in the methylcyclohexane to be due solely to isotopic contributions from *m/e* 83 (theoretically should be 6.60%), then it is obvious that the remaining 84 ions for the ethyl and dimethyl compounds are excessive in their intensities. The excess reflects a small amount of this rearrangement. The ejection of an ethylene neutral decreases with increasing chain length, as is shown by the respective M − 28 ions. (The 112 ion in the secondary butyl spectrum is due to isotopic contributions from the 111 ion.)

Also in evidence in the spectra commencing with the methyl compound are moderate to large ions at *m/e* 56. In the case of the methyl compound, a metastable ion is reported for the process

$$98^+ \xrightarrow{*} 56^+ + 42^0 \tag{5-87}$$

Since none of the other compounds showed the production of the 56 ion by a metastable peak, very little more can be said about this ion. The work of Budzikiewicz *et al.* (1967a) and Beynon *et al.* (1968c) shed some light upon possible mechanisms. The loss of a neutral olefin of mass 42 suggests itself. A metastable ion in support of this mechanism appears only in the spectrum of the methylcyclohexane. Several odd-mass even-electron ions are rather prominent and can be considered to be formed by related processes. These are the ions at *m/e* 69, 55, and 41, all in the −1 series. For all of the *n*-alkyl compounds producing the 83 ion in appreciable intensity, a metastable peak is observed which confirms the process

$$83 \xrightarrow{*} 55^{+} + C_2H_4^{0} \qquad (5\text{-}88)$$

For the 1,1-dimethyl compound, a metastable was observed for both the processes

$$97 \xrightarrow{*} 69^{+} + C_2H_4^{0} \qquad (5\text{-}89)$$

and

$$97 \xrightarrow{*} 55^{+} + C_3H_6^{0} \qquad (5\text{-}90)$$

The ions at *m/e* 55 and 69 are more prominent in the branched-chain compounds of a given carbon number. For all of the propyl and butyl compounds and the 1,1-dimethylcyclohexane, the process of

$$69 \xrightarrow{*} 41 + C_2H_4^{0} \qquad (5\text{-}91)$$

was observed. Spectra of the 1,2-, 1,3-, and 1,4-dimethylcyclohexanes (*cis* and *trans*) are in the API-44 catalog at Serial Nos. 220–227 and show additional interesting points concerning these types of molecules. They will not be discussed here.

The *m/e* 41 ion species and those ions at lower *m/e* usually are regarded as the end points of fragmentation reactions because they can usually be formed in so many ways. Further reactions are rather limited, usually confined to loss of two hydrogens, loss of a methyl or an ethyl fragment. Hence, it is of note here that the intensities of the 41 ions gradually decrease in a rather regular order for the *n*-alkyl compounds, indicating that the energy of the original molecular ion has been distributed over the additional oscillators present in the alkyl group as the chain length increases. For the branched-chain compounds this is not true because the processes leading to this ion show considerable variety.

Some additional very long chain cyclohexane derivatives have appeared in the API-44 files at a later time. Three compounds are shown in Table 5-6 to illustrate what happens with extremely long chain substitution. The compounds are 1,3-di-*n*-decylcyclohexane, 1,4-dimethyl-3-*n*-octadecylcyclohexane, and 1,3,5-trimethyl-4-*n*-octadecylcyclohexane. The small molecule ion intensities of the two branched compounds are expected in contrast to the considerably larger molecular ion intensity for the di-*n*-decyl compound. The amount of methyl substitution is immediately evident from the spectra. One should be cautioned that the replacement of the dimethyls with an ethyl or of the trimethyls with a propyl would lead to about the same result. The appearance of a small amount of *m/e* 335 in the spectra of the trimethyl compound raises

TABLE 5-6

PARTIAL MASS SPECTRA OF SOME C_{20}–C_{21} CYCLOHEXANES

m/e	Relative intensities (1,3-di-C_{10}-cyclohexane)	(1-C-3-C_{18}-4-C-cyclohexane)	(1-C_{18}-2-C-4-C-6-C-cyclohexane)
83	100.0	7.80	26.7
97	89.5	6.05	11.1
111	41.5	100.0	16.0
125	20.1		100.0
222	25.5		
223	50.3		
335	0.38	0.07	1.89
336	0.13	0.05	0.06
364	4.48 M	0.54 M	
378			0.57 M
Source:	API 1504	API 1505	API 1506

the question of a possible impurity or a chain branching so as to cause the formation of such an M − 43 ion. The most interesting event is the appearance of the *m/e* 222 peak in the di-*n*-decyl compound. Being of the relative intensity that it is indicates the strength of the hydrogen transfer from the ring (or the other decyl group) to the cleaving decyl group to produce a neutral decane moiety.

The partial mass spectra of some alkylcyclopentanes appear in Table 5-7. These spectra are obviously more complex than those of the cyclohexanes. Intuitively one might attribute this to the variation in ring form, as compared to the chair and boat forms of the cyclohexane. Regardless of the problems, a few statements can be made to assist in attempting to correlate structure with a spectrum that one has in hand. In summary these are as follows:

1. The molecular ion intensities vary similarly to the cyclohexanes, with chain branching and multiple substitution being of importance.
2. The ion at *m/e* 69 *cannot* be counted upon to identify the cyclopentane ring as the *m/e* 83 ion was used in the cyclohexane case.
3. Processes involving the ejection of neutral ethylene or propylene fragments appear to be much more important in these spectra.
4. The fragment at *m/e* 43 and/or 57 appears to increase with chain branching and/or additional methyl substitution.

TABLE 5-7

PARTIAL MASS SPECTRA OF SOME CYCLOPENTANES

	Relative intensities									
m/e	Cyclo-pentane	Methyl-cyclo-pentane	Ethyl-cyclo-pentane	Propyl-cyclo-pentane	Isopropyl-cyclo-pentane	1-Methyl-1-ethyl-cyclo-pentane	1,1,2-Tri-methyl-cyclo-pentane	1,1,3-Tri-methyl-cyclo-pentane	*n*-Butyl-cyclo-pentane	Isobutyl-cyclo-pentane
39	20.9	25.1	40.3	37.1	26.4	20.4	26.9	35.5	35.8	37.1
40	7.33	5.97	7.20	6.79	5.11	3.57	5.72	6.54	6.38	6.57
41	28.9	63.8	93.8	100.0	72.2	46.3	72.1	76.5	100.0	100.0
42	100.0	28.8	52.1	42.3	18.3	9.43	16.8	14.4	26.8	19.6
43	3.35	11.7	9.80	26.5	26.4	8.54	11.5	12.7	23.3	38.4
44		0.4	0.33	0.97	0.92	0.28	0.39	0.40		
54		4.03	7.22	7.59	2.00	2.49	2.99	3.01	10.7	7.51
55	29.0	25.1	56.1	57.2	43.6	81.7	78.2	100.0	63.4	68.9
56	1.3	100.0	52.0	59.2	31.0	12.7	100.0	90.2	59.0	77.8
57	0.07	4.91	7.79	9.50	11.6	2.31	40.7	62.7	14.2	35.3
67	1.5	2.44	10.2	13.5	16.4	8.72	4.09	3.98	14.3	20.3
68	0.4	4.13	72.8	70.0	100.0	1.42	5.73	9.21	72.3	37.7
69	1.1	32.0	100.0	90.7	85.1	13.9	80.2	62.0	86.2	90.7
70	29.3 M	1.72	54.7	38.1	12.5	34.0	34.1	22.4	40.3	20.8
82			0.61	9.43	0.36	25.3	0.47	1.17	12.0	44.4
83			7.72	37.6	3.15	100.0	8.96	59.1	34.2	70.9
84		15.9 M	1.14	25.7	5.67	9.07	26.7	7.55	20.1	13.5
97			0.76	2.44	26.0	9.72	25.0	52.8	17.2	2.64
98			16.7 M	0.13	1.87	0.73	1.91	3.82	13.8	4.74
112				16.4 M	2.88 M	0.94 M	13.7 M	5.95 M	0.65	9.50
125									0.14	0.26
126									16.7 M	6.66 M
Source:	API 182	API 183	API 184	API 190	API 191	API 192	API 194	API 195	API 206	API 207

5. The size of the *m/e* 68 peak (cyclopentene?) may offer a correlation with the monoalkyl-substituted compounds. Since only C_2–C_4-substituted compounds are available, additional data would be necessary to assist in ascertaining the validity of such a correlation.

6. It appears that the alkylcyclopentanes are more complex than the cyclohexanes. Experience has shown that more compounds are needed to ensure good correlations.

The movement of a hydrogen atom to form a *neutral saturated alkane*, which is ejected, is intriguing. This fragmentation problem should be investigated further.

In Table 5-8 are given the partial mass spectra of three highly branched,

TABLE 5-8

PARTIAL MASS SPECTRA OF BRANCHED ALKYLCYCLOPENTANE DERIVATIVES

	Relative intensities		
m/e	nC_6—C—nC_6 with C–cyclopentyl	nC_{10}—C—nC_{10} with C–(dimethylcyclopentyl)	nC_8—C—nC_8 with C_3–cyclopentyl
180	21.8	0.32	0.67
181	25.0	2.36	1.66
182	19.50	1.85	0.43
183	10.69	2.14	0.91
210	0.08	0.49	0.60
236	0.03	0.18	30.7
237	0.08	0.30	23.3
238	0.04	0.28	18.4
239	0.04	0.38	13.9
264	0.16	12.7	0.28
265	0.04	19.8	0.25
266	0.57 M	4.62	0.15
267	0.11	0.50	0.12
294		7.80	0.02
295		7.57	
296		1.49	
297		0.17	
350		0.02	0.42 M
406		3.21 M	
Source:	API 594	API 869	API 868

long-chain alkyl cyclopentanes. The relatively small M^+ ion peaks are expected, but note that the trisubstituted compound (with two methyl groups) has the highest M^+ intensity.

The most interesting point is that the alkyl chain branching controls the fragmentation paths. The molecules behave like substituted alkanes. The ions at M − 83, M − 84, M − 85 and M − 86 give four successive peaks which are characteristic for compounds of this type. Preferred cleavage at the branching point to form the secondary ion is obvious. The ion at M − 86 would indicate formation of a saturated hydrocarbon as a neutral in one of rearrangement. Metastable peaks to support direct formation from the molecular ion are reported in some of the spectra. The ion at M − 84 would also form a neutral olefin, ejected at the time of ion formation. These processes also have metastable peak support. These competing mechanisms involve a hydrogen atom transfer, apparently moving in opposite directions in the molecule. Something similar to this will be seen in connection with some aromatic compounds, but the driving forces to cause the rearrangements must be different.

The peaks at *m/e* 294–297 for compound no. 869 are puzzling. They occur at 28 mass units above the peaks expected from direct cleavage at the point of branching. They are not present in the spectra of the other molecules. While it is tempting to suggest an impurity present, with one alkyl group being C_8 instead of C_{10}, this is an impossibility since no M^+ ion at *m/e* 378 is found. It appears that some other alkane–alkene type of elimination is operating. Another minor interesting event is the apparent increase in alkane neutral formation as the point of branching is moved further from the chain. Note that the *m/e* 236 is larger than 237!

In Table 5-9 are given partial mass spectra of two long-chain *n*-alkylcyclopentanes. Note that the M^+ intensities for all the *n*-alkylcyclopentanes gradually

TABLE 5-9

PARTIAL MASS SPECTRA OF *n*-ALKYLCYCLOPENTANES

	cyclopentyl–nC_{10}		cyclopentyl–nC_{21}	
Ion	*m/e*	Relative intensity	*m/e*	Relative intensity
M^+	210	10.9	364	4.81
$(M-28)^+$	182	3.47	336	1.28
m*	158	Found	311.2	Found
	157.6	Calculated	310.6	Calculated
Source:	API 889		API 707	

decline from the lowest MW compounds to the C_{21} derivative. This is quite comparable to the cyclohexanes and can be used for correlation with *n*-alkyl substitution. Of additional interest in these spectra, however, are metastable ion peaks corresponding to the loss of a neutral ethylene fragment from the molecular ion by a direct process. The peaks found and calculated are listed in the table. A recheck of the spectra of all of the long-chain *n*-alkylcyclohexanes did not show a single metastable ion peak for this process. Hence, the process must be associated with ejection of the $C_2H_4{}^0$ from molecules containing a monosubstituted cyclopentane ring. Since the aforementioned suggestions are speculation, labeling studies should be made.

The alkylcycloalkane compounds described in the previous discussion were all obtained by Shell Oil Co. (API-44 Nos. 886–899) or Atlantic Refining Co. (API-44 Nos. 861–880). They show how to obtain data from which correlations with structure may be made. From data of this type, inroads were made on the composition of various petroleum stocks and crude oils. Without this type of information (even if "in-house") it is impossible to "guesstimate" compositions of such complex mixtures. (See Chapter 6 for application of these principles in type analysis.)

The A_3 cleavage also produces some of the primary ions in the alkynes. Cleavage beta to the triple bond does produce some of the ions in the mass spectra, but other processes, some of which are apparently quite complex and involve randomization of the hydrogens, occur according to the studies of Dolejšek *et al.* (1966). The 1-alkynes characteristically have smaller molecular ion intensities than their M − 1 ions, usually by factors of 80–100/1. As the alkyl chain gets longer, the M^+ ion intensities get smaller and are difficult to observe by the time 1-decyne is examined. Kendall and Eccleston (1966) reported spectra that show the increase in the molecular ion intensity as the triple bond moves into the chain. Their data give (see also API-44 Nos. 1809–1825):

Compound	Relative intensity (%)
1-Decyne	0.09
2-Decyne	0.29
3-Decyne	0.94
4-Decyne	2.54
5-Decyne	8.95

This is in agreement with Beynon *et al.* (1968d), who pointed to the 100-fold difference of the parent-ion peaks of these species. It should be noted that the large ion at M − 1 for the 1-alkynes might easily be mistaken for a nitrogen compound molecular ion containing an odd number of nitrogen atoms. The presence of only odd *m/e* fragment ions will quickly correct this notion.

The alkyl cycloalkenes, while producing fragment ions in the —3 (C_nH_{2n-3}) series, have been almost neglected in formal studies. A few spectra occur in the API-44 group, but generalities are difficult to develop other than the retro-Diels–Alder type of fragmentation, which will be discussed in detail in a later section.

The monoterpenes, diterpenes, and higher fused-ring systems of the terpenes have been studied and were reported by Reed (1963). These compounds are sensitive to fragmentation; many variations occur within the series and even from instrument to instrument. Much of the difficulty is now attributed to the thermal sensitivity of these compounds in the ion source. The mass spectroscopist should be looking forward to many new papers in this area of natural products which, except for a few pioneers, has been somewhat neglected. In addition to the references in Appendix I, Thomas and Willhalm (1964) and Gilchrist and Reed (1960) provide class correlation information.

4. The A_4 Type of Fragmentation

The origins of this type of cleavage arose during fragmentation studies of the alkylbenzenes. Kinney and Cook (1952) and Meyerson (1955) initiated correlation studies of the alkylbenzenes from which has stemmed much of the mass spectroscopists' knowledge of the behavior of aromatic molecules under electron impact. The chapter of Grubb and Meyerson (1963b) provides an excellent synopsis of the many facets of this work. The reader is referred to it for the many details which must be omitted here. The basic fragmentation is as shown, with a concurrent rearrangement occurring (as indicated by $\xrightarrow{\ell}$) rather than that shown in the introductory summary of fragmentations. The reason is simple.

$$\left[C_6H_5\text{—}CH_2\text{—}\wr\, CH_2\text{—}CH_2\text{—}CH_3 \right] \xrightarrow{\ell} C_7H_7^+ \text{ (tropylium)} + {}^{\bullet}R \qquad (5\text{-}92)$$

m/e 134 $\qquad$ m/e 91 (C_7H_7)

Cursory examination of the alkylbenzene spectra leads to the conclusion that a bond, beta to the phenyl group, is ruptured for the *n*-alkylbenzenes leading to the base peak ion at m/e 91. For the alkylbenzenes the product ion has been shown to be the tropylium ion (I) rather than the less symmetrical benzyl ion (II).

$$\underset{\text{I}}{C_7H_7^+ \text{ (tropylium, (CH)}_7^+)} \;\not\leftrightarrow\; \underset{\text{II}}{C_6H_5\text{—}CH_2^+} \qquad (5\text{-}93)$$

This was done by Rylander *et al.* (1957) using monodeuterotoluenes, labeled in the α, *ortho*, *meta*, and *para* positions. The mass spectra are essentially identical from which it can be concluded that the hydrogen atoms in the toluene, whether present on the ring or in the methyl group, become equivalent before the molecular ion fragments. Using the energetics observed leads to the same conclusion. The $C_7H_7^+$ fragment formed has a higher appearance potential than would be expected on the basis of the strength of the benzyl bond. In addition, the resulting ion is a perfectly good Hückel ion having six π electrons and complete delocalization of the charge.

As a further example of how the tropylium ion assisted in explaining behavior of the alkylbenzenes, the three isomeric xylenes exhibited intense peaks at the m/e 91 which was M − 15. Toluene, possessing what was at first thought to be a labile methyl group, did not lose a methyl group to any appreciable extent. If the xylene (III) first rearranges to methylcycloheptatriene (IV) and then a loss of the methyl group ensues, the stable tropylium ion is formed. If the toluene loses the methyl group, the less energetically favorable phenyl ion is formed. In determination of the size of the alkyl group cleaved, the β cleavage should be kept in mind.

$$\text{III} \rightarrow \text{IV} \xrightarrow{-CH_3} \text{I} \qquad (5\text{-}94)$$

As research continued, the excess intensity of the m/e 92 ion peak (above the normal isotopic contribution) was noted. By the time the chain length of n-octylbenzene was reached, the 92 peak was of slightly greater intensity than the 91. This is still true for the eicosylbenzene. This raised some questions about the mechanism producing the m/e 91 ion. Meanwhile, Hanuš and Čermák (1959) had extended the ring expansion concept to the thiophenes. Cook and Foster (1961) reported similar observations on a large number of alkylthiophenes and indicated that the expanded-thiophene-ring ion, thiacyclohexatrienium or thiapyrilium, at m/e 98 was formed first at low voltage with subsequent formation of the m/e 97 ion from it. The distinction between the tropylium and thiapyrilium ions was thought to be possibly due to the presence of the sulfur atom with its available 10-electron outer shell (Cilento, 1960; Price, 1964), which involves the d orbitals. Evidence for the ring expansion of pyridine types was forthcoming from La Lau (1960), who believed this occurred in his examination of the pyridines, quinolines, and indoles where two rings were present.

The chapter by Grubb and Meyerson (1963b) gives an account of the alkylbenzenes and should be consulted by all concerned with alkylbenzene spectra. Despite the mechanism problems, correlations that were useful for the identification and analysis of alkylbenzenes proved very helpful. Meyerson (1955) gave a scheme for the analysis of all alkylbenzenes in the gasoline range and which could be applied to a high degree well up into the kerosene and gas oil ranges of petroleum products. Similar correlations for the alkylthiophenes were developed by Cook and Foster (1961) and Foster *et al.* (1964, 1965, 1968). Some examples are given to illustrate fragmentation in general when an aromatic ring cleaves at the bond beta to the ring system. Merely changing the *m/e* of the aromatic nucleus as appropriate will allow the correlations to be generalized to other systems. Table 5-10 shows partial mass spectral data for some 2-*n*-alkylthiophenes.

TABLE 5-10

PARTIAL MASS SPECTRA OF SOME 2-*n*-ALKYLTHIOPHENES[a]

Thiophene	M^+	Peaks from cleavage at designated locations (from ring)				
		γ (*m/e* 111)	$\beta+1$ (*m/e* 98)	α (*m/e* 83)	$\alpha+1$ (*m/e* 84)	$\alpha+2$ (*m/e* 85)
Ethyl	38.5	7.74	6.42	0.65	2.20	2.07
Propyl	25.7	1.73	7.19	0.43	1.64	0.96
Butyl	22.9	3.66	16.37	0.37	2.59	1.66
Pentyl	20.8	4.42	21.00	0.31	3.32	2.13
Hexyl	24.6	5.87	23.11	0.33	3.33	2.53
Heptyl	21.1	7.26	24.69	0.32	3.90	3.10
Octyl	20.3	8.83	33.11	0.35	4.02	3.43
Decyl	19.3	11.19	30.65	0.57	4.21	3.76
Dodecyl	17.5	13.09	39.09	0.66	3.94	4.13

[a] Relative intensities to *m/e* 97 = 100, β cleavage.

For the alkylthiophenes out to *n*-dodecyl, the base peak was found to occur at *m/e* 97 [note this is 6 *u* (atomic mass units), above the corresponding benzene

$$\left[\text{C}_4\text{H}_3\text{S}\overset{\alpha}{-}\text{CH}_2\overset{\beta}{-}\text{CH}_2\overset{\gamma}{-}\text{CH}_2\overset{\delta}{-}\text{R}\right]^{+\cdot} \longrightarrow \left[\text{C}_4\text{H}_3\text{S}-\text{CH}_2\right]^{+} \; (m/e\ 97) \longrightarrow \text{thiopyrylium ion (HC–C(H)–CH, HC–S–CH ring)}^{+} \qquad (5\text{-}95)$$

ion fragments throughout]. The excess of the 98 peak is obvious from the table. Cleavage at the γ carbon–carbon bond from the ring does occur and increases regularly with increased chain length. The very small direct α cleavage process is established by the insignificant peaks at m/e 83, the thiophenyl ion. This is in contrast to the m/e 77 which appears with varying intensity in the alkylbenzenes. Peaks at m/e 84 and 85 correspond to the ring plus one hydrogen and plus two hydrogens, respectively, of the alkylbenzenes and may resemble the m/e 78 and 79 ions. The presence of the sulfur in these fragments is readily established on the basis of the ^{34}S isotope contribution to the $m/e + 2$. It thus provides a naturally labeled situation from which ion compositions are readily deduced; less intuition plays a part in mechanism studies than is supposed by some researchers. It is true that the label is only 4%, and thus the certainties are not what one obtains with considerably richer ^{13}C or D labeling. It should be noted that the m/e 84 and 85 peaks go through a minimum intensity for the propyl compound and then increase regularly with increased chain length. Again, this behavior is considerably more regular for these compounds than for the corresponding alkylbenzenes. These data will allow one to predict with fair assurance the approximate intensities of the various ions that should be found for these compounds.

Similar data are presented for some branched-chain 2-substituted alkylthiophenes in Table 5-11. Base peaks are formed by a β cleavage process, and

TABLE 5-11

PARTIAL MASS SPECTRA OF 2-BRANCHED ALKYLTHIOPHENES[a]

		Peaks from cleavage at designated locations (from ring)				
Alkylthiophene	M^+	γ (m/e 111)	$\beta + 1$ (m/e 98)	α (m/e 83)	$\alpha + 1$ (m/e 84)	$\alpha + 2$ (m/e 85)
2-(2-Methylpropyl-)	26.2	0.42	13.33	0.26	1.11	0.61
2-(3-Methylbutyl-)	18.8	4.37	41.06	0.36	3.64	3.69
2-(4-Methylpentyl-)	24.1	3.82	27.35	0.52	3.08	3.09
2-(2-Ethylbutyl-)	25.4	1.18	46.42	0.33	1.66	1.62
2-(2-Ethylhexyl-)	18.4	1.20	38.78	0.47	2.03	1.15

[a] Relative intensities to m/e 97 = 100, β cleavage.

the direct α cleavage route is again insignificant. Branching at the β carbon atom or beyond has a marked influence upon the intensity of the m/e 111 ion. It is quite apparent that the amount of branching has a marked influence upon the amounts of 84 and 85 ions and, therefore, upon the mechanism which must occur to form these ions. The variation in the amount of 98 ion excess is also quite apparent and suggests that branching has a noticeable effect upon the production of *m/e* 98 ion.

Data for the 2,5-di-*n*-alkylthiophenes are to be found in Table 5-12. Here one has the opportunity to observe the apparent effect of chain lengthening and competition among long chains (C_4 or greater) for the point of favored cleavage (the bond beta to the ring system). The ion of largest intensity is formed by cleavage at the bond β from the ring in the longest chain. Even

TABLE 5-12

PARTIAL MASS SPECTRA OF SOME 2,5-DIALKYLTHIOPHENES[a]

R_2—CH_2—(thiophene, S)—CH_2—R_1

R_2—CH_2	R_1—CH_2	β Cleavage: Loss of R_2	β Cleavage: Loss of R_1	$\beta + 1$ for R_1 loss	Excess intensity above isotope contribution
Ethyl	Ethyl	100	100	8.77	0.03
Ethyl	Propyl	12.2	100	9.13	0.34
Ethyl	Butyl	8.37	100	10.91	2.17
Ethyl	Hexyl	5.43	100	12.43	3.69
Ethyl	Heptyl	5.42	100	13.14	4.40
Ethyl	Octyl	5.45	100	12.46	3.72
Propyl	Propyl	100	100	9.83	−0.12
Propyl	Butyl	68.1	100	8.43	−1.52
Propyl	Hexyl	57.4	100	12.60	+2.65
Propyl	Heptyl	54.8	100	12.77	2.82
Butyl	Butyl	100	100	13.95	2.88
Butyl	Hexyl	79.3	100	13.82	2.75
Hexyl	Heptyl	90.0	100	14.02	0.61
Heptyl	Heptyl	100	100	15.58	−0.07

[a] Relative intensities.

for the hexyl–heptyl compound this is true, and hence there is established a correlation to determine the size of the side chains substituted upon the thiophene moiety. In every case, the length of each of the side chains is easily determined for these types of molecules. When the difference in chain length is large and one chain is quite short, such as in the case of the 2-ethyl-5-alkyl compounds, cleavage of the short chain seems to reach a minimum of slightly more than 5% of the longer chain (it being the base peak). Another interesting point to note from the table is the ion intensities observed for the peak at base plus one (β cleavage plus one on the longest chain). While the peak intensity increases with increasing mass of the ion (as it should), the actual excess above the isotopic contribution reaches a maximum of only 4.40% for the 2-ethyl-5-*n*-heptylthiophene. Hence, the rearrangement ion is of far less importance here. This may be due to the effect of the second substituent on the ring, which would tend to set up an inductive effect opposed to the longer alkyl chain. Cook and Foster (1961) concluded that there were no correlations existing between the ions at *m/e* 111 and 97 and the compound structures. Note that these ions must arise from further fragmentation processes from the molecular or base peak ions. The isotopic peaks associated with the β' cleavage were sometimes present in excess and sometimes present in far too small an intensity, indicating that the β' cleavage peak might not all be from the simple loss of the R_2 group.

The isopropyl-, *t*-butyl-, and neopentylthiophenes, while cleaving at a β bond as a first fragmentation event, lead to ions of different *m/e* and, as such, are distinguishable. For the isopropylthiophenes, Kinney and Cook (1952) reported the loss of a methyl group to form the base peak. This is envisioned as happening as shown [Eq. (5-96)] to form a secondary carbonium ion. The

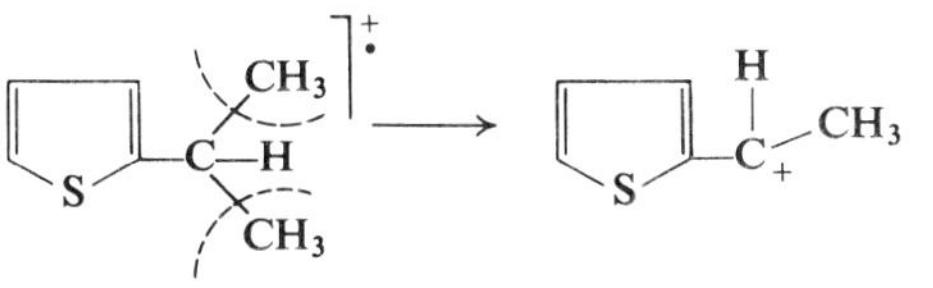

CH₃ — C—H — CH₃ (thiophene, S)]⁺· ⟶ (thiophene, S) —C⁺(H)—CH₃ (5-96)

amount of cleavage of the C–H bond on the chain is negligible, but the loss of the methyl is reminiscent of the loss of methyl groups at a branched carbon atom in the cases for the alkanes, etc., as shown earlier. For the *t*-butylthiophenes, similarly, a β bond to the ring cleaves to produce the M − 15 ion as follows:

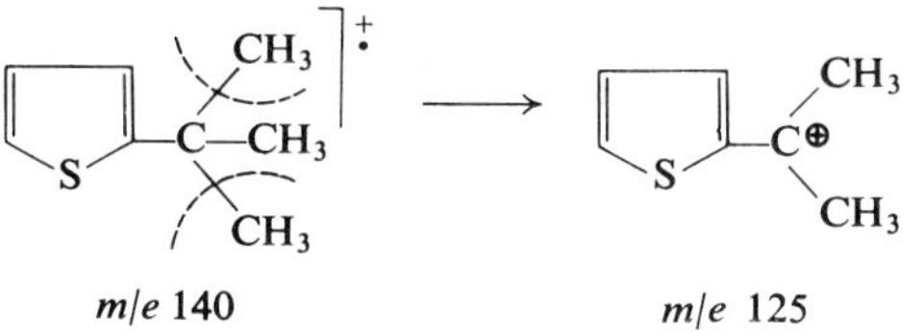

(thiophene, S) —C(CH₃)₃]⁺· ⟶ (thiophene, S) —C⊕(CH₃)₂ (5-97)

m/e 140 *m/e* 125

The subsequent events are quite interesting. A similar event occurs for the *t*-butylbenzene. For the 2-*t*- and 3-*t*-butylthiophenes this is followed by the loss of a neutral C_3H_4 particle as shown in Scheme 1. Rylander *et al.* (1956) observed this behavior for the *t*-butylbenzene and proposed the cationated cyclopropane ring as a probable ionic intermediate. In the case of the *t*-butylthiophenes the alternate structure suggested with the hydrogen on the sulfur atom is quite intriguing.

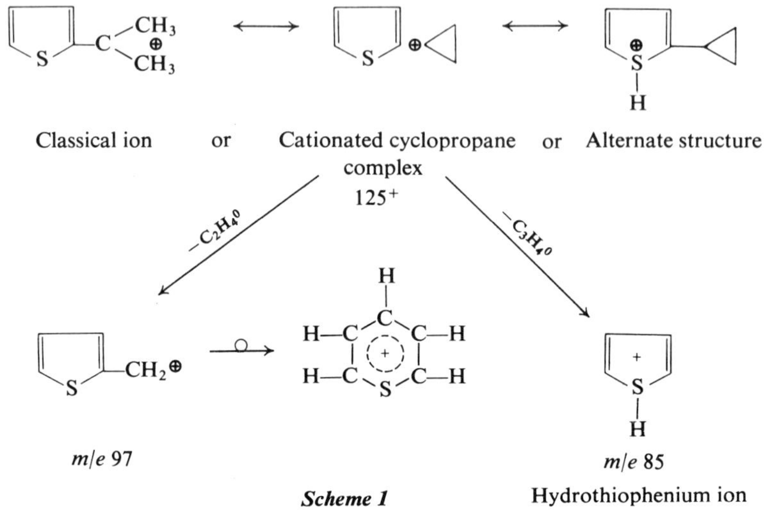

Classical ion or Cationated cyclopropane complex 125^+ or Alternate structure

$-C_2H_4O$ $-C_3H_4O$

m/e 97

m/e 85 Hydrothiophenium ion

Scheme 1

The scheme given for fragmentation is useful for correlations with other thiophenes. The ejection of the neutral C_2H_4 group with its hydrogen transfer does compete with the loss of the C_3H_4 neutral group. The m/e 97 intensities are about 18%, while the m/e 85 ion intensities are about 13% for the 2-*t*- and 3-*t*-butylthiophenes. Kinney and Cook (1952) pointed out that the 85 peak was comparatively high (10–15% of the base peak) for monoalkyl substitution when the carbon adjacent to the ring was a tertiary carbon. The m/e 85 peak was intermediate in intensity (4–10% of the base peak when the carbon adjacent to the ring was a secondary carbon atom). For all other types of substitution available in the various studies the 85 peak was less than 4% of the base peak. For the 2-*n*-alkyl- and 2-5-di-*n*-alkylthiophenes described earlier the 84 peak was more intense than the 85 peak, and therefore no confusion should arise from this source. The 84 peak was assumed to be from α cleavage with a single hydrogen transfer while the 85 peak represented two hydrogen atoms transferred. Foster (1966) reported that for the di-*t*-butylthiophenes (2,4- and 2,5-), β cleavage with the loss of a neutral methyl group was observed. Apparently the second event to occur was the loss of a second methyl group from the remaining *t*-butyl group that was intact. It should be noted, however, that the second loss of mass 15 produced an even-mass ion at $(M - 30^+)$. This could easily confuse the analyst who is attempting to identify unknown materials from the mass spectrum. More will be given on this fragmentation problem in the section on labeling studies of mechanisms.

The size (and possibly the electronegativity) of the atoms in the isopropyl- and *t*-butyl-substituted aromatic compounds has much to do with the fragmentation experienced. The aromatic fluorocarbons present a case in point. Gale (1968) has shown that the fluorinated isopropyl-substituted benzene loses the CF_3 group (69 *u*) to form the *m/e* 177 ion in much the same manner as the isopropylbenzene or thiophene would. However, the loss of a single F atom to form the 227 ion presumed to be from the single C–F bond to form a secondary ion occurs to a much larger extent than the corresponding loss of a hydrogen atom in the same location.

$$C_6H_5\text{-}C(F)(CF_3)_2^{+\cdot} \xrightarrow{-69^{\cdot}} C_6H_5\text{-}\overset{+}{C}(F)CF_3 \xrightarrow{-50^{\cdot}} C_6H_5\text{-}\overset{+}{C}F_2 \qquad (5\text{-}98)$$

m/e 246 *m/e* 177 *m/e* 127

Further, the second fragmentation of the *m/e* 177 species occurs to eject a neutral mass 50 moiety, without doubt the CF_2 group. This is considerably different from the corresponding benzene, etc. Ring expansion to form a tropylium like ion, $C_7H_6F^+$, is not claimed to occur in the spectra of the 10 fluorinated compounds examined.

Another interesting fragmentation occurs for the neopentylbenzene and thiophene which are quite parallel. Beta cleavage still occurs, but there is competition for the placement of the charge in the resulting products:

$$C_4H_3S\text{-}CH_2 \vdots C(CH_3)_3^{+\cdot} \longrightarrow C_4H_3S\text{-}\overset{+}{C}H_2 \quad \text{or} \quad {}^{+}C(CH_3)_3 \qquad (5\text{-}99)$$

m/e 97 *m/e* 57

The partial mass spectrum of the 2-neopentylthiophene is given in Table 5-13. The loss of the methyl group is now a γ-bond cleavage, and from the weakness expected at the β bond, an intensity ion this high is somewhat surprising. The loss of 29 or 43 mass units produces small ions, and this might be expected since one is dealing not only with neopentane but also with the thiophene ring. As written, the base peak is the *m/e* 97 ion. Very obviously, several processes are competing to form large ions at *m/e* 98 and 57. Since the 57 ion is presumed to be the ion formed by simple cleavage of the β bond from the ring, it must be more stable by a factor of $\frac{6}{5}$ greater than the ion at *m/e* 97. It is assumed here that only the *m/e* 98 ion is formed by a rearrangement involving a hydrogen shift to the thiophene moiety and that the 97 and 57 ions are produced by direct cleavage and charge competition. Since the

TABLE 5-13

PARTIAL MASS SPECTRUM OF 2-NEOPENTYLTHIOPHENE

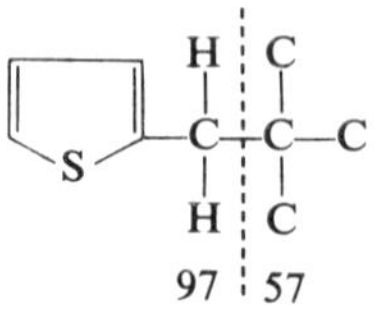

Ion	Relative intensity	Ion	Relative intensity
M^+	38.4	$(M - 71)^+$ (83)	0.65
$(M - 15)^+$	21.9	*m/e* 84	1.53
$(M - 29)^+$	0.4	*m/e* 85	1.78
$(M - 43)^+$	3.59	*m/e* 57	127.6
$(M - 57)^+$ (97)	100	*m/e* 98	91.4

m/e 97 ion may be a daughter ion of the *m/e* 98 ion, the test of charge competition is not that straightforward. Note that if the 98 ion is formed from the molecular ion, a hydrogen atom transfer from the *t*-butyl group results in the expulsion of a neutral isobutylene molecule. To establish the actual mechanisms, labeling and low ionization voltage studies had to be made. This is discussed further in Section V. It should be apparent that the presence of this type of molecule might momentarily confuse the mass spectroscopist, but a warning and a bit of experience should suffice to keep the interpreter out of difficulties such as presented by these molecules.

Many of the ideas presented for the alkylthiophenes are completely parallel with the alkylbenzenes, and it is this parallelism that leads to strongly supported correlations for many similar aromatic systems. Pioneering work on the hydrocarbons and alkylbenzenes proved to be of immense value to those who followed with correlations of other types of molecules. The effect of chain branching for the alkylbenzenes is further demonstrated by the data given in Table 5-14. Partial spectra for three cycloalkylbenzene derivatives are given and show that the cleavage at the β bond still persists to produce the *m/e* 91 and 92 peaks for both API Serial Nos. 36-m and 38-m. For the molecule of API 37-m, the β cleavage occurs, and, perforce, the α-substituted methyl group must be carried with the moiety. Even so, the presence of such a large *m/e* 106 ion indicates that the hydrogen rearrangement still accompanies the cleavage. It is also interesting to note that the hydrogen originates in a saturated-ring system and moves toward the moiety containing the highly unsaturated phenyl ring.

Table 5-14

PARTIAL MASS SPECTRA OF SOME CYCLOALKYLBENZENES

	Total ion intensity (%)		
m/e	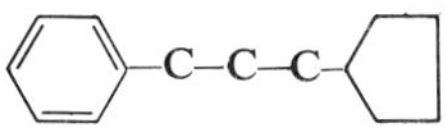 Ph—C—C—C—cyclobutyl	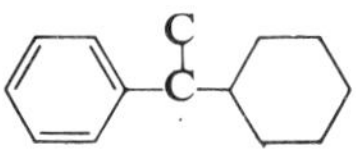 Ph—C(C)—cyclohexyl	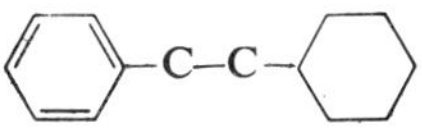 Ph—C—C—cyclohexyl
91	19.6	6.47	13.3
92	20.0	0.55	31.6
105		23.8	
106		19.5	
188 M	7.0	4.2	5.3
API No.	36-m	37-m	38-m

5. *The A_5 Type of Cleavage*

$$\left[-\overset{|}{\underset{|}{C}}-X\right]^{+\cdot} \longrightarrow -\overset{|}{\underset{|}{C}}^{+} + {}^{\cdot}X \qquad X = \text{halogen, OR, SR, NR}_2\text{, R} = \text{H or alkyl)}$$

In this type of cleavage, the bond between a carbon atom and a hetero atom is broken. Since this is more difficult than the cleavage of a carbon–carbon bond if X in the C–X bond is O, N, S, F, or Cl, the positive charge remains with the carbon atom rather than with the hetero atom. Such cleavages are usually rare, and the qualification "if it occurs" is to be considered. The ethers are probably the best examples of this sort of behavior.

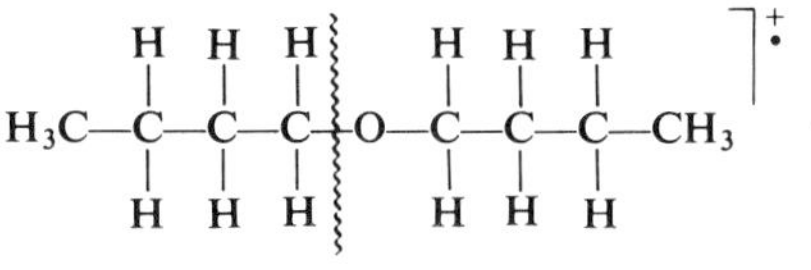

$$\left[H_3C-CH_2-CH_2-CH_2\,\wr\,O-CH_2-CH_2-CH_2-CH_3\right]^{+\cdot} \longrightarrow C_3H_7^{+} \qquad (5\text{-}100)$$

1.52% 100%

The charge remains preferentially with the alkyl ion.

In the case of the corresponding thiaalkane, the ions at *m/e* 146 and 57 are, respectively, 31.1 and 31.1%, or 1/1. Cook and Foster (1961) gave the base peak as *m/e* 61 and suggested a cyclic structure for it. While cyclic intermediates occur in many reactions and especially so in mass spectrometry, it is interesting to observe that the simple cleavage does not appear to occur to any large extent except for the oxygenated compounds; for other hetero atoms the cyclic intermediate seems to be the rule. For 2-thiapropane Cook and Foster (1961) proposed a three-membered cyclic intermediate to explain the major fragment ions observed. The *m/e* 35 ion was very large, being 35% of the *m/e* 47 peak,

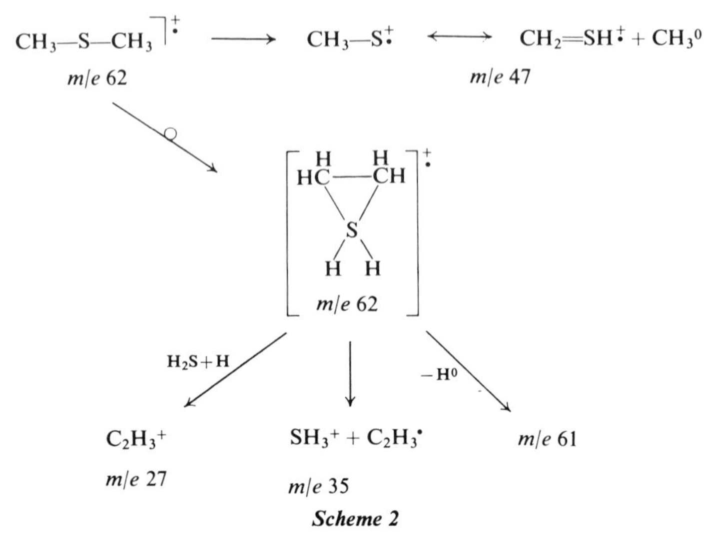

Scheme 2

which was the ion of largest intensity. Being a small molecule, the loss of a hydrogen to form the *m/e* 61 ion of 30% also was not too surprising. However, the appearance of a prominent ion at *m/e* 27, consisting of $C_2H_3^+$, definitely called for a three-ring intermediate. This is depicted in the sequences of Scheme 2, which may well account for the production of all of these ions. For the alkylthiaalkanes the intensity of the *m/e* 35 ion drops rapidly with increasing chain length. This indicates that additional lower energy paths of fragmentation are available to the excited molecular ion and, therefore, that the formation of the *m/e* 35 ion probably involves sizable energy differences from other commonly observed processes for the thiaalkanes.

Levy and Stahl (1961) observed that the prominent ions in the spectra of thiaalkanes arose from α cleavage processes, carbon–sulfur bond cleavages, and olefin elimination to form mercaptan ions of the *m/e* 47 and homologous variety. The α cleavage also occurs as an important fragmentation process for ethers and amines. Sample and Djerassi (1966) studied labeled isopropyl *n*-amyl sulfide and isopropyl *n*-butyl sulfide and showed that in addition to these processes, four-, five-, and six-membered cyclic transition states were involved. In their studies they reached the conclusion that the *m/e* 61 ion was arising from an α cleavage process followed by a McLafferty rearrangement (see Section IV,F). The presence of the sulfur atom leads to problems in molecules containing other hetero atoms as well. McFadden *et al.* (1965) reported on some anomalies in the mass spectra of the thioesters. An important rearrangement reaction produced the loss of $SCHRCH_2$– from

R′COSCHRCH$_2$ (C_nH_{2n+1}), in favorable cases contributing more than 10% of the total ion current. A labeled study of aliphatic ethers and aliphatic amines, conducted by Djerassi and Fenselau (1965a,b), showed similar results but emphasized the fact that the nature of the hetero atom exerts an effect, as judged by the quantitative differences between the ethers and amines. Additionally, in these cases and the subsequent work of Sample and Djerassi (1966), three-, four-, five-, and six-membered-ring intermediates were possible, and when a hydrogen atom was abstracted to the hetero atom, a methylene hydrogen atom was favored over the abstraction of a methyl hydrogen when both were available.

The above material could have just as well been discussed under rearrangements of the E_1 type, as will be seen later. However, it was introduced here to serve as an example and warning of the possible oversimplification of fragmentation schemes by the chemist anxious merely to utilize a correlation.

This type of cleavage is more often observed in aliphatic ethers than in most other substances. In some mercaptans, depending upon chain branching, etc., the alkyl ions do comprise a rather large percentage of the total ion intensity. The effect is even more notable if alkyl groups are attached to the carbon atom adjacent to the X grouping. Alcohols such as the tertiary variety give fragments attributable to the loss of the OH group in contrast to the primary alcohols, most of which eliminate a neutral water molecule. For the oxygen and nitrogen heteroatomic systems the detection of O–CH_3 or N–CH_3 groups in many organic molecules is very difficult as compared to the C–CH_3 groups which usually give rise to reasonable or even large M − 15 ions as described previously. The presence of S–CH_3 groups is usually possible to detect, although it is not always certain. A bit more can be said concerning the halogens; the behavior of fluorine is quite interesting and will be discussed later.

6. The B Type of Cleavage

This type of cleavage is more common for systems involving a hetero atom

$$\left[\mathrm{R{-}\overset{|}{\underset{|\underline{X}}{C}}{-}R}\right]^{+\cdot} \longrightarrow \mathrm{R^{\bullet}} + {}^{+}\mathrm{\overset{|}{\underset{|\underline{X}}{C}}{-}R} \longleftrightarrow \mathrm{\overset{|}{\underset{\|X^{+}}{C}}{-}R}$$

in the location given. The free electron pair of the hetero atom is involved in that it stabilizes the positive charge on the adjacent carbon atom through resonance forms of the type shown. Biemann (1962c) pointed out that the mass spectra of molecules containing this type of hetero atom are much less monotonous than the spectra of hydrocarbons. He also established that the effectiveness of stabilization is not in the order of electronegativities for the hetero atoms. It should be noted that the resonance stabilization is far more effective than the inductive effects of the alkyl groups. Earlier in this chapter it was shown that Cummings and Bleakney (1940) suggested the participation of the free electron pair on oxygen in the CH_2OH^+ fragment from methanol based upon the appearance potential of this ion. It should be kept in mind that the various hetero atoms behave somewhat differently since each has its own electron valence shell configuration and, further, that some, such as sulfur, may have low-lying *d* orbitals available for excited states of ions.

The classic example of this type of behavior is that of the aminoethanol

$$\underset{NH_2}{\overset{\cdot}{C}H_2} + \underset{\substack{OH \\ m/e\ 31}}{\overset{+}{C}H_2} \longleftarrow \left[\begin{matrix} CH_2 - CH_2 \\ | \quad\quad | \\ NH_2 \quad OH \end{matrix}\right]^{+\cdot} \longrightarrow \underset{\substack{NH_2 \\ m/e\ 30}}{{}^{+}CH_2} + \underset{OH}{{}^{\cdot}CH_2} \qquad (5\text{-}101)$$

molecule ion. On the basis of the ion intensities at *m/e* 30 = 57% Σ_{13} and *m/e* 31 = 3.1% Σ_{13}, one would have to conclude that the charge is preferentially held by the nitrogen-containing species by about 20/1.

In the case of the corresponding thioethanol, the ratio of the *m/e* 47 ion to the *m/e* 31 ion is about 2/1, while for 2-chloroethanol the ratio of the *m/e* 49 ion to the *m/e* 31 ion is about 1/20. In the latter case, the oxygen is far better able to stabilize the positive charge. Charge competition among fragments is a very important part of the interpretation of mass spectra and sometimes is the major key in deducing structures, just as shown above. For some molecules the presence of one group may so dominate the fragmentation pattern of a given species that it makes the detection of a second hetero atom very difficult. This simply underscores the fact that the use of every means at the disposal of the mass spectroscopist should be used to determine ion compositions, i.e. neutral particles ejected, and the presence of multiply charged ions.

7. *The C Type of Cleavage*

This cleavage is typical of carbonyl-containing compounds and others containing a hetero atom similarly situated. The distinction between this and the type B cleavage is that the hetero atom in this case carries the charge, has an unshared pair of electrons, and is triply bonded to the carbon atom. The mechanism of formation may be

$$R'^{\cdot} + R-\underset{\underset{O}{\|}}{C^{+}} \longleftarrow \left[R-\underset{\underset{|\overline{O}|}{\|}}{C}-R'\right]^{+\cdot} \longrightarrow {}^{+}\underset{\underset{|\overline{O}|}{\|}}{C}-R' \longleftrightarrow \underset{\underset{|O^{+}}{\|\|}}{C}-R' + R' \qquad (5\text{-}102)$$

The benzaldehyde molecule ion loses a hydrogen atom to produce the very stable ion at *m/e* 105 in this fashion. The 1-phenyl-1-alkanones similarly produce this ion in large intensity in their spectra.

In cleavage at the bond α to the C–X group it does not matter how one views the series of events since either

$$\underset{|X^+}{\overset{}{R-\overset{\|}{C}-R'}} \longrightarrow R-\underset{X^+}{\overset{|||}{C}} + {}^{\bullet}R' \qquad (5\text{-}103)$$

or

$$\left[R-\underset{|X|}{\overset{\|}{C}} \wr R'\right]^{+\bullet} \longrightarrow R-\underset{|X|}{\overset{\|}{C^+}} \longleftrightarrow R-\underset{|X^+}{\overset{|||}{C}} + {}^{\bullet}R' \qquad (5\text{-}104)$$

produces the same net end result. This is also true for the singly bonded hetero atom group.

In some cases atoms contain two hetero atoms of the same kind but in different bonding situations. The hydroxy alkanones are an example. If the oxygen atoms are on adjacent carbon atoms, the spectrum usually shows peaks in the *m/e* 45, 59, 73 series due to the presence of the $RCH{-}OH^+$ ions in addition to those due to the $R{-}C{=}O^+$ ions. The relative amounts vary considerably depending upon the molecular weight and structure of R. If a relatively long alkyl chain with a primary alcohol group is present, the molecule may revert to the loss of water, just as an ordinary primary alcohol. The keto group should still be detected and should cause relatively large peaks at 43, 57, 71, 85, etc., due to the $R{-}C{=}O^+$ ion.

$$\left[H_3C-\underset{O}{\overset{\|}{C}}-\underset{OH}{\overset{H}{C}}-CH_3\right]^{+\bullet}$$

$$H_3C-C{\equiv}O^+ + {}^{\bullet}\underset{OH}{\overset{H}{C}}-CH_3 \qquad\qquad H_3C-\overset{H}{\underset{+}{C}}-OH + R-\underset{O}{\overset{\|}{C^{\bullet}}} \qquad (5\text{-}105)$$

m/e 43   *m/e* 45

8. *The D Type of Cleavage—An Internal Rearrangement*

In effect this type of molecular ion electron shifting, bond breaking and making is an internal rearrangement. It has become known as a retro-Diels–Alder reaction and results in the ejection of a neutral olefinic particle. It arises from an odd-electron ion and produces an odd-electron ion fragment, supposedly less stable because of its odd-electron character. It was known to workers in the petroleum field, who regularly saw it in conjunction with the fragmentation of the tetralins.

$$\text{[tetralin]}^{+\cdot} \longrightarrow \text{[o-xylylene: } CH_2, CH_2]^{+\cdot} + C_2H_4{}^0 \qquad (5\text{-}106)$$

m/e 132 m/e 104

This type of cleavage involves the shifting or rearranging of two electrons but differs from A_2 in that two bonds are cleaved and two new bonds are made. Budzikiewicz *et al.* (1965) have published a rather extensive review of the subject. Cook and Foster (1961) described the importance of this behavior for some aromatic sulfur compounds. Data presented in Table 5-15 show the extent to which the reaction goes on for these large aromatic molecules. It

TABLE 5-15

PARTIAL MASS SPECTRA—TETRAHYDRODIBENZOTHIOPHENE AND DERIVATIVES

	Relative intensities						
	MW 188	202	238	238	298	192	190
m/e							
110							375.8
M − 56					13.4	19.4	8.5
M − 55					3.3	3.7	32.9
M − 28	100.0	100.0	81.1	69.6	33.9	100.0	0.3
M − 27	12.6	14.6	13.8	12.3	6.7	9.1	0.8
M − 1	24.3	24.7	17.8	16.1	11.6	21.7	0.6
M	76.2	90.0	100.0	100.0	100.0	66.1	100.0[a]
M + 1	11.6	15.0	18.8	19.2	21.7	11.0	11.3
Source:	API 1375	API 1307	API 1377	API 1345	API 1407	USBM[b]	USBM

[a] Parent ion used as base; actually *m/e* 110 = 375.8%.
[b] U.S. Bureau of Mines.

should be noted that the molecules which can experience two retro-Diels–Alder (RDA) reactions do so. The ions at M − 56 are not as intense as the ions at M − 28, but they are still quite intense. The rearrangement ions appeared at relatively low ionization voltages, and this led to possible analytical interference. As further proof that the rearrangement was restricted to a molecule containing a double bond in a cyclohexane ring, the mass spectrum of hexahydrodibenzo[*b*]thiophene was obtained. The compound shown in the last column of Table 5-15 was prepared by Hopkins (1959). Once the double bond was gone, the molecular ion could not undergo the RDA but did undergo the rearrangement of two hydrogen atoms to produce the ion of largest intensity at *m/e* 110. The probable mechanism is

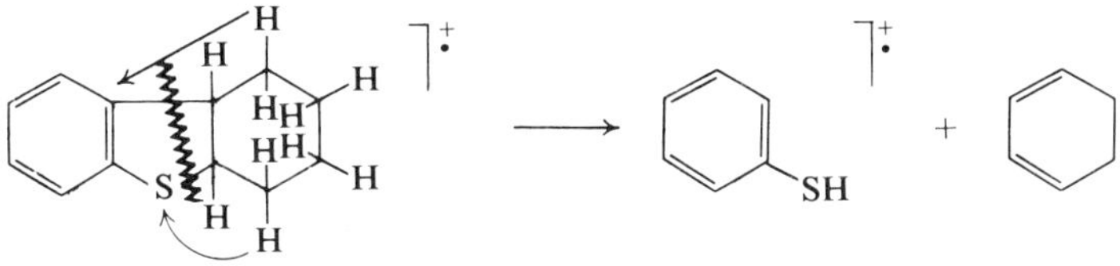

(5-107)

m/e 190 *m/e* 110 Neutral olefin

It should be noted that at low voltage the ions at 109 and 110 were prominent, but ions at *m/e* 79 and 80 were also present. Dialkyl ethers (greater than ethyl) undergo similar rearrangements to produce neutral alcohol fragments and an olefinic ion. The mass spectroscopist with scanty prior information upon examining the sample might think that the sample contained the stated molecule plus thiophenol (benzenethiol) and possibly the cyclohexadiene.

Analytical use of the RDA can be obtained for the 1- and 2-thiatetralins, which both give large molecular ions and the fragments at *m/e* 122 and 104 of varying intensities as given in Table 5-16. It should be noted that many other reactions also take part in the fragmentation pattern of the thiatetralins as contrasted with the tetrahydrodibenzothiophene derivatives where the RDA process predominates. The thiatetralins also produce ions characteristic of both the benzenes and the naphthalenes. Cook and Foster (1961) also reported that the 1- and 2-thiaindans did not follow this type of cleavage. Instead, strong M^+ ions and strong $M - 1^+$ ions were observed. The loss of a single hydrogen atom results in a very good Hückel ion if one hydrogen atom is moved to the sulfur atom leaving an ion that could be likened to naphthalene if one considers the sulfur atom (+H) as equivalent to the C_2H_2 group in the naphthalene.

Harrison *et al.* (1970) reported the operation of an RDA reaction in the thiochromanones and related compounds. The behavior was quite similar to the thiatetralins as shown in Schemes 3a and b. The position of the sulfur atom was readily determinable, and the RDA reaction preceded all further fragmentations. With the sulfur in the 2 position, the principal path involved

TABLE 5-16

THE RDA REACTION FOR THIATETRALIN

Structure		m/e	Relative intensity (%) 1-Thiatetralin	2-Thiatetralin
m/e 150 $\xrightarrow{-28}$	m/e 122 (CH_2, S)	104	14.5	74.0
m/e 150 $\xrightarrow{-46}$	m/e 104 (CH_2, CH_2)	122	28.3	25.7
		150	100 M	100 M

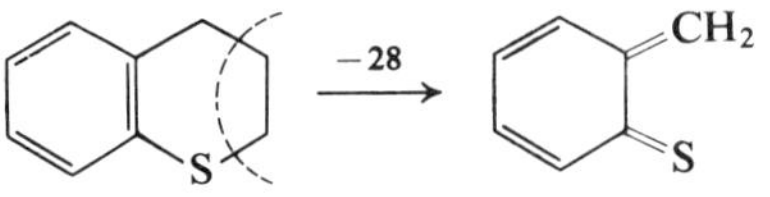

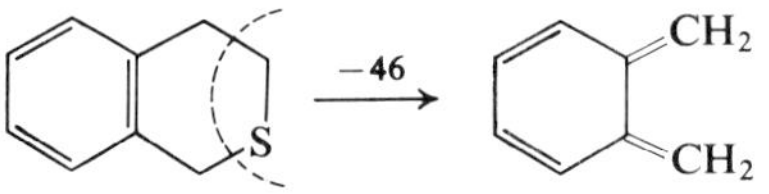

the RDA, then the loss of a neutral CO particle, followed by either the loss of a hydrogen to produce the *m/e* 89 moiety or the typical loss of C_2H_2 from a tropylium ring to produce the *m/e* 64 ion. Thus, in Scheme 3a a pathway is delineated by metastable ions for three successive processes leading to odd-electron ions. The contrasting ions that appear when the sulfur atom is in the 1 position are quite obvious from Scheme 3b. In this case a neutral CS particle can be ejected analogous to the CO.

These same workers showed that when a methyl group is present on the benzo ring of the system the methyl group is more or less simply carried along. By examining Schemes 4a and b, this can be observed to be in contrast to the compound with the methyl group on the ring with the sulfur atom and carbonyl

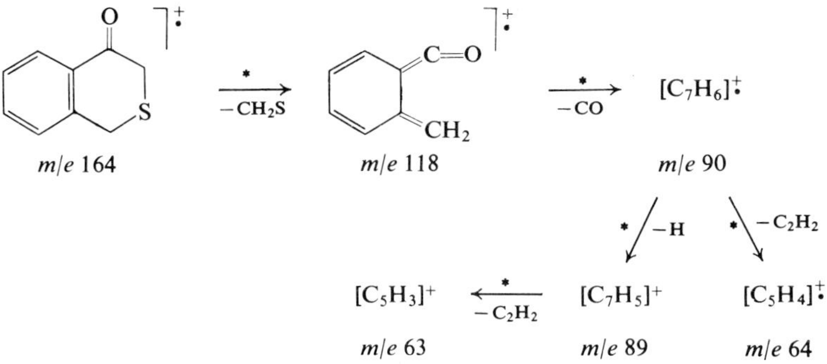

m/e 164 $\xrightarrow[-CH_2S]{*}$ m/e 118 $\xrightarrow[-CO]{*}$ $[C_7H_6]^{+\cdot}$ m/e 90

m/e 90 $\xrightarrow[-H]{*}$ $[C_7H_5]^+$ m/e 89 $\xrightarrow[-C_2H_2]{*}$ $[C_5H_3]^+$ m/e 63

m/e 90 $\xrightarrow[-C_2H_2]{*}$ $[C_5H_4]^{+\cdot}$ m/e 64

Scheme 3a

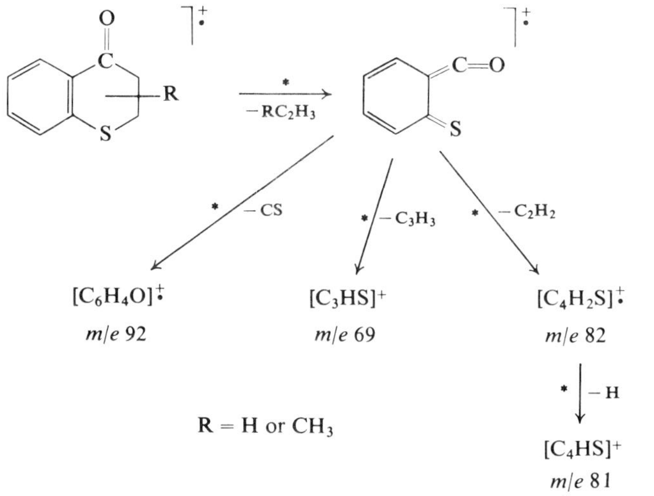

O, C, S, R (structure) $\xrightarrow[-RC_2H_3]{*}$ C=O, S (structure)

$[C_6H_4O]^{+\cdot}$ m/e 92 (* −CS)

$[C_3HS]^+$ m/e 69 (* −C_3H_3)

$[C_4H_2S]^{+\cdot}$ m/e 82 (* −C_2H_2)

↓ * −H

$[C_4HS]^+$ m/e 81

R = H or CH_3

Scheme 3b

O, C, S, CH_3 (structure) m/e 178 $\xrightarrow[-C_2H_4]{*}$ C=O, S, CH_3 (structure) m/e 150

↓ −CO

$[C_7H_5S]^+$ m/e 121 $\xleftarrow[-H]{*}$ $[C_7H_6S]^{+\cdot}$ m/e 122 $\xrightarrow[-CS]{*}$ $[C_6H_6]^+$ m/e 78

↓ * −CS (from m/e 121) ↓ * −C_2H_2 (from m/e 122)

$[C_6H_5]^+$ m/e 77

$[C_5H_4S]^{+\cdot}$ m/e 96 $\xrightarrow[-H]{*}$ $[C_5H_3S]^+$ m/e 95

↓ * −C_2H_2

$[C_3HS]^+$ m/e 69

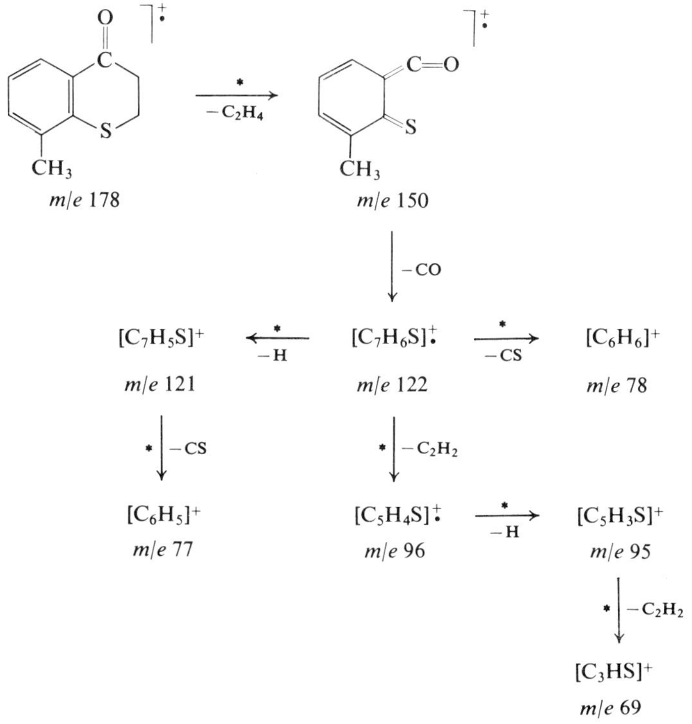

Scheme 4a

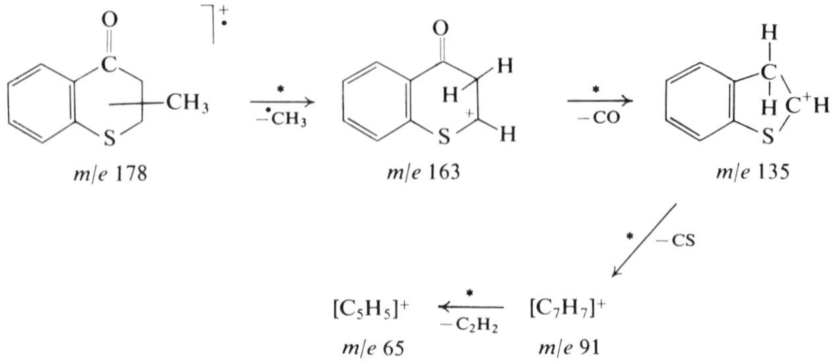

$-\dot{C}H_3$ $-CO$ $-CS$ $-C_2H_2$

m/e 178 m/e 163 m/e 135

$[C_5H_5]^+$ m/e 65 $[C_7H_7]^+$ m/e 91

Scheme 4b

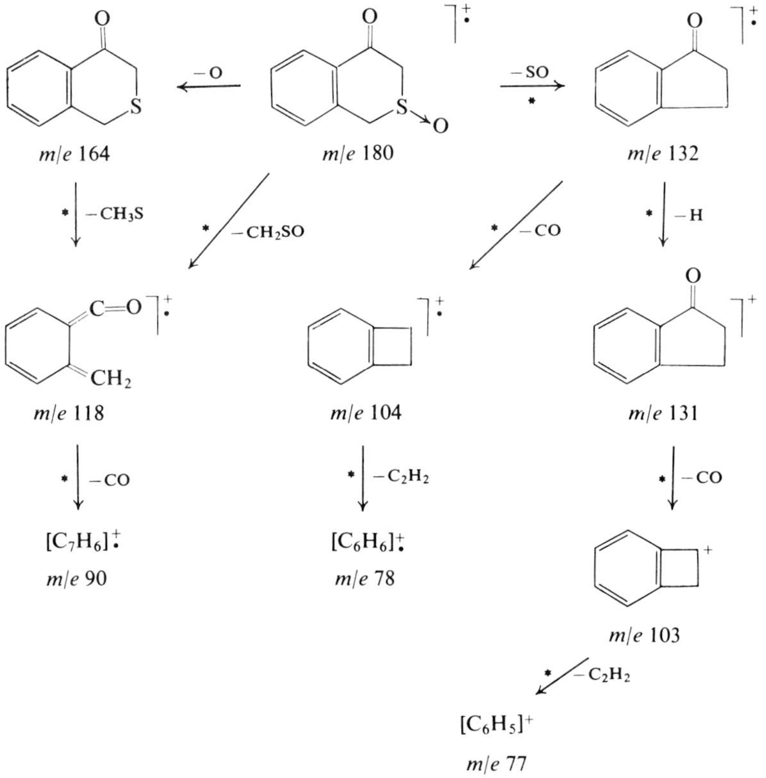

$-O$ $-SO$

m/e 164 m/e 180 m/e 132

$-CH_3S$ $-CH_2SO$ $-CO$ $-H$

m/e 118 m/e 104 m/e 131

$-CO$ $-C_2H_2$ $-CO$

$[C_7H_6]^{+\cdot}$ m/e 90

$[C_6H_6]^{+\cdot}$ m/e 78

m/e 103

$-C_2H_2$

$[C_6H_5]^+$ m/e 77

Scheme 5a

group. In the case of 4b, the neutral methyl group is lost; the second event is the loss of the neutral CO group leading to the *m/e* 135 ion, without doubt of the same structure as the 135 ion observed by Cook and Foster (1961) in the thiaindans.

The sulfoxide compounds showed the ejection of the oxygen atom quite similar to the *N*-oxide compounds. Further fragmentation then produced ions by the same mechanism as shown in Scheme 3b. Alternatively the ejection of the entire neutral SO fragment was established by the presence of a metastable peak leading to the fragmentation paths shown in Schemes 5a and b. Once again the positional isomer could be established with confidence. It is left to the reader to prove that alkyl substitution on one or both of the rings, etc., could readily be established by the appropriate fragments. For the sulfones, the RDA reactions occurred, regardless of the position of the sulfur atom, but did so in competition with other pathways. The losses of C_2H_4, SO_2, CH_2SO_2, and C_2H_4O were all established by metastable ions (m*) depending upon the positional precursor. The direct loss of SO_2 was only a very minor process, and the original work contains many additional details.

Willhalm *et al.* (1964) reported on a number of oxygenated species including the chromanones. Although of less importance than some other pathways, an RDA reaction quite similar to those just discussed is observed in this case.

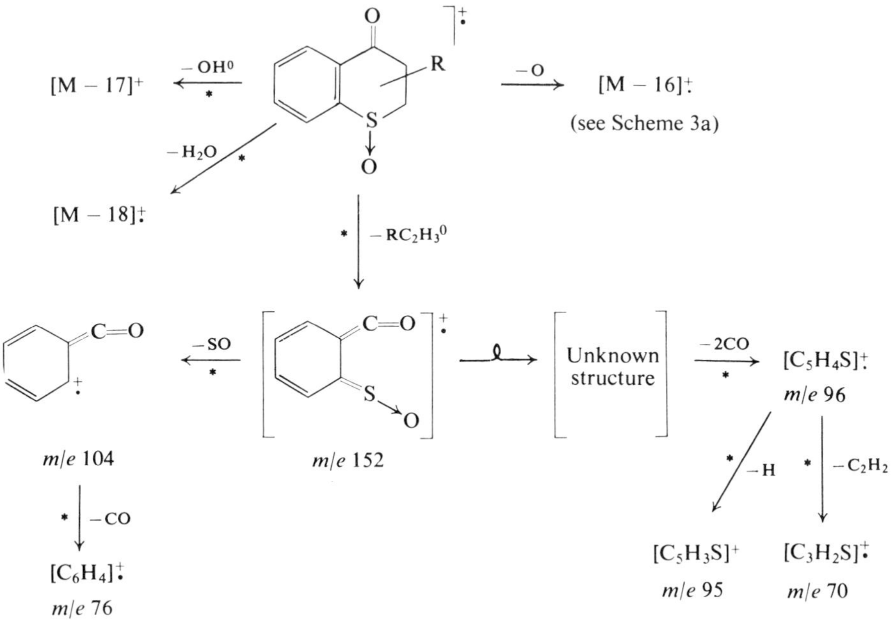

Scheme 5b

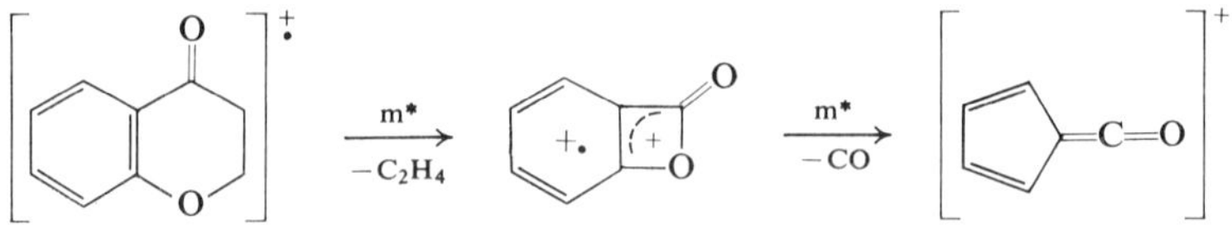

m* $-C_2H_4$ m* $-CO$ C=O

m/e 148 m/e 120 m/e 92 (5-108)

Further proof of the widespread applicability of the RDA reaction has been given by Reiser (1969) for some substituted uracils. For the two molecules shown below, the strong and approximately equal molecular ions, two mass units apart and typical of a bromine-containing moiety, are present. The M − 55 ions are strong in each case, attributable to the cleavage of the secondary butyl group. The next ions observed are from the ejection of the neutral isocyanate and isothiocyanate species.

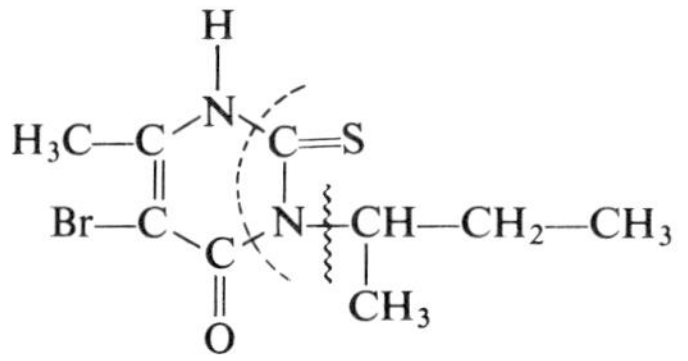

H_3C—C, Br—C, N—H, C=S, N—CH—CH_2—CH_3, CH_3, C=O and H_3C—C, Br—C, N—H, C=O, N—CH—CH_2—CH_3, CH_3, C=S

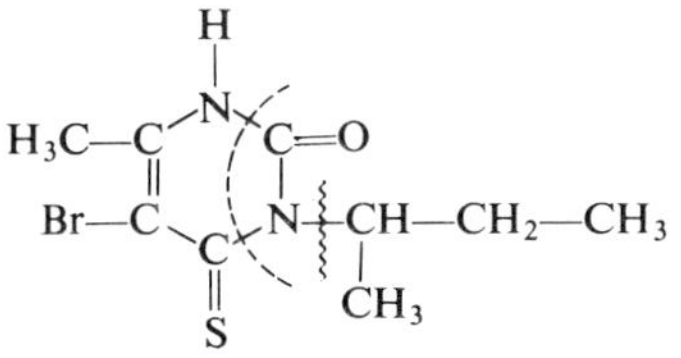

Draper and MacLean (1968) reported the mass spectra of some tetrahydroquinolines. The molecules behave quite differently depending upon the presence of the nitrogen atom in the tetrahydro ring. Thus, in this respect the nitrogen compounds behave differently than the sulfur compounds.

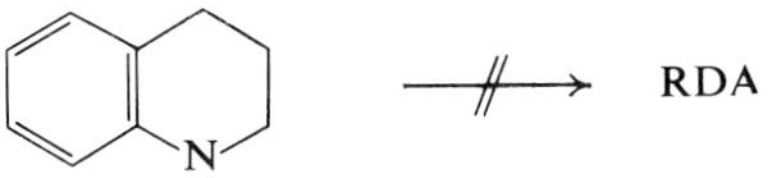

$\not\longrightarrow$ RDA (5-109)

RDA CH_2 CH_2 + $C_2H_4{}^{\bullet 0}$ (5-110)

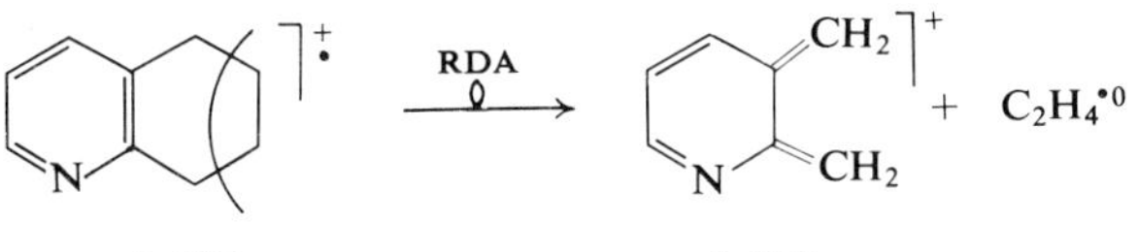

m/e 133 m/e 105

In an examination of some chlorinated pesticidal compounds, Damico *et al.* (1968) reported the operation of RDA ejections from bridged polycyclic compounds. Some HCl and also neutral H and Cl ejections were noted, so that matters were not so simple. Binks *et al.* (1969) have also reported RDA reactions for polychlorinated bridged-ring compounds based upon the norbornene skeleton. Staley and Reichard (1968) reported that the course of some RDA reactions are influenced by olefin side chains present in cyclohexene derivatives.

Based on structure, an RDA reaction can sometimes be anticipated. Yet for the analyst who is examining an unknown material, the appearance of strong peaks 28 mass units below a presumed molecular ion and persisting at low ionization voltage strongly suggests compounds in the same series two carbon numbers apart. Experience with the RDA reactions must be gained in order to complete the interpretation of the spectrum.

Some molecules eject neutral olefin fragments in conjunction with the rearrangement processes of the type shown for the alkylbenzenes, alkylthiophenes, etc. Quite naturally, the loss of mass 28 might be observed here, but little thought is given to an RDA type of mechanism for these cases. There are molecules, however, which can eject neutral ethylene fragments from a cyclic structure, and these might be confused with the RDA mechanism. Since a double bond does not exist in the correct position, the mechanism is not RDA, but the comparison is of interest. The 1,4-dithiane molecule is one such case. Condé-Caprace and Collin (1969) have described this system in great detail, including ionization potential studies. The molecule readily ejects the ethylene group from the cyclic system, and at least for some of the fragments, the ring is remade because the subsequent reactions shown do occur. The presence of the S_2 ion can be taken as evidence in support of the ring closure. The partial mechanism is shown in Scheme 6. Other events typical of aliphatic sulfur molecules also take place.

$$[\text{1,4-dithiane}]^{+\cdot} \longrightarrow C_2H_4 + [C_2H_4S_2]^{+\cdot} \longrightarrow [S_2]^{+\cdot} + C_2H_4 \quad \text{or} \quad [CH_2S]^{+\cdot} + CH_2S$$

$$\longrightarrow \text{etc.}$$

Scheme 6

These same authors also reported on the 1,4-oxathiane compound. The molecular ion exhibited less stability than the 1,4-dithiane or 1,4-dioxane, quite in contrast to general observations of the stability of molecular ions for sulfur compounds as compared to oxygenated compounds. Except for three minor ions, the oxathiane does not produce oxygen-containing ions once the molecular ion is formed. Thus, the generality of charge remaining with the sulfur-atom-containing moiety is borne out despite the discrepancy for the molecular ion. The OS^+ ion was observed, although to a lesser extent, indicating that less of the above type of rearrangement occurred. The similarities in fragmentation were considerably greater than the comparison of the ethers and thio ethers made earlier. These authors also point out that the IP of the two molecules are equal within experimental error, thus indicating in both cases that the sulfur atom carries the charge. In the case of the oxathiane, the cleavage of the M^+ ion to produce essentially only ions in which the charge is preferentially held on the sulfur atom is in agreement with Stevenson's rule (1951) because sulfur molecules and radicals have considerably lower IP's than the corresponding oxygen moieties.

F. Rearrangement Processes

From the previous discussions of the RDA and related processes, it should be obvious that a clear-cut separation of fragmentation and rearrangement reactions is not always possible. It is generally accepted that if an atom in a molecule breaks a bond with another atom and reforms a bond with some other atom, a rearrangement has taken place. On this basis, the 1,4-dithiane ejection of ethylene is a rearrangement only if the four-membered cyclic ring is reformed! In the processes to be discussed in this section, there will be little doubt of the fact of a rearrangement.

1. The E_1 Type of Elimination

The E_1 type of elimination of water from an alcohol is a rearrangement on the basis of the cyclic intermediate state involved. Friedel *et al.* (1956) gave correlations for the mass spectra of alcohols with their structures. This work showed the ejection of a water molecule from the molecular ion and pointed to the fact that once this event occurred the remaining mass spectrum strongly resembled that of an olefin. McFadden *et al.* (1958, 1963) studied labeled molecules and showed that this was essentially all 1,3 and 1,4 elimination of the water, thus involving either of the following cyclic structures as intermediates:

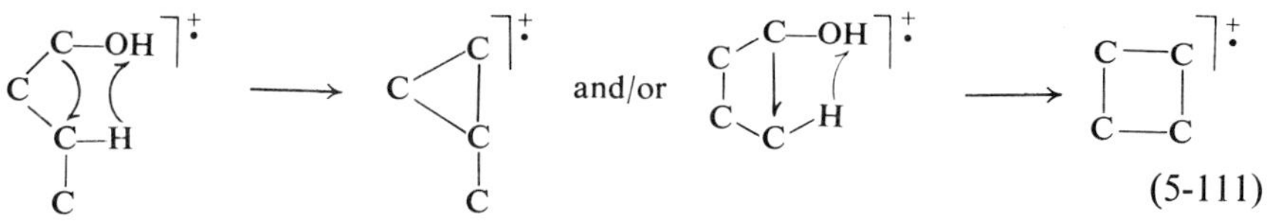

C—OH, C, C—H, C $]^{+\cdot}$ ⟶ C, C, C, C $]^{+\cdot}$ and/or C—OH, C, C, C—H $]^{+\cdot}$ ⟶ C—C, C—C $]^{+\cdot}$

(5-111)

McDonald *et al.* (1963) reached the same conclusions from their studies of alcohols. Benz and Biemann (1964) examined butyl through heptyl alcohols, including some labeled species. They claimed that elimination of the 1,4 variety was much more prominent than the 1,3 elimination. Meyerson and Leitch (1964) examined normal hexanol and labeled molecules at 4-*d*, 5-*d*, and 6-*d* and reached similar conclusions. Benz and Biemann (1964) also observed some evidence that a small contribution came from a hydrogen atom located at C-5. All in all, the result is that the elimination of water occurs, preferentially as a 1,4 elimination with some 1,3 and possibly a very small amount of 1,5 elimination. The resulting ion $(M - 18)^+$ is an odd-electron ion. As a consequence, the behavior as an alkene and the close similarity of the mass spectra to alkenes should not be surprising. In addition to the alkene type of mass spectra, there are ions at masses 31, 45, etc., which contain the oxygen atom and, therefore, distinguish the spectrum from that of a simple olefin.

Friedel *et al.* (1956) noted that the amount of ions containing oxygen increased in intensity as the hydroxy group moved into a secondary or tertiary position. For secondary alcohols, about 15% or more of the total ion current came from oxygenated ions, while for tertiary alcohols, the oxygen-containing ions accounted for over 30% of the total ion current. They also noted that the parent-ion intensity was usually less than 2% of the base peak intensity, while the tertiary alcohols had insignificant parent ions, a fact which stimulated this research group to make trimethylsilyl ether derivatives in order to determine what the molecular ion was. Friedel *et al.* (1956) also provided a very important observation. For the molecules they studied, the secondary alcohols usually had a base peak that was an oxygen-containing ion, as did the tertiary alcohols. Correspondingly, these compounds had mass spectra that resembled the alkene spectra far less than the *n*-alcohols. Beynon *et al.* (1968e) discussed additional types of alcohols, including multifunctional molecules containing an OH group.

When considering other heteroatomic systems, matters change a bit. For the primary mercaptans, the loss of H_2S is still observed but not as prominently as in the alcohols. Hence the mass spectrum of a primary mercaptan does not resemble that of an alkene as closely as its oxygenated counterpart. Instead, ions of the type $CH_2{=}SH^+$ at *m/e* 47, 59, etc., are more important. This is also in accord with the behavior of the amines, as described earlier.

2. The E_2 Type of Elimination

Elimination of a neutral molecule between adjacent groups on a *cis* double bond or an aromatic system comprises this type of elimination. Many organic chemists might like to classify the various reactions that have been shown in a different fashion, perhaps in terms of multicentered reactions, etc. The scheme used was chosen because it presents the mass spectroscopist with species to be watched for in association with a given fragmentation or rearrangement. Since an olefinic double bond is easy to visualize in the molecule, the E_2 type of elimination is rather easy to predict. Williams *et al.* (1967) have shown excellent examples of this elimination for the α- and β-unsaturated nitriles and esters. For the nitrile case, the following events occur:

$$\left[(R)(H)C{=}C(CN)(X)\right]^{+\cdot} \xrightarrow{\ \ell\ } RCN^{+\cdot} + H{-}C{\equiv}CX \qquad (5\text{-}112)$$

Note the use of the arrow with a circle on it to indicate a skeletal rearrangement. The skeletal rearrangement is facilitated by the proximity of the highly unsaturated group. The compounds examined included the cases in which R was phenyl–, methylphenyl–, furanyl–, thienyl–, and chlorophenyl–. The X group was *t*-butyl–, cyano–, and CO_2–ethyl.

Some alkoxy and OH-containing esters were examined. Williams *et al.* (1967) thought that the mechanism was best represented by a 1,3 shift. The ejection of the neutral ketenic type of molecule is a distinguishing feature of this elimination, as shown in Scheme 7. These high-resolution studies established the validity of the mechanism for both the cases given above. It was proposed

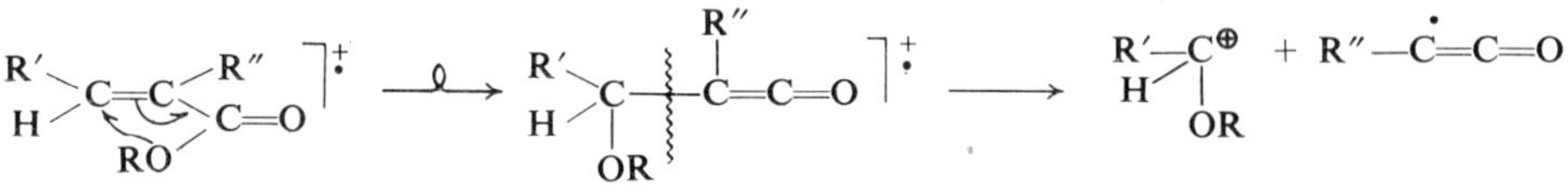

Scheme 7

that a 1,5 shift is possible if the phenyl substituent is conjugated with the double bond. In *trans*-cinnamates, the operation of a 1,5 shift requires isomerization after the primary ionization.

Also classified as E_2 elimination processes are reactions in mass spectra that are frequently referred to as "*ortho* effects" since the elimination and rearrangements involve *ortho* substitution on a phenyl group or other equivalent bonding situation. McLafferty and Gohlke (1959) showed that a neutral water molecule was lost from *ortho*-methoxy-substituted aryl acids:

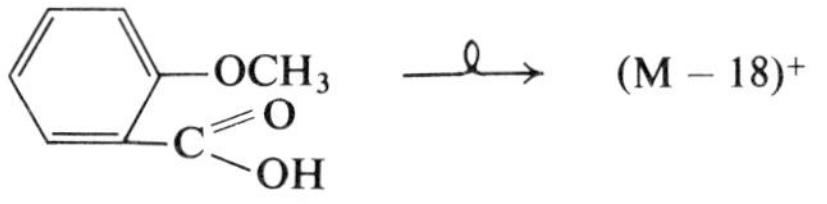

$$\longrightarrow (M - 18)^+ \qquad (5\text{-}113)$$

Aczel and Lumpkin (1960) reported prominent ions due to the loss of water from *o*-substituted aromatic alcohols. Aczel and Lumpkin (1961) also reported the loss of CH_3OH from *ortho*-substituted methyl esters of aryl acids. Presumably, these ions were produced in the manner shown in Scheme 8. Lumpkin

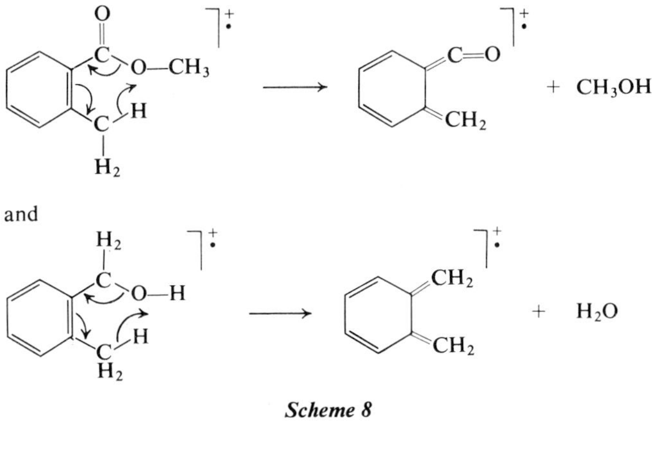

$+ \; CH_3OH$

and

$+ \; H_2O$

Scheme 8

and Nicholson (1960) and Aczel and Lumpkin (1962) followed up these studies with the examination of the *o*-xylyl-*o*-toluate, which in theory should be able to undergo this rearrangement in two directions according to Eq. (5-114):

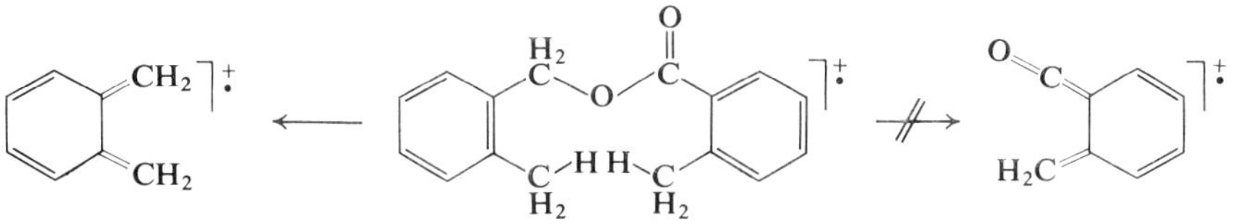

CH_2 CH_2 ← H_2 C O C O C H H C H_2 H_2 ↛ O C H_2C

m/e 104 *m/e* 118

(5-114)

They found no peak intensity at *m/e* 118. In contrast, the *m/e* 104 ion was 27.66% Σ_{73}, while the *m/e* 105 ion was 30.2% Σ_{73} and was the most intense ion in the spectrum. Structures other than the one shown may be drawn for the ion at *m/e* 104.

Meyerson *et al.* (1964, 1966) used deuterated species of diaryl methanes to prove that the *m/e* 104 ion in their studies was indeed

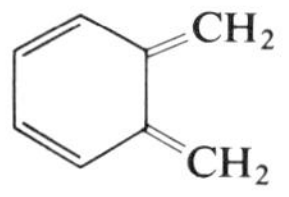

CH_2 CH_2

m/e 104

They also observed an ion at *m/e* 106 from *o*-tolyl ethers which has the structure

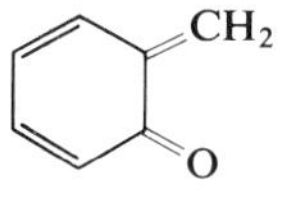

CH_2 O

m/e 106

A six-membered cyclic intermediate is involved in all of these mechanisms.

An interesting variation of the E_2 and E_1 elimination occurs in the case of benzoic acid. Meyerson and Corbin (1965) have shown that 18% of the $C_5H_7O^+$ ions produced arise by the loss of an OH group containing a hydrogen atom from the *ortho* position on the ring. This implies only a five-membered-ring transition state. From *D-ortho*-labeled species, the mechanism shown in Scheme 9 was suggested. The ion was distinct from the isomeric and more

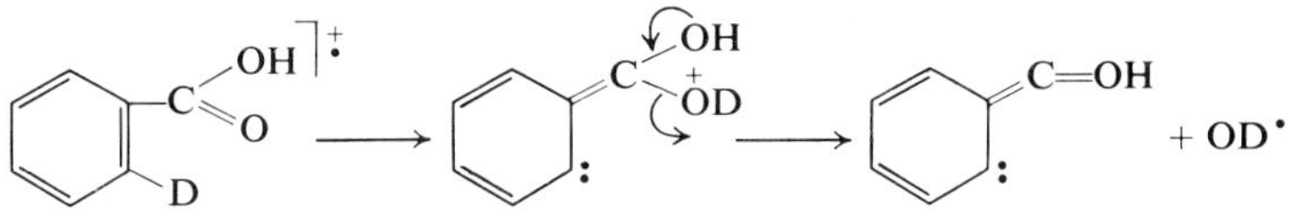

OH C O D → C OH $\overset{+}{O}D$ → C=OH $+ OD^{\bullet}$

Scheme 9

abundant benzoylium ion. There is little doubt that this type of mechanism also competes with ring expansion in any similar situation involving hetero atomic systems, etc., provided only that one of the functional groups is able to supply a transferable hydrogen, while the other supplies the remaining portion of the neutral molecule that is being eliminated.

Roberts *et al.* (1967) have shown that the "*ortho* isomer peculiarity" persists in ferrocenylbenzoate. Some of the fragmentation paths (Scheme 10) observed by these workers are quite consistent with expected processes, but the observation of a loss of the neutral CH_2O fragment is a rather unusual event. These

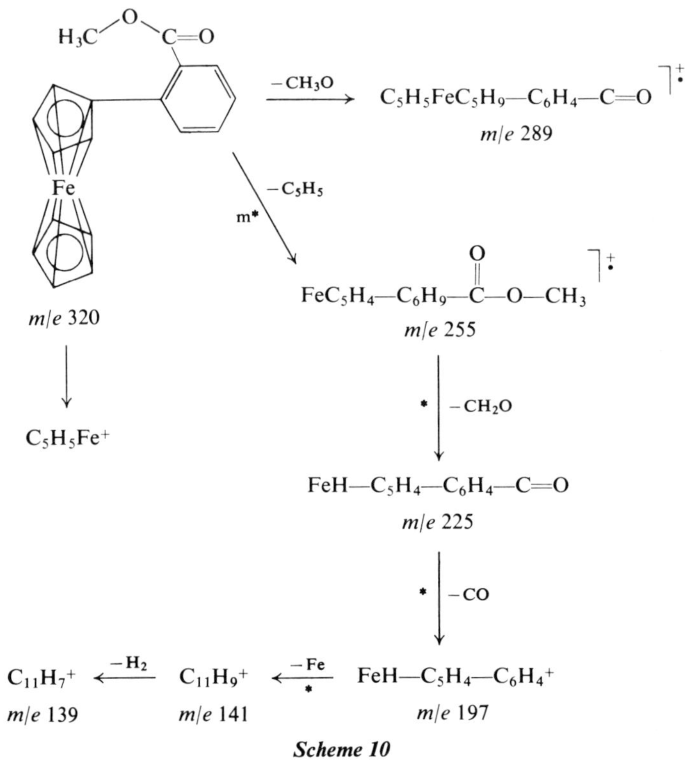

$H_3C-O-C=O$ (ferrocenyl benzoate), *m/e* 320 → $C_5H_5Fe^+$

m/e 320 —($-CH_3O$)→ $C_5H_5FeC_5H_9—C_6H_4—C{=}O]^{+\cdot}$ *m/e* 289

m/e 320 —($-C_5H_5$, m*)→ $FeC_5H_4—C_6H_9—C(=O)—O—CH_3]^{+\cdot}$ *m/e* 255

—(*, $-CH_2O$)→ $FeH—C_5H_4—C_6H_4—C{=}O$ *m/e* 225

—(*, $-CO$)→ $FeH—C_5H_4—C_6H_4^+$ *m/e* 197

—($-Fe$, *)→ $C_{11}H_9^+$ *m/e* 141 —($-H_2$)→ $C_{11}H_7^+$ *m/e* 139

Scheme 10

authors concurred with Meyerson *et al.* (1964, 1966) that a σ or π complex is formed in this instance between the ferrocene unit and the transferred hydrogen atom before the rate-determining step in the fragmentation. Since this type of rearrangement could be explained in terms of the reduced energy involved for a ring size larger than six membered in the transition state, it is tempting to assume this to be true and not consider other possibilities. Roberts and his co-workers felt that the facts are more easily accommodated by postulating the easy formation of a complex. No doubt the future literature will contain additional studies aimed at explanations of this sort of behavior. In the meantime these authors also suggested that there may be a general class of "*ortho* effect" reactions *not* predictable because a six-membered-ring transition state is not likely for the molecule in question. It appears that other factors such as nonplanar π-bonded systems, the stereochemistry of the complex, and the available orbitals of excited ionic species are among many possibilities that will have to be considered in depth.

Levi *et al.* (1969) reported another interesting "*ortho* effect" fragmentation shown in Scheme 11. In this case, α-hydroxyamides were examined. The

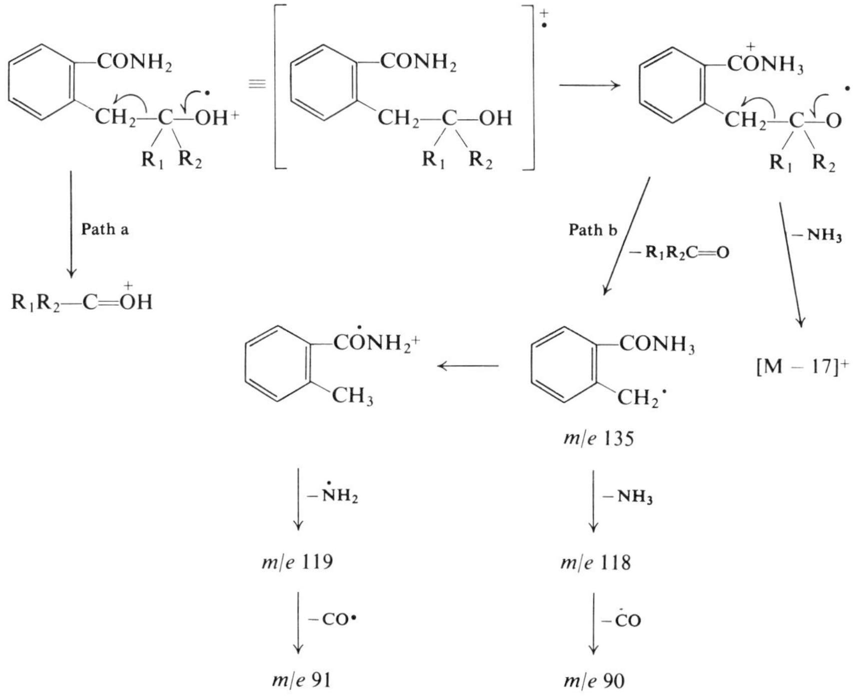

$CONH_2$; $CH_2—C—OH^+$; R_1 R_2 ≡ $[CONH_2$; $CH_2—C—OH$; R_1 $R_2]^{+\cdot}$ ⟶ $\overset{+}{C}ONH_3$; $CH_2—C—O^{\cdot}$; R_1 R_2

Path a ⟶ $R_1R_2—C=\overset{+}{O}H$

Path b, $-R_1R_2C=O$ ⟶ $CONH_3$; $CH_2^{\cdot}$ (*m*/*e* 135)

$-NH_3$ ⟶ $[M - 17]^+$

m/*e* 135 ⟶ $\dot{C}ONH_2^+$; CH_3

$-\dot{N}H_2$ ⟶ *m*/*e* 119; $-CO^{\cdot}$ ⟶ *m*/*e* 91

$-NH_3$ ⟶ *m*/*e* 118; $-\dot{C}O$ ⟶ *m*/*e* 90

Scheme 11

compound underwent a normal alcohol cleavage (path a) but, in addition, underwent an unusual path (path b). A hydrogen atom migrates from oxygen to nitrogen and was accompanied by carbon–carbon bond cleavage. The loss of 17 mass units from the M^+ ion occurred to some extent. From high-resolution data, the loss was established as NH_3 and not the loss of the hydroxyl radical. It is interesting to note in Scheme 11 that even for path b, the ejection of a neutral NH_3 fragment still occurs. The ejection of a neutral CO then follows this event. In many other examples, the ejection of CO occurs earlier in the path; hence the "*ortho* effect" must be very important in this case.

Benoit and Holmes (1969) have shown an "*ortho* effect" in the mass spectrum of *o*-nitrophenylhydrazine. After the formation of the M^+ ion at *m/e* 153, the loss of water by such an "*ortho* effect" dominates the fragmentation as

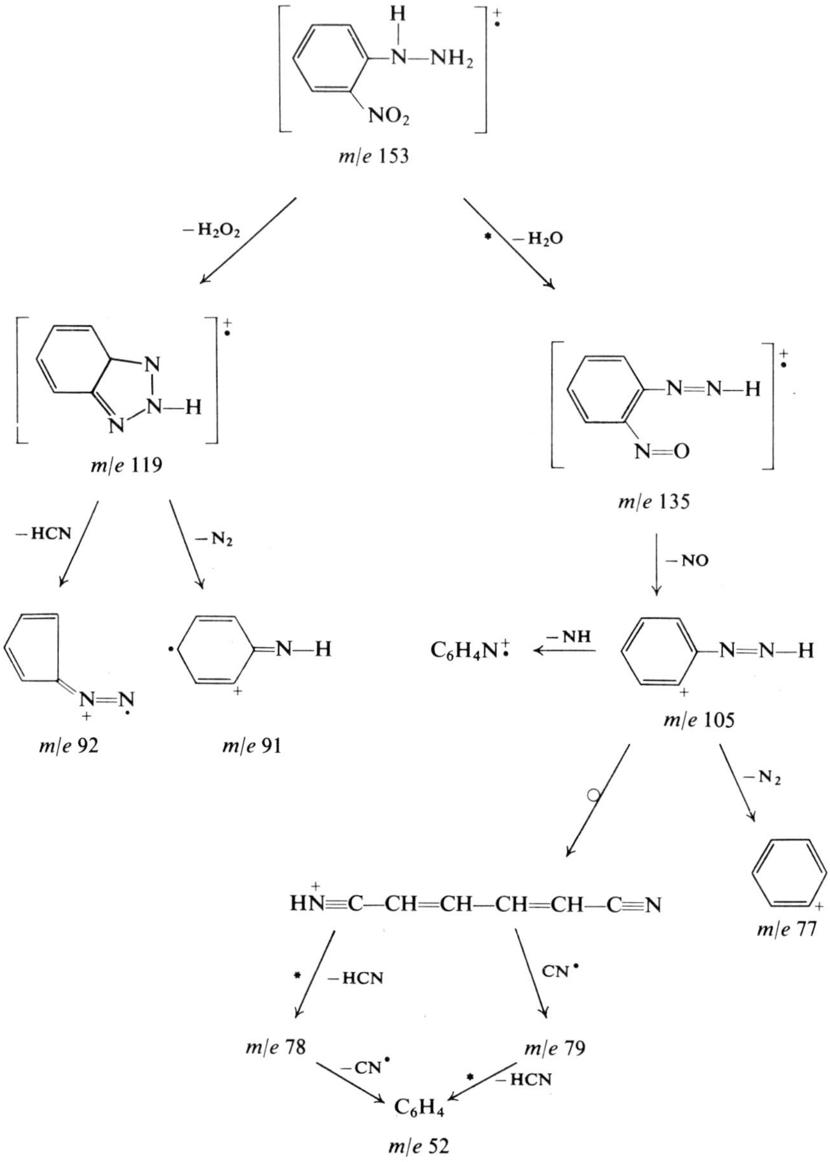

m/e 153

$-H_2O_2$ $-H_2O$

m/e 119 *m/e* 135

$-HCN$ $-N_2$ $-NO$

m/e 92 *m/e* 91 $C_6H_4N^{+\cdot}$ $\xleftarrow{-NH}$ *m/e* 105

$-N_2$

$HN^+{\equiv}C{-}CH{=}CH{-}CH{=}CH{-}C{\equiv}N$ *m/e* 77

$-HCN$ $CN^\cdot$

m/e 78 *m/e* 79

$-CN^\cdot$ $-HCN$

C_6H_4

m/e 52

Scheme 12

shown in Scheme 12. By means of deuterium labeling, metastable ions, and the high-resolution instrument, proof of the paths shown is firmly established. After the loss of H_2O, the loss of NO follows to produce an ion at *m/e* 105. This ion could lead to the species at *m/e* 91, which could even be an expanded-ring azatropylium ion, or alternatively, such an ion might undergo the major rearrangement of a ring rupture. The chain form then would fragment primarily by typical ejections of HCN and CN neutral fragments to produce the remainder of the important peaks in the mass spectrum.

Perhaps the most interesting event is the loss of a neutral hydrogen peroxide fragment. Benoit and Holmes (1969) pointed out that this is not widely encountered. When the event occurs, the rather stable benzotriazole ion is formed at *m/e* 119. The further fragmentation by the loss of either N_2 or HCN is considered typical behavior. A further point of interest for the mass spectroscopist is that with the exception of the ions *m/e* 153 (M^+) and 135, the balance of the important ions would suggest the presence of an alkylbenzene. Of course, with high resolution, there would be little doubt, but the laboratory with low to medium resolution would have to be cautious in approaching the identification of this type of molecule.

3. *The F-Type Rearrangement of Fragment Ions*

This type of fragmentation and rearrangement, really a two-step process, is the source of many ion peaks which appear in the mass spectra of compounds that normally would not be expected to produce such peaks. As an example, it was shown earlier that the primary alcohols have a large *m/e* 31 intensity, while the secondary alcohols do not but do show a large ion at *m/e* 45, etc. How then does one explain the appearance of moderate-intensity peaks at *m/e* 31 from some secondary alcohols? The explanation of this lies in the type F rearrangement:

$$\left[CH_3\text{—}CH_2\text{—}\underset{OH}{CH}\text{—}C_9H_{19}\right]^{+\cdot} \longrightarrow \underset{m/e\ 59}{H_2\overset{H}{C}\text{—}CH_2\text{—}\underset{OH}{CH^+}} \longrightarrow \underset{m/e\ 31}{CH_2\text{=}CH_2 + \underset{OH}{\overset{+}{C}H_2}} \qquad (5\text{-}115)$$

Such behavior is also to be noted for the amines and the thiols of the secondary type:

$$\left[R\text{—}\overset{H}{\underset{H}{C}}\text{—}\overset{H}{\underset{H}{C}}\text{—}\overset{H}{\underset{NH_2}{C}}\text{—}R'\right]^{+\cdot} \longrightarrow R\text{—}\overset{H}{\underset{H}{C}}\text{—}\overset{H}{\underset{H}{C}}\text{—}\overset{H}{\underset{NH_2}{C^+}} + R' \longrightarrow R\text{—}CH_2\text{=}CH_2\text{=} + \underset{\substack{NH_2 \\ m/e\ 30}}{\overset{+}{C}H_2} \qquad (5\text{-}116)$$

and

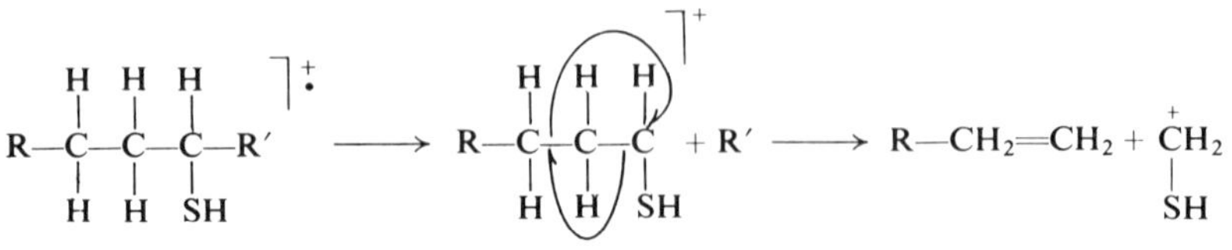

$$R-\underset{H}{\overset{H}{C}}-\underset{H}{\overset{H}{C}}-\underset{SH}{\overset{H}{C}}-R' \;]^{+\cdot} \longrightarrow R-\underset{H}{\overset{H}{C}}-\underset{H}{\overset{H}{C}}-\underset{SH}{\overset{H}{C}}{}^{+} + R' \longrightarrow R-CH_2{=}CH_2 + \underset{SH}{\overset{+}{C}H_2} \qquad (5\text{-}117)$$

where generally R′ > R.

Biemann (1962d) pointed out that it is common practice to utilize ions originating from type B fragmentation to locate functional groups in a molecule. He also noted that these type B ions are always much more intense than the ions arising by further decomposition of the type F given above. It should be noted that the loss of the neutral olefin in these cases permits a considerable drop in excess energy for the particles involved. The ion must reach the resonance form shown and, further, four-membered transition states are probably involved. The type F fragmentation is involved quite frequently in fragmentation events that sometimes lead to many other ions.

4. The G-Type Rearrangement of Fragment Ions

This type of fragmentation differs from the one given above only in that the hetero atom in this case is already in a "double-bonded" situation, i.e.,

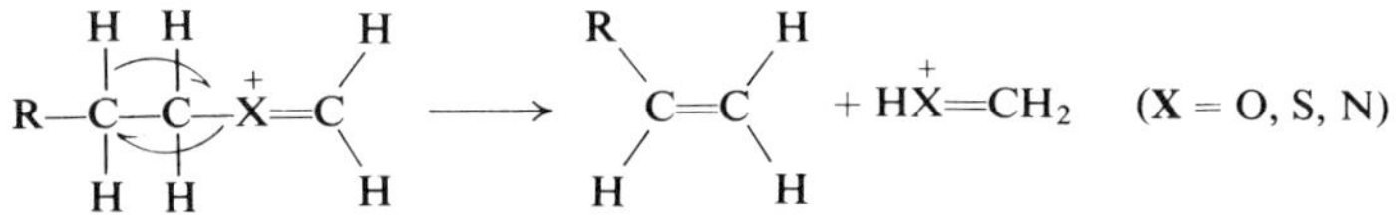

$$R-\underset{H}{\overset{H}{C}}-\underset{H}{\overset{H}{C}}-\overset{+}{X}{=}C\begin{matrix}H\\H\end{matrix} \longrightarrow \begin{matrix}R\\H\end{matrix}C{=}C\begin{matrix}H\\H\end{matrix} + H\overset{+}{X}{=}CH_2 \qquad (X = O, S, N)$$

The unshared pair of electrons of a hetero atom can participate in this rearrangement, but the type B fragment ion shown must have at least two carbon atoms in a chain with at least one hydrogen atom in the β position from the hetero atom for the olefin elimination to take place. McLafferty (1957)

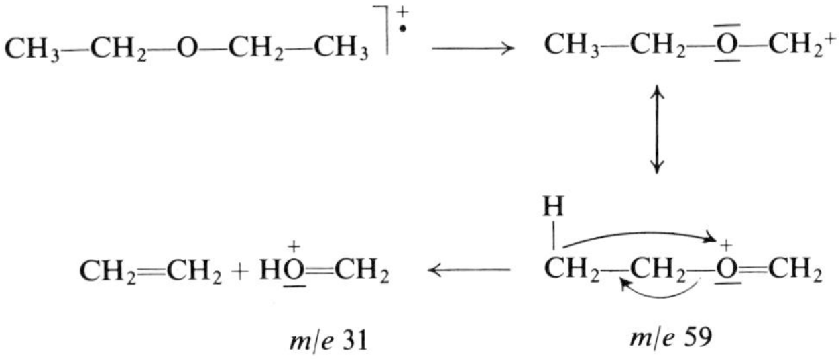

$$CH_3-CH_2-O-CH_2-CH_3\;]^{+\cdot} \longrightarrow CH_3-CH_2-\overline{\underline{O}}-CH_2^{+}$$

$$\updownarrow$$

$$CH_2{=}CH_2 + H\overset{+}{\underline{O}}{=}CH_2 \longleftarrow \overset{H}{C}H_2-CH_2-\overset{+}{\underline{O}}{=}CH_2$$

m/e 31 $\qquad\qquad$ m/e 59

Scheme 13

has shown that the *m/e* 31 ion (the ion of largest intensity) in the mass spectrum of diethyl ether is formed as shown in Scheme 13. A substituent at the carbon atom retained in the fragment may be detected by the increase in mass.

Just as in the case of the alcohols described earlier, this type of rearrangement is not restricted to a four-membered-ring transition; five- and six-membered rings are also possible. Under the E_2 elimination, the benzolyium ion is formed in competition with the isomeric $\cdot\!\!$ (cyclohexadienylidene)=C=OH$^+$ ion and probably by a different transition state. In a sense, this type of optional behavior may make the student ask why one uses an artificial system of classification of fragmentation and rearrangement reactions. The reply is that chemists have been doing this sort of construction of "crutches" as long as they proved of use. After all, the freshman chemistry student has heard only of ionic and covalent bonds in high school and sometimes is shocked when he finds out about "percent ionic character," etc. It is the authors' opinion that there is already considerable material to learn in the field of mass spectrometry for even the student with an advanced degree, and hence learning through such a classification can be useful.

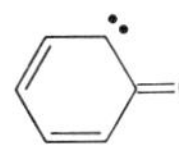

Meyerson and McCollum (1959) reported studies of the alkyl phenyl sulfoxides and discussed their fragmentation paths. The same samples were made available for some further studies at the U.S. Bureau of Mines, Bartlesville Research Center. Hirsch and Foster (1960, 1961) examined these compounds at 135°C and at a later time at 100°C in an all-glass inlet system. The problem seemed to be that thermal decompositions were rather marked and interfered with the interpretation of fragmentation events. With the all-glass system, it was established that the problem was of a thermal nature not arising from the contact of the compounds with gallium or indium then in use in the instrument of Meyerson and McCollum (1959). In addition to many other thermal problems, the basic fragmentation steps from medium-resolution and low ionization voltage data are shown in Scheme 14. The loss of mass 16 from the sulfoxide was not particularly surprising, and this event probably led to an excited state phenyl alkyl sulfide ion as depicted. This ion in turn underwent a type G fragmentation to produce the ion at 110. Alternatively, the type G fragmentation–rearrangement could occur with the molecular ion to produce the *m/e* 126 ion and a neutral olefin. From here, the *m/e* 126 ion could fragment to form the 110 by the ejection of the oxygen. However, it was thought that the bulk of the 126 ions went on to eject the SO group with the movement of the hydrogen atom producing the mass 78 ion. At 70 eV, the *m/e* 78 ion was the base peak. Regardless of the specific mechanism, the type G rearrangement–fragmentation dominates the first two steps. Note that actually all odd-electron ions are produced in reaching the *m/e* 78.

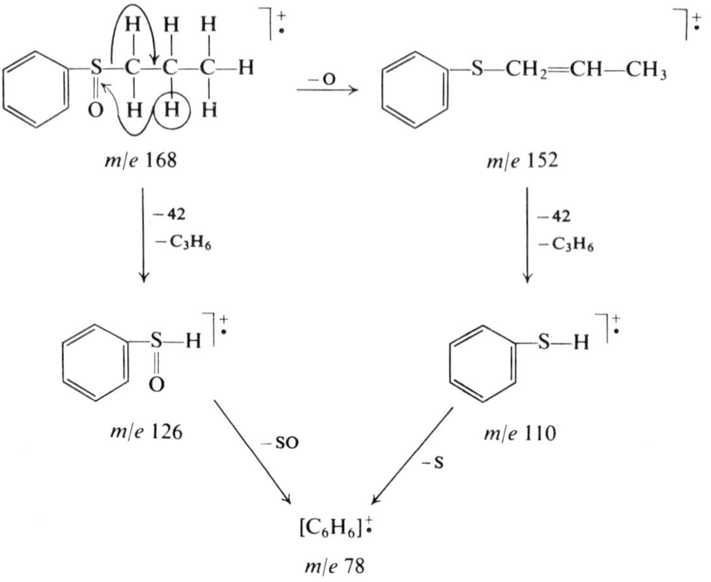

m/e 168 → (−O) → m/e 152

m/e 168 → (−42, $-C_3H_6$) → m/e 126

m/e 152 → (−42, $-C_3H_6$) → m/e 110

m/e 126 → (−SO) → $[C_6H_6]^{+\cdot}$ m/e 78

m/e 110 → (−S) → $[C_6H_6]^{+\cdot}$ m/e 78

Scheme 14

5. *The H-Type Rearrangement (The McLafferty Rearrangement)*

Because of its widespread occurrence, this rearrangement is noted early in the mass spectroscopist's training. The general reaction is

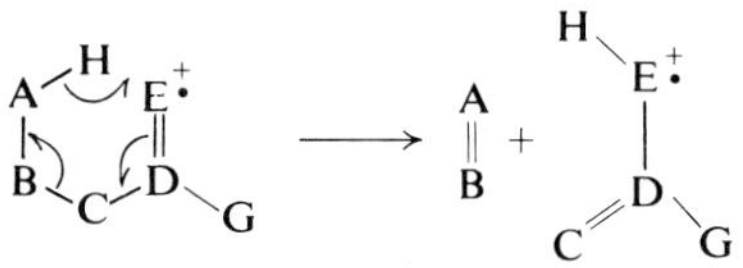

$$\text{A(H)–B–C–D(G)=E}^{+\cdot} \longrightarrow \text{A=B} + \text{H–E}^{+\cdot}\text{–D(G)=C} \tag{5-83}$$

A multiple bond must be between D and E, and a hydrogen atom (H) must be available on the atom gamma from the D–E multiple bond. A variety of atoms or groups may be at A–E and G.

As an example of this rearrangement for 2-isopropyl-2,5-dimethylcyclohexanone, the work of Kulkarni *et al.* (1970) is discussed. In their work, deuterated compounds as well as unlabeled species were used. The mass spectra showed intense peaks at M − 42 which were attributed to the loss of a C_3H_6 neutral fragment resulting from the McLafferty rearrangement. The mechanism is shown in Scheme 15. These results are quite typical for a McLafferty rearrangement. In the work of Kulkarni *et al.* (1970), a second fragmentation sequence of importance appeared resulting from 1,2 or "α" cleavage. The high-resolution mass spectrometer and metastable transitions

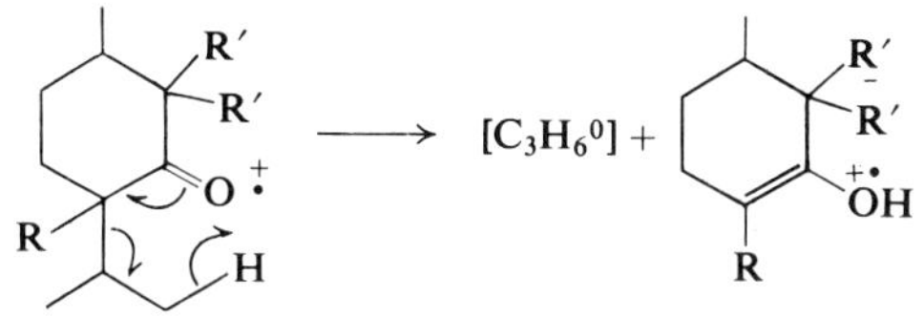

$\longrightarrow [C_3H_6O] +$

m/e 168 *m/e* 126

Scheme 15

were used to establish the paths shown in Scheme 16. From the high-resolution mass spectra of these molecules, it could be established that the fragment ion

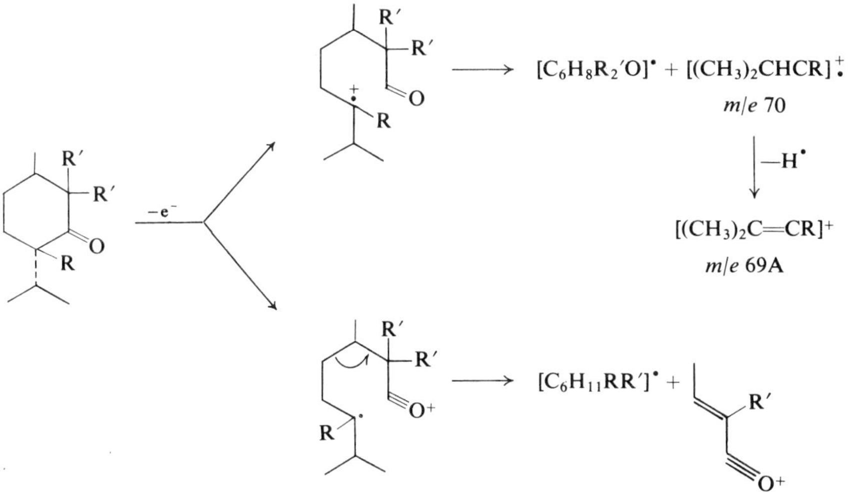

$\longrightarrow [C_6H_8R_2'O]^{\bullet} + [(CH_3)_2CHCR]^{+\bullet}$

m/e 70

$-H^{\bullet}$

$[(CH_3)_2C{=}CR]^+$

m/e 69A

$\longrightarrow [C_6H_{11}RR']^{\bullet} +$

m/e 168 *m/e* 69B

Scheme 16

at nominal *m/e* 69 was composed of $\frac{1}{3}C_4H_5O$ (see 69B) and $\frac{2}{3}C_6H_9$ (see 69A). Clearly, more than one cleavage route is involved for these molecules. Additional paths involving loss of methyl groups and ring openings are also discussed. When a 2-isopropenyl group was substituted for the 2-isopropyl group, the processes attributed to a McLafferty rearrangement disappeared. For the unsaturated ketone, cleavages "α" and "α'" predominated, but other additional differences were also noted.

McLafferty and Hamming (1958) and McLafferty (1959) reported the occurrence of a second hydrogen atom transfer in a six-membered cyclic transition state for the esters. This produced structures suggested by McLafferty (1963c) which located the second hydrogen atom as follows:

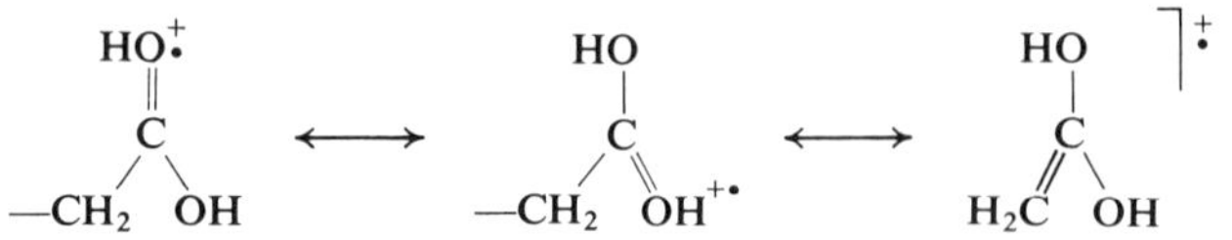

$$\begin{matrix} HO^{+\bullet} \\ \| \\ C \\ / \quad \backslash \\ -CH_2 \quad OH \end{matrix} \longleftrightarrow \begin{matrix} HO \\ | \\ C \\ / \quad \backslash\backslash \\ -CH_2 \quad OH^{+\bullet} \end{matrix} \longleftrightarrow \left[\begin{matrix} HO \\ | \\ C \\ // \quad \backslash \\ H_2C \quad OH \end{matrix}\right]^{+\bullet} \qquad (5\text{-}118)$$

It was recognized that the odd-electron ion would be stabilized by resonance, some forms of which are shown above. Because of the site specificity of this second rearranged hydrogen atom, the rearrangement has been referred to as a double McLafferty rearrangement. This rearrangement accounts for the *m/e* 75 ion that appears in the mass spectra of many esters and in some cases to ths extent that the 75 rearrangement ion is larger than the *m/e* 74 ion. Rol (1965) has observed the rearrangement in aliphatic acid.

Difficulties with nomenclature have arisen concerning this rearrangement. In addition to "double," Carpenter *et al*, (1968) have referred to it as the "protonated McLafferty rearrangement," while Haynes *et al.* (1968) have referred to it as a "super-McLafferty rearrangement." Many other rearrangements take place in which the McLafferty rearrangement is followed by nonspecific or random hydrogen transfers.

Budzikiewicz *et al.* (1967b) have presented an extensive review of the McLafferty rearrangement, while Spiteller (1966) has reinforced the claim of a double rearrangement for the aliphatic ketones. Carpenter *et al.* (1968) undertook isotopically labeled studies of the aliphatic ketones. Low ionization voltage was used, and this showed that deuterium scrambling occurs among the carbon atoms of the aliphatic chains. This work would be very convincing if NMR spectra of the variously labeled ketones were presented. The reason for this statement is that the work of the late Professor R. W. Higgins showed that a great deal of mobility of deuterium atoms existed during the preparation of labeled thiophenes from thienyl ketones. A γ-dideuterated thiophene could be prepared by a Wolff–Kishner reduction (using N_2D_4). Its isotopic purity was about 99%. The corresponding α- and β-ketones could not be converted to high positional purity by this means. Deuterium atoms were even detected on the thiophene ring by NMR, suggesting that considerable exchange was going on during the synthesis. Despite the low ionization voltage scrambling, the site specificity of the "protonated McLafferty rearrangement" was determined by Carpenter *et al.* (1968); this report will be worthy of study by those who wish to perform label studies. The importance of the McLafferty rearrangement ion plus 14 in the low-voltage mass spectra was noted. Eadon and Djerassi (1970) have continued these studies of the branched ketones and esters and the dependence of hydrogen rearrangements upon ionization voltage.

MacLeod and Djerassi (1967) have reported on competitive McLafferty rearrangements in phenylalkylmethyl ketones and keto esters. The γ hydrogen transfers preferentially to the carbonyl moiety, but the dominant fragmentation process in some of the compounds is a 1,6 elimination.

Kraft and Spiteller (1969) have discussed the McLafferty rearrangement as it occurs in 1,1-di-, tri-, and tetraalkyl-substituted ethylene derivatives. They claim that in contrast to olefins of the type A and B shown below, no hydrogen isomerization takes place, and the molecular ions of the varieties other than A and B undergo fragmentation via the McLafferty rearrangement. If the

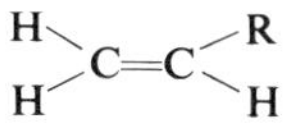

$(H)(H)C{=}C(R)(H)$ Type A $(R)(H)C{=}C(R)(H)$ Type B $(H_3C)(H_3C)C{=}C(R)(H)$ Type C

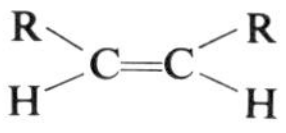

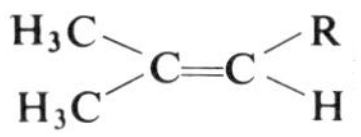

same McLafferty rearrangement product is formed, then the spectra are nearly identical. An olefin of the type C, in which the McLafferty rearrangement cannot take place, shows preferential cleavage of the allyl bond. These concepts can be added to the correlations for olefin behavior as given earlier and are examples of how the mass spectroscopist gradually expands structure-correlation studies to assist in identification of such materials.

Biemann (1962e) compared the competition between fragmentation of type B and type H in the case of the aryl ether, the thio ether, and the amine. In this particular series, the extent of rearrangement decreases while the simple cleavage increases:

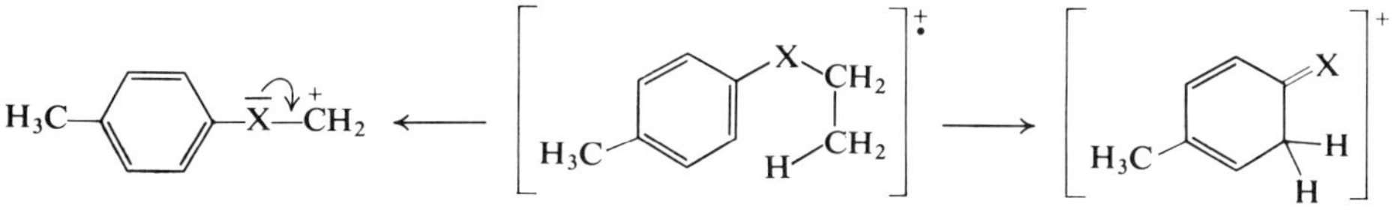

$H_3C-C_6H_4-\overset{+}{X}-CH_2$ ⟵ $[H_3C-C_6H_4-X-CH_2-CH_2-H]^{+\cdot}$ ⟶ $[H_3C-C_6H_4(=X)(H)(H)]^{+}$

X	$\%\Sigma_{27}$
—O	0.14
—S	11.3
—NH	36.6

Molecular ion

X = O, S, or NH

X	$\%\Sigma_{27}$
—O	24.2
—S	6.3
—NH	0.4

(5-119)

Budzikiewicz *et al.* (1967c) showed that the migrating γ-positioned hydrogen atom and the carbonyl oxygen atom have a rigorous proximity requirement. If the distance exceeds 1.8 Å in rigid systems, the fragmentation usually takes a different course. Also, it should be noted that while fragments formed may be similar in *m/e* value to those expected from the McLafferty rearrangement, the hydrogen atoms migrate from positions *other* than those γ to the carbonyl atom. This, in addition to the observations of hydrogen scrambling at low ionization voltages, as described later, leads to considerable difficulties in deciding whether the rearrangement took place as a McLafferty variety.

Deutsch and Mandelbaum (1969), however, have shown that the McLafferty rearrangement does not take place for an ion in which the hydrogens may be within 1.5 Å of the carbonyl oxygen. Instead, a retro-Diels–Alder fragmentation occurs with the accompaniment of a double hydrogen atom migration. The molecules studied were adducts of *p*-benzoquinone with bi-1-cycloalken-1-yls. As an example, consider Scheme 17. A normal RDA for this molecule

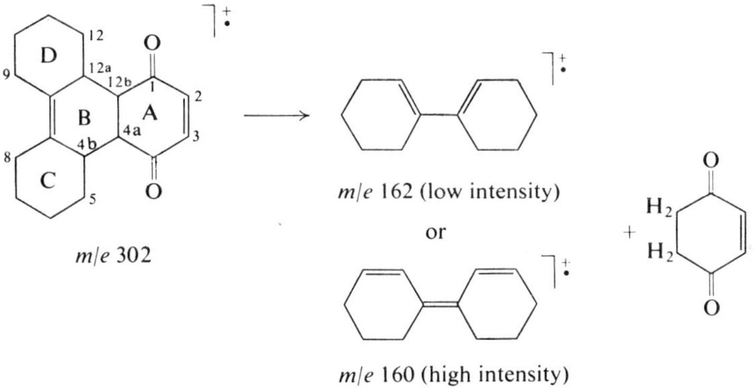

m/*e* 302

m/*e* 162 (low intensity)

or

m/*e* 160 (high intensity)

Scheme 17

would lead to the *m*/*e* 162 ion. This ion is of very low intensity as compared to the ion at *m*/*e* 160. An ion of this type may be formed only by the migration of two hydrogen atoms from rings C and D to ring A, followed by the cleavage of bonds 12a–12b and 4a–4b in this molecular ion. These authors assumed that this molecule exists in the *endo* configuration since the original Diels–Alder additions were carried out at room temperature. If this is true, H-12 and H-5 will be about 1.5 Å from the carbonyl oxygen atom. Under these conditions, the double McLafferty rearrangement should be possible. It should be noted that this behavior is quite similar to that of the hexahydrodibenzothiophene case described earlier with the RDA reaction systems. Further, it should be noted that the double RDA reaction was possible in the cases discussed earlier.

Carpenter *et al.* (1969) have investigated the effect of fluorine substitution on the McLafferty rearrangement in aliphatic ketones. One of their most interesting conclusions is that the presence of a trifluoromethyl group adjacent to the γ hydrogen atom in the molecule shown inhibits the McLafferty rearrangement. If the rearrangement were operative, an *m*/*e* 154 ion would appear. The intensity of the ion is negligible, being almost zero. The second McLafferty rearrangement occurs on the opposite side of the carbonyl group to produce the *m*/*e* 58 ion. It attains only 3% relative abundance, whereas

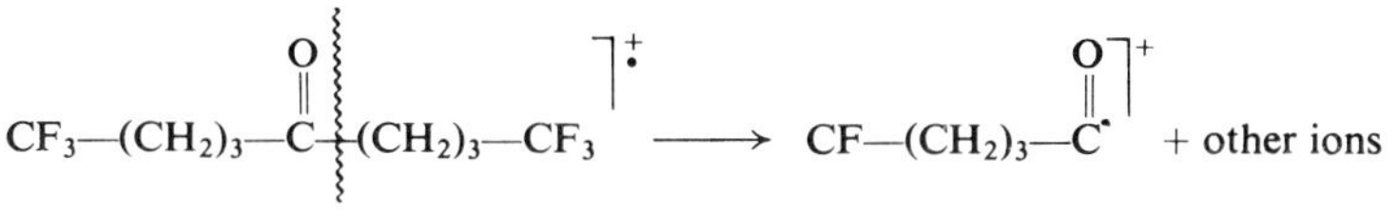

$$CF_3{-}(CH_2)_3{-}\overset{O}{\overset{\|}{C}}{\wr}(CH_2)_3{-}CF_3\ \Big]^{+\cdot} \longrightarrow CF{-}(CH_2)_3{-}\overset{O}{\overset{\|}{C}}{}^{\cdot}\ \Big]^{+} + \text{other ions} \qquad (5\text{-}120)$$

m/e 250 $\qquad\qquad$ m/e 139

in the case of 5-nonanone, this same process accounts for an ion of 70% relative abundance. Thus it appears that the suppression of the McLafferty rearrangement can occur under these conditions.

Carpenter *et al.* (1969) gave two explanations for this surprising difference. The first is based upon the work of Russell and Brown (1955), which shows that polar effects would be very significant in the transition state leading to the ion at m/e 154 because of the strong electronegativity of the oxygen atom in the molecular ion which carries the charge on the oxygen. The virtual

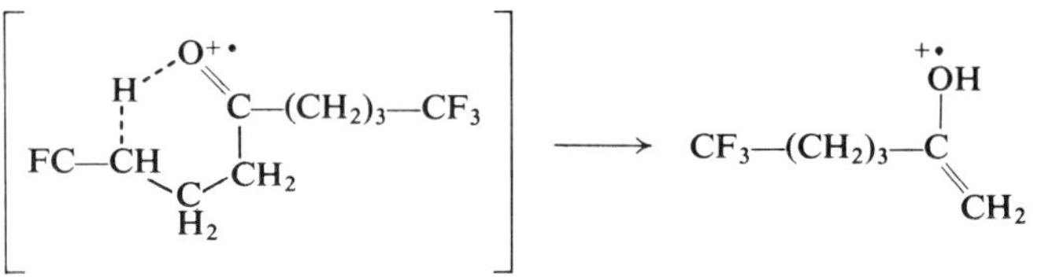

$$\left[\begin{array}{c} H\cdots O^{+\cdot}=C{-}(CH_2)_3{-}CF_3 \\ FC{-}CH{-}CH_2{-}CH_2 \end{array}\right] \longrightarrow CF_3{-}(CH_2)_3{-}\overset{\overset{+\cdot}{OH}}{\overset{|}{C}}{=}CH_2 \qquad (5\text{-}121)$$

Transition state, m/e 154 $\qquad$ m/e 154 (Russell and Brown, 1955)

absence of the 154 ion from the mass spectra of the fluorinated compound could also result from the known strengthening of the C–H bond adjacent to a trifluoromethyl group if a radical abstraction of a hydrogen atom is considered. This generalization has been established by Giles *et al.* (1967) and is shown by the fact that it is more difficult to remove hydrogen atoms by radicals from 1,1,1-trifluoroethane than from ethane.

Molecules containing other hetero atoms in a situation similar to the carbonyls have also been shown to undergo McLafferty types of rearrangements. Gamble *et al.* (1969) reported the study of a series of dialkyl sulfites. Three main fragmentation routes were observed: (a) the loss of an alkyl radical, (b) cleavage of an S–O bond to lose an alkoxy group, and (c) McLafferty rearrangements. Two different pathways are described as shown in Scheme 18, both of which ultimately lead to an ion quite analogous to the ions shown in Scheme 15, except for the change of an OH group for the methylene and the fact that the sulfur atom is now playing the role of the carbon atom. Here, as in a number of other sulfur atom bonding situations, the 10-electron sulfur atom may be involved. The use of an excited ion species that involves the *d* orbitals is shown in Scheme 18.

Another interesting problem occurs when various competitive reactions can occur. An example has been reported by Meyerson and Leitch (1966) for

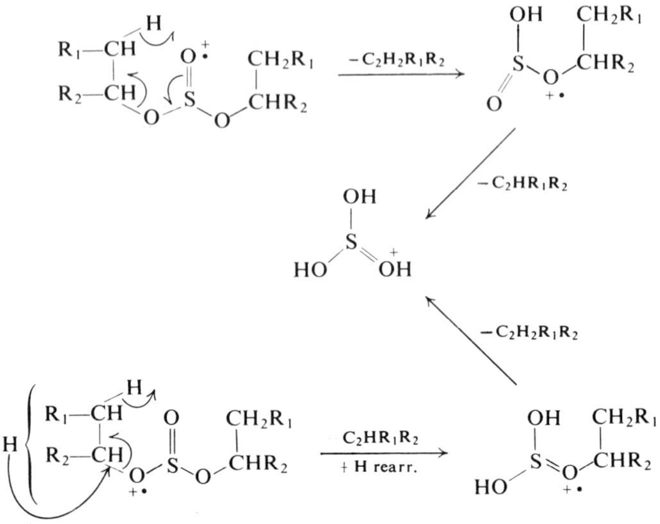

Scheme 18

ω-phenylalkylmethyl esters if the carbonyl group occurs six carbons from the ring phenyl carbon atom. In this event, instead of a McLafferty rearrangement occurring, matters proceed as shown in Scheme 19. Labeled molecules

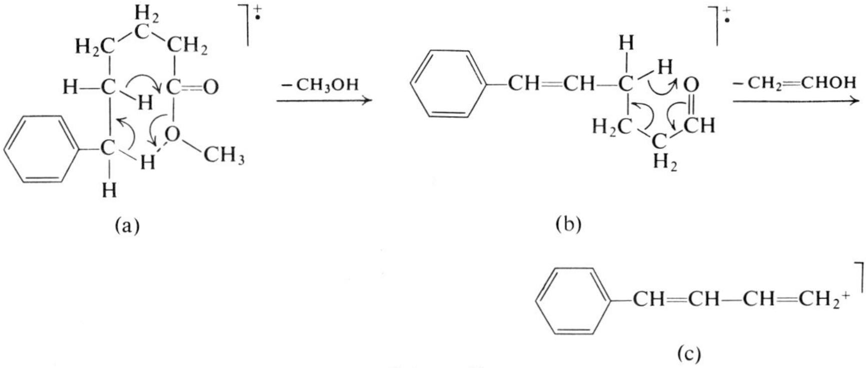

Scheme 19

were used to establish the paths indicated. Low ionizing voltage data shows that the paths to ions (b) and (c) require substantially less energy than paths that produce the $C_7H_8^+$ ion or $C_3H_6O_2^+$. It would appear that part of the increased stability of these ions is due to conjugated olefinic systems. Because of the energy differences, the McLafferty rearrangement ion intensity is very low. These authors also indicated that the *m/e* 74 rearrangement ion is present in all of the methyl esters reported in the literature. The one exception is 6-methyloctadecanoate. The 5- and 7-methyloctadecanoate compounds do show small rearrangement peaks but have strong M − 76 ion peaks.

To assist in identification work, Table 5-17 presents some commonly

TABLE 5-17

SOME REARRANGEMENT IONS

Series	*m/e*	Characteristic ion features
−10	32	CH_3OH from *n*-alkanols
	60	CH_3COOH from alkanoic acids
	74	$H_3C{-}C({=}O){-}OCH_2R$ if R = H, methyl esters; R also = CH_3–, C_2H_5, etc.
	102	Ion 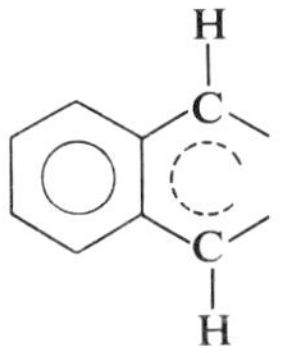 (benzene ring fused to H–C⋯C–H) system
	116	$C_4H_4S_2$ ion from thienyl thiaalkanes
−9	61	CH_2SHCH_2 ion from R—S—R′ compounds
	47 + R, 75	Double hydrogen rearrangement of esters (MW > 88)
−8	48	CH_3SH ion from mercaptans
	104	RDA ion common to benzocyclics 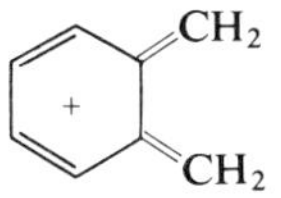 (+ ring with $=CH_2$, $=CH_2$)
	160	Rearrangement in naphthalenethiaalkanes
−7	91	Tropylium ion from aromatics
	91	C_4H_8Cl ion from terminal chloroalkanes
	147	Benzothiatropylium ion
−6	64	S_2 by rearrangement and ejection from thienyl sulfides, and benzo[*b*]thienyl sulfides
	92	C_7H_8 ion, double rearrangement ion (common to alkylbenzenes)
	92	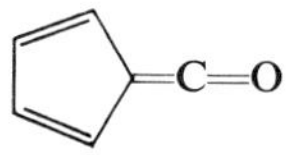 (cyclopentadienylidene)=C=O
	120	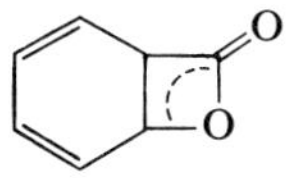 (benzo-fused ring with C=O and O)

TABLE 5-17 (*continued*)

Series	*m/e*	Characteristic ion features
−5	79	Double hydrogen rearrangement present in alkylbenzenes
	135	C_8H_7S ion
	135	C_4H_8Br ion
−4	94	C_6H_5OH ion from benzopyrans
−2	54 + R	Common to cycloalkyl derivatives
	166	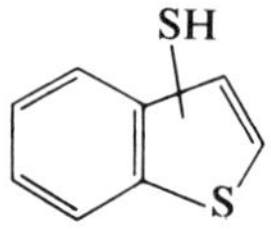 SH (benzothiophene structure)
−1	97	C_5H_5S ion in monoalkylthiophenes
	97	$C_5H_{10}C{=}NH$ in nitriles (C_{11} and above)
0	42	$CH_2{=}C{=}O$ ion from some oxygenates
	84	C_4H_4S ion from alkylthiophenes
	98	C_5H_6S ion, double rearrangement in monoalkylthiophenes (note correspondence to $C_7H_8^+$ at 92 for alkylbenzenes)
+1, −13	85	C_4H_5S ion double rearrangement in alkylthiophenes (compare to 79 for alkylbenzenes)
	141	$C_{11}H_9^+$ from alkylnaphthalenes (if with 91^+)
+2, −12	30 + R	Rearrangement ions of $R{-}NH_2$
	58	CH_3COCH_3 from alkylalkanones; prominent in methyl derivatives
	142	$C_{10}H_8N^+$ from quinolines
+3, −11	115	C_9H_7 ion from naphthalenes; also produces strong *m/e* 57.5 from 115^{2+}

encountered rearrangement ions. For the aromatic heterocyclic compounds one can utilize the table by making the substitution of N for CH (as in pyridine) and increasing or decreasing the net mass of the ion appropriately (+1 in this case). For aromatic ring systems a sulfur analog amounts to substituting the sulfur atom (mass = 32) for the C_2H_2 group (mass = 26); hence the net increase in the mass of the ion is 6 *u*. Rearrangement ions from condensed-ring systems wherein a benzo group (or several) has been added can quickly be determined because the addition of each group adds exactly 50 *u* to the mass of the ion. For the series benzene, naphthalene, and anthracene, the molecular ions are 78, 128, and 178, respectively. Similarly for pyridine, quinoline, etc. the molecular ions will be at 79, 129, and 179, all differing by 50 *u*. Lastly, for thiophene, benzothiophene, and dibenzothiophene, the molecular ions would occur at 84, 134, and 184. For any related ion, the same relationship holds. The table is not comprehensive but will prove of value to the beginner. Additions should be made as soon as one begins to observe specific rearrangement ions in the types of materials being investigated.

G. Multistep and Concerted Processes

The mass spectroscopist is now well aware of the fact that competition exists among decomposition paths for an excited-state ion. Even small differences in the energies of activation and the stabilities of daughter ions can be important in this competition. Sometimes it appears that a so-called multistep process or concerted process takes place. In these cases difunctional molecules interact to produce rather surprising results when first viewed by the mass spectroscopist.

Barnes and Occolowitz (1963) reported that the loss of the lactonic carbonyl function from coumarins was a significant and sequential process. Shapiro and Djerassi (1965) considered the methoxylated coumarins, and using 6,7-dimethoxycoumarin showed peaks corresponding to M–CH_3, M–CO, and M–(CO + CH_3). These are assignable to a process of the multistep and/or concerted variety.

Nibbering and de Boer (1968a) have produced a series of studies showing reactions that are multistep in character and provide excellent examples. The consequences of a series of events for the 3-phenyl-1-propanol resemble in part the behavior reported by Meyerson and Leitch (1966) in Scheme 19. Nibbering and de Boer (1968a) showed by means of deuterated species that water is eliminated involving the terminal OH group and the *ortho* hydrogen atom and/or the hydrogen on the alkyl chain adjacent to the ring. It is note-

(5-122)

worthy that these authors referred to the carbon atom adjacent to the ring as γ (which it is with respect to the OH group), whereas other workers might refer to this as the α carbon (to the ring). Hence, in reporting spectra or in describing a system, great care should be taken to adequately define the hydrogen atom in which one is interested. There appears to be no set standard of nomenclature in this area, and hence literature will be found using both methods given above.

In a second article Nibbering and de Boer (1968b) showed the operation of a quite similar six-membered cyclic intermediate for the equivalent bromo compound. This is shown in Scheme 20. These authors claimed that for the hydroxy compound a mutual exchange takes place between the hydrogen atoms from the OH group, the γ methylene group, and in the *ortho* positions in the ring. This is limited to the ring *ortho* positions for the bromo compound shown.

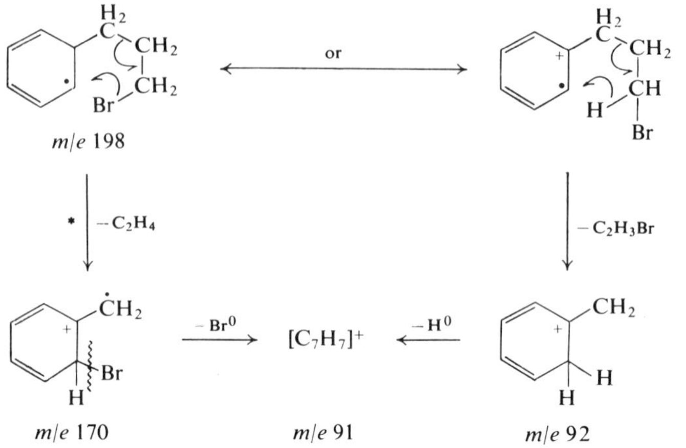

Scheme 20

In a more recent article (Nibbering and de Boer, 1970), the 3-phenylnitropropane and 3-phenylpropylnitrite molecules were investigated. From the spectra of the 3-phenylnitropropane and variously deuterated compounds, three major fragmentation paths were discernible: (a) successive elimination of a water molecule and an oxygen atom (as shown in Scheme 21a), (b) γ hydrogen transfer and water elimination as shown in Scheme 21b, and (c) formation of an $(M - NO)^+$ ion (*m/e* 135) and its further degradation by either the loss of water to form *m/e* 117 or loss of ethylene and formaldehyde from the $(M - NO)^+$ ion. The multistep and/or concerted process nature of these schemes is quite obvious.

In the case of the 3-phenylpropylnitrite molecule, a number of the processes described in this section obviously are operating. A partial fragmentation is given in Scheme 22.

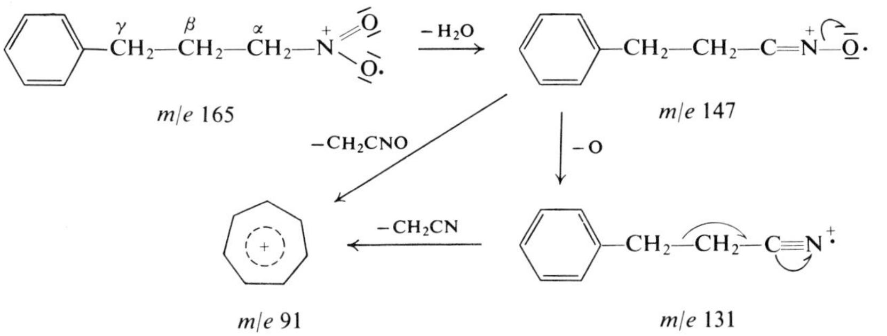

Scheme 21a

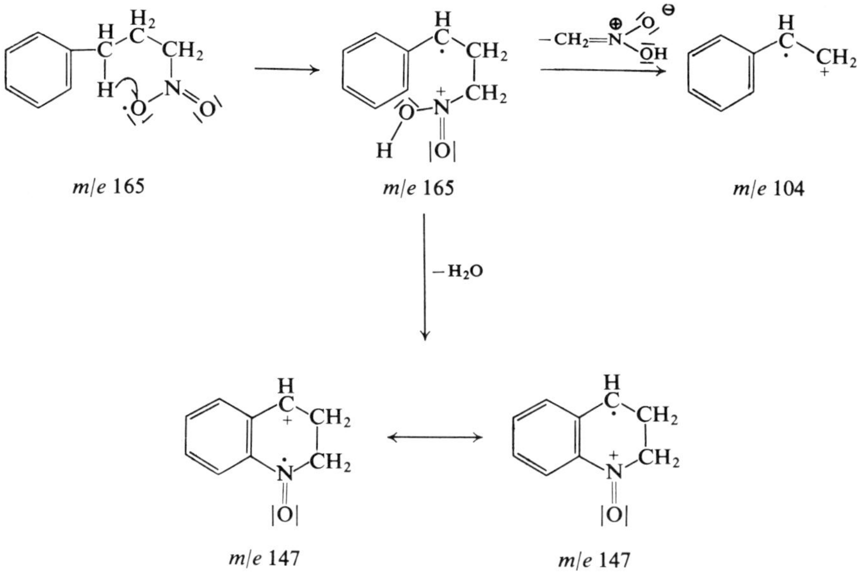

C_6H_5–CH–CH_2–CH_2–N(=O)–O· (H transfer) → C_6H_5–$\dot{C}H$–CH_2–CH_2–$\overset{+}{N}$(=O)–OH —($-CH_2{=}\overset{\oplus}{N}(O^{\ominus})OH$)→ C_6H_5–$\dot{C}H$–$\overset{+}{C}H_2$

m/e 165 m/e 165 m/e 104

m/e 165 ↓ $-H_2O$

H, C, CH_2, CH_2, $\dot{N}$, |O| ⟷ H, C·, CH_2, CH_2, $\overset{+}{N}$, |O|

m/e 147 m/e 147

Scheme 21b

$C_6H_5—CH_2—CH_2—CH_2—O—NO^{+\bullet}$ or C_6H_5–CH_2–CH_2–C(H)(H)–O–$\overset{+}{N}$·=O

m/e 165

$-CH_2ONO$ → $[C_8H_9]^{+}$ m/e 105

$-C_8H_9$ → $H_2C{=}\overset{+}{O}—N{=}O$ m/e 60

$-NO$ → $C_9H_{11}O^+$ m/e 135

$-HNO$ → C_6H_5–CH_2–CH_2–H–C·–$\overset{+}{O}$ m/e 134

* $-C_2H_4$ → $[C_7H_7O]^+$ m/e 107

* $-C_2H_2O$ → $[C_7H_8]^{+\bullet}$ m/e 92

* $-CO$ → $[C_6H_7]^+$ m/e 79

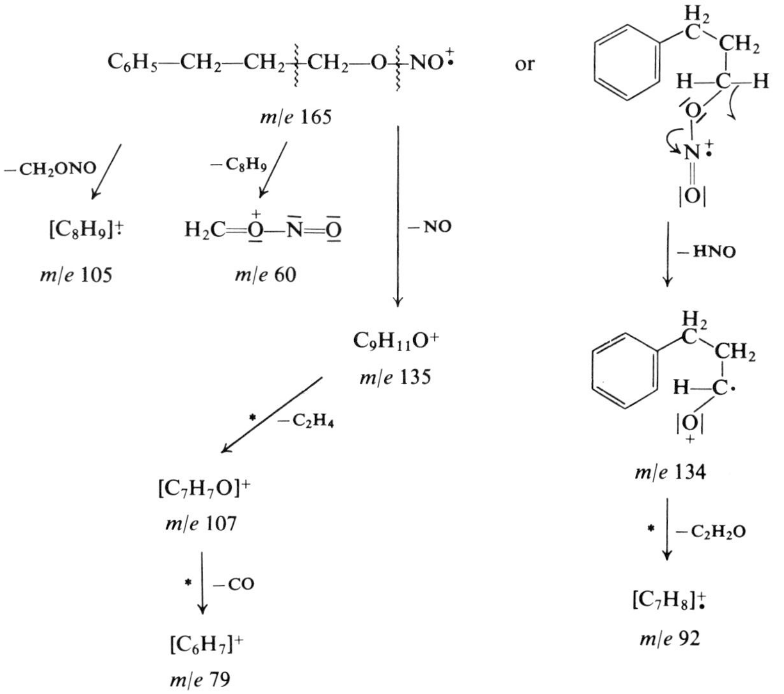

Scheme 22

An example of a transannular amide–amide interaction in the mass spectrometer for the cyclopeptides has been given by Denisov *et al.* (1968). In Scheme 23, it is shown that the proximity of the interacting groups is critical

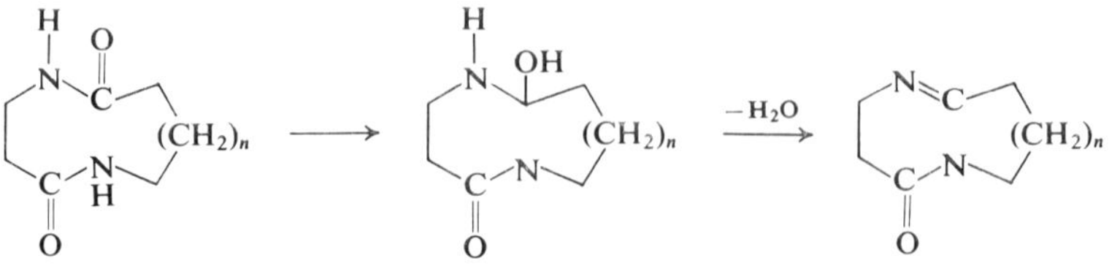

Scheme 23

to the process so as to eliminate the water molecule and make the new C–N bond. When $n = 1$, the nine-membered ring gives a good fit and the elimination occurs. When $n = 2$, the reaction still occurs, but when $n = 3$, the interaction is apparently not possible. Thus, for at least several of these types of molecules, the elimination of water might be expected to occur, producing M − 18 ion peaks of note and probably interfering with the interpretation of a mass spectrum of such a substance.

Quite recently Lengyel *et al.* (1970a) have shown a novel skeletal rearrangement in difunctional bromamides. Within the framework of a broad program undertaken to investigate effects of stereochemistry on mass spectra, a number of bis-α-bromoamides were examined. The compounds included the two given below. The *t*-butyl derivative compound (I) provided a singlet peak at *m/e* 210 of 20% relative intensity, which in a high-resolution measurement was shown to have the composition of $C_{13}H_{24}NO$. A 90% intensity shift to *m/e* 212 was provided if both N–H groups were deuterated. Exchange of the *t*-butyl group for an adamantyl group ($C_{10}H_{15}$) in I resulted in the formation of a bromine-free, even-electron ion at *m/e* 288. The bisamide (II) as might be predicted

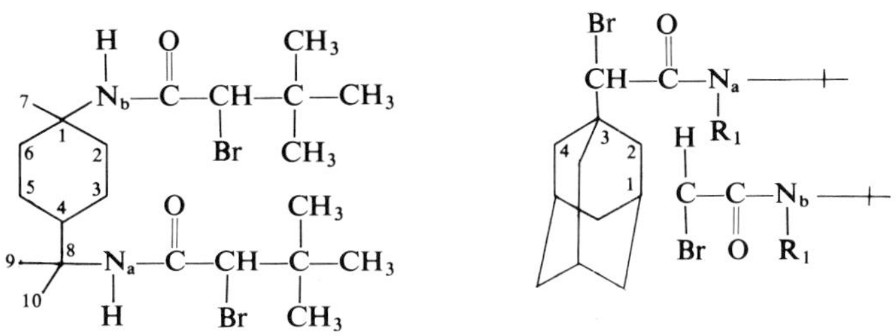

I

II

showed no evidence for a similar remote-group interaction since this molecule is skeletally and conformationally rigid.

In Scheme 24, the suggested rearrangement and fragmentation are given. It should be noted that two particles, the bromine radical and a neutral

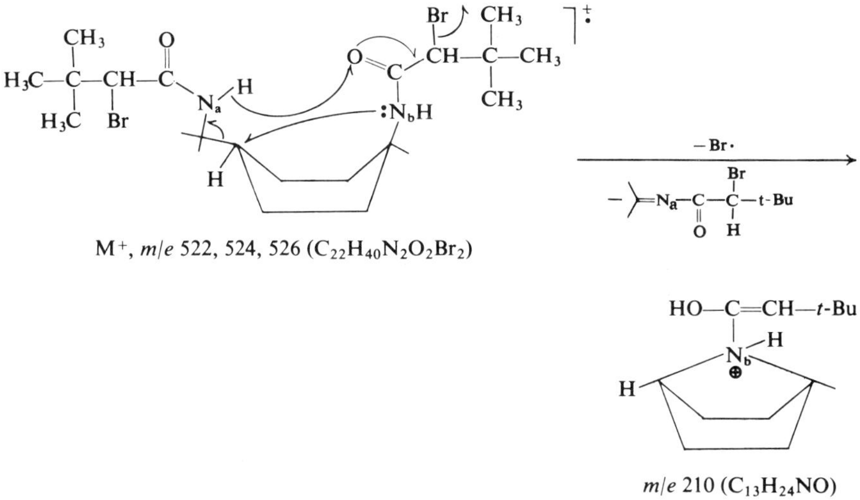

M^+, *m/e* 522, 524, 526 ($C_{22}H_{40}N_2O_2Br_2$)

$-Br\cdot$

m/e 210 ($C_{13}H_{24}NO$)

Scheme 24

fragment containing the second bromine atom, are ejected concurrently. This is a multistep fragmentation, fully dependent upon the spatial conformation of the molecular ion, and possibly can be considered a concerted reaction, depending upon the viewpoint taken. To the working mass spectroscopist, the actual process is often of little concern, but the possibility of such a fragmenta- taking place is of interest; hence the question of stereochemistry and conformation becomes quite important.

H. Stereochemical Effects upon Fragmentation

An excellent review of stereoisomeric effects on mass spectra has been prepared by Meyerson and Weitkamp (1968). Important discrepancies have appeared in the literature, and these authors attempted to assess and resolve these problems so that they could consider the results obtained on a series of methyldecalins which Meyerson and Weitkamp (1969) reported separately.

Das *et al.* (1970) have examined some epimeric 7-substituted norbornenes for differences in their fragmentation modes due to stereochemical differences. Little influence was found during labeling and high-resolution studies. Rather, they concluded that the spectra could be rationalized in terms of the formation of common intermediates or transition states by ring-opening processes. It was difficult to distinguish between the *syn*- and *anti*-7-substituted norbornenes examined by means of mass spectrometry. The problem of a symmetrical carbonium ion intermediate versus a rapidly equilibrating set of ions which simulate such an intermediate is not solved. No information is available on the structure of the norbornyl ion, and it is not known if the "norbornyl" type of ion exists, as such, in the mass spectrometer. It is conceded that the ion may be a monocyclic species, an open-chain species, or even a mixture of all these possible structures.

Bunton and Del Pasco (1969) have discussed the mass spectra of some methylnorbornyl chlorides observed from 12 to 80 eV. Wagner–Meerwein rearrangements were thought to occur very readily in the ion source, and compounds relatable by these rearrangements proved to give very similar fragmentation patterns. As might be anticipated, thermal decompositions with the tertiary chlorides were observed to be important at higher source temperatures. Thus the problems presented for these types of molecules will require considerably more study in the future.

The *cis*- and *trans*-1,2-diphenylcyclobutane isomers have been examined for stereochemical effects by Gross and Wilkins (1969). Differences in the spectra of the two compounds were noted, and low ionization voltage studies were also conducted. Complications do occur in these studies. Similarly, Bursey and Hoffman (1969) studied the loss of NO by substituted nitrobenzene ions and reported differences in the low-ionization spectra which were not anticipated and which caused complications. Some of the problems were thought to be associated with ring expansions. The study was restricted to substituent effects of methyl groups *ortho* to resonance donors. *ortho*-Methyl groups seem to prevent coplanarity of the dimethylamino group but not the methoxy group. This conclusion was supported by metastable ion peaks and by the comparison of peak intensities. Because the methyl group is not completely inert as a hindering group for this reaction (particularly at low voltage), the suggestion is made that it may be preferable to use halogen substituents in studying steric effects.

Several studies of water elimination from *t*-butylcyclohexanols have been made. For the studies different sources and instruments were used. Brion and Hall (1966) reported the photoionization mass spectra of *cis*- and *trans*-4-*t*-butylcyclohexanols. They showed that the elimination of water occurs more readily from the *trans* derivative and concluded that the *trans* isomer reacts in the boat form in which the hydroxyl group and the hydrogen from position 4 come close together and are eliminated. Such a 1,4 elimination would be much less likely to occur in the *cis* isomer. Dolejš and Hanuš (1968) agreed that the ratio of the ions $(M - 18^+)/(M^+)$ is significantly higher for the *trans* derivative when the molecules are the isomeric pairs of 2-methyl-4-*t*-butylcyclohexanols and 2,2-dimethyl-4-*t*-butylcyclohexanols. These authors also claimed that it is the equatorial hydroxyl group that is more readily eliminated if the molecules are in electronically unexcited states.

Of the isomeric pairs of 3-*t*-butylcyclohexanols and their homologs, the 2-methyl-5-*t*-butylcyclohexanols, and the 2,2-dimethyl-5-*t*-butylcyclohexanols studied, the *trans* compound again had an enhanced tendency to split out water. Dolejš and Hanuš (1968) pointed out that the *trans* compound cannot preferentially lose water by a 1,4 elimination. They suggested that the elimination would occur from the *cis* compound and, therefore, the explanation for the *trans* compound is formulated on the basis of the chair conformation as a 1,3 elimination of the axial hydroxyl group and the axial tertiary hydrogen

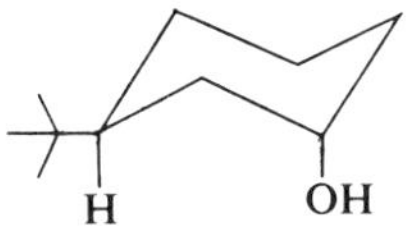

H OH

atom. From the deuterated molecules studied, they concluded that 1,2 elimination did not occur to any significant extent. They concluded, also, that the tertiary butyl group seemed to preserve the equatorial conformation in the ionized state and that the elimination of the tertiary hydrogen atom from the carbon atom carrying the butyl group was preferred over the elimination of the other hydrogens present.

The general question of conformation seems to be under more scrutiny. Eadon and Djerassi (1970) have studied the effect of ring size in reciprocal hydrogen transfer using 2-(1′-octyl)-cyclohexanone as a model. Deuterium labeling was used to confirm that the ring size is crucial to the ease of migration of the hydrogen atoms.

I. Group Migration

In most of the rearrangements discussed so far, the hydrogen atom is the principal atom that migrates in the molecule. Sometimes the fluorine atom moves somewhat similarly to the hydrogen. Larger groups are known to migrate within the molecule; perhaps one of the earliest of these to be reported is that by Funke *et al.* (1964). In this case, the aryl group migrates from a

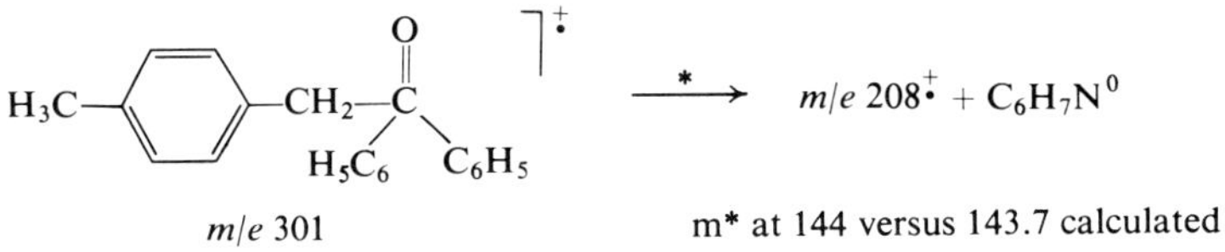

H_3C–C_6H_4–CH_2–C(=O)... H_5C_6 C_6H_5 $]^{+\cdot}$ $\xrightarrow{*}$ *m/e* $208^{+\cdot}$ + $C_6H_7N^0$

m/e 301 m* at 144 versus 143.7 calculated

Scheme 25

nitrogen atom to a carbon atom as shown in Scheme 25. The structure of the ion at *m/e* 208 is thought to be as follows:

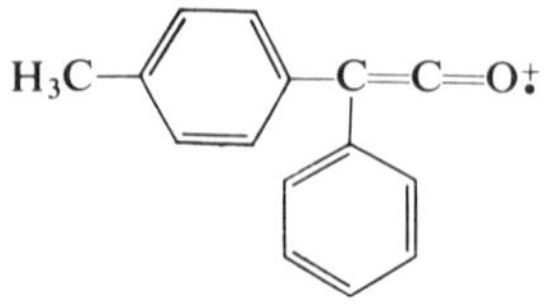

H_3C–C₆H₄–C(C₆H₅)=C=O⁺·

Scheppele *et al.* (1968, 1970) have reported the migration of a phenyl group, a methyl, and a hydrogen in the mass spectrum of acetophenone azine. Gray and Djerassi (1970) have reported a phenyl migration in *trans*-10-phenyl-Δ^3-2-octalone. Numerous reports of methyl migration have now appeared. The transfer of a hydrogen atom from a side chain to the nitrogen atom in a triazine ring has been reported by Preston *et al.* (1970).

Considerable skeletal bond changes are also noted in compounds containing hetero atoms other than N, O, and S. Cooks and Gerrard (1968) have shown that compounds with the structure –P(:S)O–R can be shown to undergo molecular ion rearrangements to produce –P(:O)S–R molecules. In addition, if an aryl group is present, one is likely to observe bond formation with the phosphorus. They also reported hydrogen rearrangement which gave the phenol molecular ion (or an isomer) in compounds having a phenoxy substituent. Cavell and Dobbie (1968) reported the rearrangement of compounds of the type $(CF_3)_3$–P–P–(CF_3) to give ions with P–F bonds. This also occurred if the phosphorus atom were separated by P–O–P, P–S–P, and in compounds of the type $(CF_3P)_4S$.

Organometallic compounds often exhibit processes in which new bonds are formed concurrently with the abstraction of a fluorine atom by a metal. Miller (1969) reported the formation of perfluorophenylene ions from $(C_6F_5)_4M$ compounds with the abstraction of fluorides of the elements Si, Ge, Sn, and Pb. It is interesting to note that the SnF^+ and PbF^+ ions become the base peak for such compounds, whereas if M is a carbon atom [for the compounds $(C_6F_5)_3OH$ and $(C_6F_5)_2CO$], there is little evidence for rearrangements or transfer of fluorine to the central atom. Clobes *et al.* (1969) similarly showed the loss of neutral metal fluorides in the mass spectra of tris(1,1,1,5,5,5-hexafluoro-2,4-pentanedionato) metal complexes.

Lawrence and Waight (1970) showed cases in which concerted migrations occur for some substituted 1,2,3-benzotriazoles resulting in elimination of neutral nitrogen and formation of a new ring. The 9,10-dihydrophenanthridine and carbazole ions show up in the mass spectra of these types of compounds. Thus the resulting spectra may appear to be a mixture to the newcomer to the mass spectrometry field. Low-voltage spectra usually provide a means of differentiation.

J. Ring-Expansion Processes

The expanded-ring processes that occur in mass spectrometry frequently have a good deal to do with the course of fragmentation paths. Rylander *et*

TABLE 5-18

Some Expanded-Ring Ions Observed in Mass Spectrometry

Precursor	Ion	*m/e*	Reference
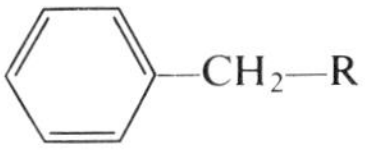 (benzene ring)—CH_2—R ⟶	$C_7H_7^+$, tropylium	91	Rylander *et al.* (1957)
	$C_7H_8^+$	92	
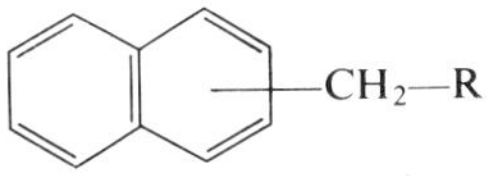 (naphthalene ring)—CH_2—R ⟶	$C_{11}H_9^+$, benzotropylium	141	Beynon *et al.* (1959)
Three or more fused rings	$C_{13}H_{11}^+$, dibenzotropylium	191	Extrapolation; observed by many workers
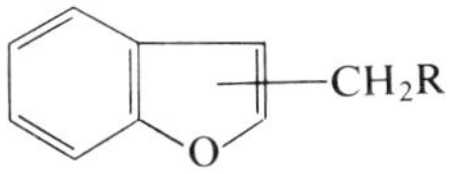 (benzofuran ring)—CH_2R ⟶	$C_9H_7O^+$	131	Willhalm *et al.* (1964), Reed *et al.* (1963)
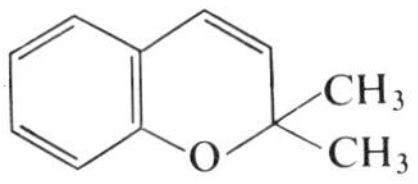 (2,2-dimethylchromene, CH_3, CH_3) ⟶	$C_{10}H_9O^+$	145	Barnes and Occolowitz (1963, 1964)
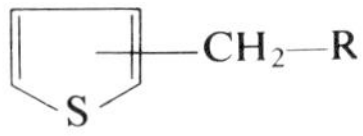 (thiophene ring)—CH_2—R ⟶	$C_5H_5S^+$	97	Hanuš and Čermák (1959)
	$C_5H_6S^+$	98	Cook and Foster (1961)
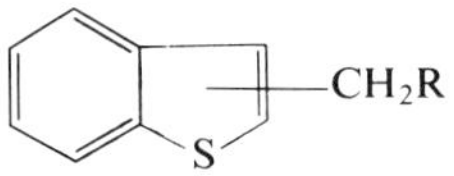 (benzothiophene ring)—CH_2R ⟶	$C_9H_7S^+$	147	Cook and Foster (1961), Porter (1967), Ritter *et al.* (1969)
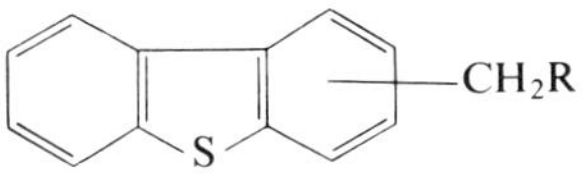 (dibenzothiophene ring)—CH_2R ⟶	$C_{13}H_9S^+$	197	Extrapolation
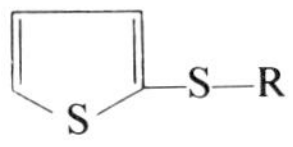 (thiophene ring)—S—R ⟶	$C_4H_4S_2^+$	116	Foster *et al.* (1968)
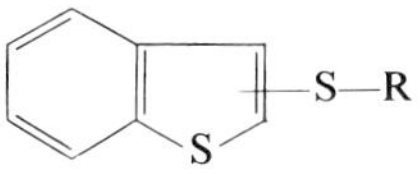 (benzothiophene ring)—S—R ⟶	$C_8H_6S_2^+$	166	Foster *et al.* (1969)

TABLE 5-18 (*continued*)

Precursor	Ion	*m/e*	Reference
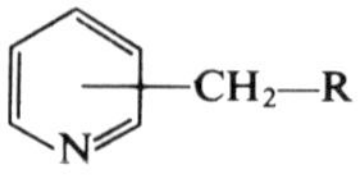 (N-ring)—CH_2—R →	$C_5H_5N^+$	79	Marx and Djerassi (1968)
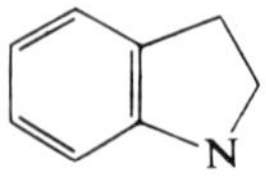 (N-ring) →	$C_5H_6N^+$	80	La Lau (1960)
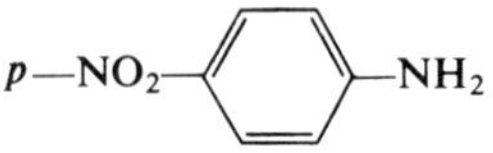 *p*—NO_2—(ring)—NH_2 →	$C_5H_6N^+$	80	Robertson and Djerassi (1968)
	$C_6H_6N^+$ (even electron)	92	
	$C_6H_7N^{+\cdot}$	93	
Quinolines		92	Marx and Djerassi (1968)
Isoquinolines	$C_6H_7N^+$, azatropylium		
Methylindole			
N-methylindole	Pyridiunium, quinolinium (var.) and other expanded-ring types		Forkey (1969)
N-Methylpyrrole			
Aniline			
Acetonitrile	$C_6H_6N^+$, azepinium	92	
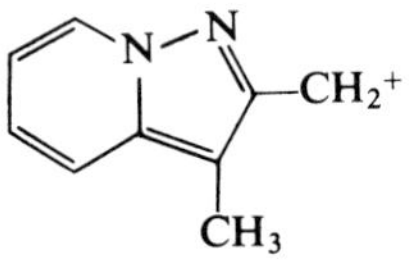 (N—N ring)—CH_2^+, CH_3 →	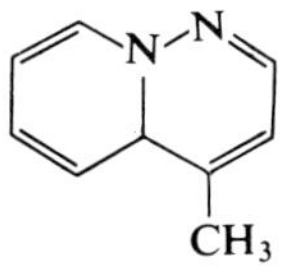 (N—N ring), CH_3	145	Potts and Singh (1970)
↘	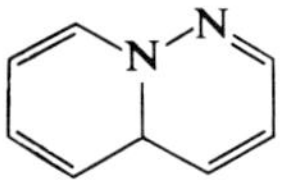 (N—N ring)	131	Potts and Singh (1970)
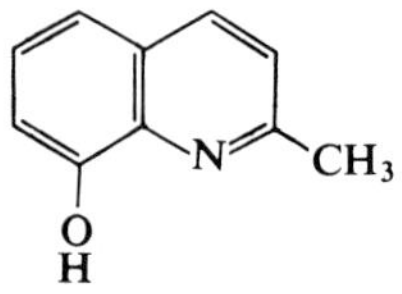 (ring, N)—CH_3, OH →	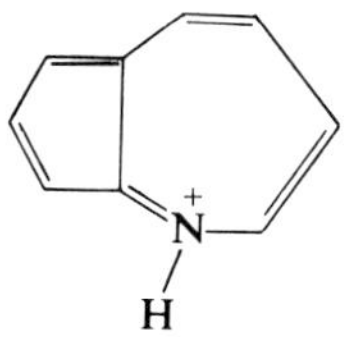 ($\overset{+}{N}$—H ring)	130	Stevenson *et al.* (1969)

al. (1957) proposed the tropylium ion to explain many of the problems with the fragmentation of alkylbenzenes. An excellent summary and review by Grubb and Meyerson (1963b) covered the complete background of this proposal and the accompanying labeling experiments that were done to establish firmly the seven-membered ring structure for this species. It should be remembered that Hückel (1931) had predicted that the tropylium ion should be a stable cation because it has six π electrons. In any event, the $C_7H_7^+$ ion is commonly found in the spectra of many compounds in addition to the alkylbenzenes. Given the opportunity to form, at least some of the ion appears in the mass spectra of aromatic compounds. As was indicated earlier in the discussion of fragmentation schemes, the ion is often accompanied by the $C_7H_8^+$ ion, a rearrangement ion plus an expanded-ring ion (*m/e* 92). This rearrangement was found to be analytically useful in the case of the monosubstituted alkylbenzenes. The subsequent breakdown of these rings also contributes important peaks to the mass spectra, but discussion is deferred until consideration of the heteroatomic aromatic ions can be added.

Expanded-ring ions for the thiophenes have been described—$C_5H_5S^+$ by Hanuš and Čermák (1959) and $C_5H_5S^+$ and $C_5H_6S^+$ by Cook and Foster (1961)—while La Lau (1960) demonstrated that the $C_5H_6N^+$ ion appeared from the alkylindoles. Subsequently, many other such rings have been observed, and a listing is provided in Table 5-18. The table is not complete, but sufficient examples are given to illustrate what the mass spectroscopist may expect to find in the spectrum from aromatic-ring-containing compounds. The nitrogen cases are still open to considerable question as to mechanism and extent of ring expansion.

An oxygenated material will be used as an example of a case in which the hetero atom oxygen is ejected as a neutral carbon monoxide moiety with the resulting ion going on to form a benzotropylium ion plus a hydrogen (mass 142) followed by the ejection of the neutral hydrogen to form the benzotropylium ion at mass 141. Beynon *et al.* (1959) have proposed a mechanism of the type shown in Scheme 26. It should be noted that the scheme involves

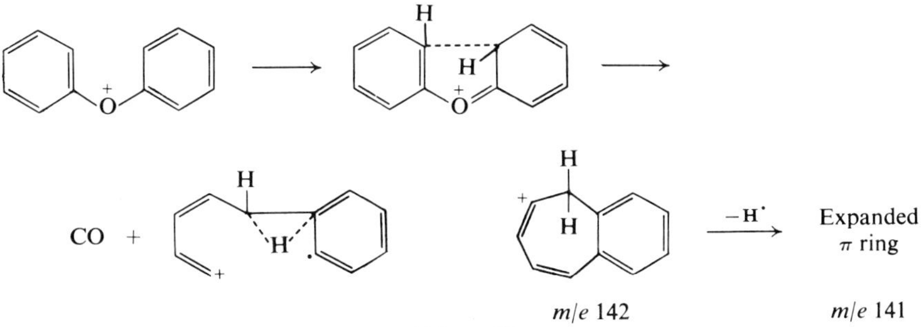

CO + ... $\xrightarrow{-H^{\cdot}}$ Expanded π ring

m/e 142 *m/e* 141

Scheme 26

simply diphenyl ether, not a fused-aromatic-ring compound. Eland and Danby (1965) have reported similar mechanisms for diphenylmethane, while Liao *et al.* (1968) have verified it for the 1-phenyl-1-(2-thienyl)methane. Beynon *et al.* (1959) indicated the principal process for the dibenzofuran to be the following:

$$\text{dibenzofuran}^{+\bullet} \xrightarrow{-CO} [m/e\ 140]^+ \begin{cases} \xrightarrow{-H^\bullet} 139^+ \\ \xrightarrow{-C_2H_2} 114^+ \xrightarrow{-H^\bullet} 113^+ \end{cases} \quad (5\text{-}123)$$

The exact structures of these ions were not assured, but it is apparent that the aromatic-ring systems still prevail, and the further fragmentation processes are characteristic of moieties that are still π-bonded systems and aromatic rings. Beynon *et al.* (1959) claimed that the ring is opened when the ion at *m/e* 113 is formed. For the diphenyl ether given above, the *m/e* 141 ion loses two neutral hydrogen atoms (or a molecule) to form the 139 ion. Then it, in turn, loses a neutral C_2H_2 (mass 26) to form the ion of mass 113. These workers, Eland and Danby (1965), and Foster *et al.* (1969), reported the elimination of HCN and CS neutrals, essentially analogous to the benzofurans for carbazole, dibenzothiophenes, and benzothienyl thiaalkanes, respectively. Other mechanisms of fragmentation are also operating for all of these types of compounds.

The expanded-ring ions form for all of the heteroatomic O, S, and N compounds, but matters do seem a good deal more involved. Barnes and Occolowitz (1964) reported that the compounds derived from coumarin lose oxygen atoms as CO and usually form a stable benzofuran ion. By contrast, 2,2-dimethylchromenes lose a methyl radical to give a stable benzopyrilium ion. For other oxygenated compounds of these general types, additional cleavages also enter the picture. Reed *et al.* (1963), working with methoxy-substituted chromenes, reported the probable mechanism to be the following:

$$\text{(CH}_3\text{O)}_2\text{-chromene } (m/e\ 192) \xrightarrow{-CH_3^\bullet} \text{O=, CH}_3\text{O-benzopyrylium } (m/e\ 177) \xrightarrow{-CH_3^\bullet} \text{etc.} \quad (5\text{-}124)$$

This was supported by Barnes and Occolowitz (1963) and Willhalm *et al.* (1964). The complication of the methoxy groups was not present in the molecule considered by Willhalm *et al.* (1964), who showed the simple loss of a hydrogen atom to produce the expanded-ring species, which then underwent the loss of neutral CO and a neutral H_2 to produce the very stable *m/e* 115 hydrocarbon aromatic ion:

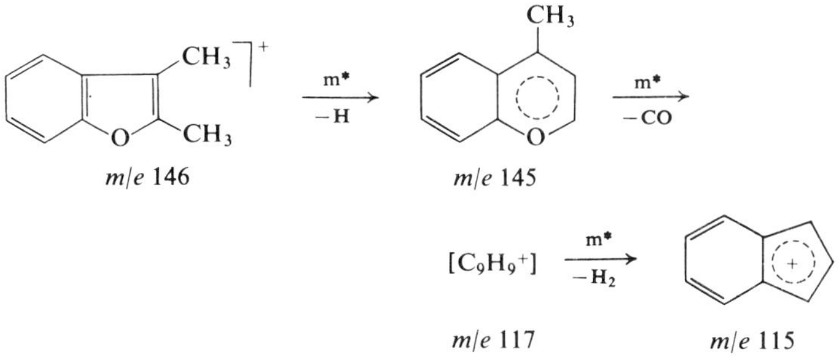

$$[C_9H_9^+] \xrightarrow[-H_2]{m^*} \quad (5\text{-}125)$$

m/e 146 —(m*, −H)→ *m/e* 145 —(m*, −CO)→ *m/e* 117 —(m*, $-H_2$)→ *m/e* 115

Among the processes also occurring were those described earlier under "*ortho* effects." Eland and Danby (1965) pointed out that for the hydrocarbons and oxygenated species, at least, the neutrals ejected, and their order might be predicted on the basis of the IP of the neutrals. Thus the series expected would be $CO > C_2H_2 > CH_3 > C_3H_3$ where the IP's are, respectively, 14.1, 11.6, 9.0, and 3-4 eV.

Cook and Foster (1961) predicted the ring expansion for the alkylbenzothiophenes, while Porter (1967) concluded that the thienotropylium ion was formed in the fragmentation of 4-, 5-, and 7-methylbenzo[*b*]thiophenes because the mass spectra of these compounds were identical, indicating a common intermediate ion. Van Brunt and Wacks (1964) reported a similar situation for the mass spectra of naphthalene and azulene wherein it appeared that a common transition state was present which led to the *m/e* 115 ion as a first fragmentation step for both substances. For a series of 2- and 3-thiaalkyl-substituted benzo[*b*]thiophenes, Foster *et al.* (1969) showed that a rearrangement and ring expansion occurred to produce ions of greatest

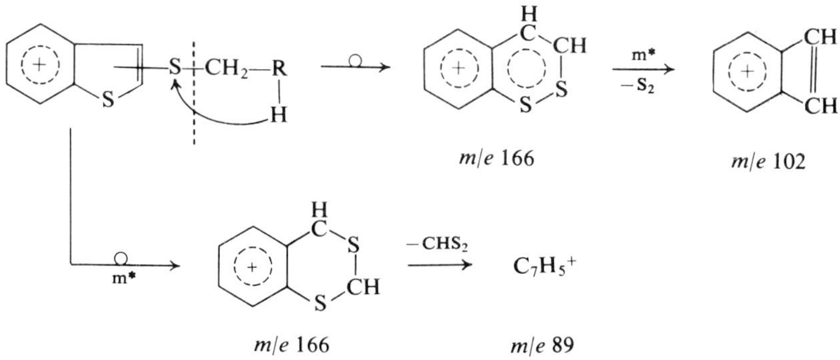

S—CH_2—R, H; *m/e* 166 —(m*, $-S_2$)→ *m/e* 102; m* → *m/e* 166 —($-CHS_2$)→ $C_7H_5^+$ *m/e* 89

Scheme 27

intensity at *m/e* 166. At least some of these ions underwent a ring expansion as shown in Scheme 27. The presence of peaks at *m/e* 64 and 66 verified the presence of an ion S_2^+. Ions at *m/e* 76 and 77 consisting of CS_2^+ and CS_2H were observed.

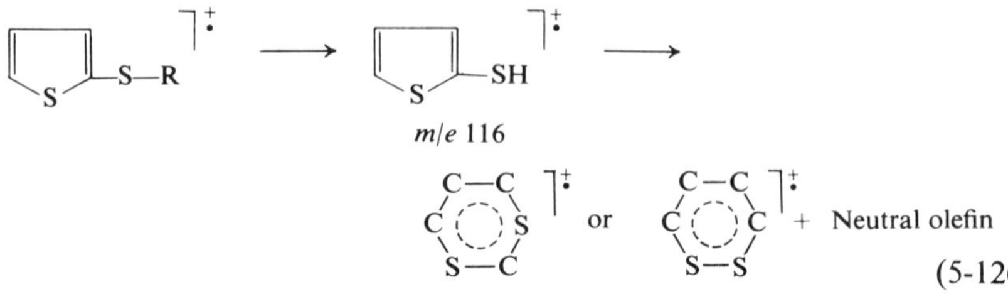

S—R → SH → *m/e* 116

C—C, C S, S—C or C—C, C C, S—S + Neutral olefin (5-126)

It should be noted that since naturally occurring sulfur has a little over 4% of the ^{34}S isotope present, the mass spectroscopist is able to utilize the naturally labeled molecules as a guide to *some* fragmentation reactions. The percentage of label is quite low but also quite constant. Therefore, the results obtained are of a little better degree of certainty and not strictly in the area of judicious speculation. A similar rearrangement occurred in the case of the thienyl thiaalkanes as shown by Foster *et al.* (1968). In this case the ring expansion may occur, although it was expected that the presence of the benzo group in the previous case would lead to a more stable ion. For these compounds and the benzo derivatives, in addition to the major rearrangement process, a process occurred to produce ions at mass 129 (and 179 in the benzo case) and fragmentation products from this ion. These ions were not large in intensity but the daughter ions were. There was some evidence of expansion to include the methyl group, followed by the loss of C_2H_2 as a neutral. This process is quite typical of the breakdown of aromatic rings. Without ^{13}C labeling the specifics of the process could not be further delineated.

For the 1-(2-thienyl)-1-butanone molecule Foster and Higgins (1969) reported no evidence for a ring expansion on the basis of both ^{13}C and deuterium labeling. For this molecule, the McLafferty type of rearrangement dominated the fragmentation. This behavior is also noted for the 1-phenyl alkyl ketones.

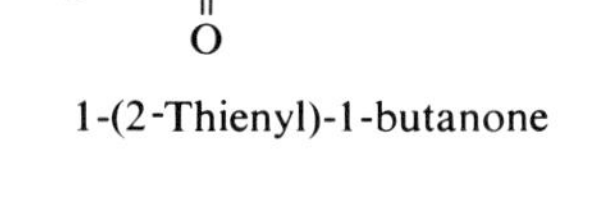

S C(=O)—CH_2—CH_2—CH_3

1-(2-Thienyl)-1-butanone

For 1-(2-thienyl)-3-thiaalkanes and -4-thiaalkanes, Foster *et al.* (1968) showed that the presence of the sulfur atom in the λ position allowed only a negligible amount of the *m/e* 98 rearrangement ion to form, but an *m/e* 97

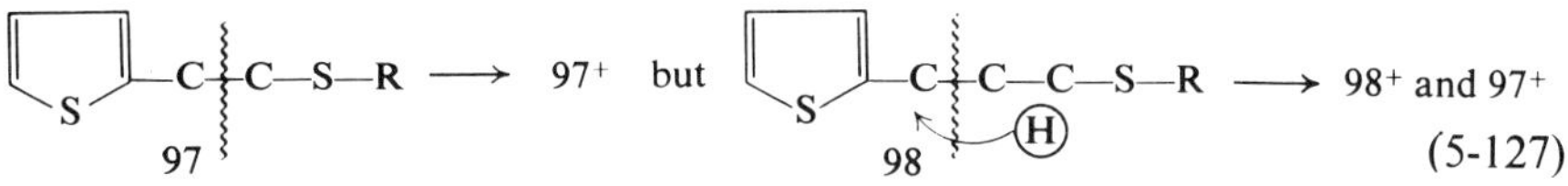

$$\text{Th—C}\,\vdots\,\text{C—S—R} \longrightarrow 97^+ \quad \text{but} \quad \text{Th—C}\,\vdots\,\text{C—C—S—R} \longrightarrow 98^+ \text{ and } 97^+ \qquad (5\text{-}127)$$

ion was formed. In contrast, the 4-thiaalkane (δ position) again permitted the formation of the *m/e* 98 ion. It cannot be determined from these data alone that a McLafferty type of rearrangement occurred or whether a β hydrogen shift was involved. Also, one cannot decide whether a four-centered transition state involving the sulfur atom of the thiophene ring and the γ-carbon-atom hydrogens leads to the production of the *m/e* 98 ion. There is some evidence that specific *and* random rearrangements are occurring, based upon the work of Foster and Higgins (1966).

Beynon and Williams (1959, 1960) reported a rearrangement ion at *m/e* 103 for a series of eleven alkylindoles. This ion proved to be $C_8H_7^+$ and arose from the $(M - H)^+$ ion by a process established by a metastable ion:

$$\underset{m/e\ 131}{C_9H_9N^+} \xrightarrow[-H]{} \underset{m/e\ 130}{C_9H_8N^+} \xrightarrow{m^*} \underset{m/e\ 103}{C_8H_7^+} + HCN \qquad (5\text{-}128)$$

The *m/e* 103 ion fragmented further by typical aromatic processes, losing mass mass 26 $(C_2H_2)^0$ twice to produce ions at *m/e* 77 and subsequently at *m/e* 51. These processes were also supported by metastable peaks. Speculation about the styryl structure for the 103 ion was given in Beynon *et al.* (1968f), but a further important point was noted. For the case of 2,3-dimethylindole, which these authors indicated should form a resonance-stabilized system by the loss of one of the methyl groups, the reaction pathway may be the following:

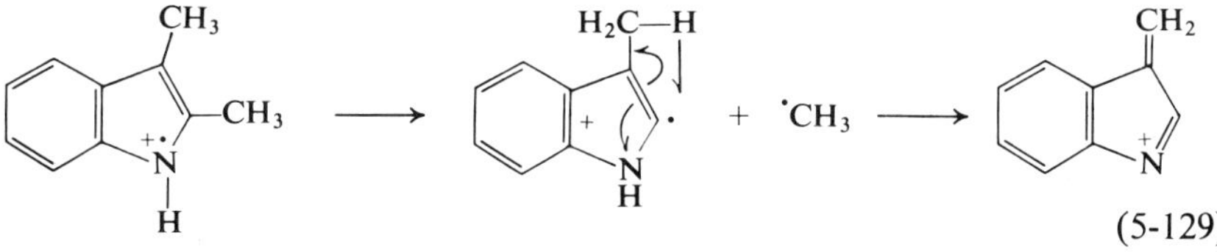

$$\longrightarrow \quad + \ {}^{\bullet}CH_3 \longrightarrow \qquad (5\text{-}129)$$

This is a resonance-stabilized ion. It may also undergo a further rearrangement to give the quinolinium ion, quite analogous to the formation of the tropylium ion formed from many alkylbenzene molecular or fragment ions.

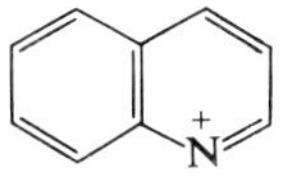

Subsequent to La Lau's (1960) report of an expanded-ring $C_5H_6N^+$ ion in the alkylindoles, the nitrogen heterocycles did not receive too much attention.

Studies of unlabeled aniline and three labeled compounds, the 3-*d*, 2,4,6-d_3, and ^{15}N, by Rylander *et al.* (1963) showed that aniline does not behave like toluene in forming the tropylium structure but rather undergoes the loss of HCN as the major fragmentation route. The very strong point was made that the aniline ion which forms is more stable due to the number of resonance

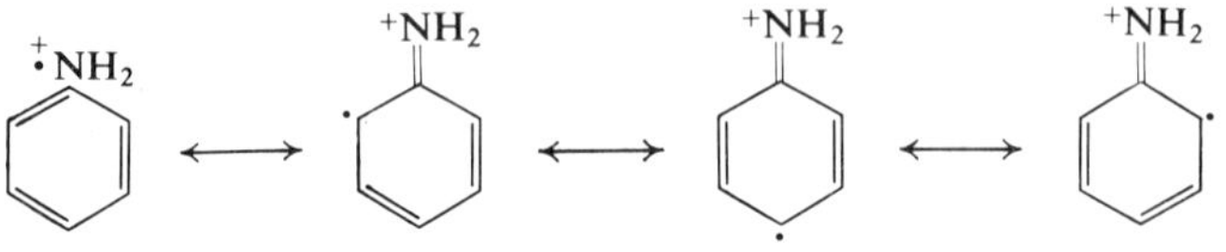

$\overset{+}{\cdot}NH_2$ ⟷ $^{+}NH_2$ ⟷ $^{+}NH_2$ ⟷ $^{+}NH_2$

forms it can assume. Thus no appreciable additional stabilization should occur if a ring expansion took place. Further, the asymmetry introduced by the nitrogen atom in the ring does not make the proposal as attractive as the hydrocarbon case. It should be noted that the nitrogen atom and, to some extent, the oxygen atom differ appreciably from the sulfur atom, which can employ the 3*d* electrons in hybridized orbital forms.

Marx and Djerassi (1968) undertook studies of ^{13}C-labeled nitrogen heterocycles to promote further understanding of the ring expansion in such molecules. Two isoquinolines, three indoles, and a pyrrole were studied. A large degree of skeletal rearrangement was observed in each compound. The migration involved the exocyclic methyl group and was found to accompany the fragmentation leading to an $[M-(H+HCN)]$ ion. These authors pointed out that their data were consistent with the intermediacy of quinolinium, pyridinium, azatropylium, or other ring-expanded ions along some of the fragmentation paths.

Robertson and Djerassi (1968) used ^{13}C-labeled aniline, acetanilide, sulfanilamide, and *p*-nitroaniline to further study rearrangements and ring expansions. In these molecules the azepinium ($C_6H_6N^+$) ion, an even-electron ion, appeared in each mass spectrum. For the *p*-nitroaniline-1-^{13}C the ion $C_5H_6N^+$ was formed by successive losses of NO· and CO. This ion contained only $\frac{5}{6}$ of the label, suggesting that the structure for the ion $C_6H_5NO^+$ from which the carbon monoxide was expelled had the label uniformly randomized. Thus, a ring expansion took place somewhere along the reaction pathway. Interestingly, in addition to the azepinium ion and the above ion, an odd-electron ion, $C_6H_7N^{+\cdot}$, was observed in both the *p*-nitroaniline and the sulfanilamide spectra.

Potts and Singh (1970) have considered a series of bridgehead nitrogen heterocycles and have reported ring-expansion ions still containing both nitrogen atoms in the mass spectra of a number of pyrazolo[1,5-*a*]pyridines. An example of portions of the processes involved is shown in Scheme 28. Attention is called to the fact that these molecules behave similarly to the

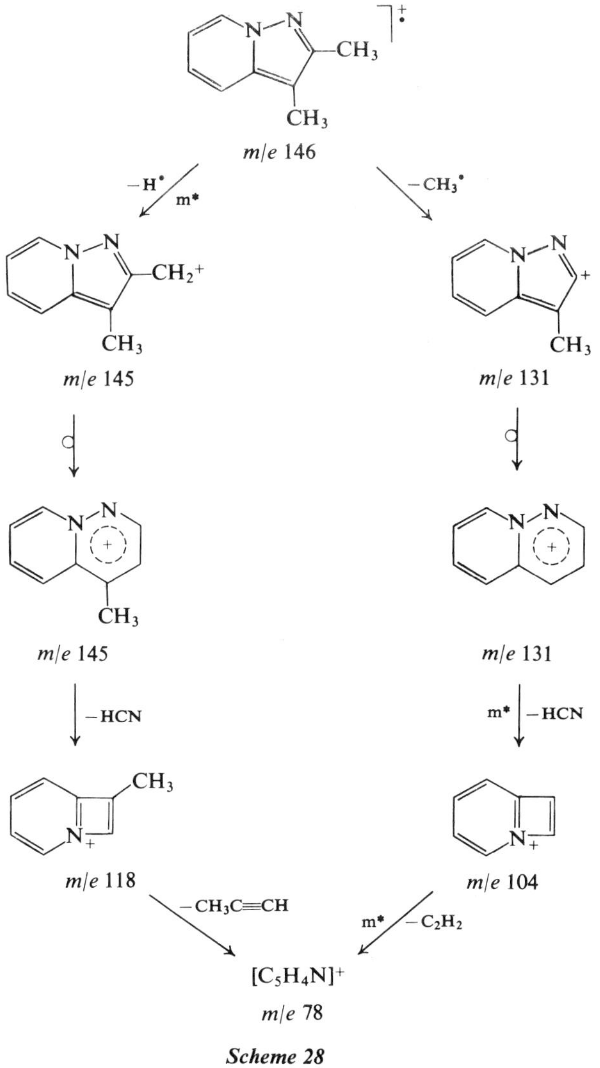

m/e 146

$-H^{\bullet}$ m* $-CH_3^{\bullet}$

m/e 145 m/e 131

m/e 145 m/e 131

$-HCN$ m* $-HCN$

m/e 118 m/e 104

$-CH_3C{\equiv}CH$ m* $-C_2H_2$

$[C_5H_4N]^+$

m/e 78

Scheme 28

2,3-dimethylindoles discussed above. The presence of two substitutional groups (in this case methyl) in the 2,3 positions aids the formation of species such as m/e 131 and 145, which then undergo a ring expansion. Once the ring expansion has occurred, the loss of HCN follows, leading to ions which are identical except for the methyl substitution in the one. The m/e 104 ion is probably analogous in structure to the m/e 102 ion observed in a number of processes involving hydrocarbon aromatic systems that have been shown in this chapter. The processes of ring degradation and fragmentation are quite similar to other aromatic systems shown in the work of Potts and Singh (1970).

In addition to considering the behavior of heterocyclic compounds with the possibilities of ring expansion, one should also consider those molecules possessing a hetero atom in the α chain position. The case for aniline was described above. Apparently for phenol and phenol-substituted compounds (and methoxy in some instances) the stabilization of the six-membered cyclic ring by many resonance forms also makes the ring expansion unattractive. For additional details on the specific processes see Beynon *et al.* (1968g). Occolowitz (1964) reported that the number of hydroxyl groups (or methoxy) substituted on the aromatic ring can be determined directly from the most abundant ion, which thus includes all of the resonance-stabilized forms of the types shown for an alkyl dihydroxy compound. The mass spectroscopist

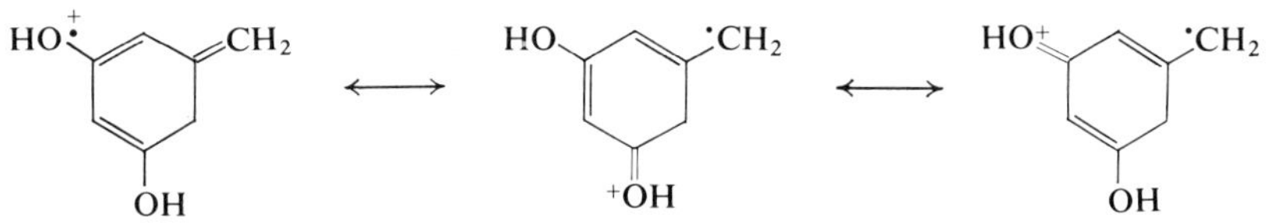

HO: CH_2 ⟷ HO $\dot{C}H_2$ ⟷ HO^+ $\dot{C}H_2$
OH ^+OH OH

working with a series of phenols, cresols, resorcinols, etc., can perform a reasonably good analysis when armed with this type of information.

In the case of the sulfur compounds, the rearrangement of the aromatic thiaalkanes was established by Foster *et al.* (1968, 1969). For the case of benzenethiol (thiophenol), Earnshaw *et al.* (1964) reported that at least some of the benzenethiol molecular ion undergoes a ring expansion to a seven-membered structure, but a large portion of the molecular ion does not undergo such a ring expansion. This work involved the study of a deuterated species and is presented in more detail in the next section of this chapter.

The tropylium ion also appears in the mass spectra of compounds which could not produce it without a rearrangement. The $C_7H_7^+$ ion may become a very prominent ion even though it cannot be formed by a simple cleavage process. Djerassi and his co-workers [see Bach *et al.* (1965)] by means of a metastable ion have established the formation of the mass 91 ion from the mass 119 ion in the spectrum of 1-phenylpropylisothiocyanate. The events are depicted as follows:

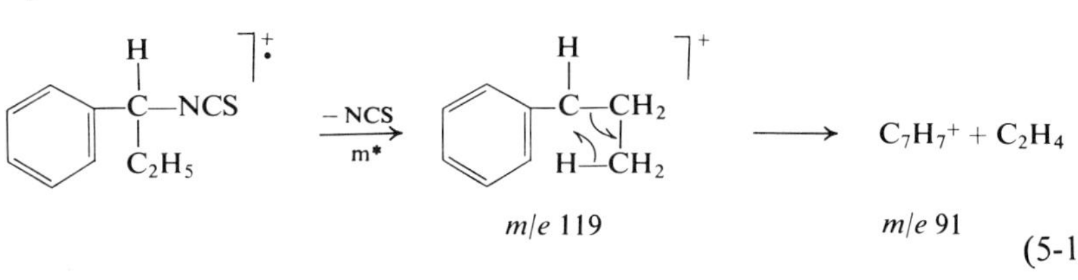

$$C_6H_5\text{-}CH(C_2H_5)\text{-}NCS^{+\cdot} \xrightarrow[m^*]{-NCS} [C_6H_5\text{-}CH(CH_2)\text{-}CH_2\text{-}H]^+ \longrightarrow C_7H_7^+ + C_2H_4 \quad (5\text{-}130)$$

m/e 119 *m/e* 91

The tropylium structure for the *m/e* 91 ion is assumed.

The favorable change in entropy of activation (ΔS_a) terms may often be a factor in the formation of a given expanded-ring ion. The randomness of atoms in the expanded-ring configuration is consistent with the entropy change being positive and, therefore, may contribute sufficiently to the overall ΔG_a (activation energy) along the reaction coordinate to make the reaction spontaneous, that is, energetically favored. McLafferty (1963d) cited this in terms of the well-known difference in reactivities of phenyl and benzyl halides. In the mass spectrum of chlorobenzene, for example, the phenyl ion produced by the cleavage of the C–Cl bond is rather small. In the spectrum of benzyl chloride, the largest ion is the $C_7H_7^+$ ion, assumed to be the tropylium structure. The entropy change is obvious. Such generalities can lead to problems, however. Some mass spectroscopists have overgeneralized to the extent that the $(M - Cl)^+$ ion is claimed to be prominent when the Cl atom is attached to the ring, in direct contrast to the above observations. The analytical mass spectroscopist must be armed with the above information when approaching any interpretive problem. It is quite true that $(M - Cl)^+$ ions are quite prominent in some cases where the chlorine is on the ring, but the case of *p*-chloromonochlorotoluene can become confusing. The presence of the two chlorine atoms will readily be noted on the basis of the isotopic peaks associated with the molecular ion. From this point on, the problem is made more difficult. Frequently, different instruments will produce substantial differences in intensities of ions involving the loss of the chlorine from the ring or from a short side chain. A few known compounds and calibration will be of much greater help than general correlations. The work of Grubb and Meyerson (1963b) should be consulted for additional details on some labeled studies of the chloroalkylbenzenes.

Substituted tropylium ions are known. The above-cited work of Grubb and Meyerson should be consulted for the details of formation of the methyltropylium ion, which is claimed to be a common intermediate in the spectra of the various xylenes and which thus prevents a distinction of these compounds by mass spectrometry. Bowie and White (1969) have suggested the formation of an interesting ion in the fragmentation pathways of the aryl dithianes. An ion of moderate intensity appears at *m/e* 131 in the dithiane, as shown below, supported by a metastable ion for the loss of S_2H from the molecular ion at *m/e* 196.

$$\text{(2-phenyl-1,3-dithiane)}^{+\cdot} \xrightarrow[m^*]{-S_2H^{\cdot}} (C_{10}H_{11})^+ \xrightarrow[m^*]{-C_3H_4} C_7H_7^+ \qquad (5\text{-}131)$$

m/e 196   *m/e* 131   *m/e* 91

The structure suggested resembles those proposed for the cationated cyclopropane complex and was also suggested by Foster (1966) as a possible structure of the *m/e* 125 ion shown in Scheme 1 for the *t*-butylthiophenes. Also shown was the loss of C_3H_4 as a neutral to produce the *m/e* 85 ion commonly found in the mass spectra of the *t*-butylthiophenes. It is presumed that the 125 ion can be classified by a charitable reader as a substituted tropylium ion. Bruce and Thomas (1968) discussed penta- and heptafluorotropylium ions from polyfluoro aromatic compounds. Nibbering and de Boer (1968c) in labeling experiments suggested the formation of a cyanotropylium ion as a part of a fragmentation pathway. Nibbering and de Boer (1968b) reported the presence of the $C_7H_6Br^+$ ion. The hydroxytropylium ion was suggested to be found in the mass spectrum of benzyl alcohol by Shannon (1962). Nibbering and de Boer (1968a,d) have confirmed its presence in labeled studies involving 1-phenylethanol-1. Kadentsev *et al.* (1968) showed the presence of an *m/e* 107 ion identified as the hydroxytropylium ion arising from an oxygen rearrangement to the phenyl ring in aromatic esters. There are, doubtless, many other claims of substituted ions to be found in the literature.

K. The Ejection of Neutral Particles

The ejection of neutral species from ions in mass spectrometry has proven to be of great importance in interpretation. It was pointed out earlier during the cleavage, RDA, and rearrangement discussions that when such ejection leads to ions in the same series as the molecular ion, there are difficulties with the interpretation. In fact, it can be said that this is probably one of the prime blocks to completely computer interpreted mass spectra. If one knows the class or classes of compounds present and these types of rearrangements are not of major importance, the computer interpretation can most likely lead to a successful analysis. For those working with high-resolution and element mapping, Bhati *et al.* (1966) have shown how neutral-fragment eliminations from internal segments of an ion can put a serious limitation upon the techniques of elemental mapping. In any case, with or without the aid of the computer, the mass spectroscopist must use his knowledge of eliminations of neutral fragments to interpret the spectrum.

Many of the neutrals (but not all) that are eliminated are small fragments, comprising from one to four atoms. Wacks and his co-workers have started a series of papers entitled "The Elimination of Small Neutral Fragments." An example from this work will be discussed in the section to follow. Typical of the small pieces ejected are such neutrals as CO, N_2, NO, S, H_2O, H_2S, and HCN. To aid the analyst, some of the common neutral fragments observed are given in Table 5-19. The table is not complete, but rather is suggestive of the many types of neutrals that may be formed. The working mass spectro-

TABLE 5-19

Some Common Neutral Fragments

Parent or fragment minus	Fragment lost
1	$H\cdot$
15	$CH_3\cdot$
16	O (from *N*-oxides), NH_2 (aromatic amides)
17	$HO\cdot$ (simple aromatic acids), NH_3 (from diamino compounds)
18	H_2O
19	$F\cdot$
20	HF
26	$CH{\equiv}CH$ (aromatic-ring fragment), $C{\equiv}N$
27	$CH_2{=}CH\cdot$, $HC{\equiv}N$ (*N*-heterocyclics)
28	$CH_2{=}CH_2$, CO(*O*-heterocyclics), N_2 $H_2C{\equiv}N$ (*N*-heterocyclics)
29	$CH_3CH_2\cdot$, $\cdot CHO$, $CH_2{=}NH\cdot$ (purines, etc.)
30	$CH_2O\cdot$, $\cdot NO$ (aromatic nitro compounds), $NH_2CH_2\cdot$
31	$CH_3O\cdot$, $\cdot CH_2OH$, CH_3NH_2
32	CH_3OH, S (usually aromatic)
33	$HS\cdot$, $[—CH_3 + H_2O]$
34	H_2S (usually aliphatic)
35	$Cl\cdot$
36	HCl, [$H_2O + H_2O$ from di- and polyhydroxy compounds]
37	H_2Cl
38	$C_3H_2\cdot$, C_2N, F_2
39	C_3H_3, HC_2N
40	$CH_3—C{\equiv}CH$, C_2O, C_3H_4
41	$CH_2{=}CHCH_2\cdot$, $CH_3—CN$
42	$CH_2{=}CHCH_3$, $CH_2—CH_2$, $CH_2{=}C{=}O$, $NC—NH_2$ (purines), NCO \ / CH_2
43	$C_3H_7\cdot$, $CH_3C{=}O\cdot$, $CH_2{=}CH—O\cdot$, $[CH_3\cdot + CH_2{=}CH_2]$, HCNO
44	O ‖ $CH_2{=}CHOH$, CO_2, CS, N_2O, $C—NH_2$ (aromatic amides), $NH_2—CH—CH_3$
45	CH_3CHOH, $CH_3CH_2O\cdot$, CSH, CO_2H, $CH_3—CH_2—NH_2$
46	CH_3CH_2OH, $[H_2O + CH_2{=}CH_2]$, CSH_2, $\cdot NO_2$ (aliphatic nitro compounds)
47	$CH_2{=}SH\cdot$
48	CH_3SH, SO, O_3
49	$\cdot CH_2Cl$
50	$\cdot CF_2$, C_3N
51	$\cdot CHF_2$
52	C_4H_4, C_2N_2

TABLE 5-19 (continued)

Parent or fragment minus	Fragment lost
53	C_4H_5
54	$CH_2{=}CH{-}CH{=}CH_2$
55	$CH_2{=}CH{-}\dot{C}HCH_3$
56	$CH_2{=}CH{-}CH_2{-}CH_3$, $CH_3CH{=}CH{-}CH_3$, [CO + CO, quinones] C_2S
57	$C_4H_9\cdot$, C_2SH
58	$^{\cdot}NCS$, C_2SH_2, [NO + CO], 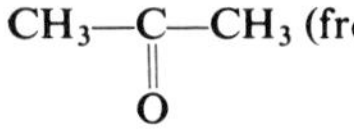 $CH_3{-}C({=}O){-}CH_3$ (from ketones)
59	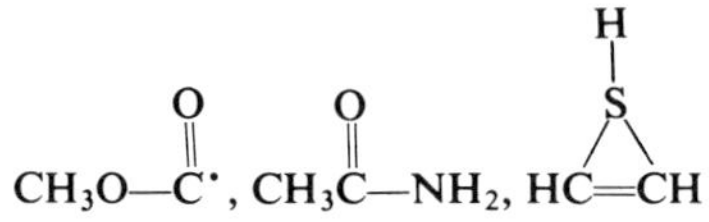 $CH_3O{-}\dot{C}{=}O$, $CH_3C({=}O){-}NH_2$, $HC{=}CH$ (bridged by S–H)
60	C_3H_7OH
61	$CH_3CH_2S^{\cdot}$, 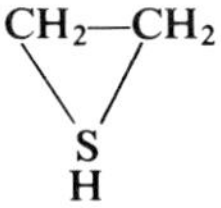 $CH_2{-}CH_2$ (bridged by S–H)
62	$[H_2S + CH_2{=}CH_2]$
63	$^{\cdot}CH_2CH_2Cl$
64	C_5H_4, S_2, SO_2
68	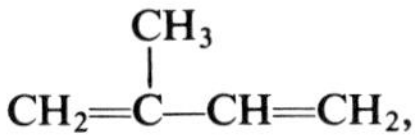 $CH_2{=}C(CH_3){-}CH{=}CH_2$, C_3S
69	$CF_3^{\cdot}$, $C_5H_9^{\cdot}$, C_3SH
71	$C_5H_{11}^{\cdot}$, C_3H_3S
73	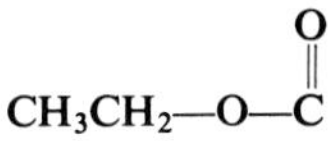 $CH_3CH_2{-}O{-}\dot{C}{=}O$
74	C_4H_9OH (ester rearrangement)
75	C_6H_3
76	C_6H_4, CS_2
77	C_6H_5, CS_2H
78	C_6H_6, CS_2H_2, C_5H_4N
79	$Br\cdot$, C_5H_5N
80	HBr

TABLE 5-19 (*continued*)

Parent or fragment minus	Fragment lost
85	$\cdot CClF_2$
100	$CF_2{=}CF_2$
119	$CF_3{-}CF_2\cdot$
122	C_6H_5COOH
127	$I\cdot$
128	HI

scopist will again develop his own addenda to this table as he moves into a given research area and new fragments are observed.

Throughout the previous sections, the reader has probably become quite conscious of ejections of neutral fragments from the various aromatic moieties. Thus, only a general view is given here. For details of the breakdown of the tropylium ion, other alkylbenzene ions, etc., the work of Meyerson (1955) and Grubb and Meyerson (1963a) and the texts of Budzikiewicz *et al.* (1967c), Beynon (1960a), and Beynon *et al.* (1968a) should be consulted.

The tropylium ion was shown to undergo a first ring fragmentation of

$$\underset{m/e\ 91}{C_7H_7^+} \rightarrow \underset{m/e\ 65}{C_5H_5^+} + C_2H_2 \tag{5-132}$$

The *m/e* 65 ion, in turn, could lose a second neutral C_2H_2 species to form the stable $C_3H_3^+$ ion at *m/e* 39. Ions associated with the ring fragmentation in alkylbenzenes in general also include *m/e* 63, 54, 53, 52, and 51 and small amounts of 27, 26, and 25, depending upon the original molecular ion. Many of these ions are formed by the ejection of H_2, H, CH_3, C_2H_2, etc., in successive processes. In other words, a group of fragmentations which produces quantities of the above-mentioned *m/e* ions coupled with other evidence leads the mass spectroscopist to conclude that at least the benzene ring is present. These same neutral fragments are also found to be ejected from fused-ring systems and from heterocyclic aromatic compounds.

Ions from the loss of these same fragments are found in the mass spectra of diphenyl, diphenylacetylene, and phenanthrene as reported by Natalis and Franklin (1965). These authors also called attention to doubly and triply charged ions from the loss of these neutrals. They noted that in almost all cases, the electron energy required to bring about a specific process increases with the resonance stabilization in the neutral parent molecule. With the exception of the molecular ions, all ions of the same composition have similar heats of formation. Natalis and Franklin (1965) also concluded that all reactions studied involving energies appreciably above the ionization potential occur with ring opening. Energetics have been the best guide to the question of an open-chain or cyclic ion form and probably will continue to be for some time to come. Detailed studies are beginning to appear in the literature.

Meyerson and Fields (1969a) have used deuterated polymethylbenzenes to confirm and extend the model derived from the labeled *p*-xylenes (Grubb and Meyerson, 1963b) for the loss of a methyl radical under electron impact. They concluded that a methyl group flanked by two methyl groups undergoes no exchange of hydrogens before it is ejected. However, if an unsubstituted position was adjacent to the methyl group, a hydrogen appeared to exchange with the *ortho* hydrogen atom through a methyl–methylene–methyl cycle, somewhat as shown:

CH_3, CH_2, H_x, H_y $\longrightarrow$ CH_3, CH_2, H_y, H_x (5-133)

The point was made that such a similar concerted coupling of the methylene group and hydrogen and loss of the methyl group reformed could account for an apparent inconsistency existing between the mass spectra of methyl alkyl and phenyl alkyl ketones. With deuteration on the methyls, it was shown that hydrogen was present, but not to the exact statistical expectations, in the methyl radical ejected. Johnstone and Millard (1966) have shown the loss of nonterminal carbon atoms as methyl radicals from species such as diphenylmethyl, 1,2-diphenylmethyl, and stilbene ions. Study of these articles in depth will convince the mass spectroscopist just entering the field that there is much to be done concerning mechanisms in the second (or third) generation of organic mass spectrometry. The complicated processes will require a considerable effort to learn, and in view of the massive number of literature entries, the reading alone will be quite difficult.

Much of what has been learned about the fragmentation of the hydrocarbons has been translated to the corresponding heteroatomic species. There are many special cases as will be seen in the following material, but often the presence of the hetero atom serves to accentuate the differences between aromatic and linear fragmentation processes and thus assists the mass spectroscopist in some generalizations. Earlier in this chapter we have seen, for example, that when a sulfur atom is involved in an aromatic ring system, decomposition occurs by routes that tend to eject the neutral sulfur, a neutral CS, and only rarely a neutral CHS or RCS. Conversely, these latter two plus the ejection of HS and H_2S are more typical of aliphatic processes. Gillis and Occolowitz (1966) have shown that only a few aliphatic compounds have an $(M - S)^+$ ion greater than 1% of the largest ion in the spectra. Hence, one may expect such differences to show up in oxygen- and/or nitrogen-containing species. Such is the case, as we proceed to demonstrate.

In addition to the ejection of a methyl radical from the middle of a molecule, there are some fine examples of such ejections involving heteroatomic molecules. Perhaps the best known of these is that described by Beynon and Williams (1960) and discussed at length in Beynon *et al.* (1968h). These authors noted that ionization of the anthraquinone molecule occurs preferentially by loss of one of the unpaired electrons on the oxygen atom. Since the O^+ ion is trivalent, one of the carbon–carbon bonds next to the carbonyl group may become separated to provide an orbital which can give a π-bonding effect on the lone-pair oxygen orbital. A complete π-bonding effect cannot initially be realized, since the direction of this bond orbital is not parallel to the orbital available on the oxygen. Beynon *et al.* (1968h) claimed the changes can be represented by the following:

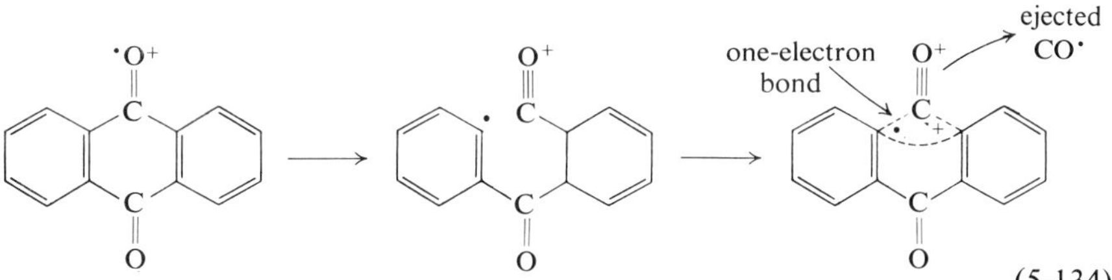

one-electron bond
ejected CO·

(5-134)

The positive charge on the oxygen atom induces a polarity on the adjacent bonds, the maximum occurring in the remaining carbon–carbon bond. A weakening of this bond occurs to the point where the electron pair may be considered to be essentially entirely in the diagonal orbital. The excess internal energy produced at ionization is sufficient to increase the vibrational energy beyond the limit of dissociation. The neutral CO is ejected, and a *one-electron bond* is set up as shown. This is a highly excited fluorenone ion. To reach the ground state, one of the unpaired oxygen electrons migrates to this bond, closing the bond to a normal covalent distance and producing the fluorenone

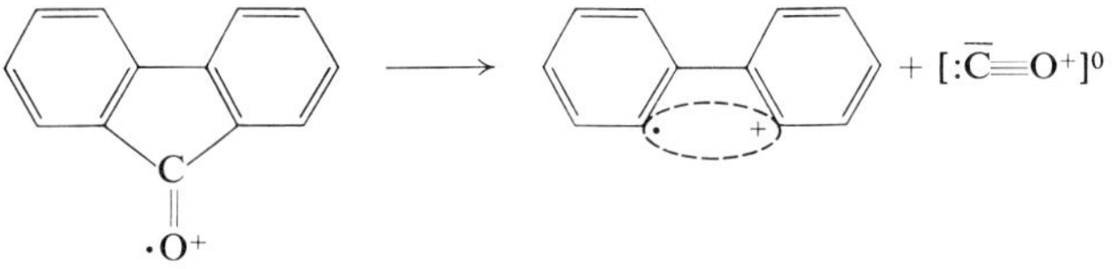

$$\longrightarrow \quad + [:\bar{C}{\equiv}O^+]^0 \qquad (5\text{-}135)$$

ion. This, in turn, can undergo the same processes to give the highly excited *o*-biphenylene ion and a second neutral CO. Accompanying this fragmentation is a metastable ion peak for the transition of a doubly charged ion into a second doubly charged ion with the ejection of a neutral carbon monoxide particle. Beynon (1960b) described the reaction as

$$(M - CO)^{2+} \xrightarrow{m^* = 64.2} (M - C_2O_2)^{2+} + CO \qquad (5\text{-}136)$$

This was the first transition between doubly charged ions to have been detected and established by means of a metastable ion peak.

The ejection of the CO continues even when other functional groups are present, but to a much lesser extent. Elwood *et al.* (1970) demonstrated this with isolapachol, as shown in Scheme 29. The above is only a partial description

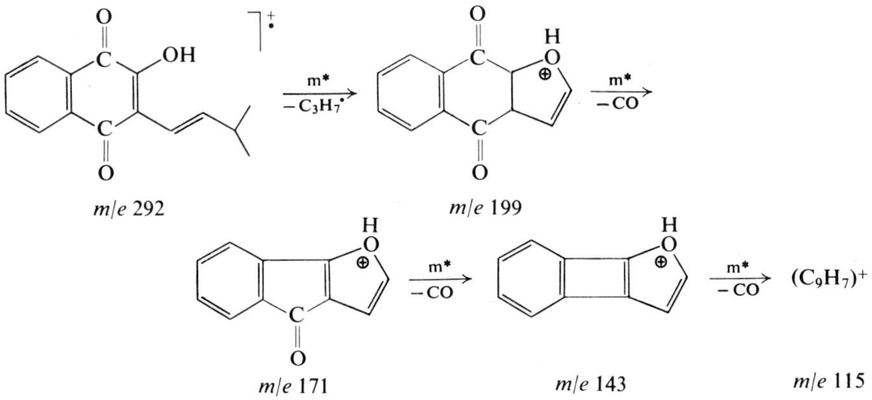

m/e 292 $\xrightarrow[-C_3H_7^{\cdot}]{m^*}$ m/e 199 $\xrightarrow[-CO]{m^*}$ m/e 171 $\xrightarrow[-CO]{m^*}$ m/e 143 $\xrightarrow[-CO]{m^*}$ $(C_9H_7)^+$ m/e 115

Scheme **29**

of the processes going on. The molecular ion also loses a methyl radical to produce an ion at *m/e* 227 which, in turn, loses a neutral water molecule to produce the *m/e* 209 ion. This ion successively loses two neutral CO particles to arrive at *m/e* 153. Ions at *m/e* 105 and 104 are also produced from the molecular ion, the 105 being the structure one would expect from earlier fragmentations:

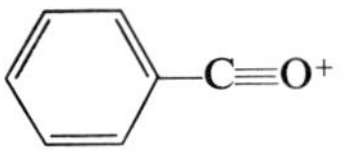

$C_6H_5-C{\equiv}O^+$

Heiss *et al.* (1968) described some dibenzothiophene and related sulfur- and oxygen-containing molecules, which permit one to make a comparison to the Beynon mechanism above. For the first molecule (a), the intensities of the ions ejected initially from the molecular ion follow the values of $CS > S > C_2H_2$. For compound (c) the loss of CO was almost as great as the loss of S. Subsequently, it appeared that the moiety which had lost the sulfur atom next

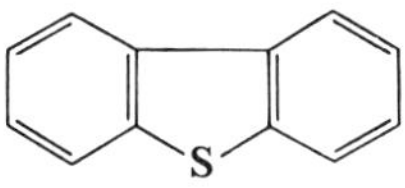

(a) 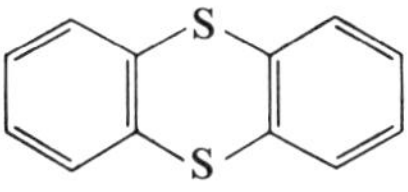 (b) 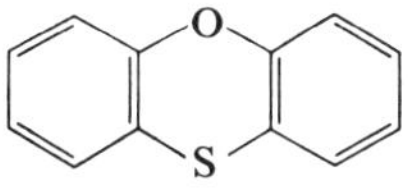 (c)

lost a COH group, while the moiety that lost the CO then lost a hydrogen atom, followed by the loss of CS rather than the sulfur atom. The thianthrene (b) first lost a CS, followed by a hydrogen, both supported with metastable peaks. Once the first sulfur was lost from (b), the moiety remaining was identical to (a) and hence a CS was ejected.

If the sulfoxides of (a)–(c) are considered, [a′] showed the loss of an oxygen atom to give the largest ion in the spectrum. This has been observed in previous cases. A neutral CO was also lost giving an ion of only about 10% the intensity of the $(M - O)^+$ ion. Subsequently, the $(M - O)^+$ ion decomposed by losing CSH > S > CS. The (b) sulfoxide [b′] molecular ion showed the loss of SO > COH > O. Subsequently, the $(M - SO)^+$ ion lost CSH, the intensity being twice that of the loss of S. The ion formed by $(M - CHO)^+$ subsequently lost a sulfur neutral atom to form an ion about equal in intensity to the $(M - SO - CSH)^+$ above. The (c) sulfoxide [c′] ion showed losses of SO > CO > –O. Subsequent events were quite complex and included the loss of a second CO, an S, a CS, and a COH. These workers also gave details on the corresponding fragmentations of the sulfones. In addition to the above processes, the ejection of a neutral SO_2 fragment was observed. In all cases, the ejection of these neutrals dominated the fragmentation, and for the sulfone of (c), three separate ejections of CO could be observed.

The ejection of a neutral SO_2 is not restricted to the above type of molecules. Grostic *et al.* (1966) reported the ejection of such a fragment from the middle of a molecule of the following type:

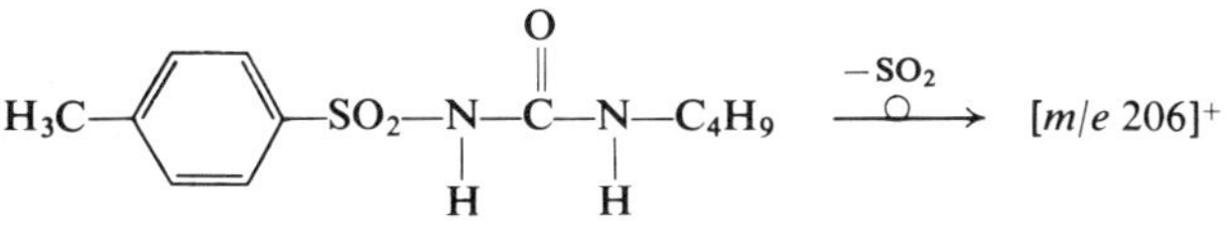

$$H_3C-C_6H_4-SO_2-\underset{H}{N}-\overset{O}{\overset{\|}{C}}-\underset{H}{N}-C_4H_9 \xrightarrow{-SO_2} [m/e\ 206]^+ \qquad (5\text{-}137)$$

At 19 eV, the *m/e* 206 ion was the most intense ion.

Another example of an ejection of a neutral from the middle of a molecule or a fragment ion was reported by Foster *et al.* (1965). The presence of alkylbenzenelike sulfur-free ions in the mass spectra of the alkylthiophenes was shown by Kinney and Cook (1952). The process or processes by which these arose were not known. The low ionizing voltage data of Foster *et al.* (1965) showed that these ions did not appear in any appreciable quantities at low voltages but became more evident in spectra taken at voltages of 20–25 eV and higher. The interesting point is that despite the uncertainty of the mechanism of formation, the net effect was to form ions devoid of sulfur and correlatable with the presence of an ethyl or propyl substitution in a dialkylthiophene. The probable events were as follows:

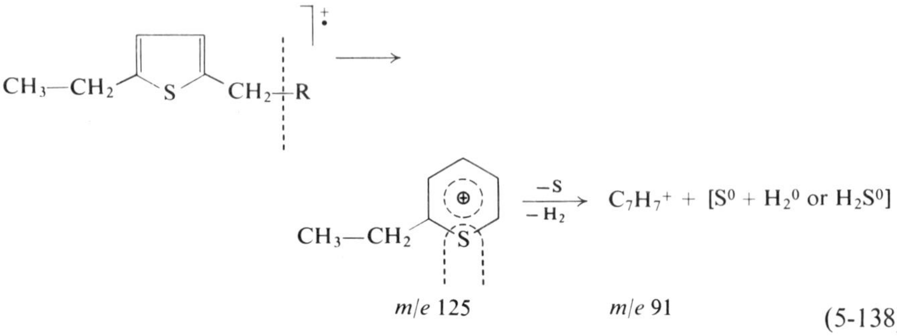

$CH_3—CH_2$ S CH_2 ⫶ R $]^{+\cdot}$ ⟶

$CH_3—CH_2$ S ⊕ $\xrightarrow[-H_2]{-S}$ $C_7H_7^+ + [S^0 + H_2^0 \text{ or } H_2S^0]$

m/e 125 *m/e* 91 (5-138)

The correct number of carbon atoms would be present, but the process of ejection of S or H_2S was not verifiable by a metastable peak. The ion intensities were low. The intensity of the *m/e* 91 ion was greater than 4 % of the base peak if the compound was an ethyl alkyl derivative, while it was from about 2.1 to 3.8 % of the base peak if the compound was a propyl alkyl derivative. Higher derivatives were of lower intensity than 2.1 %. In each case the alkyl group had to be equal to or larger than the ethyl or propyl group. In some work done on these molecules by the Shell (Houston) high-resolution mass spectrometer (Bailey, 1961), a portion of the data indicated the formation of an $(M - 6)^+$ ion at high voltages. This sort of hydrogen stripping reaction was dismaying at first because it suggested that a highly unsaturated species was formed before the ejection of the neutral sulfur moiety, an obvious contradiction to the suggestions above. In any event, this problem was never finally solved. Such stripping was not observed to occur in the ^{13}C-labeled 2-*n*-hexylthiophene reported by Foster and Higgins (1968).

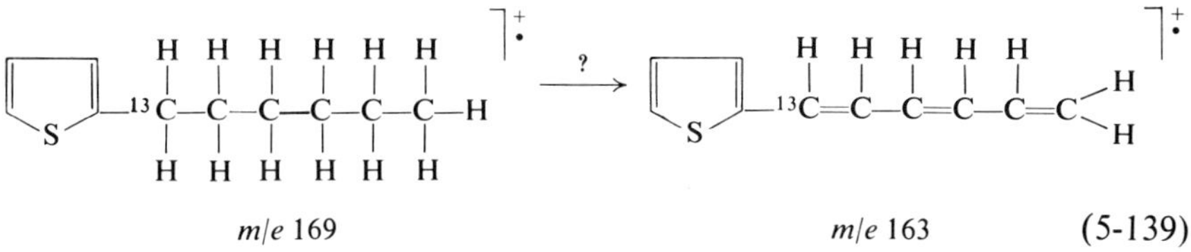

S —$^{13}C(H_2)$—$(CH_2)_4$—CH_3 $]^{+\cdot}$ $\xrightarrow{?}$ S —$^{13}CH{=}CH—CH{=}CH—CH{=}CH_2$ $]^{+\cdot}$

m/e 169 *m/e* 163 (5-139)

Eliminations of neutral species from two different parts of a molecular ion or a fragment ion have been shown in other examples given throughout this chapter. Probably most notable were the successive losses of CO and, in the RDA case, the loss of neutral C_2H_4 from quite separate parts of the ion. A further example of this sort of behavior is provided by examination of the partial mass spectrum of *d,l*-4-*exo*-methyl-7,7-dimethyl-6-thiabicyclo[3.2.1]-octane. The compound was provided by the courtesy of Dr. A. W. Weitkamp,

TABLE 5-20

PARTIAL MASS SPECTRUM OF *d,l*-4-*exo*-METHYL-7,7-DIMETHYL-6-THIABICYCLO[3.2.1]OCTANE[a]

m/e	RI	*m/e*	RI
93	58.5	121	66.9
95	37.5	136	100
99	12.9	155	58.0
107	12.7	170	51.0
113	60.3		

[a] Compound provided by courtesy A. W. Weitkamp, American Oil Co.

American Oil Co. The partial mass spectrum is presented in Table 5-20. The strong parent and the isotopes (not shown) for the molecular ion established the presence of the sulfur atom. The ion at *m/e* 155 clearly retained the sulfur atom plus its built-in isotope label. The largest ion at *m/e* 136 did not contain sulfur. It was thought that the loss of the neutral H_2S would lead to the limonene ion. The complexity of the spectra was reminiscent of the terpenes. Note that the loss of methyl could occur at two different places, but successive

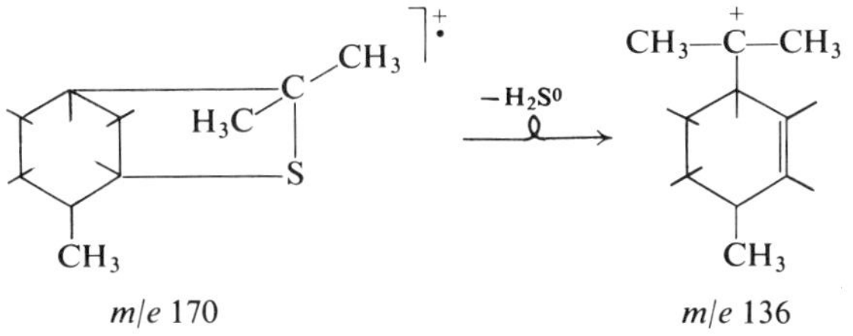

$$m/e\ 170 \xrightarrow{-H_2S^0} m/e\ 136 \quad (5\text{-}140)$$

losses of methyl were not observed. The pathways are difficult to construct, let alone speculate about. Some unpublished speculation is given in the mechanism shown in Scheme 30. The 155–113 process could have been the loss of

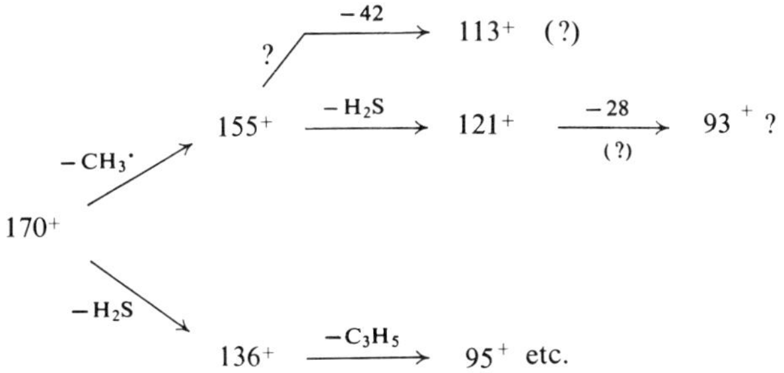

$$170^+ \xrightarrow{-CH_3^\cdot} 155^+ \xrightarrow{-H_2S} 121^+ \xrightarrow[(?)]{-28} 93^+ ?$$
$$155^+ \xrightarrow[?]{-42} 113^+ \ (?)$$
$$170^+ \xrightarrow{-H_2S} 136^+ \xrightarrow{-C_3H_5} 95^+ \text{ etc.}$$

Scheme 30

a methyl from the six-membered ring of the molecular ion followed by the ejection of the isopropyl moiety. If true, where did the sulfur go? It was possible that it formed an *m/e* 79 ion with the ejection of H_2S, but this could not be established. Indeed the terpene types of compounds lead to many fragmentations.

Often in molecules where neutrals are ejected, there are obvious cases of competition in the primary reactions. This was true in the previous case, but the work of Meyerson and Vander Haar (1968) provides a very concise example of how erroneous conclusions can be drawn in regard to the relative importance of two reaction paths. The type of compound examined was the 1-tetralone, and the method was to consider the high-resolution mass spectrum. These authors gave a partial decomposition scheme, the major pathways of which arise from competitive decomposition of the molecular ion to lead to doublet ions at *m/e* 118. The neutral ejections were of CO and C_2H_4:

$-C_2H_4$ $\longleftarrow$ $\longrightarrow$ $-CO$ (5-141)

m/e 118 I — *m/e* 146 — *m/e* 118 II

The structures of ions I and II are justified in the work. By considering the high resolution for identification of the atomic content of the ions, the conclusion was reached that 74% of the fragmentation occurred through branch I while 26% followed from species II. In contrast the low-resolution mass spectrum of 1-tetralone shows evidence for the existence of several ions that are readily explained only in terms of the primary loss of CO. In the high-resolution–high-voltage spectrum, the *m/e* 118 ions give the ion of type I 96.4% of the total intensity at nominal mass 118. Species II accounts for the remaining 3.6%. How is this reconcilable with the 26% given above, which does make the ion quite significant in the pathways? The relative intensities of I and II were shown to vary markedly by lowering the ionization voltage sufficiently to substantially limit the extent of further reactions of the ions. The *m/e* 118 ion of species II rises from the low value given to about 25% of the *m/e* 118 doublet at 10 eV. The agreement with the 26% obtained from the proposed decomposition scheme was excellent. The most important conclusion in this work is that "equating the intensity of an ion with its importance in the underlying chemistry may well lead to serious oversights."

Spiteller-Friedmann and Spiteller (1967) have stated that the $C_6H_{13}OH_2^+$ ion from di-*n*-hexyl ether appears to be a significant intermediate leading to the $C_6H_{14}^+$ ion even though it is present in very low intensity in the 70 eV spectrum.

A number of reports have appeared dealing with the ejection of neutrals from various fused-ring nitrogen heterocycles. In the previous sections, we have observed some behavior of oxygen and sulfur mixed heteroaromatics. The 1,2,3-benzotriazoles have been described by Lawrence and Waight (1970). Typical fragmentation paths are shown in Scheme 31a. The (b) path leads to an opening of the triazole ring and thence to the loss of a neutral N_2CH_3 species. The *m*/*e* 90 species then undergoes the loss of a neutral HCN leaving

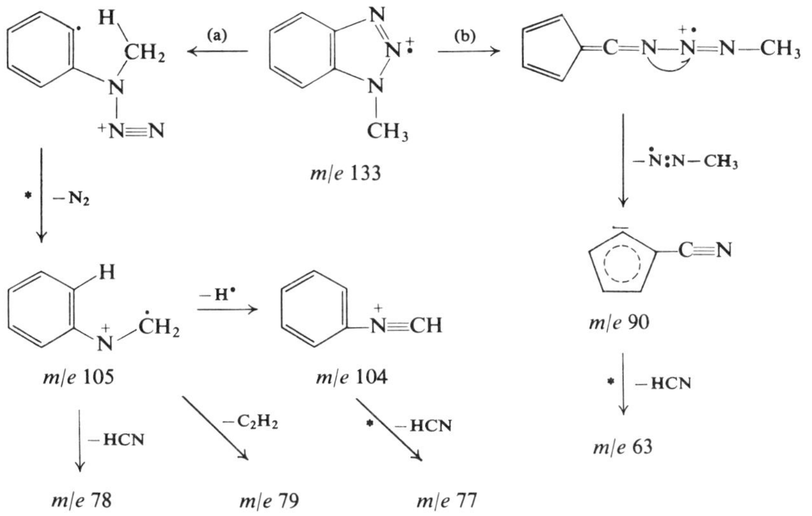

(a) (b) *m*/*e* 133 $-N_2$ $-\dot{N}{:}N-CH_3$ $-H^\bullet$ *m*/*e* 105 *m*/*e* 104 *m*/*e* 90 $-HCN$ $-C_2H_2$ $-HCN$ $-HCN$ *m*/*e* 63 *m*/*e* 78 *m*/*e* 79 *m*/*e* 77

Scheme 31a

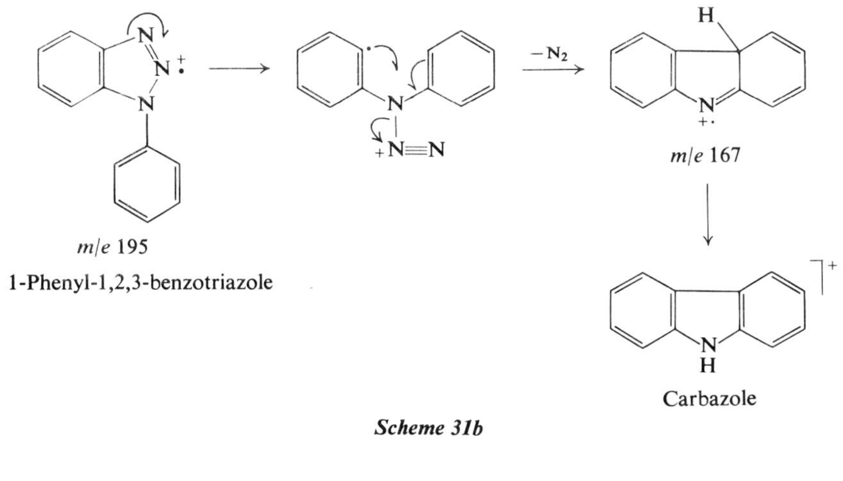

$-N_2$ *m*/*e* 167 *m*/*e* 195

1-Phenyl-1,2,3-benzotriazole

Carbazole

Scheme 31b

a moiety consisting of C_5H_3 at *m/e* 63. By contrast the (a) type of cleavage leads to an opening of the triazole ring and the transfer of a hydrogen atom to the ring from a spatially favored orientation of the methyl group. The transition state must at this point be favorably organized for the ejection of a neutral nitrogen molecule, as shown. The *m/e* 105 ion then has at least three subpaths involving the ejection of a neutral HCN, a neutral C_2H_2, and lastly a neutral hydrogen followed by a neutral HCN. The remaining ions, in both paths (a) and (b), are reminiscent of the alkylbenzenes and are nitrogen free.

For the 1-phenyl-2,3-benzotriazole shown in Scheme 31b, matters are a little different. The favored process is the ring opening followed by the loss of neutral nitrogen but leading to a more condensed ring structure, the carbazole ion, through the loss of a neutral hydrogen atom. The similarities between the (a) process of Schemes 31a and b are obvious. These authors also pointed out that for similar molecules, the 1,2,3-thiadiazole, and the 2,1,3-thiadiazole, the intensities of the ions from the loss of a neutral NS fragment are essentially nonexistent. Hence, they concluded that the ejection of the neutral NS must be a difficult, if not impossible, process.

Gelling *et al.* (1969) described the fragmentation processes occurring in some pyridopyrimidin-4(3*H*)-ones and pyridopyrimidine-2,4(1*H*,3*H*)-diones. When the CO group is present, its ejection is the initial reaction, followed by two successive losses of one molecule of HCN. A minor but competing pathway involves the initial loss of HCN while a second involves the ejection of either HCNO and HNCO, followed by a second loss of HNCO, followed by a loss of HCN to produce an ion at *m/e* 77, which still contains a nitrogen atom.

The mass spectra of the nonbenzenoid aromatic 2*H*- and 2-methyl-2*H*-cyclopenta[*d*]pyridazines have been studied by Forkey (1969). Even in these cases, the majority of the nitrogen ejected is lost in the form of HCN or H_2CN. The molecular ion is shown to have at least six possible paths of initial loss,

CH_3 $]^{+\cdot}$ N N *m/e* 132 → N^+ N *m/e* 117; $-N_2$ → 89^+ → $-C_2H_4$ → 63^+; $-HCN$ → 90^+ → $-HCN$ → 63^+ (5-142)

some being the neutral H, CH_3 radical, HCN, H_2CN, and, a rare event, the loss of CH_4. The path started by the loss of hydrogen is followed by two successive losses of HCN and finally by the loss of C_2H_2 to form the *m/e* 51 ion. If the methyl radical initial loss occurs, then the choice is between the loss of N_2 followed by the loss of C_2H_2 arriving at the *m/e* 63 ion, while the second branch from the *m/e* 117 ion contains two successive losses of HCN to form the same m/e 63 ion. The initial loss of HCN pathway employs a second loss of HCN followed by the loss of C_2H_2 to arrive at an ion of m/e 52. When the similar phenyl derivative is considered, much the same processes occur except, of course, the loss of a methyl radical is impossible, while the loss of a phenyl radical does not occur. The loss of a nitrogen molecule was also not observed.

Desmarchelier and Johns (1969) have examined some indazolones and pyrazolones and have shown some parallels and deviations from the above types of behavior. In Scheme 32 are shown data for a methyl-substituted

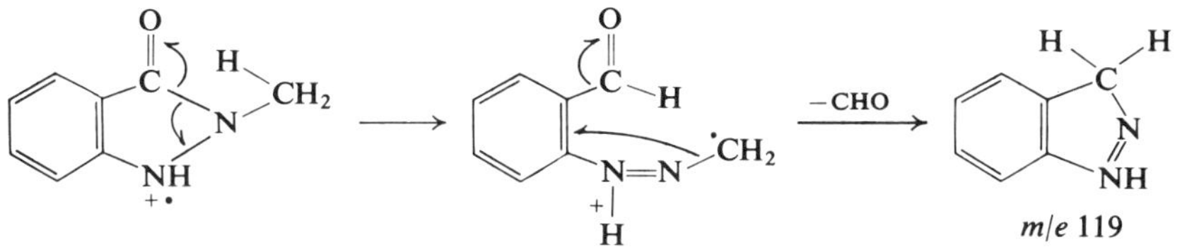

m/e 119

Scheme 32

indazolone. The unsubstituted molecular ion undergoes primarily the loss of N_2H to form the m/e 105 ion which then undergoes rearrangement to become the simple phenyl–CO ion. This ion then undergoes its usual decomposition by the loss of a neutral CO to form the m/e 77 ion, and it, in turn, loses the neutral C_2H_2 particle resulting in the m/e 51 ion so commonly found in aromatic spectra. As can be seen from the scheme for the methyl compound, the loss of a neutral CHO (57%) and the loss of the methyl radical itself compete with the processes described for the unsubstituted ion. These workers were striving for a means of distinguishing the two *N*-methyl isomers from the *O*-methylindazole. Each of these compounds provided a triplet peak at m/e 119, which proved to correspond to the loss of CHO, N_2H, and $CH_2{=}NH$. The marked difference in the intensities of ions formed by the loss of CHO may provide a means of differentiation between the isomers. It also reflects the difference in the ground-state structure. Thus, it can be seen from this work that a high-resolution instrument is essential to determination of such compounds.

The four-membered-ring systems containing nitrogen, the azetidines, and 2-azetidinones or β-lactams have been reported by Jackson *et al.* (1968). These are not aromatic-ring compounds, but frequently small-ring compounds tend to fragment in unusual ways. The mass spectra of the azetidines with three (or more) substituents are generally simple and involve ring cleavage. If alkyl groups larger than methyl are present, additional fragmentation of the side chain is also noted. Neutral losses detected, in addition to HCN and C_2H_4 and C_2H_2, include $CH_2{=}NH$, C_2H_4O, CH_3, C_2H_5N, C_8H_9N, and C_7H_7 (when a phenyl group is available). The 2-azetidinones are characterized by low intensity (if any) of the molecular ions. The prominent processes involve the ejection of CO and CHNO as neutrals, but other common ejections are also observed. The spectra can become quite complex, depending upon substitution.

Two studies have recently reported on the three-membered heterocyclic rings containing nitrogen—the aziridinones (α-lactams). The work of Baumgarten *et al.* (1969) involved molecules in which the ring was stabilized by a *t*-butyl group. These molecules were dominated by the presence of the *t*-butyl ion at *m/e* 57. The remaining major processes involved the loss of CO and also the loss of *t*-butyl isocyanate [$(CH_3)_3C{-}N{=}C{=}O$]. In the work of Lengyel and Uliss (1969), a high-resolution instrument was used, establishing that the primary fragmentation step was the ejection of carbon monoxide from the molecular ion, as shown in Scheme 33. This process was still predominant

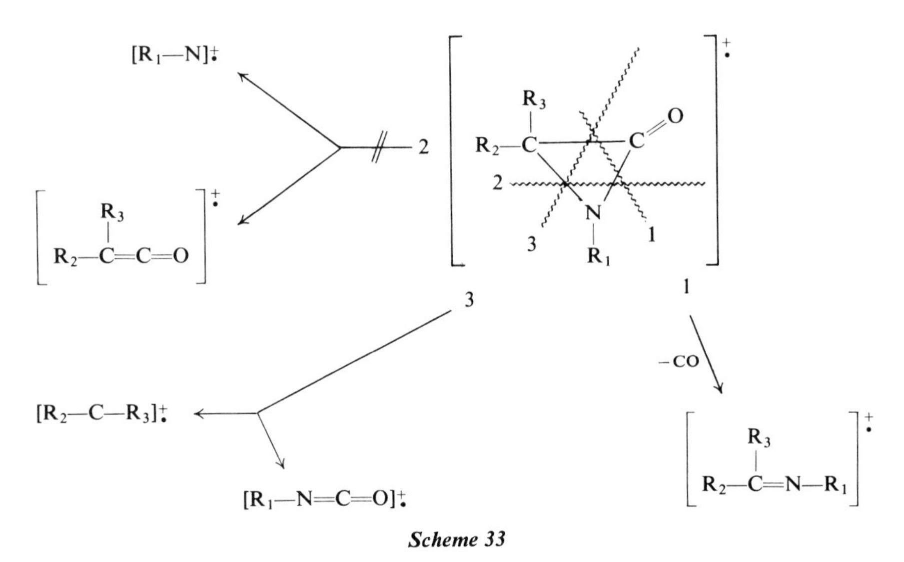

Scheme 33

at 12 eV. The other processes shown to occur are less important. Note that cleavage 2 does not occur to produce ions or neutrals of the type shown. In addition to the *t*-butyl-stabilized compounds, these workers also gave data on compounds in which the adamantyl group was used as a stabilizer. The major effects noticed were that the reduction of intensity of ions from N–C bond cleavage and the formation of an isocyanate type of ion, $R_1{-}N{=}C{=}O$, become prominent. The tendency of the adamantyl type of moiety to be retained is attributed to the partial double-bond character of the *N*-adamantane bond.

For phenyl-substituted linear molecules containing two nitrogen atoms, Scheppele *et al.* (1970) showed the loss of a neutral CH_3CN fragment concurrent with the migration of a phenyl group. This is depicted as a 1,3 phenyl shift:

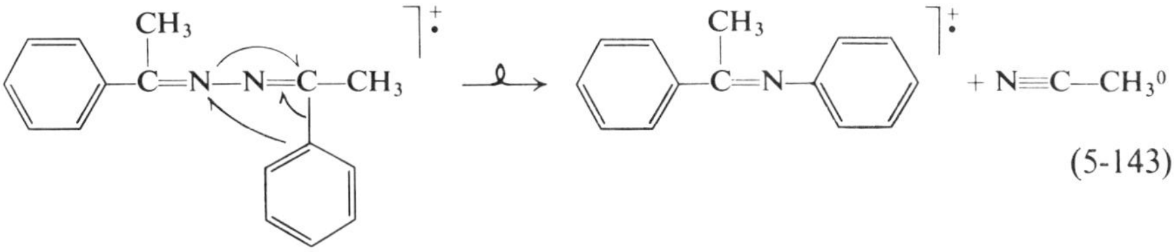

$$C_6H_5\text{—}C(CH_3)\text{=N—N=C(C}_6H_5)\text{—}CH_3{}^{+\cdot} \rightarrow C_6H_5\text{—}C(CH_3)\text{=N—}C_6H_5{}^{+\cdot} + N\equiv C\text{—}CH_3{}^0$$

(5-143)

Stevenson *et al.* (1969) reported the spectral pathways for the 2-substituted 8-hydroxyquinolines. Typical compounds were 8-quinolinol, 2-methyl-8-quinolinol and 4-methyl-8-quinolinol, 2-hydroxymethyl-8-quinolinol and 2-aminomethyl-8-quinolinol. All of these compounds exhibited a major fragmentation process in which the neutral CO was lost. For the aminomethyl compound, a prominent competing process involving the loss of a neutral NH_3 fragment was observed. Metastable peaks were observed that indicated that the loss of CO could occur before or after the loss of the ammonia. These workers also made an interesting observation on the presence of an *m/e* 89 ion to be found in all eight of the compounds they examined. Its composition was $C_7H_5{}^+$, and it was claimed that it was characteristic of compounds of the type

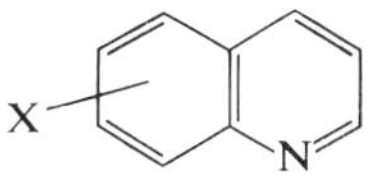

X—(quinoline)

where X is OR, SR, NH_2, etc. Because the ion is present in all the substituted methoxyl isomers, a cyclic structure is suggested, although a lower-energy open-chain species is considered a possibility. Foster *et al.* (1969, 1970) have observed the *m/e* 89 ion present in the mass spectra of compounds of the following types:

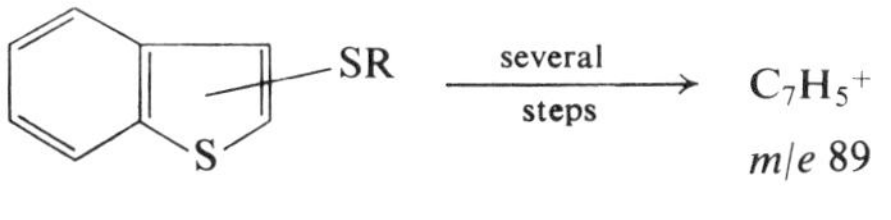

$$\text{(benzothiophene)—SR} \xrightarrow[\text{steps}]{\text{several}} C_7H_5{}^+ \quad m/e\ 89 \qquad (5\text{-}144)$$

and

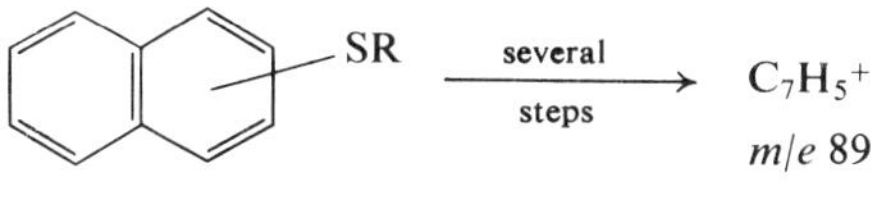

$$\text{(naphthalene)—SR} \xrightarrow[\text{steps}]{\text{several}} C_7H_5{}^+ \quad m/e\ 89 \qquad (5\text{-}145)$$

In both cases the fragmentation scheme is quite involved, and yet the arrival at an 89 ion instead of the possible *m/e* 91 suggests that this is characteristic of a fused-ring system carrying the X groups as suggested by Stevenson *et al.*

(1969). The hetero atoms probably are the explanation of this ion's formation, and the fact that the alkylquinolines do not show it is simply because the tropylium ion will be favored to form since sufficient hydrogen is present on the carbon atom alpha to the ring. For the heteroatomic species, it appears that only one hydrogen atom can be carried in by rearrangement (if it occurs), and thus the *m/e* 90 or 91 ions do not form. In each molecule involved in the above cases, it should be noted that the inclusion of the α carbon plus hydrogen atoms from the second ring can be included. The hetero atom cleaves from the benzo ring in all cases of Stevenson *et al.* (1969), whereas the naphthalenethiols do not meet this requirement. Further study as to the more general cause of this behavior should be made.

Fraas *et al.* (1969) have reported the eliminations of CO and C_2H_4 neutrals from metal acetylacetonates. Scheme 34 depicts how this is thought to occur. Note that the ultimate fragment ion is the metal ion itself.

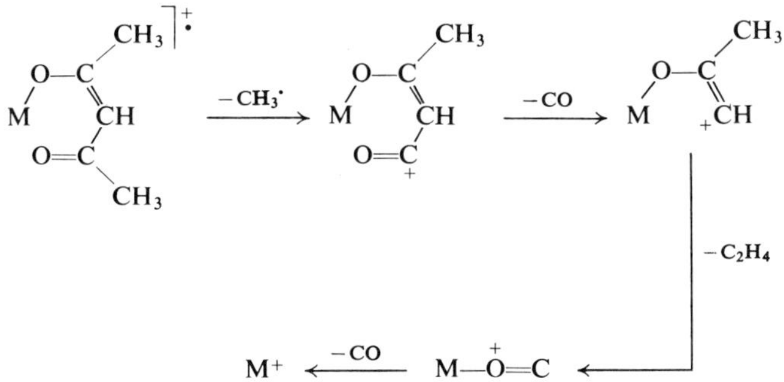

$-CH_3^{\cdot}$ $-CO$ $-C_2H_4$ $-CO$

M^+ $M—\overset{+}{O}=C$

Scheme 34

An interesting loss of neutral water occurs in the case of alkyl ketones possessing an alkyl chain of seven or more carbon atoms. Yeo and Williams (1969) also examined ketones in which the alkyl chain was six or less carbons and did not observe the loss of water. Metastable ion peaks suggest that the process occurs in one step. These workers believe a 10-membered transition state may be involved, with the ketone being in its enol form as shown below:

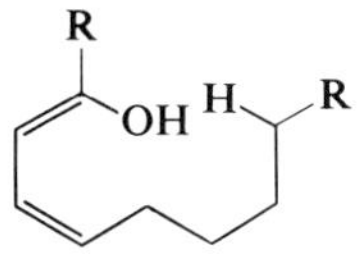

R = H or alkyl

The curling back of hydrocarbons upon themselves has been known for a considerable length of time and was first considered to occur in mass spectrometry by Ryhage and Stenhagen (1960). The structure of the ion resulting above was not given; however, it must be a highly unsaturated species.

Among the neutrals ejected and considered so far, the loss of CO_2 has not been mentioned. Cava *et al.* (1966) reported such a loss from phthalic anhydride in which the carbon dioxide is first lost, followed by the expected ejection of a neutral CO. These processes also take place for the pyridine-2,3-dicarboxylic anhydride, the ultimate ion being the pyridine ion at *m/e* 77. In the phthalic anhydride case, the ion would be the benzyne ion at *m/e* 76. Johnstone *et al.* (1966) reported that the loss of CO_2 occurred from *N*-methylphthalimide. This requires some rearrangement of the molecular ion before cleavage occurs. In the work of Desmarchelier and Johns (1969) examination of an ethyl ester of indazolone having only a hydrogen atom on the second nitrogen atom (see Scheme 32) showed the direct ejection of the CO_2 moiety and the attachment of the ethyl group to the nitrogen atom where the ester group was originally bound.

Sometimes neutral-fragment losses are not to be observed, and several examples may alert the mass spectroscopist to the types of molecules that do not produce loss of such neutrals as obvious characteristics of the mass spectrum.

Ungnade and Loughran (1964) reported on the mass spectra of the furoxans and indicate they have large NO^+ (*m/e* 30) peaks because of the fragmentation of the ring system at C–N and N–O bonds. Minor peaks arise from cleavage of substituents and ring fission between carbons 3 and 4. For those interested in analytical work, the spectra are characteristic enough to permit identification of these compounds. The related furazans give different fragmentation patterns because carbon–carbon cleavages predominate.

Even more drastic are the compounds benzotrifurazan (I) and benzotrifuroxan (II) reported by Bailey *et al.* (1969). The mass spectrum of I contains only two intense ions, both of about the same intensity. The ions are NO^+ and the molecular ion of *m/e* 206. In contrast the compound II has only very small $(M - NO)^+$, $(M - 2NO)^+$, and $(M - O)^+$ peaks. Most of the peaks observed are formed by fission of the carbon–carbon bonds to split out the furazan units (III). These further fragment by another cleavage of the carbon–carbon bond. The major ions formed are NO^+, $HCNO^+$, and HCN^+. Neutral losses are only inferred from the events described above.

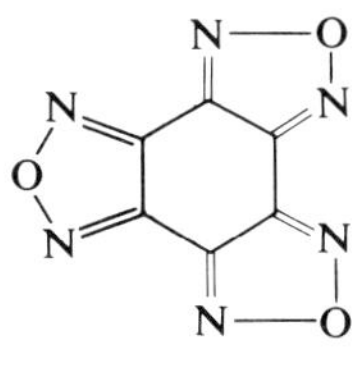

I

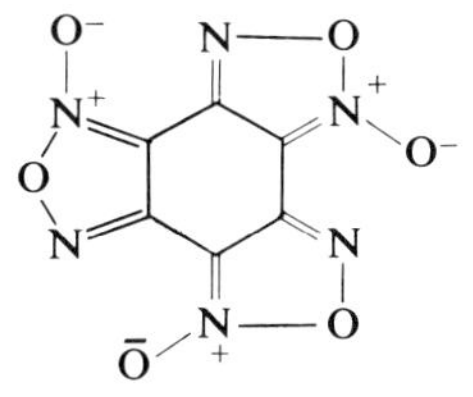

II

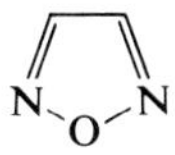

III

L. Charge Localization and Competition in Fragmentation

The location (or localization) of the charge in an ion formed in the mass spectrometer can be accomplished in some cases, while in many others it cannot. When a hetero atom is found in the molecular, rearrangement, or fragment ion, the charge is considered to be localized since one of the unshared pairs of electrons is lost in forming the ion for an odd-electron moiety. The positive sign has been placed upon the atom considered to carry the charge. If the charge cannot be localized, the ion has usually been shown in brackets or partial brackets $\urcorner^{+}$ to indicate this fact. Examination of the literature often discloses that some authors have claimed ionization at the silicon atom or the nitrogen atom, etc. What is really meant is that an electron has been lost from somewhere in the moiety during the electron impact process, and in the matter of a very few atomic vibrations, the electrons have adjusted so that the missing electron appears to be from a particular atom or group which is now shown to carry the charge. When the excited-state molecular ion fragments, the positive charge should reside on the fragment of lowest ionization potential after the bond cleavage. This is known as Stevenson's rule (1951). A corollary (and inversion) of this rule is that the neutral fragment first ejected should be the one with the highest ionization potential. In the previous section, this was shown (Eland and Danby, 1965) to be approximately obeyed, but often competing decompositions became important.

Thus, it can be seen that ordinarily one should be able to predict which fragment will carry the charge if a fragmentation scheme is set up. McLafferty (1963a) reemphasized that care must be taken since even for a molecule of only moderate size, a great many competing decomposition pathways can exist. He also pointed out that the even-electron fragment, whether an ion or neutral, usually has much greater stability than the odd-electron ion. As a consequence the even-electron ion has a much greater influence on the fragmentation paths than the odd-electron ion. It was noted in some of the examples that despite these generalizations, odd-electron ions may be the largest ions in the spectrum. When this happens, other dominant factors are usually found to be controlling the decompositions observed.

An interesting point has been raised by Lengyel *et al.* (1970b). If charge migration in ions can occur, they ask, "how does it occur?" Does the charge migrate through space or through chemical bonds and, if the latter, can it be transmitted by σ bonds alone? Many of the successive neutral ejections discussed certainly follow the corollary to Stevenson's rule. Careful examination of the examples given shows that many of these involve π-bonded systems which can readily transmit the charge by delocalization. The ejection of the successive carbon monoxides from anthraquinones required a charge migration through an aromatic system. The RDA ejection of two successive (or possibly concomitant) neutral ethylene particles follows the corollary of Stevenson's rule. No other "ejectable" neutral groups had a higher ionization potential. The charge was delocalized in all probability but did not have to be, nor did it have to migrate. Mandelbaum and Biemann (1968) presented evidence for charge migration with fragment ions. The compounds were *p,p'*-disubstituted 1,3-diphenylcyclopentanes of unknown stereochemistry. Kinstle and Oliver (1969) reported the loss of two molecules of propylene in successive steps by McLafferty rearrangements from *p,p'*-bisvalerylphenyl ether. In both cases the charge migrated within an odd-electron fragment ion. Foster (1966) reported successive loss of methyl groups from di-*t*-butylthiophenes. It was thought that this occurred from the separated *t*-butyl groups. Fowler *et al.* (1969) used deuterated molecules to prove that this was indeed the case. It should be noted that only one σ bond was between the charge locale and a strong π-bonded system.

Lengyel *et al.* (1970b) devised a set of six prerequisites for a "model" compound that would provide a test of charge migration. Meeting these requirements would insure that the charge was transferred through only σ bonds and not through space by interaction with another part of the molecule. Other interferences were also considered. The compound bisaziridinone-2 was selected with *t*-butyl groups chosen as the *N*-substituent because of the known effect on thermal stability. The two lactam rings, which can eject CO fragments from widely separated parts of the molecule, are connected by the 1,3-adamantylene group through C_3–C_3'. Thus, a rigid inflexible molecule possessing a plane of symmetry is obtained.

Lengyel *et al.* (1970b) gave Scheme 35 for fragmentation and charge migration for the molecular ion. The subsequent fragmentation after the formation of the *m/e* 302 ion is the elimination of a methyl radical, a typical fragmentation for a *t*-butyl group. The loss of the second carbon monoxide requires the charge to migrate from the N_a atom to the N_b atom. This would be through six σ bonds. These authors concluded that charge migration in fragment ions is possible even if the two sites are separated exclusively by σ bonds. Further, stereochemical considerations prohibited nonbonded interactions through space.

Their results support the conclusion of Mandelbaum and Biemann (1968) that the fixation of the charge in fragment ions is "unwarranted." They also tend to agree with Mandelbaum and Biemann, who visualize a "dynamic distribution of charge...statistically maximized at the site of the lowest ionization potential" within fragment ions. Thus, "charge migration" is influenced by factors other than the ionization potentials at various sites. These include stereochemistry, ion structure, release of ring strain, π-bonded character in the molecule, etc. While it is obvious that knowledge in these

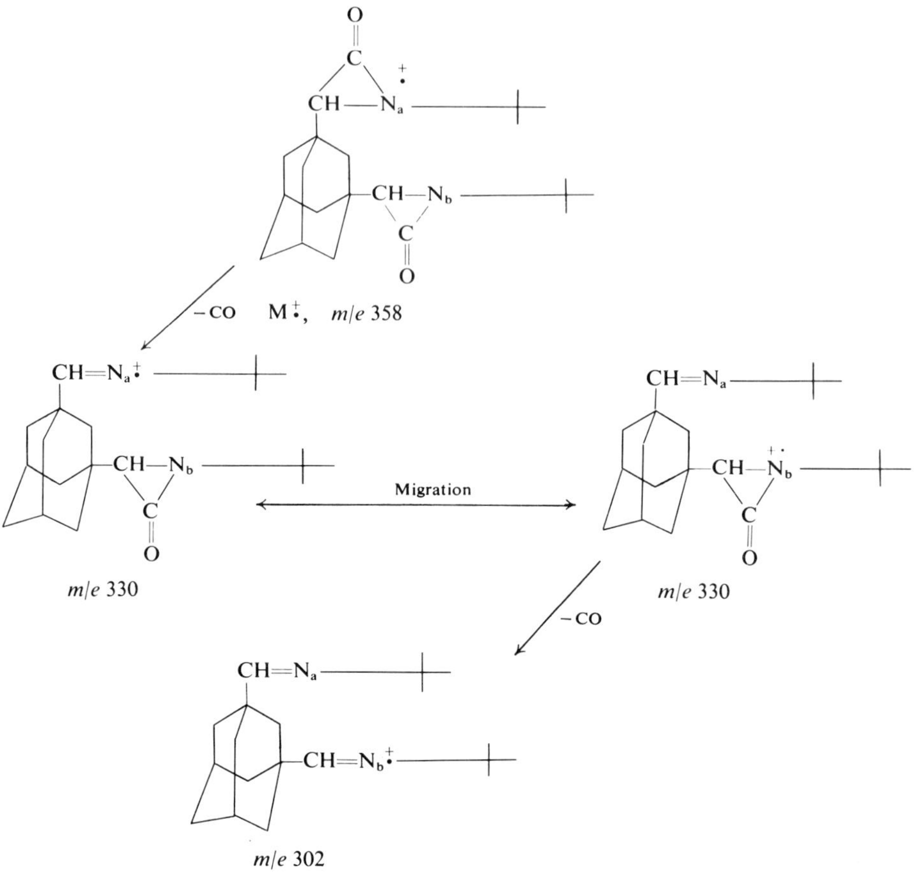

$-CO$ $M^{+\cdot}$, *m/e* 358

Migration

m/e 330

m/e 330

$-CO$

m/e 302

Scheme 35

areas should be extended, the reader must be reminded that the charge is simply a point of electron vacancy. Future studies may shed considerable light on the problem.

V. Obtaining Supporting Evidence for Fragmentation Mechanisms

It is quite obvious from an inspection of a given mass spectrum that the major ions formed can be determined simply by considering their abundances of the total ion intensity. The mechanism of formation of those ions may be very well hidden. While it is quite tempting to use a simple approach and state that the molecular ion obviously fragments into species A, B, and C, for example, and that these ions in turn produce the daughter and subsequent ions of moderate to large intensity by what appear to be reasonable processes, this is not a safe assumption. One must consider the entire spectrum, the presence of metastable ions, and the respective appearance potentials of the various fragments. In the example above it is important to know if ion C is produced by a direct process from the molecular ion or by fragmentation of either species A or B, or possibly by all three processes. The general procedures used to determine this are a necessary part of the mass spectroscopist's working knowledge.

A. Comparison to Ordinary Organic Mechanisms

Comparison of the suggested fragmentations should always be made to known organic mechanisms. A great many mechanisms that are neither first order nor possible in the high-vacuum system can be discarded as highly unlikely. The justification for a maximum probability of collisions occurring after ionization of only one ion in 10,000 at 10^{-6} torr was given earlier. In many instances it can be found that the usual driving forces (and inhibitions) are operating in mass spectrometry much as they do in organic chemistry. Steric hindrance, release of strains, formation of a more stable carbonium ion, etc., can all be observed. The apparent ease of release of a *t*-butyl ion can be seen in Fig. 5-5 where the *m/e* 57 ion appeared at low voltages from a neopentyl alkyl group. The *m/e* 57 ion did not appear at the same potential from the corresponding *n*-pentyl compound. This could be regarded as one of H. C. Brown's famous I-strain releases, since in the process bond hybridization goes from sp^3 to sp^2, allowing a change in bond angles [see Brown *et al.* (1944) and Brown and Fletcher (1949)]. It is interesting to note that, as usually occurs

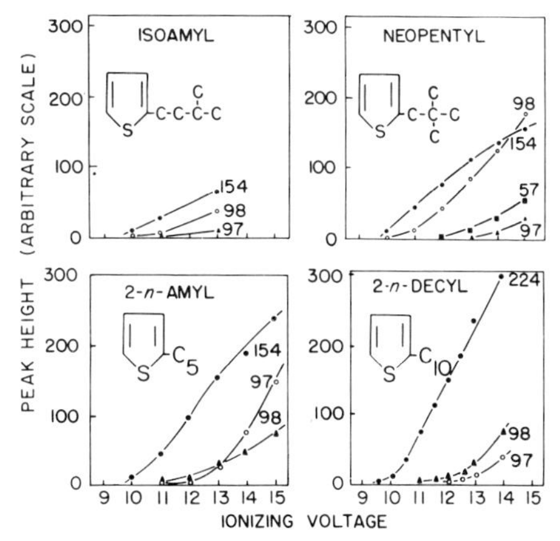

Fig. 5-5. Typical low ionization voltage data for selected alkylthiophenes. [From Cook and Foster (1961). Courtesy of American Petroleum Institute.]

in science, someone quite early predicts the fundamental cause of future general behavior in explaining his research. The preferential formation of the series of carbonium ions, tertiary > secondary > primary, may be found in the predictions of Demjanow [for details see Fuson (1943)] in his explanation of rearrangements in the ring expansion of cycloalkylmethylamines. Whitmore (1948) has formulated some rules associated with carbonium ion formation in solutions; an understanding of these will be helpful to the mass spectroscopist. For an introduction to this area, the text of Alexander (1950) is recommended. Additional current reference books were cited at the end of Section III of this chapter. Without a consideration of the "regular" organic mechanisms operating, even though in solution, the mass spectroscopist is hindered in his predictions.

B. The Use of Energetics to Define Processes

Knowledge of the ionization potential of the molecule is very helpful and can usually be obtained by one of the methods given in Section III. One can consider this potential as the ground state of the molecular ion for the fragmentation events that follow. The appearance potentials of the ions that are prominent in the spectra will then assist directly. For the example given above, C may have its appearance potential closest to the IP of the M^+ ion. Hence, it would be foolish to continue to entertain the possibility of the process $M^+ \rightarrow A^+ \rightarrow C^+$ occurring in the ranges examined. It is always possible that the process given could occur but not so as to produce C^+ fragments at their AP. They would instead be produced with considerably higher energy, according to Stevenson's rule. It should be noted that throughout this discussion the processes are being considered because the ions A, B, and C are produced in large quantities. If A^+ were only a very minor ion, it is possible that it could be an intermediate in the process of M–A–C and be of such low intensity as to be undetectable, even under the more favored conditions of examination at low ionization voltages.

A system which, at least a number of years ago, proved to be somewhat elusive can be used to illustrate the point. Low ionization voltage data for four alkylthiophenes are shown in Fig. 5-5, taken from Cook and Foster (1961). The ionization voltage scale was calibrated against the appearance potential of argon, run as a mixture with the pure compound. In each case the M^+ ion appears first and steadily increases in intensity with increased voltage. The second ion to appear is mass 98, followed closely by mass 97. For the earlier type of instruments, the voltages reported are, at best, good to about ± 0.3 V. Hence, one cannot argue that the *m/e* 98 ion is simply the isotopic ion from *m/e* 97, since it would be expected that the 98 isotope ion AP would be closer to the 97 ion than this. Further, the intensity of the 98 and its growth rate are seen to be dependent upon the molecular structure and not upon the size of the 97 ion intensity. Thus, in setting up the fragmentation mechanisms of the alkylthiophenes one must consider that the *m/e* 98 ion (a rearrangement ion since a hydrogen atom must be transferred in its formation) may be a precursor to the *m/e* 97 ion, which eventually goes on to be the largest ion in the mass spectrum. Certainly the order of appearance is M^+, 98^+, 97^+, and, in the case of the neopentyl compound, the ion at *m/e* 57 (most probably the *t*-butyl carbonium ion).

From the thermodynamic viewpoint this leaves the situation depicted in Fig. 5-6 wherein further matters are considered. Hanuš and Čermák (1959),

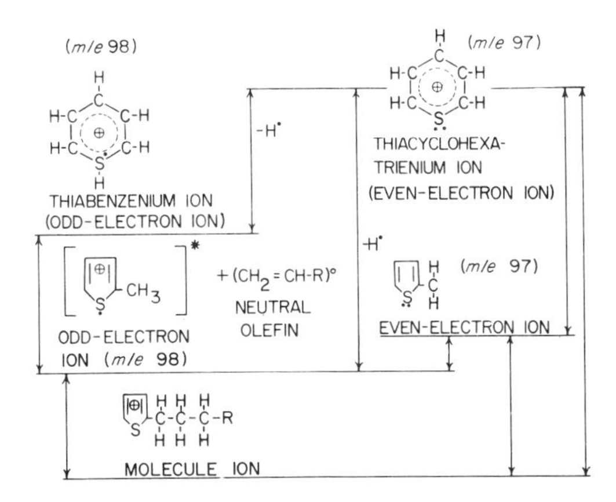

Fig. 5-6. Some possible reaction paths for base peak formation of a typical 2-*n*-alkylthiophene. [From Cook and Foster (1961). Courtesy of the American Petroleum Institute.]

utilizing the similarity of the thiophenes to the alkylbenzenes, postulated the expansion of the five-membered thiophene ring into a six-membered ring—thiapyrilium, or the thiacyclohexatrienium ion (an even-electron ion)—to account for the large abundance of the *m/e* 97 ion in the mass spectrum of the thiophenes. Thus it can be seen that the introduction of the low-voltage data by Cook and Foster (1961) complicated the picture of the mechanism to some extent. It should be noted that every possibility of ion structure should be considered. Open-chain structures were considered to require considerably more energy than the closed-ring structures shown here. The odd-electron at 98 was starred to show that the methylthiophene form is considered highly unlikely. Rather, the direct transition by a concerted means to the *m/e* 98 expanded-ring species is more probable. Attaching the hydrogen to the ring sulfur is possible if the ten-electron sulfur atom involving hybridized *d* orbitals is used. There was no evidence for the molecules studied that the *m/e* 97 ion was formed directly from the molecule ion. Low-ionization data plus the energetics involved would probably rule this out. The loss of a hydrogen from *m/e* 98 to form the *m/e* 97 ion would probably require relatively little energy. It should be noted that the spacings shown for the energy scale are for convenience in presentation and do not represent actual energy differences.

The investigation of the thiophenes at low ionization voltages was done only after the Bartlesville research laboratories of the Bureau of Mines turned up indications of spurious molecular ion peaks at *m/e* 98 in the mass spectra of a reference mixture of thiophenes received for use in a cooperative testing of an analytical method for thiophenes in gasoline and naphtha fractions. Interference from olefins had been negated by removal of the aromatics and sulfur compounds as a concentrate by means of a gel percolation using cyclopentane as a "spacer." In running the reference calibrations, each of the thiophenes still gave a contribution to the *m/e* 98 peak, even at the low ionization voltages specified by the method. The methylthiophenes were not present as contaminants in the pure reference materials; hence, the 98 peak became an interference and required explanation. Throughout this section we shall employ this problem as an example of what may be done to verify mechanisms.

It should be noted that many workers still employ energetics and metastable data to provide studies of specific types of compounds. Franklin and Carroll (1969) have studied 1,3- and 1,4-cyclohexadiene and 1,3,5-hexatriene by this approach and have produced very useful information. Li and McGee (1969) have similarly examined cyclobutadiene. Gowenlock *et al.* (1963) have considered the sulfides and disulfides in the same manner. Zamir *et al.* (1969) have used both varying energies and inlet temperatures to study keto–enol tautomeric equilibria by means of mass spectrometry.

C. The Use of Metastable Ion Data

It should be recalled (see Chapter 3, Section V,5) that the presence of the metastable ion proves that the process is taking place, while the absence of the metastable peak proves nothing. It may well be that the investigator is unable to "see" the metastable ion for various reasons; however, the process may still be occurring. The obvious metastable peak to look for in the thiophene case, to prove that the *m/e* 98 was producing the *m/e* 97, occurs at *m/e* 96. This peak could not be seen by the CEC-102 instrument operating at Bartlesville but was observed in a sample of 2-*n*-hexylthiophene by Meyerson (API No. 1876); it was subsequently seen many times on the CEC-103 at Continental Oil Co., and on the CEC-104 at Texas Woman's University. Hence, the presence of the m* peak proved that at least *some* of the *m/e* 97 was produced directly from the *m/e* 98 precursor. Metastable peaks for the direct production of the 97 ion from M^+ have not been observed. It should be noted that *some* was emphasized concerning the *m/e* 97 from 98. Both processes may be occurring, but proof of only one has been obtained.

When structure correlation studies are being conducted today, it is customary to report metastable peaks associated with a given process for as many of the compounds as are studied. If one studies eight different isomers of a given species and finds the same metastable peak in each, statistically at least he is on quite firm ground. If, however, he observes only one metastable transition for the eight compounds, he should pause and ask why only that single metastable peak is observed. Consideration of processes other than the one postulated may lead to a process that could be found only for the compound in question.

The work of Beynon *et al.* (1965) is recommended as a source of relatively easy discovery of which processes are feasible for the ion in question. One may consider the principal ions present and examine the metastable possibilities that occur. Then, using the above reference, a list of the likely metastable peaks may be made for the various processes simply by looking up the various initial and final ions which cause a metastable to appear at the observed value. A little experience will permit the determination of m*'s for the reverse processes. Alternatively, a number of texts, Beynon (1960a) and Kiser (1965e) in particular, contain nomograghs which are of great value in finding the various possible ions. Lastly, one may employ the computer (Rhodes *et al.*, 1966) to calculate the various m*'s to be found for the set of processes that one, perhaps intuitively, believes to be present.

The use of metastable data has increased markedly in the last few years. In part this is because of the fact that many instruments observe these peaks in the course of routine work with only minor adjustments in instrumental conditions of operation. However, refinements in the techniques are being constantly made. Kiser *et al.* (1969) have described the use of electric sector variation for double-focusing mass spectrometers. There are advantages and limitations, but overall this represents a good step forward. Beynon *et al.* (1970a) have shown that the metastable ions observed in methane provide a good means of calibration in the low mass range. Beynon *et al.* (1970d) have also reported a new method to determine the percent deuteration from metastable data. This was developed during a study of deuterated toluenes. Hills *et al.* (1969) have reported consecutive metastable peaks in mass spectra and described calculations for the abundance of these species. Primary and secondary rates for a toluene consecutive dissociation were calculated. Considerable discussion is given for the intercomparison of mass spectra from double-focusing instruments.

A very significant contribution from Beynon *et al.* (1970b) is a discussion of ion kinetic energy spectra (IKES) of organic molecules observed on a double-focusing instrument. Peaks are commonly observed which possess energies in excess of that of the main beam. While considerably more work is planned for investigating many of the aspects of the problem, these workers have concluded that using the technique permits detailed studies of reactions occurring in the field-free region in front of the electric sector of the double-focusing mass spectrometer. Unimolecular decompositions of singly (or multiply) charged species can be studied. By introducing a collision gas, intermolecular processes, including ion–molecule reactions and collision-induced fragmentations, can be studied. Because of the resolution of the instrument the multiply charged ions can be separated from the intense overlapping peaks of the normal mass spectra. Hence, their characteristics are studied more easily. It is possible, because of the greater precision of measurement, to investigate charge localization in multiply charged ions and, particularly, to determine the importance of hetero atoms such as oxygen, nitrogen, and sulfur. Examples are given of the detail surrounding metastable peaks and how, for some cases, these may be resolved in terms of excess energies. The calculation of energy released is shown.

Subsequently, Beynon *et al.* (1970c) utilized this technique to detect the transition of a triply charged ion into a doubly charged ion and a methyl ion:

$$C_{12}H_{10}^{3+} \rightarrow C_{11}H_7^{2+} + CH_3^+$$

The charge distribution in the triply charged ion was discussed.

D. The Use of Doubly Charged Ions

Although doubly charged ions from polyatomic species have been known since the work of Conrad (1930), progress in their use has been relatively slow. Mohler *et al.* (1954) attributed the low abundance of doubly charged ions in the mass spectra of aliphatic hydrocarbons in part to the possibility that fragmentation of the doubly charged ion is rapid and forms two singly charged ions. Higgins and Jennings (1965) verified by means of metastable peaks the dissociation of the doubly charged benzene ion into CH_3^+ and $C_5H_3^+$. The decomposition of doubly charged ions into a neutral particle and a doubly charged fragment ion has been verified by Beynon *et al.* (1959), Meyerson and Vander Haar (1962), and McLafferty and Bursey (1968). It is notable that often the fragmentation schemes of doubly charged ions do not correspond to those of the singly charged species (see Foster *et al.*, 1968, 1969). It is generally recognized that olefinic bonds and hetero atoms enhance the stability and abundance of doubly charged ions. The presence of π-bonded systems also tends to assist the formation of these ions. Vouros and Biemann (1969) have emphasized the structural significance of doubly charged ion spectra. Their studies were made on phenylenediamine derivatives. These authors supplied some ground rules concerning identification of doubly charged ions and the calculation of their abundances. A warning was given concerning the problem of isotopic ions containing ^{13}C, D, and perhaps just the ion at the same nominal mass containing one H atom more than the natural or unlabeled ion. Examples of stabilization of doubly charged ions by ease of charge separation were shown by some of the compounds they studied (I, II).

$$\underset{\textbf{I}}{CH_2{=}\overset{+}{HN}{-}C_6H_4{-}\overset{+}{N}{\equiv}CH} \quad \text{and} \quad \underset{\textbf{II}}{CH_3{-}CH{=}\overset{+}{HN}{-}C_6H_4{-}\overset{+}{N}H{=}CH{-}CH_3}$$

From Vouros and Biemann (1969), the factors contributing to the formation and stabilization of doubly charged ions were found to be maximum separation of charges, the presence of nonbonding or π electrons, which are easily lost upon electron impact or shared after ionization and cleavage, the formation of even-electron fragments, and the formation of a more conjugated system. The spectra of the doubly charged ion spectrum may permit differentiation between structures not distinguishable from just the singly charged ion spectra. They also concluded that a doubly charged ion may often be much more abundant than a singly charged ion of the same mass and corresponding structure. The nitrogen compounds exhibit a much greater tendency to form doubly charged ions than oxygen. In the various mass spectra considered, Foster and his co-workers have observed that the aromatic sulfur compounds tend to exhibit substantial amounts of doubly charged ions. There is little doubt that those possessing high-resolution equipment will utilize this technique to a much greater extent in mechanism determination. Hence, to the trio of energetics, metastables, and isotopic labeling, one must add the use of multiply charged ions.

E. The Use of Isotopic Labeling

The use of isotopically labeled species often assists in mechanism studies. The presence of natural isotopes associated with the various ionic species is readily recognized and put to use, even by the novice. He easily learns that he can determine whether the m/e 57 ion observed is that of $C_4H_9^+$ or C_2SH^+ based upon the appearance of the accompanying 59 peak due to the presence of the ^{34}S isotope. This is often referred to as the "isotope profile" of a peak (see Chapter 6 for a full discussion). Essentially no contribution to the 59 peak would be expected for a butyl ion, whereas the sulfur-containing ion would contribute about 4.4% to that m/e. If high enough resolution (about 2000) is available, the difference in the two ions at m/e 57 may be detected directly because of the mass defect of sulfur. In a study of the fragmentation of sulfur compounds where hydrocarbon fragments are also present, the ^{34}S isotope almost makes a perfect means of suggesting which fragment contains the sulfur atom. Thus, this effect alone allows one to find fragmentation routes with quite a bit more confidence than does simple intuition. The chlorides and bromides also may assist in this manner quite directly. Because of the low abundances of natural isotopes of nitrogen and oxygen, compounds containing these labeled atoms are expensive to study by the use of natural isotopes. Of course, high-resolution instruments are a great help today, but a good deal was learned with just the presence of the sulfur isotope on compounds where this could be studied. Literally, the molecule under study had a built-in label.

Quite often one is forced to employ isotopic labeling of species to prove what is taking place during the fragmentation in question. Most of the labeling studies today involve deuterium because of the relatively low cost, even though often the synthetic-organic chemist is challenged to come up with a high-purity compound with the label in the desired position. More difficulties may arise here than are anticipated.

The appearance of a mechanism not observed in organic chemistry is relatively rare, and will be indicated to the reader in passing. It is also interesting to note that some times the occurrence of a rearrangement or mechanism in mass spectrometry has prompted the "true" organic chemist to do a little research to prove the possible existence of such a species. In line with our discussion, after the prediction of the appearance of the thiapyrilium ion in the mass spectrometer from alkylthiophenes, Koutecký (1959) calculated the delocalization energy of the ion species and pronounced that it should be more stable than the tropylium ion. Subsequently, Pettit (1960) synthesized thiapyrilium bromide, a salt and a perfectly stable compound. Turnbo *et al.* (1964) described the preparation of the salts of additional expanded-ring compounds containing the sulfur atom. It is well to remember that fragmentations which occur in the mass spectrometer may be completely inhibited in a solvent system or, at best, only produce minor products. In the absence of solvation, other possible reaction paths are available and sometimes become populated. Carbon-13 labeling is often done, but the costs of compound preparation are usually quite high. Often times only a compound enriched up to 60% of ^{13}C in the desired position can be obtained. Usually the carbon labeling is done to determine skeletal rearrangements as opposed to the use of deuterium to determine where those labile, elusive, and otherwise nefarious hydrogen atoms are located. An excellent review of the accomplishments of ^{13}C labeling has been prepared by Meyerson and Fields (1969b).

The first examination of a labeled hydrocarbon was that of Delfosse and Hipple (1938). In 1950, Dibeler and Mohler reported the mass spectra of all the possible deuteromethanes and their findings of differences in the probability of dissociation of the C–H and C–D bonds. From these beginnings, labeling studies have yearly increased in number until today they form a large portion of the literature in mass spectrometry. In view of the high cost of many natural isotopes and the current lack of research money, the number of such studies may well diminish in the coming years despite the general increase in mass spectrometry literature.

In making such studies, in addition to the cost, the mass spectroscopist is faced with two major new problems: (a) how to synthesize and purify the correctly labeled substance with a satisfactorily high ratio of labeled-to-unlabeled material and (b) how to treat the mass spectral data once they are obtained.

The first problem is best solved by association with an excellent synthetic-organic chemist and probably a "separations specialist." Even here the organic chemist will be in for some surprises and will probably broaden considerably his knowledge of mechanisms. It is quite true that many labeled materials may be obtained by reductions using D_2 (gas) instead of H_2, N_2D_4 instead of N_2H_4, $LiAlD_4$ instead of $LiAlH_4$, etc., but matters are not always simple. Foster and Higgins (1968) found deuterium all along the chain and in the ring when a Wolff–Kishner reduction of 1-(2-thienyl)-2-butanone was performed with the intent of producing β-dideutero-2-*n*-butylthiophene (see also Lee, 1964). The lithium aluminum deuteride did produce the desired molecule with about 95+ % di-D, 4+ % mono-D, and less than 1 % of unlabeled material. This is considered a satisfactory preparation since 95+ % labeled material is adequate for most purposes. Matters can become quite complicated when reactions are run in solution because of the possibility of ion formation and improved opportunities for exchange. One doesn't change the pH but uses D_2SO_4 to change the pD. Also, some step in the proposed synthesis may lead to some equilibration leading to a randomization of the label.

Biemann (1962f) provided an excellent chapter on the entire subject, describing in sufficient detail most of the facets involved in the general problem. For the synthesis side, Biemann gave the following general methods for the the incorporation of stable isotopes for use in mass spectrometry:

1. There can be exchange with D_2, D_2O, D_2SO_4, deuterated solvents, such as ROD, ozonation followed by exchange with ^{18}O. Exchange with ^{13}C or ^{15}N is rarely possible.
2. Replacement of a functional group can be accomplished by $LiAlD_4$, or decomposition of Grignard reagents with DCl, among other reactions.
3. Addition of deuterium to multiple bonds aids in locating double bonds.

Weinberg and Scoggins (1969) have reported an example for dideuterio-octanes.

In addition to these common problems, others develop because of multiple functional groups in the molecule to be labeled. Some synthetic-organic chemists have had as much fun with this as computer programmers have had trying to reduce a program of 52 steps to one of 50. The literature is full of such reports of synthesis, and a rather unusual new journal the *Journal of Labeled Compounds*, is also available.

The separations specialist alluded to above must be considerably more than a technician with a GC instrument. Microtechniques must be used, since the quantity of labeled material to be prepared is quite small. In addition, the specialist should have considerable experience with GC (capillary and prep-scale included), TLC, paper chromatography, electrophoresis, etc., to achieve the desired end results. It should be noted that *usually* separations of the labeled isotopic compounds from their unlabeled (or rather, naturally labeled) counterparts cannot be achieved.

Biemann (1962g) described the selection of operating conditions for the instrument, sources of error, and listed seven assumptions that are normally made in arriving at a calculated isotopic distribution in a labeled sample. More important, Biemann (1962h) gave two completely worked-out examples of the mass spectra obtained from a sample in which partial labeling is involved. The importance of this problem should not be underestimated in light of the incompleteness of many organic reactions. The results given for the thienyl butanone above appear simple. In fact, a rather simple but tedious calculation is necessary to obtain the mole percent of unlabeled, monodeuterated, and dideuterated species. The reason for the problem is that the unlabeled compound produces a molecular ion at m/e 154 and has sizeable $M + 1$, $M + 2$ and small $M + 3$ and $M + 4$ peaks arising from the naturally occurring C, H, and S isotopes. The $M + 1$ contributes to the $M_D{}^+$ at m/e 155 of the monodeuterated species. The $M_D{}^+$ also produces $M_D + 1$ ions at m/e 156 and $M_D + 2$ ions at m/e 157, etc. Similarly, the $M + 2$ and $M_D + 1$ ions contribute to the intensity of the M_{D_2} molecular ion at m/e 156, etc. The problem is readily solvable and computer usage is certainly indicated. If one wishes to estimate the isotopic composition when the molecular ion intensities are small, a trial-and-error calculation on other peaks in the spectrum may be desirable. Biemann's chapter is essential reading for any mass spectroscopist planning to work with labeled materials.

In the fragmentation and rearrangement discussion given to this point, many of the mechanisms have been established from labeled studies, but many have been arrived at by other means. Deuterium labeling was used on many, an example already given being the work of Meyerson and Corbin (1965). The label revealed that 18% of the $C_7H_5O^+$ ions in the mass spectrum of benzoic acid came by loss of a hydroxyl group containing hydrogen from the *ortho* position. The product was chemically distinct from the isomeric and more abundant benzoylium ion. Benz and Biemann (1964) used a series of aliphatic alcohols (*n*-butyl through *n*-heptyl) labeled with a CD_2 group in consecutive positions (C-1 to C-5) from the hydroxyl group to examine the elimination process of water from alcohols in the mass spectrometer. Their data indicated that elimination of water, leading to formation of the $(M - 18)^+$ ion in the undeuterated alcohols, involves a hydrogen from C-4 (90%) with the remainder coming from C-3 and C-5. In tertiary alcohols the elimination of HOH is less specific. As a further example of labeled studies, Gerrard and Djerassi (1969) examined a series of isomeric 1-phenylheptenes. The use of D and ^{13}C labeling provided evidence for extensive hydrogen and phenyl rearrangements. They concluded that mass spectrometry, therefore, is not a very useful tool for differentiating between double-bond isomers. Further, they attribute this to the ease of double-bond migration after ionization.

It may appear that by using isotopic labeling, the mass spectroscopists have opened Pandora's box. For each problem that appears to be solved, at least two new ones arise! Berlin and Shupe (1969) have shown the elimination of a ^{12}C-atom-containing fragment from trityl compounds of the type $(C_6H_5)_3{}^{13}CH$. In the mechanism proposed, the labeled carbon is included in a seven-membered expanded ring. The resulting ion then undergoes a phenyl shift to this seven-membered ring, attaching itself to the ^{12}C adjacent to the labeled carbon atom. With some further shifting of bonds, hydrogen atoms, and charges, an ion is arrived at which can eject the methyl radical containing a ^{12}C atom. Four ion species of different structures, but all isomeric, are involved as resonance forms. The fourth one ejects the methyl radical. This excellent work provides many additional ideas about similar processes in multiply fused ring systems which have been annoying mass spectroscopists for some time (since the advent of labeling and the discovery that matters were far more complicated than originally envisioned).

Williams and his co-workers have reached the conclusion that many molecules can suffer not only carbon scrambling but hydrogen scrambling after ionization occurs but before fragmentation occurs. Williams *et al.* (1968a) show this type of behavior for the diphenylmethanol and diphenylmethyl chloride. Williams *et al.* (1968b) also reported similar problems occurring for diphenyl ether and diphenyl carbonate. Johnstone and Ward (1968) reported evidence for cyclization processes in the elimination of methyl from stilbene analogs. Regardless of the impact of this type of scrambling in the mass spectroscopist's mechanism studies, the sum total may contribute in a very important fashion to settling the "carbonium ion controversy," as reported and discussed by Bernhard (1969). Equilibration is involved and, in fact, mandatory for all four of the studies made above. The difficulty is that some of the species that are described as intermediate resonance structures are nonclassical, yet the equilibration goes on, as Brown describes it, according to Bernhard (1969). Meyerson and Fields (1968) confirmed that labeled 2- and

3-phenylthiophenes exhibit fragmentations indicating that "deep-seated reorganizations of the molecular ion" must be occurring. This particular work also contains many fine details as to further ring breakdown and even differences due to the 2 or 3 substitution.

Earnshaw *et al.* (1964) studied selected ions in the mass spectra of labeled and unlabeled benzenethiol rather early in the game. An examination of a small part of their data shows some of the problems with which the mass spectroscopist was faced. In Table 5-21 are relative intensity data for selected

TABLE 5-21

PARTIAL MASS SPECTRA OF BENZENETHIOL AND DEUTERATED BENZENETHIOL (MONOISOTOPIC)

		Relative intensity	
m/e	Formula	Unlabeled	Labeled
33	SH^+	2.1	0.2
34	SD^+		1.9
66	$C_5H_6{}^+$	27.3	
67	$C_5H_5D^+$		27.3
109	$C_6H_5S^+$	23.3	10.2
110	$C_6H_4DS^+$		13.1
110	$C_6H_5SH^+$	98.2	
111	$C_6H_5SD^+$		98.2

ions. The molecular ion is the ion of largest intensity, but the 98.2 values are from monoisotopic spectra for each molecular species. The $(M - 1)^+$ distribution shows that more than half of the ions of the labeled species contain the D atom associated with ring carbons. This would indicate that some sort of equilibration had occurred in the molecular ion before a hydrogen (or D) was lost. Other decomposition processes must be operating, however, since the amount of SD^+ ion formed represents 90% ejection of the label directly. Apparently the equilibration between structures does lead back to a six-membered ring, this time with an H instead of D. Fragmentation then occurs to produce the SH^+ species at *m/e* 33. Earnshaw *et al.* (1964) proposed that the molecular ion consisted of two forms as shown. The I form would be a six-membered ring, while the II form would be a seven-membered (expanded-ring) form. It should be noted that the ions at *m/e* 66 and 67 must arise only from fragmentation of the seven-membered ring. These authors also considered the energetics of the various ions and, thus, an examination of the report will provide a good example of the combination of energetics with isotopic labeling

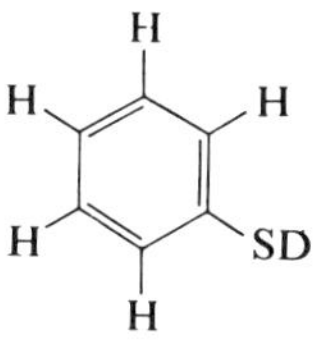

I

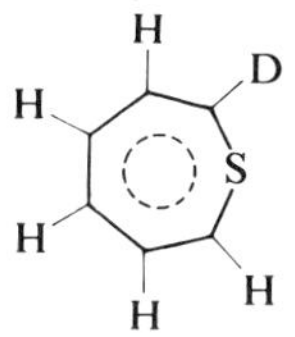

II

to study fragmentation mechanisms. These authors concluded that some of the ions from C_6H_5SD issue from the six-membered ring structure (I), but other ions arise from another structure such as the seven-membered ring (II) where the hydrogens and deuterium are equivalent.

In the work of Foster and Higgins (1968) α ^{13}C-labeled 2-hexylthiophene was studied, and further complications arose. This work permitted the proof that, again, at least some of the thiophene underwent ring expansion after initial β cleavage of the neutral olefin:

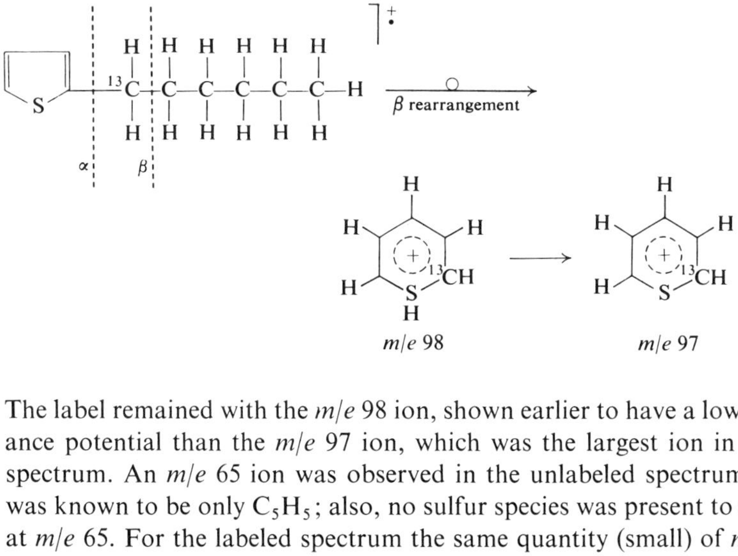

β rearrangement

m/*e* 98 → *m*/*e* 97 (5-146)

The label remained with the *m*/*e* 98 ion, shown earlier to have a lower appearance potential than the *m*/*e* 97 ion, which was the largest ion in the 70-eV spectrum. An *m*/*e* 65 ion was observed in the unlabeled spectrum, and this was known to be only C_5H_5; also, no sulfur species was present to any extent at *m*/*e* 65. For the labeled spectrum the same quantity (small) of *m*/*e* 66 was observed, indicating ring expansion. These workers further showed that the *m*/*e* 84 and 85 peaks came from α cleavage since the label was not present in either species, and the relative intensities of the ions agreed within experimental error. Establishing the next processes to occur, however, was difficult because fragmentation of either the five- or six-membered rings occurred to produce similar ions. The 84 ion could produce the CSH^+ ion; so could the *m*/*e* 85 ion, the *m*/*e* 97 ion, and even the *m*/*e* 98 ion. The other fragments normally found could be produced from any of these ions.

Additional evidence for the expanded-ring six-membered ion was found for the small ion at *m/e* 71. For the labeled molecule this appeared at *m/e* 72 and again indicated that at least part of the ring had expanded. The cleavage,

$$\text{HC}\overset{\text{H}}{\text{C}}\text{CH}\ (+)\ \text{HC}\,\text{S}\,\text{CH} \longrightarrow C_3H_3S^+ + C_2H_2 \qquad (5\text{-}147)$$

m/e 97 → *m/e* 71

is quite comparable to that of the tropylium ion. For additional details the original work should be consulted.

In retrospect, it also appears that perhaps equilibration between ion forms was going on, at least to some extent. This was evident by the intensities of the ions at *m/e* 125 and 126 as shown in Table 5-22. The label distributions for the minor amounts of ions produced by cleavage at the gamma C–C bond and beyond in the chain were in good agreement with a simple cleavage process. The discrepancy of the *m/e* 125 and 126 ions was noted but was thought to be of minor importance. It now appears that the α ^{13}C label could have been removed if the terminal part of the chain interacted with the molecular ion and the original ring C–^{13}C bond was broken. This is typical of the type of minor, yet disturbing, event that can be observed. The interference from fragmentation to form the same ion by several different pathways is extensive and, in the case presented, prevented more information being obtained from labeled mass spectral studies.

Labeling studies may be done on molecules that cannot be labeled by any of the ordinary chemical reactions done in the laboratory. Waller *et al.* (1966)

TABLE 5-22

PARTIAL MASS SPECTRA OF 1-(2-THIENYL)-1-HEXANE AND 1-^{13}C-(2-THIENYL)-1-HEXANE

	Relative intensity			Relative intensity	
m/e	Unlabeled	Labeled	*m/e*	Unlabeled	Labeled
110	1.66	0.88	153	0.27	0.06
111	5.54	1.81	154	0.04	0.26
112	0.82	5.00	168	19.52 M	0.05
125	1.09	0.79	169	2.18	19.64 M
126	0.28	0.42	170	0.96	1.58
139	1.40	0.11	171	0.12	0.94
140	0.28	1.28	172	0.01	0.15

reported studies of the mass spectrometry of biosynthetically labeled ricinine. This work proves that labeling in this fashion is most satisfactory for those interested in labeling natural products. It may also suggest approaches to problems that the new-generation mass spectroscopist may have to face in connection with pollution studies, environmental problems, etc.

Normally, the labeled studies are conducted with the intention of disclosing the mass spectral fragmentation of the species of interest. Karabatsos *et al.* (1964) provided an interesting and classical use of labeled studies to prove the behavior of organic molecules in solution reactions. The deamination of neopentyl-1-^{13}C and neopentyl-1-1-d_2-amines was conducted and the resulting *t*-butyl alcohols examined by both NMR and mass spectrometry (these authors also reported other neopentyl substituted compounds similarly labeled). The label originally present at C-1 of the neopentyl compounds always ended up at C-3 of the *t*-amyl compounds. These workers provided a thorough discussion of this interesting application to mechanism studies.

Before leaving this section it should be mentioned that isotopic labeling is not necessarily the only approach to mechanism studies involving organic substances. Bursey (1968) has developed a method referred to as the *p*-fluoro labeling technique to aid in mechanism research. Employing the technique, Bursey *et al.* (1968) have reported a cyclobutadiene cation radical in the mass spectrometer. Bursey and Elwood (1970) have also described the fragmentation without rearrangement of the *para*-fluoro label in the mass spectra of some six-membered heterocycles. These workers recognized that these results are discordant with statistical randomization of the molecular ions from similar six-membered aromatic compounds found by deuterium labeling studies. They have tried to advance a number of possible explanations. A further report by Hoffman *et al.* (1970) suggests that operation of a double-focusing mass spectrometer in the defocused mode shows that the integrity of the molecular ion in different systems is affected to varying degrees. This was observed to range from total lack of rearrangement to essentially complete scrambling.

VI. Predicting Fragmentation Mechanisms

A. *A priori* Comparison

It should be quite obvious to the reader by this time that one can begin rudimentary prediction of some fragmentation mechanisms based upon the acquired knowledge obtained in this treatise. Further, the method of comparison to a similar, known compound will provide considerable help. If the type of molecule in question contains alkyl substitutions of various sizes on a benzene ring, a little study of the chapter of Grubb and Meyerson (1963b) should provide enough information for even the novice to predict the major ions that should be observed in the mass spectrum. Similarly, for molecules containing other common structural features and differing only in the type of hetero atom present, consideration of the generalized differences among the various hetero atom species should allow reasonable prediction of the mass spectral pattern. From this the fragmentation mechanisms may be extrapolated. With a bit of experience one can even carry the prediction to the point of estimating the intensity of the molecular ion, what the base peak ion will be, and also some of the major fragment ions.

B. Methods of Predicting Mechanisms by Calculations

A molecular orbital (MO) approach to the interpretation of organic mass spectra of mesoionic compounds has been reported by Dougherty *et al.* (1970). In the form used, the MO interpretation is an extension of the usual resonance approach to the explanation of mass spectral reactions. This approach should prove quite popular for some time to come, especially with computer facilities available to the MO-oriented researcher.

Hirota *et al.* (1970) have used the CNDO method to calculate electronic distributions in *n*-propylamine and ethylamine. From this it was hoped that the probability of bond scission could be determined. The electron density of the highest occupied orbital in the skeletal bonds of both amines is found to be the largest on the α C–C bond adjacent to the CN bond. If equivalent orbital calculations are made, the electron density is the largest on the CN bond. If one assumes that there are fast and slow processes in the fragmentation of the alkylamines, then the change of scission probability at low ionizing energies can be explained. These authors also showed that the calculated dipole moment agrees with the observed, a fact which supports the reliability of the calculated values. Knoop *et al.* (1969) have discussed the low-energy electron impact excitation spectra of 1,3,5-cycloheptatriene and 1,3,5,7-cyclooctatetraene on the basis of quantum chemical calculations. Schiller and Jacobi (1969) have presented an interrelation discussion of mass spectra and classical dynamics.

C. The Kinetic Approach to Mass Spectrometry

Howe and Williams and their co-workers have presented a number of reports concerning the kinetic approach to mass spectrometry. Using this approach, they have shown that the relative ease of fragmentation of a substituent can be predicted from daughter–parent-ion ratios $(A^+)/(M^{+\cdot})$ at 18 eV. Howe and Williams (1968) proved that the ratio $(A^+)/(M^{+\cdot})$ is a function of the rates of competing unimolecular decompositions. Howe and Williams

(1969a) related the size of the metastable peak to the activation energy required for the decomposition in question. A total of 17 monosubstituted benzenes were classified according to the daughter–parent-ion ratios and their ease of fragmentation. These results also show that the process of lowest appearance potential gives the most abundant metastable peak. The use of this classification also predicts the more abundant daughter ion. Howe and Williams (1969b) consider substituent effects of some γ- and β-substituted methyl butyrates. The variation in the parent–daughter ratios at different energies is explained in terms of relative AP's and frequency factors. The mass spectra of substituted phenyl benzyl ethers have been examined similarly by Ward *et al.* (1969). Qualitative agreement with the theory and the earlier work was obtained, but some molecular ion abundances were much greater than the calculated value.

Paralleling this approach, but also in some conflict with the above approach, is the "physical-organic chemist's" treatment *via* the famous Hammett equation (Hammett, 1940):

$$\log(k/k_0) = \rho\sigma \tag{5-148}$$

Arising from solution chemistry, the equation relates the rate, k_0, for the reaction of an unsubstituted compound with rate k for the substituted compound. It has been most successful in applications comparing and predicting reactions of *meta*- and *para*-substituted aromatic compounds.

The σ is a substituent constant which reflects the electron-donating or -withdrawing nature of the substituent being considered. The constant, ρ, reflects the degree to which the influence of the substituent is transmitted to the reaction site for the particular reaction. The equation can also be written

$$(\Delta G^{+} - \Delta G_0{}^{+}) = -RT\rho\sigma \tag{5-149}$$

a perhaps more familiar form. Recalling that $\Delta G^{+} = \Delta H^{+} - T\Delta S^{+}$, the relationship to activation energies and entropies is obvious.

Bursey (1968) has reviewed the literature concerning its application to mass spectrometry prior to that time. He has also shown that for any precursor ion (P) leading to a daughter ion (A) the concentration ratio $Z = (A)/(P)$ is equal to the rate of formation of the precursor ion divided by the sum of the various rates of removal (further fragmentation) of the daughter ion. In the steady-state approximation of kinetics, the rate of change of (A) with time is zero, and this justifies the above statement. Hence, the ratio Z/Z_0 is numerically equal to k/k_0 so that one may employ concentrations of ions rather than the rate constants. Bursey discussed the assumptions made in the use of the Hammett equation in considerable detail.

McLafferty and Bursey (1968) developed the use of substituent effects as a probe for studying reaction rates and ion structures. In the equation they used $Z_0 = (A)^+/(M_0^{+\cdot})$ for the reaction for a substituted aromatic compound wherein the fragmentation of $M_0^{+\cdot} \rightarrow A^+$ was of interest. The Z_0 represented the recorded ratio of the unsubstituted parent compound. Hence, the Z/Z_0 value would be the rate of the $M^+ \rightarrow A^+$ reaction for a substituted M^+ relative to the rate of the same reaction for an unsubstituted M_0^+. A plot of the $\log(Z/Z_0)$ versus Hammett's σ (or σ^+), a substituent constant, gives, in some cases, a straight line; the relationship was thought to be analogous to the Hammett linear free-energy relationship used in solution chemistry. Admittedly this may be a rather far stretch of usage, but at least it sounds like a physical-organic chemist or even a physical chemist because the plotting of the data suggests a typical approach used by these types of chemists.

In the work of Williams and his co-workers given above, some aspects of the kinetic approach have been questioned. They provided experimental evidence from a group of monosubstituted diphenyl ethers in which the fraction of total ion current carried by the molecular ions decreased only a very small amount with the doubling of decay time. Hence, it appears that almost all of the molecular ions possessing enough energy to fragment did so in the source while those ions which did not have sufficient energy to fragment reached the collector and were recorded. Since these ions were not precursors for (A^+) they cannot be related to the rate of formation of (A^+) by the expression of

$$d(A^+)/dt = k_1(M)^{+\cdot} \tag{5-150}$$

where $(M)^{+\cdot}$ is the recorded abundance of the ion $M^{+\cdot}$. Howe and Williams (1968) have devoted a publication to elaborating on some of the other problems associated with both the qualitative and quantitative assumptions and aspects that are inherent in the approach employed by McLafferty and Bursey (1968).

Shapiro and Tomer (1969) reported a study concerned with the effects of substituents on the reactions (a) expulsion of C_2H_2O, a rearrangement, and (b) simple cleavage of CH_3CO^+ in the molecular ions of a series of *m*- and *p*-substituted acetanilides and the corresponding series of phenyl acetates. In terms of Z_m/Z_p, identical values of Z/Z_0 were obtained for the *m*- and *p*-phenyl acetates and almost identical Z/Z_0 values for the *m*- and *p*-methoxyphenyl acetate. The $(M - C_2H_2O)^+$ ions, however, underwent different secondary decompositions. For some of the isomers examined, identical Z/Z_0 values were obtained in competing simple cleavages. Shapiro and Tomer (1969) concluded that the substituents do not lose their positional identity in the molecular ion since the first fragment ions decompose by completely different secondary pathways. They further concluded that identical Z/Z_0 values can be misleading criteria for determining whether substituents retain or lose positional identities. Brown (1968a,b,c) has published work in this area while Lum and Smith (1969) have also described a linear free-energy relationship involving *ortho* substituents in mass spectrometry.

Bursey and Hoffman (1969) studied effects of methyl groups *ortho* to resonance donors for the loss of NO by substituted nitrobenzene ions in the mass spectrometer. The *o*-methyl groups seem to prevent coplanarity of the dimethylamino group in this specific system, but they do not prevent coplanarity for the methoxy group. At low voltage a complication appears, possibly associated with a ring expansion. Nagai *et al.* (1970) have considered the migratory aptitude of the aryl group in the diaryl sulfones from the viewpoint of substituent effects. Bursey and Kissinger (1970) reported an unexpected correlation of ion intensities with σ constants for the loss of substituents from biphenyl molecular ions. They discussed the significance of Hammett correlations in mass spectral decompositions. They conceded that their observations are more easily reconciled with an argument based upon energy distributions in molecular ions than on interpretation in terms of rates. This report faces the problem very squarely, pointing to the many problems and yet showing that, in some cases, the correlations can be useful.

The authors of this treatise were tempted to include the above discussion under the heading of Section V of this chapter, and will predict that, after many studies and much detailed effort, a link will be forged to permit satisfactory use of the kinetic approach as a means of determining or at least gaining some information about ion structures and fragmentation processes.

REFERENCES

Aczel, T., and Lumpkin, H. E. (1960). *Anal. Chem.* **32**, 1819.

Aczel, T., and Lumpkin, H. E. (1961). *Anal. Chem.* **33**, 386.

Aczel, T., and Lumpkin, H. E. (1962). *Anal. Chem.* **34**, 33.

Alexander, E. R. (1950). "Principles of Ionic Organic Reactions," 318 pp. Wiley, New York.

Bach, E. A., Kjaer, A., Shapiro, R. H., and Djerassi, C. (1965). *Acta Chem. Scand.* **19**, 2438.

Bailey, A. S., Gutch, C. J. W., Peach, J. M., and Waters, W. A. (1969). *J. Chem. Soc. B*, 681.

Bailey, W. A., Jr. (1961). Private communication.

Barnes, C. S., and Occolowitz, J. L. (1963). *Aust. J. Chem.* **16**, 219.

Barnes, C. S., and Occolowitz, J. L. (1964). *Aust. J. Chem.* **17**, 975.

Basson, R. A., and van der Linde, H. J. (1969). *J. Chem. Soc. A*, **11**, 1618.

Baumgarten, H. E., Parker, R. G., and von Minden, D. L. (1969). *Org. Mass Spectrom.* **2**, 1221.

Bendoraitis, J. G., Brown, B. L., and Hepner, L. S. (1962). *Anal. Chem.* **34**, 49.

Benoit, F., and Holmes, J. L. (1969). *Can. J. Chem.* **47**, 3611.

Benz, W., and Biemann, K. (1964). *J. Amer. Chem. Soc.* **86**, 2375.

Berlin, K. D., and Shupe, R. D. (1969). *Org. Mass Spectrom.* **2**, 447.

Bernhard, H. (1969). *Sci. Res.* August 18, 26.

Beynon, J. H. (1960a). "Mass Spectrometry and Its Applications to Organic Chemistry," p. 259. Elsevier, Amsterdam.

Beynon, J. H. (1960b). "Mass Spectrometry and Its Applications to Organic Chemistry," p. 271. Elsevier, Amsterdam.
Beynon, J. H., and Williams, A. E. (1959). *Appl. Spectrosc.* **13**, 101.
Beynon, J. H., and Williams, A. E. (1960). *Appl. Spectrosc.* **14**, 27.
Beynon, J. H., Lester, G. R., and Williams, A. E. (1959). *J. Phys. Chem.* **63**, 1861.
Beynon, J. H., Saunders, R. A., Topham, A., and Williams, A. E. (1961). *J. Phys. Chem.* **65**, 114.
Beynon, J. H., Saunders, R. A., and Williams, A. E. (1965). "Table of Meta-stable Transitions." Amer. Elsevier, New York.
Beynon, J. H., Saunders, R. A., and Williams, A. E. (1968a). "The Mass Spectra of Organic Molecules," p. 96, Amer. Elsevier, New York.
Beynon, J. H., Saunders, R. A., and Williams, A. E. (1968b). "The Mass Spectra of Organic Molecules," p. 105. Amer. Elsevier, New York.
Beynon, J. H., Saunders, R. A., and Williams, A. E. (1968c). "The Mass Spectra of Organic Molecules," pp. 108–112. Amer. Elsevier, New York.
Beynon, J. H., Saunders, R. A., and Williams, A. E. (1968d). "The Mass Spectra of Organic Molecules," p. 107. Amer. Elsevier, New York.
Beynon, J. H., Saunders, R. A., and Williams, A. E. (1968e). "The Mass Spectra of Organic Molecules," pp. 145–148. Amer. Elsevier, New York.
Beynon, J. H., Saunders, R. A., and Williams, A. E. (1968f). "The Mass Spectra of Organic Molecules," p. 302. Amer. Elsevier, New York.
Beynon, J. H., Saunders, R. A., and Williams, A. E. (1968g). "The Mass Spectra of Organic Molecules," p. 153, Amer. Elsevier, New York.
Beynon, J. H., Saunders, R. A., and Williams, A. E. (1968h). "The Mass Spectra of Organic Molecules," pp. 205–210. Amer. Elsevier, New York.
Beynon, J. H., Caprioli, R. M., Baitinger, W. E., and Amy, J. W. (1970a). *Org. Mass Spectrom.* **3**, 479.
Beynon, J. H., Caprioli, R. M., Baitinger, W. E., and Amy, J. W. (1970b). *Org. Mass Spectrom.* **3**, 455.
Beynon, J. H., Caprioli, R. M., Baitinger, W. E., and Amy, J. W. (1970c). *Org. Mass Spectrom.* **3**, 661.
Beynon, J. H., Corn, J. E., Baitinger, W. E., Amy, J. W., and Benkesser, R. A. (1970d). *Org. Mass Spectrom.* **3**, 192.
Bhati, A., Johnstone, R. A. W., and Millard, B. J. (1966). *J. Chem. Soc. C*, 358.
Biemann, K. (1962a). "Mass Spectrometry: Organic Chemical Applications," p. 51. McGraw-Hill, New York.
Biemann, K. (1962b). "Mass Spectrometry: Organic Chemical Applications," 370 pp. McGraw-Hill, New York.
Biemann, K. (1962c). "Mass Spectrometry: Organic Chemical Applications," pp. 74–87. McGraw-Hill, New York.
Biemann, K. (1962d). "Mass Spectrometry: Organic Chemical Applications," p. 116. McGraw-Hill, New York.
Biemann, K. (1962e). "Mass Spectrometry: Organic Chemical Applications," p. 124. McGraw-Hill, New York.
Biemann, K. (1962f). "Mass Spectrometry: Organic Chemical Applications," pp. 205–250. McGraw-Hill, New York.
Biemann, K. (1962g). "Mass Spectrometry: Organic Chemical Applications," p. 209. McGraw-Hill, New York.
Biemann, K. (1962h). "Mass Spectrometry: Organic Chemical Applications," pp. 223–227. McGraw-Hill, New York.

Binks, R., Mackenzie, K., and Williams-Smith, D. L. (1969). *J. Chem. Soc. C*, 1528.
Blades, A. T., and Gilderson, P. W. (1960). *Can. J. Chem.* **38**, 1401.
Bleakney, W., Condon, E. U., and Smith, L. G. (1937). *J. Phys. Chem.* **41**, 197.
Bowie, J. H., and White, P. Y. (1969). *Org. Mass Spectrom.* **2**, 611.
Brion, C. E., and Hall, L. D. (1966). *J. Amer. Chem. Soc.* **88**, 3661.
Brown, H. C., and Fletcher, R. S. (1949). *J. Amer. Chem. Soc.* **71**, 1845.
Brown, H. C., Bartholomay, H., Jr., and Taylor, M. D. (1944). *J. Amer. Chem. Soc.* **66**, 435.
Brown, P. (1968a). *J. Amer. Chem. Soc.* **90**, 2694.
Brown, P. (1968b). *J. Amer. Chem. Soc.* **90**, 4459.
Brown, P. (1968c). *J. Amer. Chem. Soc.* **90**, 4461.
Bruce, M. I., and Thomas, M. A. (1968). *Org. Mass Spectrom.* **1**, 417.
Budzikiewicz, H. (1969). *Org. Mass Spectrom.* **2**, 249.
Budzikiewicz, H., Brauman, J. I., and Djerassi, C. (1965). *Tetrahedron* **21**, 1855.
Budzikiewicz, H., Djerassi, C., and Williams, D. H. (1967a). "Mass Spectrometry of Organic Compounds," pp. 60–64. Holden-Day, San Francisco, California.
Budzikiewicz, H., Djerassi, C., and Williams, D. H. (1967b). "Mass Spectrometry of Organic Compounds," pp. 155–162, 174–213. Holden-Day, San Francisco, California.
Budzikiewicz, H., Djerassi, C., and Williams, D. H. (1967c). "Mass Spectrometry of Organic Compounds," p. 115. Holden-Day, San Francisco, California.
Bunton, C. A., and Del Pasco, T. W. (1969). *Org. Mass Spectrom.* **2**, 81.
Bursey, M. M. (1968). *Org. Mass Spectrom.* **1**, 31.
Bursey, M. M., and Elwood, T. A. (1970). *J. Org. Chem.* **35**, 793.
Bursey, M. M., and Hoffman, M. K. (1969). *J. Amer. Chem. Soc.* **91**, 5023.
Bursey, M. M., and Kissinger, P. T. (1970). *Org. Mass Spectrom.* **3**, 395.
Bursey, M. M., Rieke, R. D., Elwood, T. A., and Dusold, L. R. (1968). *J. Amer. Chem. Soc.* **90**, 1557.
Capellen, J., Svec, H. J., Jordan, J. R., and Sun, R. (1970a). "Bibliography of Mass Spectroscopy Literature for the First Half of 1968 Compiled by a Computer Method," 463 pp. Division of Technical Information, U.S. Atomic Energy Commission, Oak Ridge, Tennessee.
Capellen, J., Svec, H. J., Jordan, J. R., and Watkins, W. J. (1970b). "Bibliography of Mass Spectroscopy Literature for the Last Half of 1967 Compiled by a Computer Method," 458 pp. Division of Technical Information, U.S. Atomic Energy Commission, Oak Ridge, Tennessee.
Carpenter, W., Duffield, A. M., and Djerassi, C. (1967). *J. Amer. Chem. Soc.* **89**, 6164.
Carpenter, W., Duffield, A. M., and Djerassi, C. (1968). *J. Amer. Chem. Soc.* **90**, 160.
Carpenter, W., Duffield, A. M., and Djerassi, C. (1969). *Org. Mass Spectrom.* **2**, 317.
Cava, M. P., Mitchell, M. J., DeJongh, D. C., and Van Fossen, R. Y. (1966). *Tetrahedron Lett.* **26**, 2947.
Cavell, R. G., and Dobbie, R. C. (1968). *Inorg. Chem.* **7**, 690.
Chupka, W. A. (1959). *J. Chem. Phys.* **30**, 191.
Chupka, W. A., and Berkowitz, J. (1960). *J. Chem. Phys.* **32**, 1546.
Cilento, G. (1960). *Chem. Rev.* **60**, 147.
Clobes, A. L., Morris, M. L., and Kolb, R. D. (1969). *J. Amer. Chem. Soc.* **91**, 3087.
Coggeshall, N. D. (1960). *J. Chem. Phys.* **33**, 1247.
Coggeshall, N. D. (1963). *J. Chem. Phys.* **47**, 183.
Collin, J. E. (1965). *In* "Mass Spectrometry; A NATO Advanced Study Institute on Theory, Design and Applications held in Glasgow, August 1964" (R. I. Reed, ed.), pp. 183–201. Academic Press, New York.

Condé-Caprace, G., and Collin, J. E. (1969). *Org. Mass Spectrom.* **2**, 1277.
Conrad, R. (1930). *Phys. Z.* **31**, 888.
Cook, G. L., and Foster, N. G. (1961). *Proc. Amer. Petrol. Inst.* **41**, (III), 199.
Cooks, R. G., and Gerrard, A. F. (1968). *J. Chem. Soc. B*, 1327.
Cottrell, T. L. (1958). "The Strengths of Chemical Bonds," 2nd Ed., p. 271. Butterworth, London.
Cummings, C. S., and Bleakney, W. (1940). *Phys. Rev.* **58**, 787.
Damico, J. N., Barron, R. P., and Ruth, J. M. (1968). *Org. Mass Spectrom.* **1**, 331.
Das, K. G., Nayar, M. S. B., and Chinchwadkar, C. A. (1970). *Org. Mass Spectrom.* **3**, 303.
Dean, R. A., and Whitehead, E. V. (1961). *Tetrahedron Lett.* **21**, 768.
Delfosse, J., and Bleakney, W. (1939). *Phys. Rev.* **56**, 256.
Delfosse, J., and Hipple, J. A. (1938). *Phys. Rev.* **54**, 1060.
Denisov, Yu. V., Puchkov, V. A., Vulfson, N. S., Agadzhanyan, T. E., Antonov, V. K., and Shemyakin, M. M. (1968). *Zh. Obshch. Khim.* **38**, 770.
Desmarchelier, J. M., and Johns, R. B. (1969). *Org. Mass Spectrom.* **2**, 37.
Deutsch, J., and Mandelbaum, A. (1969). *J. Amer. Chem. Soc.* **91**, 4809.
Dibeler, V. H., and Mohler, F. L. (1950). *J. Res. Nat. Bur. Stand. A* **45**, 441.
Dibeler, V. H., Reese, R. M., and Franklin, J. L. (1961). *J. Amer. Chem. Soc.* **83**, 1813.
Djerassi, C., and Fenselau, C. (1965a). *J. Amer. Chem. Soc.* **87**, 5747.
Djerassi, C., and Fenselau, C. (1965b). *J. Amer. Chem. Soc.* **87**, 5752.
Dolejš, L., and Hanuš, V. (1968). *Collect. Czech. Chem. Commun.* **33**, 332.
Dolejšek, Z., Hanuš, V., and Vokáč, K. (1966). *In* "Advances in Mass Spectrometry" (W. L. Mead, ed.), Vol. III, pp. 503–514. Institute of Petroleum, London.
Dooley, J. E., Kendall, R. F., and Hirsch, D. E. (1970). *U.S. Bur. Mines Rep. Invest.* No. 7351.
Dorman, F. H., and Morrison, J. D. (1959a). *J. Chem. Phys.* **31**, 1320.
Dorman, F. H., and Morrison, J. D. (1959b). *J. Chem. Phys.* **31**, 1335.
Dorman, F. H., and Morrison, J. D. (1960). *J. Chem. Phys.* **32**, 378.
Dorman, F. H., and Morrison, J. D. (1961a). *J. Chem. Phys.* **34**, 578.
Dorman, F. H., and Morrison, J. D. (1961b). *J. Chem. Phys.* **34**, 1407.
Dorman, F. H., and Morrison, J. D. (1961c). *J. Chem. Phys.* **35**, 575.
Dougherty, R. C., Foltz, R. L., and Kier, L. B. (1970). *Tetrahedron* **26**, 1989.
Draper, P. M., and MacLean, D. B. (1968). *Can. J. Chem.* **46**, 1499.
Eadon, G., and Djerassi, C. (1969). *J. Amer. Chem. Soc.* **91**, 2724.
Eadon, G., and Djerassi, C. (1970). *J. Amer. Chem. Soc.* **92**, 3084.
Earnshaw, D. G., Cook, G. L., and Dinneen, G. U. (1964). *J. Phys. Chem.* **68**, 296.
Eland, J. H. D., and Danby, C. J. (1965). *J. Chem. Soc.* 5935.
Elwood, T. A., Dudley, K. H., Tesarek, J. M., Rogerson, P. F., and Bursey, M. M. (1970). *Org. Mass Spectrom.* **3**, 841.
Eyring, E. M., and Wahrhaftig, A. L. (1961). *J. Chem. Phys.* **34**, 23.
Eyring, H., Walter, J., and Kimball, G. E. (1944). "Quantum Chemistry," Chap. 16. Wiley, New York.
Fano, U. (1961). *Phys. Rev.* **124**, 1866.
Fenselau, C., and Wang, S. Y. (1969). *Tetrahedron* **25**, 2853.
Field, F. H., and Franklin, J. L. (1957). "Electron Impact Phenomena and the Properties of Gaseous Ions," pp. 36, 64, 382. Academic Press, New York.
Forkey, D. M. (1969). *Org. Mass Spectrom.* **2**, 309.
Foster, N. G. (1966). *U.S. Bur. Mines Rep. Invest.* No. 6741.
Foster, N. G., and Higgins, R. W. (1966). Paper No. 97. Fourteenth Annual Conference on Mass Spectrometry, ASTM Committee E-14, Dallas, Texas.

Foster, N. G., and Higgins, R. W. (1968). *Org. Mass Spectrom.* **1**, 191.
Foster, N. G., and Higgins, R. W. (1969). *Org. Mass Spectrom.* **2**, 1005.
Foster, N. G., Hirsch, D. E., Kendall, R. F., and Eccleston, B. H. (1964). *U.S. Bur. Mines Rep. Invest.* No. 6433.
Foster, N. G., Hirsch, D. E., Kendall, R. F., and Eccleston, B. H. (1965). *U.S. Bur. Mines Rep. Invest.* No. 6671.
Foster, N. G., Shiu, D. W.-K., and Higgins, R. W. (1968). Paper No. 71. Sixteenth Annual Conference on Mass Spectrometry, ASTM Committee E-14, Pittsburgh, Pennsylvania.
Foster, N. G., Liao, J. P., and Higgins, R. W. (1969). Paper No. 162. Seventeenth Annual Conference on Mass Spectrometry, ASTM Committee E-14, Dallas, Texas.
Foster, N. G., Higgins, R. W., and Hamming, M. C. (1970). Paper No. G-9. Eighteenth Annual Conference on Mass Spectrometry, ASMS with ASTM Committee E-14, San Francisco, California.
Fowler, R. G., Foster, N. G., and Higgins, R. W. (1969). Paper No. 144. Seventeenth Annual Conference on Mass Spectrometry, ASTM Committee E-14, Dallas, Texas.
Fox, R. E. (1960a). *J. Chem. Phys.* **32**, 385.
Fox, R. E. (1960b). *J. Chem. Phys.* **33**, 200.
Fox, R. E. (1961). *J. Chem. Phys.* **35**, 1379.
Fox, R. E., Hickam, W. M., Grove, D. J., and Kjeldaas, T. (1951). *Phys. Rev.* **84**, 859.
Fox, R. E., Hickam, W. M., Grove, D. J., and Kjeldaas, T. (1955). *Rev. Sci. Instrum.* **26**, 1101.
Fraas, R. E., Kiser, R. W., and Chaney, G. L. (1969). *Org. Mass Spectrom.* **2**, 1171.
Franklin, J. L. (1949). *Ind. Eng. Chem.* **41**, 1070.
Franklin, J. L. (1953). *J. Chem. Phys.* **21**, 2029.
Franklin, J. L. (1954). *J. Chem. Phys.* **22**, 1034.
Franklin, J. L. (1963). *J. Chem. Ed.* **40**, 284.
Franklin, J. L., and Carroll, S. R. (1969). *J. Amer. Chem. Soc.* **91**, 6564.
Franklin, J. L., Dillard, J. G., Rosenstock, H. M., Herron, J. T., and Draxl, K. (1969). "Ionization Potentials, Appearance Potentials and Heats of Formation of Gaseous Positive Ions," p. 285. U.S. Department of Commerce, NSRDS-NBS, U.S. Government Printing Office, Washington, D.C.
Friedel, R. A., Schultz, J. L., and Sharkey, A. G., Jr. (1956). *Anal. Chem.* **28**, 926.
Friedman, L., Long, F. A., and Wolfsberg, M. (1957). *J. Chem. Phys.* **26**, 714.
Funke, P., Das, K. G., and Bose, A. K. (1964). *J. Amer. Chem. Soc.* **86**, 2527.
Fuson, N. (1943). *In* "Organic Chemistry, an Advanced Treatise" (H. Gilman, ed.), p. 96. Wiley, New York.
Gale, D. M. (1968). *Tetrahedron* **24**, 1811.
Gallegos, E. J., and Kiser, R. W. (1961). *J. Amer. Chem. Soc.* **83**, 773.
Gamble, A. A., Gilbert, J. R., Tillett, J. G., Coombs, R. E., and Wilkinson, A. J. (1969). *J. Chem. Soc. B*, 655.
Gelling, I. R., Irwin, W. J., and Wibberley, D. G. (1969). *J. Chem. Soc. B*, 513.
George, W. O. (1962). *Nature* **194**, 672.
Gerrard, A. F., and Djerassi, C. (1969). *J. Amer. Chem. Soc.* **91**, 6808.
Gilchrist, T., and Reed, R. I. (1960). *Experimentia* **16**, 134.
Giles, R. D., Quick, L. M., and Whittle, E. (1967). *Trans. Faraday Soc.* **63**, 662.
Gillis, R. G., and Occolowitz, J. L. (1966). *Tetrahedron Lett.* **18**, 1997.
Glasstone, S., Laidler, K. J., and Eyring, H. (1941). "The Theory of Rate Processes." McGraw-Hill, New York.
Gohlke, R. S., and McLafferty, F. W. (1962). *Anal. Chem.* **34**, 1281.
Gowenlock, B. G., Kay, J., and Majer, J. R. (1963). *J. Chem. Soc.*, 2463.

Gray, R. T., and Djerassi, C. (1970). *J. Org. Chem.* **35**, 753.
Gross, M. L., and Wilkins, C. L. (1969). *Tetrahedron Lett.* **44**, 3875.
Grostic, M. F., Wnuk, R. J., and MacKellar, F. A. (1966). *J. Amer. Chem. Soc.* **88**, 4664.
Grubb, H. M., and Meyerson, S. (1963a). *In* "Mass Spectrometry of Organic Ions" (F. W. McLafferty, ed.), p. 518. Academic Press, New York.
Grubb, H. M., and Meyerson, S. (1963b). *In* "Mass Spectrometry of Organic Ions" (F. W. McLafferty, ed.), pp. 453–527. Academic Press, New York.
Hall, G. G. (1951). *Proc. Roy. Soc. London* **205*A***, 541.
Hall, G. G. (1953). *Trans. Faraday Soc.* **49**, 113.
Hammett, L. P. (1940). "Physical Organic Chemistry," 1st Ed., 405 pp. McGraw-Hill, New York.
Haney, M. A., and Franklin, J. L. (1969). *J. Phys. Chem.* **73**, 4328.
Hanuš, V., and Čermák, V. (1959). *Collect. Czech. Chem. Commun.* **24**, 1602.
Harrison, A. G. (1963). *In* "Mass Spectrometry of Organic Ions" (F. W. McLafferty, ed.), pp. 207–251. Academic Press, New York.
Harrison, A. G., Thomas, M. T., and Still, I. W. J. (1970). *Org. Mass Spectrom.* **3**, 899.
Haynes, L. J., Kirkien-Konasiewicz, A., Loudon, A. G., and Maccoll, A. (1968). *Org. Mass Spectrom.* **1**, 743.
Heiss, J., Zeller, K.-P., and Zeeh, B. (1968). *Tetrahedron* **24**, 3255.
Henley, E. J., and Johnson, E. R. (1969). "The Chemistry and Physics of High Energy Reactions," 475 pp. Washington Univ. Press, Washington, D.C.
Herzberg, G. (1950). "Molecular Spectra and Molecular Structure. Vol. I. Spectra of Diatomic Molecules," 2nd Ed., p. 194. Van Nostrand-Reinhold, Princeton, New Jersey.
Higgins, W., and Jennings, K. R. (1965). *Chem. Commun.* 99.
Hills, L. P., Futrell, J. H., and Wahrhaftig, A. L. (1969). *J. Chem. Phys.* **51**, 5255.
Hirota, K., Fujita, I., Yamamoto, M., and Niwa, Y. Y. (1970). *J. Phys. Chem.* **74**, 410.
Hirsch, D. E., and Foster, N. G. (1960). Unpublished work. U.S. Bureau of Mines, Bartlesville Petroleum Research Center, Bartlesville, Oklahoma.
Hirsch, D. E., and Foster, N. G. (1961). Unpublished work. U.S. Bureau of Mines, Bartlesville Petroleum Research Center, Bartlesville, Oklahoma.
Hoffman, M. K., Elwood, T. A., Rogerson, P. F., Tesarek, J. M., Bursey, M. M., and Rosenthal, D. (1970). *Org. Mass Spectrom.* **3**, 891.
Honig, R. E. (1948). *J. Chem. Phys.* **16**, 105.
Hood, A. (1963). *In* "Mass Spectrometry of Organic Ions" (F. W. McLafferty, ed.), p. 597. Academic Press, New York.
Hopkins, R. L. (1959). Private communications and American Petroleum Institute Project 48 reports, U.S. Bureau of Mines, Petroleum Experiment Station, Bartlesville, Oklahoma.
Howe, I., and Williams, D. H. (1968). *J. Chem. Soc. B*, 1213.
Howe, I., and Williams, D. H. (1969a). *J. Amer. Chem. Soc.* **91**, 7137.
Howe, I., and Williams, D. H. (1969b). *J. Chem. Soc. B*, 439.
Hückel, E. (1931). *Z. Phys.* **70**, 204.
Jackson, M. B., Spotswood, T. M., and Bowie, J. H. (1968). *Org. Mass Spectrom.* **1**, 857.
Jesse, W. P., and Platzman, R. L. (1962). *Nature* **195**, 790.
Johnstone, R. A. W., and Millard, B. J. (1966). *Z. Naturforsch.* **21a,** 604.
Johnstone, R. A. W., and Ward, S. D. (1968). *J. Chem. Soc. C*, 2540.
Johnstone, R. A. W., Millard, B. J., and Millington, D. S. (1966). *Chem. Commun.* 600.
Junk, G. A., and Svec, H. J. (1963). *J. Amer. Chem. Soc.* **85**, 839.
Kadentsev, V. I., Zolotarev, B. M., Chizov, O. S., Shachidayatov, C., Yanovskaya, L. A., and Kucherov, V. F. (1968). *Org. Mass Spectrom.* **1**, 899.

Kaminsky, M. S., and Chupka, W. A. (1959). *Bull. Amer. Phys. Soc.* **4**, 235.

Karabatsos, G. J., Orzech, C. E., Jr., and Meyerson, S. (1964). *J. Amer. Chem. Soc.* **86**, 1994.

Kaufman, J. J., and Koski, W. S. (1960). *J. Amer. Chem. Soc.* **82**, 3262.

Kendall, R. F., and Eccleston, B. H. (1966). *U.S. Bur. Mines Rep. Invest.* No. 6854.

King, A. B., and Long, F. A. (1958). *J. Chem. Phys.* **29**, 374.

Kinney, I. W., Jr., and Cook, G. L. (1952). *Anal. Chem.* **21**, 1391.

Kinstle, T. H., and Oliver, W. R. (1969). *J. Amer. Chem. Soc.* **91**, 1864.

Kiser, R. W. (1960). "Tables of Ionization Potentials," TID-6142, June 20 and supplements. U.S. Department of Commerce, Washington, D.C.

Kiser, R. W. (1965a). "Introduction to Mass Spectrometry and Its Applications," p. 144. Prentice-Hall, Englewood Cliffs, New Jersey.

Kiser, R. W. (1965b). "Introduction to Mass Spectrometry and Its Applications," pp. 142–149. Prentice-Hall, Englewood Cliffs, New Jersey.

Kiser, R. W. (1965c). "Introduction to Mass Spectrometry and Its Applications," p. 155. Prentice-Hall, Englewood Cliffs, New Jersey.

Kiser, R. W. (1965d). "Introduction to Mass Spectrometry and Its Applications," p. 196. Prentice-Hall, Englewood Cliffs, New Jersey.

Kiser, R. W. (1965e). "Introduction to Mass Spectrometry and Its Applications," pp. 162–206. Prentice-Hall, Englewood Cliffs, New Jersey.

Kiser, R. W. (1965f). "Introduction to Mass Spectrometry and Its Applications," p. 170. Prentice-Hall, Englewood Cliffs, New Jersey.

Kiser, R. W. (1965g). "Introduction to Mass Spectrometry and Its Applications," p. 177. Prentice-Hall, Englewood Cliffs, New Jersey.

Kiser, R. W. (1965h). "Introduction to Mass Spectrometry and Its Applications," pp. 179–182. Prentice-Hall, Englewood Cliffs, New Jersey.

Kiser, R. W., and Gallegos, E. J. (1962). *J. Phys. Chem.* **66**, 947.

Kiser, R. W., and Hobrock, B. G. (1962). *J. Phys. Chem.* **66**, 957.

Kiser, R. W., Sullivan, R. E., and Lupin, M. S. (1969). *Anal. Chem.* **41**, 1958.

Knoop, F. W. E., Kistemaker, J., and Oosterhoff, L. J. (1969). *Chem. Phys. Lett.* **3**, 41.

Koutecký, J. (1959). *Collect. Czech. Chem. Commun.* **24**, 1608.

Kraft, M., and Spiteller, G. (1969). *Org. Mass Spectrom.* **2**, 865.

Krauss, M., and Dibeler, V. H. (1963a). *In* "Mass Spectrometry of Organic Ions" (F. W. McLafferty, ed.), pp. 120–123. Academic Press, New York.

Krauss, M., and Dibeler, V. H. (1963b). *In* "Mass Spectrometry of Organic Ions" (F. W. McLafferty, ed.), pp. 117–161. Academic Press, New York.

Kulkarni, M. V., Eisenbraun, E. J., and Hamming, M. C. (1970). *J. Org. Chem.* **35**, 686.

La Lau, C. (1960). *Anal. Chim. Acta* **22**, 239.

La Londe, R. T., and Davis, C. B. (1969). *Can. J. Chem.* **47**, 3250.

Lampe, F. W., Franklin, J. L., and Field, F. H. (1961). "Kinetics of the Reactions of Ions with Molecules" (G. Porter, ed.), Vol. 1, pp. 67–103. Pergamon, Oxford.

Lawrence, R., and Waight, E. S. (1970). *Org. Mass Spectrom.* **3**, 367.

Lee, F. C. (1964). "Synthesis of 2-(Butyl-1,1d_2)thiophene," M.S. thesis. Texas Woman's Univ., Denton, Texas.

Lengyel, I., and Uliss, D. B. (1969). *Org. Mass Spectrom.* **2**, 1239.

Lengyel, I., Uliss, D. B., and Mark, R. V. (1970a). Paper No. G-11. Eighteenth Annual Conference on Mass Spectrometry, ASMS with ASTM Committee E-14, San Francisco, California.

Lengyel, I., Uliss, D. B., and Mark, R. V. (1970b). *J. Org. Chem.* **35**, 4077.

Lennard-Jones, J., and Hall, G. G. (1952a). *Proc. Roy. Soc. London* **213*A***, 102.

Lennard-Jones, J., and Hall, G. G. (1952b). *Trans. Faraday Soc.* **48**, 581.
Levi, E. M., Mao, C. L., and Hauser, C. R. (1969). *Can. J. Chem.* **47**, 3671.
Levy, E. J., and Stahl, W. A. (1961). *Anal. Chem.* **33**, 707.
Li, P. H., and McGee, H. A., Jr. (1969). *Chem. Commun.* 592.
Liao, J. P., Foster, N. G., and Higgins, R. W. (1968). Southwest Regional Meeting of the American Chemical Society, Austin, Texas.
Liberles, A. (1968). "Introduction to Theoretical Organic Chemistry," 722 pp. Macmillan, New York.
Lorquet, J. C. (1960). *Rev. Mod. Phys.* **32**, 312.
Lossing, F. P., Tickner, A. W., and Bryce, W. A. (1951). *J. Chem. Phys.* **19**, 1254.
Lum, K. K., and Smith, G. G. (1969). *J. Org. Chem.* **34**, 2095.
Lumpkin, H. E., and Nicholson, D. E. (1960). *Anal. Chem.* **32**, 74.
McAllister, T., and Lossing, F. P. (1969). *J. Phys. Chem.* **73**, 2996.
McDonald, C. G., Shannon, J. S., and Sugowdz, G. (1963). *Tetrahedron Lett.* **13**, 807.
McDowell, C. A. (1963). *In* "Mass Spectrometry" (C. A. McDowell, ed.), pp. 506–588. McGraw-Hill, New York.
McDowell, C. A., Lossing, F. P., Henderson, I. H. S., and Farmer, J. B. (1956). *Can. J. Chem.* **34**, 345.
McFadden, W. H., Lounsbury, M., and Wahrhaftig, A. L. (1958). *Can. J. Chem.* **36**, 990.
McFadden, W. H., Black, D. R., and Corse, J. W. (1963). *Phys. Chem.* **67**, 1517.
McFadden, W. H., Seifert, R. M., and Wasserman, J. (1965). *Anal. Chem.* **37**, 560.
McLafferty, F. W. (1957). *Anal. Chem.* **29**, 1782.
McLafferty, F. W. (1959). *Anal. Chem.* **31**, 82.
McLafferty, F. W. (1963a). *In* "Mass Spectrometry of Organic Ions" (F. W. McLafferty, ed.), pp. 316–329. Academic Press, New York.
McLafferty, F. W., ed. (1963b). "Mass Spectrometry of Organic Ions," p. 324. Academic Press, New York.
McLafferty, F. W., ed. (1963c). "Mass Spectrometry of Organic Ions," p. 312. Academic Press, New York.
McLafferty, F. W., ed. (1963d). "Mass Spectrometry of Organic Ions," p. 316. Academic Press, New York.
McLafferty, F. W., and Bursey, M. M. (1968). *J. Amer. Chem. Soc.* **90**, 5299.
McLafferty, F. W., and Gohlke, R. S. (1959). *Anal. Chem.* **31**, 2076.
McLafferty, F. W., and Hamming, M. C. (1958). *Chem. Ind.* (*London*), p. 1366.
MacLeod, J. K., and Djerassi, C. (1967). *J. Org. Chem.* **32**, 3485.
Magat, M. (1951). *Discuss. Faraday Soc.* **10**, 113.
Magat, M., and Viallard, R. (1951). *J. Chim. Phys.* **48**, 385.
Magee, J. L. (1952). *Proc. Nat. Acad. Sci. U. S.* **38**, 764.
Mandelbaum, A., and Biemann, K. (1968). *J. Amer. Chem. Soc.* **90**, 2975.
Mann, M. M., Hustrulid, A., and Tate, J. T. (1940). *Phys. Rev.* **58**, 340.
March, J. (1968). "Advanced Organic Chemistry: Reactions, Mechanisms, and Structure," 1008 pp. McGraw-Hill, New York.
Martin, R. H., Lampe, F. W., and Taft, R. W. (1966). *J. Amer. Chem. Soc.* **88**, 1353.
Marx, M., and Djerassi, C. (1968). *J. Amer. Chem. Soc.* **90**, 678.
Matsuura, T., and Kitaura, Y. (1969). *Tetrahedron* **25**, 4487.
Melton, C. E. (1970). "Principles of Mass Spectrometry and Negative Ions," 328 pp. Dekker, New York.
Meyerson, S. (1955). *Appl. Spectrosc.* **9**, 120.
Meyerson, S. (1964). *J. Phys. Chem.* **68**, 968.

Meyerson, S. (1965). *J. Chem. Phys.* **42**, 2181.
Meyerson, S., and Corbin, J. L. (1965). *J. Amer. Chem. Soc.* **87**, 3045.
Meyerson, S., and Fields, E. K. (1968). *Org. Mass Spectrom.* **1**, 263.
Meyerson, S., and Fields, E. K. (1969a). *Org. Mass Spectrom.* **2**, 1309.
Meyerson, S., and Fields, E. K. (1969b). *Science*, **166**, 325.
Meyerson, S., and Leitch, L. C. (1964). *J. Amer. Chem. Soc.* **86**, 2555.
Meyerson, S., and Leitch, L. C. (1966). *J. Amer. Chem. Soc.* **88**, 56.
Meyerson, S., and McCollum, J. D. (1959). Division of Physical Chemistry, 136th Meeting of the American Chemical Society.
Meyerson, S., and Vander Haar, R. W. (1962). *J. Chem. Phys.* **37**, 2458.
Meyerson, S., and Vander Haar, R. W. (1968). *Org. Mass Spectrom.* **1**, 397.
Meyerson, S., and Weitkamp, A. W. (1968). *Org. Mass Spectrom.* **1**, 659.
Meyerson, S., and Weitkamp, A. W. (1969). *Org. Mass Spectrom.* **2**, 603.
Meyerson, S., Nevitt, T. D., and Rylander, P. N. (1963). *In* "Advances in Mass Spectrometry" (R. M. Elliott, ed.), Vol. II, pp. 313–336. Macmillan, New York.
Meyerson, S., Drews, H., and Fields, E. K. (1964). *J. Amer. Chem. Soc.* **86**, 4964.
Meyerson, S., Puskas, I., and Fields, E. K. (1966). *J. Amer. Chem. Soc.* **88**, 4974.
Miller, J. M. (1969). *Can. J. Chem.* **47**, 1613.
Mohler, F. L. (1948). *J. Wash. Acad. Sci.* **38**, 193.
Mohler, F. L., Dibeler, V. H., and Reese, R. M. (1954). *J. Chem. Phys.* **22**, 394.
Morrison, J. D. (1953). *J. Chem. Phys.* **21**, 1767.
Nagai, T., Maeno, T., and Tokura, N. (1970). *Bull. Chem. Soc. Jap.* **43**, 462.
Natalis, P., and Franklin, J. L. (1965). *J. Phys. Chem.* **69**, 2935.
Newton, A. S., and Strom, P. D. (1958). *J. Phys. Chem.* **62**, 24.
Nibbering, N. M. M., and de Boer, T. J. (1968a). *Tetrahedron* **24**, 1415.
Nibbering, N. M. M., and de Boer, T. J. (1968b). *Tetrahedron* **24**, 1427.
Nibbering, N. M. M., and de Boer, T. J. (1968c). *Tetrahedron* **24**, 1435.
Nibbering, N. M. M., and de Boer, T. J. (1968d). *Org. Mass Spectrom.* **1**, 365.
Nibbering, N. M. M., and de Boer, T. J. (1970). *Org. Mass Spectrom.* **3**, 487.
Nicholls, R. W. (1961). *J. Res. Nat. Bur. Stand. Sect. A* **65**, 451.
Nicholls, R. W., and Stewart, A. L. (1962). "Allowed Transitions in Atomic and Molecular Processes" (D. R. Bates, ed.), pp. 47–78. Academic Press, New York.
Nicholson, A. J. C. (1958). *J. Chem. Phys.* **29**, 1312.
Occolowitz, J. L. (1964). *Anal. Chem.* **36**, 2177.
Oro, J., Nooner, D. W., Zlatkis, A., Wikström, S. A., and Barghoorn, E. S. (1965). *Science* **148**, 77.
Pettit, R. (1960). *Tetrahedron Lett.* **23**, 11.
Platzman, R. L. (1960). *J. Phys. Radium* **21**, 853.
Platzman, R. L. (1962). *Radiat. Res.* **17**, 419.
Plimmer, J. R., Kearney, P. C., and Klingebiel, U. I. (1969). *Tetrahedron Lett.* **44**, 3891.
Polyakova, A. A., and Khmel'nitskii, R. A. (1961). *Zh. Obshch. Khim.* **31**, 4059.
Porter, Q. N. (1967). *Aust. J. Chem.* **20**, 103.
Pottie, R. F., and Lossing, F. P. (1961). *J. Amer. Chem. Soc.* **83**, 4737.
Potts, K. T., and Singh, U. P. (1970). *Org. Mass Spectrom.* **3**, 433.
Prasil, Z., and Forst, W. (1968). *J. Amer. Chem. Soc.* **90**, 3344.
Preston, P. N., Steedman, W., Palmer, M. H., Mackenzie, S. M., and Stevens, M. F. G. (1970). *Org. Mass Spectrom.* **3**, 863.
Price, C. C. (1964). *Chem. Eng. News*, November 30, 58.
Reed, R. I. (1963). *In* "Mass Spectrometry of Organic Ions" (F. W. McLafferty, ed.), pp. 637–699. Academic Press, New York.

Reed, R. I., Reid, W. K., and Wilson, J. M. (1963). *In* "Advances in Mass Spectrometry" (R. M. Elliott, ed.), Vol. II, pp. 416–439. Pergamon, Oxford.
Reiser, R. W. (1969). *Org. Mass Spectrom.* **2**, 467.
Rhodes, R. E., Barber, M., and Anderson, R. L. (1966). *Anal. Chem.* **38**, 48.
Ritter, M. L., McElrath, E. N., and Meisels, G. G. (1969). Southwest Regional American Chemical Society Meeting, Tulsa, Oklahoma.
Roberts, D. T., Jr., Little, W. G., and Bursey, M. M. (1967). *J. Amer. Chem. Soc.* **89**, 4917.
Robertson, A. V., and Djerassi, C. (1968). *J. Amer. Chem. Soc.* **90**, 6992.
Robinson, W. E., Cummins, J. J., and Dinneen, G. U. (1965). *Geochim. Cosmochim. Acta* **29**, 249.
Rol, N. C. (1965). *Rec. Trav. Chim. Pays-Bas* **84**, 413.
Rosenstock, H. M. (1952). Ph.D. thesis, University of Utah, Salt Lake City.
Rosenstock, H. M., and Krauss, M. (1963). *In* "Mass Spectrometry of Organic Ions" (F. W. McLafferty, ed.), pp. 1–64. Academic Press, New York.
Rosenstock, H. M., Wallenstein, M. B., Wahrhaftig, A. L., and Eyring, H. (1952). *Proc. Nat. Acad. Sci. U. S.* **38**, 667.
Russell, G. A., and Brown, H. C. (1955). *J. Amer. Chem. Soc.* **77**, 4578.
Ryhage, R., and Stenhagen, E. (1960). *J. Lipid Res.* **1**, 361.
Rylander, P. N., Meyerson, S., and Grubb, H. M. (1956). *J. Amer. Chem. Soc.* **78**, 5799.
Rylander, P. N., Meyerson, S., and Grubb, H. M. (1957). *J. Amer. Chem. Soc.* **79**, 842.
Rylander, P. N., Meyerson, S., Eliel, E. L., and McCollum, J. D. (1963). *J. Amer. Chem. Soc.* **85**, 2723.
Sample, S., and Djerassi, C. (1966). *J. Amer. Chem. Soc.* **88**, 1937.
Scheppele, S. E., Grigsby, R. D., Mitchell, E. D., Miller, D. W., and Waller, G. R. (1968). *J. Amer. Chem. Soc.* **90**, 3521.
Scheppele, S. E., Grigsby, R. D., Whitaker, D. W., Hinds, S. D., Kinneberg, K. F., and Mitchum, R. K. (1970). *Org. Mass Spectrom.* **3**, 571.
Schiller, R., and Jacobi, L. (1969). *Phys. Rev.* **186**, 186.
Scott, D. W., Finke, H. L., Gross, M. E., Guthrie, G. B., and Huffman, H. M. (1950). *J. Amer. Chem. Soc.* **72**, 424.
Shannon, J. S. (1962). *Aust. J. Chem.* **15**, 265.
Shapiro, R. H., and Djerassi, C. (1965). *J. Org. Chem.* **30**, 955.
Shapiro, R. H., and Tomer, K. B. (1969). *Org. Mass Spectrom.* **2**, 579.
Shapiro, R. H., and Turk, J. (1969). *Org. Mass Spectrom.* **2**, 1067.
Sharkey, A. G., Jr., Schultz, J. L., and Friedel, R. A. (1959). *Anal. Chem.* **31**, 87.
Silverstein, R. M., and Bassler, G. C. (1967). "Spectrometric Identification of Organic Compounds," 2nd Ed., pp. 18–32. Wiley, New York.
Spiteller, G. (1966). "Massenspektrometrische Strukturanalyse Organischer Verbindungen," p. 127. Verlag Chemie, Weinheim.
Spiteller-Friedmann, M., and Spiteller, G. (1967). *Chem. Ber.* **100**, 79.
Staley, S., and Reichard, D. W. (1968). *J. Amer. Chem. Soc.* **90**, 816.
Stevenson, D. P. (1951). *Discuss. Faraday Soc.* **10**, 35.
Stevenson, D. P. (1958). *J. Amer. Chem. Soc.* **80**, 1571.
Stevenson, D. P. (1963). *In* "Mass Spectrometry" (C. A. McDowell, ed.), pp. 589–615. McGraw-Hill, New York.
Stevenson, D. P., and Hipple, J. A. (1942a). *J. Amer. Chem. Soc.* **64**, 1588.
Stevenson, D. P., and Hipple, J. A. (1942b). *J. Amer. Chem. Soc.* **64**, 2766.
Stevenson, R. L., Wacks, M. E., and Scott, W. M. (1969). *Org. Mass Spectrom.* **2**, 261.
Streitwieser, A., Jr. (1960). *J. Amer. Chem. Soc.* **82**, 4123.

Streitwieser, A., Jr. (1961). "Molecular Orbital Theory for Organic Chemists." Wiley, New York.
Streitwieser, A., Jr., and Nair, P. M. (1959). *Tetrahedron* **5**, 149.
Svec, H. J. (1970). Private communication.
Svec, H. J., and Junk, G. A. (1967). *J. Amer. Chem. Soc.* **89**, 790.
Swarc, M. (1950). *Chem. Rev.* **47**, 75.
Thomas, A. F., and Willhalm, B. (1964). *Helv. Chim. Acta* **47**, 475.
Turnbo, R. G., Sullivan, D. L., Pettit, R. (1964). *J. Amer. Chem. Soc.* **86**, 5630.
Ungnade, H. E., and Loughran, E. D. (1964). *J. Heterocycl. Chem.* **1**, 61.
Van Brunt, R. J., and Wacks, M. E. (1964). *J. Chem. Phys.* **41**, 3195.
Viallard, R., and Magat, M. (1949). *C. R. Acad. Sci.* **228**, 1118.
Vought, R. H. (1947). *Phys. Rev.* **71**, 93.
Vouros, P., and Biemann, K. (1969). *Org. Mass Spectrom.* **2**, 375.
Wacks, M. E., and Dibeler, V. H. (1959). *J. Chem. Phys.* **31**, 1557.
Waldron, J. D., and Wood, K. (1952). "Mass Spectrometry." Institute of Petroleum, London.
Waller, G. R., Ryhage, R., and Meyerson, S. (1966). *Anal. Biochem.* **16**, 277.
Ward, R. S., Cooks, R. G., and Williams, D. H. (1969). *J. Amer. Chem. Soc.* **91**, 2727.
Warren, J. W. (1950). *Nature* **165**, 810.
Washburn, H. W., Wiley, H. F., Rock, S., and Berry, C. E. (1945). *Ind. Eng. Chem. Anal. Ed.* **17**, 74.
Weinberg, D. S., and Scoggins, M. W. (1969). *Org. Mass Spectrom.* **2**, 553.
Weissler, G. L. (1956). "Handbuch der Physik," Vol. 21, pp. 304–382. Springer-Verlag, Berlin.
Whitmore, F. C. (1948). *Ind. Eng. Chem. News Ed.* **26**, 669.
Willhalm, B., Thomas, A. F., and Gautschi, F. (1964). *Tetrahedron* **20**, 1185.
Williams, D. H., Cooks, R. G., Bowie, J. H., Madsen, P., Schroll, G., and Lawesson, S.-O. (1967). *Tetrahedron* **23**, 3173.
Williams, D. H., Ward, R. S., and Cooks, R. G. (1968a). *J. Chem. Soc. B*, 522.
Williams, D. H., Tam, S. W., and Cooks, R. G. (1968b). *J. Amer. Chem. Soc.* **90**, 2150.
Yeo, A. N. H., and Williams, D. H. (1968). *J. Chem. Soc. C*, 2666.
Yeo, A. N. H., and Williams, D. H. (1969). *Org. Mass Spectrom.* **2**, 331.
Zamir, L., Jenson, B. S., and Larsen, E. (1969). *Org. Mass Spectrom.* **2**, 49.

CHAPTER 6

PROCEEDING WITH AN INTERPRETATION

Because of the high cost usually associated with the operation of a mass spectrometry laboratory, the procedure of interpretation is of much consequence. In a research organization, a great deal of individual judgment is usually exercised in obtaining spectra data and examining it. The objective of this chapter is to supply general procedures which have been found by experience to be workable, without directly furnishing a set of instructions.

I. Analytical Objectives

When a mass spectrometric examination of a sample is desired, it may appear that the problems involved should be only those associated with the technique. However, experience shows that too often when objectives are not clearly defined the organic researcher does not receive the desired information on the "first try." This is because mass spectrometry can often be "fitted" to specific analytical applications. A discussion of objectives is thus desirable in an effort to blend the extent of need with the limits of capability. In many cases, a discussion leads to a mutual benefit beyond that originally "planned."

The wide variety of applications of mass spectrometry can, however, become an expensive pitfall. Scientific personnel knowing of its widespread application are likely to request that samples be examined by mass spectrometry without any clear-cut objectives. While an examination of research samples by mass spectrometry has often provided unexpected results, an agreed-upon objective is desirable. Several well-established objectives for analyzing samples will be discussed in the hope of conserving time and effort when practical problems are undertaken. These objectives are intended as ground rules for researchers using mass spectrometry in a modern research center. At the same time, a person with a new interest in this field will find these objectives instructive in a general manner.

General objectives usually become more specific with use. For example, the researcher may at first be concerned only with determining the average molecular weight of a mixture. Next he may desire a molecular weight distribution of the compound types, followed by accurate mass measurements of the species involved. Often, the analytical accuracy can be tailored to the needs of analyses, presuming that the manpower cost can be justified.

A. Molecular Weight Determination

One of the more common analytical objectives is determination of molecular weight. For the objective to be fully realized, it should be clearly understood what information is supplied in a molecular weight determination.

Normally, the molecular ion is taken to be the ion produced by the molecule in question assuming a composition of the lowest atomic mass unit isotopes and summing these to find the integral atomic mass units, or mass of the molecular ion. For example, the organic chemist ordinarily sums up each carbon atom mass as 12, hydrogen as 1, oxygen as 16, nitrogen as 14, sulfur as 32, fluorine as 19, chlorine as 35 (not the 35.45 atomic weight of $^{35}Cl + {}^{37}Cl$ due to the abundance of the chlorine isotope). Hence, a molecular ion peak may consist of several different elemental compositions, all of which occur at the same nominal mass. Often the elemental composition can be determined by examination of the spectrum for the abundances of isotopic peaks or the "isotope profile." Accurate mass measurements by high-resolution mass spectrometry also permit determining elemental compositions of ions.

Molecular ions depicted on the mass spectrum are individually associated with the corresponding isotope peaks. For example, the spectrum of carbon monoxide (API-44 No. 156) shows a molecular ion peak for the $^{12}CO^{16}$ isotope at m/e 28. The ^{13}C isotope is shown at m/e 29 for the $^{13}C^{16}O$ (and a trace from $^{12}C^{17}O$), and the $^{12}C^{18}O$ ion is shown at m/e 30. The intensities of the peaks at m/e 29 and 30 are in the approximate ratio of the natural abundance of these elements. Thus, the intensities of the isotopic peaks can be used in determining the elemental compositions.

Mass spectrometric molecular weight determinations of synthetic products can often provide rapid preliminary results. If the product is pure, the determination of the molecular weight is a rather simple task. For example, a sample of 2-methyl-4-heptanone would be expected to have a molecular ion peak at *m/e* 128. A structure determination would require an additional investigation, such as comparing the sample fragmentation pattern with a known reference standard (API-44–TRC No. 109-m).

If the sample is a mixture, several different molecular ions may be present on a single mass spectrum. Each of the molecular ions denotes the molecular weight of the compound which it represents. For example, a molecular ion peak at *m/e* 190 denotes only those compounds of molecular weight 190. This is an advantage in analyzing many mixtures because isomers are thus grouped together. For example, when several isomers make up a single molecular ion peak, accurate average molecular weight determinations can be made, as illustrated by Boyer *et al.* (1963) for mixtures of alkylbenzenes.

Some substances do not give a detectable molecular ion peak on a 70-eV spectrum. Such compounds often may still be analyzed for molecular weights by applying fragmentation rules. For example, the molecular weight of a primary unbranched aliphatic alcohol without a detectable molecular ion peak can be determined by adding 18 mass units to the olefinic peak (C_nH_{2n} ion) which predominates in the high-mass region of the spectrum. Obviously, caution has to be used to avoid errors in applying such rules. Other methods for establishing molecular weights by chemical conversions to either give strong molecular ions or remove the ions in question from interferences by raising them to a considerably higher *m/e* range are discussed in Chapter 4.

B. Functional Group Determinations

Functional group determinations are usually of secondary importance to compound or compound-type analyses in mass spectrometric analysis. The reason for this becomes clear with consideration of the typical fragmentation of an alcohol and a formate. The functional groups are different, but predominate fragment peaks are similar. A primary saturated aliphatic alcohol with a hydroxyl group (true also of bifunctional molecules) gives an intense peak at *m/e* 31 (CH_2OH^+). It does not follow, however, that a peak at *m/e* 31 is exclusive evidence of a hydroxyl group. Consider as an example the spectrum of *n*-propyl formate (API-44 No. 386), which has its most intense peak at *m/e* 31. For this reason, in mass spectrometry, classes of compounds are usually discussed in more detail than functional groups. In cases where the objective is the detection of a functional group, the class of compounds is usually determined in the process.

A functional group greatly influences the fragmentation under electron impact, and its presence is often detectable by characteristic peaks. Identifying a functional group in a sample may require that several known spectra of different classes of compounds be examined for common characteristic spectral features. By observing if these common spectral features appear on an unknown spectrum, it then becomes possible to deduce the presence of a functional group. In many cases, correlation studies reported in the literature are used for the detection of functional groups. Consider the detection of the mercaptan group in a mixture of hydrocarbons. Spectra of mercaptans are known to show evidence of the SH or SH_2 group by peaks at *m/e* 33 and 34 (and at $P - H_2S$). These ions are not greatly interfered with by hydrocarbons. Assume that a hydrocarbon sample is being examined for cyclohexenethiols and that no reference standards are available for cyclohexenethiols at the time of the analyses. It then can be assumed that cyclohexene-

TABLE 6-1

INDICATORS FOR FUNCTIONAL GROUPS AND ATOMS

Group	Spectral indicators
$—CH_3$, $—CH_2—CH_3$, etc. *t*-Butyl and isopropyl groups	Peaks at 15, 29, 43, etc.
—CH=CH— Alkene groups	Peaks at 28, 42, 56, etc.
H—C≡C—H Vinyl groups	Peaks at 26, 40, 54, etc.
Aldehyde groups Alkyl Aryl	Usually exhibit a loss of the CHO group if alkyl types H atom is lost, usually followed by loss of the neutral CO from the very stable "oylium" ion formed
(benzene ring) Phenyl groups	Peaks at 77, 91, 105, etc.
—OH (primary)	Peak at *m/e* 31 characteristic; also M − 18 in an *n*-propyl or longer chain
—OH (secondary)	Peaks at 45, 59, 73, or higher *m/e* in this series
—OH (phenol)	Peaks at 94, 108, 122, or higher *m/e* in this series
$CH_3—C(=O)—OH$	Peak at *m/e* 60 characteristic of monocarboxylic acids with four carbons or more (and a few related compounds)

TABLE 6-1 (*continued*)

Group	Spectral indicators
C=O	Peaks at 58, 72, 86, or higher *m/e* in this series, also M − CO ion prominent
—SH, —S—	Sulfur isotope profile is easily recognizable, mercaptans lose 34 or 33 as a neutral
R—S—S—R	Shows ions at R—S—S—H, H—S—S—H
R—SH	Ions appear at *m/e* 47 from CH_2SH in alkanethiols (mercaptans)
R—S—R	Thiaalkanes usually show an ion at *m/e* 61
 [thiophene ring] and other aryl S compounds	Ions or neutrals of mass 44 (CS) and 45 (CSH) are characteristic of an aromatic sulfur structure, *m/e* 84, 85, 97, 98, etc.
N, odd number of atoms	Odd MW ion—nitrogen rule
$—NH_2$, NH, N—	Peaks at 30 and 44 are prime indicators
—NO	Usually a direct loss of the O shown by $(M - 16)^+$ in the nitroxides
NO_2, $—ONO_2$	Sometimes a loss of mass 30; ions at *m/e* 30, sometimes at *m/e* 46
Fluorine	Usually indicated by M − 19 or *m/e* 19, sometimes by HF species; perfluorocarbons show a prominent *m/e* 69 due to CF_3^+; also look for multiples of CF (31) and CF_2 (50); fluorine is monoisotopic
Chlorine	Usually indicated by its isotopic profile on parent ions; also, individual ions still containing the chlorine will appear as M − 35 and M − 37 in proportion; chlorines on an aromatic ring usually are fairly easily ejected
Bromine	Usually indicated by its isotopic profile, two roughly equal peaks 2 mass units apart. Similarly, fragment ions should show M − 79 and M − 81 in proportion until bromine atom is lost
Iodine	If present, usually gives a large MW ion, and the isotopic profile will indicate considerably fewer M + 1 ions than if the chain of $(CH_2)_n$ were present to account for the large MW; M − 127 ions present; iodine is monoisotopic
Silicon	A characteristic isotopic profile with all ions containing Si
Boron	Characteristic isotopic profile due to ^{10}B and ^{11}B; fragments are atypical of the usual mass 14 differences observed in hydrocarbons

thiols fragment to produce the characteristic peaks at *m/e* 33 and 34 similar to the fragmentation exhibited by known standards with a mercaptan group. If the sample spectra show the characteristic peaks at *m/e* 33 and 34, an indication is then given of the presence of a mercaptan group. The identification of many functional groups can be made using the interpretation maps discussed in Chapter 7. A few examples of functional groups that may be detected are given in Table 6-1. Experience with interpretation of mass spectra will lead to many more for the spectroscopist, but he will also understand the many "if's, and's and but's" that qualify the use of the mass spectrometer as a means of detecting functional groups.

C. Component Analysis

An important objective of mass spectrometry is quantitative analysis of the individual components in a mixture. Before such an analysis can be made, each component must be known. Procedures for identifying components in a mixture will be discussed in later sections of this volume. In some cases, it is possible to quantitatively determine only one component without the complete identification of the remaining components in the sample, but, in general, all components are identified before doing a quantitative analysis.

When the objective is a quantitative analysis, careful consideration should be given to obtaining the mass spectral data. Reference standards are often scanned at the same time as the sample spectrum. In other cases, a test mixture with a composition similar to that of the sample is scanned either just before or just after the sample in order to achieve a high degree of accuracy. Consideration should also be given to changes in the sensitivity of the mass spectrometer, unexpected background interference from previous samples, and several other factors. It is advisable to check the mass spectrometric results with other analytical techniques when accuracy cannot be otherwise established. However, in complex mixtures, the absolute accuracy often cannot be evaluated because no similar sample of known composition is available for comparison, nor is another analytical technique known by which results can be compared.

Usually associated with a quantitative analysis is a statement of the expected accuracy. Accuracy depends greatly upon the type of sample and the instrument adjustments used in obtaining the spectral data. In general, accuracy is within 1% of the amount of a component present in most mixtures of 20 or 25 components but is not the same for all components in a mixture. Interference with analytical peak intensities varies among different components and greatly affects the accuracy of analysis for any given component. For example, toluene may be determined accurately to $\pm 0.01\%$ in a mixture of alkanes where its identifying peaks are nearly free of peak intensity interference from the alkanes, but it is determined with an accuracy of less than $\pm 0.01\%$ in a mixture of aromatics where interference is largely from other aromatics.

For some types of mixtures, a component analysis may be impossible from a single mass spectrum of the mixture. The fractionation of the original sample may be required and each fraction individually analyzed. In a sense, using a gas chromatograph–mass spectrometer is a means of making component identifications by individual analysis of the GC eluates.

An important analytical objective is the characterization of complex mixtures. Such analyses often group together *compound types* such as alkanes, cycloalkanes of various ring sizes, and aromatics. A carbon-number distribution is often possible for the compound types determined. There are two forms of type analysis. One is based upon parent or molecular ion peaks as obtained by low voltages. The peak intensities of these ions are converted into percentages by sensitivities (Lumpkin and Aczel, 1963, 1964; Lumpkin and Johnson, 1954). The other method of characterization is use of high-voltage (70 eV) summation analysis.

Type analyses are of great importance to oil companies because the composition dictates conversion processes and the refining equipment requirements involving considerable financial support. The usefulness of type analysis is realized by chemical companies, as well as by petroleum companies. In a sense, analysis of isomeric mixtures is type analysis because no single structure is determined.

The methods of characterization can be extended by available knowledge of fragmentation. For example, consider a request to characterize samples of chloroalkanes as to differences in the content of terminal and internal chlorinated alkanes. If one knows that there is a mass spectral difference between these two classes of compounds, it becomes a simple task to set up a calculation routine. The terminal chloroalkanes give an intense peak at m/e 91, while the internal chloroalkanes give intense olefinic ions in the C_nH_{2n} series of peaks. (Some examples of component analysis with calculations are provided in the last section of this chapter.)

D. Trace Analysis

When the objective is trace analysis, it should be clearly understood that the detection, identification, and quantitative determination of very low concentrations of certain components are difficult to achieve. In this type of work the components to be detected are often present in the parts per billion–parts per million range, and tedious isolation procedures must be utilized to obtain fractions of a microliter of sample.

Literature references to trace analysis are numerous even in view of the difficulties involved. Volatile components at trace levels have been identified in roasted peanuts (Mason *et al.*, 1966, 1967). Components in coffee have been identified by Merritt *et al.* (1963) and by Gianturco and Giammarino (1966). Flavor components of soybean oil were identified by Smouse and Chang (1967). Volatiles from strawberries have been determined by McFadden *et al.* (1965). Potato flavor components were examined by Deck and Chang (1965). Day and co-workers (Day and Anderson, 1965; Day and Libbey, 1964) investigated cheese flavors. Pineapple flavors were investigated by Silverstein *et al.* (1965). Buttery *et al.* (1965) examined hop oil, and Boldingh and Taylor (1962) investigated butter, for the volatile components. The aroma of black currants was investigated by Andersson and von Sydow (1964). Trace components in cream were the subject of investigation by Begemann and Koster (1964). Concentrating air pollutants has been an objective of several mass spectrometric investigations in trace analyses. One example is shown by Hoshino *et al.* (1964) using a modified freezeout technique. These are only a few of the many workers who have illustrated the care and detail in doing careful trace analysis by mass spectrometry.

E. Identification from the Spectrum Itself

A mass spectrum has both a "singing" and a "silent" feature. The "singing" feature applies when the objective of an analysis is, for example, the well-recognized use of a spectrum as supporting evidence for the identification of an expected molecular structure. This is a common objective of a mass spectrometric analysis. Usually a definitive choice between proposed structures can be made by applying fragmentation rules (Chapter 5). If the fragmentation pattern can be interpreted readily in terms of the structure proposed, mass spectrometry is then usable as confirming evidence. For example, if a sample of 2-heptanol is submitted for structure confirmation, the spectrum would be expected to have an intense peak at m/e 45 due to the characteristic cleavage of the C–C alpha to the hydroxyl group. If the experimental spectrum from the sample did indeed have an *interpretive* peak (Chapter 7) at m/e 45 and other fragment peaks typical of an alcohol, the spectrum obtained from this sample would be considered as evidence of the proposed structure. In this case, a reference standard is available for the direct comparison (API–TRC No. 37-m).

Structure verification, by comparison with a known reference standard, can be impressive when one considers that for a simple compound such as 2-heptanol, over 70 individual peaks of various intensities are available as supporting evidence. The spectrum of 2-heptanol is shown in Fig. 6-1. Support that the sample submitted for identification was 2-heptanol is given by the close agreement of intensities with a reference spectrum, as shown in Fig. 6-2. Such mass spectrometric identification is impressive when used alone. When

RELATIVE INTENSITIES

The ratio of $\frac{\text{mass}}{\text{charge}}$ is indicated by the superscript in the upper left hand corner of each block.

Magnetic Field: 2398 gauss to m/e 90; 4138 gauss from m/e 91
Electron Energy: 70 volts

									^{12} 21	^{13} 33	^{14} 173	^{15} 1074	^{16} 52
^{17} 25	^{18} 62	^{19} 445	^{20}	^{21}	^{22}	^{23}	^{24}	^{25} 13	^{26} 767	^{27} **7019**	^{28} 1483	^{29} **5315**	^{30} 177
^{31} 1274	^{32} 83	^{33} 15	^{34}	^{35} 9	^{36} 25	^{37} 60	^{38} 258	^{39} **2796**	^{40} 502	^{41} **5185**	^{42} 2031	^{43} **5704**	^{44} **3414**
^{45} **40927**	^{46} 941	^{47} 132	^{48} 15	^{49} 28	^{50} 117	^{51} 183	^{52} 75	^{53} 351	^{54} 210	^{55} **5235**	^{56} **2037**	^{57} 1554	^{58} 432
^{59} 283	^{60} 13	^{61} 11	^{62} 21	^{63} 38	^{64} 8	^{65} 47	^{66} 21	^{67} 114	^{68} 92	^{69} 848	^{70} 1459	^{71} 357	^{72} 77
^{73} 45	^{74} 9	^{75}	^{76}	^{77} 23	^{78} 9	^{79} 17	^{80}	^{81} 55	^{82} 38	^{83} **2747**	^{84} 201	^{85} 19	^{86}
^{87} 17	^{88}	^{89}	^{90}	^{91} 8	^{92}	^{93}	^{94}	^{95}	^{96} 9	^{97} 60	^{98} 1296	^{99} 120	^{100} 12
^{101} 1311	^{102} 91	^{103}	^{104}	^{105}	^{106}	^{107}	^{108}	^{109}	^{110}	^{111}	^{112} 10	^{113}	^{114} 17
^{115} 128	^{116} p 11	^{117}	^{118}	^{119}	^{120}	^{121}	^{122}	^{123}	^{124}	^{125}	^{126}	^{127}	^{128}
^{129}	^{130}	^{131}	^{132}	^{133}	^{134}	^{135}	^{136}	^{137}	^{138}	^{139}	^{140}	^{141}	^{142}
^{143}	^{144}	^{145}	^{146}	^{147}	^{148}	^{149}	^{150}	^{151}	^{152}	^{153}	^{154}	^{155}	^{156}
^{157}	^{158}	^{159}	^{160}	^{161}	^{162}	^{163}	^{164}	^{165}	^{166}	^{167}	^{168}	^{169}	^{170}
^{171}	^{172}	^{173}	^{174}	^{175}	^{176}	^{177}	^{178}	^{179}	^{180}	^{181}	^{182}	^{183}	^{184}
^{185}	^{186}	^{187}	^{188}	^{189}	^{190}	^{191}	^{192}	^{193}	^{194}	^{195}	^{196}	^{197}	^{198}
^{199}	^{200}	^{201}	^{202}	^{203}	^{204}	^{205}	^{206}	^{207}	^{208}	^{209}	^{210}	^{211}	^{212}
^{213}	^{214}	^{215}	^{216}	^{217}	^{218}	^{219}	^{220}	^{221}	^{222}	^{223}	^{224}	^{225}	^{226}
^{227}	^{228}	^{229}	^{230}	^{231}	^{232}	^{233}	^{234}	^{235}	^{236}	^{237}	^{238}	^{239}	^{240}

IONS WITH FRACTIONAL RATIO OF MASS TO CHARGE

mass/charge	Relative Intensity	mass/charge	Relative Intensity	mass/charge	Relative Intensity	mass/charge	Relative Intensity
37.5	2398 gauss 105						

Total Ionization for Compound: div/micron/microampere; div/lambda/microampere

n-BUTANE	n-HEXADECANE
Magnetic Field: 2398 gauss	Magnetic Field: 2398 / 4138 gauss
Sensitivity at $\frac{\text{mass}}{\text{charge}}$ 43:	Total Ionization:
10.37 div/micron/microampere	div/micron/microampere; div/lambda/microampere

mass/charge	Relative Intensity
15	8.68
29	33.94
43	100.00
58	11.92

mass/charge	Relative Intensity	mass/charge	Relative Intensity
57	100.00	141	2.49
71	56.72	155	2.15
85	35.21	169	1.79
99	5.70	183	1.24
113	3.81	197	0.63
127	2.99	226	3.94

COMPOUND

Name: 2-Heptanol

Molecular Weight	Molecular Formula	Approximate Boiling Point	Approximate Freezing Point	Approximate Density
116.1197	$C_7H_{16}O$	°C	°C	g/ml at °C

Semi-structural Formula:

```
 OH
 |
C-C-C-C-C-C-C
```

Source: Gallard-Schlessinger Chemical Manufacturing Corporation, Carle Place, New York

Purity: Sample purified by G.C.

LABORATORY: Continental Oil Company, Research and Development Department, Ponca City, Oklahoma 74601

MASS SPECTROMETER

Maker and Model: Consolidated Model No. 21-103C

Ionizing Current	Ion Chamber Temperature	Vapor Temperature	Collector Slit Width
10.0 microamperes	230 °C	130 °C	30 mils to m/e 90; 7 mils from m/e 91

Sample Pressure and Basis of Measurement:

Additional Information:
Sensitivity for n-C_4, I/ΣI 43: 0.298
Sensitivity for n-C_{16}, I/ΣI 57: 0.175
Ion beam: Focused at high sensitivity
Recording system: Consolidated Mascot
Mass spectral data edited by: Mynard C. Hamming

Note: The intensities set in bold face type in the matrix are the ten most intense peaks in this spectrum.

Date of Measurement: February 25, 1964

Fig. 6-1. Complete mass spectrum as published in the "Catalog of Mass Spectral Data," Serial No. 37-m, Thermodynamics Research Center Data Project, Thermodynamics Research Center, Texas A & M University, College Station, Texas. [Contributed by the Continental Oil Company, Ponca City, Oklahoma.]

UNKNOWN SAMPLE: NO. 44-37-M 12/8/69 LLH 15

INSTRUMENT: CEC 103C ION SOURCE: 250C EV: 70

POLYISOTOPIC PEAKS IN PERCENT TOTAL IONIZATION M.S. NO. LLH 15

M/E	INT	M/E	INT	M/E	INT	M/E	INT	M/E	INT	M/E	INT	M/E	INT	M/E	INT	M/E	INT	M/E	INT	M/E	INT	M/E	INT	M/E	INT	M/E	INT
3		4		5		6		7		8		9		10		11		12	2	13	3	14	17	15	111	16	5
17	2	18	9	19	51	20		21		22		23		24		25	1	26	74	27	681	28	142	29	533	30	18
31	124	32	6	33	1	34		35	1	36	1	37	4	38	26	39	279	40	50	41	527	42	198	43	570	44	346
45	4113	46	93	47	14	48	1	49	3	50	12	51	17	52	7	53	36	54	21	55	527	56	204	57	155	58	45
59	32	60	1	61	1	62	2	63	4	64	1	65	5	66	2	67	13	68	8	69	85	70	137	71	35	72	8
73	5	74	1	75		76		77	2	78	1	79	2	80		81	6	82	4	83	273	84	19	85	2	86	
87	2	88		89		90		91	1	92		93		94		95		96	1	97	6	98	127	99	11	100	1
101	137	102	9	103		104		105		106		107		108		109		110		111		112	1	113		114	2
115	12	116	1																								
SUBTOTALS																											
	4427		120		67		3		11		15		28		35		335		160		2102		845		1417		425
																								TOTAL			9990

GRAND TOTAL INCL. FRACTIONAL PEAKS, 10001

Fig. 6-2. Complete mass spectrum as obtained from a sample submitted for identification. Notice the close comparison of intensities expressed as a percent total ionization with the known reference spectrum shown in Fig. 6-1.

used in combination with synthesis, GLC, IR, NMR, etc., it becomes conclusive proof of structures, as often is required, for example, by drug firms.

The "silent" or the ordinarily dormant source of information in a mass spectrum includes those features which attract additional attention. The fragmentation pattern is now recognized as an important source of information. Organic compounds fragment in the mass spectrometer by pathways analogous to those in chemical reactions. Thus, it is possible to find support for a postulated scheme of chemical degradative conditions from the fragmentation pattern produced by electron impact. Yates (1967) has illustrated this for the conversion of xanthone to the benzophenone derivate. McLafferty (1966a) has cited several examples of the relationship between fragmentation patterns and chemical processes. Meyerson and McCollum (1963) have shown that mass spectra can be correlated with conventional chemical systems.

A mass spectrometric study of nonylphenolethylene oxide adducts (Harless and Crabb, 1969) has been used to speculate that certain biodegradation processes are similar to mass spectral fragmentation processes. Such work has pointed to the value of mass spectra in developing novel reactions, possibly an important future objective of mass spectrometry. However, the comparison of photochemical reactions with those of mass spectra must be carefully considered, as expressed by Meyerson (1964). It should be kept in mind that the extrapolation from results in the gas phase at 10^{-5} torr to liquid-phase primary reaction steps may be a possibility, but the probability is far less for higher-energy processes.

Bond formation in mass spectrometry has been discussed by Cooks (1969) because of the practical importance it has in structure determinations. The occurrence of rearrangements is becoming more predictable based upon a better understanding of bond formation and bond breaking as derived from mass spectra. Another example of the "silent" information found in a mass spectrum is given by Seibl (1969) for the significance of doubly charged ions. This trend in examination of mass spectra is expected to continue.

F. Structure Determination

Determination of molecular structures by mass spectrometry has become an important objective to the organic research chemist. Organic structure determination by mass spectrometry had meager beginnings in the 1920's and 1930's, but concerted programs at the National Bureau of Standards to obtain reference hydrocarbon spectra permitted the consideration of structure determination in addition to simply providing spectra that gave matrix coefficients. During the postwar years, spectroscopists began studying a number of isomers from a series of compounds, many of which are referred to in

Chapter 5. A natural outgrowth of these studies were correlations with structure. Predictions based upon mechanisms of fragmentation followed as a logical consequence.

Structure determination can be one of the most intriguing objectives in mass spectrometry. Studies usually consist of the use of known fragmentation processes, the detailed mechanisms which have been verified by isotopically labeled compounds, metastable ions, accurate mass measurements, and low-voltage data. Ingenious use of these devices has been widely applied in individual laboratories. Proof of the migration of a hydrogen atom can be investigated by replacing the hydrogen in question with a deuterium atom. In an as yet incomplete examination of xylylmethylbutyric acids (Grigsby and Hamming, 1966), it was observed that when the hydrogen atom on the carboxyl group of γ-(*p*-xylyl-2)-α-methylbutyric acid was replaced with deuterium, the deuterium was by some means transferred from the carboxyl group to the β carbon atom during initial fragmentation. Such observations often remain unpublished because of lack of labeled data or compounds. The experimental procedure for deuterium labeling may be rather simple, some examples of which have been discussed in detail as related to mass spectrometry by Budzikiewicz *et al.* (1964). However, in application, labeling is often not straightforward.

McLafferty *et al.* (1969) have reemphasized the importance of metastable ion data for deriving structure information. Oversimplified, this is possible because metastable decompositions are formed from low-energy precursor ions, causing the products to be strongly influenced by activation energies. Beynon (1968) has presented a summary on the applications of metastable peaks including analytical, deriving mechanisms, determining formulas of fragment ions, and others.

The objective of a structure determination can be realized by rather simple correlations between fragmentation patterns and molecular structures. For example, from an examination of a series of saturated straight-chain acids, it can be observed that an intense peak is formed at m/e 60 for compounds above *n*-butanoic acid. Such a peak is formed by the McLafferty rearrangement, which requires transfer of a γ hydrogen. Thus, the formation of an intense peak at m/e 60 requires the presence of a hydrogen atom on C-4 (position four of carbon chain). The spectrum of *n*-butanoic acid shows such a peak at m/e 60, together with a molecular ion at m/e 88. Consider as an unknown spectrum an aliphatic acid with a molecular ion peak at m/e 88 but without an intense peak at m/e 60. It could then be concluded that this spectrum had no C-4 with an available hydrogen atom and thus would be expected to be isobutyric acid.

Low voltage has been used for several years as a useful approach to structure elucidation. The technique was applied for various purposes by a number of workers, including Stevenson and Wagner (1950), Tickner *et al.* (1951), Lumpkin and Johnson (1954), Field and Hastings (1956), Lumpkin (1958), Sharkey *et al.* (1959), Thompson *et al.* (1966), and Budzikiewicz *et al.* (1967a). (See also Chapter 5 of this text.)

II. Miscellaneous Objectives

Mass spectrometry is today enormously diversified in its areas of applications and, thus, objectives. The analytical objectives so far discussed have been those commonly encountered by the "nonspecialist." Anyone would be well advised to seek expert advice before embarking upon specialized objectives. Most of the objectives listed below are too detailed to be discussed here, but the wide scope is at least indicative of the versatility of mass spectrometry:

Atmospheric pollution
Biochemistry studies
Chemical bonding studies
Chemical ionization reactions
Chemical processing
Collision reactions
Drug metabolites
Energetics of ion formation and decomposition
Food contaminants
Free-radical studies
Geology and upper-atmosphere problems
Heats of hydration, solvation
High-temperature chemical and physics studies
High-temperature studies of Geo and cosmo applications
Ionization cross sections
Ionization efficiency curves
Ionization processes related to photoionization
Ionization studies—surface
Ion–molecule reactions
Ion optics and production
Isotopic ratio—precise measurements
Medical research studies
Molecular beam studies
Negative-ion impact experiments
Nuclear research studies
Quality control
Reaction completion studies
Reaction kinetics
Reaction studies—monomolecular and bimolecular
Solvent quality
Space physics
Theory of mass spectra
Thermal analysis
Toxicology

Any one of the above-listed items has within its scope several objectives. Many of these have been under investigation for several years. For example, Kebarle and Hogg (1965) have shown that mass spectrometric data permit the possibility of measuring heats and entropies of hydration.

By considering additional objectives, new applications of spectral data can often be devised. For example, Natalis (1965) has shown that *cis* and *trans* geometrical isomers can often be differentiated despite similarities in their mass spectra and that the observed mass spectra relate to the thermodynamic stabilities of these isomers and their ions. The more thermodynamically stable *trans* isomers exhibit, in general, more intense molecular ion peaks than the corresponding *cis* isomers. A brief discussion of this work resulted in Hamming and Keen (1969) showing that the geometrical isomers of the *anti* and *syn* type could also be differentiated for a fluoranthene derivative. By such discussions, many additional objectives may be undertaken.

The time-of-flight mass spectrometer has been used in several specific applications. One of the most useful is for kinetic studies of fast reactions. Such a study is possible because spectra can be continuously recorded on an oscilloscope of reaction intermediates in fast, gas-phase chemical reactions. Other applications have included the investigation of explosion phenomena and flash photochemical reactions. A review of these and other specific applications has been made by Damoth (1965). A current progress series of the dynamic applications of mass spectrometry is provided. The first volume of this series is edited by Price and Williams (1969).

A review of the prolific applications of mass spectrometry has also been presented by White (1968). Many of these are nonanalytical and are thus different from the ones discussed in this chapter.

It was of interest to note that about one-fourth of the original principal investigators involved in the analysis of lunar materials brought back by Apollo 11 in 1969 used mass spectrometry. This illustrates the importance of miscellaneous objectives of mass spectrometry. Simmonds (1970) pointed out that the gas chromatograph–mass spectrometry combination is potentially one of the most useful for the comprehensive analysis of multicomponent, unknown planetary atmospheres.

III. Assembling Information about a Sample

All available information about a sample should be collected and carefully examined at the time the sample is received for analysis. Usually some information is known or assumed concerning the origin of the sample. After all previously determined analytical data is collected, it is important to know which are the firmly established facts, the indicated or probable facts, and the assumed facts or conclusions. The assembling and evaluation of this information will influence how the mass spectral experimental data is obtained.

The information collected should be compatible with the objectives. If a sample is to be examined to determine if the desired reaction product has been formed, it may be necessary only to know the molecular weight of the sample. Eisenbraun *et al.* (1968) and many others have used this approach for checking the organic synthesis of pure compounds. In cases where the molecular weight determined by mass spectrometry does not agree with the expected results, mass spectrometry becomes a more exciting technique because it supplies auxiliary information without additional effort. The person submitting the sample then usually realizes the uniqueness of mass spectrometry and finds that it can do much more than determine molecular weights.

The information obtained should be made available to all those concerned with the sample. When the operator and the interpreter of the data are the same person, no real problem is presented. However, in large industrial laboratories nonprofessional personnel often operate mass spectrometers, and it is advisable to *write* all pertinent information supplied by the originator of the sample on a request form. The information given to the operator should be limited to the needs of obtaining the desired spectral data, such as the inlet temperature, the method of loading, the scan range to cover, and any special instructions. Also it may be desirable to inform the operator of expected intense peaks. This practice prevents misunderstandings and forces the interpreter to have a more clearly formulated plan of obtaining the desired mass spectral data. This includes a selection of a mass spectrometer best suited to the problem, the operating conditions to be used, as well as the mass range to be scanned, the temperature of the inlet, and the means of loading the samples. Equipment available in any given laboratory will, of course, govern the routing of information. A great deal of expense can be saved if a few simple procedures are followed in obtaining the known information about a sample submitted for examination.

IV. Preinterpretation of Raw Data

The spectral data, as obtained from the mass spectrometer, should first be carefully checked for possible errors. On a single digitized spectrum a random error may be impossible to detect in either the mass number or the intensity. However, if a single value is in error, usually this will become evident when the spectrum is edited. If the spectrum is to be hand-processed, care should be taken that peaks are assigned correct m/e values and that intensities are carefully measured. If a photographic plate is used, several steps must be considered, depending upon the process involved.

The objective of preinterpretation is to put the raw data into a form that will facilitate interpretation. Too often not enough care is given to this phase of the work, resulting in much of the costly and valuable information on the spectra not being fully utilized. For example, if simple gross errors are made in peak measurements, an identification of the complete sample may be in error. For such reasons, preliminary data processing is similar to defining objectives; both can carelessly block the usefulness of mass spectrometry.

Unexpected peaks should be viewed with caution. If a solvent has been used, peaks from the solvent may have to be edited out of the spectrum. Peaks which may have originated from substrates used in chromatography may appear on a spectrum. For example, peaks at *m/e* 207 and 355 may be silicon-containing peaks (Beynon, 1960a).

The first real problem encountered in examining an unknown spectrum is to determine if the sample loaded was a pure substance or a mixture. If the spectrum is from a pure substance, references, indexes, and fragmentation rules will aid in the identification. If the spectrum is from a mixture, these same indexes and rules may only add confusion to the interpretation. Even purification by gas–liquid chromatography or other separation techniques does not always yield a monocomponent sample. However, the use of previously known information can be helpful in such cases. Consideration of the retention times, for example, can do much to help reach a conclusion because it eliminates certain compounds and even some isomers.

In many cases when the spectrum is believed to be from a pure sample, peaks which could be due to impurities should be considered. The detection of impurities in a mass spectrum may be a rather simple task from an examination of the general appearance of a spectrum or it may be nearly impossible. For example, it would be a simple matter to detect impurities in a sample of pentanes (different isomers of pentane) which showed detectable peaks at *m/e* 85, 99, and 113. These peaks are all at mass numbers greater than the isotopic peaks of pentanes (pentanes have molecular ion peaks at *m/e* 72 with isotopic peaks usually detectable at *m/e* 73 and 74). The peaks above *m/e* 74 thus could not originate from the fragmentation of pentanes and for this reason would be interpretated as due to impurities (assuming no background). However, had this sample of pentanes also contained small amounts of alkanes of molecular weights less than pentanes, identification of the impurities would have been difficult or impossible. This is because intense fragments of butanes, for example, overlap with the major fragments of pentanes (*m/e* 29, 43, and 57). In this case, the detection of butanes would depend mainly upon the detection of molecular ion peaks, which could be very difficult or impossible if the butanes were in very low concentration and only a 70-eV mass spectrum had been obtained.

A mixture spectrum is usually evident by a series of homologous peaks which differ by 14 mass units. For example, a mixture of alkylbenzenes shows molecular ions in the *m/e* 92, 106, 120, etc., series with intense fragment ions in the *m/e* 91, 105, 119, etc., series. A general knowledge of fragmentation processes for different classes of compounds is helpful. Interpretation maps discussed in Chapter 7 provide empirical information on the structural significance of series of peaks.

In view of the uncertainty of detecting minor impurities in a sample and the general complexity of a mass spectrum, it may seem as if the general appearance of a spectrum has little value. However, with experience, it is usually possible to rationalize which peaks are logical for a given type of molecule. The localized-charge concept, discussed by Budzikiewicz *et al.* (1967b), is sometimes useful for predicting the triggering of fragmentation mechanisms and thus provides information on formation of intense peaks from a pure substance. Other concepts, such as rearrangement processes which produce even-mass-numbered peaks from odd-electron ions (McLafferty, 1963a), have to be kept in mind in the gross examination of a mass spectrum.

The degree of preinterpretation of raw data varies greatly among laboratories. There is also some question as to where the preinterpretation stops and the interpretation starts. In a well-computerized laboratory the preinterpretation may stop at the time an element map (Biemann *et al.*, 1964) is formulated. In other laboratories the peaks may be measured and recorded by laboratory personnel doing the interpretation with no clear dividing line. As the volume of spectra in any one laboratory becomes large, the preinterpretation step becomes more defined. It can, if effectively applied, save much time of highly trained personnel.

V. The Molecular Ion as a Starting Point

The molecular ion provides a starting point for many interpretations, as well as providing the molecular weight of the substance being examined. To know that a ketone has a molecular weight of 156 from a molecular ion peak at *m/e* 156 can in itself be useful information. It can also be the point at which a structural determination is started. Several examples are provided in later chapters of this volume using the most intense peak as a starting point of interpretation.

Consider as one example an unknown spectrum which has a rather large peak at *m/e* 188 in the high region of the spectrum and a peak of lower intensity at *m/e* 190. It could be assumed that both peaks are molecular ion peaks. In going toward lower mass numbers an intense peak at *m/e* 153 is observed. The difference of 35 mass units between 188 and 153 brings to mind the mass of a chlorine atom. Upon reexamination of the ratio of the intensities of the peaks at *m/e* 188 and 190, it is observed that the peak at *m/e* 190 is about one-third as intense as the 188 peak. It could then be considered that the peak at *m/e* 190 is due to the ^{37}Cl isotope. Absolute mass measurements could then be determined (depending upon an availability of instruments). The point here is that the molecular ion is usually one of the *most diagnostic* peaks in a spectrum, if not *the most* diagnostic peak.

A. Determination of the m/e of the Molecular Ion from the Spectrum

In practice it may at times be difficult to determine which of the peaks in the high-mass region are molecular ion peaks, as in the case of a mixture. One approach is the use of low voltage. Low voltage has for a number of years been applied to enhance the intensity of molecular ion peaks relative to the fragment ions. It has the disadvantage of producing an overall decreased sensitivity of all ion peaks per sample size loaded. The technique of low-voltage mass spectrometry reduces fragmentation and, as such, is a means for differentiating between fragment ions and molecular ions in a mixture spectrum. Usually the instrument is adjusted to an ionizing electron current of 10–15 eV rather than 70 eV. Those ions which increase in relative intensity as a result of reducing the ionizing voltage can usually be considered as molecular ions. Low-voltage spectra can be obtained when the electron energy used is slightly (0.5–1) above the ionization potential of a given molecule, as indicated in Chapter 5. As such, the energy is just above that required to ionize the molecule but usually not enough to fragment it. As demonstrated a number of years ago by Lumpkin (1958), low voltage simplifies a mass spectrum of a complex mixture because it gives a spectrum which consists mainly of molecular ions (a few fragment ions of low intensities are usually present). This, of course, aids greatly in examining an unknown spectrum when molecular ions are to be used for the identification of components in a mixture.

Several methods of molecular ion detection have been established. McLafferty (1962) has discussed the use of the inlet system as an effusiometer and the "ion-molecular" method. Also, chemical conversion from a substance not giving a detectable molecular ion to a substance giving a molecular ion peak is often used for molecular weight determinations as described in Chapter 4.

In laboratories where field ion mass spectrometry (Beckey *et al.*, 1966) is is applied, the detection of molecular ions has not been found to be a serious problem. Most field ion spectra show intense molecular ion peaks, even when molecular ions for the same components are not detectable by electron-impact-produced spectra (70-eV spectra).

As pointed out by Yeo and Williams (1968), considerable effort has been made to avoid thermal decomposition in molecular ion formation, while little has been done to enhance intensities of molecular ions of compounds which undergo very fast unimolecular decay. For such compounds it should be possible, in principle, to obtain a molecular ion by the formation of suitable derivatives and use of very low ionization potentials.

Molecular ions are more readily detected if the complexity of the mixture is reduced. Today, many mass spectrometers are designed for probe loading in the ion source. By varying the source temperature, samples can be "probe fractionated" within the source to provide a series of spectra, each representing a different fraction of a total sample.

B. Determination of the Exact Composition of the Molecular Ion

Once the mass spectroscopist has located the molecular ion and determined the nominal mass of the species, his next question will be: What is the exact formula for the ion appearing at mass *i*? For high-resolution instruments, this is done according to the procedure given later in this section. For low- and medium-resolution instruments, the use of isotopic peaks and the *isotopic profile* come into use. For the molecular ion, the isotopic profile is the appearance of the molecular ion intensity relative to those of the naturally occurring isotopes accompanying that given molecule. The relative intensities of isotopic peaks can be important in the determination of the elemental composition of *both* molecular ions and fragment ions. On a mass spectrum, individual isotopes appear as individual peaks (Chapter 3 and Appendix VI B). Such elements as chlorine and bromine, with large isotopes, are easily recognized by two different peaks 2 mass units apart. Naturally occurring chlorine has two stable isotopes in the abundance of 75.8% ^{35}Cl and 24.2% ^{37}Cl. Thus, peaks 2 mass units apart in the ratio of 3:1 can be expected to be found for any ion species with one chlorine atom. This does not mean that the substance being analyzed contains only one chlorine atom. This means that the ion species observed contains one chlorine atom, and it is possible that this ion species may be a molecular ion or fragment ion. If the isotopic peak due to ^{37}Cl is 2/3 as intense as the first peak (^{35}Cl), then two chlorine atoms are present in the ion species observed. Ions containing one bromine atom are characterized by peaks 2 mass units apart in the ratio of 1:1. If two bromine atoms are present in the ion, three peaks appear, each 2 mass units apart, in the ratio of 1:2:1. Isotopic ratios for combinations of chlorine atoms, bromine atoms, and both chlorine and bromine atoms are shown in Fig. 6-3. The figure shows the characteristic intensities of the isotopic profile of each combination.

An application of the use of isotopic peaks is given by the fragmentation of terminal chloroalkanes and alkylbenzenes. Both compound types give an intense characteristic peak at m/e 91. Consideration of the isotopic peak can make it possible to differentiate between these two compound types. The peak at m/e 91 from terminal chloroalkanes has the composition of C_7H_7. Since low-resolution mass spectrometry does not differentiate between these different elemental compositions, the differentiation is dependent upon isotopes. The terminal chloroalkanes give an isotopic peak at m/e 93 one-third as intense as the peak at m/e 91. The 93 peak is due to the ^{37}Cl isotope of the C_4H_8Cl ion species. The alkylbenzenes give an isotopic peak of a very minor intensity at m/e 93 from the C_7H_7 ion at m/e 91 (the isotopic peak at m/e 93 is in the ratio

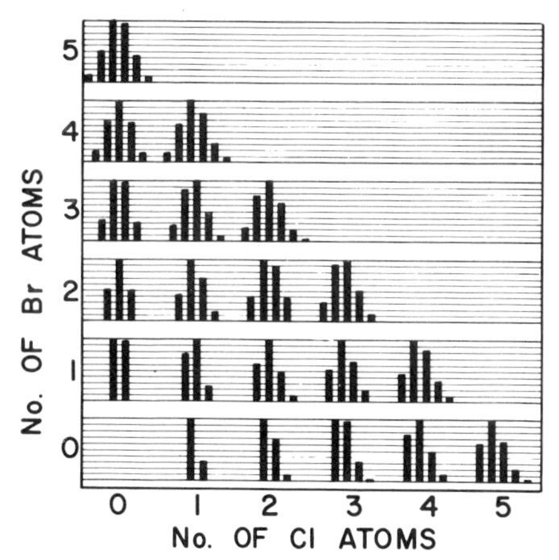

Fig. 6-3. Isotopic ratios of various combinations of chlorine and bromine atoms. Note that the peaks are 2 mass units apart.

0.3:100.0). Thus these two classes of compounds can be differentiated by a difference in intensities caused by isotopic peaks.

The isotope profile may be interfered with if the $(P - 1)^+$ ion peak is of more than a few percent of the intensity of the P^+ peak. It may cause a contribution to the P, P + 1, P + 2 peaks from its isotopes. The use of low voltage to minimize fragmentation and thereby reduce the P − 1 peak to small intensity allows the determination of the true isotopic profile of the molecular ion.

C. Further Uses of the Isotopic Profile

One must be cautious in using the isotope profile. With experience a good bit of "eyeballing" is resorted to, but when this is done, one can very easily be misled. The following example is a clear-cut case of this problem. The partial spectrum is shown in the tabulation below.

m/e	Relative intensity
108	26.4
134	13.2
169	100.
170	8.82
171	36.6
172	3.23

At a first glance two facts are apparent: The odd molecular weight ion indicates an odd number of nitrogen atoms, while the P and P + 2 intensities show the profile of a chlorine atom being present. The loss of 35 (169 → 134) and 37 (171 → 134) immediately confirms the presence of chlorine. If no further consideration of the isotope profile were given, the mass spectroscopist would begin his interpretation by including a chlorine and nitrogen atom in his first thoughts. Careful calculation of the chlorine isotope contribution shows that the intensity of *m/e* 171 is too high. Only 32.6% of the observed 36.6% can be accounted for by the chlorine. The remaining 4% is large enough to suggest the presence of a sulfur atom, allowing for a small experimental error in the determination of the peak intensity. If the sulfur is present, about 0.7% contribution to the P + 1 peak will be made by the naturally occurring ^{33}S isotope. The remainder of the P + 1 peak then suggests only about seven carbon atoms. This provides us with considerable information. To summarize,

169 P^+
−14 (assuming one nitrogen atom)
155
−35 (assuming one chlorine atom)
120
−32 (assuming one sulfur atom)
88
−84 (assuming seven carbon atoms)
4 (and we now assume these are H atoms)

The formula of this molecule must then be C_7H_4SClN. This is highly deficient in hydrogen atoms and hence must be aromatic. The large parent-ion intensity supports this idea. The chlorine atom must fragment off quite easily to leave a moiety of mass 134, still containing the sulfur and nitrogen atoms. The next loss of mass 26 could be due to either C_2H_2 (typical of benzenoid systems) or to CN, in this case still typical of an aromatic system. At this point, the compound 2-chlorobenzothiazole is strongly suggested. Placing the chlorine atom in this position is consistent with the fragmentation wherein the loss of HCN might be expected if the chloro group were on the benzo ring. Comparison with the spectra for 2-methylbenzothiazole (API 1760) shows that this compound is probably 2-chlorobenzothiazole.

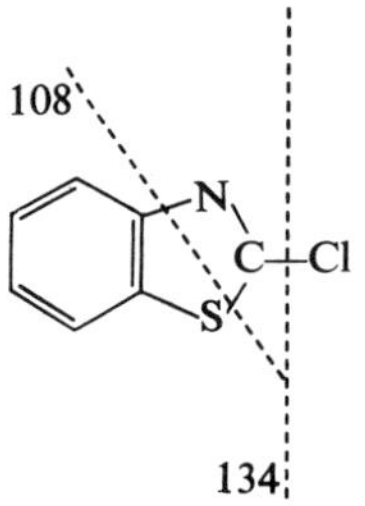

108
N
C—Cl = C_7H_4SClN (169)
S
134

The further use of isotope profile data could assist in ascertaining if the *m/e* 108 moiety still contains the sulfur atom. Such an ion is known to be rather stable and thus is also consistent with fragmentation knowledge. This example illustrates the depth of information that can be obtained from just a little "bit of data." It further illustrates the fact that "eyeballing" isotopic data can lead to trouble. In this case, the mass spectroscopist would have lost his S (sulfur atom).

Another example of utilizing the observations made possible from isotopic peaks on a low-resolution spectrum is found in the examination of the spectrum of 2-chloro-5-trifluoromethylaniline (*Archives* 1, No. 1, 1970, pp. 178–179). A partial spectrum of this compound is shown in the tabulation below:

m/e	Intensities (percent total ion yield)
160	2.6
162	0.2
176	3.8
178	1.2
195 ($^{12}C_7H_5{}^{14}NF_3{}^{35}Cl$)	19.0
196	1.7
197	6.1

NH2
Cl
CF3

Again this compound has more than one detectable molecular ion because of individual isotopes. The peak at *m/e* 195 consists mainly of the ion with contribution of the isotopes $^{12}C_7H_5{}^{14}NF_3{}^{35}Cl$. The peak at *m/e* 196 is comprised mainly of the isotopes ^{13}C or ^{15}N in the same ion, and the peak at *m/e* 197 is comprised mainly of the isotope ^{37}Cl. The peak 176 was formed by removal of one of the monoisotopic fluorine atoms (195 − 19). Because fluorine is monoisotopic, the ratio of the 178 peak to the 176 peak should remain the same as that of the 197 peak to the 195 peak (approximately correct). The chlorine atom must have been removed to form the ion at *m/e* 160 (195 − 35 and 197 − 37) as is evident by the small intensity of *m/e* 162 (the presence of a chlorine atom in the 160 moiety would require a 3:1 ratio between 160 and 162). Other isotopic peaks, as they relate to mass spectral data, have been discussed in detail by several authors referred to in Chapter 5.

Isotopic tables should be frequently referred to in attempting to identify unknown substances from mass spectral data. Appendix IV gives a set of isotopic mass and abundance tables for atoms commonly found in mass spectrometry applications. The values shown are those readily detectable on the usual mass spectrum. The arrangement of the data is similar to that of Benz (1968) and works well for finding approximate isotopic values from combinations of atoms. For example, an isotopic contribution from a $C_{16}H_{16}$ ion to an ion one higher mass number (designated in Appendix IV as $m + 1$) is 18.0% and 0.2% from the 16 carbons and 16 hydrogens, respectively, or a total isotopic contribution of 18.2% of the intensity of the $C_{16}H_{16}$ peak. At two higher mass numbers ($m + 2$), only the contribution of the 16 carbons is meaningful. In the case of 16 carbons this contribution is 1.5% of the intensity of the monoisotopic $C_{16}H_{16}$ peak. More accurate values can be found in Beynon and Williams (1963).

A commonly encountered problem in mass spectrometry is to identify the composition of a molecular ion with very low isotopic profile peaks except for the P + 1. If a check of the P/(P + 1) ratio indicates a deficiency from that expected for a molecule consisting of C_xH_y and of a molecular weight equal to the *m/e* of P^+, oxygen and/or nitrogen are probably present. If a high-resolution mass spectrometer is available the procedure is to obtain the exact mass of the molecule ion. By following the method as outlined by Beynon *et al.* (1968b) and Beynon and Williams (1963) the composition is usually limited to perhaps a half dozen possible ions, depending upon the uncertainty in the absolute mass value. The high-resolution approach may limit the use of isotopic peaks for the determination of elemental compositions. Since not all laboratories are gifted with a high-resolution instrument the mass spectroscopist having data from a low or medium instrument still faces a problem. The calculation of the total number of rings plus double bonds will be of assistance if the possible structures of an ion of the formula $C_xH_yN_zO_n$ are to be considered.

D. Calculation of Rings Plus Double Bonds

For the case of the problem posed above, consider that a molecular ion of *m/e* 128 is observed in the mass spectrum. If the formula is C_9H_{20}, then the intensity of the P + 1 peak should be about 10.08% of the intensity of the molecular ion. However, if the formula is $C_{10}H_8$ (the −12 series) then the P + 1 intensity should be 11.2% of the *m/e* 128 ion intensity. The measured value of the P + 1 peak is only 5.6% of the 128 ion, suggesting that only five carbon atoms are present. Since the molecular ion is of even mass, the presence of nitrogen atoms in even numbers is possible, but one, three, or five nitrogen atoms cannot be present. This is because nitrogen is the only common element with an even atomic mass and an odd valence. Therefore if an odd number of nitrogen atoms are present in the molecule one observes an odd molecular weight ion. The assumption of a formula of $C_5H_8N_2O_2$ is a reasonable place to start, since the alternative of $C_5H_4O_4$ will lead to few realistic organic structures that would be immediately obvious to the spectroscopist. One is given below. This possible compound is quite highly oxidized and its presence would be known to the mass spectroscopist on the basis of its very characteristic fragmentation.

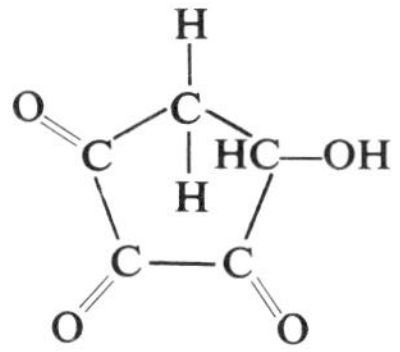

No doubt other structures having this same formula can be visualized. But let us return to the compound wherein there are two nitrogens and two oxygens, not discernible on the basis of the isotopic ions. An aid may be the use of the formula for calculating the number of rings plus double bonds. This formula, which arises because of the limited valences of the atoms involved and their constant character, is as follows:

$$R\text{ (rings + double bonds)} = x - (y/2) + (z/2) + 1$$

where x, y, and z are the number of carbon, hydrogen, and nitrogen atoms, respectively, in the molecular formula. For the nitrogen-containing formula this leads to the following calculation:

$$R = 5 - 8/2 + 2/2 + 1 \quad \text{or} \quad 5 - 4 + 1 + 1 = 3\text{ (rings plus double bonds)}$$

Let us construct several molecules of this formula that meet this requirement. Two linear molecules are

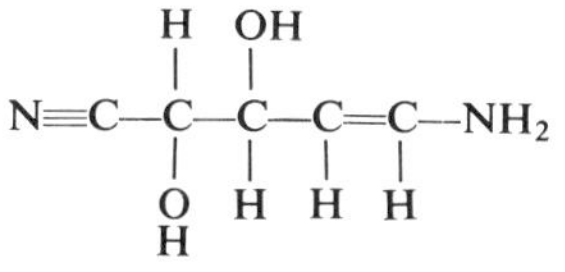

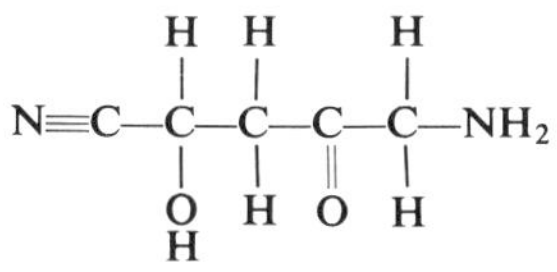

while at least four structures involving rings can be written quite quickly:

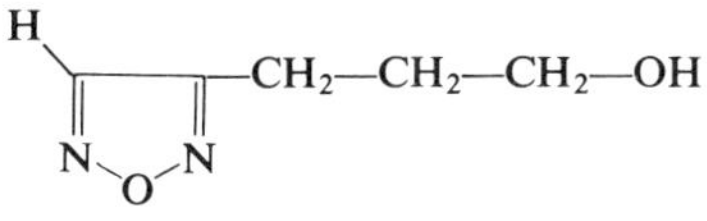

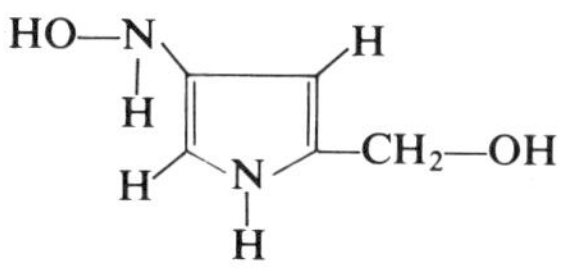

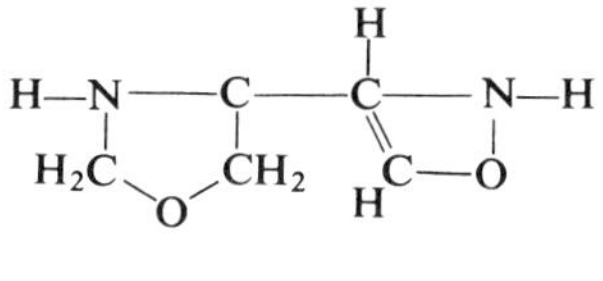

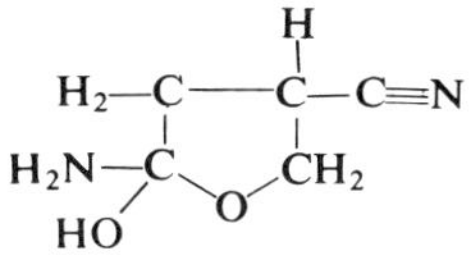

The organic chemist can consider the likelihood of existence of these various species, and if the person submitting the sample considers them, a number may immediately be ruled out as impossibilities, while others may be added. These suggested structures all have one thing in common, they do have a total of rings plus double bonds equal to 3. Note that the cyano group represents two double bonds for the triple bond.

To exercise the reader a bit further, suppose now that the molecular ion is 129 instead of 128. By the nitrogen rule this should contain an odd number of nitrogen atoms, say one, three, five, etc. For a first trial assume only one nitrogen. Subtracting 14 from the 129 leaves 115 mass units. If a hydrocarbon is a possibility, we could have either C_9H_7 or C_8H_{19} (which is an impossibility) to cover this mass 115. Here the degree of unsaturation is obvious, so the rings plus double bonds may not be of much assistance. Quinoline is an obvious possibility. The calculated $R = 9 - 7/2 + 1/2 + 1 = 7$, and there are two rings and five double bonds present in the system. Note that the carbon isotope ratio would lead to about 11% on the P + 1, not in agreement with the presumption of a low percentage of P + 1 intensity. Trying the case for three nitrogen atoms leads to $129 - 42 = 87$ for the hydrocarbon balance. Clearly $C_7H_3N_3$ (I) and $C_6H_{15}N_3$ (II) are the only reasonable formulas. Thus for the former (I),

$$R = 7 - 3/2 + 3/2 + 1 = 8$$

while for the latter (II),

$$R = 6 - 15/2 + 3/2 + 1 = 1$$

Using the R of 8 leads to structures for formula I of the types shown:

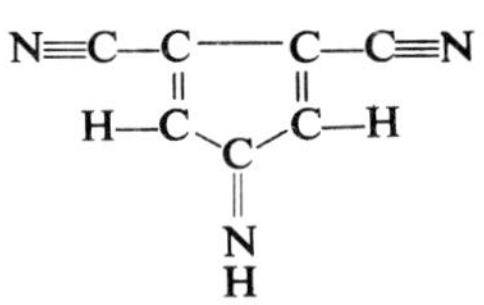

Ia or Ib or Ic

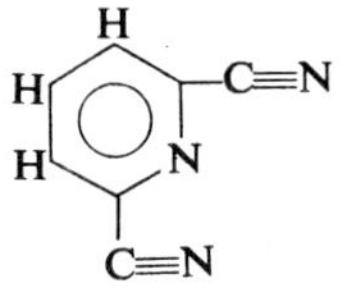

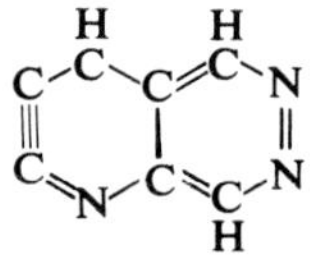

Again, reality is to be judged by the organic chemist, not by what can be drawn. For the case in which R is 1 the formula of $C_6H_{15}N_3$ (II) being obviously rather saturated leads immediately to the following very reasonable structures. Lastly, still using the MW 129 and assuming one nitrogen and two oxygen

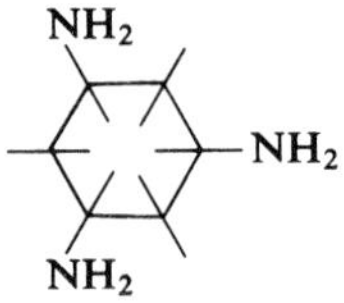

IIa or IIb

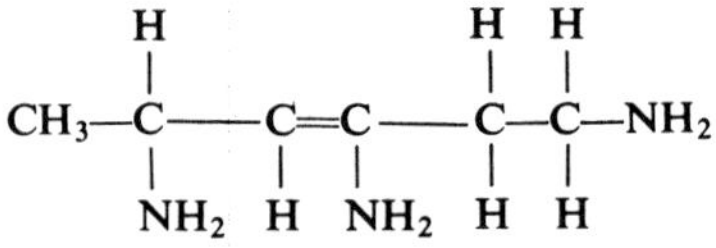

atoms, assume the molecular formula to be $C_xH_yN_1O_2$. Since $NO_2 = 46$, subtract this from 129 and obtain 83. The hydrocarbon portion can then be $C_6H_{11} = 83$, but not C_5H_{23} (which is impossibly supersaturated) or C_7H_{-1} (which is not allowable in the chemists' game). So the formula is $C_6H_{11}NO_2$. Applying the calculation of $R = 6 - 11/2 + 1/2 + 1 = 7 - 5$ or 2 leads one to consider the formula,

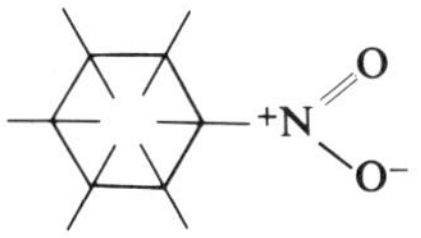

The behavior of the nitro group should be noted. One could also have a structure such as

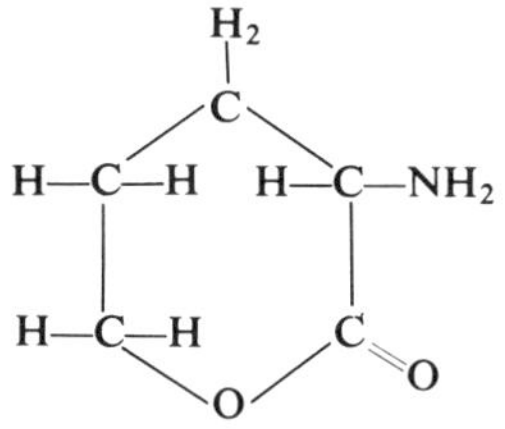

There are many other possibilities and one should always remember that someone else can add an additional structure to those that have been accumulated.

McLafferty (1966c) pointed out that this formula leads to a value ending in 1/2 for some ions such as $C_6H_5{-}C{\equiv}O^+$ and hence the true value of R is found by subtracting the 1/2. The use of the formula is based upon the lowest valence states of each element. It does not count double bonds formed to elements in higher valence states. A general formula may be written as $\mathrm{I}_a\,\mathrm{II}_b\,\mathrm{III}_c\,\mathrm{IV}_d$, where I = H, F, Cl, Br, etc. (any element of valence 1), II = O, S, Se, etc., III = N, P, As, etc., IV = C, Si, etc., and a, b, c, and d are the integer numbers of each atom in the formula. Note that this allows one to add the number of atoms of the same valence in a molecule, i.e., sulfur and oxygen or carbon and silicon. Beynon (1960b) has shown examples of the R calculation carried to rings as large as coronene and metal-free phthalocyanine.

VI. Deriving Qualitative Information

Each spectrum has a few vital peaks which are of greatest diagnostic value. However, the complete spectrum has to be considered in deriving qualitative information. The information provided by the experimental spectrum must be carefully combined with the information previously known about the sample. For many samples the mass spectral data will not provide an unequivocal assignment of a specific structure. The following discussion will provide a general guide for reporting qualitative information from the interpretation of an unknown spectrum. Great care should be taken in examining an unknown spectrum so that gross errors are not made in identifications. The identification of a completely unexpected component in a mixture or an unexpected structure can become an exciting analytical aspect of mass spectrometry.

A. The Use of Collections of Spectra and Indexes

Several current catalogs and compilations of mass spectral data are available (see Appendix III). These are useful in the identification of unknown substances for which spectra have been published and when the approach to identification is to be empirical. The empirical approach is used if little is known about the composition of the sample and the mass spectrometrist has a spectrum consisting of several intense peaks. The use of these indexes will vary with the problem. To illustrate their uses, examples will be given. For the most part these examples are from actual problems, modified only slightly to enhance their value in illustrating the use of indexes and the collections of mass spectra.

Consider a spectrum of an unknown mixture with two intense peaks above mass 65, one at m/e 104 and another at m/e 76 (of nearly equal intensities). The reason for examining peaks of higher mass numbers is because such peaks are usually of more diagnostic significance than peaks of lower mass numbers. Several starting points are possible, and there is usually no clear-cut selection of which index to use first. Thus the order of the steps taken here should not be considered as necessarily the fastest route to identification.

For the present case the compilations of Cornu and Massot (1966, 1967) are first investigated in the identification of a substance giving intense peaks at m/e 104 and 76. The fragment-ion section of this index lists five spectra as having the most intense peaks at m/e 104 and 76 (four in the 1966 volume and one in the 1967 supplement). It should be recalled that in this example the substance is known to be a mixture, and thus the spectrum may have peaks other than m/e 104 and 76 of higher intensities. The reverse is not true, however. For example, one of the five possible spectra listed shows a peak at m/e 31 of an intensity half that of the 104 peak. If the mixture spectrum does not have a peak at m/e 31 of such an intensity, the compound listed has to be ruled out as a possibility.

It may appear that the selection of reference spectra has narrowed to three in the identification of the 104 and 76 peaks. However, additional probing of the data will often produce additional valuable information. In this case the spectra of phthalic acid and phthalic acid anhydride show peaks of nearly the same relative intensities at *m/e* 104, 76, 74, and 60, and these ratios are nearly the same for the sample spectrum. This could lead to the "conclusion" that the sample contained either phthalic acid or phthalic acid anhydride and that these two compounds gave very similar fragmentation patterns. This possible "conclusion" is stated to illustrate the point that gross errors can be made when only the empirical approach is taken to identification of an unknown spectrum. The spectrum of phthalic acid was taken from "uncertified" mass spectral data Gohlke (1963), while the spectrum of its anhydride was found in the collection of selected high-quality mass spectra in the catalogs of the American Petroleum Institute Research Project 44 and the Thermodynamics Research Center Data Project (formerly known as the Manufacturing Chemist Association Research Project). In checking the literature, Budzikiewicz *et al.* (1967d) discussed the fragmentation of phthalic acid, and an inconsistency was found between the intense peaks listed in the "uncertified" spectra and that of another spectrum of phthalic acid (Beynon *et al.*, 1965). The Cornu and Massot compilations (1966, 1967) list "uncertified" spectra, as well as spectra from other sources. (An attempt had been made to remove spectra which were not high quality in the "uncertified" list, but this becomes very difficult). In this case, the phthalic acid dehydrated in the inlet system of the mass spectrum to the anhydride, and the spectrum listed as phthalic acid in the "uncertified" spectra was primarily that of the anhydride. The point here is that not all of the spectra listed in indexes are reliable because often the data indexed is in error. Errors of this type have been a difficult problem to resolve in the release of reference mass spectral data. Because of this problem, some of the new sources of spectra are now being carefully checked by experienced mass spectrometrists. The corrected data are published in journals such as the *Archives* (quarterly publication, editors: E. Stenhagen, S. Abrahamsson, and F. W. McLafferty). An index published by the American Society for Testing and Materials, Philadelphia, Pennsylvania (ASTM Special Technical Publication No. 356), lists only phthalic acid anhydride as having the most intense peak at *m/e* 104 and a peak of nearly the same intensity at *m/e* 76. A "conclusion" is easier to arrive at when one compound is listed rather than several, but it should be remembered that smaller indexes may leave out many possible structures.

Another useful index of structurally significant peaks and common elemental compositions is an American Chemical Society publication authored by McLafferty (1963b). Had that index been used and had the composition of the 104 peak been determined by high-resolution mass spectrometry to be C_7H_4O, a compound having a CO group attached to a phenyl group would have been found.

Spectral indexes and collections of mass spectra will very likely be greatly enlarged and improved upon in the future with the use of computers. Individual approaches as to the method of indexing may then become possible. One example would be indexing by the last few peaks of a spectrum above a given percent of the total ion yield.

Collections of spectral data are also useful for the identification of unknown substances. The usefulness depends upon the organization of the spectral index within any given laboratory. For example, when a sample of branched-chain alkanols (alcohols) is examined, it is convenient to have all available spectra grouped together and not in several different notebooks. Several of the classes of compounds listed in the American Petroleum Institute (API) and the Thermodynamics Research Center (TRC) have been submitted at one time to increase their usefulness as well as to group similar classes of compounds together. Nineteen branched-chain alcohols, all of a molecular weight of 130, are listed in a series, API-44–TRC Serial Nos. 44-m through 61-m. The identification of an unknown isomer or compound can be made directly from such a series of spectra more readily than can be arrived at from a correlation study. Because it is not always convenient to group all spectra of one type of compound together when publishing, it becomes desirable to devise an indexing system within a given laboratory to accomplish this. For example, 13 alkyl indanones were published in API–TRC in December 1966; three additional indanones were published in 1968. It is convenient if all alkyl indanones are grouped together in the files of spectral data within a given laboratory when identification work is undertaken.

B. The Empirical Approach and Fragmentation Principles

One approach to an identification of an unknown substance from its mass spectrum is a combination of empirical information (Capellen *et al.*, 1970a, 1970b; Stenhagen *et al.*, 1969; Tatematsu and Tsuchiya, 1966; Zwolinski and Wilhoit, 1968) and a knowledge of fragmentation. Usually a few intense peaks are considered on an empirical basis with the remaining structure features established by fragmentation rules. A few examples will illustrate this type of procedure.

Consider a mass spectrum with an intense molecular ion peak equal to 10% of the total ion intensity. From spectral compilations an aromatic or a compound with a conjugated system or cyclic structure is indicated (not an alkane or a compound with a long alkyl chain). Several structures given by the index should be possible within the fragmentation principles. A spectrum of a hetero compound which has a moderately intense molecular ion peak is more likely to contain a double bond attached to the hetero atom than a single bond attached to the hetero atom. For example, 2-heptanone has a molecular ion with an intensity of about 2% (defined here as moderately intense) of the total ionization, while the molecular ion of 2-heptanol is not more than 0.2% of the total ion yield. The fact that a double bond in a molecule has the effect of producing a more intense molecular ion peak should be one of the many rules to keep in mind when combining fragmentation knowledge with information empirically derived.

Rationalization of the observed ion fragments combined with the empirical approach can be illustrated by considering a spectrum known to be a hydrocarbon and with the most intense peak at *m/e* 57. The peak at *m/e* 57 can be assumed to have the formula C_4H_9. Why is the peak at *m/e* 57 more intense than any of the others in the C_nH_{2n+1} series? The answer may be that the unknown substance has a *t*-butyl structure or is branched at a carbon (such as 3,5-dimethylheptane), which would give a C_4H_9 fragment. Mass spectrometrically, the *t*-butyl ion has a high degree of stability and is thus an intense peak in the C_nH_{2n+1} series.

The prominent peaks should, if possible, be considered as a group rather than individually. For example, *the three most intense peaks* for 2-butyl acetate occur at *m/e* 43, 56, and 87. From the indicated molecular weight of 116 and isotopic peaks indicative of only CH and O, then with the difference between it (116) and 87 being 29, the structure probably contains an ethyl group. If the structure is simple, such as 2-butyl acetate, the location of branching becomes rather simple to deduce. In this case, it is highly probable that branching is at the second carbon on the four-carbon chain. Note that the site of branching aids in the formation of an intense peak at *m/e* 87, which in turn points to the ethyl group.

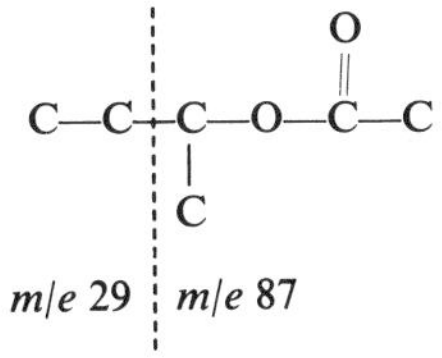

```
             ¦         O
             ¦         ‖
      C—C ¦ C—O—C—C
             ¦  |
             ¦  C
  m/e 29 ¦ m/e 87
```

C. The Use of the Ejection of Neutral Species

The most intense peaks in a spectrum are the most conspicuous, and there often exists a desire to note the mass differences between these intense peaks. Differences between intense peaks which correspond to ejection of small neutral species, if evident, can be important in identifying an unknown substance. Beynon *et al.* (1968a) have tabulated values of neutral species lost from the molecular ion for typical classes of compounds. A "neutral loss" table is given in Chapter 5 of this text.

One rather simple example to illustrate the use of ejection of a small neutral species can be made with the spectrum of 6-methyl-5-hepten-2-ol (API-44–TRC No. 70-m), which has intense peaks in the high-mass region of the spectrum at *m/e* 128, 110, and 95. The difference between these mass peaks are 18 (128 – 110) and 15 (110 – 95). A difference of 18 mass units suggests the loss of water from the molecular ion at *m/e* 128. The difference of 15 mass numbers suggests the loss of a methyl group which was readily cleaved during electron impact. It could then be tentatively concluded that the unknown has an OH group (such as an alcohol) and at least one methyl group. Several similar examples have been cited by McLafferty (1966b).

It should be remembered that not all organic compounds have been examined by mass spectrometry and that not all recorded spectra have been published or included in the various compilations of mass spectra. Spectra of unrelated compound types can show similarities, and these can lead to serious errors in identification of unknown substances. For example, a mass spectrum may consist of what appears to be solely fragment ions of hydrocarbons and actually may not be from a hydrocarbon compound at all. Mass spectra of 1-chlorooctane and 2-chlorooctane (API-44–TRC Nos. 75-m and 76-m, respectively) illustrate this point. The spectrum of the 1-chlorooctane has its most intense peak at *m/e* 91. This peak could bring to mind an aromatic compound having a benzene ring with at least one carbon in an alkyl group. If such a spectrum also has intense peaks at *m/e* 41, 43, 27, 55, 29, 57, 39, and 69, one could use them as supporting evidence for an identification of a hydrocarbon compound. The fact that the tenth most intense peak in this spectrum is at *m/e* 93 could easily be overlooked in an attempt to make a rapid identification. (The peak at *m/e* 93 is the isotopic peak of ^{37}Cl.) It is clear from this example that a check of the "isotope profile" should always be made for the major ions found in an unknown spectrum.

It should be emphasized that because of the large amount of information in a single mass spectrum, there exists a temptation to select only those peaks which will fit a preconceived identification (or those which will fit into the scope of knowledge of the interpreter). By yielding to such temptations, mass spectrometry is denied one of its most valuable assets, i.e., the identification of unexpected components. High-resolution mass measurements would have shown, in the previous example, that the peak at *m/e* 91 had the elemental composition of C_4H_8Cl (typical of terminal chloroparaffins) rather than C_7H_7 (typical of aromatics).

The second illustrative case (API-44–TRC No. 76-m), the spectrum of 2-chlorooctane, has most of its intense peaks in the C_nH_{2n} and C_nH_{2n-1} series of ions. Such a spectrum would thus have the general appearance of an olefinic type of hydrocarbon (alkene) or a cycloparaffin. In fact, the intense peak at *m/e* 112 has the appearance of the molecular ion of a C_8H_{16} compound.

Indeed, the elemental composition of the 112 peak is that of a hydrocarbon, and high-resolution mass spectrometry would not have aided in the identification of a chloroalkane by an accurate mass measurement of the 112 peak. It is of interest to note that these two spectra illustrate the fragmentation difference between a terminally substituted chloroalkane (intense peak at m/e 91) and an internally substituted chloroalkane (intense olefinic ions). Of importance also is the fact that because of the large amount of information given by an individual peak, there existed the possibility of using part of the information upon which to base a conclusion.

This approach to identifications is not infallible, as shown by Budzikiewicz *et al.* (1967c). It is not intended here to imply that the use of masses resulting from ejection of neutral species removes the need for understanding fragmentation reactions based upon the mechanistic approach.

D. Confidence in Identification

Today mass spectrometry is recognized as a powerful tool for organic structure elucidation, along with nuclear magnetic resonance and infrared spectroscopy. The pioneering work of the group at the Dow Chemical Co. in the 1950's, inspired by McLafferty, did much to show the usefulness of mass spectrometry for structure determinations. Stenhagen and Ryhage in Sweden did impressive work in structure identifications, as well as Biemann and his collaborators at M.I.T., Cambridge, Massachusetts. Reed's group in Glasgow and Beynon's associates at Manchester plus several other hard-working individuals in various laboratories developed much confidence among scientists in the technique.

Some of the lack of confidence in the early years by those not knowledgeable in fragmentation of molecules under electron impact may have arisen because empirical rules of fragmentation processes were often not sufficiently specific or well organized. Rules were generalized to a point that a remark was once made that "to make a structural identification was like finding which buckshot from a shotgun killed a certain bird." Confidence in identifications was too often more a case of earned confidence in the individual doing the interpretation rather than in the technique.

The future of acceptance of structural identifications using mass spectrometry is bright. Fragmentation routes are becoming better understood and applied. Applications are widely published today. High-resolution mass spectrometry has added a measure of confidence which is easily accepted because of the experimentally determined accurate mass measurements. Confirmation is often more easily gained of a ketone structure from the experimentally determined 57 peak of 57.0215 with an elemental composition of

C_2H_3O than from the intense characteristic 58 peak resulting from transfer of a hydrogen atom.

Additional confidence in structural identification by mass spectrometry is gained from agreement with other techniques when very complex mixtures are examined. One example is an investigation undertaken to identify nitrogen, oxygen, and sulfur compounds found in catalytically cracked distillate fuel (Davis *et al.*, 1963). The objective of this investigation at the Bureau of Mines at Bartlesville, Oklahoma, was the identification of classes of hetero compounds for the satisfactory treatment for their removal in these distillate fuels. In this investigation, nitrogen-containing concentrates gave mass spectra with peaks at m/e 130, 144, and 158, which indicated the possible presence of naphthyridines, cinnolines, or phthalazines. This information was supported by infrared data and by the combination of the two techniques. It was concluded that naphthyridines were more likely present than cinnolines and phthalazines because of the lack of strong N=N bands in the infrared spectrum.

In practical applications identifications are often made by mass spectrometry which are not expected. Confidence in the identification is often questioned. How could that be in these samples? For example, consider an identification of silicon tetrafluoride in a sample expected to be free of silicon. It may require that the original laboratory equipment be investigated before confidence is gained and then only after finding that the stoppers used in the apparatus were partly silicon.

VII. Analysis of Mixtures

The applicability of mass spectrometry to the analysis of gas mixtures was demonstrated by Hoover and Washburn (1940) with equipment in the laboratories of Consolidated Engineering Corp. It was indicated then that such instruments would not be generally available for routine purchase for several years, and the method of interpretation of mass spectra was complicated. Other methods then known for gas analysis were, however, time-consuming and difficult to carry out. A national defense program at that time made the possibility of rapid analysis of hydrocarbon mixtures attractive. Progress in analytical mass spectrometry was based upon the remarkable reproducibility shown by mass spectra of volatile components in complex mixtures (Washburn *et al.*, 1943, 1945). Methods of calculation were being computerized in the early 1950's. Today, quantitative analysis of complex hydrocarbon mixtures continues to be a supporting aspect of many mass spectrometry industrial laboratories. For this reason, calculation details will be presented with illustrations.

It is assumed that the mass spectrometer to be used has been evaluated and found to meet operational standards. A general evaluation procedure is presented in the ASTM Standards (ASTM, 1968). Similar considerations also have been covered by Reed (1966).

A. Quantitative Analysis without a Matrix

Qualitative analysis has to precede quantitative analysis. Often the mass spectrum provides the qualitative information and, in fact, that stepwise method of calculation (without a matrix) is in itself a series of qualitative analyses.

Mass spectrometry requires, as do other spectrometric procedures, that some feature or combination of features shown by the complete spectrum be unique to each of the components in a mixture. In a simple mixture, this could merely mean the selection of a peak unique to a given component, sometimes referred to as the "analytical peak." The peak height from the sample spectrum then has to be adjusted with the peak response per unit of sample material. This response factor is usually referred to as the sensitivity of a given compound. Different compounds have different sensitivities. For example, in a mixture of methane and nitrogen, the peak at m/e 16 is unique to methane and the peak at m/e 28 is unique to nitrogen; however, equal peak heights do not mean equal amounts of these two components in a mixture. Relative sensitivities are adequate (absolute sensitivities can be made relative). Consider, for example, a mixture of methane and nitrogen where the peak at m/e 16 is from methane and the peak at m/e 28 is from nitrogen and each measures on the sample spectrum 100 scale divisions. On the same mass spectrometer on the same day, it was found that a pure standard sample of methane gave 48.7 divisions of peak height for each micron of sample pressure, while a pure standard sample of nitrogen gave 61.8 divisions per micron of pressure. The measured sample peak heights for each component identified in the mixture are divided by the sensitivity obtained from a pure standard. These adjusted partial pressures are normalized to a given percent, which is usually 100%. Thus, the percent methane from 100 scale divisions is 47.7%, while the percent nitrogen from the same number of divisions is 52.3%.

Most mixtures are too complicated to expect to find a single peak unique to each given component. Overlapping contributions to analytical peaks from different components in a mixture are almost always found in practical problems. This overlapping has to be removed in the calculation. This can be done in a variety of ways. One approach to a quantitative analysis is to use a stepwise calculation method. Another common method is to set up a series of linear simultaneous equations relating peak heights with sensitivities for selected peaks. Stepwise calculations are usually carried out on a desk calculator and, because of simplicity, will be discussed first. The stepwise

TABLE 6-2

STEPWISE CALCULATION OF A MASS SPECTRUM[a]

		1-Dodecene		1-Undecene			1-Decene	
m/e	Gross intensities from sample spectrum	Reference intensities from API Serial No. 15-m	Calculated intensities using API Serial No. 15-m	Reference intensities from API Serial No. 14-m	Calculated intensities using API Serial No. 14-m	Reference intensities from API Serial No. 13-m	Calculated intensities using API Serial No. 13-m	Calculated intensities
125	47.8	0.374	27.7	0.328	21.5	0.027	1.8	51.0
126	64.4	0.485	35.9	0.497	32.6	—	—	68.5
140	100.7	0.382	28.3	—	—	1.093	72.4	100.7
141	11.3	0.045	3.3	—	—	0.124	8.2	11.5
154	58.3	—	—	0.890	58.3	—	—	58.3
155	6.8	—	—	0.104	6.8	—	—	6.8
168	59.0	0.797	59.0	—	—	—	—	59.0
Ratio factors			74.0		65.5		66.2	
Percentages			36.0		31.8		32.2	

[a] Underlined peak is used as the analytical peak for that substance.

method aids in understanding the more complex linear equation method or other computation methods. Today, in larger oil and chemical companies, most of the mass spectrometric calculations are carried out by various computer techniques.

The stepwise calculation is easily described by a simple example. Consider the mixture to consist of 1-olefins in the C_{10}, C_{11}, and C_{12} range. A partial spectrum of a mixture is shown in columns 1 and 2 of Table 6-2 as it is tabulated from an experimental mass spectrum. The procedure consists of the following steps:

1. Peaks to be used in the calculation are selected in advance. In this case, the molecular ions of the C_{10}, C_{11}, and C_{12} α-olefins were selected (*m/e* 140, 154, and 168, respectively) as the *analytical peaks*, as well as certain fragment ions which are useful as check peaks (125, 126, 141, and 155).
2. Reference standards for each of the compounds present in the sample are obtained. Spectra for the compounds in this mixture can be found in the API "Catalog of Mass Spectral Data" under Serial Nos. 13-m, 14-m, and 15-m. However, whenever possible, a laboratory should calibrate from reference spectra obtained on the same mass spectrometer used to obtain the sample spectral data.
3. The calculations are started with the component having the highest *m/e* selected as unique to that compound. One component at a time is removed from the mixture spectrum by subtracting contributing amounts from the heavier components to the higher components. This is sometimes referred to as the "artichoke" method because one "peels" out the peaks associated with a given compound.

The peak selected for 1-dodecene is *m/e* 168, the highest of the three molecular ions (140, 154, and 168). Its intensity is 59.0 units. The 59.0 units are recorded opposite *m/e* 168 and under column 4 in Table 6-2. The reference standard (API Serial No. 15-m) provides the coefficients for calculation of the relative intensity values of fragments of 1-dodecene (*m/e* 168). This is done through calculating a ratio factor by dividing 59.0 (gross intensity) by 0.797 (reference intensities expressed in percent total ionization), which is, in this case, 74.0. The value obtained is called a *ratio factor*, which is used later to calculate the percentages of each component present. Each of the other reference intensities for 1-dodecene (from API Serial No. 15-m) is then multiplied by this ratio factor to get the calculated intensities (3.3 for *m/e* 141, 28.3 for *m/e* 140, etc.). No overlap exists between the fragmentation of 1-dodecene and 1-undecene; therefore, the intensity of 58.3 is transferred to the *calculated intensity* column, and ratio intensities of fragments are calculated as done for 1-dodecene (58.3/0.890) to obtain a ratio factor of 65.5. For the component 1-decene, with a gross intensity value of 100.7, a value of 28.3 is subtracted because of the overlap from 1-dodecene, leaving 72.4 (underlined) as the component value of 1-decene. Again, ratio intensities of fragments are calculated from a standard reference with 72.4/1.093. The last column shows the calculated intensities which can be compared to the experimental values listed as gross intensities. The value of 47.8 was recorded on the experimental spectrum for *m/e* 125, which compared fairly well with the calculated intensity of 51.0 (sum of intensities of the calculated fragments from 1-dodecene and 1-undecene).

4. The ratio factors calculated in step 3 give the relative percentage of each of the components when they are normalized. For the example shown, the derived percentages are shown below the ratio factors in Table 6-2.

The above example is perhaps oversimplified, but this means of calculation has been applied to hydrocarbon mixtures containing up to 30 components. The mathematical manipulations are, however, cumbersome using a desk calculator. The ratio factors used to calculate percentages were somewhat of a special case. Percent total ion yield of reference standards was used as sensitivity factors. Other means of sensitivities could have been used. Judgment must also be applied in making approximations in certain of the calculations. A considerable amount of qualitative work is done by the artichoke method, since it can be often done easily on a desk calculator and involves simple tasks such as subtracting the contribution to peaks of a solvent. Often one component is easily recognized, and it then becomes a simple matter to determine if other components are present by finding those peaks which have intensities in excess of a standard reference spectrum.

B. Quantitative Analysis with a Matrix

Methods for doing quantitative calculations have been presented by Barnard (1953), Dibeler (1963), Reed (1966) and Tunnicliff and Wadsworth (1965). These methods all involve a set of n simultaneous linear equations which require knowing the components in the mixture (c), the peak height (H) of an individual component peak (i), and sensitivity coefficients (S_{ic}) for the components in the mixtures. The unknowns to be determined are the partial pressures (p_c) of each of the components. Assuming Dalton's law, the sum of the individual partial pressures is considered as the total pressure of the sample loaded. (This usually can be experimentally checked if a particular sample causes problems.) The series of linear equations becomes

$$\begin{aligned} S_{11}p_1 + S_{12}p_2 + \ldots + S_{1n}p_n &= H_1 \\ S_{21}p_1 + S_{22}p_2 + \ldots + S_{2n}p_n &= H_2 \\ \cdots\cdots\cdots\cdots\cdots\cdots\cdots\cdots\cdots\cdots\cdots \\ S_{n1}p_1 + S_{n2}p_2 + \ldots + S_{nn}p_n &= H_n \end{aligned}$$

The above calculations imply that reservoir pressure measurements were made (at least in the determination of sensitivity coefficients for the pure standards) or that some other measurements of sample size were used, such as total ionization for high molecular weight hydrocarbons, milligram of sample or microliter of liquid material charged. Ruth (1968) has derived a general solution applying the standard mixture method in cases where samples are introduced directly into the ion source by means of a probe and a standard mixture method of calculation can be applied. However, in many research problems requiring probe loading, pure substances for blending a standard mixture are nonexistent, thus reducing the analysis to a semiquantitative determination.

The selection of peaks from a mass spectrum which should be included in a matrix is important. Consider the quantitative analysis for five isomeric hexanes in the fashion shown later in this chapter; the question of *what peaks could be used* arises. The partial spectrum presented in Table 5-4 (Chapter 5) shows all of the largest ion peaks in the original spectra. A selection of suitable peaks would be difficult in light of the contributions to several of the large peaks. The probable selection would be m/e 86 for *n*-hexane, 72 for 2,2-dimethylbutane, 42 for 2,3-dimethylbutane, 56 for 3-methylpentane, and 43 for 2-methylpentane, despite the objection that the other components all make sizable contributions to the 43 peak. This, in a sense, is a good example of how the 2-methylpentane could be discriminated against during a component analysis. One would probably find that component in the largest error from a known mixture, while the 2,2-dimethylbutane would come out to the best value. It is assumed here that calibrations on the pure compounds would be used in obtaining the matrix coefficients.

In practice, the desk calculator manipulations for inversion and solution become too cumbersone and costly for matrices greater than five or six components. Today computer programs are in common use in mass spectrometer laboratories. These are often set up for the person doing mass spectrometer calculations. Computer programs are discussed in much more detail in Chapter 8. However, the effective use of computerized calculations is dependent upon an understanding of the method of calculations. The matrix calculations are given by a somewhat more general approach in the next section.

C. Type Analyses—Semiquantitative by Summation of 70-eV Peaks

Type analysis methods are used on complex mixtures consisting of various types of classes of organic compounds. These methods require the selection of peaks which are characteristic for a given class of compounds. For example, the C_nH_{2n+1} series of peaks (*m/e* 71, 85, 99, 113, etc.) are among the most intense peaks of pure alkanes. These peaks are therefore referred to as *characteristic peaks of alkanes*. Pure alkanes have, however, peaks of smaller intensities in other series, such as the C_nH_{2n-1} series (*m/e* 69, 83, 97, etc.), which are characteristic of cycloparaffins and olefins. The overlapping which exists between the different classes of compounds has to be removed in the calculations. The selection of characteristic peaks for any given set of samples should be made by a careful examination of mass spectral data of pure compounds which relate to the samples being analyzed.

Sensitivity coefficients are required to relate intensities sums of characteristic peaks to each of the types of compounds making up a mixture. Such sensitivities are the peak responses per unit of material loaded into the mass spectrometer. Sensitivity values are applied to the summation of characteristic peaks after removal of the overlap of intensities between different classes of compounds in the same way as used when a unique peak is selected for a single component. Often, in practical problems, relative sensitivities published in the current literature can be used in type analysis calculations; an example is the calibrated matrix inverses published by Gallegos *et al.* (1967). In cases where literature values are not acceptable, sensitivities for hydrocarbons can be derived by total ionization measurements (Chapter 3). The molar volume relationships for high molecular weight hydrocarbons are such that the total ion yield is approximately proportional to the liquid volume of the sample charged to the mass spectrometer. Meyerson *et al.* (1963) have pointed out that concern should be given to the validity of the use of total ionization.

A typical example will illustrate a practical method for type analysis calculations. Consider a complex hydrocarbon mixture for which the following compound types are expected to be present. The characteristic peaks selected for each compound type are those used by Lumpkin (1956), who claims the method can be used to a molecular weight of about 450–500 (see tabulation below).

Types of compounds	Selected characteristic peaks
Alkanes	Σ 71, 83, 99, 113
Noncondensed naphthenes	Σ 69, 83, 97, 111, 125, 139
Condensed naphthenes	
2-Ring molecule	Σ 109, 123, 137, 151, 165, 179, 193
3-Ring molecule	Σ 149, 163, 177, 191, 205, 219, 233, 247
4-Ring molecule	Σ 189, 203, 217, 231, 245, 259, 273, 287, 301
5-Ring molecule	Σ 229, 243, 257, 271, 285, 299, 313, 327, 341, 355
6-Ring molecule	Σ 269, 283, 297, 311, 325, 339, 353, 367, 381, 395, 409

The sum of the characteristic peaks is treated as a single peak in these calculations. The sum of the characteristic peaks of alkanes is referred to as the $\sum$ 71 peak and is the sum of the intensities of the peaks at *m/e* 71, 85, 99, and 113. In addition to knowing the sum of measured peak heights of the characteristic peaks, it is necessary to correct these values by sensitivity coefficients, as well as to remove overlapping intensities between the different characteristic peaks. The values given in Table 6-3 are typical of coefficients

TABLE 6-3

MATRIX FORMAT FOR HYDROCARBON TYPE ANALYSIS

$\sum$ Peak	Alkanes	Noncondensed naphthenes	Condensed naphthenes				
			2-Ring	3-Ring	4-Ring	5-Ring	6-Ring
71	20.15	7.93	0.77	1.37	0.63	0.33	0.21
69	6.15	22.04	9.18	5.68	3.40	2.20	1.66
109	0.08	0.83	19.40	6.16	3.55	2.30	1.80
149	0.02	0.03	0.24	11.60	4.16	2.35	1.56
189	—	0.02	0.04	0.36	7.95	2.37	1.32
229	—	0.01	0.01	0.02	0.13	6.02	1.66
269	—	0.01	0.01	—	—	0.09	4.83

which have been adjusted for differences of sensitivities (underlined values); they also provide coefficients for removing overlap between the summation intensities. The horizontal values represent interference to the peak height from the other components in the mixture. At first this may appear as an impossible task in a stepwise calculation procedure, as illustrated by the previous example. However, judgment in making approximations will give a surprisingly rapid calculation.

The stepwise method will be illustrated for a sample spectrum found to have summation peaks ($\sum$ peak) with measured intensities shown in the tabulation below.

Components	$\sum$ Peak	Intensities measured in scale divisions
Alkenes	71	256.0
Noncondensed naphthenes	69	377.0
2-Ring condensed naphthenes	109	152.0
3-Ring condensed naphthenes	149	61.2

It is now possible in this calculation to set up an individual matrix calculation format. This format is shown in Table 6-4. Only four of the seven possible types shown in Table 6-3 had detectable intensities on the experimental sample spectrum. That part of Table 6-3 which is usable in this example has been transferred to Table 6-4 together with the matrix coefficients to be used. In such a calculation, it is important to realize that for each compound type the calculated values in the sample calculations format have to remain in the same ratios as the coefficients in Table 6-4. Also, the sum of the calculated vertical values should equal the measured values found on the spectrum. The reason for this is that a mass spectrum obtained from scanning a multi-component mixture is the result of the sum or linear superposition of the spectra for all of the individual components present.

TABLE 6-4

SAMPLE SPECTRUM AND MATRIX COEFFICIENTS

				Condensed naphthenes	
$\sum$ Peak	Peak height of sample spectrum	Alkanes	Noncondensed naphthenes	2-Ring	3-Ring
71	256.0	20.15	7.93	0.77	1.37
69	377.0	6.15	22.04	9.18	5.68
109	152.0	0.08	0.83	19.40	6.16
149	61.2	0.02	0.03	0.24	11.60

In Table 6-4, the values on the diagonal will be called the base coefficients ($b_{i,j}$), where $i = j$ and the value of i and j ranges from 1 to 4 (i.e., the base coefficient of the three-ring components is $b_{4,4}$, of the two-ring is $b_{3,3}$, etc.). The peaks sums are designated S_j, $j = 1$–4, where $\sum$ 71 is S_1, $\sum$ 69 is S_2, etc. These definitions apply in the following calculation steps:

1. Start calculations with the height, S_j, listed (the peak 149 of 61.2 intensity units with a coefficient of 11.60 in the above example) and divide it by the appropriate base coefficient found in the above format.

$$S_j/b_{i,j} = t_j$$

Consider the example of the highest S_j as S_4 where the $\sum$ 149 is 61.2 units. Thus,

$$\begin{matrix} i = 4 \\ j = 4 \end{matrix} \qquad \frac{S_4}{b_{4,4}} = t_4 \qquad \frac{61.2}{11.60} = 5.3534$$

2. Multiply each of the three-ring coefficients in the above format, including the three-ring base coefficient, by the t_j value ($t_j \times$ coefficient $= a_{i,j}$). Thus,

$$j = 4 \qquad t_4 = 5.3534$$
$$i = 1\text{–}4$$
$$t_4 \times 1.37 = a_{1,4} \qquad 5.3534 \times 1.37 = 7.2$$
$$t_4 \times 5.68 = a_{2,4} \qquad 5.3534 \times 5.68 = 30.0$$
$$t_4 \times 6.16 = a_{3,4} \qquad 5.3534 \times 6.16 = 32.5$$
$$t_4 \times 11.60 = a_{4,4} \qquad 5.3534 \times 11.60 = 61.2$$

The values at the right above (7.2, 30.0, 32.5, 61.2) are corrected first-order approximations, i.e., second-order approximations.

3. Using S_3 ($\sum$ 109 = 152.0), calculate t_j.

$$t_j = \frac{S_j - (a_{i,j+1})}{b_{i,j}}$$

Thus,

$$i = 3, \; j = 3 \qquad t_3 = \frac{S_3 - a_{3,4}}{b_{3,3}} \qquad \frac{152 - 32.5}{19.40} = 6.1597$$

4. Multiply the two-ring coefficients by the t_j value.

$$j = 3 \qquad t_3 = 6.1597$$
$$i = 1\text{–}4$$
$$t_3 \times 0.77 = a_{1,3} \qquad 6.1597 \times 0.77 = 4.7$$
$$t_3 \times 9.18 = a_{2,3} \qquad 6.1597 \times 9.18 = 56.5$$
$$t_3 \times 19.40 = a_{3,3} \qquad 6.1597 \times 19.40 = 119.5$$
$$t_3 \times 0.24 = a_{4,3} \qquad 6.1597 \times 0.24 = 1.5$$

5. Making successive approximations using the above scheme, it becomes possible to recalculate base coefficients for each of the four compound types. This process of approximations leads to the following values of corrected base coefficients (underlined values on the diagonal) shown in the tabulation below.

$\sum$ Peak	Measured intensity units	Paraffins	Noncondensed naphthenes	Condensed naphthenes	
				2-Ring	3-Ring
71	256.0	<u>155.3</u>	89.3	4.4	7.0
69	377.0	47.4	<u>248.3</u>	52.3	29.0
109	152.0	0.6	9.4	<u>110.5</u>	31.5
149	61.2	0.2	0.3	1.4	<u>59.3</u>

These corrected base coefficient values (underlined) are now summed and normalized to a given percent (usually 100%). The result to be reported for this analysis is shown in the tabulation below.

Component	Percent
Paraffins	27.1
Noncondensed naphthenes	43.3
Two-ring condensed naphthenes	19.3
Three-ring condensed naphthenes	10.3

The example given is concluded with these results. It is important to realize that today computer programs are widely used. One illustration is a simple matrix inversion program (Etter, 1968) which obtains the roots of the simultaneous equations. These unnormalized factors (roots) are then normalized to a given percent expressing the normalized components in a mixture as shown by the inverted matrix in Table 6-5. Computer techniques are covered in greater detail in Chapter 8.

TABLE 6-5

MATRIX COEFFICIENTS FOR AN INVERTED MATRIX

	Summation peaks			
Components	71	69	109	149
Alkanes	1.1246	−0.4101	0.1495	−0.0116
Noncondensed naphthenes	−0.3472	1.1450	−0.5251	−0.2401
Two-ring condensed naphthenes	0.0095	−0.0417	1.0260	−0.5257
Three-ring condensed naphthenes	−0.0012	−0.0003	−0.0126	1.0070

D. CHARACTERIZATION OF COMPLEX MIXTURES

In the characterization of complex mixtures, the mass spectroscopist is seeking to aid the investigator who has no idea which types of compounds may predominate in his material and primarily wants to find directions for further separations that will lead to identification of individual compounds or, at the very least, classes of compounds. In this type of analysis, fragmentation is minimized by using low ionization voltage, thereby emphasizing the molecular ions of the various species present. Stevenson and Wagner (1950) were early workers in this field who pointed to possible analytical applications. Lumpkin and Johnson (1954), Field and Hastings (1956), Lumpkin (1958), and Sharkey *et al.* (1959) all contributed to the development of the technique. Revised and additional sensitivity data have been provided by Reid *et al.* (1966) and Lumpkin and Aczel (1963, 1964).

As an example of the application of LVMS (low-voltage mass spectrometry) to a specific but broad problem, consider the work of Thompson *et al.* (1966) on sulfur compound characterization studies of high-boiling petroleum fractions. This work is only a small portion of the general study of separation and identification of sulfur compounds in petroleum which was conducted by the U.S. Bureau of Mines, Bartlesville Petroleum Research Center, in cooperation with the American Petroleum Institute Research Project 48. In this work sulfur compound concentrates boiling from 225° to 400°C were prepared, and the various fractions from several separational techniques were examined by LVMS to lead to a report of the sulfur compound types detected in these concentrates. The present interest is in the application of LVMS rather than in the laborious process of preliminary separations necessary to reach the stage of having the sulfur concentrates. The details of the separations appear in the above reference and also in the article of Thompson *et al.* (1955). For the purpose here it suffices that a complete and representative distillate boiling from 225° to approximately 400°C from the Wasson, Texas, crude oil was available.

A Consolidated Vacuum Corp. (CVC) brush still was employed to separate the wide-boiling distillates into a series of fractions. Two preliminary distillations were made as a guide to establish operating conditions. The ASTM D158 distillation data and sulfur determinations were used to indicate that reasonable fractionation was being accomplished using the brush still. Although some overlap existed between the fractions, no significant improvement was achieved by conducting the distillation at a slower rate. Despite the desirability of more efficient fractionation, no better equipment or methods are known for the separation, by boiling point, of such high-boiling material needed in a short time and with less danger of thermal decomposition. A total of 18 batch distillations was made, each using a 1.6-liter charge and obtaining seven 200-ml fractions and a residue. Corresponding fractions were recombined to yield about 1 gallon of each fraction. The distillation conditions were a takeoff rate of 120 ml/hour at about 100 μ pressure for the first three fractions and at 5 μ or less for the final four fractions. No evolution of hydrogen sulfide or other signs of decomposition were noted. In the interest of conserving sample, no ASTM D158 distillations were made on the final distillates.

Table 6-6 shows the sulfur–aromatic and paraffin–naphthene content of the seven brush still fractions. Thompson *et al.* (1966) reported gas–liquid chromatograms in their original work, which shows the considerable degree

TABLE 6-6

ANALYSIS OF SEVEN BRUSH STILL FRACTIONS

Fraction	Crude oil (wt %)	Sulfur (wt %)	Fluorescent indicator adsorption analysis (ASTM D1319): Aromatics (vol %)	Fluorescent indicator adsorption analysis (ASTM D1319): Paraffin–naphthene (vol %)
1	3.869	0.974	29	71
2	3.846	1.21	31	69
3	3.919	1.50	34	66
4	4.098	1.72	39	61
5	4.132	1.97	45	55
6	4.122	2.20	57	43
7	4.144	2.14	—	[a]
Residue	4.445	2.34	—	[a]

[a] Paraffin solidified in column—no accurate measurement made.

of fractionation achieved by the brush still separation described above. It remained to make sulfur compound concentrates from these fractions obtained from the brush still because the sulfur content in most crude oil distillates is too low to permit easy identification of the sulfur compounds. The use of liquid–solid chromatography is one of the most effective methods for concentrating sulfur compounds, as has been described by Thompson *et al.* (1955). This was done as shown on Fig. 6-4 to fraction 1 of the brush still distillation. The aromatic–sulfur concentrates were examined by LVMS techniques as described below and found to require further separation. Fractions 2–7 were treated similarly.

The aromatic–sulfur compound concentrates were repercolated through alumina to further concentrate the sulfur compounds. Figure 6-4 also shows a typical treatment for brush still fraction 1. The details of the alumina gel percolation were as follows: Fluid grade H-41 gel of 200 mesh was used with a gel-to-sample ratio of about 20/1. The aromatic–sulfur concentrate was diluted with pentane about 5/1, charged to the column, and eluted with *n*-pentane until no product was obtained. This produced fractions 1–3. Next 8–9 liters of 5% benzene in pentane was used until no product was produced. This yielded fractions 4 and 5. Six liters of ethyl ether in *n*-pentane was used next to produce fractions 6 and 7. Finally, methanol was used to displace the remaining sample, which gave fraction 8. The percent sulfur and the percent original crude oil are given for the appropriate fraction in Table 6-6. These

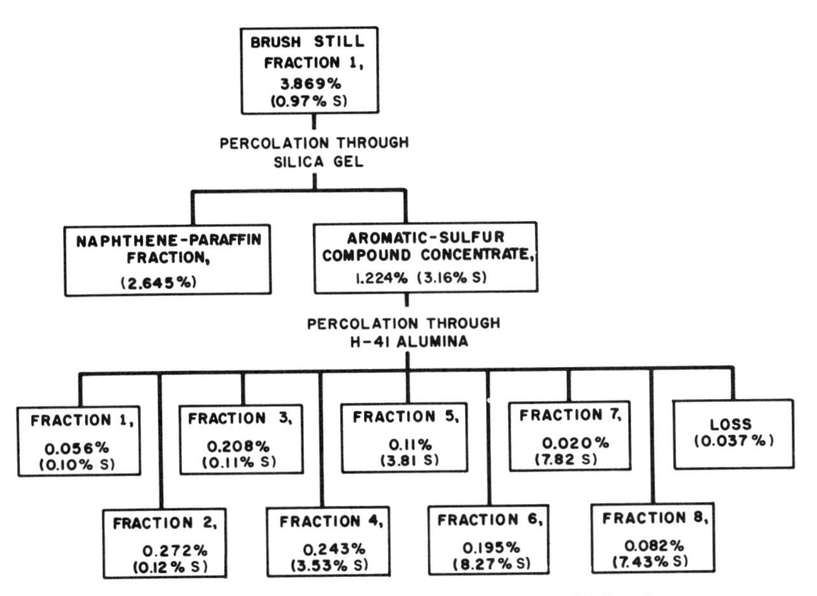

Fig. 6-4. Typical treatment for a brush still fraction.

aromatic–sulfur concentrates were similarly examined by LVMS techniques as described below.

In LVMS characterization work it is well to establish a listing of probable types of compounds thought to be present and thought to be absent, either on the basis of their likely nonexistence or removal to insignificant concentrations by preliminary separations. In Table 6-7 a partial list of the compound types expected to be present is given. This limited list is based in part upon the work of Lumpkin and Johnson (1954) and Clerc and O'Neal (1961) and has been extended by the experience of the Bureau of Mines API Project 48. Some of the compound types believed to be present in each series were considered, but this study was purposely directed to emphasize sulfur compounds. This does not mean that hydrocarbons of types other than the few mentioned are not present. The selected sulfur types listed are given to indicate some of the structural types expected in each series. The LVMS characterization was aided by the knowledge of sulfur concentration data and boiling range for each fraction, once again establishing the value of the use of all available knowledge about samples.

The use of LVMS to characterize petroleum fractions has been reported by others, in particular, Lumpkin and Johnson (1954), Clerc and O'Neal (1961), and Reid *et al.* (1966), and parent peak sensitivities of the aromatic classes in the gas–oil range are available. Relative sensitivities of the parent peaks of compounds are obtained on the basis of liquid volume charges

TABLE 6-7

POSSIBLE COMPOUND TYPES IN THE 225°–400°C FRACTION OF WASSON PETROLEUM

	Hydrocarbon types		Sulfur types	
Series	Name	Typical structure	Name	Typical structure
+2, −12	Naphthalenes	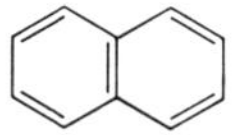	Dibenzothiophenes	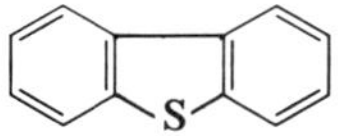
			Bicyclic sulfides	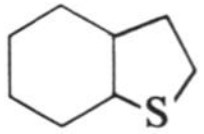
0, −14	Biphenyls	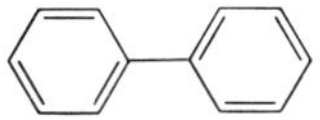	Tricyclic sulfides, thiophenes	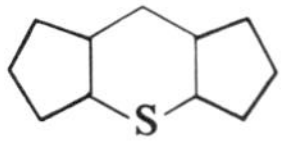 
	Acenaphthenes	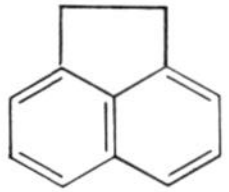	Acenaphthenothiophenes	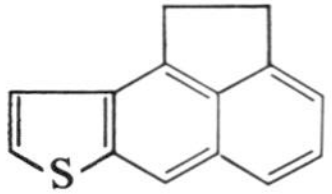
−2, −16	Fluorenes	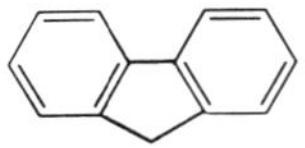	Alkyl aryl sulfides and aryl thiols	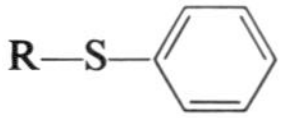 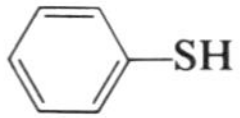
	Acenaphthalenes	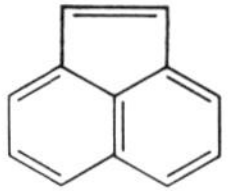	Acenaphthalenothiophenes	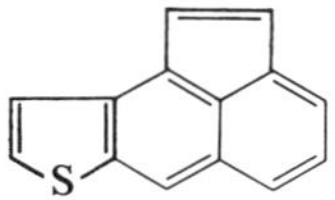

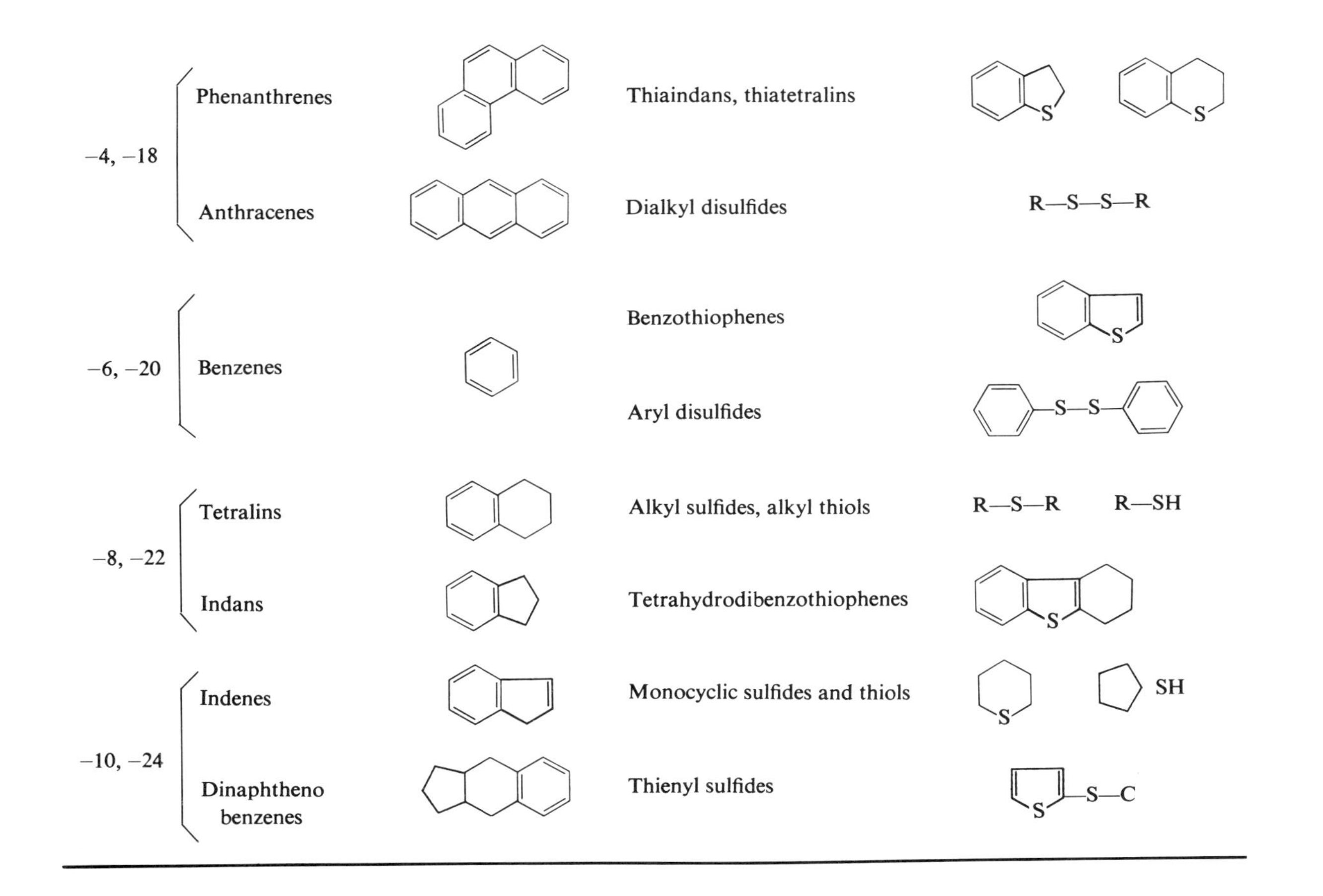

−4, −18
Phenanthrenes
Thiaindans, thiatetralins
S
S
Anthracenes
Dialkyl disulfides
R—S—S—R
−6, −20
Benzenes
Benzothiophenes
S
Aryl disulfides
S—S
−8, −22
Tetralins
Alkyl sulfides, alkyl thiols
R—S—R
R—SH
Indans
Tetrahydrodibenzothiophenes
S
−10, −24
Indenes
Monocyclic sulfides and thiols
S
SH
Dinaphtheno benzenes
Thienyl sulfides
S—C
S

compared to ethylbenzene as an arbitrary standard having a relative sensitivity of 1.0 for the parent peak height obtained on a given instrument for a 1-μliter charge and with the ionizing voltage being specified for the comparison. Published and unpublished relative sensitivity data on aromatic and sulfur compounds were accumulated. These can be adjusted to literature data by comparing key compounds of a given series on the instrument to be used with the literature or those obtained from unpublished sources. These data combined

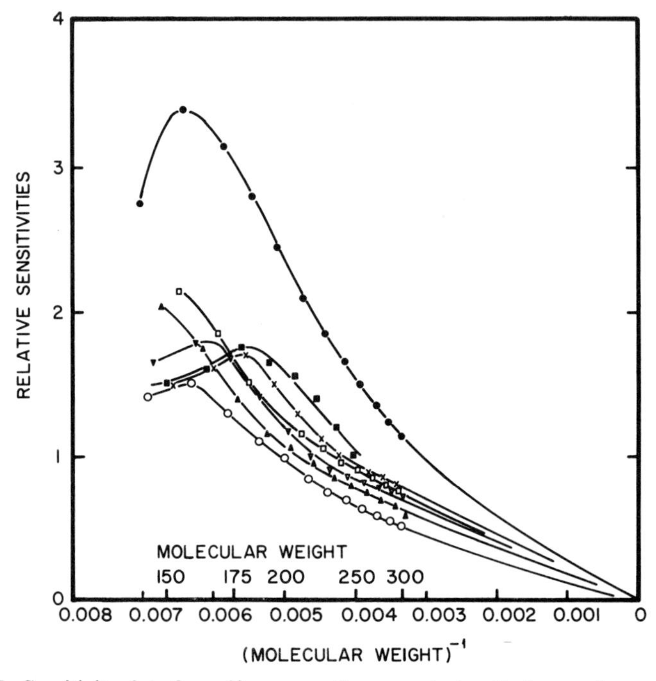

Fig. 6-5. Sensitivity data for sulfur–aromatic concentrates. Series numbers: ●, +2, −12; □, 0, −14; ×, −2, −16; ■, −4, −18; ▲, −6, −20; ▽, −8, −22; ○, −10, −24.

with some sensitivities determined in the Bartlesville Petroleum Research Center led to the determination of relative sensitivity curves for each series of compounds in the aromatic–sulfur system under investigation. The data is shown in Figure 6-5.

Peak heights from the 7-V (meter) scans on the mass spectrometer were too small to allow differentiation of compound types. Almost linear sensitivity increases were observed as the scanning voltage was increased from 9 to 11 V on the meter. Because of the near absence of nonaromatic hydrocarbons (resulting from the preliminary percolation), the use of the 11-V spectra was attractive, since only aromatic rearrangement ions could interfere with the use of the partial fragmentation of the sulfur type of molecular ions to support class identifications. If the nonaromatic hydrocarbons were present to any appreciable extent, the fragmentation resulting from molecular ions of these compounds would be expected to be quite appreciable. It was elected to employ the 11 V (meter) in scanning; this corresponded to 15.0 V (absolute) when the meter was calibrated, using the appearance potential of argon. The LVMS patterns were obtained on a CEC Model 21-102 spectrometer, modified to the equivalent of the 21-103. A CEC heated inlet system was employed at 135°C.

In processing the spectral data, the peak heights of the various *m/e*'s were determined and tabulated by increasing MW within a given series. Each peak height of a given *m/e* within a series was divided by the relative sensitivity shown for the series in Fig. 6-5. For example, a peak at *m/e* 156 would correspond to the total of C_3-substituted naphthalenes, if an aromatic, and hence the peak height would be divided by 3.4 so as to produce a peak height that could be related to the ethylbenzene standard of 1 and thus normalized for volume percent present in the sample. Similarly, the sensitivities shown for the −6, −20 series are for the alkylbenzenes. In cases where sulfur compounds were evident in the series (as shown by a bimodal distribution of peak heights within a given series and isotopic distributions), a sulfur compound sensitivity was used. At the higher molecular weights, these sensitivities did not vary greatly, as can be seen from Fig. 6-5. Hence, it was necessary to apply these special-case factors with care only when one was at the first few carbon numbers of the series. It should be obvious that these sensitivities do not account for all the possible isomers present at a given carbon number. It is an impossible task to obtain information on all the compounds. Hence the concept of class sensitivities, while useful, has limited validity.

Once each observed ion peak has been assigned to a series and corrected by the sensitivity factor, the corrected peak height is entered in the tabulation. The intensities of all of the peaks are summed to give a total peak intensity. Then to determine the percentages of each series member at a given MW, the peak height, p_i, is multiplied by 100 and divided by the sum of the peaks.

The appearance of any odd-integer peaks at these low voltages should be an indication of nitrogen-containing compounds, sensitivities for which were not available at the time this particular investigation was carried out. Despite this deficiency, considerable information was obtained, as is shown later. For such gross mixtures, this form of "characterization" is very useful and, while not semiquantitative in the true sense, deserves a name which indicates the type of information that can be forthcoming from the analysis. In LVMS work, the name characterization is used in this manner. One should note that as the molecular weight increases, so does the molecular volume. Hence all

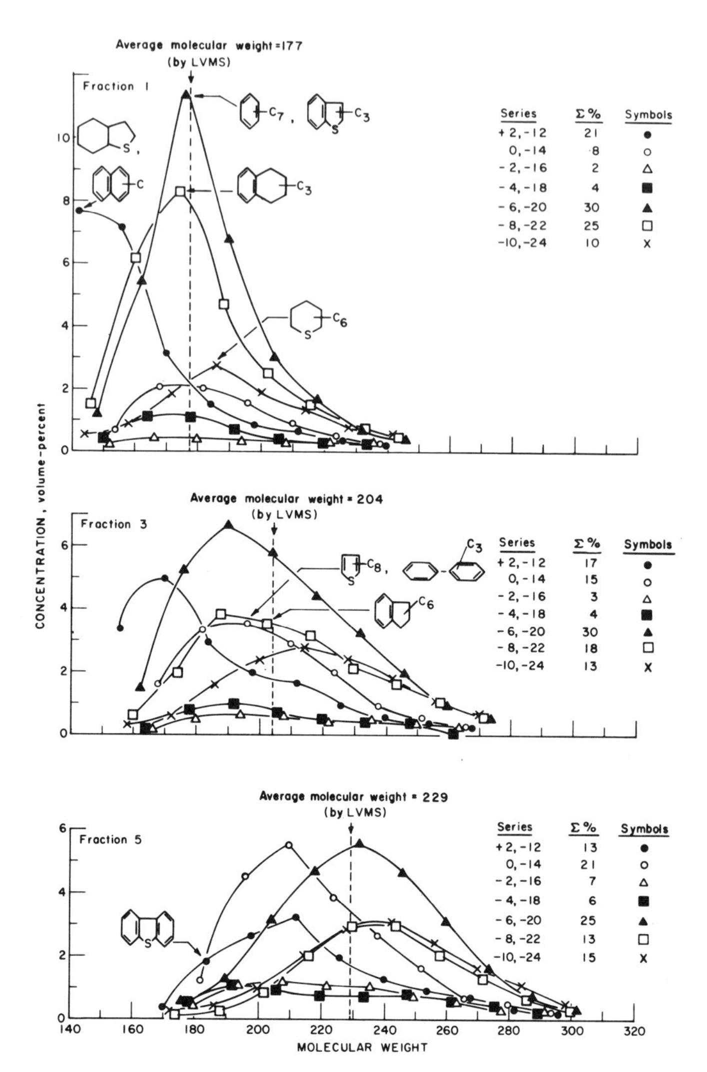

Fig. 6-6. Low-voltage mass spectrometer data on the aromatic–sulfur concentrates from brush still fractions 1, 3, and 5. Structural formulas indicate plausible components. (Courtesy of U.S. Bureau of Mines, Bartlesville, Oklahoma.)

liquid–volume relative sensitivities should approach zero at infinite molecular weight.

The characterization data (sometimes referred to as series profile data) shown in Fig. 6-6 were obtained in this fashion. The percentages found for each carbon number of a series were summed to illustrate the approximate amounts of various classes present. In fraction 1 the alkyl bicyclic sulfides would fall in the +2, −12 series and can be seen as occurring at the higher molecular weight range. The black circle at *m/e* 212 represents a sulfur-containing moiety as the most concentrated class. This bimodal distribution is emphasized even more for the materials observed in fractions 3 and 5. It is not reasonable that even with a distillation such as the brush still gave one should find a distribution of over 12 carbon numbers of the same material. This would call for a boiling range of over 200°C for each fraction, and this was simply not observed in the ASTM 0158 distillations of the fractions. It is more plausible that alkyl sulfides of an average molecular weight of 212 would be boiling in about the same range as naphthalenes of molecular weights 156, 170, and 184. These data reflect the need for further concentration of the sulfur compounds, and hence the alumina gel percolations were made.

One important direct piece of information is also available from these curves. Table 6-8 gives the molecular weight range and the average molecular weight,

TABLE 6-8

MOLECULAR WEIGHTS AND SULFUR CONTENTS OF CONCENTRATES FROM SELECTED BRUSH STILL FRACTIONS

Fraction	Sulfur (%)	Molecular weight	
		Range	Average
1	3.16	142–246	177
3	4.17	156–274	204
5	4.69	170–302	229
7	4.65	184–330	255

as determined by LVMS, of the concentrates produced in the silica gel percolation of brush still fractions 1, 3, 5, and 7.

The average molecular weights of fractions 1, 3, and 5 are calculated values, but the value for fraction 7 is estimated from the −6, −20 and +2, −12 series parent-ion distributions. Fraction 7 was not examined extensively because it was not volatile enough to insure a representative sample at the 135°C inlet temperature employed.

To obtain additional detailed information, brush still fraction 1 was re-chromatographed using alumina gel as described earlier. The alumina gel fractions were examined by both LVMS and GLC. In the original work of Thompson *et al.* (1966), the gas–liquid chromatograms were presented and showed the extent of separation achieved on what was still a very complex mixture. The LVMS data showed this also. In this work the type separations followed the usual alumina gel separation pattern, with aromatics eluting first, followed by increasing concentrations of sulfur compounds. Sulfur concentration data presented with other quantitative data in Table 6-9 support this observation.*

TABLE 6-9

LOW-VOLTAGE MASS SPECTRAL CHARACTERIZATION OF ALUMINA GEL FRACTIONS 1–8

	Concentration (liquid vol %)							
	Fraction							
Series[a]	1	2	3	4	5	6	7	8
+2, −12	—	—	0.6	35.3	59.8	40.3	34.6	14.6
0, −14	2.7	0.7	0.8	4.5	14.7	16.1	13.2	34.5
−2, −16	4.0	—	—	—	—	—	2.1	9.8
−4, −18	2.7	—	—	—	—	4.9	10.5	29.7
−6, −20	63.9	48.4	27.9	18.4	18.9	31.1	38.5	9.5
−8, −22	22.3	42.9	54.3	25.7	2.2	—	1.1	1.9
−10, −24	4.4	8.0	16.4	16.1	4.4	7.6	—	—
Total	100.0	100.0	100.0	100.0	100.0	100.0	100.0	100.0
Sulfur (wt %)	0.10	0.12	0.11	3.53	3.81	8.27	7.82	7.43
Crude oil (wt %)	0.0199	0.0959	0.0734	0.0860	0.0393	0.0687	0.0069	0.0294

[a] Hydrocarbon series (see Table 6-7 for representative compound types).

Based upon the series profiles of Fig. 6-7 and the data of Table 6-9, naphthalenes are to be found in fractions 4–8, along with the bicyclic sulfides in fractions 5–8 in the +2, −12 series. In the 0, −14 series, biphenyls are present in fractions 4–7, and the tricyclic sulfides and/or thienothiophenes occur in

* The mass spectrometric portions of this work were presented at the Ninth Annual ASTM E-14 Meeting in Chicago, Illinois, June 1961. The session chairman was Mr. Earl Lumpkin (Humble Oil Co., Baytown, Texas), who raised a question concerning the alumina gel order of elution. Representatives of Humble Oil Co., Shell Oil Co., and the American Oil Co. all confirmed that the observed appearance of the lowest MW naphthalenes ahead of the higher MW material was contrary to their experience. This is the reverse of the case presented above for the alkylbenzenes and alkyltetralin–indans. A representative of California Research Corp. suggested that the H-41 alumina gel might contain some silica gel, which would probably account for this difference. The F-20 alumina gel was in use by all of the other workers. The U.S. Bureau of Mines still had the H-41 gel on hand and desired to continue to use it for self-consistency in the API-48 project work. The H-41 was no longer available from the source, and no more was to be prepared. The reader can easily see that reproducing conditions for preliminary separations can be a problem. It may not always be possible to reproduce the work of others because of factors of this sort.

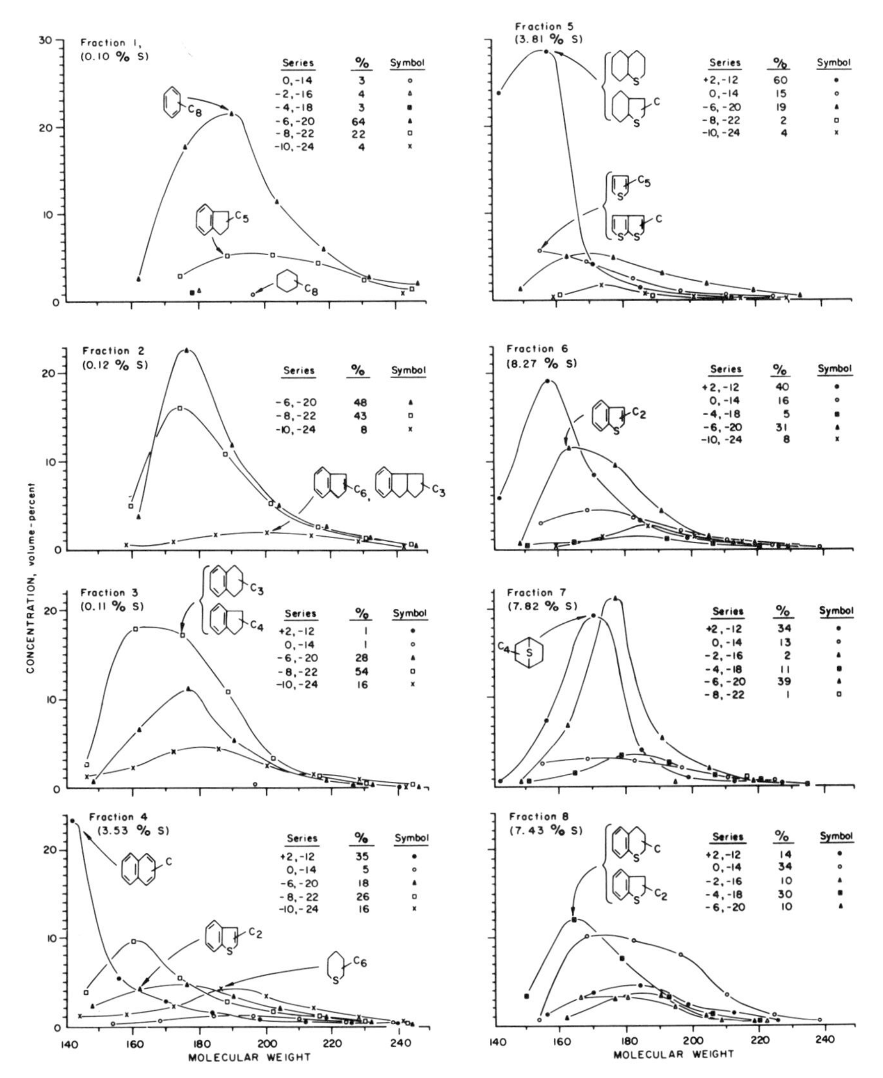

Fig. 6-7. Low-voltage mass spectrometer data on alumina adsorption fractions 1–8 on brush still fraction 1. Structural formulas indicate plausible compounds. (Courtesy of U.S. Bureau of Mines, Bartlesville, Oklahoma.)

fractions 7 and 8. Benzo[*b*]thiophene and the 2-methyl- and 3-methylbenzo-[*b*]thiophenes have been identified by Coleman *et al.* (1961) in the Wasson 200°–250°C boiling range distillate, as have biphenyl and methylthieno-thiophene by Hopkins *et al.* (1966).

The −4, −18 series appears to contain thiatetralins and thiaindans in fractions 6, 7, and 8. In the −2, −16 series, tetrahydrobenzothiophenes and/or dithienyls are possible types in fractions 7 and 8. Alkylbenzenes dominate the −6, −20 series of fractions 1–5, while the alkylbenzothiophenes occur in concentrations in fractions 5–8. Tetralins and indans dominate the −8, −22 series of fractions 1–5, although this same series contains some chain sulfides in fractions 2–5. In the −10, −24 series, the dinaphthenobenzenes and/or indenes are indicated in fractions 1–6, but in lower concentration than the tetralins and indans. The cyclic sulfides are present in much greater concentration than the chain sulfides. In support of these series indicated as present, one should also utilize the regular high-voltage spectrum to find strong fragment peaks associated with the various classes. For example, in the −6, −20 series, a distribution of peaks at m/e 91, 105, 119, and 133 is noticeable and supports the presence of the alkylbenzenes. In the same series, peaks at m/e 147, 161, 175, etc., support the presence of the alkylbenzothiophenes class. The presence of the sulfur atom is readily determinable from the isotopic profiles of these ions and the molecular ion peaks, in addition to the fact that the bimodal distribution of ion intensities is reflected in the fragment peaks as it was in the molecular ion peaks. Thus, in effect, the LVMS method supplies the best characterization when coupled with examination of the regular high-voltage spectrum. It also shows the researcher where concentrations of possibly identifiable material may occur.

Since an analysis of the brush still fraction 1 was available, it was possible to make a comparison with the summed volume percents obtained for the various series found to be present in the eight alumina gel fractions. Table 6-10 compares the results found for fraction 1 with the summed percents after normalization for the amount present in each of the alumina gel fractions, as shown in Table 6-6. If the sensitivities had been in gross error, then the

TABLE 6-10

Class Analysis of Fraction 1 of the Original Concentrate and of Alumina Gel Fractions Thereof

Series	Silica gel concentrate (vol %)	Eight alumina fractions (summed vol %)
+2, −12	20.9	21.0
0, −14	7.6	7.9
−2, −16	1.7	0.9
−4, −18	3.8	3.1
−6, −20	30.0	31.0
−8, −22	25.6	26.2
−10, −24	10.4	9.9
Total	100.0	100.0

application of these sensitivities to the separated fractions obtained from the alumina gel would have shown considerable error in the analysis. It is of interest to note that the −2, −16 series shows the only appreciable error. It is believed to be due to the fact that with such a small percentage of the total material falling in the series, it is probable that the guess as to composition is in error; therefore, the application of the wrong sensitivity factor was made. The data generally shows good enough agreement and testifies to some extent to the validity of the sensitivity data and the application of the method of LVMS.

Nine classes of sulfur compounds were identified in this study of Wasson crude oil. Homologs and isomers of benzothiophenes, dibenzothiophenes, naphthanothiophenes, and naphthenobenzothiophenes were readily identified from the mass spectral data. Bicyclic, cyclic, and chain sulfides comprised a significant portion of the gas–oil range sulfur compound types. The presence of some classes of sulfur compounds containing two sulfur atoms per molecule was indicated.

Similar analytical information has been obtained by Levy *et al.* (1961); prefractionation was done by GLC, followed by type analysis methods on waxes from petroleum sources. Robinson and Cook (1969) have reported low-resolution mass spectrometric determination of aromatic fractions from petroleum. Seifert and Teeter (1970) used high-resolution mass spectrometry for the identification of polycyclic aromatic and heterocyclic crude oil carboxylic acids. They have expanded by about 40 classes the conventional view of the "naphthenic acids" in petroleum. Kajdas and Tummler (1969) used reaction with urea to simplify the materials from slack-wax and followed this with electron attachment (EA) mass spectrography with negative ions to analyze solid-petroleum hydrocarbons. Mead (1968) has used field ionization mass spectrometry to advantage in analyzing petroleum waxes. Thus the reader should recognize that analysis and characterization of complex mixtures will continue to occupy the interest of at least a substantial number of mass spectroscopists.

REFERENCES

Andersson, J., and von Sydow, E. (1964). *Acta Chem. Scand.* **18**, 1105.

ASTM (1968). "ASTM Publications on Mass Spectrometry," pp. 3–7. American Society for Testing and Materials Standards, Philadelphia.

Barnard, G. P. (1953). "Modern Mass Spectrometry," pp. 214–229. Institute of Physics, London.

Beckey, H. D., Knoppel, H., Metzinger, G., and Schulze, P. (1966). *In* "Advances in Mass Spectrometry" (W. L. Mead, ed.), Vol. III, p. 35. Institute of Petroleum, London.

Begemann, P. H., and Koster, J. C. (1964). *Nature* **202**, 552.

Benz, W. (1968). "Massenspektrometric Organischer Verbindungen," pp. 385–391. Akad. Verlagsgesellschaft, Frankfurt a. M.

Beynon, J. H. (1960a). "Mass Spectrometry and Its Applications to Organic Chemistry," pp. 421–422. Elsevier, Amsterdam.

Beynon, J. H. (1960b). "Mass Spectrometry and Its Applications to Organic Chemistry," pp. 312–319. Elsevier, Amsterdam.

Beynon, J. H. (1968). *In* "Advances in Mass Spectrometry" (E. Kendrick, ed.), Vol. 4, pp. 123–138. Institute of Petroleum, London.

Beynon, J. H., and Williams, A. E. (1963). "Mass and Abundance Tables for Use in Mass Spectrometry," 570 pp. Amer. Elsevier, New York.

Beynon, J. H., Job, B. E., and Williams, A. E. (1965). *Z. Naturforsch. A* **20**, 883.

Beynon, J. H., Saunders, R. A., and Williams, A. E. (1968a). "The Mass Spectra of Organic Molecules," 510 pp. Amer. Elsevier, New York.

Beynon, J. H., Saunders, R. A., and Williams, A. E. (1968b). "The Mass Spectra of Organic Molecules," p. 21. Amer. Elsevier, New York.

Biemann, K., Bommer, P., and Desiderio, D. M. (1964). *Tetrahedron Lett.*, 1725.

Boldingh, J., and Taylor, R. S. (1962). *Nature* **194**, 909.

Boyer, E. W., Hamming, M. C., and Ford, H. T. (1963). *Anal. Chem.* **35**, 1168.

Budzikiewicz, H., Djerassi, C., and Williams, D. H. (1964). "Structure Elucidation of Natural Products by Mass Spectrometry," Vol. I, pp. 17–40. Holden-Day, San Francisco, California.

Budzikiewicz, H., Djerassi, C., and Williams, D. H. (1967a). "Mass Spectrometry of Organic Compounds," p. 100. Holden-Day, San Francisco, California.

Budzikiewicz, H., Djerassi, C., and Williams, D. H. (1967b). "Mass Spectrometry of Organic Compounds," 690 pp. Holden-Day, San Francisco, California.

Budzikiewicz, H., Djerassi, C., and Williams, D. H. (1967c). "Mass Spectrometry of Organic Compounds," p. 5. Holden-Day, San Francisco, California.

Budzikiewicz, H., Djerassi, C., and Williams, D. H. (1967d). "Mass Spectrometry of Organic Compounds," pp. 221–225. Holden-Day, San Francisco, California.

Buttery, R. G., Black, D. R., and Kealy, M. P. (1965). *J. Chromatogr.* **18**, 399.

Capellen, J., Svec, H. J., Jordan, J. R., and Sun, R. (1970a). "Bibliography of Mass Spectroscopy Literature for the First Half of 1968 Compiled by a Computer Method," 463 pp. Division of Technical Information, U.S. Atomic Energy Commission, Oak Ridge, Tennessee.

Capellen, J., Svec, H. J., Jordan, J. R., and Watkins, W. J. (1970b). "Bibliography of Mass Spectroscopy Literature for the Last Half of 1967 Compiled by a Computer Method," 458 pp. Division of Technical Information, U.S. Atomic Energy Commission, Oak Ridge, Tennessee.

Clerc, R. J., and O'Neal, M. J. (1961). *Anal. Chem.* **33**, 380.

Coleman, H. J., Thompson, C. J., Hopkins, R. L., Foster, N. G., Whisman, M. L., and Richardson, D. M. (1961). *J. Chem. Eng. Data* **6**, 464.

Cooks, R. G. (1969). *Org. Mass Spectrom.* **2**, 481.

Cornu, A., and Massot, R. (1966). "Compilation of Mass Spectral Data." Heydon, London, and Presses Universitaires de France.

Cornu, A., and Massot, R. (1967). "First Supplement to Compilation of Mass Spectral Data." Heydon, London, and Presses Universitaires de France.

Damoth, D. C. (1965). *In* "Advances in Analytical Chemistry and Instrumentation" (C. N. Reilley, ed.), Vol. 4, pp. 317–410. Wiley, New York.

Davis, J. W., Hirsch, D. E., Foster, N. G., and Schwartz, F. G. (1963). *U.S. Bur. Mines Rep. Invest.* No. 6298.

Day, E. A., and Anderson, D. F. (1965). *J. Agr. Food Chem.* **13**, 2.

Day, E. A., and Libbey, L. M. (1964). *J. Food Sci.* **29**, 583.

Deck, R. E., and Chang, S. S. (1965). *Chem. Ind.* (*London*), 1343.
Dibeler, V. H. (1963). *In* "Mass Spectrometry" (C. A. McDowell, ed.), pp. 358–365. McGraw-Hill, New York.
Eisenbraun, E. J., Mattox, J. R., Bansal, R. C., Wilhelm, M. S., Flanagan, P. W. K., Carel, A. B., Laramy, R. E., and Hamming, M. C. (1968). *J. Org. Chem.* **33**, 2000.
Etter, D. M. (1968). Personal communications. Continental Oil Co., Ponca City, Oklahoma.
Field, F. H., and Hastings, S. H. (1956). *Anal. Chem.* **28**, 1248.
Gallegos, E. J., Green, J. W., Lindeman, L. P., Le Tourneau, R. L., and Teeter, R. M. (1967). *Anal. Chem.* **39**, 1833.
Gianturco, M. A., and Giammarino, A. S. (1966). *Nature* **210**, 1358.
Gohlke, R. S. (1963). "Uncertified Mass Spectral Data." Dow Chemical Co., Midland, Michigan.
Grigsby, R. D., and Hamming, M. C. (1966). Unpublished results. Continental Oil Co., Ponca City, Oklahoma.
Hamming, M. C., and Keen, G. W. (1969). *Int. J. Mass Spectrom. Ion Phys.* **3**, App. 1.
Harless, H. R., and Crabb, N. T. (1969). *J. Amer. Oil Chem. Soc.* **46**, 238.
Hoover, H. W., Jr., and Washburn, H. W. (1940). *Amer. Inst. Met. Eng. Technol. Publ.* No. 1205.
Hopkins, R. L., Thompson, C. J., Coleman, H. J., and Rall, H. T. (1966). *U.S. Bur. Mines Rep. Invest.* No. 6795.
Hoshino, H., Wasada, N., and Tsuchiya, T. (1964). *Bull. Chem. Soc. Jap.* **37**, 1310.
Kajdas, C., and Tummler, R. (1969). *Org. Mass Spectrom.* **2**, 1049.
Kebarle, P., and Hogg, A. M. (1965). *J. Chem. Phys.* **42**, 798.
Lee, F. C. (1969). Private communications.
Levy, E. J., Doyle, R. R., Brown, R. A., and Melpolder, F. W. (1961). *Anal. Chem.* **33**, 639.
Lumpkin, H. E. (1956). *Anal. Chem.* **28**, 1946.
Lumpkin, H. E. (1958). *Anal. Chem.* **30**, 321.
Lumpkin, H. E., and Aczel, T. (1963). Paper No. 47. Eleventh Annual Conference on Mass Spectrometry, ASTM Committee E-14, San Francisco, California.
Lumpkin, H. E., and Aczel, T. (1964). *Anal. Chem.* **36**, 181.
Lumpkin, H. E., and Johnson, B. H. (1954). *Anal. Chem.* **26**, 1719.
McFadden, W. H., Teranishi, R., Corse, J. W., Black, D. R., and Mon, T. R. (1965). *J. Chromatogr.* **18**, 10.
McLafferty, F. W. (1962). *In* "Determination of Organic Structures by Physical Methods" (F. C. Nachod and W. D. Phillips, eds.), Vol. 2, pp. 103–104. Academic Press, New York.
McLafferty, F. W. (1963a). *In* "Mass Spectrometry of Organic Ions" (F. W. McLafferty, ed.), pp. 309–342. Academic Press, New York.
McLafferty, F. W. (1963b). "Mass Spectral Correlations," 117 pp. American Chemical Society, Washington, D.C.
McLafferty, F. W. (1966a). "Interpretation of Mass Spectra," pp. 78–149. Benjamin, New York.
McLafferty, F. W. (1966b). "Interpretation of Mass Spectra," p. 63. Benjamin, New York.
McLafferty, F. W. (1966c). "Interpretation of Mass Spectra," p. 20. Benjamin, New York.
McLafferty, F. W., McAdoo, D. J., Schuddemage, H. D. R., and Smith, J. S. (1969). Paper No. 31. Seventeenth Annual Conference on Mass Spectrometry, ASTM Committee E-14, Dallas, Texas.
Mason, M. E., Johnson, B., and Hamming, M. C. (1966). *J. Agr. Food Chem.* **14**, 454.
Mason, M. E., Johnson, B., and Hamming, M. C. (1967). *J. Agr. Food Chem.* **15**, 66.

Mead, W. L. (1968). *Anal. Chem.* **40**, 743.
Merritt, C., Jr., Bazinet, M. L., Sullivan, J. H., and Robertson, D. H. (1963). *J. Agr. Food Chem.* **11**, 152.
Meyerson, S. (1964). *J. Phys. Chem.* **68**, 968.
Meyerson, S., and McCollum, J. D. (1963). *Advan. Phys. Org. Chem.* **2**, 179.
Meyerson, S., Grubb, H. M., and van der Haar, R. W. (1963). *J. Chem. Phys.* **39**, 1445.
Meyerson, S., Drews, H., and Fields, E. K. (1964). *Anal. Chem.* **36**, 1294.
Natalis, P. (1965). *In* "Mass Spectrometry" (R. I. Reed, ed.), pp. 379–399. Academic Press, New York.
Price, D., and Williams, J. E., eds. (1969). "Dynamic Mass Spectrometry," Vol. I. Heyden, London.
Reed, R. I. (1966). "Applications of Mass Spectrometry to Organic Chemistry," p. 128. Academic Press, New York.
Reid, W. K., Mead, W. L., and Bowen, K. M. (1966). *In* "Advances In Mass Spectrometry" (W. L. Mead, ed.), Vol. 3, p. 731–745. Institute of Petroleum, London.
Robinson, C. J., and Cook, G. L. (1969). *Anal. Chem.* **41**, 1548.
Ruth, J. M. (1968). *Anal. Chem.* **40**, 747.
Seibl, J. (1969). *Org. Mass Spectrom.* **2**, 1033.
Seifert, W. K., and Teeter, R. M. (1970). *Anal. Chem.* **42**, 750.
Sharkey, A. G., Jr., Wood, G., Shultz, J. L., Wender, I. L., and Friedel, R. A. (1959). *Fuel* **38**, 315.
Silverstein, R. M., Rodin, J. O., Himel, C. M., and Leeper, R. W. (1965). *J. Food Sci.* **30**, 668.
Simmonds, P. G. (1970). *Amer. Lab.* Oct., 8.
Smouse, T. H., and Chang, S. S. (1967). *J. Amer. Oil Chem. Soc.* **44**, 509.
Stenhagen, E., Abrahamsson, S., and McLafferty, F. W. (1969). "Atlas of Mass Spectral Data." Vols. I, II, III, 2,266 pp. Wiley (Interscience), New York.
Stevenson, D. P., and Wagner, C. C. (1950). *J. Amer. Chem. Soc.* **72**, 5612.
Tatematsu, A., and Tsuchiya, T. (1966). "Structure Indexed Literature of Organic Mass Spectra," 275 pp. Academic Press of Japan, Japan.
Thompson, C. J., Coleman, H. J., Rall, H. T., and Smith, H. M. (1955). *Anal. Chem.* **27**, 175.
Thompson, C. J., Foster, N. G., Coleman, H. J., and Rall, H. T. (1966). *U.S. Bur. Mines Rep. Invest.* No. 6879.
Tickner, A. W., Bryce, W. A., and Lossing, F. P. (1951). *J. Amer. Chem. Soc.* **73**, 5001.
Tunnicliff, D. D., and Wadsworth, P. A. (1965). *Anal. Chem.* **37**, 1082.
Washburn, H. W., Wiley, H. F., and Rock, S. M. (1943). *Ind. Eng. Chem. Anal. Ed.* **15**, 541.
Washburn, H. W., Wiley, H. F., Rock, S. M., and Berry, C. E. (1945). *Ind. Eng. Chem. Anal. Ed.* **17**, 74.
White, F. A. (1968). "Mass Spectrometry in Science and Technology," pp. 296–316. Wiley, New York.
Yates, P. (1967). "Structure Determination," pp. 47–49. Benjamin, New York.
Yeo, A. N. H., and Williams, D. H. (1968). *J. Chem. Soc.* (*C*) **21**, 2666.
Zwolinski, B. J., and Wilhoit, R. C. (1968). "Comprehensive Index of API 44–TRC Selected Data on Thermodynamics and Spectroscopy," 507 pp. Thermodynamics Research Center, Texas A & M University, College Station, Texas.

CHAPTER 7

THE RECTANGULAR ARRAY AND INTERPRETATION MAPS

I. Introduction

A practical procedure for identifying an unknown from its mass spectrum is to first arrange the spectrum in the form of a rectangular array (Hamming and Grigsby, 1967) and then to determine the structural significance of peaks or series of peaks from interpretation maps. Spectra can be arranged in the form of the rectangular array by computers or by hand recording of intensities. The peaks used for interpretation are indexed by series numbers which refer the interpreter to a set of interpretation maps showing either the probable compound type (if the class of substance is known) or the ion composition for a completely unknown material. These maps are from a collection and coalition of a great number of mass spectra. Each map summarizes empirical fragmentation data in a form which can be referred to quickly. The interpretation maps are given in Appendix II of this volume. Much of the intuition often common to interpretation problems has been removed by the use of interpretative peaks (to be defined later).

The matrix (array) format was first brought to attention by Danti (1960) in a set of directions for contributing mass spectra to API-44. Contributions to its formation in the API were made by several interested investigators on the Advisory Committee of API Research Project 44 (G. F. Crable and N. D. Coggeshall were largely responsible for the matrix form in API). However, publication of spectra using the array was not widely accepted for several years. The slow acceptance of the array may have been due to the failure to realize the interpretation aid provided by such an arrangement of the data. The use of "grids" had been employed by Clerc *et al.* (1955) as an aid in interpreting spectral data of high molecular weight saturated hydrocarbons.

The rectangular array consists of a sequence of rows, each containing 14 ascending, integral, consecutive mass numbers (shown as *m/e*). This arrangement produces columns having elements which differ by 14 mass units from adjacent elements in that column. The significance of the 14 mass difference for hydrocarbons is related to the mass of one methylene group (CH_2). One such array is shown in Table 7-1.

TABLE 7-1

PEAKS (*m/e*) ARRANGED IN THE RECTANGULAR ARRAY

m/e (−11)	*m/e* (−10)	*m/e* (−9)	*m/e* (−8)	*m/e* (−7)	*m/e* (−6)	*m/e* (−5)	*m/e* (−4)	*m/e* (−3)	*m/e* (−2)	*m/e* (−1)	*m/e* (+0)	*m/e* (+1)	*m/e* (+2)
							10	11	12	13	14	15	16
17	18	19	20	21	22	23	24	25	26	27	28	29	30
31	32	33	34	35	36	37	38	39	40	41	42	43	44
45	46	47	48	49	50	51	etc.						

An outstanding feature of the rectangular array is that the molecular weights of any homologous series fall within a single column. For example, molecular weights (molecular ions) of alkanes are found (Table 7-1) in the column that begins with *m/e* 16 (the molecular weight of methane). Directly below *m/e* 16 in the same column is *m/e* 30, the molecular weight of ethane. Of equal importance in interpretation is the fact that intense fragment ions are listed in columns characteristic of the various classes of organic compounds. Spectra of alkanes, for example, are dominated by peaks of relatively high intensities in the column starting at *m/e* 15 or the +1 column (Table 7-1). The peaks in this series thus have diagnostic value for identifying alkanes (or an alkane moiety).

All of the mass spectra used to develop the interpretation maps in Appendix II were first arranged in an array. Therefore the array is inseparable from both the development and use of the interpretative aid herein discussed.

Since organic chemistry has long dealt with homologous series of compounds in which the difference was the methylene group ($-CH_2-$) in the general formula, it is natural to expect the members of the series to repeat every 14 mass units. The numbers shown in parentheses at the top of Table 7-1 are the numbers for the series associated with each of the columns of the array. These series numbers are the "name" used to designate the peaks within that column. This classification finds its origin in the number z in the general formula for hydrocarbons, namely,

$$C_nH_{2n+z}$$

When the number z is +2, the formula represents all of the alkanes; when the number is 0, the formula represents all of the cycloalkanes and/or the alkenes. For $z = -6$, we have the general formula for the alkylbenzenes. If only integers are used in the formula, all possible hydrocarbon molecules or ions may be represented. It is apparent from Table 7-1 that an integral step listing of all atomic or molecular weights of any ions from ^{10}B upwards is possible. Indexing by this means has many advantages when dealing with organic compounds. For convenience, all other substances are indexed by this system for interpretive purposes. The utility of the system will become apparent with a little use. Note that essentially the indexing is by atomic mass sum of the given moiety, whether molecule, atom, or ion. Thus for substances containing atoms other than carbon and hydrogen, the use of a general formula is prohibited. Instead, one simply finds the sum of the atomic masses making up the moiety in question. For example, ethanol with a formula of C_2H_6O has a mass of 46. This ion falls in the mass series −10 in Table 7-1. An enlarged version of this index table is to be found as Map A, Appendix II, and its use will prove essential to the reader from this point on. Note that the mass 46 would not be found in the +2 series column if one attempted to use the fact that the C_2H_6 portion of the formula yields a +2. A similar situation arises if the formula is C_2H_6S (MW = 62) or C_2H_6N (MW = 44). The hydrocarbon portion is of little help in locating the series number for the substance, but it must be added in obtaining the MW and hence the mass at which the molecular ion would appear. Note that isotopes have been neglected in this discussion.

The array presentation of a mass spectrum aids in interpretative problems because it displays in visual form the fragmentation reactions taking place by electron impact in the mass spectrometer. When a molecular ion dissociates into fragment ions, the intense fragment ions often differ from one another by 14 mass units or a methylene group (CH_2). These ions are grouped in one column by the array rather than being distributed over the entire spectrum, as is the case when spectra are tabulated by increasing mass numbers. The peaks which stand out because of higher intensities in a given column usually can be related to structural features of the molecule. Specific decomposition routes and rearrangement processes cause certain ions in a series to be prominent.

Two different sets of interpretation maps are provided in Appendix II (Maps 1–25 and Maps 26–39). One set is referred to as compound class or "specific maps" and a second set as "ion composition maps." The specific maps are intended to be used in structural identifications *when the class of compounds is known*. For example, if the sample is known to be a hydrocarbon, the hydrocarbon set of maps should be used. The second set of interpretation maps or "ion composition maps" is intended to be used if the unknown sample could be one of several different classes of compounds.

Specific maps will be discussed first. These maps include the following sets: CH, hydrocarbons (Maps 1–5); CHN and CHNO compounds (Maps 6–10); CHS and CHSO compounds (Maps 11–15); CHNS compounds (Maps 16–18); CHO, oxygen-containing compounds (Maps 19–25). The arrangement of the maps for any one class of compounds is similar to the peaks shown in Table 7-1. The first map of a set of maps starts with the column of m/e on the left and moves by columns to the right. In this way, several maps are included in each set of maps. For example, peaks in the −11 and −10 series make up the first map of the set of specific interpretation maps for hydrocarbons.

Mass spectral data used in the formation of these maps were obtained over a period of several years from a wide variety of sources. These sources include selected spectra from the American Petroleum Institute Research Project 44, the Chemical Thermodynamics Properties Center, the American Society for Testing and Materials Committee E-14 spectral data index (ASTM, 1963), and spectral data published by the Dow Chemical Co. (Gohlke, 1963) and the Manufacturing Chemists Association. Also included are unpublished mass spectra, as well as selected mass spectral data compiled by Cornu and Massot (1966, 1967) and miscellaneous spectra discussed in the literature. In the formation of these maps, an added benefit became evident. These maps provide a convenient means by which newly published information can be compared and assimilated with older knowledge of fragmentation, and thus such bits of information are less likely to become isolated facts.

While these maps cover the fragmentation of a large number of compound classes, they do not include *all* classes of organic compounds. Several groups of compounds have not been examined by mass spectrometry in enough detail to establish a correlation between fragmentation and structural features and thus could not be included. Other classes of compounds with a well-established correlation were not included because of a lack of general interest, while still others may well have been omitted by error or poor judgment. Admittedly, the types of compounds listed are only a small part of the total number of organic compounds known. However, the wide number of sources examined should provide a representation of the common classes of organic compounds.

These maps are not intended to duplicate the outstanding tabulation made by McLafferty (1963a) listing the structural significance of individual peaks. The main point of similarity between the two pieces of work is in the desire to aid the inexperienced person attempting to identify an unknown compound from its mass spectrum. The interpretation maps include both molecular ions and fragment ions for use in aiding the identification of classes of compounds. Indanes, for example, can be identified by molecular ions in the −8 series where the series starts at m/e 118 and increases by 14 mass numbers. Thus a molecular ion of an alkyl-substituted indane is included on the maps, even if such a given spectrum has not been published or included with the data used to originate these maps.

The combined uses of the rectangular array and interpretation maps will be illustrated first for hydrocarbons. A more general discussion involving compounds with hetero atoms will follow. Some brief remarks will be made on the fragmentation processes involved, which will serve to introduce the reader to a more detailed discussion of fragmentation as applied to interpretation presented in Chapter 9.

II. Columnar Grouping of Characteristic Peaks

One advantage of the array presentation of a mass spectrum is that characteristic ions for different classes of compounds are grouped in columns. This advantage applies to spectra of both pure substances and complexes mixtures. Mass spectra of complex mixtures will have series of characteristic peaks and molecular ions of homologs in individual columns; only spectra of pure compounds will be considered in this chapter. Mixtures are discussed in Chapter 6 and elsewhere.

A. The General Formula Approach—Applied to Hydrocarbons

In mass spectrometry advantage is taken of the fact that all hydrocarbons can be classified by the general formula, C_nH_{2n+z}. As the numerical value of z changes, so does the compound type. Alkanes have a z value of +2 (C_nH_{2n+2}), while cycloalkanes have a z value of zero (+0). The z value may change, however, within a given series column. For example, a molecular ion peak at m/e 128 (coming from a compound of molecular weight 128) could be an alkane (C_nH_{2n+2}) or naphthalene (C_nH_{2n+12}). The actual formulas would be C_9H_{20} or $C_{10}H_8$, both with nominal masses of 128. Thus a molecular ion at m/e 128 could be one of these two different z numbers, +2 or −12. It could not have the z number −26 (another 14 or CH_2 removed) because for C_nH_{2n-26}

one would have to have $n = 14$ before the ion could theoretically exist, i.e., $C_{14}H_2$. This ion would occur at m/e 174, but such a highly hydrogen deficient ion is unlikely in any case. Hence another variable enters the picture in the form of the cycling of z numbers, which takes place in the fashion as shown in Table 7-2. For numbers beyond this table, cycling continues, and 14 should

TABLE 7-2

CYCLING OF z NUMBERS IN THE ARRAY

													Start ↓
−11	−10	−9	−8	−7	−6	−5	−4	−3	−2	−1	+0	+1	+2
−25	−24	−23	−22	−21	−20	−19	−18	−17	−16	−15	−14	−13	−12
−39	−38	−37	−36	−35	−34	−33	−32	−31	−30	−29	−28	−27	−26
→ Continues													

be subtracted to get the next column entry for any series column. It should be remembered that all series of peaks shown on the interpretation maps in Appendix II are designated by this scheme. In such cases where the compound type changes within a given series, it may be necessary to know the elemental composition of the species before a correct empirical formula can be determined. Medium- or high-resolution mass spectrometry can be used to make accurate mass-to-charge measurements for empirical formulas. For example, if the 128 peak had the empirical formula of C_9H_{20}, it would have an absolute mass measurement of 128.1565, which could easily be resolved from the $C_{10}H_8$ compound of absolute mass 128.0626. Absolute mass measurement values are given in a reference of Beynon and Williams (1963). The reason for the overlap between classes of hydrocarbons such as alkanes and naphthalenes is that the general formula of naphthalenes differs by 14 mass units from alkanes (C_nH_{2n-12} for naphthalenes as compared with C_nH_{2n+2} for alkanes).

It is pertinent to understanding fragmentation to know that if the z number is an even number, the peak on the spectrum is also an even-numbered m/e peak. For hydrocarbons even-numbered peaks are odd-electron ions (molecular ions or rearrangement ions). If the z number is an odd number, the peak on the spectrum is also at an odd-numbered peak (for hydrocarbons these are even-electron ions). Odd-electron ions and even-electron ions are discussed in detail by McLafferty (1966a).

The utilization of z numbers is common to analytical applications of mass spectrometry in hydrocarbon characterization and the analysis of very complex mixtures, as shown in Chapter 6. Most of these methods provide quantitative estimates of the major types of compounds present in crude oils or refinery samples. Analysis need not be confined to these types of samples alone, as illustrated by Ford (1964), who developed a complete analytical method for the determination of the phenyl position of straight-chain biodegradable detergent alkylates using only the –6 and –7 series of z numbers.

B. THE +2 SERIES OF PEAKS—SATURATED HYDROCARBONS

An example of a partial mass spectrum (70 eV) in the array format is shown in Table 7-3 for *n*-decane. This is an *n*-alkane with a C_nH_{2n+2} formula type,

TABLE 7-3

PARTIAL MASS SPECTRUM OF *n*-DECANE[a]

m/e (−1)	Int	*m/e* (+0)	Int	*m/e* (+1)	Int	*m/e* (+2)	Int
13	—	14	0.1	15	0.5	16	—
27	6.3	28	1.4	29	8.1	30	0.2
41	9.5	42	3.3	43	20.7	44	0.7
55	3.0	56	3.4	57	16.7	58	0.8
69	0.8	70	2.3	71	6.2	72	0.3
83	0.2	84	1.3	85	4.1	86	0.3
97	0.1	98	0.7	99	0.9	100	—
111	—	112	0.3	113	0.6	114	—
125	—	126	—	127	—	128	—
139	—	140	—	141	—	142	1.2

[a] In the tables of partial mass spectral data in this chapter the following designations will apply: *m/e* is peak of mass/charge ratio; Int denotes intensities expressed as a fraction of total ion current produced. The significance of the underlined values is explained in the text.

which will therefore be designated as a +2 hydrocarbon. The fragmentation shown for *n*-decane is typical of long-chain hydrocarbons. It should be noted that the peaks with odd mass numbers are usually more intense than peaks of even mass numbers (Friedman and Long, 1953). The column beginning with *m/e* 15 contains +1 ions, and many of the most intense peaks in the spectrum are in this series. Note that the +1 series ions are 1 mass unit less than the +2 series in which the parent ion occurs. This does not mean that only a hydrogen atom is lost, and, in fact, it probably will not be lost unless it is very labile. This behavior is typical for many compounds containing alkyl groups in which the losses are usually initially a moiety of mass 29 or

$29 + (CH_2)_n$ with the ion intensities decreasing rapidly as n increases. The fragmentation shown is thus characteristic for spectra of alkanes. In this example, the most intense peak in this series is at *m/e* 43 (a C_3H_7 ion having an intensity of 20.7% of the total ion yield). This peak is also the most intense peak in the spectrum (referred to as the base peak).

A branched alkane gives a similar, however more complex, spectrum as illustrated for the compound 3-ethyl-2-methylheptane shown in Table 7-4. The

TABLE 7-4

PARTIAL MASS SPECTRUM OF 3-ETHYL-2-METHYLHEPTANE

m/e (−1)	Int	*m/e* (+0)	Int	*m/e* (+1)	Int	*m/e* (+2)	Int
13	—	14	0.1	15	0.7	16	—
27	5.8	28	0.9	29	4.9	30	0.1
41	8.9	42	2.2	43	12.8	44	0.5
55	3.3	56	4.1	57	30.9	58	1.4
69	2.4	70	2.1	71	1.2	72	0.1
83	0.4	84	0.1	85	0.4	86	—
97	0.1	98	7.4	99	2.1	100	0.1
96	—	112	—	113	0.1	114	—
125	—	126	—	127	—	128	—
139	—	140	—	141	—	142	0·2

spectra of *n*-decane and 3-ethyl-2-methylheptane are similar because the most intense fragment ions are in the +1 series of peaks and both compounds have a molecular ion at *m/e* 142 in the +2 series. The most striking difference is in the intensities of the peaks in the +1 series. For *n*-decane this series of peaks has one maximum (at *m/e* 43), but for 3-ethyl-2-methylheptane the +1 series has two maxima (one at *m/e* 57 and one at *m/e* 99). An examination of the structure of 3-ethyl-2-methylheptane will show a reason for the two prominent peaks at *m/e* 57 and 99.

Cleavage at point of branching → $C_7H_{15}^+$ or $C_4H_9^+$

m/e 142

m/e 99 $(-C_3H_7)$ or *m/e* 57 $(-C_6H_{13})$

The relative intensity of a molecular ion of the above branched compound is much less than for an unbranched compound of the same molecular weight (0.2% compared to 1.2%). A reduction in intensities of molecular ions with branching is characteristic of most classes of organic compounds. The most pertinent point in presenting these two spectra of +2 hydrocarbons is that they illustrate the grouping of peaks having the greatest diagnostic value in one column on the array format. The grouping of such peaks allows spectral differences to be easily detected, which can often be rationalized with differences in molecular structures.

C. THE −12 SERIES OF PEAKS—NAPHTHALENES

Compounds in the −12 series of peaks cannot have fewer than 10 carbon atoms ($C_{10}H_8$). Thus the series starts with naphthalene itself. For each methyl substitution the molecular weight (and thus the molecular ion) is increased

TABLE 7-5

PARTIAL MASS SPECTRUM OF 1-METHYLNAPHTHALENE

m/e	Int	*m/e*	Int	*m/e*	Int	*m/e* (−12)	Int
13	—	14	0.1	15	0.4	16	—
27	1.4	28	0.2	29	0.1	30	—
41	0.1	42	—	43	0.1	44	—
55	0.1	56	0.1	57	0.3	58	—
69	0.9	70	1.6	71	2.8	72	–
83	—	84	—	85	0.2	86	0.1
97	—	98	0.3	99	0.2	100	—
111	—	112	—	113	0.6	114	0.4
125	—	126	—	127	—	128	—
139	2.7	140	1.6	141	20.1	142	26.3

by 14 mass units. The fragmentation of 1-methylnaphthalene is shown in Table 7-5.

It should be noted that the appearance of the *n*-decane spectrum (of a typical alkane) differs greatly from a methyl-substituted naphthalene. The intense peaks typical of alkanes at *m/e* 29, 43, and 57 are essentially absent for naphthalenes. This difference is easily observed in the array format presentation of these two spectra, because both the +1 and −13 series of peaks are in the column to the left of the molecular ion peaks. This difference becomes a key to differentiating between alkanes and naphthalenes. Thus, a compound of molecular weight 142 (molecular ion at *m/e* 142) with intense fragment peaks at *m/e* 15, 29, 43, 57, and 71 would be denoted as a C_{10} alkane, while, if these peaks were essentially absent, a methyl-substituted naphthalene would be indicated. Additional evidence could be obtained by medium- or high-resolution mass spectrometry, which could resolve the mass difference between $C_{10}H_{22}$ and $C_{11}H_{10}$.

The fragmentation of naphthalenes as contrasted with alkanes illustrates the differences between mass spectra of different compound types. On the array format the most diagnostic peaks are grouped in the same corresponding column for both compounds. The most characteristic peaks would have been distributed over the complete range of the spectrum if the spectra had been tabulated by increasing mass numbers or presented on line drawings.

D. The Interpretative Peak Concept

A visually interpretative peak (VIP) *is a peak which has an intensity stronger than that of another peak* 14 *mass units higher and* 14 *mass units lower and also an intensity stronger than that of another peak* 1 *mass unit higher and* 1 *mass unit lower*. The relationship between such neighboring peaks is easily visualized when a mass spectrum is presented on the array format. Any difference equal to or greater than 0.2% is convenient to use in making the selection for most classes of compounds.

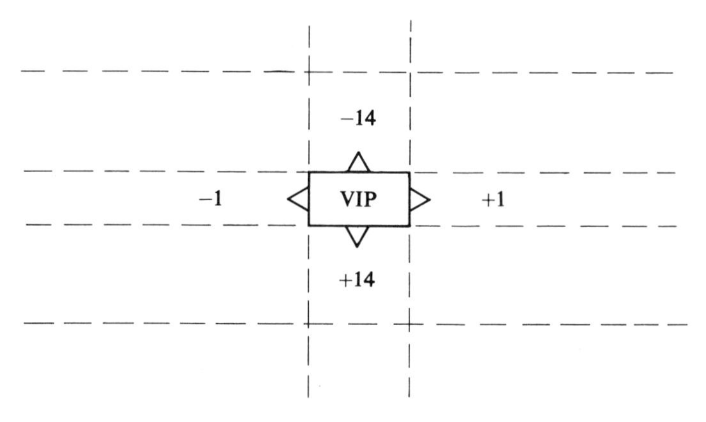

Because interpretative peaks are well defined, they can be selected by computer routines. The number of such peaks selected from a single spectrum is often quite small. It is about the same number arbitrarily set when tabulating the most intense peaks (10 peaks). This is because usually many of the most intense peaks on a spectrum differ by 14 mass units. For example, the three most intense peaks for *n*-hexadecane are *m*/*e* 43, 57, and 71. Only one of these peaks would be selected as an interpretative peak. However, the 226 peak, which is the molecular ion peak or parent peak (PIP) and would thus have identifying significance, would not be selected unless the 18 most intense peaks were compiled (the value of 18 depends somewhat on the mass spectrometer used).

III. Interpretation Maps of Hydrocarbons

Hydrocarbons were among the first classes of compounds to be examined in depth by mass spectrometry. As the number of mass spectra of hydrocarbons grew, it was observed that certain empirical rules could be set up relating the fragmentation with structural features of the molecules. The fragmentation appeared to be complicated by rearrangement reactions. However, often the early workers observing mass spectra were not concerned with understanding fragmentation reactions but only with the application to a given analytical problem. The specific applications were often not organized into any easy-to-apply form. Many usable applications in the pioneering days of mass spectrometry were not published and remained "in-house know-how."

A. Illustrating Interpretation Maps with the −6 Series

Once interpretative peaks are selected, the question arises as to how they can be applied to obtain an identification or interpretation of the sample. (The same question arises if interpretation peaks are not selected.) The utilization of interpretative peaks is intended to be coupled with interpretation maps provided in Appendix II. The maps provide a type of correlation which exists between mass spectra and the molecular structures of organic compounds. Their use allows the interpreter to restrict the probable class of compound(s) present to a small number out of the myriad of "all possible compounds," thus speeding the course of the identification. For example, an examination of a large number of spectra from alkylbenzene compounds has shown that these spectra exhibit intense peaks at one or more *m/e* in the −7 series (77, 91, 105, 119, etc.). Such fragment ions are characteristic of alkylbenzenes. Because this is the type of information found on interpretation maps, they can be used in making at least qualitative type analysis.

If an unknown mass spectrum has an interpretative peak at any one of the mass numbers in the −7 series, an indication is given that an alkylbenzene type of compound is present or, perhaps in a more general sense, aromatic compounds. It does not follow that proof of the presence of alkylbenzenes or aromatics is given. The maps can provide *additional* evidence for or against a given class of compounds. Consider a spectrum with two interpretative peaks, one at *m/e* 91 and another at *m/e* 120. The maps indicate that both interpretative peaks are characteristic of an alkylbenzene. This can be seen from a section of the interpretation map taken from Appendix II which is shown in Table 7-6. The molecular ion (M^+) has nine carbon atoms (start with six at *m/e* 78 and count down the column by one carbon until *m/e* 120 is reached). The empirical formula C_nH_{2n-6} is an alkylbenzene.

It should be noted in Table 7-6 that the circled number −7 is used to designate

TABLE 7-6

SECTION FROM INTERPRETATION MAP FOR HYDROCARBONS (MAP 2 OF APPENDIX II-B)

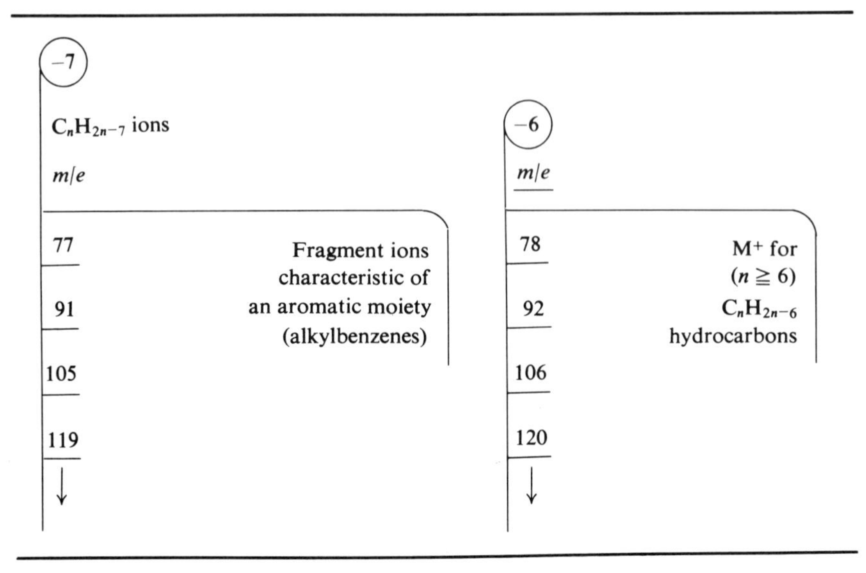

the series of ions shown below it. The *m/e* values in this series increase by 14 mass numbers. To the right, the words "fragment ions characteristic of an aromatic moiety" give the structural meaning, which is denoted by this series of peaks. Likewise, the circled number −6 designates the peaks below it. The remark to the right of this series provides the meaning of these peaks. In this case, these are molecular or parent ion peaks (PIP) of C_nH_{2n-6} hydrocarbons where n is six or more carbon atoms.

TABLE 7-7

PARTIAL MASS SPECTRUM OF 1-PHENYLOCTANE

m/e (−7)	Int	*m/e* (−6)	Int
91	29.3	92	24.6
105	3.1	106	0.3
119	0.6	120	0.1
133	1.4	134	0.2
147	0·3	148	0.1
161	—	162	0.1
175	—	176	—
189	—	190	7.5

The fragmentation of alkylbenzenes is well established (Grubb and Meyerson, 1963). These compounds give intense molecular ion peaks, as would be expected by their chemical stability (Pahl, 1954). The aromatic nucleus becomes a center for the stabilization of the positive charge during electron impact. Alkylbenzenes undergo cleavage of the bond β to the aromatic ring, which gives intense fragment ions shown in the −7 series on the maps. A typical spectrum of a monoalkylbenzene undergoing β cleavage, as illustrated in the monoisotopic partial spectrum of 1-phenyloctane, is shown in Table 7-7. The peak at *m/e* 91 is a tropylium ion (Rylander *et al.*, 1957) and is depicted below as being formed from 1-phenyloctane.

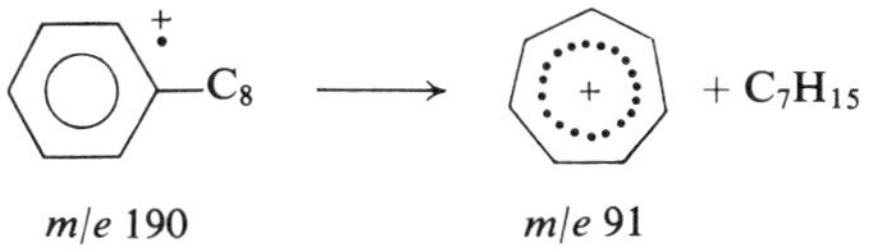

C_8 ⟶ $+ C_7H_{15}$

m/e 190 *m/e* 91

The monoisotopic spectrum of pure 1-phenyloctane has a peak at *m/e* 92 of nearly the same intensity as the tropylium ion at *m/e* 91. The peak at *m/e* 92 is in the same column as the molecular ion at *m/e* 190 and could easily be confused with the molecular ion of toluene (which is at *m/e* 92). The peak shown in the spectrum at *m/e* 92 is a rearrangement ion peak (RIP) formed by transfer of a hydrogen atom. It should be noted here that the array format has listed this rearrangement peak in the same column as the molecular ion peak, where such peaks can easily be observed.

B. The C_nH_{2n} Series—The +0 Series

The C_nH_{2n} series of peaks is designated as the +0 series of hydrocarbons. Typical compounds in this series are shown on the interpretation map under the +0 column (Map 4). These compounds have intense fragment ions in the −1 series of peaks, as illustrated by the partial spectrum for 2,6-dimethyl-2-octene (an alkene), shown in Table 7-8. It should be noted from this spectrum that the intense peaks have been grouped in the −1 series of peaks by the array format. The interpretative map in Appendix II presents this information in the form of a simple correlation for all alkenes and cycloalkenes. For the compound 2,6-dimethyl-2-octene the most intense peak in the −1 series is at *m/e* 69. The formation of the peak at *m/e* 69 can be rationalized as depicted below.

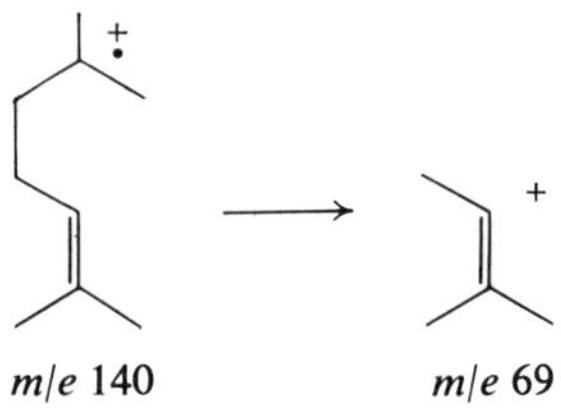

⟶

m/e 140 *m/e* 69

TABLE 7-8

PARTIAL MASS SPECTRUM OF 2,6-DIMETHYL-2-OCTENE

m/e (−1)	Int	*m/e* (+0)	Int	*m/e* (+1)	Int
13	—	14	0.1	15	0.8
27	5.5	28	1.1	29	5.9
41	14.7	42	1.4	43	4.8
55	8.3	56	5.6	57	3.0
69	15.6	70	8.0	71	1.2
83	2.5	84	1.9	85	—
97	0.3	98	0.1	99	—
111	0.9	112	0.1	113	—
125	0.1	126	—	127	—
139	—	140	2.8	141	—

Another compound also with a molecular ion at *m/e* 140 in the +0 series is 1-isopropyl-4-methylcyclohexane (Table 7-9). Again the molecular ion peak (PIP) is an interpretative peak with the most intense peaks of the spectrum in

TABLE 7-9

PARTIAL MASS SPECTRUM OF 1-ISOPROPYL-4-METHYLCYCLOHEXANE

m/e (−1)	Int	*m/e* (+0)	Int	*m/e* (+1)	Int
13	—	14	0.2	15	1.1
27	5.3	28	1.6	29	3.8
41	8.6	42	1.1	43	3.6
55	23.6	56	0.9	57	0.8
69	3.9	70	0.4	71	0.1
83	1.3	84	0.1	85	—
97	14.4	98	0.2	99	—
111	—	112	—	113	—
125	0.1	126	—	127	—
139	—	140	1.8	141	—

the −1 series. The interpretation map (Map 4) shows that this type of fragmentation is typical for cycloalkanes. The fragmentation is depicted below (other schemes of fragmentation are possible).

m/e 140 → *m/e* 97

These two compounds illustrate the typical fragmentation of hydrocarbons of the +0 series. It should be kept in mind that the identification of unknown spectra is the reverse of observing typical fragmentation processes. Thus intense peaks in the −1 series for compounds of molecular weights in the +0 series are identifying clues of hydrocarbons such as alkenes and cycloalkanes. This is typical of identifying clues given by interpretation maps.

C. The −2 Series and −4 Series

The −2 series of hydrocarbons includes several classes of compounds. Among these classes are bicycloalkanes, cycloalkanes, dialkenes, and alkynes or acetylenes. These compounds are listed for rapid reference on the interpretative map under the series designated by a circled number 2 (Map A).

The typical fragmentation of the −2 hydrocarbons produces molecular ions in the −2 series and intense fragment ions in the −3 series. This fragmentation can be illustrated with the partial spectra of two different −2 compounds, each of molecular weight 96. A bicycloalkane structure is shown in Table 7-10, and a dialkene structure is shown in Table 7-11. It is of interest to note that the

TABLE 7-10

Partial Mass Spectrum of Bicyclo(2,2,1)heptane

m/e (−4)	Int	*m/e* (−3)	Int	*m/e* (−2)	Int	*m/e* (−1)	Int
24	—	25	0.1	26	1.0	27	5.6
40	0.9	39	8.3	40	1.6	41	5.1
52	0.5	53	3.0	54	7.7	55	4.9
66	2.2	<u>67</u>	<u>16.3</u>	68	13.3	69	0.9
80	0.3	81	11.1	82	0.7	83	—
94	—	95	0.6	<u>96</u>	<u>3.0</u>	97	0.2

rearrangement peak (RIP) at *m/e* 68 of bicycloheptane is nearly as intense as the fragment peak at *m/e* 67. Fragmentation is depicted below.

TABLE 7-11

PARTIAL MASS SPECTRUM OF 5-METHYL-1,2-HEXADIENE

m/e (−4)	Int	*m/e* (−3)	Int	*m/e* (−2)	Int	*m/e* (−1)	Int
24	—	25	0.1	26	1.0	27	9.6
38	1.0	38	8.3	40	1.1	41	9.4
52	1.4	53	6.7	54	7.7	55	3.0
66	0.4	67	1.6	68	0.9	69	0.5
80	0.4	81	13.4	82	0.9	83	0.1
94	0.1	95	0.4	96	1.1	97	0.1

$$C_5H_9^+ \xleftarrow{-C_2H_5} [\text{structure}]^{+\cdot} \xrightarrow{-C_2H_4} C_5H_8^{+\cdot}$$

m/e 67 *m/e* 96 *m/e* 68

The spectrum of 5-methyl-1,2-hexadiene shown in Table 7-11 has an interpretative peak (PIP) at the molecular ion. The interpretative peak (VIP) at *m/e* 81 is attributed to removal of the methyl group at C-5, as depicted below.

$$H_2C{=}C{=}CH{-}CH_2{-}\overset{CH_3^{+\cdot}}{\overset{|}{C}H}{-}CH_3 \longrightarrow C_6H_9^+ + CH_3$$

m/e 96 *m/e* 81

Both −2 compounds have common spectral features. The intense fragment peaks are, as previously noted, in the column to the left of the molecular ion peaks. This correlation is expressed on Map 3 (Appendix II), which states that the −3 series of fragment ions is characteristic of an olefinic compound (alkene) or a cyclic moiety. Had these been spectra of unknown compounds, information on the interpretative map would have provided a clue to their identification.

Compounds of the −4 series of hydrocarbons include alkenynes, tetradecahydrophenanthrenes, dodecahydrofluorenes, tricycloalkylalkanes, cycloalkyldecahydronaphthylalkanes, and dicycloalkylcycloalkanes. The mass spectrum of 1,3-cyclooctadiene illustrates the fragmentation of a −4 hydrocarbon (Table 7-12). Again, the intense fragment peak, VIP (*m/e* 79), is in the column to the left of the molecular ion (*m/e* 108).

TABLE 7-12

PARTIAL MASS SPECTRUM OF 1,3-CYCLOOCTADIENE

m/e (−6)	Int	*m/e* (−5)	Int	*m/e* (−4)	Int	*m/e* (−3)	Int
36	—	37	0.4	38	1.0	39	7.7
50	1.3	51	2.5	52	1.5	53	1.8
64	0.2	65	2.1	66	3.0	67	8.1
78	2.1	79	15.7	80	8.3	81	0.8
92	0.4	93	7.8	94	0.6	95	—
106	0.1	107	0.7	108	6.9	109	0.5

D. THE −8 SERIES

Compounds in the −8 series include indans and tetralins, as well as several other classes of compounds. Because of the aromatic character of the −8 compounds, their mass spectra show abundant molecular ion peaks. The intense molecular ions serve to identify these compounds by molecular formula. For example, the partial monoisotopic spectrum of an indane is shown in Table 7-13. The molecular ion peak at *m/e* 188 establishes the molecular formula

TABLE 7-13

PARTIAL MASS SPECTRUM OF 1,1,4,5,6-PENTAMETHYLINDAN

m/e (−10)	Int	*m/e* (−9)	Int	*m/e* (−8)	Int	*m/e* (−7)	Int
88	—	89	0.2	90	0.6	91	1.2
102	0.2	103	0.3	104	0.1	105	0.5
116	0.4	117	0.4	118	0.1	119	0.4
130	0.3	131	0.5	132	0.1	133	0.4
144	0.6	145	0.8	146	0.1	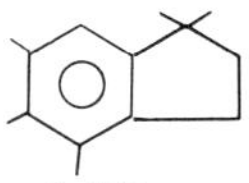	
158	3.6	159	0.3	160	0.1		
172	0.8	173	33.4	174	—	*m/e* 118	
186	0.1	187	0.5	188	5.8	*m/e*	

to be $C_{14}H_{20}$. This compound is substituted with five methyl groups. The "*gem*-dimethyl rule" is operating, as evident by the M − 15 peak (loss of a methyl group), which is the most intense peak (BLIP—base, largest ion peak) in the spectrum. Again the interpretative peak at *m/e* 173 (a VIP) is in the column to the left of the molecular ion peak (*m/e* 188). This information is presented as a generalization on the interpretative Map 2 for −8 compounds. It should be noted from the interpretative map that the −9 peaks starting at *m/e* 117 are characteristic of all −8 compounds, as indicated by the horizontal line extending beyond the vertical line listing characteristic fragment ions of −22 hydrocarbons. The −22 hydrocarbons are in the same series of peaks as the −8 hydrocarbons (8 + 14 = 22).

Table 7-14 shows a spectrum of another −8 compound. Because of the

TABLE 7-14

PARTIAL MASS SPECTRUM OF 1-PHENYL-3-CYCLOPENTYLPROPANE

m/e (−9)	Int	*m/e* (−8)	Int	*m/e* (−7)	Int	*m/e* (−6)	Int
89	0.4	90	0.2	<u>91</u>	<u>21.5</u>	92	20.3
103	0.9	104	3.3	105	2.5	93	2.2
117	1.4	118	0.4	119	0.7	120	0.2
131	1.2	132	0.1	133	0.1	134	0.2
145	0.4	146	0.2	147	0.1	148	—
159	0.2	160	0.1	161	—	162	—
173	0.1	174	—	175	—	176	—
187	—	<u>188</u>	<u>7.7</u>	189	—	190	—

structure, the β-cleavage rule is free to operate, and thus an intense peak (a VIP) is formed at *m/e* 91 as depicted below.

C—C—C (β cleavage) ⟶ + C_7H_{13}

m/e 188 *m/e* 91 (Tropylium ion)

The contrast between the spectra of these two −8 compounds is great. The methyl-substituted indane gives an intense interpretative peak in the column to the left of the molecular ion peak (−9 series), while the phenylcycloalkane compound gives an intense interpretative peak in the column to the right of the molecular ion peak (−7 series). The difference is clearly shown on the array format and demonstrates the great advantage of the array format presentation for interpretation. From such observations, correlations can often be developed which would have otherwise not been defined.

Many other classes of hydrocarbons are mass spectrometrically identifiable using the *z*-number approach. While such compounds are important, further discussion at this point would not appear to greatly enhance the understanding of the use of interpretation maps. The interpretation maps are intended to be used as a desk reference for doing identification work in an analytical mass spectrometry laboratory. For this reason, the maps should be carefully examined. Map A lists mass numbers which are grouped by series with each series designated by a circled number at the top of the series.

E. "Do It Yourself" Examples

With the fragmentation of several classes of hydrocarbons in mind and the availability of identification maps, it becomes possible for the reader to test his skill by attempting to identify hydrocarbons from their mass spectra. One means of doing this is to examine API mass spectral data sheets of the "m" series, which are published in the catalog of the Thermodynamics Research Center, Texas A & M University, College Station, Texas. These spectra are presented in the matrix or array format. By considering peaks and their intensities *only* (names and structures being covered), each spectrum serves as an "unknown." For example, Serial No. 31-m has a molecular ion at m/e 168 with the most intense peak in the spectrum at m/e 41. A clue to the identification of this compound, presuming it to be a hydrocarbon, could quickly be obtained from the interpretative maps (Map 4). The compound is shown to be a +0 compound (alkene or cycloalkane). Another example is the spectrum shown on Serial No. 42-m. The molecular ion at m/e 338, with intense fragment peaks at m/e 29, 43, 57, 71, and 85, shows this compound to be an alkane with 28 carbon atoms (Map 5).

If the API catalog is not presently available to the reader, he can test his knowledge by attempting to identify the hydrocarbons from partial spectra presented in the following tables.

Table 7-15 shows a partial spectrum from the API catalog Serial No. 135-m.

TABLE 7-15

Partial Mass Spectrum API 135-*m*

m/e (−3)	Int	m/e (−1)	Int	m/e (−2)	Int
52	0.1	53	0.1	54	—
66	—	67	—	68	—
80	—	81	0.1	82	0.4
178	3.2	179	5.0	180	10.7
192	0.4	193	0.3	194	0.1
206	0.4	207	4.4	<u>208</u>	<u>16.1</u>

The interpretative peak at *m/e* 208 is assumed to be a molecular ion peak (PIP). The column in which this peak will appear on the interpretative maps can be found from the list of mass numbers given in Map 1. The 208 peak is listed as in the −2 column. The interpretative map with peaks in the −2 column is then referred to in Appendix II. This interpretation map shows that for the

TABLE 7-16

PARTIAL MASS SPECTRUM API 109-*m*

m/e (−8)	Int	*m/e* (−7)	Int	*m/e* (−6)	Int	*m/e* (−5)	Int
76	0.3	77	2.8	78	1.0	79	1.1
90	0.2	91	4.5	92	0.6	93	0.4
104	0.9	105	1.3	106	0.1	107	0.1
118	0.8	119	33.0	120	3.2	121	0.1
132	0.3	133	3.0	134	16.5	135	1.8

molecular ion peak to be a −2 hydrocarbon, the spectrum should have intense peaks at *m/e* 53, 67, and 81, but the spectrum being examined shows that these peaks are of low intensities. For this reason, a −16 hydrocarbon is more probable. We could thus conclude that this unknown spectrum is an aromatic hydrocarbon of molecular weight 208.

The partial spectrum shown in Table 7-16 appears to be a compound of molecular weight 134. The spectrum is polyisotopic; therefore, the peak at

TABLE 7-17

PARTIAL MASS SPECTRUM API 23-*m*

m/e (+0)	Int	*m/e* (+1)	Int	*m/e* (+2)	Int
42	2.7	43	17.0	44	0.6
56	3.1	57	17.5	58	0.8
70	2.3	71	9.5	72	0.5
85	1.5	85	6.4	86	0.4
98	1.1	99	1.0	100	0.1
112	0.6	113	0.6	114	0.1
126	0.3	127	0.6	128	0.1
140	0.1	141	0.3	142	—
154	—	155	—	156	—
168	—	169	—	170	1.4

m/e 135 is due to the contributions of the isotopes ^{13}C and ^{2}H. This molecular ion is shown on the interpretation map (Map 2) to be a –6 aromatic hydrocarbon. The interpretative peak at *m/e* 119 (VIP) is a characteristic fragment ion of aromatics. Thus, it can be concluded from this brief examination of the unknown spectrum that it is an aromatic compound of molecular weight 134 and undergoes a fragmentation to produce quantities of P – 15 ions at *m/e* 119.

Another partial spectrum is shown in Table 7-17. The interpretation map shown as Map 4 is used to aid the identification. The intense peaks in the +1 ion series (*m/e* 43, 57, etc.) and a peak which appears to be a molecular ion peak at *m/e* 170 indicate a +2 hydrocarbon (an alkane). From the molecular ion peak, the molecular formula has to be $C_{12}H_{26}$.

IV. Detection of Functional Groups

Compounds with functional groups and/or hetero atoms are more interesting to examine by mass spectrometry than are hydrocarbons. Functional groups give rise to intense peaks or series of peaks for ions which are either of much higher intensities than found in spectra of hydrocarbons or not possible from any reasonable combination of carbon and hydrogen atoms. For example, the 31 peak is characteristic of many oxygenated compounds, but such an ion cannot be formed by any reasonable combination of carbon and hydrogen atoms. The ion at *m/e* 31 has the elemental composition of CH_3O (*not* C_2H_7). Another example is the peaks in the series *m/e* 47, 61, 75, 89, etc., which are found in small quantities on spectra of hydrocarbons but are much more intense for aliphatic sulfur compounds. Thus, these peaks have significant diagnostic value in the detection of sulfur compounds.

The reason an organic compound with a functional group gives characteristic peaks under electron impact is that the functional group provides a center of preferred localization of the positive charge (McLafferty, 1963b; Budzikiewicz *et al.*, 1967). The localization center acts as a promoter for specific bond cleavages and rearrangement processes. The result is often a well-defined spectrum of which the fragmentation can be rationalized in accordance with the molecular structure of the molecule. To gain an understanding of the use of interpretation maps for compounds with different functional groups, several classes of compounds will be discussed. The discussion will start with nitrogen-containing compounds and will be followed by various classes of sulfur, sulfoxyl, and CHSN compounds. After that, oxygenated compounds will be discussed. Compounds containing a combination of hetero atoms will be discussed last in the examination of the general interpretative maps. Which set of interpretation maps to use for aiding in the identification of any given spectrum will depend upon the information known about the sample. The discussion covering CHN and CHNO identification maps is based on the assumption that the unknown sample falls within these classes of compounds. The same is true for the other specific sets of interpretative maps. If the sample is completely unknown, the ion composition maps should be applied (Section V). Examples have been selected to illustrate the following:

1. The procedure for using interpretative maps of hetero compounds
2. The fragmentation typical of heteroatomic compounds
3. The terminology used in describing the fragmentation processes by electron impact

A. Interpretation Maps for CHN and CHNO Compounds

The fragmentation of nitrogen compounds serves to illustrate several points in the use of interpretative maps. The interpretative maps for CHN and CHNO (Maps 6-10) are arranged similarly to the hydrocarbon maps. Several typical fragmentation features are shown on these maps. Many of the CHN and CHNO compounds give interpretative peaks in the series starting at m/e 30 and increasing by 14 mass numbers. Aliphatic amines and amides give intense rearrangement ions in the 31, 45, 59, etc., series. Compounds of the general formula $C_nH_{2n+1}N$ and $C_nH_{2n-1}NO$ usually give peaks in the 42, 56, 70, etc., series, which are more intense than peaks in the 41, 55, 69, etc., or 43, 57, 85, etc., series. The fragmentation of amines has been studied (Gohlke and McLafferty, 1962) and found to exhibit both α and β cleavage of the C–C bond adjacent to the nitrogen atom, depending upon the type of amine and the length of the alkyl chain.

A sense of reality is added to identifying an unknown substance from its mass spectrum if the spectrum is considered by the reader as truly unknown. For this reason, many of the partial spectra are designated by numbers. These numbers are, whenever possible, the same as the API or TRC series numbers. The reader can thus locate the complete spectra in the mass spectral data catalogs of API and TRC. Numbers starting with C are unpublished spectra in the files at Continental Oil Co.

Table 7-18 shows a mass spectrum of a compound which, for the purpose of this illustration, has been restricted to a compound containing the elements carbon, hydrogen, and nitrogen (CHN compound). Because the spectrum is a CHN compound, the 101 peak can be assumed to be a molecular ion peak and, as such, is the starting point for the interpretation. This peak is designated as being in the −11 series from the listing on Map A.

The −11 series of peaks is found on Map 6 of the interpretative maps for

TABLE 7-18

PARTIAL MASS SPECTRUM C-9660

m/e (+0)	Int	m/e (+1)	Int	m/e (+2)	Int	m/e (−11)	Int
28	6.2	29	6.6	30	9.8	31	—
42	5.4	43	0.7	44	3.9	45	—
56	2.6	57	0.6	58	7.1	59	—
70	1.2	71	0.4	72	0.9	73	—
84	0.2	85	0.2	86	30.2	87	—
98	—	99	0.1	100	2.0	101	4.4

CHN and CHNO compounds (recall Chapter 6). The reader should now refer to details on Map 6 (Appendix II). Lines are used to confine classes of compounds within a designated series. For example, this map indicates that quinolines and isoquinolines have molecular ions (M^+) of an m/e 129 or greater. These classes of compounds cannot exist with less than nine carbon atoms in their structures. For this reason, in the example given of a CHN compound with a molecular ion of 101, a quinoline or isoquinoline structure is impossible. (These maps do not, of course, exclude other possible classes of compounds.) If classes of compounds containing oxygen are excluded, the maps leave only the possibility of amines. Starting with m/e 31 as an amine with one carbon atom ($C_nH_{2n+1}NH_2$ with $n \geqq 1$), by counting down the −11 column by one carbon for each m/e shown, it is found that an amine of molecular weight 101 has six carbons or a formula $C_6H_{13}NH_2$. This compound is known to be triethylamine.

The degree of saturation of a compound has to be kept in mind in using these interpretation maps. Thus, a cyclic structure or a monoolefinic structure having the same number of carbon atoms would have a molecular ion at two mass numbers less than that of a saturated structure. For example, both

TABLE 7-19

PARTIAL MASS SPECTRUM OF HEXAMETHYLENIMINE

m/e (−1)	Int	m/e (+0)	Int	m/e (+1)	Int	m/e (+2)	Int
55	1.6	56	8.0	57	5.6	58	0.3
69	0.6	70	10.6	71	1.1	72	0.1
83	0.1	84	1.4	85	0.2	86	—
97	0.2	98	1.7	99	3.8	100	0.3

triethylamine and hexamethylenimine (Tables 7-18 and 7-19, respectively) contain six carbon atoms but differ in the number of hydrogen atoms, and thus they are in columns separated by two mass numbers on the array (*m/e* 101 and 99). The interpretative maps for hetero compounds are set up mainly for saturated structures and may not always include the corresponding unsaturated or cyclic class of compounds. In this case, the corresponding cyclic structures are included, as illustrated for the compound hexamethylenimine (Table 7-19). The molecular ion is at *m/e* 99 (again at an odd mass number). The molecular ion peak at *m/e* 99 is found in the +1 series in Appendix II. The maps indicate that this "unknown" compound has a cyclic amine structure and has an empirical formula of $C_6H_{13}N$ (CHNO compounds are excluded in this example).

The spectrum of another "unknown" nitrogen-containing compound is shown in Table 7-20. If this spectrum represented an unknown sample

TABLE 7-20

PARTIAL MASS SPECTRUM C-9529

m/e (−5)	Int	*m/e* (−4)	Int	*m/e* (−3)	Int	*m/e* (−2)	Int
23	—	24	0.1	25	0.3	26	1.9
37	1.5	38	2.4	39	7.7	40	2.0
51	2.1	52	1.5	53	6.2	54	2.8
65	0.1	66	0.8	67	0.1	68	0.1
79	0.6	80	15.4	81	18.8	82	1.1

submitted for identification, it would be best to examine the −3 series of peaks on Map 8 because of the interpretative peak at *m/e* 81. Compounds of the $C_nH_{2n-5}NO$ type, such as pyridones, are excluded because their molecular ions do not exist below *m/e* 95 (the molecular weight of 2-pyridone itself is 95). Pyrroles are thus indicated as a probable compound type. It should be remembered that the general formula $C_nH_{2n-3}N$ may include classes of compounds other than pyrroles, and for that reason, the identification should be worded with this possibility in mind. This spectrum is known to be *N*-methylpyrrole.

The spectrum shown in Table 7-21 has many even-numbered peaks, more intense than the odd-numbered peaks. The first three peaks listed in the +0 series are more intense than either of the first three peaks in the −1 or +1 series. This spectral feature is characteristic of nitrogen-containing compounds. The peak at *m/e* 100 is assumed to be a molecular ion peak and is an *even*-numbered peak. If this spectrum represents a nitrogen-containing compound,

TABLE 7-21

PARTIAL MASS SPECTRUM API 175-*m*

m/e (−1)	Int	*m/e* (+0)	Int	*m/e* (+1)	Int	*m/e* (+2)	Int
41	4.8	42	11.3	43	7.9	44	21.8
55	0.6	56	9.9	57	8.7	58	5.6
69	0.6	70	2.0	71	1.8	72	0.1
83	0.5	84	0.5	85	10.2	86	0.5
97	0.2	98	0.1	99	0.8	100	2.1

it has to contain an *even* number of nitrogen atoms. The +2 series of peaks shown on Map 10 gives the general formula $C_nH_{2n+2}N_2$ for this type of compound and denotes piperazines as a possible structure. To fit the general formula, the peak at *m/e* 100 has to be a $C_5H_{12}N_2$ compound. The spectrum is known to be 2-methylpiperazine.

B. INTERPRETATION MAPS FOR SULFUR, SULFOXYL, AND CHNS COMPOUNDS

Sulfur-containing organic compounds fragment under electron impact producing characteristic series of peaks which provide clues to their identification. These are expressed on interpretative maps (Maps 11–15). The two most characteristic series of peaks for aliphatic sulfur-containing compounds are at the $C_nH_{2n+1}S^+$ ions (*m/e* 47, 61, 75, etc.) and $C_nH_{2n-1}S^+$ ions (*m/e* 45, 59, 73, etc.). The first series of ions starts with the elemental composition of CH_3S, and the second series starts with CHS ions. These two series of peaks are designated as −9 and −11 on the interpretative maps shown on Maps 11 and 15, respectively. Aromatic sulfur ions have prominent peaks at various *m/e*'s including the 97 series (thiophenes), the 110 and 123 series (aromatic thiols), and the 147 series (benzothiophenes).

Several examples will illustrate the use of CHS and CHSO maps. The partial spectrum of an "unknown" is shown in Table 7-22. The sample is known to be a CHS compound. The objective is to identify this substance from its mass spectrum by using the interpretative maps. The 126 peak is in the +0 series shown on interpretative Map 15. Again it is convenient to use Map A in locating the interpretative map for any given peak. From the +0 series of peaks, one sees that the peak at *m/e* 126 could be one of three general formulas, $C_nH_{2n-4}S$, $C_nH_{2n+2}S_3$, or $C_nH_{2n+2}SO_4$, but not $C_nH_{2n-6}SO$. The line confining the $C_nH_{2n-6}SO$ compounds excludes the possibility of compounds of this general formula being of molecular weights below 140. Medium- or high-

TABLE 7-22

PARTIAL MASS SPECTRUM C-859

m/e (−11)	Int	m/e (−9)	Int	m/e (−1)	Int	m/e (+0)	Int
45	9.2	47	0.8	55	0.1	56	0.1
59	0.6	61	0.3	69	1.7	70	0.6
73	0.1	75	0.1	83	0.2	84	0.9
87	0.1	89	0.1	97	26.4	98	3.3
101	—	103	—	111	0.8	112	0.1
129	—	117	—	125	0.3	126	11.0

resolution mass spectrometry could resolve which of the four compound types is present. If high-resolution mass spectrometry is not available, the maps provide, in this case, a means of selecting one of these compound types. If we can narrow our selection to only CHS compounds (assumed at the start), we have only two general formulas, $C_nH_{2n-4}S$ and $C_nH_{2n+2}S_3$. The latter requires that three sulfur atoms be present, leaving the hydrocarbon moiety to be 30 [$126 - (3 \times 32) = 30$]. The value of 30 agrees with the C_nH_{2n+2} part of the $C_nH_{2n+2}S_3$ formula; however, only two carbons are possible for mass of 30 (C_2H_6). The value of two could be ruled out because such a structure as $C_2H_6S_3$, while it might exist in one of the forms shown below, would demonstrate its presence by other interpretative peaks.

$$HS—CH_2—CH_2—S—SH \quad \text{or} \quad HS—CH_2—S—CH_2—SH$$

The VIP = 34 from H_2S^+; perhaps even an H_3S^+ ion at m/e 35, and $H_2S_2^+$ at m/e 66 are other interpretative peaks expected from $C_2H_6S_3$.

It should be noted that the two peaks at m/e 97 and 126 could be confused with those of a cycloalkane or alkene except that in this spectrum, interpretative peaks appear at m/e 45 and m/e 47. These peaks would not have been as intense had this spectrum been a hydrocarbon. Another indication that the peak at m/e 126 is not from a hydrocarbon can be deduced from the complete spectrum by inspecting the isotopic peaks above the molecular ion peak. The sulfur atom has two principal stable isotopes, one of mass 32 and another of mass 34. The ^{34}S isotope, 4.2% of naturally occurring ^{32}S, is useful for identifying sulfur-containing compounds. Thus, a sulfur-containing ion of an intensity of 100 divisions gives an isotopic peak of about four divisions at a peak 2 mass units greater. It can then be concluded that alkylthiophenes are the most probable identification for this spectrum. Other structures are, or course, possible. In this case, note from the map that the general formula $C_nH_{2n-4}S$ exists only for compounds starting with four carbon atoms ($n \geqq 4$). The m/e 126 is *three* numbers below m/e 84 or four carbons in the +0 series; therefore, the compound is indicated to be a C_3 alkyl-substituted thiophene. The spectrum is known to be 2-propylthiophene ($C_7H_{10}S$).

The fragmentation of 2-propylthiophene illustrates carbon–carbon cleavage at the bond β to the aromatic ring to form the base peak (most intense peak in the spectrum). This is the same type of cleavage characteristic of alkylbenzenes. For this compound, β-bond cleavage produced the interpretative peak at m/e 97, as depicted below. The methylethylthiophene could also cleave

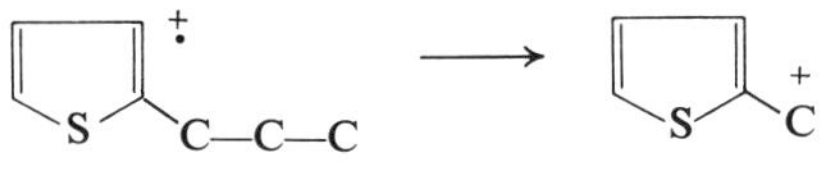

S C—C—C → S C

m/e 126 m/e 97

β to the ring on the longest chain. Its molecular ion would be at m/e 126, but its base peak would be at m/e 111 rather than at m/e 97.

Another "unknown" compound is shown in the partial spectrum in Table 7-23. The molecular ion at m/e 160 is in the −8 series as shown from Map A.

TABLE 7-23

PARTIAL MASS SPECTRUM OF 3-PHENYLTHIOPHENE

m/e (−11)	Int	m/e (−10)	Int	m/e (−9)	Int	m/e (−8)	Int
45	3.0	46	0.1	47	0.2	48	—
59	0.1	60	0.1	61	0.5	62	1.1
73	0.2	74	0.9	75	0.8	76	0.7
87	0.5	88	0.3	89	2.0	90	0.5
101	0.2	102	1.2	103	0.2	104	—
115	9.0	116	2.0	117	0.2	118	—
129	0.2	130	—	131	—	132	0.2
143	—	144	—	145	0.2	146	—
157	0.1	158	1.0	159	2.0	160	32.0

The interpretative map (Map 11) shows that several classes of compounds are possible for the peak at m/e 160. In such cases, it is a wise procedure to examine other peaks from the mass spectrum so as to narrow down the possibilities. The peak at m/e 115 is shown on the hydrocarbon maps (Map 1) to be characteristic of compounds having an aromatic moiety. Only one of the empirical formulas from the −8 series suggests an aromatic type of compound; therefore, it can be assumed that the compound has a general formula of $C_nH_{2n-12}S$, and the specific formula is thus $C_{10}H_8S$. The compound is known to be 3-phenylthiophene, for which the fragmentation with rearrangement is shown below.

$$\xrightarrow{-CHS\cdot} \qquad \xrightarrow{-C_2H_2}$$

$M^+ = 160$ *m/e* 115 *m/e* 89

The loss of the CHS radical going from M^+ 160 to *m/e* 115 is equal to 45 mass units. The ion at *m/e* 115 further degrades by loss at ethylene to give the interpretive peak at *m/e* 89. It should be noted that the β-bond cleavage rule which operates for alkyl substitution cannot apply to this compound. The interpretative peak at *m/e* 89 is shown above to be a hydrocarbon ion containing seven carbon atoms, bearing the same relationship to a tropylium ion that the benzyne ion bears to the benzene ion.

The last two ions described, namely, at *m/e* 115 and 89, occur rather commonly from condensed-ring aromatic systems, including heteroatomic aromatics. They provide doubly charged ions, and hence contribute peaks at *m/e* 57.5 and 44.5. For the alkylnaphthalenes series of compounds, the peak at 57.5 exceeds in intensity one at either *m/e* 57 or 58 and hence calls attention to an unusual "doubly charged interpretative peak" (DIP). These double charged ions are listed separately on the API-44 matrix format, permitting a rapid check on such a spectrum for this type of ion peak. These peaks are also included in the summation of peaks for total ion intensity data in the computer printout as developed at Continental Oil Co.

The partial spectrum shown in Table 7-24 is another "unknown" containing sulfur. The molecular ion is in the same series of peaks (−8) as the previous unknown. The possibility of a $C_nH_{2n-12}S$ compound is ruled out because the

TABLE 7-24

PARTIAL MASS SPECTRUM OF 3-METHYL-3-PENTANETHIOL

m/e (+1)	Int	*m/e* (−10)	Int	*m/e* (−9)	Int	*m/e* (−8)	Int
43	17.4	46	—	47	1.6	48	—
57	2.5	60	0.2	61	1.4	62	—
71	0.3	74	0.3	75	—	76	—
85	9.3	88	0.4	89	3.9	90	0.3
99	—	102	—	103	—	104	—
113	—	116	—	117	—	118	2.7

interpretative peak of highest mass number is at *m/e* 118. However, remaining possibilities (Map 11) include an aliphatic thiol (R–SH), an aliphatic thio ether (R–S–R), and compounds of the general formula $C_nH_{2n}SO$. It should be noted that peaks in the sample spectrum at *m/e* 43 and 85 in the +1 series are indicated on Map 15 to be characteristic of thiols. The peak at *m/e* 85 is 33 mass units (SH) below the molecular ion (118 − 33 = 85), which is an empirical characteristic of the fragmentation of thiols. In such cases, it may be necessary to report only that the compound probably has an elemental composition of $C_6H_{14}S$, for which a thiol structure is possible. Positive proof of identification requires a more detailed examination of known reference standards.

Interpretative maps are also included in the appendix for CHNS compounds (Maps 16–18). Consider the partial spectrum shown in Table 7-25 as an

TABLE 7-25

PARTIAL MASS SPECTRUM API 1759

m/e (+2)	Int	*m/e* (−11)	Int	*m/e* (−7)	Int	*m/e* (−5)	Int
44	0.4	45	3.6	49	0.6	51	0.9
58	1.8	59	0.1	63	3.8	65	0.1
72	0.1	73	0.3	77	0.3	79	—
86	0.1	87	0.1	91	1.8	93	0.6
100	—	101	—	105	0.2	107	0.5
114	—	115	—	133	0.1	<u>135</u>	<u>29.4</u>

"unknown" CHNS compound. Again we start with the interpretative peak in the high-mass region of the spectrum, which is the 135 peak. This peak is in the −5 series and can be found on Map 17. Both compound types shown on the maps are of the general formula $C_nH_{2n-9}NS$. A more detailed identification depends upon fragmentation differences between these two classes of compounds. In such cases, it is necessary to examine the various collections of published mass spectral data.

C. INTERPRETATION MAPS FOR OXYGEN-CONTAINING COMPOUNDS

Oxygen-containing compounds have been extensively examined by mass spectrometry. Budzikiewicz *et al.* (1967) have reviewed and discussed many fragmentation processes. The interpretative maps in Appendix II reflect the extensive fragmentation knowledge by listing ionic structures for some of the more common fragmentation ions containing oxygen. The classes of compounds and ions shown are those which are commonly encountered in a research analytical mass spectrometry laboratory.

The purpose of this chapter has been to introduce the reader to the use of interpretative maps as an aid to making identifications of substances from their mass spectra. At this point, the reader should be quite well informed in the use of these maps. For this reason, the examples of CHO compounds will be confined to spectra which are difficult to identify and for which the maps may do no more than to direct the procedure to follow in arriving at a compound class identification. In many practical problems, a small amount of additional information will often make it possible to deduce an unambiguous identification. For example, if an unknown is indicated by the maps to be an ester, a more extensive examination of the fragmentation of esters can be undertaken. It is hoped that with the use of interpretative maps, many samples

TABLE 7-26

PARTIAL MASS SPECTRUM C-9254

m/e (−1)	Int	*m/e* (+0)	Int	*m/e* (−1)	Int	*m/e* (+2)	Int
41	10.4	42	1.3	43	18.6	44	0.9
55	5.8	56	1.4	57	0.3	<u>58</u>	<u>2.6</u>
69	5.1	70	0.4	<u>71</u>	<u>1.8</u>	72	0.1
83	1.2	84	0.2	85	0.1	86	—
97	0.1	98	—	99	—	100	—
<u>111</u>	<u>1.8</u>	112	0.2	113	—	114	—
125	—	<u>126</u>	<u>1.1</u>	127	0.1	128	—

which had fallen by the wayside and were reported as "no structural information could be obtained from the mass spectral data" will be moved into a probable class identification at least. The reader should not be led to believe that by selecting oxygenated compounds to illustrate some of the difficulties in identifications, this class of compound is, in general, more difficult to identify than many of the other classes of organic compounds.

Consider the spectrum in Table 7-26 as an "unknown." The peaks at *m/e* 58, 71, 111, and 126 are interpretative peaks. The odd-numbered peaks could be attributed to fragment ions from the even-numbered molecular ion peak at *m/e* 126. The peak at *m/e* 58 would be expected to be a rearrangement peak in view of the fact that it is an even-mass-numbered peak from an even-numbered molecular ion peak. Such rearrangement peaks often are important clues to identifications of unknown compounds. From Map 25 the 58 peak is designated as the first of the rearrangement ions in the +2 series of peaks, which are characteristic of ketones. Saturated aliphatic ketones have molecular ions in the same series of peaks as these rearrangement peaks. At first this may appear to rule out identifying ketones, because the molecular ion appears to be in the +0 series rather than the +2 series. It should be kept in mind, however, that a double bond or a cyclic structure has the effect of transferring the molecular ion to a column of two lower mass numbers. If this bit of information is applied to this sample, the peak at *m/e* 126 could be an unsaturated ketone or a cyclic ketone. Such an incomplete identification can be reported as a probable $C_nH_{2n-2}O$ compound of molecular weight 126. If a known reference spectrum for this compound is available, proof of identification would, of course, be possible. This compound is known to be 6-methyl-5-hepten-2-one.

TABLE 7-27

PARTIAL MASS SPECTRUM C-5953

m/e (+1)	Int	*m/e* (+2)	Int	*m/e* (−11)	Int	*m/e* (−10)	Int
43	6.9	44	0.3	45	3.6	46	0.1
57	0.7	58	0.1	59	0.7	60	18.9
71	2.4	72	0.6	73	10.0	74	1.2
85	0.2	86	—	87	3.2	88	0.2
99	—	100	—	101	0.1	102	0.1
127	—	128	—	129	—	130	0.1

Another class of oxygenated compound is illustrated in the spectrum shown in Table 7-27. The even-numbered interpretative peak at *m/e* 60 appears to be a key peak to the identification. This peak is shown on Map 19 to be characteristic of acids. Again, the other information shown has to be, if possible, excluded from consideration. In practice it is not always possible to exclude all of the bits of information supplied by the interpretative map or to quickly select the peak(s) which will lead to the correct identification. The low intensity of peaks shown on the spectrum in the series *m/e* 58, 72, 86, etc., serves to exclude ketones (previously discussed). Likewise, the low intensities for peaks at *m/e* greater than 74 in the −10 series (Map 19) help to exclude esters. The compound shown in Table 7-27 is known to be heptanoic acid. While aliphatic acids are given on Map 19, it may be difficult to select from the maps alone the correct identification. Identification may require auxiliary information about the sample or the availability of known reference standards for direct comparison. In this case, the selection of an acid is somewhat selectively indicated by the direct line from the peak at 60 to the listing of saturated aliphatic acids.

Another spectrum is shown in Table 7-28 which could be difficult to identify even by one experienced in interpretation. The peak at *m/e* 31 is listed on Map 19 as characteristic of alcohols, ethers, and esters. The peak at *m/e* 44 on Map 25 is shown as a $CH_2{=}CHOH^+$ ion (Meyerson and McCollum, 1963) and is a characteristic rearrangement ion of aldehydes and/or vinyl ethers. A compound identification may be difficult to make from this mass spectrum, and even a compound class identification is not unambiguous. This spectrum

TABLE 7-28

PARTIAL MASS SPECTRUM C-4770

m/e (−11)	Int	*m/e* (+0)	Int	*m/e* (+1)	Int	*m/e* (+2)	Int
31	2.0	42	3.7	43	29.1	44	2.8
45	1.1	56	5.0	57	3.6	58	0.2
59	0.3	70	0.2	71	0.2	72	—
73	—	84	1.4	85	0.9	86	0.1
101	—	112	—	113	1.9	114	0.2
115	—	126	—	127	—	128	0.1

is known to be hexyl vinyl ether. Some merit is realized, however, by an examination of such a spectrum if no more than that this sample is not well suited to identification by mass spectrometry. Other analytical techniques can then be considered.

Since the use of a computer for aid in interpretation is the ultimate aim in mass spectrometry today, the use of the abbreviations for ion peaks such as have been employed should not be surprising. One needs abbreviations for feeding the computer data and marking on spectra, and some such designations are helpful. To summarize the suggested usage for various peaks, the following is offered:

PIP Parent interpretative peak
BLIP Base, largest intensity peak
VIP Visually interpretative peak
RIP Rearrangement ion peak
CLIP Clustered isotope peak(s)
DIP Doubly charged interpretative peak
MIP Metastable ion peak
SIP Small isolated interpretable peak

DRIP Double rearrangement interpretable peak
TRIP Triple rearrangement interpretable peak

Handling numerical data and "picking peaks" can be tedious, and hence any form of lightening the load for the processer helps keep the team rolling along.

V. Ion Composition Maps

A. Introduction

The ion composition maps, Maps 26–39, are intended to aid in the identification of completely unknown substances. One of the several practical applications is found when unknown substances are collected by gas chromatography and submitted to mass spectrometry for identification.

These maps were formulated from intense peaks which appear on mass spectra and not based upon mathematical possibilities of combinations of ions. Spectral *data* made available over the period of several years have been used. The ion composition maps are arranged similarly to specific maps. One different series of *m/e* occurs on each of the 14 maps. The numbers circled at the top of each series represent those from the hydrocarbon *z* numbers −11 to +2 (C_nH_{2n-11} to C_nH_{2n+2}), just as employed earlier. To locate a map for a given *m/e*, it is convenient to first find the series in which the *m/e* appears from those listed on Map A. By knowing the series, the desired interpretation map becomes rather easy to locate since these are in the order of the series numbers (−11 to +2).

Lines are drawn on the maps to confine certain combinations of hetero atoms to their correct *m/e* values in each series. For example, on Map 26 lines are used to confine peaks starting at *m/e* 227 in this series to ions with a bromine atom, a chlorine atom, and four fluorine atoms combined with a C_nH_{2n-5} alkyl group when *n* is equal to three or more carbon atoms ($79 + 35 + (4 \times 19) + C_3H = 227$). The *z* value in the formula $C_nH_{2n+z}BrClF_4$ is shown in parentheses below the hetero atom(s). As *n* is increased by one, the *m/e* in the series is increased by 14. Both the moieties of the hetero atoms and the hydrocarbons are considered in the confining lines; thus in this example only a C_nH_{2n-5} hydrocarbon can be combined with $BrClF_4$ and be in the series starting at *m/e* 227. When a combination of hetero atoms does not include both a hydrogen atom and a carbon atom, no *z* number in parentheses is shown with the hetero atoms. Such hetero atoms are confined to the *m/e* which they represent. For example, HO and H_3N are confined to *m/e* 17 (Map 26). In cases where the number of carbon atoms would appear to aid in the interpretation, that number is supplied below the *z* number in parentheses as C_4 to designate four carbons in the general formula C_nH_{2n+z}.

B. Using Ion Composition Maps

A few examples will illustrate the use of the maps. Consider a mass spectrum (API Serial No. 1227) with interpretative peaks at 47, 54, 67, 82, 87, and 116. The peak at *m/e* 116 (VIP) is the highest *m/e* of these interpretative peaks and therefore is used as the starting point. Appendix II shows the 116 peak to be in the −10 series. The next step is to select from the remaining interpretative peaks those peaks which are in columns nearest the −10 column. The peak at *m/e* 87 is in the −11 series, and the peak at *m/e* 47 is in the −9 series. The mass difference between the 116 peak and the 87 peak is 29 mass units, indicating cleavage of a hydrocarbon moiety, probably C_2H_5. If this is true, then both the 87 and 116 ions have the same hetero atoms present (assuming the organic compound is not a hydrocarbon). The hydrocarbon moiety combined with the same hetero atom would thus contain one less hydrogen atom in the −11 series than in the −10 series. Likewise, the hydrocarbon moiety in the −10 series contains one less hydrogen atom than an ion with the same hetero atom in the −9 series. If we now take the lowest of these three *m/e*'s (47, 87, and 116), we can arrive at the fewest possibilities for identifying the hydrocarbons. For example, the 47 peak is shown on Map 28 to be combined with the atoms S, O_2, and F. (The 116 peak would have given a much greater number of possibilities.) Further, at least C_2H_5 is included somewhere in the 116 ion moiety. The hydrocarbon moieties decrease by one hydrogen atom for each of these hetero atoms going from the −9 series to the −11 series. The atoms listed directly across from the 47 peak are O_2 and S and, as such, are somewhat more probable than the F atom. The point being illustrated is that we have narrowed the composition of the unknown to a rather small number of possibilities. Additional work can often further narrow down the number of possibilities and lead to an identification which would not otherwise have been achieved.

Consider the example of an unknown spectrum with interpretative peaks at *m/e* 102 and 104 in the high-mass region of the spectrum. The spectrum being considered is dichloromonofluoromethane (ASTM No. 1093), in which the peak at *m/e* 102 is one of the interpretative peaks. From Map A the 102 peak is shown to be in the series designated as −10 and thus is found on ion composition Map 27. Two combinations of hetero atoms are shown directly across from *m/e* 102, Cl_2F and the F_3S, each of which is combined with a C_nH_{2n-1} moiety where *n* is one or greater. At this point in the interpretation of such a spectrum, one could report that either or both identifications are possible. More information can, however, be obtained. If the ratio between 102 and 104 peaks is 3:2, two chlorine atoms are indicated. If the ratio between the 102 and 104 peaks is 100:4, a single atom of sulfur would be indicated. In this example, the 102 and 104 ratio is 3:2, indicating a dichloro compound. It should be kept in mind that peaks in the 102 series at mass numbers greater than 102 could be higher homologs of the type of compound identified from the 102 peak. For example, a peak at *m/e* 116 (14 mass units greater than 102) could be dichloromonofluoromethane.

Consider another spectrum (API Serial No. 91-m) as an unknown. This spectrum has interpretative peaks at *m/e* 27, 49, 51, 62, 64, 98, and 100. The peaks at *m/e* 98 and 100 are in the ratio of 3:2, suggesting that the two chlorine atoms are present in the peak at *m/e* 98. Map A is used to locate the series of peaks in which *m/e* 98 appears. Thus, reference is made to the +0 series in Map 37, where across from peak 84 are listed two chlorine atoms with a C_nH_{2n} hydrocarbon moiety. For the ion at *m/e* 98, the empirical formula would then be $C_2H_4Cl_2$. This agrees with the spectrum which is known to be 1,2-dichloroethane. However, care should be taken that the ease of an identification does not lead to serious errors. The ion composition maps aid in reducing such errors by showing other possibilities. For example, Map 37 shows another characteristic ion in the +0 series also containing two chlorine atoms for ions at *m/e* 140 and greater. These ions are shown to contain two oxygen atoms in addition to two chlorine atoms. Consideration should be given, in such cases, to the possibility that the fragment peak at *m/e* 98 could be an ion containing two chlorine atoms which were formed from a more complicated compound not giving a detectable molecular ion peak.

VI. Some Additional Benefits of the Array

A. Rearrangement Peaks and Other Odd-Electron Ion Peaks on the Array

Many classes of compounds are known to give intense rearrangement peaks. These ions are sometimes referred to as being formed by the "McLafferty rearrangement" (McLafferty, 1956, 1959). The detection of rearrangement ions is expedited by the array format because such ions are usually grouped in the same column as molecular ions. In the partial spectrum of 4-octanone shown in Table 7-29, peaks at *m/e* 58 and 86 are prominent rearrangement ions (RIP) from the molecular ion at *m/e* 128.

Similar observations can be made from spectral data when odd-electron ions are formed by other ion reaction processes, such as the "retro-Diels–Alder cleavage" (Biemann, 1962). Hamming and Eisenbraun (1966) have shown that this type of fragmentation is common for alkyldihydronaphthalenones. One of these compounds is shown in Table 7-30, in which the RIP ion at *m/e*

TABLE 7-29

PARTIAL MASS SPECTRUM OF 4-OCTANONE

m/e (+1)	Int	m/e (+2)	Int	m/e (−11)	Int
43	13.9	44	0.5	45	0.1
57	11.0	58	6.7	59	0.3
71	10.4	72	0.5	73	0.1
85	7.5	86	2.4	86	0.2
99	0.4	100	0.1	101	—
113	0.1	114	—	115	—
127	—	128	2.1	129	0.2

TABLE 7-30

PARTIAL MASS SPECTRUM OF 3,4-DIHYDRO-2,5,8-TRIMETHYL-1(2*H*)-NAPHTHALENONE

m/e (−9)	Int	m/e (−8)	Int	m/e (−7)	Int
131	0.3	132	0.2	133	0.1
145	1.4	146	18.2	147	1.9
159	2.5	160	0.7	161	0.1
173	1.4	174	0.2	175	—
187	0.2	188	7.5	189	1.1

146 is 42 mass numbers less than the molecular ion at m/e 188. The fragmentation is shown below.

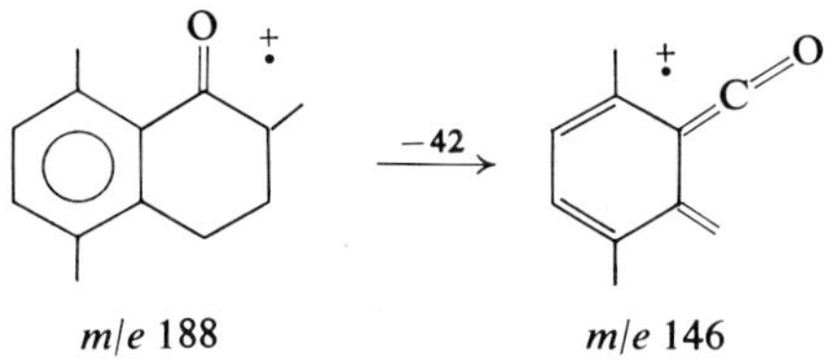

−42

m/e 188 m/e 146

B. ISOTOPIC PEAKS ON THE ARRAY

Mass spectra of certain classes of organic compounds contain intense isotopic peaks. McLafferty (1966b) has described peaks of natural isotopic abundances as being in "isotopic clusters" (CLIP). The rectangular array provides a visual display of such peaks, which often are in sets of "isotopic clusters" which differ by 14 or 28 mass units. An example of the isotopic peaks presented on the array is shown in Table 7-31. The 164–166 peaks (PIP) are

TABLE 7-31

PARTIAL MASS SPECTRUM OF 1-BROMOHEXANE

m/e (−5)	Int	m/e (−4)	Int	m/e (−3)	Int	m/e (−2)	Int
93	0.3	94	—	95	0.3	96	—
107	1.0	108	0.1	109	1.0	110	—
121	0.1	122	—	123	0.1	124	—
135	10.4	136	0.5	137	10.2	138	0.5
149	—	150	—	151	—	152	—
163	—	164	1.0	165	—	166	0.9

molecular ions which contain the naturally occurring isotopes ^{79}Br and ^{81}Br. The ratio 1:1 is also shown in the other two "isotopic clusters" (CLIP) at *m/e* 135–137 and 107–109. Therefore these ions still contain the bromine atom.

C. COLUMNAR SUMMATION OF TYPE ANALYSIS

Mass spectrometry has been used extensively to provide hydrocarbon type analyses. Such methods are usually based upon a summation of characteristic peaks which is then used in place of a single peak. In the development of type analyses methods, large amounts of calibration data usually have to be carefully examined in order to find which set of peaks is most characteristic of a given class of compounds. Lumpkin (1956) has provided an excellent illustration of the development of a hydrocarbon type analyses method.

It is often desirable to develop individual type analyses methods. Instrumental differences between mass spectrometers cause production of varying fragmentation patterns which, in turn, can cause rather large differences in analytical results obtained on a given sample. For this reason, calibration data obtained in the same laboratory in which the samples are being analyzed are more desirable than calibration data found in the literature. The main objection to a development of an individual hydrocarbon type method is the time and cost involved. The array format, together with computer summation of a selected set of peaks, can greatly expedite development of an individual laboratory method. Type analyses are discussed in more detail in Chapter 6.

D. Columnar Summation for Structural Elucidations

Columnar summations of peaks in the array format can be used to make "spot correlations." A striking example is that C_nH_{2n-8} compounds with a single nucleus (condensed structure) have a much higher sum of −9 series peaks than −8 compounds with two nuclei per molecule (noncondensed or dinuclear), which are characterized by an intense summation of −7 fragment ions. Several mass spectra were examined in the formation of this "spot correlation." Two typical compounds are shown with summation intensities in Table 7-32. The

TABLE 7-32

Homologous Series Summations for C_nH_{2n-8} Compounds

Structure	Reference	C_nH_{2n-9}	C_nH_{2n-8}	C_nH_{2n-7}
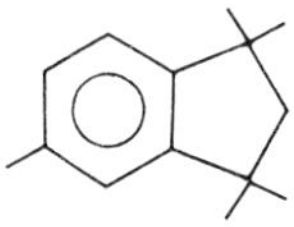	API 210-m	37.8	10.8	7.1
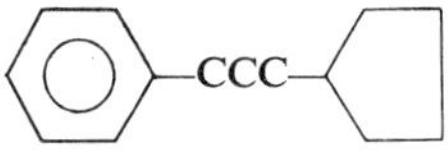 CCC	API 36-m	2.0	7.5	17.8

examination consisted of finding the sum of the series −9, −8, and −7 and relating these summations to the different molecular structures. For the C_nH_{2n-8} compounds examined, it was found that compounds with a single nucleus (condensed structure) had much more intense −9 series peaks than the −8 or −7 series. However, C_nH_{2n-8} compounds with two nuclei per molecule (noncondensed or dinuclear) were found to have a −7 series that was much more intense than the −8 or −9 series.

Identifications can also be made from columnar summations. A computer program has been developed for the identification of aliphatic hydrocarbons which utilize column summations (Pettersson and Ryhage, 1967). Such work illustrates the concept in making identification in which peak summations are used in place of single peaks. It should be possible to extend the concept to other types of organic compounds.

REFERENCES

ASTM (1963). "Index of Mass Spectral Data," ASTM Special Technical Publication No. 356. American Society for Testing and Materials Committee E-14, Philadelphia, Pennsylvania.

Beynon, J. H., and Williams, A. E. (1963). "Mass and Abundance Tables for Use in Mass Spectrometry," 570 pp. Amer. Elsevier, New York.

Biemann, K. (1962). "Mass Spectrometry: Organic Chemical Applications," p. 102. McGraw-Hill, New York.

Budzikiewicz, H., Djerassi, C., and Williams, D. H. (1967). "Mass Spectrometry of Organic Compounds," 690 pp. Holden-Day, San Francisco, California.

Clerc, R. J., Hood, A., and O'Neal, M. J., Jr. (1955). *Anal. Chem.* **27**, 868.

Cornu, A., and Massot, R. (1966). "Complication of Mass Spectral Data." Heyden, London, and Presses Universitaires de France, Paris.

Cornu, A., and Massot, R. (1967). "First Supplement to Compilation of Mass Spectral Data." Heyden, London, and Presses Universitaires de France, Paris.

Danti, A. (1960). "Report of Investigation of the American Petroleum Institute Research Project 44 and the Manufacturing Chemists Association Research Project." Carnegie Institute of Technology, Pittsburgh, Pennsylvania.

Ford, H. T. (1964). Paper No. 83. Twelfth Annual Conference on Mass Spectrometry, ASTM Committee E-14, Montreal, Canada.

Friedman, L., and Long, F. A. (1953). *J. Amer. Chem. Soc.* **75**, 2832.

Gohlke, R. S. (1963). "Uncertified Mass Spectral Data." Dow Chemical Co., Midland, Michigan.

Gohlke, R. S., and McLafferty, F. W. (1962). *Anal. Chem.* **34**, 1281.

Grubb, H. M., and Meyerson, S. (1963). *In* "Mass Spectrometry of Organic Ions" (F. W. McLafferty, ed.), pp. 453–527. Academic Press, New York.

Hamming, M. C., and Eisenbraun, E. J. (1966). Paper No. 90. Fourteenth Annual Conference on Mass Spectrometry, ASTM Committee E-14, Dallas, Texas.

Hamming, M. C., and Grigsby, R. D. (1967). Paper No. 37. Fifteenth Annual Conference on Mass Spectrometry, ASTM Committee E-14, Denver, Colorado.

Lumpkin, H. E. (1956). *Anal. Chem.* **28**, 1946.

McLafferty, F. W. (1956). *Anal. Chem.* **28**, 306.

McLafferty, F. W. (1959). *Anal. Chem.* **31**, 82.

McLafferty, F. W. (1963a). "Mass Spectral Correlations." American Chemical Society, Washington, D.C.

McLafferty, F. W., ed. (1963b). *In* "Mass Spectrometry of Organic Ions," pp. 309–342. Academic Press, New York.

McLafferty, F. W. (1966a). "Interpretation of Mass Spectra," p. 73. Benjamin, New York.

McLafferty, F. W. (1966b). "Interpretation of Mass Spectra," p. 20. Benjamin, New York.

Meyerson, S., and McCollum, J. D. (1963). *In* "Advances in Analytical Chemistry and Instrumentation" (C. N. Reilley, ed.), Vol. 2, p. 212. Wiley, New York.

Pahl, M. (1954). *Z. Naturforsch. B* **9**, 188.

Pettersson, B., and Ryhage, R. (1967). *Anal. Chem.* **39**, 790.

Rylander, P. N., Meyerson, S., and Grubb, H. M. (1957). *J. Amer. Chem. Soc.* **79**, 842.

CHAPTER 8

COMPUTERIZING MASS SPECTRAL DATA

I. Introduction

The transformation of experimentally obtained spectra to a readily interpretable format can be a tedious and time-consuming task. The problem has been accelerated by fast-scanning mass spectrometers and the combination of mass spectrometry with gas chromatography. In some instances, as many as 180 different mass spectra are obtained in an hour from a single complex mixture (Li *et al.*, 1968). Such problems have emphasized the need for computerizing the processing of mass spectral data. At the same time the data from the computer need to be presented in a format which allows visual grasping of the salient features of a spectrum. If the computer does not achieve a means of presenting the data which aids mental assimilation, it has done no more than given a different set of problems and not aided in the subsequent interpretation or application of the data.

The complete process of computerizing mass spectral data includes several steps, such as interfacing scanning, circuitry, and the conversion of analog signals to digital printout, as well as the processing of the digitized data. However, only the processing of data in the digitized form will be dealt with here in any detail. At such a point the human effort can be substituted by computers.

Computers have given the mass spectroscopist a new degree of freedom. One has only to consider that a simple compound such as *n*-hexane contains over 50 peaks, each portraying the mass and abundance of specific ion types formed when the hexane molecule is fragmented, to realize the laborious mathematical or spectral manipulations which are required. Only those who have used desk calculators day after day to resolve mass spectra can fully appreciate the efficiency that computers have brought to mass spectrometry. Many operations that were impossible by reason of the time required are now both possible and practical with computers. It is becoming generally accepted that the large quantity of data obtainable in a modern mass spectrometer laboratory can no longer be economically processed by nonautomated techniques. Data processing and interpretation are often the bottleneck preventing increased instrument usefulness and thus a wider application of mass spectrometry.

Large numbers of authentic spectra can be tabulated, examined and indexed by computers, as well as stored in computer-accessible form. The assembling of the large volume of data associated with correlation studies in searching for characteristic fragmentation modes was illustrated by Hamming and Eisenbraun (1966). While these are rather simple and unsophisticated tasks, such work has often severely hampered the full utilization of mass spectral data.

Computers can be used in a wide variety of applications involving the various steps of data processing. The use of computers in any one laboratory should be coordinated or keyed to the objectives of the laboratory. The main consideration should be the specific steps to be taken in applying computers to mass spectrometry. These steps are often more difficult than selecting a heated inlet system or designing new equipment to be attached to a mass spectrometer. Several examples will be given in illustrating the application of computers. Programming will not be covered in detail. It is assumed that a mass spectrometry laboratory will have the cooperation of a computer laboratory. The American Society of Mass Spectrometry has organized a committee involved in sharing computer programs (Hamming, 1972).

Eventually the computer should be able to print out the structural formula from an unknown spectrum without human intervention. However, this will require a much greater understanding of fragmentation than is now available. Many of the applications today are unsophisticated, but the use of computers may prompt a greater understanding of fragmentation, and with this knowledge computers will find greater applications.

Mass spectrometer–computer systems of the type described by Reynolds *et al.* (1970) are expected to become common. In such systems the computer queries the users for parameters to control the operation of the mass spectrometer and makes spectral information available to the chemist within minutes.

II. Objectives of Computerizing

Discussions of sophisticated data acquisition and reduction facilities can sometimes lead to the belief that such systems are all that is required for setting up an effective mass spectrometry laboratory. It must be realized that the basic knowledge of fragmentation mechanisms and application techniques remain at the root of computerization. It is rather easy to list computerization as an objective, but for computerization to be effectively applied, several questions should be considered:

1. Have the proposed computer operations been manipulated without a computer? The simple task of identifying an unknown spectrum by making a comparison with a library of earlier recorded reference standards should be considered. Spectra of similar compounds have similar indicators, as illustrated in the previous chapter. Spectra obtained on instruments at different laboratories are different. Certain limits should be considered to be included in a program. This becomes a question of judgment based upon observations of a large number of spectra when as many relevant spectral features as possible are taken into consideration. Utilization of other such operations should be equally well established before being computerized. In fact, something that is fairly clear in a given case has to be well defined for the computer. Some haphazard and unsystematic approaches to interpretation may require a considerable amount of refinement.
2. Does the computerization fit the operation of the laboratory? With ever-changing techniques and corresponding changes in the operation of individual laboratories, a balance between computerization and individual efforts is difficult. It is important to establish such a balance only in that it prevents a large computer from doing the work of a desk calculator. On the other hand, replacing a desk calculator with a computer can lead to new and improved methods. It is difficult to realize the future value of the computer to mass spectrometry without using it as an alternative method to the older methods of calculations. High-resolution mass spectrometry fragmentation patterns have made clear the need for computerization, and this has become a vast new aspect of mass spectrometry which is somewhat beyond the scope of this volume.
3. How much time and expense are required for computerization? It should be realized that not only the original cost is involved but computerization usually requires constant attention if it is to be kept up to date and in a reasonable balance with other changes in the laboratory. The computer program must

be updated frequently as new knowledge emerges. For example, a library of spectral data requires constant changes to be current with available spectral data. Matrix calculations of mixtures continues to change with improved identification of components.

4. Will the data output from the computer become unwieldy? One of the purposes of the computer is to present the data in a more compact form which can be easily mentally assimilated for interpretation. The format of printout can do much to accomplish this goal. One example is the approach termed "topographic element mapping" proposed by Venkataraghavan *et al.* (1968). Less sophisticated presentations of data are also important. The data presented from the computer should be easily read by those receiving the data. The computer program which does not meet such needs may even add confusion to the understanding of results.

5. Can computerization be easily modified? With the rapid advance in mass spectrometry, one important consideration is the ease of computer program modification. A program designed with changes in mind can remain useful for several years. A computer program having several subroutines can often be easily changed.

III. Processing Batched Data

Computers can be used in processing batched data acquired by conventional recorders. The computer, thus used, can alleviate the arduous chore of manually listing masses of peaks and their intensities, together with calculations usually associated with these values.

A. Computer Acquisition of Mass Spectra

In actual practice, the interpreter of mass spectra will usually find that conclusions concerning molecular structures can be reached more efficiently and accurately from computerized display spectra than from those manually derived.

There are several approaches to computer acquisition of mass spectral data. The approach used depends upon the needs and the availability of equipment in any given laboratory. Individual accomplishments are a continuous source of topics in the literature of mass spectrometry. The computer program discussed by Hamming *et al.* (1967) used in the Analytical Research Section at Continental Oil Co. serves as an illustration of a flexible computer program for the acquisition of mass spectral data. This program is suited to the diverse research activities which require continuing modification. The program consists of one main system with several subroutines. Subroutines can easily be added,

TABLE 8-1

COMPUTER PRINTOUT OF THE MONO- AND POLYISOTOPIC SPECTRUM OF *n*-BUTANE IN TABULATED FORM

```
DATE     7/ 7/69     MS NO.  6710     ANAL NO.                 PAGE   1

N-BUTANE     C4 H10          CH3-CH2-CH2-CH3               MW 58

INLET TEMP: 100   ION TEMP: 250   EV: 70

BREAKING POINT
```

		POLYISOTOPIC			MONOISOTOPIC	
M/E	GROSS PEAK	PERCENT TOTAL IONIZATION	RELATIVE ABUNDANCE	CORRECTED PEAK	PERCENT TOTAL IONIZATION	RELATIVE ABUNDANCE
12.0	12.7	0.066	0.22	12.7	0.068	0.22
13.0	33.0	0.171	0.57	32.9	0.175	0.57
14.0	114.0	0.590	1.97	113.6	0.606	1.97
15.0	558.0	2.889	9.62	556.7	2.967	9.64
16.0	15.0	0.078	0.26	8.6	0.046	0.15
24.0	5.0	0.026	0.09	5.0	0.027	0.09
25.0	40.0	0.207	0.69	39.9	0.213	0.69
25.5	24.0	0.124	0.41	24.0	0.128	0.42
26.0	470.0	2.433	8.10	469.1	2.500	8.12
26.5	4.7	0.024	0.08	4.2	0.022	0.07
27.0	2400.0	12.426	41.38	2389.6	12.734	41.37
28.0	1818.0	9.412	31.34	1764.8	9.405	30.55
29.0	2430.0	12.581	41.90	2390.2	12.737	41.37
30.0	57.7	0.299	0.99	3.6	0.019	0.06
36.0	9.8	0.051	0.17	9.8	0.052	0.17
37.0	92.0	0.476	1.59	91.7	0.489	1.59
38.0	168.9	0.874	2.91	165.9	0.884	2.87
39.0	987.0	5.110	17.02	981.5	5.230	16.99
40.0	150.0	0.777	2.59	117.5	0.626	2.03
41.0	1857.0	9.614	32.02	1852.7	9.873	32.07
42.0	728.0	3.769	12.55	666.2	3.550	11.53
43.0	5800.0	30.029	100.00	5776.9	30.785	100.00
44.0	203.1	1.052	3.50	8.5	0.045	0.15
48.0	7.0	0.036	0.12	7.0	0.037	0.12
49.0	35.9	0.186	0.62	35.6	0.190	0.62
50.0	112.5	0.582	1.94	110.9	0.591	1.92
51.0	84.3	0.436	1.45	79.4	0.423	1.37
52.0	20.8	0.108	0.36	17.2	0.092	0.30
53.0	57.6	0.298	0.99	56.8	0.303	0.98
54.0	16.5	0.085	0.28	14.0	0.075	0.24
55.0	69.4	0.359	1.20	68.7	0.366	1.19
56.0	55.0	0.285	0.95	51.9	0.277	0.90
57.0	171.0	0.885	2.95	168.6	0.899	2.92
58.0	677.0	3.505	11.67	669.4	3.567	11.59
59.0	30.0	0.155	0.52			
	19314.9			18765.38		

deleted, or changed without disturbing the main system. This program can be a replacement for the laborious manual tabulation of reference standards and sample spectra generated in a research laboratory. An example of the output for *n*-butane as tabulated by the computer is shown in Table 8-1. The column on the left lists *m*/*e*'s. The second column lists peaks in scale divisions as recorded by the mass spectrometer. These divisions are summed, and from the sum, the percent total ionization is calculated for each of the individual polyisotopic peaks as given in the third column. The next column lists the relative abundance of each peak as compared to the most intense peak which is selected by the computer. If another peak is to be used as the base peak in calculating relative abundances, the computer can be so instructed. Monoisotopic peaks for hydrocarbons are calculated, and the deisotope peaks are then tabulated in the same manner as done for polyisotopic peaks. The calculation of monoisotopic hydrocarbon peaks may be omitted if desired.

All input to this computer program is from IBM cards. Several forms of printed input on cards are acceptable. Examples of three forms are shown below. Peaks and intensities from unattenuated digitized tape

03802563 03903987 04002500 04104619
04203728 04305580 04402677 04801070

Peaks and intensities from a "slave" key punch coupled to a mass spectrometer

03802563 03903981 04002500 04104619
04203728 04305588 04402677

Peaks and intensities from literature sources expressed in relative intensities

40.0 2.6 41.0 32.0 42.0 12.6 43.0 100.0 44.0 3.5

A computer program which can be used to correct low-resolution mass spectra for naturally occurring heavy isotopes ^{13}C, ^{2}H, ^{15}N, ^{17}O, and ^{18}O has

TABLE 8-2

COMPUTER PRINTOUT OF SELECTED PEAKS IN THE SPECTRUM OF *n*-BUTANE

CATE 7/ 7/69 MS NO. 6710 ANAL NO. PAGE 1

N-BUTANE C4 H10 CH3-CH2-CH2-CH3 MW 58

INLET TEMP: 100 ION TEMP: 250 EV: 70

BREAKING PCINT

MOST INTENSE PEAKS

ORDER	M/E	INTENSITY
1	43.0	30.029
2	29.0	12.581
3	27.0	12.426
4	41.0	9.614
5	28.0	9.412
6	39.0	5.110
7	42.0	3.769
8	58.0	3.505
9	15.0	2.889
10	26.0	2.433

FRACTIONAL PEAKS

25.5	0.124
26.5	0.024
TOTAL-	0.149

LAST PEAKS GREATER THAN 0.5

ORDER	M/E	INTENSITY
1	58.0	3.505
2	57.0	0.885
3	50.0	0.582
4	44.0	1.052
5	43.0	30.029
6	42.0	3.769

TABLE 8-3

Computer Printout of the Mono- and Polyisotopic Spectrum of *n*-Butane Arranged in the Array Format

N-BLTANE C4 H10 CH3-CH2-CH2-CH3 MW 58

INLET TEMP: 100 ION TEMP: 250 EV: 70

MONOISOTCPIC PEAKS IN PERCENT TOTAL IONIZATION M.S. NO. 6710

M/E	INT	M/E	INT	M/E	INT	M/E	INT	M/E	INT	M/E	INT	M/E	INT	M/E	INT	M/E	INT	M/E	INT	M/E	INT	M/E	INT	M/E	INT	M/E	INT
3		4		5		6		7		8		9		10		11		12	7	13	18	14	61	15	297	16	5
17		18		19		20		21		22		23		24	3	25	21	26	250	27	1273	28	940	29	1274	30	2
31		32		33		34		35		36	5	37	49	38	88	39	523	40	63	41	987	42	355	43	3079	44	5
45		46		47		48	4	49	19	50	59	51	42	52	9	53	30	54	7	55	37	56	28	57	90	58	357
SUBTOTALS																											
							4		19		64		91		100		574		327		2315		1384		4740		369
																									TOTAL		9987

N-BUTANE C4 H10 CH3-CH2-CH2-CH3 MW 58

INLET TEMP: 100 ION TEMP: 250 EV: 7C

POLYISCTCPIC PEAKS IN PERCENT TOTAL IONIZATION M.S. NO. 6710

M/E	INT	M/E	INT	M/E	INT	M/E	INT	M/E	INT	M/E	INT	M/E	INT	M/E	INT	M/E	INT	M/E	INT	M/E	INT	M/E	INT	M/E	INT	M/E	INT
3		4		5		6		7		8		9		10		11		12	7	13	17	14	59	15	289	16	8
17		18		19		20		21		22		23		24	3	25	21	26	243	27	1243	28	941	29	1258	30	30
31		32		33		34		35		36	5	37	48	38	87	39	511	40	78	41	961	42	377	43	3003	44	105
45		46		47		48	4	49	19	50	58	51	44	52	11	53	30	54	9	55	36	56	28	57	89	58	351
59	16																										
SUBTOTALS																											
	16						4		19		63		92		101		562		337		2257		1405		4639		494
																									TOTAL		9989

GRAND TOTAL INCL. FRACTIONAL PEAKS, 10004

been described by Boone *et al.* (1970). Copies of this program are available from S. E. Scheppele of the Department of Chemistry, Oklahoma State University, Stillwater, Oklahoma. The computer printout of isotopic correction factors from another program has been made available for several years from D. McAdams at Humble Oil, Baton Rouge, Louisiana.

The Hamming *et al.* (1967) program selects the 10 most intense peaks in a spectrum, the last six peaks which have intensities greater than 0.5% of the total ionization, and fractional peaks (Table 8-2). The program presents data in both relative abundance and percent total ionization, printed either in tabular form or as a block array (rectangular array).

In addition to the above-tabulated data, this computer program displays the data in the form of the rectangular array (Chapter 7) as shown in Table 8-3. Both the polyisotopic and monoisotopic data are given with the array series of peaks which are summed. While this phase of data processing is rather simple, it is a tedious and time-consuming task with many possibilities for human error. It should be realized that a large number of standares (50–60) can be tabulated by computers in a few minutes, while the same work would require several man-hours and too often would fall short of being finished.

Several data acquisition developments have been discussed in the literature. Among those described are those by Hites and Biemann (1967, 1968a) for data acquisition on-line to a computer. Hedfjäll *et al.* (1969) discussed the use of a computer for processing recorded digitized mass spectra. Desiderio and Mead (1968) discussed mass spectral data acquired on a photoplate and reduced with a real-time remote time-shared digital computer. Obtaining data from a photoplate with use of a computer has been reported previously by Biemann and Desiderio (1964), McLafferty *et al.* (1967), Olsen (1965), and Cone *et al.* (1967). A small high-speed computer connected to a magnetically scanned mass spectrometer has been discussed by Bowen *et al.* (1967) and Burlingame *et al.* (1968).

Ultimately, completely computerized quantitative analyses with the mass spectrometer and the computer will be realized. This will involve computer control of the mass spectrometer operator's console in obtaining a mass spectrum. Matrix multiplications on specified peaks or sum of peaks will allow analyses to be printed on a typewriter in the laboratory. These aims and others are being realized and reported on by Lumpkin and Harding (1970) from low-resolution mass spectral data.

B. Matrix Computations

Mass spectrometric analysis of mixtures are possible by the solution of a system of simultaneous linear equations (Barnard, 1953). The formation of such a matrix solution is most useful for repetitive samples. A computer program is

TABLE 8-4

COEFFICIENTS FOR A MATRIX CALCULATION

Peak	Paraffins	Noncondensed naphthalenes	Condensed naphthalenes 2-Ring	Condensed naphthalenes 3-Ring	Sample intensities
71	155.3	89.3	4.4	7.0	256.0
69	47.4	248.3	52.3	29.0	377.0
109	0.6	9.4	110.5	31.5	152.0
149	0.2	0.3	1.4	59.3	61.2

given (Appendix V) to illustrate the execution of a system of simultaneous linear equations using the same example as discussed in Chapter 6, Section VI,B. It should be kept in mind that matrix computations require that components in the sample mixture be known and that suitable calibration standards be available. The calibration standards provide the coefficients consisting of peaks (m/e) and corresponding intensities.

This program (Etter, 1968) is only one of several possible variations. Loeffler and Goodman (1962) made available their program, which was developed for the automatic assembly of matrices. The dictionary of calibration data consists of m/e values with corresponding intensities (pattern coefficients and sensitivities). Such a program can be updated readily for recalibration of pattern coefficients. These and other similar programs have removed much of the burden associated with the nonrepetitive quantitative analysis.

The computer program given as an illustration will be referred to as INVER (Appendix V). To be used, the coefficients in Table 8-4 must be kept in the

TABLE 8-5

COEFFICIENTS PUT INTO RATIO FOR COMPUTER PROGRAM

Paraffins	Noncondensed naphthalenes	Condensed naphthalenes 2-Ring	Condensed naphthalenes 3-Ring	Sample intensities
1.000	0.360	0.040	0.118	2.560
0.305	1.000	0.473	0.489	3.770
0.004	0.038	1.000	0.531	1.520
0.001	0.001	0.013	1.000	0.612

original ratio but adjusted as shown in Table 8-5. This program is designed as an illustration and has been kept simple for that reason. Information related to the INVER program is given below.

Dictionary of Variables

A	Array of coefficients of matrix
B	Sample identification
I	Counter used in inverting matrix
J	Counter used in inverting matrix
JL	Counter used in building identity matrix
K	Counter used in inverting matrix
N	Number of rows of matrix
NJ	Number of columns in augmented matrix
MN	Number of columns in augmented matrix plus identity matrix
N2	Column number in which matrix inverse begins

Coding Instructions

First and second cards	Sample identification may be put in all 80 columns of both cards.
Third card	Matrix rank (number of rows) must be right justified in columns 1 and 2.
Data cards	Each coefficient is allowed 10 columns (right justified), with six coefficients to a card. A data card then has an element in columns 1–10, 11–20, 21–30, 31–40, 41–50, 51–60. The total number of elements for each set of data must be NN where N is the matrix rank.

Restrictions

Matrix rank is restricted to 20 or less by the dimension statement. Double-precision computation will give fairly accurate solutions for matrices of this size (20 × 20) and smaller.

TABLE 8-6

EXAMPLE OF ACTUAL CARD INPUT FOR INVER PROGRAM

	Column spaces on card					
Card no.	1–10	11–20	21–30	31–40	41–50	51–60
1	1.000	0.360	0.040	0.118	2.560	0.305
2	1.000	0.473	0.489	3.770	0.004	0.038
3	1.000	0.531	1.520	0.001	0.001	0.013
4	1.000	0.612				

Output in the program is written for matrices of rank less than 10. Only the output statements need to be modified for matrices with rank from 10 to 20.

The first data point cannot be zero. The program will not work if one of the diagonal points is changed to zero in the compilation of the matrix inversion.

Computer Input and Printout

The actual card input to the computer for this example is shown in Table 8-6. The flow chart for the program INVER is shown in Fig. 8-1. The printout for this example with names of components added is shown in Table 8-7.

C. Computer-Aided Correlation Studies

Several examples have been given to illustrate that one practical analytical approach to identifying unknown substances by mass spectrometry is finding common trends in the fragmentation of a group of organic compounds containing a single functional group (Chapter 7). Such studies are useful for structure identification when no reference spectra are available. However, such studies usually involve 50 or more spectra, each containing over 150 peaks of various intensities. Such an examination is prone to human errors, and often there is an inclination to "find" spectra correlations which exist only for part of the available spectra.

Hamming and Eisenbraun (1966) have done preliminary work on a correlation of several cyclic ketones with a tetralin and indan structure using a computer. The computer selected intense peaks when the data was arranged in various forms. By so doing, certain relationships became evident between the fragment peaks formed and the known structures. As pointed out by Barber and Wolstenholme (1967), the logical extension of such simple and unsophisticated routines are computer searches to find the portion of the molecule which gives rise to intense fragment peaks and thus should take much of the detail data examination from the scientist.

Almost any group of mass spectra of a single class of organic compounds give spectra features which can be related to their molecular structures. A computer examination of such data provides a rapid means of finding correlations which may well not have been published or if published are sometimes difficult to tabulate. For example, several spectra of alkyl indanones included in the catalog of spectra of the Thermodynamics Research Center Data Project (API-44–TRC Nos. 1-m–13-m and 32-m–34-m) were examined. If the six most intense peaks in each spectrum are selected and if the computer routine is used to select the two most intense peaks in each homologous series (peaks differing by 14 mass numbers), it can be seen that the structures which are

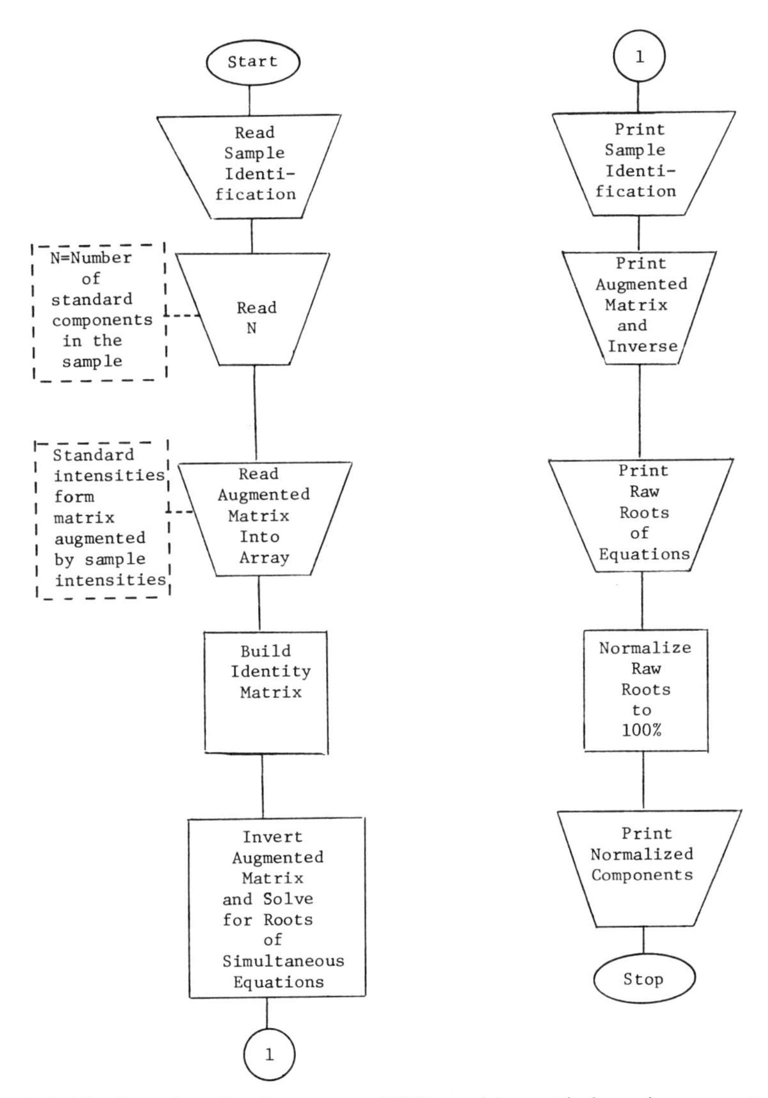

Fig. 8-1. The flow chart for the program IVER used in matrix inversion computations of mixtures.

not methyl substituted on the saturated moiety in positions 2 and 3 (API-44–TRC Nos. 1-m and 4-m–6-m) have peaks in this series which differ by 28 mass units. Such information could also be the basis for development of a fragmentation mechanism for alkyl indanones.

Usually several different subroutines can be made to operate on a class of compounds at the same time in searching for common trends in the fragmenta-

TABLE 8-7

COMPUTER PRINTOUT SHOWING INVERTED MATRIX, COMPONENTS, AND CALCULATED PERCENTAGES

EXAMPLE OF COMPOUND-TYPE ANALYSIS
INVERTED MATRIX PROGRAM FOR MASS SPECTROMETRIC CALCULATIONS

MATRIX FOR ANALYSIS

1.000	0.360	0.040	0.118	0.079	0.055	0.043	1.284
0.305	1.000	0.473	0.490	0.428	0.365	0.344	2.330
0.004	0.038	1.000	0.531	0.447	0.382	0.373	1.104
0.001	0.001	0.012	1.000	0.523	0.390	0.323	0.488
0.000	0.001	0.002	0.031	1.000	0.394	0.273	0.352
0.000	0.000	0.000	0.002	0.016	1.000	0.344	0.204
0.000	0.000	0.000	0.000	0.000	0.015	1.000	0.123

INVERTED MATRIX

1.1246	-0.4103	0.1496	-0.0122	0.0256	0.0255	0.0245
-0.3471	1.1452	-0.5249	-0.2378	-0.1025	-0.0649	-0.0558
0.0090	-0.0409	1.0259	-0.5204	-0.1672	-0.1068	-0.1182
-0.0008	-0.0001	-0.0114	1.0228	-0.5269	-0.1857	-0.1182
0.0001	-0.0006	-0.0012	-0.0301	1.0229	-0.3883	-0.1357
0.0000	-0.0003	-0.0001	-0.0010	-0.0159	1.0119	-0.3430
0.0001	-0.0005	-0.0003	0.0003	0.0003	-0.0149	1.0052

UN-NORMALIZED FACTORS

0.66

1.47

0.70

0.25

0.25

0.16

0.12

NORMALIZED RESULTS

COMPONENTS	PERCENT
Alkanes	18.4
Noncondensed naphthenes	40.8
Two - ring condensed naphthenes	19.4
Three - ring condensed naphthenes	6.9
Four - ring condensed naphthenes	6.8
Five - ring condensed naphthenes	4.4
Six - ring condensed naphthenes	3.3

tion. For example, one subroutine can select interpretative peaks and arrange these starting with the molecular ion to show mass differences which then can be related to the decomposition of the molecule. Another subroutine can be set up to examine all peaks above a given intensity which are between the most intense peak and the molecular ion peak. Such an examination becomes most useful when the most intense peaks are near the center of the spectrum. Another subroutine found useful at the Continental Oil Co. is called the decomposition routine. This routine searches a spectrum starting with the molecular ion for the first peak of an intensity greater than a given input value. The search then proceeds toward lower mass numbers for peaks of increasing intensities until the most intense peak is reached and recorded. Then it selects the next eight peaks of lower mass numbers which are of intensities greater than a given input value.

Another useful subroutine is the summation routine. This routine has quantitative application and it aids in correlation studies. The quantitative application is the familiar procedure of selecting peaks in a mixture and dividing each peak by the sensitivity coefficient and normalizing the values to a given percent (100%). This routine has the advantage of using peak sums of various selected peaks in the same way as individual peaks are used in structural diagnosis. For example, with the alkylbenzenes, as the number of substituted alkyl groups increases so do the sums of the C_nH_{2n-7} and C_nH_{2n-9} ions. Spectral data of two alkylbenzenes serve as an illustration. Dineopentyldurene (API-44 No. 86-m) has alkyl groups attached at all carbons of the benzene ring while 1-phenyltetradecane (API-44 No. 73-m) has only one alkyl group. Both compounds give an intense molecular ion peak at *m/e* 274 but exhibit marked differences in the −7 and −9 series ions.

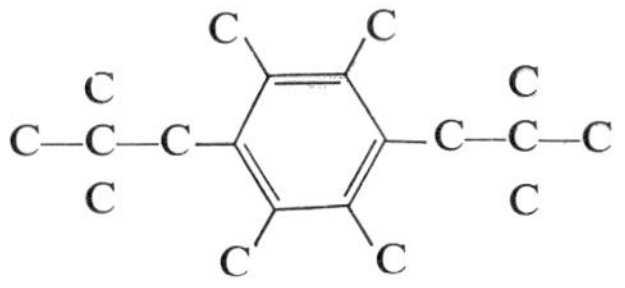

C_nH_{2n-7} ions = 40.2%
C_nH_{2n-9} ions = 4.8 %

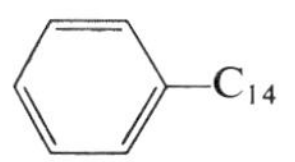

C_nH_{2n-7} ions = 29.7%
C_nH_{2n-9} ions = 2.0%

IV. Computer Identifications by Direct Comparison

A means of identifying an unknown substance is by computer comparing its mass spectrum to that of a known reference standard. Today there are tens of thousands of reference spectra available (see Chapter 6, Section V,B). Tabulated *m/e* values and intensities of reference spectra may be stored on magnetic tape or disks. Such large libraries of spectra can then be searched by means of a computer for the "best fit" of an unknown spectrum with the reference spectra. Usually some form of peak selection is made from the complete spectrum prior to recording the data into a library. A well-defined method for the "abbreviation" of computer-indexing spectra has been discussed by Hites and Biemann (1968b), together with a searching technique as a means of identifying an unknown spectrum. Libraries of reference spectra include as many as 7000 spectra (Abrahamsson *et al.*, 1966). The information retrieval systems of Crawford and Morrison (1968) and Abrahamsson (1967) have done much to speed up identifications by comparison of known spectra. Pettersson and Ryhage (1967) used a computer program storing about 2000 spectra of diterpenes and various low molecular weight acids, esters, aldehydes, ketones, alcohols, and amines. Grotch (1970) studied a statistical examination of matching large groups of mass spectra in a library matching search technique. This study was from the viewpoint of statistics and information theory. The results obtained indicated that low-resolution spectra are highly specific signatures even when encoded to only one bit.

Several tests have been carried out on how well such a program can identify a substance. The program is usually given several known standards in the form of unknowns. As different methods are evaluated, improvements are to be expected, and this should become an important means of identification. One problem is that the reference standards put into the program must be rather free of errors and impurities.

The task of setting up a large library of mass spectra on magnetic tape or disks by an individual laboratory can be very time-consuming. Libraries of spectral data may be purchased from the Mass Spectrometry Data Centre, AWRE Aldermaston, England. Identifications by direct comparison are covered in detail by Ridley (1971).

V. Computer Identifications without a Reference Library

Many of the samples received for identification in a modern mass spectrometry laboratory are compounds for which previous spectra are unavailable for comparison. In such cases, identification if done by computers has to be *ab initio*. This means that ideally a computer program would embody all rules for determining a molecular structure from an unknown spectrum for which no reference standard is available for direct comparison. Several approaches have been taken in this type of computer identifications. Crawford and Morrison (1968) have devised a means of making class identifications from specific peak probabilities and "hypersphere cluster centers" by computer techniques. By applying the nearest neighbor matrix concept for a structure,

Crawford and Morrison (1968) have shown that their computer program is able to store known or partially known structures and then recover printouts of molecular structures. Duffield *et al.* (1969) showed the application of artificial intelligence to computer interpretation using aliphatic ketones. It appears that this approach will have merit in future applications, as illustrated by Schroll *et al.* (1969) and Duffield *et al.* (1970).

The mass spectrometrist is faced first with the problem of finding specific fragment rules and, second, of putting these rules into a computer program. Such fragmentation rules often are not readily available. The type of rules required must be well defined and amenable to programming. In many cases, this calls for a new approach to a correlation study. An example of the type of fragmentation statements is given in the tabulation below for aliphatic acids having five or more carbons.

Check for interpretative peak at *m/e*[a]	Conclusion possible from "yes" answer
45, 60, 73, 115 + ($n \times 14$)	Saturated straight-chain aliphatic acid
45 and 74	Methyl branched at C-2
45 and 88 + ($n \times 14$)	A two-alkyl carbon substitution at C-2; one additional alkyl carbon substitution for each increase of 14 mass units from *m/e* 88 (Select highest *m/e* in this series, subtract 19, then divide by 14 to find number of carbons substituted at C-2)
45 and 87	Two methyl groups on C-2 (pivalic acids)
45, 60, 87	Methyl branched at C-3
73 and 101	Methyl branched at C-4
45 and 66 or 66 + ($n \times 14$)	$C_nH_{2n-4}O_2$ unsaturated (Select highest *m/e* in this series and add 18 to find molecular weight of alcohol)

[a] If none of the above combinations of interpretative peaks is found, the acid is branched at more than one carbon atom or is not an aliphatic acid with a single functional group.

Such an automated approach is practical, as shown by Morrison (1969), who reported that a computer can be instructed to trace the McLafferty rearrangement and to identify structural groups on both sides of the double bond.

There remains considerable additional work to provide improved computer applications and improved spectral structure correlations. These are rapidly under way. One of the phases of applications having merit is the computerized learning machine method being applied to chemical problems. Jurs *et al.* (1970) has applied the learning machine method to elucidation of a large number of molecular structures without recourse to theory. It must be realized that such a method is not a means of solving complex structure problems from low-resolution mass spectra of unknown compounds. It does provide an important gain in obtaining useful information of ever-improving predictive ability. This feature alone will provide an important place for the computer in the future achievements of mass spectrometry.

REFERENCES

Abrahamsson, S. (1967). *Sci. Tools* **14**, 29.

Abrahamsson, S., Haggstrom, G., and Stenhagen, E. (1966). Paper No. 105. Fourteenth Annual Conference on Mass Spectrometry, ASTM Committee E-14, Dallas, Texas.

Barber, M., and Wolstenholme, W. A. (1967). *Sci. J.* Nov., 76.

Barnard, G. P. (1953). "Modern Mass Spectrometry." Institute of Physics. London.

Biemann, K., and Desiderio, D. M. (1964). Paper No. 69. Twelfth Annual Conference on Mass Spectrometry, ASTM Committee E-14, Montreal, Canada.

Boone, B., Mitchum, R. K., and Scheppele, S. E. (1970). *Int. J. Mass Spectrom. Ion Phys.* **5**, 21.

Bowen, H., Chenevix-Trench, T., Drackley, S., Faust, R., and Saunders, R. A. (1967). *J. Sci. Instrum.* **44**, 343.

Burlingame, A. L., Smith, D. H., and Alsen, R. W. (1968). *Anal. Chem.* **40**, 13.

Cone, C., Fennessey, P., Hites, R. A., Mancuso, N., and Biemann, K. (1967). Paper No. 39. Fifteenth Annual Conference on Mass Spectrometry, ASTM Committee E-14, Denver, Colorado.

Crawford, L. R., and Morrison, J. D. (1968). *Anal. Chem.* **40**, 1464, 1469.

Desiderio, D. M., Jr., and Mead, T. E. (1968). *Anal. Chem.* **40**, 2090.

Duffield, A. M., Robertson, A. V., Djerassi, C., Buchanan, B. G., Sutherland, G. L., Feigenbaum, E. A., and Lederberg, J. (1969). *J. Amer. Chem. Soc.* **91**, 2977.

Duffield, A. M., Buchs, A., Delfino, A. B., and Djerassi, C. (1970). Paper No. G2. Eighteenth Annual Conference on Mass Spectrometry, ASMS with ASTM Committee E-14, San Francisco, California.

Etter, D. M. (1968). Private communications.

Grotch, S. L. (1970). *Anal. Chem.* **42**, 1214.

Hamming, M. C. (1972). *In* "Biochemical Applications of Mass Spectrometry" (G. R. Waller, ed.), Appendix 3. Wiley (Interscience), New York.

Hamming, M. C., and Eisenbraun, E. J. (1966). Paper No. 90. Fourteenth Annual Conference on Mass Spectrometry, ASTM Committee E-14, Dallas, Texas.

Hamming, M. C., Wright, W. M., Gartside, H. G., Ford, H. T., and Haley, J. (1967). Paper No. 16. Fifteenth Annual Conference on Mass Spectrometry, ASTM Committee E-14, Denver, Colorado.

Hedfjäll, B., Jansson, P. A., Marde, Y., Ryhage, R., and Wikström, S. (1969). *J. Sci. Instrum.* E-2, 1031.

Hites, R. A., and Biemann, K. (1967). *Anal. Chem.* **39**, 965.

Hites, R. A., and Biemann, K. (1968a). *Anal. Chem.* **40**, 1217.

Hites, R. A., and Biemann, K. (1968b). *In* "Advances in Mass Spectrometry" (E. Kendrick, ed.), Vol. IV, pp. 37–54. Institute of Petroleum, London.

Jurs, P. C., Kowalski, B. R., Isenhour, T. L., and Reilley, C. N. (1970). *Anal. Chem.* **42**, 1387.

Li, H. Y., Walden, J., Etter, D. M., and Waller, G. R. (1968). Fourteenth Tetrasectional Meeting of the Oklahoma Section, American Chemical Society, Ponca City, Oklahoma.

Loeffler, M. H., and Goodman, R. (1962). *Anal. Chem.* **34**, 713.
Lumpkin, H. E., and Harding, J. H. (1970). Paper No. T12. Eighteenth Annual Conference on Mass Spectrometry, ASMS with ASTM Committee E-14, San Francisco, California.
McLafferty, F. W., Venkataraghavan, R., and Amy, J. W. (1967). *Anal. Chem.* **39**, 178.
Morrison, J. D. (1969). Paper No. 17. Seventeenth Annual Conference on Mass Spectrometry, ASTM Committee E-14, Dallas, Texas.
Olsen, R. W. (1965). Paper No. 36. Thirteenth Annual Conference on Mass Spectrometry ASTM Committee E-14, St. Louis, Missouri.
Pettersson, B., and Ryage, R. (1967). *Ark. Kemi* **26**, 293.
Reynolds, W. E., Bacon, V. A., Bridges, J. C., Coburn, T. C., Halpern, B., Lederberg, J., Levinthal, E. C., Steed, E., and Tucker, R. B. (1970). *Anal. Chem.* **42**, 1122.
Ridley, R. G. (1971). *In* "Biochemical Applications of Mass Spectrometry" (G. R. Waller, ed.), Chapter 7. Wiley (Interscience), New York.
Schroll, G., Duffield, A. M., Djerassi, C., Buchanan, B. G., Sutherland, G. L., Feigenbaum, E. A., and Lederberg, J. (1969). *J. Amer. Chem. Soc.* **91**, 7440.
Venkataraghavan, R., Board, R. D., Klimowski, R., Amy, J. W., and McLafferty, F. W. (1968). *In* "Advances in Mass Spectrometry" (E. Kendrick, ed.), pp. 65–76. Institute of Petroleum, London.

CHAPTER 9

CORRELATIONS APPLIED TO INTERPRETATIONS

I. Introduction

Correlation studies are of great interest, and fragmentation of organic compounds should continue to be the subject of many future publications. While the interest is high, it is important to clearly realize that correlations must be usefully applied if the interest is to be supported. It is hoped that the reader has been able to find specific as well as general applications. Such applications are useful with and without the help of large computers. It is the purpose here to provide a few examples of relating an observed fragmentation process to practical analytical problems.

The approach taken in this section of relating mass spectrometric correlations to interpretations or identifications differs from the usual approach. Usually spectra from a single class of related organic compounds are selected, and correlations for that one class of compounds are presented. These correlations are not then related to other classes of compounds in detail. In this way, one presents an overall view, at the risk of perhaps making some overgeneralized statements. The practical value of this approach will be appreciated by those puzzled over a spectrum from a completely unknown substance (either pure or a simple mixture).

Ideally, an all-inclusive, practical set of rules would be used, in the same way a slide rule is used, to meet the needs of those who would desire to interpret spectral data. However, even the most enthusiastic mass spectroscopist finds it is a surprisingly difficult task to list simple rules which can be applied in a foolproof manner. In this chapter an attempt will be made to summarize some of the common principles of fragmentation which have been established. Too often a gap exists between understanding a fragmentation scheme for the decomposition of a known compound and the interpretation problem associated with an unknown spectrum. An attempt will be made in this chapter to close that gap and to inspire additional work in this direction.

The reader should be aware that theoretical approaches have been made toward understanding the decomposition of molecules in the mass spectrometer. Molecules receive an excess energy by collision with energetic electrons (usually 70 eV). Intellectual challenge is one stimulus for gaining an understanding of these complicated dissociation processes. One theoretical approach is the quasiequilibrium theory, which assumes that a molecular ion does not dissociate within the time of one molecular vibration after having received excess energy. Another approach is the molecular orbital theory, which assumes dissociation within the time of one molecular vibration after receiving the excess energy. Modified versions of the molecular orbital theory have predicted, with reasonable success, the spectra of alkanes (Fueki and Hirota, 1959), alkyl ketones, (Hatada and Hirota, 1965), alkylamines (Hirota and Itoh, 1966), and cycloalkanes (Hirota and Niwa, 1968). Any real analytical practical utility from such theories is difficult to visualize in the foreseeable future. Thus the usefulness of empirical research in the formation of fragmentation rules seems undeniable.

Fragmentation schemes have often not shown analytically useful applications. It should be kept in mind that a fragmentation process which leads to the formation of major fragment peaks is also a means to structural diagnosis. For example, 2-ethylthiophene (API Serial No. 241-m) shows an intense peak 15 mass units below the molecular ion, and 2-*n*-propylthiophene (API Serial No. 242-m) shows an intense peak 29 mass units below the molecular ion. In both cases the prominent cleavage is at the bond β to the ring. If either of the spectra were from unknown substances, the β-bond cleavage rule could be applied to aid in structural diagnosis. Such applications move mass spectrometry from an area of mere interest to an area of practical application which has merit to many department managers in research organizations or to others concerned with the cost of research.

A wider application of computerized interpretation of mass spectra should be possible as fragmentation rules are simplified and assembled. When computer usage is not feasible, the same set of rules can also be used by individuals. Simplified rules should allow researchers to often complete their own interpretation of mass spectral data, thus greatly extending mass spectrometry.

The empirical rules of breakdown cannot be expected, except in certain cases, to be sufficiently specific to ensure a detailed structural identification of many of the complex molecules. Empirical rules often can be used only as a general indicator for identifying a research sample which is completely unknown. Several examples will follow to provide the reader with a knowledge of how to apply correlations to typical analytical problems requiring some degree of structural diagnosis. Not all rules will be given. Those selected have been found by experience to apply to several different types of identifications.

Correlation rules in qualitative identifications should be applied with caution. This essentially involves careful consideration of all possible cases within a given rule. For example, β-bond cleavage with a hydrogen transfer results in a rearrangement peak (RIP) at m/e 58 for 2-octanone (API–TRC No. 98-m) in the 58, 72, 86, etc., series of peaks. It would thus be a simple matter to use the β-bond cleavage rule to deduce methyl substitution on the α carbon from observation of an increase in the RIP peak from m/e 58 to m/e 72. This identification would be correct for the compound 3-methyl-2-heptanone, and the same cleavage rule could also apply to 3-octanone (API–TRC No. 99-m) and other structures. Likewise an RIP peak in the same homologous series at m/e 86 could be due to 3,3-dimethyl-2-hexanone (API–TRC No. 106-m). In fact, it should be kept in mind that some compounds are apparently "outside" the rules. For example, 2,2-dimethyl-3-hexanone (API–TRC No. 114-m) gives a maximum at m/e 58 rather than m/e 100, as the β-bond cleavage rule would indicate. These rules will become clear to the reader in the discussion to follow.

II. Applications of the Branching Rule

One of the earliest and simplest observations concerning fragmentation of organic compounds under electron impact (70 eV) was the tendency for preferential fragmentation at branched positions on an alkyl chain. The rule may be expressed as follows:

> *Fragmentation is favored at a branched carbon atom position on an alkyl chain resulting in more than one maximum in a characteristic series of homologous peaks differing by* 14 *mass units* (CH_2 *group*).

The branch positions are distinctive sites for the localization of the positive

TABLE 9-1

COMPARISON OF INTENSITIES OF C_nH_{2n+1} IONS FOR SETS OF BRANCHED AND NORMAL ALKANES[a]

	Intensities of monoisotopic peaks (% total ion yield)					
m/e	3-Methylpentane (branched)	*n*-Hexane (unbranched)	3,3-Dimethyl-pentane (branched)	*n*-Heptane (unbranched)	3-Methylheptane (branched)	*n*-Octane (unbranched)
15	2.5	1.7	1.7	1.1	1.3	0.9
29	5.0	10.6	6.5	9.5	6.7	9.3
43	28.0	13.5	27.9	21.4	22.3	26.2
57	3.1	17.3	3.1	9.6	16.8	8.8
71	8.0*	0.8	16.5*	9.0	2.8	6.0
85	0.2	0.1	4.6	0.4	0.3	5.3
99	—	—	0.1	—	2.3*	—
113	—	—	—	—	—	—

[a] In the tables of this chapter, an asterisk signifies an interpretative peak.

charge, as described in Chapter 5, and, as such, the breaking of a given C–C bond is enhanced.

For this simple rule to be analytically applied to structural diagnoses, a comparison is implied between a branched structure and a straight-chain structure of a corresponding compound of the same MW. This comparison need not be necessarily directly applied, as illustrated by the intensities of the C_nH_{2n+1} of branched and unbranched hydrocarbons shown in Table 9-1. The maximum alkyl peak intensity for the normal compounds occurs at *m/e* 57 or 43, and the C_nH_{2n+1} peak intensities decrease from this point on up toward the molecular ion. Two or more maxima are noted for branched alkyl compounds. It should be noted that applying the branching rule is not a case of making a direct comparison between any one of the C_nH_{2n+1} ion series of peaks. If that were the case, *n*-hexane would be concluded to be more branched than 3-methylpentane, based upon the fact that the 57 peak is more intense in the spectrum of *n*-hexane than the 71 peak in the spectrum of 3-methylpentane. The peak at *m/e* 71 is 10 times more intense for the branched compound than for the unbranched compound (8.0 compared to 0.8). If the two structures are examined, the reason for selecting a comparison of their 71 peaks becomes evident. The problem here, for the mass spectroscopist, is to use correlation principles, *not structures* which he is trying to determine, to account for the formation of intense peaks. The 71 peak was selected from the mass spectrum of 3-methylpentane because in the intensities shown it breaks from the decreasing intensities of the maximum peak at *m/e* 43 in this homologous series (two maxima instead of one occur). This break in the intensity

TABLE 9-2

COMPARISON OF INTENSITIES OF C_nH_{2n-1} IONS FOR BRANCHED AND NORMAL ALKENES

	Intensities of monoisotopic peaks (% total ion yield)		
m/e	3,6,8-Trimethylnonene-1	2-Ethyldecene-1	Dodecene-1 (unbranched)
27	4.1	4.9	5.3
41	12.1*	11.0	11.9*
55	4.8	13.1*	8.9
69	5.9*	4.0	5.7
83	2.5	4.3*	4.3
97	0.5	0.8	2.1
111	1.7*	0.2	0.7
125	0.9	—	0.2
139	—	0.5*	0.1

values is a means of selecting a peak of diagnostic value and to point to the indicated branching in the structure. The same type of application is illustrated for the other sets of hydrocarbons shown in Table 9-2.

Another illustration of the branching rule for hydrocarbons is provided in the data shown in Table 9-2. The position of the olefinic bond remains the same for the compounds shown and, therefore, differences in the intensities of the C_nH_{2n-1} ions can be attributed to an alkyl chain rather than the influence of a double bond or other structural features. It must be kept in mind that an attempt is being made here to apply rules to spectral data for the purpose of deducing structural information. The rule being discussed is the branching rule and, the data in Table 9-2 are of concern. The first two series of C_nH_{2n-1}

TABLE 9-3

COMPARISON OF INTENSITIES OF C_nH_{2n+1} IONS FOR BRANCHED AND NORMAL SATURATED ALIPHATIC ACIDS

	Intensities (% total ion yield)		
m/e	Pivalic acid [$(CH_3)_3$ branched]	Isovaleric acid (CH_3 branched at C-3)	Valeric acid (unbranched)
15	2.6	2.4	1.7
29	11.0*	6.2	7.6*
43	1.7	8.7*	4.3
57	30.7*	3.9	1.0
71	0.1	0.1	0.1

ions have three maxima, while the last series (from an unbranched alkene) has but one maximum. The rule applied is as follows. *If no other functional group is present to cause this difference*, *then* the three maxima are evidence of branching of an alkyl chain. The location of the points of branching requires additional examination of a complete spectrum (in many cases, the point of branching cannot be readily determined).

The data given in Table 9-3 illustrate that cleavage of C–C bonds adjacent to a quaternary-bonded carbon atom is readily evident for pivalic acids as compared to two other C_5 aliphatic acids. From these data it should be noted that the C_nH_{2n+1} ions show only one maximum for isovaleric acid, and thus branching would not be detected by these ions alone. The *complete* spectrum should always be examined before drawing conclusions about molecular structure.

Fragmentation rules should be used with caution and good judgment. An individual may note that the peaks at *m/e* 15, 29, 43, 57, 71, 85, and 99 in the spectrum of 2-octane (TRC Serial No. 98-m) have two maxima, one at *m/e* 43 and another at *m/e* 71. This does not permit the conclusion that the structure has an alkyl branched chain. The peak at *m/e* 71 in this spectrum has been shown (Carpenter *et al.*, 1968) to be mainly an oxygen-containing fragment and thus is not a characteristic hydrocarbon ion.

III. Applications of Beta-Bond Cleavages

Preferential cleavage at the bond β to a functional group is a common phenomenon in the fragmentation of organic compounds by electron impact. Several different classes of compounds illustrate this type of fragmentation, both with and without rearrangement. The object here is to present analytical applications of β-bond cleavage as defined below:

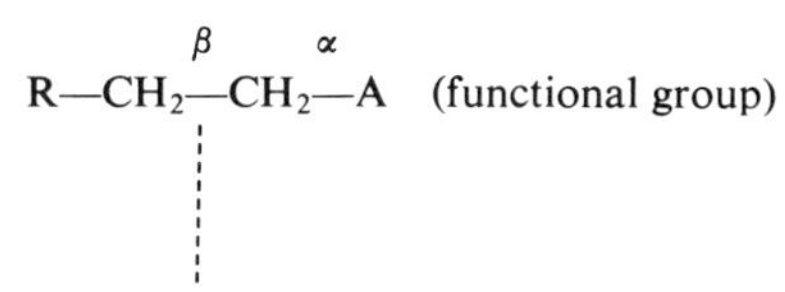

Cleavage to yield a charged species and a neutral species

The functional group A could be one of several, such as C=O, –OR, –OH, –SH, and $-NH_2$. This discussion will stem more from the viewpoint of the type of cleavage than of any one given class of compounds. This approach will be taken because considerably more than one type of cleavage takes place in the fragmentation of any one class of compounds.

A. Beta-Bond Cleavage with Rearrangement

A basic discussion of *β-bond cleavage* with rearrangement presented by McLafferty (1966) reviews many of his contributions leading to the well-known McLafferty rearrangement. The illustrations presented here are intended to show analytical applications of this widely observed type of decomposition. The β-bond cleavage rule with rearrangement may be expressed as follows:

> *Cleavage of the bond β to a functional group with hydrogen transfer gives an interpretative rearrangement peak which can be used to provide structural information such as branching or the location of a double bond.*

This β-bond rule with γ-hydrogen transfer can be applied to a number of different classes of organic compounds including acids, esters, aldehydes, ketones, sulfites, amides, nitriles, phenylalkanes, olefins, etc., and some compounds with combinations of these functional groups.

For this rule to apply, the compound must first meet the basic structural requirements. If a γ hydrogen is not available, the rule cannot apply; in fact, this in itself is of diagnostic value. For example, consider spectral data shown in Table 9-4. These data are illustrative of the β-bond cleavage rule with

TABLE 9-4

SPECTRAL DATA ILLUSTRATING THE β-BOND CLEAVAGE RULE ACCOMPANIED BY HYDROGEN REARRANGEMENT FOR SATURATED ALIPHATIC ACIDS

	Intensities (% total ion yield)				
m/e	Pivalic acid	Valeric acid	Isovaleric acid	2-Methyl-nonanoic acid	2-Ethyliso-hexanoic acid
60	0.2	26.5 (RIP)	17.7 (RIP)	0.5	0.8
74	0.1	1.0	5.2	10.5 (RIP)	0.7
88	0.1		0.2	0.5	10.6 (RIP)

rearrangement (the same three acids were used to illustrate the branching rule). The peaks selected (60, 74, and 88) are the same homologous series as molecular ions of $C_nH_{2n}O_2$ acids. Rearrangement peaks of diagnostic value are designated by RIP. These peaks are those of highest intensities in the $C_nH_{2n}O_2$ series. Pivalic acid does not meet the structural requirement of an available γ hydrogen for transfer as do the other acids shown in Table 9-4 and thus has no intense rearrangement peak. Cleavage of the bond β to the C=O group with γ-hydrogen transfer is shown for two acids with a γ hydrogen atom.

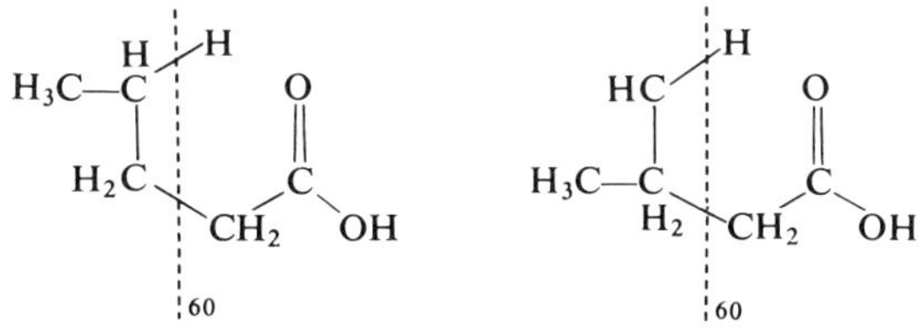

60

Valeric acid

60

Isovaleric acid

From this rule, alkyl substitution at C-2 could be detected by an increase in the rearrangement peak by 14 mass numbers for each additional methylene group. This is evident by intense peaks at *m/e* 74 and 88 for the compounds 2-methylnonanoic acid and 2-ethylisohexanoic acid, respectively (Table 9-4). The β-bond cleavage is shown below.

$$C_5H_{11}-\overset{H}{C}H-CH_2-CH(CH_3)-COOH \quad (\text{cleavage at } 74)$$

2-Methylnonanoic acid

$$(H_3C)_2C(H)-CH_2-CH(C_2H_5)-COOH \quad (\text{cleavage at } 88)$$

2-Ethylisohexanoic acid

B. Beta-Bond Cleavage without Rearrangement

Cleavage of the bond β to any one of several functional groups without hydrogen transfer gives an intense even-electron fragment ion (at an odd mass number except in the case of nitrogen compounds having an odd number of nitrogen atoms).

One of the examples of C–C bond cleavage widely applied analytically is illustrated by mass spectra of alkylbenzenes (Ford, 1964) and other aromatic compounds (Martin and Hamming, 1963). This type of fragmentation is given in a comprehensive review by Grubb and Meyerson (1963). Spectral data shown in Table 9-5 illustrate how interpretative peaks in the C_nH_{2n-7} series (m/e 91, 105, etc.) can be used in structural diagnosis of alkylbenzenes. Intensities of peaks marked with an asterisk are used in structural diagnosis (it should

TABLE 9-5

Spectral Data Showing Interpretative Peaks of Alkylbenzenes[a]

	Intensities of monoisotopic peaks (% total ion yield)			
m/e	2-Phenyldodecane	2-Methyl-2-phenylundecane	3-Phenyldodecane	4-Phenyldodecane
91	0.7	9.2	34.9	44.3
105	57.5*	3.2	3.3	3.2
119	0.5	46.8*	18.6*	1.1
133	0.2	0.1	1.0	11.8*
147	0.2	—	0.8	0.6
203	—	—	—	4.5*
217	—	—	5.2*	—
246 (M⁺)	<u>4.5</u>	<u>1.1</u>	<u>3.6</u>	<u>3.4</u>

[a] In the tables in this chapter an underlined value is the intensity of the molecular ion peak.

be noted that the peak at *m/e* 91 is common to three of the four structures and thus is of limited use in deducing structures). The β-bond cleavage rule is depicted below for 3-phenyldodecane and 4-phenyldodecane (the mechanism of the formation of these ions is more complicated than shown).

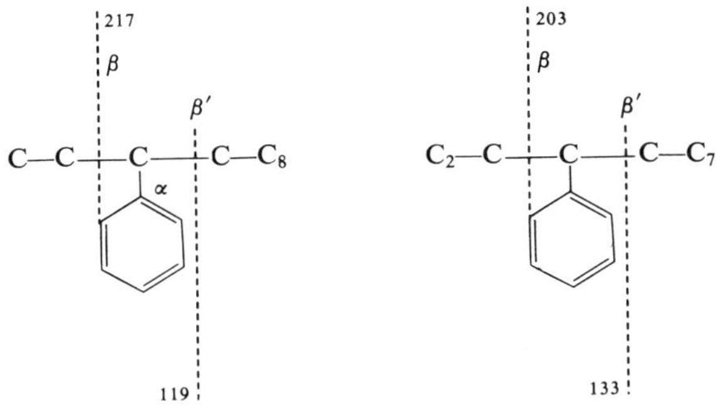

217 β β′ C—C—C—C—C_8 α 119

203 β β′ C_2—C—C—C—C_7 133

Both compounds have intense molecular ion peaks (underlined in Table 9-5), and thus the interpretative peak at *m/e* 217 is supporting evidence of 3-phenyldodecane rather than 2-methyl-2-phenylundecane.

The compounds shown in Table 9-6 illustrate the β-bond cleavage rule as

TABLE 9-6

STRUCTURAL DIAGNOSIS USING THE β-BOND CLEAVAGE RULE

	Intensities of monoiostopic peaks (% total ion yield)			
m/e	1,5-Cyclo-octadiene	1,3-Cyclo-octadiene	1-Methyl-4-isopropenyl-1-cyclohexene	1-Methyl-4-isopropyl-1-cyclohexene
54	19.6*	2.2	0.3	0.5
68	—	0.1	17.7*	8.9*
79	7.1	16.5*	3.7	2.3
108	1.2	7.3	0.7	—
136	—	—	3.0	—
138	—	—	—	4.2

applied to these cyclic hydrocarbons. Molecular ion peaks are underlined, and interpretative peaks are designated by an asterisk. Consider the spectral data shown for the two cyclooctadienes, one of these isomers having an interpretive peak at *m/e* 54 and the other at *m/e* 79. What can be deduced about the structure from these two peaks and from cleavage by the β-bond rule? The 54 peak is half the molecular ion peak, leaving a possibility of a structure with the same two C–C bonds β to the two double bonds. This possibility is correct, as shown by the spectrum of 1,5-cyclooctadiene (API No. 240-m). The interpretative peak at m/e 79 must involve the loss of two carbons from the molecular ion peak at m/e 108. Under β-bond cleavage rule operation, the structure of 1,3-cyclooctadiene is deduced (API No. 239-m). Both 1-cyclohexenes shown in Table 9-6 have an interpretative peak (marked with an asterisk) at m/e 68, with one of the intensities nearly double that of the other. The cyclohexene of molecular weight 136 has two double bonds, as compared to one double bond for the compound of molecular weight 138. The fragmentation is depicted below, together with the compound 1,3-*p*-methadiene (no m/e 68 exists in its spectrum).

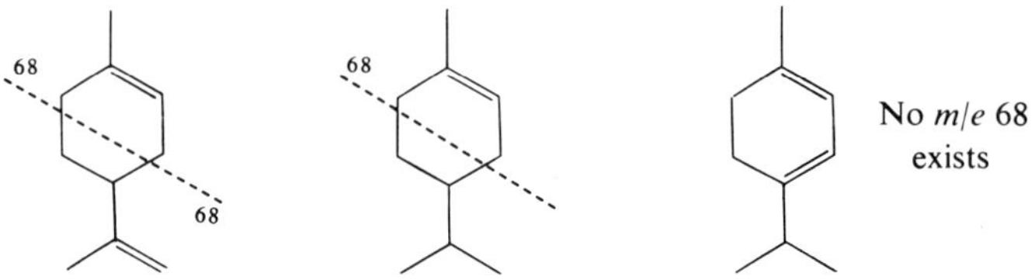

68 68 68

No m/e 68 exists

This discussion of the fragmentation of 1,3-*p*-menthadiene illustrates that a shift in the position of the double bonds can completely alter the fragmentation pattern. This change is in accordance with the rule of β-bond cleavage without hydrogen transfer.

From the data presented in Table 9-7, it is seen that the structural difference

TABLE 9-7

SPECTRAL DATA ILLUSTRATING THE β-BOND CLEAVAGE RULE ACCOMPANIED BY HYDROGEN REARRANGEMENT FOR SATURATED ESTERS[a]

	Intensities (% total ion yield)			
m/e	Methyl heptanoate	Ethyl *n*-caproate	Ethyl β-methylvalerate	Ethyl α-methylvalerate
60	0.2	5.1*	6.3	—
74	23.2**	0.5	—	9.4*
88	1.0	11.4**	19.4**	—
102	0.2	0.4	—	19.0**
144 (M^+)	0.1	0.1	1.0	0.1

[a] In the tables of this chapter, a double asterisk signifies an interpretative peak which is also a rearrangement peak.

between ethyl *n*-caproate and ethyl β-methylvalerate does not produce different interpretative peaks in the $C_nH_{2n}O_2$ series (*m/e* 60, 74, etc.), as is true of the two other compounds shown. Consider as an analytical problem the structural analysis of these spectra. Ethyl *n*-caproate and ethyl β-methylvalerate can be differentiated based upon other types of cleavage (γ bond on the alkyl chain). It must be realized that a rearrangement ion at *m/e* 88 is not absolute evidence of an ethyl ester. A methyl ester with a methyl group substituted at the α carbon would also give a rearrangement peak at *m/e* 88. Similar conclusions can be drawn for other such rearrangement ions. It should be noted that the cleavage is at the C–C bond β to the carbonyl group in such cases. Thus, it becomes possible to extend structural diagnosis. For example, the rearrangement peak at *m/e* 102 observed for ethyl α-methylvalerate could also have originated from β-bond cleavage of the C–C bond for the structure methyl α-ethylvalerate. The rearrangement ion at *m/e* 74 was useful to establish the methyl substitution at the α carbon. It should be noted that where the α carbon was not methyl substituted (ethyl β-methylvalerate), a rearrangement peak was found at *m/e* 60 as compared to *m/e* 74 from the methyl-substituted compound (ethyl α-methylvalerate). This double β cleavage will be further discussed for aliphatic ketones.

The shift of rearrangement interpretive peaks (RIP) in the structures ethyl β-methylvalerate and ethyl α-methylvalerate from *m/e* 60 to 74 and 88 to 102 is depicted below.

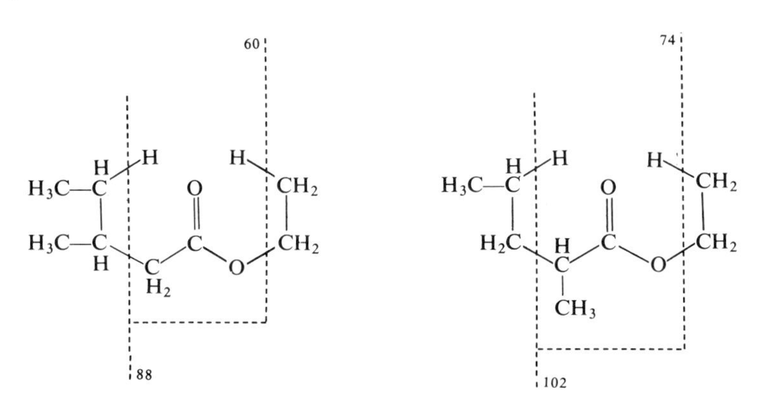

Ethyl β-methylvalerate   Ethyl α-methylvalerate

C. Beta-Bond Cleavages Involving Functional Groups

Many compounds with two or more functional groups often undergo cleavages at different bonds, each β to a different functional group. Because of such cleavages, correlations can often effectively be applied to structural determinations. An example of such cleavages is gained from three spectra published in the catalogs of the Thermodynamics Research Center Data Project (API-44–TRC Nos. 121-m, 122-m, and 123-m). These structures are shown below with β bond cleavage to the benzene ring and β bond cleavage to the carbonyl groups.

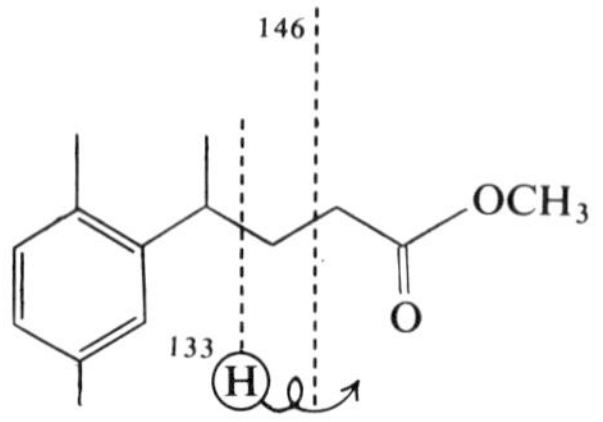

146 133 H OCH_3 O TRC 121-m

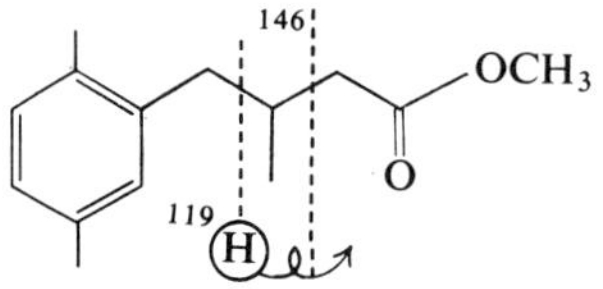

146 119 H OCH_3 O TRC 123-m

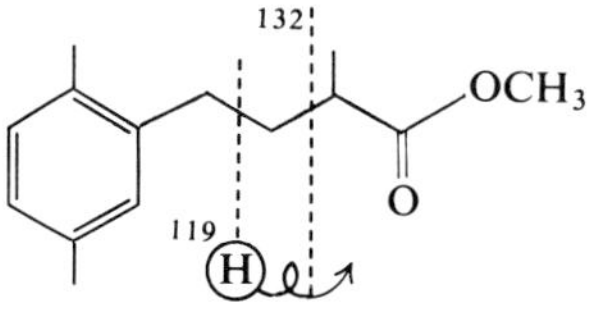

132 119 H OCH_3 O TRC 122-m

An interpretative peak at *m/e* 133 found on the TRC 121-m spectrum locates the methyl group on the carbon adjacent to the benzene ring, as opposed to an interpretative peak at *m/e* 119 for spectra TRC 123-m and TRC 122-m. This is the characteristic cleavage of the bond β to the benzene ring. This cleavage rule, however, does not help when locating the methyl or "ring methyl groups" for compounds TRC 123-m and TRC-122-m. Compounds TRC 122-m and TRC 123-m both give an interpretative peak at *m/e* 146, meaning the TRC 123-m is substituted as shown as compared to TRC 122-m, which has an interpretative peak at *m/e* 132. This cleavage is β to the carbonyl group with transfer of a hydrogen away from the charged species. Thus, a structure determination has been made from the well-defined interpretative peaks and the application of the β-bond cleavage rule to two different functional groups.

As a corollary to the above discussion, what happens to ketone fragmentations when two functional groups are adjacent? The thienyl alkyl alkanones provide information for comparison. In this case the cleavage at the bond β to the thiophene ring would appear to dominate to produce the BLIP at *m/e* 111. But the work of Foster and Higgins (1969) shows that the actual mechanism

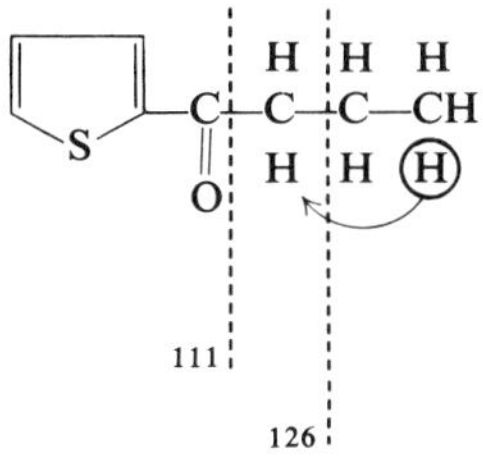

H H H
C—C—C—CH
O H H (H)
111
126

is something different (see Chapter 5). At low voltages a β cleavage with rearrangement occurs with respect to the keto grouping to produce ions at *m/e* 126. At higher ionization voltages, the *m/e* 111 becomes predominant, resulting in the BLIP at 70 eV. Thus if a π-bonded ring system is adjacent to the carbonyl group, β cleavage in the alkyl group with respect to the ring and hence α with respect to the carbonyl group should produce the interpretative ions. The phenyl ketones behave similarly from the interpretative viewpoint.

IV. Alpha-Bond Cleavage

It should be clear that several types of cleavages take place during the decomposition of a molecule under electron impact. A classification is used to provide an understanding of different types of cleavages as well as combinations for analytical applications. For example, the most intense peak in the spectrum of 3-*n*-propyl-1-cyclopentene is at *m/e* 67 (API No. 223-m). The cleavage may be depicted as shown:

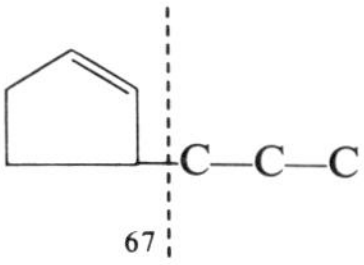

C—C—C
67

The bond being cleaved is β to the double bond of the ring (recall previous section). This same bond is also α to the cyclopentene ring. These two forces in this case support each other with the resulting fragment being the most intense peak in the spectrum.

The data in Table 9-8 show that C_nH_{2n-1} ion peaks for three cyclohexanes of molecular weight 126 (API Nos. 1226, 1227, and 1422). If these were spectra from unknown compounds, the α-bond rule could be applied to structural diagnosis. It would be well to first examine the different maxima in these three series. From the data shown the maximum at *m/e* 55 appears to be characteristic of cyclohexanes because all of these compounds show an interpretative peak at *m/e* 55. (In general, hydrocarbons have interpretative peaks in a homologous series of fragment peaks with one less hydrogen atom than the corresponding molecular ion peaks; thus C_nH_{2n} compounds have interpretative peaks in the C_nH_{2n-1} series.) The single second maximum in the first series of peaks can be rationalized easily for an *n*-propyl group when α-bond cleavage is considered. Two maxima in the second column of peaks still suggest cleavage of three carbon atoms (M − 43), but with a somewhat different arrangement of the carbon atoms. The isopropyl arrangement would be in keeping with this diagnosis, recalling that fragmentation is enhanced at the point of branching. In the third column of peaks in Table 9-8, a shift of the maximum *m/e* 97 from *m/e* 83, for the same molecular weight compound (known from the molecular ion peaks which are not shown) suggests that one of the alkyl chains has two (rather than three) carbon atoms. This could be true for a structure

TABLE 9-8

STRUCTURAL DIAGNOSIS USING THE α-BOND CLEAVAGE RULE AND THE BRANCHING RULE

	Intensities of monoisotopic peaks (% total ion yield)		
m/e	*n*-Propylcyclohexane	Isopropylcyclohexane	1-Methyl-1-ethyl-cyclohexane
41	10.5	10.7	8.3
55	16.1*	16.2*	21.7*
69	2.1	3.4	4.6
83	22.2*	18.0*	1.2
97	0.5	—	24.7*
111	—	0.4*	1.7

such as 1-methyl-1-ethylcyclohexane. Here it should be noted that several other methyl and ethyl alkyl-substituted isomers would be expected to give a similar fragmentation pattern. This type of examination illustrates applying analytically the α-bond cleavage rule.

Another example of α-bond cleavage is found in the spectra of aliphatic ketones. Recall that these compounds also exhibit β-bond cleavage to the carbonyl group. Also note that one class of compounds may exhibit more than one type of prominent cleavage. Usually a different type of cleavage will result in different interpretative peaks. The structural information determined should agree. In some cases the interpretative peaks formed will not be as informative as from another type of cleavage. For example, for aliphatic ketones α-bond cleavage is, in general, of less use than β-bond cleavage in deriving structural information and also can be misleading.

The data shown in Table 9-9 illustrate the usefulness and some pitfalls in

TABLE 9-9

SPECTRAL DATA ILLUSTRATING α-BOND CLEAVAGES FOR ALIPHATIC KETONES

	Intensities (% total ion yield)			
m/e	3-Heptanone	2-Octanone	3-Octanone	4-Octanone
29	17.6*	3.8	14.7	6.9
43	3.2	31.3*	17.4*	13.9*
57	25.7*	1.5	13.7	11.0
71	0.6	3.9*	6.4	10.4
85	7.7*	1.4	1.1	7.5
99	0.1	0.1	5.4*	0.4
114	3.2	—	—	—
128	—	1.2	1.4	2.1

the use of α-bond cleavage in obtaining structural information on aliphatic ketones (also implying the usefulness in understanding several types of cleavages). The fragmentation of 3-heptanone ideally illustrates the establishment of structural information from α-bond cleavage to the carbonyl group.

57
29
O
‖
C—C—C—C—C—C—C
85

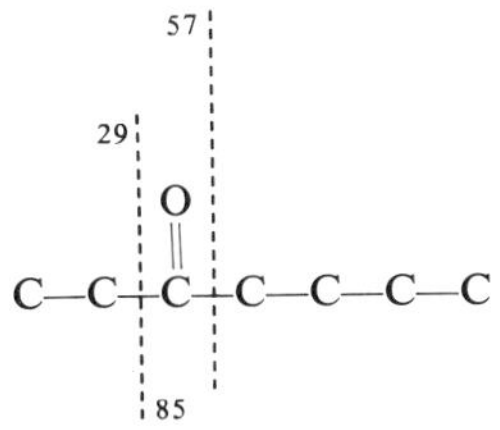

The peaks at *m/e* 29, 57, and 85 are maximum peaks in the 29, 43, 57 series, as designated by a single asterisk in Table 9-8 for 3-heptanone. The peak at *m/e* 43 (CH_3CO), shown for 2-octanone, could readily be attributed to α-bond cleavage to the carbonyl group; however, the second maximum at *m/e* 71 could not be rationalized in the same way as is 3-heptanone (a propyl group attached to a carbonyl group would be suggested by the peak at *m/e* 71). The peak at *m/e* for 3-octanone is much more intense than for 2-octanone, as would be expected by cleavage α to the carbonyl group with the charge going to the C_2H_5CO ion. However, the 43 peak is more intense than the 57 peak for 3-octanone, which would cause uncertainty if this spectrum were being examined as an unknown ketone. The maximum at *m/e* 99 may be rationalized as the $C_5H_{11}CO$ ion formed by α-bond cleavage to the carbonyl group with the charge going to the ion with the longer alkyl group. The fragmentation of 4-octanone has one maximum at *m/e* 43 which makes it appear more like the "expected" or predicted fragmentation of 2-octanone than the observed spectrum of 2-octanone. Structural information for these octanones could be readily deduced using the β-bond cleavage rule and their complete mass spectra published in the "Catalog of Mass Spectra Data" (1970).

V. Cyclic Transitions or Rearrangements Useful in Identifications

Certain cyclic ions have been found to act as identifiers of structural features. These cyclic transitions or rearrangement ions aid in structural diagnosis because they require certain structural features in order to form or to operate. The examples which follow will illustrate applying a knowledge of the formation to analytical identifications.

A cyclic ion at m/e 91 is useful to distinguish a terminal alkyl chloride from an internal chloride. The m/e 91 ion is a five-membered cyclic ion with a divalent hydrogen, as postulated by McLafferty (1962a). Because this ion contains four carbon atoms, it cannot be formed if the original structure has less than four straight-chain carbons. Spectra of internal chlorides do not give an identifying

m/e 91

TABLE 9-10

SPECTRAL DATA ON THE CYCLIC ION AT MASS 97 OF NITRILES

	Intensities (% total ion yield)		
m/e	Pelargononitrile ($C_8H_{17}CN$)	Hendecanonitrile ($C_{10}H_{21}CN$)	Myristonitrile ($C_{14}H_{29}CN$)
41	12.0	11.0	11.2
55	5.0	5.1	5.2
69	5.6	3.8	2.9
83	4.8	3.7	2.8
97	3.5	5.0*	5.8*
111	1.0	1.7	2.5
125	0.1	0.4	0.8

peak at *m/e* 91, as illustrated by the spectrum of 2-chlorooctane (API–TRC No. 76-m). A similar process of forming a cyclic transition is observed for terminal alkyl bromides having more than four carbon atoms. The cyclic ion thus formed is at *m/e* 135 rather than *m/e* 91, as is the case for terminal chlorides.

Another example of an identifying cyclic ion is the $C_5H_{10}CNH$ ion at *m/e* 97 found in spectra of nitriles. This ion does not stand out from other peaks in the 41, 55, 69, etc., series as shown in Table 9-10 for pelargononitrile but becomes an interpretive peak for hendecanonitrile and higher nitriles (*Archives* **1**, No. 2, pp. 196–213). The formation of this ion as shown by McLafferty (1962b) is depicted below.

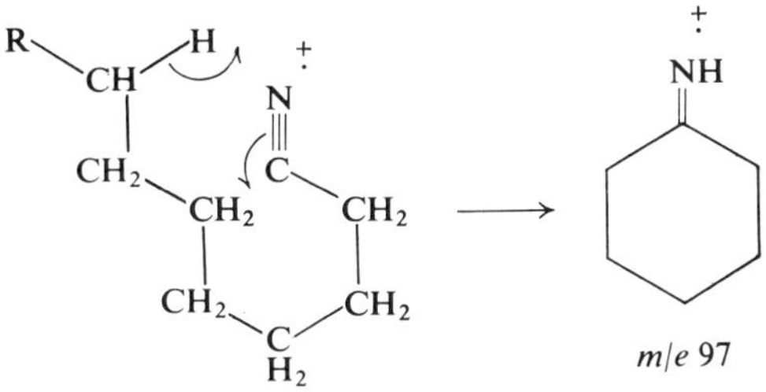

R CH H N C CH₂ CH₂ CH₂ CH₂ CH₂ C H₂ ⟶ NH *m/e* 97

$C_nH_{2n+1}CN$ compounds

The peak at *m/e* 97 is useful for the detection of nitriles because these compounds have most of their ion yield (35% or more) in the 41, 55, 69, etc., series of peaks as do many classes of organic compounds. Recalling the formation of this cyclic ion can be very useful when examining an unknown spectrum.

Draper and MacLean (1970) presented a study on ring-expanded ions of methylquinolines. In this case, the ring-expanded species was considered to be formed prior to decomposition. In the case of monomethylquinolines, one scheme of fragmentation is shown as follows:

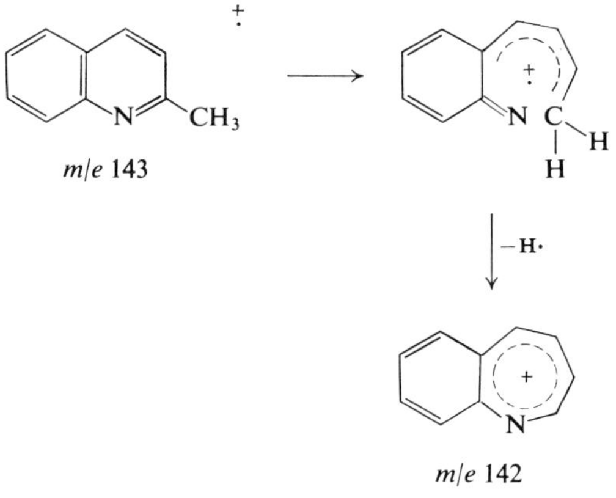

N CH₃ *m/e* 143 ⟶ N C H H

$-H\cdot$

N *m/e* 142

The purpose of citing the above work, beyond the interest it has within itself, is to enable the reader working in a research laboratory to take such a study and relate it in a practical way to the identification of unknown structures from mass spectral data. If this thinking process is developed, it can greatly aid the advancement of mass spectrometry in a way supervisors, company managers, research funding agencies, etc., can readily understand. The practical side of mass spectrometry becomes evident by applying the technique to identifying unknown reaction products. In this case, an intense peak 1 mass unit below the molecular ion of an odd mass number indicates a compound having a nitrogen atom and a methyl group. The identification is supported by formation of a ring-expanded species.

VI. Fragmentation Behavior in Polyfunctional Compounds

Fragmentation of organic molecules in the mass spectrometer usually occurs by several competing pathways. These cannot be quantitatively defined but are usually evident to various degrees in spectra of polyfunctional compounds. (The degree may be so small that it is almost obscured.) The examples provided illustrate the detection of different functional groups within a molecule.

The real test of practicality is whether the accumulated knowledge of fragmentation rules can be translated to structure elucidation. There is a great difference between following the formation of intense peaks from a known structure and identifying an unknown structure from a group of numbers on a spectrum. Consider, for example, a spectrum having its three most intense peaks at *m/e* 91, 145, and 236. Consider the peak at *m/e* 236 ($C_{17}H_{16}O$) as the molecular ion peak. If the structure is known to be 2-benzyl-3,4-dihydro-1(2*H*)-naphthalenone (API–TRC No. 218-m), the formation of these fragment peaks can be rationalized through a combination of fragmentation rules. The

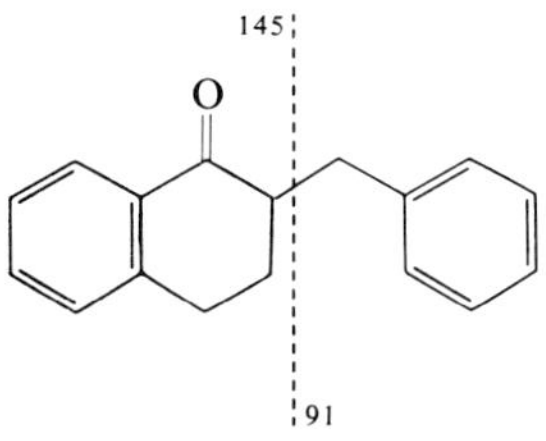

bond shown being cleaved is β to the single benzene ring and β to the carbonyl group. This same bond is also α to the saturated moiety of the molecule. These *three factors* promote the cleavage of this bond. The competition for the positive charge is shown by the intensities of the peaks at *m/e* 91 (20.4 %) and *m/e* 145 (9.7 %).

A similar example is provided by the spectrum of 2,2′,5′-trimethylacrylophenone (API–TRC No. 26-m). In the high-mass region, the area of the spectrum where the intense peaks are more efficacious for identification, peaks appear at *m/e* 174 (molecular ion, $C_{12}H_{14}O$), 159, 133, and 105. A simple combination of two fragmentation rules aids in rationalizing the formation of the peaks at *m/e* 159 and 133. The fragmentation of 2,2′,5′-trimethylacrylophenone may be depicted as I (later to be compared to II).

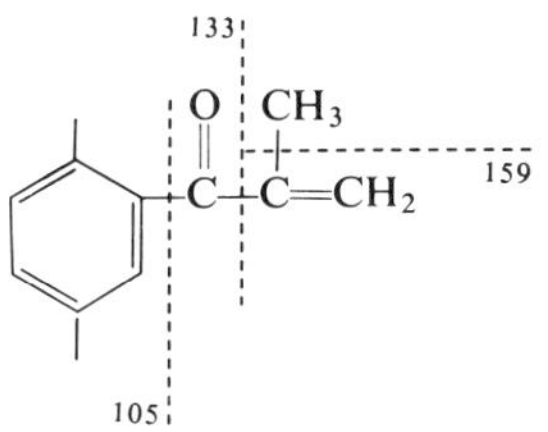

I

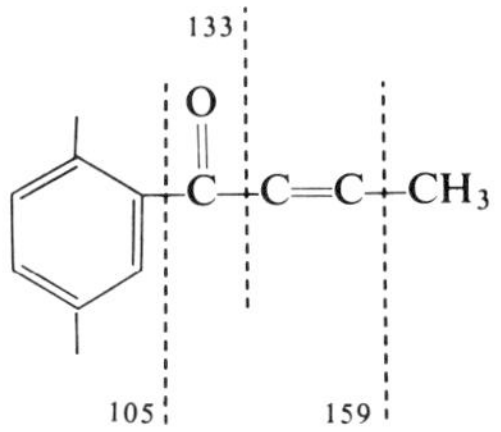

II (2′,5′-dimethylcrotonophenone)

The peak at *m/e* 159 is attributed to cleavage of the methyl group shown at a branched position. The peak at *m/e* 133 is the result of cleavage of the C–C bond β to the benzene ring. The peak at *m/e* 105 is then attributed to removal of CO from the fragment peak at *m/e* 133. What would have been the fragmentation had the structure not been branched on the alkyl group adjacent to the carbonyl group? The answer is given in the spectrum of 2′,5′-dimethylcrotonophenone (API–TRC No. 25-m). The intense peak at *m/e* 159 may be rationalized due to the ease of cleavage of the C–C bond adjacent to the double bond, rather than as a result of cleavage enhanced at the point of branching, as is the case for 2,2′,5′-trimethylacrylophenone. When the structure is known, the application of fragmentation rules is usually obvious.

The reader should examine the fragmentation of other such aromatic ketones published in the "Catalog of the Mass Spectral Data" (1968). It is well to keep in mind in the examination of these fragmentation patterns, as Reed (1966) has pointed out, that unless the functional groups are separated by three or more carbon atoms, the various functional groups affect each other differently than if present as a single group in separate molecules.

In the two previous samples, the different functional groups enhance cleavage of the same C–C bond in the molecule. The cleavage of two different C–C bonds is illustrated in the fragmentation study of Meyerson and Leitch (1966) in the decomposition of 6-substituted alkanoic acids and esters. While their results clearly showed expected fragmentation, an unexpected rearrangement–dissociation path was also observed. A compound to be examined is assumed to be known only as an alkanoic ester. The unexpected rearrangement–dissociation found by Meyerson and Leitch (1966) is such that the hydrogen atom at C-6 apparently triggers the mechanism shown in part below.

CH_2 CH_2 CH_2 H—C—H $C^{+\cdot}{=}O$ H_5C_6—C—H OCH_3 H $\xrightarrow{-CH_3OH}$ $C_6H_5CH{=}CH{-}CH_2$ O^{+} CH_2 CH CH_2

m/e 206 *m/e* 174

To apply this fragmentation route to the unknown structure of an alkanoic ester, it must be remembered that it is the hydrogen at C-6 which triggers the mechanism. Thus, without a C-6 hydrogen, an alkanoic ester would not be expected to give an intense fragment ion in the series $174 + (n \times 14)$. This is supported by spectra published in the "Catalog of Mass Spectral Data" of the Thermodynamics Research Center Data Project (API-44–TRC Nos. 117-m, 118-m, 121-m, and 123-m). Had any one of these fragmentation patterns been obtained as an unknown structure of an alkanoic ester, the conclusion could have been drawn at once that the structure did not have a C-6 carbon. This alone, of course, does not provide a complete structural identification. Several rules have to be considered for any one rather complicated structure. Such additions to fragmentation routes can lead to new applications in mass spectrometry. In fact, it is hoped that the reader has reached the point where a discussion of a new fragmentation route brings to mind applications in structural analysis.

Another problem that frequently occurs in mass spectrometry is the observation of a spectrum that presents several choices of fragmentation and several rearrangement possibilities. The peaks observed, if from an unknown compound, may mislead the interpreter into considering a mixture of several compounds or possibly major impurities present in the sample. The compound 4,5,6,7-tetrahydro-1-(2-benzo[*b*]thienyl)-1-thiapentane is used to illustrate some difficulties that may be encountered. The molecule shown presents a number of cleavage possibilities after the formation of the molecular ion has occurred. The partial mass spectrum is presented in Table 9-11. The molecular ion peak (PIP) at 226 is also the BLIP. This might be expected in view of the aromatic nature of the molecule. However, the several more saturated groups tend to offer fragmentation sites, and thus reduce the PIP intensity to only 10.1 % of the total ion intensity. The ion at *m/e* 170, an obvious RIP, is also quite large and gives an intensity of 9.94% of the total ion intensity. One immediately associates this (provided the structure is known) with a β-bond

TABLE 9-11

PARTIAL MASS SPECTRUM OF 4,5,6,7-TETRAHYDRO-1-(2-BENZO[*b*]THIENYL)-1-THIAPENTANE

m/e	$\%\Sigma_i$	*m/e*	$\%\Sigma_i$
39	3.04	134	1.09
41	3.73	135	1.85
45	2.92	136	1.76
57	1.28	137	4.57
65	1.31	142	7.16 (RIP)
71	1.34	169	2.98
77	2.15	170	9.94 (RIP)
93	0.67	198	0.86 (RIP)
125	1.15	226	10.10 (PIP and BLIP)
131	0.08		

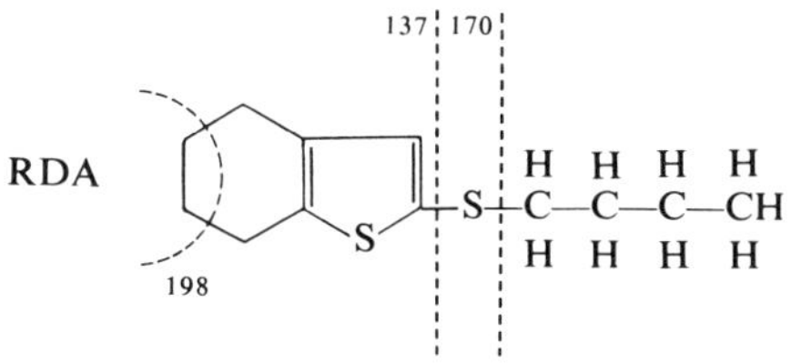

137 170

RDA

H H H H

S—C—C—C—CH

H H H H

198

$M^+ = 226$

cleavage with rearrangement leading to the *m/e* 170 ion. Further inspection shows either two more RIP's at 198 and 142 or else the presence of molecular ions of compounds in the same series, each two carbon numbers apart, commencing with 142, then 170, 198, and finally 226. Are the ions molecule ions of four different compounds, of three different compounds and a trace impurity at *m/e* 198, or just what is hapenning here? Low voltage would ultimately show only the 226 ion persisting; therefore, it is a single compound possessing three RIP's. The RIP at 198 probably comes from the RDA ejection of a neutral C_2H_4 particle from the tetrahydro portion of the molecule. Since this cleavage would be in direct competition with the β cleavage with rearrangement, it is quite clear that the β rearrangement process is favored by a little better than 11:1 (9.94% versus 0.86% of *m/e* 198). The *m/e* 170 species may lose a hydrogen atom to form the *m/e* 169 ion, but competing with this is the occurrence of the RDA ejection from the 170 species to produce the 142 RIP. In the work of Foster and Higgins (1969), it was shown from low-voltage data that the 170 ion appeared ahead of all of the other fragment ions, suggesting that the 142 did arise from the 170. A concerted process to produce this ion directly from the molecular ion could not be definitely ruled out, although no evidence for it could be found. Similarly, no evidence could be found that the *m/e* 169 ion was produced by direct β cleavage without a hydrogen transfer. Metastable ions were observed for the two RDA processes, $226^{+\cdot} \rightarrow 198^{+\cdot}$ ($m^* = 173.4$) and $170^{+} \rightarrow 142^{+}$ ($m^* = 118.6$), but not for the β cleavage with rearrangement. This is quite puzzling but reminds the mass spectroscopist that often all metastable peaks will not be seen and with no apparent reason (see also Chapter 3). Somewhat surprising in light of this was the observance of the direct α cleavage (to the fused-ring system) by the process $226^{+} \rightarrow 137^{+} + C_4H_9S^0$ ($m^* = 83.2$). The ions at P − 57 and 198 − 57 (131) were insignificant, indicating the expected retention of the charge by the hetero atom moieties. It appears that the *m/e* 169 ion must be primarily produced from the 170 but, once formed, the 169 ion does not undergo an RDA ejection of $C_2H_4{}^0$ to produce an *m/e* 141 ion. Instead, it seems to account for the ion at *m/e* 125 by the loss of a CS^0, and the 125, in turn, may form the 93 ion by the loss of S^0. The interesting group of ions in the group from 137 to 134 probably arises by one or more processes involving the ejection of S, H, and SH as neutral fragments. The 137 may arise from the loss of S^0 from the 169 ion.

At this point one may well say that this is simply a verification of Murphy's second law, namely, that everything that is likely to happen will happen. So be it. The mass spectroscopist faced with the interpretation of a spectrum presenting these types of RIP and the attendant problems should (a) obtain low voltage data, (b) check all ions for isotopic contributions in an effort to (c) determine the ion composition of all major fragments, (d) reconsult the sample submitter for the various possibilities of impurities, multiproduct reactions, or just plain "things didn't go right." Next, (e) he should take a break and reorganize his thoughts, looking at the problem afresh after putting it out of mind, as it were, for some period of time. (f) DON'T GIVE UP. TRY AGAIN!

The mass spectroscopist should recognize that this occurs enough times that the array format is his biggest aid in starting the interpretation. It allows him to find the presence of such a problem at the outset, and less time is wasted. As a convincing case which produces rearrangement ions not in the series of the molecular ion, consider the phenyl ethyl sulfoxide molecule. At 70 V, the parent ion is relatively small at *m/e* 154. Large RIP's are observed at *m/e* 126, 110, and 78, the 78 ion in fact becoming the BLIP. The probable fragmentations leading to this series of ions are as follows: 154 loses 28 ($C_2H_4{}^0$) to form the 126 ion, structure undetermined; the 126 ion loses an oxygen atom (16) to form the ion at *m/e* 110, structure not ascertained, finally, this ion ejects a neutral sulfur to produce the C_6H_6 ion at *m/e* 78.

Hence it should be no surprise to find the interpreter thinking in terms of a mixture of thiophenol (110), benzene (78), and one or two molecules having ions at *m/e* 154 and 126! Note that the 110 ion is in the −2 series and the 78

$-C_2H_4O$ → $-O$ →

(m/e 154) (m/e 126)

$-S$ →

m/e 110 m/e 78

Structure not ascertained

ion is in the −6 series. Both the 154 and 126 ions are in the 0 series. The major way out for the interpreter in this case is to use low ionization voltage, and when he does, the fragmentations and rearrangements occurring line up rather nicely, permitting the original structure to be deduced with some degree of confidence.

The rules presented here are far from all-inclusive. It is hoped that the reader will develop his own set of rules. When well formulated they should provide a new and exciting application of mass spectrometry—one in which the computer can become a working partner on nonroutine samples submitted to the mass spectroscopist.

REFERENCES

Carpenter, W., Duffield, A. M., and Djerassi, C. (1968). *J. Amer. Chem. Soc.* **90**, 160.

"Catalog of Mass Spectral Data" (1968). Loose-leaf data sheets Serial Nos. 25-m–31-m. Thermodynamics Research Center Data Project, Thermodynamics Research Center, Texas A & M University, College Station, Texas.

"Catalog of Mass Spectral Data" (1970). Loose-leaf data sheets Serial Nos. 101-m–116-m. Thermodynamics Research Center Data Project, Thermodynamics Research Center, Texas A & M University, College Station, Texas.

Draper, P. M., and MacLean, D. B. (1970). *Can. J. Chem.* **48**, 476.

Ford, H. T. (1964). Paper No. 83. Twelfth Annual Conference on Mass Spectrometry, ASTM Committee E-14, Montreal, Canada.

Foster, N. G., and Higgins, R. W. (1969). *Org. Mass Spectrom.* **2**, 1005.

Fueki, K., and Hirota, K. (1959). *Nippon Kagaku Zasshi* **80**, 1202.

Gohlke, R. S., and McLafferty, F. W. (1962). *Anal. Chem.* **34**, 1281.

Grubb, H. M., and Meyerson, S. (1963). *In* "Mass Spectrometry of Organic Ions" (F. W. McLafferty, ed.), pp. 453–527. Academic Press, New York.

Hatada, M., and Hirota, K. (1965). *Bull. Chem. Soc. Jap.* **38**, 599.

Hirota, K., and Itoh, M. (1966). *Bull. Chem. Soc. Jap.* **39**, 1406.
Hirota, K., and Niwa, Y. (1968). *J. Phys. Chem.* **72**, 5.
McLafferty, F. W. (1962a). *Anal. Chem.* **34**, 2.
McLafferty, F. W. (1962b). *Anal. Chem.* **34**, 26.
McLafferty, F. W. (1966). "Interpretation of Mass Spectra," pp. 120–150. Benjamin, New York.
Martin, T. T., and Hamming, M. C. (1963). Unpublished results. Continental Oil Co., Ponca City, Oklahoma.
Meyerson, S., and Leitch, L. C. (1966). *J. Amer. Chem. Soc.* **88**, 56.
Reed, R. I. (1966). "Applications of Mass Spectrometry to Organic Chemistry," p. 96. Academic Press, New York.

GENERAL BIBLIOGRAPHY AND COMPREHENSIVE REFERENCES

Aston, F. W. (1922). "Isotopes." Arnold, London.

Aston, F. W. (1933). "Mass Spectra and Isotopes." Arnold, London.

Aston, F. W. (1942). "Mass Spectra and Isotopes," 2nd Ed., 276 pp. Arnold, London.

Ausloos, P. J. (1966). "Ion–Molecule Reactions in the Gas Phase," *Advances in Chemistry* series no. 58, 336 pp. American Chemical Society, Washington, D.C.

Barnard, G. P. (1953). "Modern Mass Spectrometry." Institute of Physics, London.

Benz, W. (1968). "Massenspektrometrie Organischer Verbindungen," 425 pp. Akad. Verlagsges., Frankfurt a.M.

Beynon, J. H. (1960). "Mass Spectrometry and Its Applications to Organic Chemistry," 640 pp. Amer. Elsevier, New York.

Beynon, J. H., and Fontaine, A. E. (1967). *In* "Some Newer Physical Methods in Structural Chemistry" (R. Bonnett and J. G. Davis, eds.). United Trade Press, London.

Beynon, J. H., and Williams, A. E. (1963). "Mass and Abundance Tables for Use in Mass Spectrometry," 570 pp. Amer. Elsevier, New York.

Beynon, J. H., Saunders, R. A., and Williams, A. E. (1965). "Table of Meta-stable Transitions." Amer. Elsevier, New York.

Beynon, J. H., Saunders, R. A., and Williams, A. E. (1968). "The Mass Spectra of Organic Molecules," 510 pp. Amer. Elsevier, New York.

Biemann, K. (1962). "Mass Spectrometry: Organic Chemical Applications," 370 pp. McGraw-Hill, New York.

Biemann, K. (1963). *Ann. Rev. Biochem.* **32**, 755–80.

Birkenfeld, H., Haase, G., and Zahn, H. (1962). "Massenspektrometrische Isotopenanalyse." VEB Deutscher Verlag der Wissenschaften, Berlin.

Blanth, E. W. (1966). "Dynamic Mass Spectrometer," 185 pp. Amer. Elsevier, New York.

Bondarovich, H. A., and Freeman, S. K. (1965). *In* "Interpretive Spectroscopy" (S. F. Freemen, ed.), pp. 170–209. Van Nostrand-Reinhold, Princeton, New Jersey.

Bowie, J. H. (1969). *In* "Chemistry of the Carbonyl Group" (J. Zabicky, ed.), Vol. II. Wiley, New York.

Branscomb, L. M. (1957). *In* "Advances in Electronics and Electron Physics" (L. Marton, ed.), Vol. 9, pp. 43–94. Academic Press, New York.

Brunnée, C., and Voshage, H. (1964). "Massenspektrometrie." Verlag Karl Thiemig, Munich.

Budzikiewicz, H., Djerassi, C., and Williams, D. H. (1964). "Structure Elucidation of Natural Products by Mass Spectrometry," Vols. I and II, 539 pp. Holden-Day, San Francisco.

Budzikiewicz, H., Djerassi, C., and Williams, D. H. (1967). "Mass Spectrometry of Organic Compounds," 690 pp. Holden-Day, San Francisco.

Burlingame, A. L., ed. (1970). "Advances in Analytical Chemistry and Instrumentation," Vol. 8, 472 pp. Wiley, New York.

Cameron, A. E. (1961). *In* "Physical Methods in Chemical Analysis" (W. G. Berl, ed.), Vol. IV, pp. 119–32. Academic Press, New York.

Capellen, J., Svec, H. J., and Sage, C. R. (1966). "Bibliography of Mass Spectroscopy Literature, Compiled by Computer Method," IS-1335, 399 pp. Division of Technical Information, U.S. Atomic Energy Commission, Oak Ridge, Tennessee.

Capellen, J., Svec, H. J., Jordan, J. R., and Watkins, W. J. (1970). "Bibliography of Mass Spectroscopy Literature for the Last Half of 1967,Compiled by a Computer Method," IS-2058, 458 pp. Ames Laboratory of the U.S. Atomic Energy Commission.

Capellen, J., Svec, H. J., Jordan, J. R., and Sun, R. (1970). "Bibliography of Mass Spectroscopy Literature for the First Half of 1968, Compiled by a Computer Method," IS-2059, 463 pp. Ames Laboratory of the U.S. Atomic Energy Commission.

Cornu, A., and Massot, R. (1964). "Compilation of Exact Masses of Organic Ions for Use in High Resolution Mass Spectrometry," 364 pp. Heyden, London.

Cornu, A., and Massot, R. (1966). "Compilation of Mass Spectral Data." Heyden, London.

Cornu, A., and Massot, R. (1967). "First Supplement to Compilation of Mass Spectral Data." Heyden, London.

Craggs, J. D., and McDowell, C. A. (1955). *Rep. Progr. Phys.* **18**, 374–422.

Duckworth, H. E. (1958). "Mass Spectroscopy." Cambridge Univ. Press, London and New York.

Duckworth, H. E. (1960). "Mass Spectroscopy," 206 pp. Cambridge Univ. Press, London and New York.

Duckworth, H. E., ed. (1960). "Proceedings of the International Conference of Nuclidic Masses," 540 pp. Univ. of Toronto Press, Toronto.

Elliott, R. M., ed. (1963). "Advances in Mass Spectrometry," Vol. II. Pergamon, Oxford.

Elliott, R. M., Craig, R. D., and Errock, G. A. (1961). *In* "Instruments and Measurements, Proceedings of the Fifth International Instruments and Measurements Conference, Stockholm, 1960 (H. von Koch and G. Ljungberg, eds.), Vol. 1. Academic Press, New York.

Ewald, H., and Hintenberger, H. (1953). "Methoden und Anwendungen der Massenspektroskopie," 288 pp. Verlag Chemie, Weinheim.

Ewing, G. W. (1960). "Instrumental Methods of Chemical Analysis," pp. 278–295. McGraw-Hill, New York.

Field, F. H., and Franklin, J. L. (1957). "Electron Impact Phenomena and the Properties of Gaseous Ions," 350 pp. Academic Press, New York.

Fields, E. K., and Meyerson, S. (1967). *In* "Organosulfur Chemistry" (M. J. Janssen, ed.). Wiley, New York.

Fleming, I., and Williams, D. (1968). "Spectroscopic Problems in Organic Chemistry." McGraw-Hill, New York.

Franklin, J. L., Dillard, J. G., Rosenstock, H. M., Herron, J. T., Draxl, K., and Field, F. H. (1969). "Ionization Potentials, Appearance Potentials, and Heats of Formation of Gaseous Positive Ions," 289 pp. U.S. Department of Commerce National Bureau of Standards, Washington, D.C.

Geltman, S. (1969). "Topics in Atomic Collision Theory," 247 pp. Academic Press, New York.

Gohlke, R. S. (1963). "Uncertified Mass Spectral Data." Dow Chemical Co., Midland, Michigan.

Good, R. J., Jr., and Muller, E. W. (1956). "Handbuch der Physik," 2nd Ed., Vol. 21, p. 176. Springer-Verlag, Berlin and New York.

Harrington, D. B. (1960). *In* "Encyclopedia of Spectroscopy" (C. L. Clark, ed.), pp. 628–647. Van Nonstrand-Reinhold, Princeton, New Jersey.

Herzog, L. F. (1970). *In* "Analytical Chemistry in Space" (R. E. Wainerdi, ed.), pp. 109–163. Pergamon, Oxford.

Higatsberger, M. J., and Viehbock, F. P. (1961). "Electromagnetic Separation of Radioactive Isotopes," 318 pp. Springer-Verlag, Berlin and New York.

Hill, H. C. (1966). "Introduction to Mass Spectrometry," 135 pp. Heyden, London.

Hintenberger, H., ed. (1957). "Nuclear Masses and Their Determination." Pergamon, Oxford.

Hintenberger, H. (1962). *Ann. Rev. Nucl. Sci.* **12**, 435–506.

Horning, E. C., Brooks, C. J. W., and Vanden Heuvel, W. J. A. (1968). "Gas Phase Analytical Methods for the Study of Steroids," Vol. 6, pp. 237–392. Academic Press, New York.

Hutter, R. G. E. (1948). *In* "Advances in Electronics" (L. Marton, ed.), Vol .1, pp. 167–218. Academic Press, New York.

Inghram, M. G. (1948). *In* "Advances in Electronics" (L. Marton, ed.), Vol. 1, pp. 219–68 Academic Press, New York.

Inghram, M. G., and Hayden, R. J. (1954). "A Handbook of Mass Spectrometry," *Nuclear Science* series, rep. 14. Nat. Acad. Sci.-Nat. Research Council, Washington, D.C.

Jayaram, R. (1966). "Mass Spectrometry," 225 pp. Plenum, New York.

Kendrick, E., ed. (1968). "Advances in Mass Spectrometry," Vol. 4, 971 pp. Institute of Petroleum, London.

Kerwin, L. (1956). *In* "Advances in Electronics and Electron Physics" (L. Marton, ed.), Vol. 8, pp. 187–253. Academic Press, New York.

Kienitz, H., Aulinger, F., Franke, G., Habfast, K., and Spiteller, G. (1968). "Mass Spectrometry," 884 pp. Verlag Chemie, Weinheim.

Kiser, R. W. (1965). "Introduction to Mass Spectrometry and Its Applications," 356 pp. Prentice-Hall, Englewood Cliffs, New Jersey.

Knewstubb, P. F. (1969). "Mass Spectrometry and Ion-molecule Reactions," 136 pp. Cambridge Univ. Press, London and New York.

Krauss, M., Wahrhaftig, A. L., and Eyring, H. (1955). *Ann. Rev. Nucl. Sci.* **5**, 241–68.

Lampe, F. W., Franklin, J. L., and Field, F. H. (1961). *In* "Progress in Reaction Kinetics" (G. Porter, ed.), pp. 67–103. Pergamon, Oxford.

Lederberg, J. (1964). "Computation of Molecular Formulas for Mass Spectrometry," 69 pp. Holden-Day, San Francisco.

Loeb, L. B. (1955). "Basic Processes of Gaseous Electronics," 1012 pp. Univ. of California Press, Berkeley.

Loudon, A. G., and Maccoll, A. (1970). *In* "The Chemistry of Alkenes" (J. Zabicky, ed.), pp. 327–358. Wiley (Interscience), New York.

McDaniel, E. W. (1964). "Collisional Phenomena in Ionized Gases," 775 pp. Wiley, New York.

McDowell, C. A., ed. (1963). "Mass Spectrometry," 639 pp. McGraw-Hill, New York.

McLafferty, F. W. (1962). *In* "Determination of Organic Structures by Physical Methods" (F. C. Nachod and W. D. Phillips, eds.), Vol. II, pp. 93–179. Academic Press, New York.

McLafferty, F. W., ed. (1963). "Mass Spectrometry of Organic Ions," 730 pp. Academic Press, New York.

McLafferty, F. W. (1963). "Mass Spectral Correlations," *Advances in Chemistry* series no. 40, 117 pp. American Chemical Society, Washington, D.C.

McLafferty, F. W. (1966). "Interpretation of Mass Spectra," 229 pp. Benjamin, New York.

McLafferty, F. W., and Pinzelik, J. (1967). "Index and Bibliography of Mass Spectrometry." Wiley (Interscience), New York.

Massey, H. S. W. (1950). "Negative Ions," 2nd Ed., 136 pp. Cambridge Univ. Press, London and New York.

Massey, H. S. W., and Burhop, E. H. S. (1952). "Electronic and Ionic Impact Phenomena," 670 pp. Oxford Univ. Press (Clarendon), London and New York.

Massey, H. S. W., and Burhop, E. H. S. (1969). "Electronic and Ionic Impact Phenomena," Vols. 1 and 2, 1294 pp. Oxford Univ. Press, London.

Mayne, K. I. (1952). *Rep. Progr. Phys.* **15**, 24–48.

Mead, W. L., ed. (1966). "Advances in Mass Spectrometry," Vol. 3, 1064 pp. Institute of Petroleum, London.

Melton, C. E. (1970). "Principles of Mass Spectrometry and Negative Ions," 328 pp. Dekker, New York.

Meyerson, S., and McCollum, J. D. (1963). *In* "Advances in Analytical Chemistry and Instrumentation" (C. N. Reilley, ed.), Vol. 2, pp. 179–218. Wiley (Interscience), New York.

Mitchell, J. J. (1950). *In* "Physical Chemistry of the Hydrocarbons" (A. Farkas, ed.), Vol. I. Academic Press, New York.

Müller, E. W. (1960). *In* "Advances in Electronics and Electron Physics" (L. Marton, ed.), p. 83. Academic Press, New York.

Ogata, K., and Hayakawa, T., eds. (1970). "Recent Developments in Mass Spectrometry," 1328 pp. Univ. Park Press, Baltimore, Maryland.

Pasto, D. J., and Johnson, C. R. (1969). "Organic Structure Determination," 513 pp. Prentice-Hall, Englewood Cliffs, New Jersey.

Pecsok, R. L., and Shields, L. D., eds. (1968). *In* "Modern Methods of Chemical Analysis," pp. 257–289. Wiley, New York.

Polyakova, A. A., and Khmel'nitskii, R. A. (1966). "Introduction to Mass Spectroscopy of Organic Compounds," 203 pp. Khimiya, Leningrad.

Price, D., and Williams, J. D., eds. (1969). "Dynamic Mass Spectrometry," Vol. I, 247 pp. Heyden, London.

Quayle, A. (1959). "Advances in Mass Spectrometry." Pergamon, Oxford.

Reed, R. I. (1962). "Ion Production by Electron Impact," 242 pp. Academic Press, New York.

Reed, R. I., ed. (1965). "Mass Spectrometry," 463 pp. Academic Press, New York.

Reed, R. I. (1966). "Applications of Mass Spectrometry to Organic Chemistry," 256 pp. Academic Press, New York.

Reed, R. I. (1968). "Modern Aspects of Mass Spectrometry," 389 pp. Plenum, New York.

Robertson, A. J. B. (1954). "Mass Spectrometry," 135 pp. Methuen, London.

Robinson, C. F. (1960). *In* "Physical Methods in Chemical Analysis" (W. G. Berl, ed.), 2nd Rev. Ed. Vol. 1, pp. 474–545. Academic Press, New York.

Roboz, J. (1968). "Introduction to Mass Spectrometry: Instrumentation Techniques," 539 pp. Wiley, New York.

Rosenstock, H. M., Wahrhaftig, A. L., and Eyring, H. (1952). "The Mass Spectra of Large Molecules. Part II. The Application of Absolute Rate Theory," Tech. Rep. No. 2. Institute for the Study of Rate Processes, University of Utah, Salt Lake City.

Sharkey, A. G., Jr., Shultz, J. L., Kessler, T., and Friedel, R. A. (1970). *In* "Spectrometry of Fuels" (R. A. Friedel, ed.), pp. 1–14. Plenum, New York.

Shuler, K. E., and Fenn, J. B., ed. (1963). "Progress in Astronautics and Aeronautics," 409 pp. Academic Press, New York.

Shumulovskii, N. N., and Stakhovskii, R. I. (1966). "Mass Spectral Methods," 160 pp. Energiya, Moscow.

Siggia, S. (1968). "Survey of Analytical Chemistry," 304 pp. McGraw-Hill, New York.

Silverstein, R. M., and Bassler, G. C. (1967). "Spectrometric Identification of Organic Compounds," 2nd Ed., 256 pp. Wiley, New York.

Smith, M. L., ed. (1956). "Electromagnetically Enriched Isotopes and Mass Spectrometry," Proceedings of the Harwell Conferences, Sept. 13–16, 1955, 272 pp. Butterworth, London.

Snedden, W. (1963). "Advances in Mass Spectrometry," Vol. 2, 456 pp. Pergamon, Oxford.

Spiteller, G. (1966). *In* "Advances in Heterocyclic Chemistry" (A. R. Katritzky and A. J. Boulton, eds.), Vol. 7, pp. 301–376. Academic Press, New York.

Spiteller, G. (1966). "Massenspektrometrische Strukturanalyze Organischen Verbindunger," 354 pp. Verlag Chemie, Weinheim.

Stenhagen, E., Abrahamsson, S., and McLafferty, F. W. (1969). "Atlas of Mass Spectral Data," Vols. I, II, and III, 2266 pp. Wiley (Interscience), New York.

Stenhagen, E., Abrahamsson, S., and McLafferty, F. W., eds. (1970). "The Archives of Mass Spectral Data," Vol. 1, 784 pp. Wiley (Interscience), New York.

Stevenson, D. P., and Schissler, D. O. (1961). *In* "Actions Chimiques et Biologiques des Radiations" (M. Haissinsky, ed.), cinquieme serie pp. 167–271. Masson, Paris.

Streitwieser, A., Jr. (1963). *In* "Progress in Physical Organic Chemistry" (S. G. Cohen, A. Streitwieser, Jr., and R. W. Taft, eds.), Vol. 1, pp. 1–30. Wiley (Interscience), New York.

Tatematsu, A., and Tsuchiya, T. (1968). "Structure Indexed Literature of Organic Mass Spectra—1966," 275 pp. Academic Press of Japan, Tokyo, Japan.

Tatematsu, A., and Tsuchiya, T. (1969). "Structure Indexed Literature of Organic Mass Spectra—1967," 496 pp. Academic Press of Japan, Tokyo, Japan.

Thode, H. G., and Shields, R. B. (1949). *Rep. Progr. Phys.* **12**, 1–21.

Thode, H. G., McMullen, C. C., and Fritze, K. (1960). *In* "Advances in Inorganic Chemistry and Radiochemistry" (H. J. Emelus and A. G. Sharpe, eds.), Vol. II, pp. 315–363. Academic Press, New York.

Thompson, R. (1954). "Applied Mass Spectrometry." Institute of Petroleum, London.

Thomson, J. J. (1913). "Rays of Positive Electricity and Their Application to Chemical Analyses." Longmans, Green, New York.

Vastola, F. J., Pirone, A. L., Given, P. H., and Dutcher, R. R. (1970). *In* "Spectrometry of Fuels" (R. A. Friedel, ed.), pp. 29–36. Plenum, New York.

Vedeneev, V. I., Gurvich, L. V., Kondrat'ev, V. N., Medvedev, V. A., and Frankevich, E. L. (1962). "Dissociation Energies of Chemical Bonds, Ionization Potentials and Electron Affinities Handbook," 216 pp. Acad. Sci. U.S.S.R., Chemical Physics Institute, Moscow.

Von Hock, H., and Ljungberg, G., eds. (1961). "Instruments and Measurements, Proceedings of the Fifth International Instruments and Measurements Conference, Stockholm, 1960," Vol. 2. Academic Press, New York.

Waldron, J. D., ed. (1959). "Advances in Mass Spectrometry," Vol. 1, 704 pp. Pergamon, Oxford.

Wallenstein, M. B., Wahrhaftig, A. L., and Eyring, H. (1951). "The Mass Spectra of Large Molecules. I. Saturated Hydrocarbons." Univ. of Utah, Salt Lake City.

Waller, G. R., ed. (1972). "Biochemical Applications of Mass Spectrometry," 900 pp. Wiley, New York.

Wanless, G. G. (1970). *In* "Spectrometry of Fuels" (R. A. Friedel, ed.), pp. 15–28. Plenum, New York.

Washburn, H. W. (1950). *In* "Physical Methods in Chemical Analysis" (W. G. Berl, ed.), Vol. 1, pp. 587–637. Academic Press, New York.

White, F. A. (1968). "Mass Spectrometry in Science and Technology," 352 pp. Wiley, New York.

Williams, D. H., and Fleming, I. (1967). "Spectroscopic Problems in Organic Chemistry," 142 pp. McGraw-Hill, New York.

Appendixes **I–V**

APPENDIX I

STRUCTURE-CORRELATION REFERENCES FOR SOME ORGANIC AND RELATED COMPOUNDS

Organization of the Index

With so many disciplines utilizing the field of mass spectrometry it is almost impossible to set up a single, all-purpose, all-inclusive index. Often paper titles do not reflect the presence of mass spectral data. When they do, the compounds, exact derivatives, molecular weight range, etc., may be omitted and can be ascertained only by perusal of the paper. Therefore, to assist the reader of this book, this appendix provides an index with a relationship to compounds in a series. Insofar as possible, the system used is as follows:

SECTIONS

A	HC, hydrocarbon compounds only
B	CHO, compounds containing those elements only
C	CHN, compounds containing those elements, with oxygen if present
D	CHS, compounds containing those elements, with oxygen and/or nitrogen if present
E	CHX, compounds containing halogen atoms plus any of the above elements
F	CHM, compounds containing metallic atoms plus any of the above elements
G	CHB, P, Si, Se, Te, etc., essentially the nonmetals plus any of the above elements
H	Substances of biological interest, lipids, proteins, alkaloids, pesticides, etc. (this section is organized by subtitle rather than by series numbers)

In general, within Sections A–G, the reader will see that logical organization by series numbers is replaced when the series numbers alone are no longer helpful, i.e., compounds with one nitrogen atom and two oxygen atoms.

COLUMN ITEMS

I	Name of class (series numbers in parentheses)
II	Number of compounds discussed
III	Number of complete or partial spectra given
IV	Contains D, ^{13}C, ^{18}O, X = halogen, etc.
V	Coded as to type of article, i.e., M—mechanism, E—energetics, C—correlations, O—only spectra, D—discussion of type, G—general discussion
VI	Molecular weight range of compounds reported
VII	References are numbered within each section. When a reference is duplicated within a section, the second entry shows only the reference number and lists the author(s) and year.

For further discussion of the use of this appendix see text pages 267–268.

SECTION A: HYDROCARBON COMPOUNDS

Name of class (series no.)	II	III	IV	V	MW range	Reference
Alkanes (+2)				C		1
	23	13		G	16–86	2
				C		3
				C		4
				C		5
			D	M	156–170	6
			^{13}CD	M		7
Cycloalkanes (0) (naphthenes)						
1,2-Dialkylcyclopentanes	2			E		8
C_3–C_8 Cycloalkanes	6	6		M	42–112	9
1,2-Dialkylcyclohexanes				E		10
	13	13	D	M	84–98	11
Alkenes (0)						12
	23	10			28–56	13
				G		14
				M		15
Coda compounds (−2) (Cyclic olefins, diolefins, and acetylenes)						
Alkynes	23	23		O	68–138	16
	12	12		C	54–124	17
		10	D	M	54–110	18
Alkadienes	4	4		O	96–110	16
Mono- and bicyclic C_6H_{10} molecules	6	6		M	82	19
Exocyclic olefins	7	7	D	M	96	20
Endocyclic olefins	7	7		O	82–110	16
Dicycloalkanes (−2) (dinaphthenes)						
Methyldecalins	8	8		M		21
Terpenes (−4)						
Monoterpenes				G		22
	17	21		M	136	23
Cycloalkadienes (−4)						
Cyclooctadiene	3	3		M	108	24
Cyclobutane derivatives	5	5		C	108	25

SECTION A: HYDROCARBON COMPOUNDS—CONT.

Key to references

1. H. W. Washburn, H. F. Wiley, S. M. Rock, and C. E. Berry, *Ind. Eng. Chem. Anal. Ed.* **17**, 74 (1945).
2. F. L. Mohler, L. Williamson, C. E. Wise, E. J. Wells, H. M. Dean, and E. G. Bloom, *J. Res. Nat. Bur. Stand.* **44**, 291 (1950).
3. M. J. O'Neal, Jr., and T. P. Wier, *Anal. Chem.* **31**, 2072 (1951).
4. J. P. Wibaut and H. Brand, *Rec. Trav. Chim. Pays-Bas* **80**, 97 (1961).
5. A. Herlan, *Brennst. Chem.* **45**, 244 (1964).
6. S. Meyerson, *J. Chem. Phys.* **42**, 2181 (1965).
7. I. V. Goldenfeld and I. Z. Korostyshevsky, *Int. J. Mass Spectrom. Ion Phys.* **3**, 404 (1970).

8. J. Momigny and P. Natalis, *Bull. Soc. Chim. Belg.* **66**, 26 (1957).
9. P. Natalis, *Bull. Soc. Roy. Sci. Liege* **29**, 94 (1960).
10. P. Natalis, *Bull. Soc. Chim. Belg.* **72**, 416 (1963).
11. S. Meyerson, T. D. Nevitt, and P. N. Rylander, *in* "Advances in Mass Spectrometry" (R. M. Elliott, ed.), Vol. 2, pp. 313–326. Macmillan, New York, 1963.
12. B. J. Millard and D. F. Shaw, *J. Chem. Soc. B*, 664 (1966).
13. F. L. Mohler, *J. Wash. Acad. Sci.* **38**, 1933 (1948).
14. F. W. McLafferty, *Anal. Chem.* **31**, 2072 (1959).
15. M. Magat and R. Viallard, *J. Chem. Phys.* **48**, 385 (1951).

16. R. F. Kendall and B. H. Eccleston, *U.S. Bur. Mines Rep. Invest.* No. 6854 (1966).
17. R. E. Rondeau and L. A. Harrah, *J. Chem. Eng. Data* **13**, 109 (1968).
18. Z. Dolejšek, V. Hanuš, and K. Vokáč, *in* "Advances in Mass Spectrometry" (W. L. Mead, ed.), Vol. 3, pp. 503–514. Institute of Petroleum, London, 1966.
16. Kendall and Eccleston (1966).

19. R. E. Winters and J. H. Collins, *Org. Mass Spectrom.* **2**, 299 (1969).
20. T. H. Kinstle and R. E. Stark, *J. Org. Chem.* **32**, 1318 (1967).
16. Kendall and Eccleston (1966).

21. S. Meyerson and A. W. Weitkamp, *Org. Mass Spectrom.* **2**, 603 (1969).

22. R. I. Reed, *in* "Mass Spectrometry of Organic Ions" (F. W. McLafferty, ed.), pp. 637. Academic Press, New York, 1963.
23. R. Ryhage and E. von Sydow, *Acta Chem. Scand.* **17**, 225 (1963).

24. E. F. H. Brittain, C. H. J. Wells, and H. M. Paisley, *J. Chem. Soc. B*, 503 (1969).
25. H. M. Paisley, E. F. H. Brittain, and C. H. J. Wells, *J. Chem. Soc. B*, 185 (1969).

Section A: Hydrocarbon Compounds—cont.

Name of class (series no.)	II	III	IV	V	MW range	Reference
Cycloalkadienes (−4)—*cont.*						
Cyclobutanes and cyclohexenes	5	5		C	136	26
Cyclohexadiene and hexatriene	2	2		E	80	27
Cycloalkadienes and alkenynes	3	3		O	80–108	16
Alkadiynes (−6)						
1,5-Diynes						28
	3	3		O	78–106	16
Alkylbenzenes (−6)	114	114		C	78–540	29, 30
	35	35		C	78	31
	18	18		M		32
Alkenylbenzenes (−8)						
1-Phenylheptenes	6	6	D	M	174	33
C_8H_8 Isomers	7	7		ME	104	34
Alkyltetralins, alkylindans (−8)						
Diterpenes (−8)						
Carbodicyclic diterpenes	44	18	D	M	290–364	35
Podocarpa-8,11,13-trienes						36
	1	1		D		37
Alkylnaphthalenes (−12)						
Azulene and naphthalene	2	2		ME	128	38
Diaryl compounds (−14)						
Biphenyls	10	10			154–182	39
o-Substituted diaryl methanes	28	12		M		40
−16 Series						
1,2-Diphenylcyclobutanes	2	2		M	180	41
9,10-Dihydrophenanthrenes	20	10	X	M	180–292	42
−18 Series						
Phenanthrene, methyl phenanthrene, and dimethyl phenanthrene	3	3		M	178–206	43
−22 Series						
Fluoranthenes	2	2		M	262	44
−24 Series						
Triaryl ethylenes	8	8	X	M	258–336	45
Dibenzo compounds	2	2			186	39

SECTION A: HYDROCARBON COMPOUNDS—CONT.

Key to references

26. E. F. H. Brittain and C. H. J. Wells, *J. Chem. Soc. B* 3, 304 (1968).
27. J. L. Franklin and S. R. Carroll, *J. Amer. Chem. Soc.* **91**, 6564 (1969).
16. Kendall and Eccleston (1966).

28. A. A. Polyakova, R. A. Khmel'nitskii, and A. A. Petrov, *Zh. Obshch. Khim.* **34**, 3296 (1964).
16. Kendall and Eccleston (1966).
29. H. M. Grubb and S. Meyerson, *in* "Mass Spectrometry of Organic Ions" (F. W. McLafferty, ed.), pp. 457–467. Academic Press, New York, 1963.
30. S. Meyerson, *Appl. Spectrosc.* **9**, 120 (1955).
31. I. W. Kinney, Jr. and G. L. Cook, *Anal. Chem.* **24**, 1391 (1952).
32. A. B. King, *J. Chem. Phys.* **42**, 3526 (1965).

33. A. F. Gerrard and C. Djerassi, *J. Amer. Chem. Soc.* **91**, 6808 (1969).
34. J. L. Franklin and S. R. Carroll, *J. Amer. Chem. Soc.* **91**, 5940 (1969).

35. C. R. Enzell and R. Ryhage, *Ark. Kemi* **23**, 367 (1965).
36. C. R. Enzell, *Ark. Kemi* **26**, 87 (1966).
37. C. R. Enzell, *Tetrahedron* **19**, 2135 (1966).

38. R. J. Van Brunt and M. E. Wacks, *J. Chem. Phys.* **41**, 3195 (1964).

39. J. H. D. Eland and C. J. Danby, *J. Chem. Soc.* 5935 (1965).
40. S. Meyerson, H. Drews, and E. K. Fields, *J. Amer. Chem. Soc.* **86**, 4964 (1964).

41. M. L. Gross and C. L. Wilkins, *Tetrahedron Lett.* 3875 (1969).
42. E. Dynesen, S.-O. Lawesson, G. Schroll, J. H. Bowie, and R. G. Cooks, *Ark. Kemi* **26**, 379 (1967).

43. P. Nounou, *J. Chim. Phys. Physicochim. Biol.* **65**, 700 (1968).

44. M. C. Hamming and G. W. Keen, *Int. J. Mass Spectrom. Ion Phys.* **3**, App. 1 (1969).

45. J. Möller and C. T. Pedersen, *Acta Chem. Scand.* **22**, 706 (1968).
39. Eland and Danby (1965).

SECTION A: HYDROCARBON COMPOUNDS—CONT.

Name of class (series no.)	II	III	IV	V	MW range	Reference
Miscellaneous						
Naphthalene, naphthaphene						46
Chrysene, triphenylene, and pyrene (quadricondensed aromatics)		10		M		47
Triptycene, tri- and diphenylmethanes	7	7	D	M		48
Phenylmethyl derivatives	25	25	X	M		49
Dibenzofulvene derivatives	6	6		ME		50
Polyphenyls (ter–hexa)	10	10		M	230–610	51
Triterpenes						52
Pentacyclodecane derivatives	15	15		M	132–540	53
Pentacyclic triterpenes	27			M		54
Sterols					356–384	55
Pentacyclic triterpene						56
Sesquiterpene of hop oil						57

SECTION B: CHO COMPOUNDS

Name of class (series no.)	II	III	IV	V	MW range	Reference
Compounds with one O atom						
Alkanols (−10)	69			C	32–158	1
Primary		20				
Secondary		27				
Tertiary		22				
Heptanols—analysis						2
Aliphatic ethers (−10)	25	25		C	46–298	3
	15	15		M		4
				M		5
	8	8	D	EM	46–60	6
Isopropyl *n*-butyl ether (−10)	3	3	D	M	130	7
Methoxynaphthalenes (−10)	46	46		C	158–316	8
Methoxyphenanthrenes	44	44		M		9
						10
α-Arylidene cyclic ketones (−10)	24	24			200–394	11
trans-10-Phenyl-2-decalone (−10)	3	3	D	M	214–228	12

SECTION A: HYDROCARBON COMPOUNDS—CONT.

Key to references

46. M. E. Wacks and V. H. Dibeler, *J. Chem. Phys.* **31**, 1557 (1959).
47. M. E. Wacks, *J. Chem. Phys.* **41**, 1661 (1964).
48. S. Meyerson, *Org. Mass Spectrom.* **3**, 119 (1970).
49. K. D. Berlin and R. D. Shupe, *Org. Mass Spectrom.* **2**, 447 (1969).
50. C. Lifshitz, E. D. Bergmann, M. Rabinovitz, and I. Agranat, *J. Chem. Soc. B*, 732 (1968).
51. E. J. Gallegos, *J. Phys. Chem.* **71**, 1647 (1967).
52. J. P. Kutney, G. Eigendorf, and I. H. Rogers, *Tetrahedron* **25**, 3753 (1969).
53. W. L. Dilling and M. L. Dilling, *Tetrahedron* **23**, 1225 (1967).
54. H. Budzikiewicz, J. M. Wilson, and C. Djerassi, *J. Amer. Chem. Soc.* **85**, 3688 (1963).
55. S. G. Wyllie and C. Djerassi, *J. Org. Chem.* **33**, 305 (1968).
56. J. Karliner and C. Djerassi, *J. Org. Chem.* **31**, 1945 (1966).
57. A. T. McPhail, R. I. Reed, and G. A. Sim, *Chem. Ind. (London)*, 976 (1964).

SECTION B: CHO COMPOUNDS

Key to references

1. R. A. Friedel, J. L. Schultz, and A. G. Sharkey, Jr., *Anal. Chem.* **28**, 926 (1956).
2. T. I. Popova, A. A. Polyakova, and M. I. Khotimskaya, *Neftekhimiya* **5**, 149 (1965).
3. F. W. McLafferty, *Anal. Chem.* **29**, 1782 (1957).
4. C. Djerassi and C. Fenselau, *J. Amer. Chem. Soc.* **87**, 5747 (1965).
5. W. Carpenter, A. M. Duffield, and C. Djerassi, *J. Amer. Chem. Soc.* **89**, 6164 (1967).
6. C. W. Tsang and A. G. Harrison, *Org. Mass Spectrom.* **3**, 647 (1970).
7. G. A. Smith and D. H. Williams, *J. Amer. Chem. Soc.* **91**, 5254 (1969).
8. C. S. Barnes, D. J. Collins, J. H. Hobbs, P. I. Mortimer, and W. H. F. Sasse, *Aust. J. Chem.* **20**, 699 (1967).
9. F. W. McLafferty and M. M. Bursey, *J. Org. Chem.* **33**, 124 (1968).
10. C. S. Barnes and J. L. Occolowitz, *Aust. J. Chem.* **16**, 219 (1963).
11. R. L. N. Harris, F. Komitsky, Jr., and C. Djerassi, *J. Amer. Chem. Soc.* **89**, 4775 (1967).
12. R. T. Gray and C. Djerassi, *Org. Mass Spectrom.* **3**, 245 (1970).

SECTION B: CHO COMPOUNDS—CONT.

Name of class (series no.)	II	III	IV	V	MW range	Reference
Compounds with one O atom—*cont.*						
Cyclic ethers (+2)						
Ethylene and propylene oxide	2	2		E	44–58	13
Trimethylene oxide and tetrahydrofuran	2	2		E	58–72	14
Tetrahydropyran and 1,3-dioxepane	3	3			72–86	15
5-, 6-, 7-Membered rings	3	3	D	M	72–100	16
Methyl cyclohexyl ethers	4	4	D	M	114	17
Cyclobutyl- and cyclopropyl-carbinyl methyl ethers	3	3	D	M	86	18
Aliphatic aldehydes (+2)	20	20		C	30–212	19
Hexanal and deuterated analog	11	4	D	M	100	20
A cyclobutanol intermediate in aldehydes	8	8	D	M	114	21
	13	13	D	M	100–142	22
Aliphatic ketones (+2)	35	35		C	58–198	23
7-Membered ring, loss of H_2O			D	M	128–170	24
						25
Branched (H rearrangements)				M		26
Isomeric hexen-1-ols (+2)	7	7		M	100	27
Enols, enals, enones, cyclic ethers, ynols, etc.	22	22		O	56–114	28
Diphenyl ethers	3	3	D	M	170	29
Phenyl benzyl ethers (−12)	4	4		E	184	30
2-Methoxytropone–cycloheptatriene (−12)	5	5	X	M	198–235	31
Cyclic alcohols (+2)						
Menthols and carvomenthols	8	8			156–184	32
0-Series						
Acetylenic ethers	12	12		M	112–168	33
α,β-Unsaturated ketones and aldehydes	4	4		M	56–84	34
Oxiranes, butene epoxide	1	1		E	70	35
Cyclic ketones, decalones	5	5		M	84–152	36
Dimethylcyclohexanones	6	6	D	M	166–168	37
2-Norbonanols	16	16	D	M	112–168	38
Unsaturated carbonyl compounds	18	18	D	M	126–218	39
Cyclobutanones	12	12		M	98–126	40
Methone, isomenthone, and carvomenthone	7	2	D	M	154	41

Key to references

13. E. J. Gallegos and R. W. Kiser, *J. Amer. Chem. Soc.* **83**, 773 (1961).

14. E. J. Gallegos and R. W. Kiser, *J. Phys. Chem.* **66**, 136 (1962).

15. J. E. Collin and G. Condé-Caprace, *Int. J. Mass Spectrom. Ion Phys.* **1**, 213 (1968).
16. R. Smakman and T. J. de Boer, *Org. Mass Spectrom.* **1**, 403 (1968).
17. G. W. Klein and V. F. Smith, Jr., *J. Org. Chem.* **35**, 52 (1970).

18. W. G. Dauben, J. H. Smith, and J. Saltiel, *J. Org. Chem.* **34**, 261 (1969).
19. J. A. Gilpin and F. W. McLafferty, *Anal. Chem.* **29**, 990 (1957).
20. R. J. Liedtke and C. Djerassi, *J. Amer. Chem. Soc.* **91**, 6814 (1969).

21. C. Fenselau, J. L. Young, S. Meyerson, W. R. Landis, E. Selke, and L. C. Leitch, *J. Amer. Chem. Soc.* **91**, 6847 (1969).
22. S. Meyerson, C. Fenselau, J. L. Young, W. R. Landis, E. Selke, and L. C. Leitch, *Org. Mass Spectrom.* **3**, 689 (1970).
23. A. G. Sharkey, Jr., J. L. Shultz, and R. A. Friedel, *Anal. Chem.* **28**, 934 (1956).
24. A. N. H. Yeo and D. H. Williams, *Org. Mass Spectrom.* **2**, 331 (1969).
25. W. Carpenter, A. M. Duffield, and C. Djerassi, *Org. Mass Spectrom.* **2**, 317 (1969).
26. G. Eadon and C. Djerassi, *J. Amer. Chem. Soc.* **92**, 3084 (1970).
27. E. Honkanen and T. Moisio, *Acta Chem. Scand.* **17**, 2051 (1963).

28. R. F. Kendall and B. H. Eccleston, *U.S. Bur. Mines Rep. Invest.* No. 6854 (1966).
29. J. A. Ballantine and C. T. Pillinger, *Org. Mass Spectrom.* **1**, 447 (1968).
30. R. S. Ward, R. G. Cooks, and D. H. Williams, *J. Amer. Chem. Soc.* **91**, 2727 (1969).

31. S. Ito, Y. Fujise, and M. Sato, *Tetrahedron*, **25**, 691 (1969).

32. A. F. Thomas and B. Willhalm, *J. Chem. Soc. B*, 219 (1966).

33. P. E. Butler, *J. Org. Chem.* **29**, 3024 (1964).

34. A. J. Bowles, E. F. H. Brittain, and W. O. George, *Org. Mass Spectrom.* **2**, 809 (1969).
35. Y. Wada and R. W. Kiser, *J. Phys. Chem.* **66**, 1652 (1962).
36. J. H. Beynon, R. A. Saunders, and A. E. Williams, *Appl. Spectrosc.* **14**, 95 (1960).
37. M. V. Kulkarni, E. J. Eisenbraun, and M. C. Hamming, *J. Org. Chem.* **35**, 686 (1970).
38. D. R. Dimmel and J. Wolinsky, *J. Org. Chem.* **32**, 2735 (1967).

39. A. F. Thomas, B. Willhalm, and R. Müller, *Org. Mass Spectrom.* **2**, 223 (1969).
40. D. S. Weiss, R. B. Gagosian, and N. J. Turro, *Org. Mass Spectrom.* **3**, 145 (1970).

41. B. Willhalm and A. F. Thomas, *J. Chem. Soc.* 6478 (1965).

SECTION B: CHO COMPOUNDS—CONT.

Name of class (series no.)	II	III	IV	V	MW range	Reference
Compounds with one O atom—*cont.*						
0-Series —*cont.*						
Monocyclic ketones (0)	11	11	^{18}O	M		42
Bicyclic alcohols (0)	11	11	D	M	94–142	43
Monoterpene alcohols (0)	19	19		M	154–170	44
Benzophenones	2	2	D	M	196	45
–2 Series						
Unsaturated cyclic ketones	12	12		M	96–218	46
Piperitones (–2)	18	18	D	M	152+	47
Terpenes—citronellal, geranial, and ketones	12	12		M	138–154	48
Terpenes—menthenes and camphenes (–2)	2	2		M	138–152	49
Alkyl furan	18	18		M	68–138	50
Alkenyl furans	6	6		M	94–122	50
Furan	2	2	D	M	68	51
Alkenyl phenols	9	9		M	320–402	52
–4 Series						
Phenyl methyl ethers	11	11		M	108–138	10
	4	4	D	M	122	53
				M		54
Aromatic alcohols	8	8		C	108–150	55
Monoalkylphenols	14	14		C	94–220	55
	9	9		M	164–248	56
Monoalkylanisoles	9	9		M	178–262	56
–6 Series						
Triphenyl methyl ethers and cycloalkyl ethers	9	9		M	302–558	57
Aromatic aldehydes	6	6		C	106–148	58
4-Substituted cyclohexanes (–6, –10, and –12)	8	8		M	134–218	59
Bicyclo[2.2.1]heptane and bicyclo[2.2.2]octane derivatives (–6)	33	33		C	92–254	60
Acetophenone	2	2	D	M	120	61
–8 Series						
Benzofurans	4	4		M	118	62
Derivatives	11	11	XN	M	132–196	63
Chromenes, chromanols, etc.	49	49	X	M	132–288	64
7-Substituted norbornenes	2	2		M	160	65
4-Benzocycloalkenones	3	3	D	C	146–177	66

SECTION B: CHO COMPOUNDS—CONT.

Key to references

42. J. Seibl and T. Galmann, *Z. Anal. Chem.* **197**, 33 (1963).
43. H. Kwart and T. A. Blazer, *J. Org. Chem.* **35**, 2726 (1970).
44. E. von Sydow, *Acta Chem. Scand.* **17**, 2504 (1963).
45. J. A. Ballantine and C. T. Pillinger, *Org. Mass Spectrom.* **1**, 425 (1968).

46. J. H. Bowie, *Aust. J. Chem.* **19**, 1619 (1966).
47. A. F. Thomas, B. Willhalm, and J. H. Bowie, *J. Chem. Soc. B*, 392 (1967).

48. E. von Sydow, *Acta Chem. Scand.* **18**, 1099 (1964).

49. D. S. Weinberg and C. Djerassi, *J. Org. Chem.* **31**, 115 (1966).
50. K. Heyns, R. Stute, and H. Scharmann, *Tetrahedron* **22**, 2223 (1966).

51. D. H. Williams, R. G. Cooks, J. Ronayne, and S. W. Tam, *Tetrahedron Lett.*, 1777 (1968).
52. J. L. Occolowitz, *Anal. Chem.* **36**, 2177 (1964).

10. Barnes and Occolowitz (1963).
53. F. Meyer and A. G. Harrison, *Can. J. Chem.* **42**, 2008 (1964).
54. F. W. McLafferty and M. M. Bursey, *J. Org. Chem.* **33**, 124 (1967).
55. T. Aczel and H. E. Lumpkin, *Anal. Chem.* **32**, 1819 (1960).

56. M. I. Gorfinkel, L. Yu. Ivanovshaia, and V. A. Koptyug, *Org. Mass Spectrom.* **2**, 273 (1969).

57. Y. M. Sheikh, A. M. Duffield, and C. Djerassi, *Org. Mass Spectrom.* **1**, 251 (1968).
58. T. Aczel and H. E. Lumpkin, *Anal. Chem.* **33**, 386 (1961).

59. R. T. Gray, R. J. Spangler, and C. Djerassi, *J. Org. Chem.* **35**, 1525 (1970).

60. T. Goto, A. Tatematsu, Y. Hata, R. Muneyuki, H. Tanida, and K. Tori, *Tetrahedron* **22**, 2213 (1966).
61. S. E. Scheppele, R. D. Grigsby, E. D. Mitchell, D. W. Miller, and G. R. Waller, *J. Amer. Chem. Soc.* **90**, 3521 (1968).

62. R. I. Reed and W. K. Reid, *J. Chem. Soc.* 5933 (1963).
63. E. N. Givens, L. G. Alexakos, and P. B. Venuto, *Tetrahedron* **25**, 2407 (1969).
64. B. Willhalm, A. F. Thomas, and F. Gautschi, *Tetrahedron* **20**, 1185 (1964).
65. K. G. Das, M. S. B. Nayar, and C. A. Chinchwadkar, *Org. Mass Spectrom.* **3**, 303 (1970).
66. D. G. B. Boocock and E. S. Waight, *J. Chem. Soc. B*, 258 (1968).

SECTION B: CHO COMPOUNDS—CONT.

Name of class (series no.)	II	III	IV	V	MW range	Reference
Compounds with two O atoms						
−10 Series						
Alkyltetrahydropyranyl ethers	4	4		M	130–158	67
Cycloalkane-α-glycols	6	6		M	116–172	68
1,2-Cyclohexanediol						69
Acetal type of compounds	14	14		C	74–230	70
Esters	31	31		C	60–200	71
Aliphatic esters	27	27		C	60	72
Formate esters	6	6	D	M	60	73
	14	14				74
Methyl esters	9	9	D	M		75
Methyl butyrates	21	28	X	D		76
Ethyl and isopropyl esters	6	6	D	M	102–130	77
						78
1,4-Dicarbonyl compounds	7	7		M	130–248	79
Pentyl acetates	9	9		M	130	80
Butyl hexanoates	32	32	D	M	172	81
Aliphatic acids	14	14		M		82
	5	5			158–172	74
Stearates—long-chain esters	5	5	D	M	286–806	83
	14	14		M	102–158	84
+2 Series						
Lactones	17	17		C	86–284	85
δ-Lactones	6	6		O	114–144	86
α-Diketones (nonenolized)	9	9		M	86–198	87
β-Diketones	5	5		M	100–224	88
Hydroxy- and methoxy-benzophenones	7	7		M	198–226	45
Alkoxy-alkene-ones	13	13		M	100	89
Cyclobutane carboxylates	6	6			128–184	90
Cycloalkyl acetates	5	5		M	184	91
Unsaturated acetates (0)	4	4		M	182	91
Cyclopropane fatty acid esters	5	5		M	296–374	92
0 Series						
Unsaturated esters	3	3		M	190–232	93

SECTION B: CHO COMPOUNDS—CONT.

Key to references

67. S. J. Isser, A. M. Duffield, and C. Djerassi, *J. Org. Chem.* **33**, 2266 (1968).
68. S. Sasaki, Y. Itagaki, H. Abe, K. Nakanishi, T. Suga, T. Shishihori, and T. Matsuura, *Org. Mass Spectrom.* **1**, 61 (1968).
69. M. K. Strong and C. Djerassi, *Org. Mass Spectrom.* **2**, 631 (1969).
70. R. A. Friedel and A. G. Sharkey, Jr., *Anal. Chem.* **28**, 940 (1956).
71. A. G. Sharkey, Jr., J. L. Shultz, and R. A. Friedel, *Anal. Chem.* **31**, 87 (1959).
72. J. H. Beynon, R. A. Saunders, and A. E. Williams, *Anal. Chem.* **33**, 221 (1961).
73. D. Van Raalte and A. G. Harrison, *Can. J. Chem.* **41**, 2054 (1963).
74. F. M. Trent and F. D. Miller, *Appl. Spectrosc.* **15**, 64 (1961).
75. Ng. Dinh-Nguyen, R. Ryhage, S. Ställberg-Stenhagen, and E. Stenhagen, *Ark. Kemi* **18**, 393 (1961).
76. I. Howe, D. H. Williams, D. G. I. Kingston, and H. P. Tannenbaum, *J. Chem. Soc. B*, 439 (1969).
77. A. G. Harrison and E. G. Jones, *Can. J. Chem.* **43**, 960 (1965).
78. J. H. Bowie, R. G. Cooks, P. Jakobsen, S.-O. Lawesson, and G. Schroll, *Aust. J. Chem.* **20**, 689 (1967).
79. S.-O. Lawesson, J. O. Madsen, G. Schroll, J. H. Bowie, R. Grigg, and D. H. Williams, *Acta Chem. Scand.* **20**, 1129 (1966).
80. R. Teranishi, R. A. Flath, D. G. Guadagni, R. E. Lundin, T. R. Mon, K. L. Stevens, *Agr. Food Chem.* **14**, 253 (1966).
81. W. H. McFadden, L. E. Boggs, and R. G. Buttery, *J. Phys. Chem.* **40**, 3516 (1966).
82. G. P. Happ and D. W. Stewart, *J. Amer. Chem. Soc.* **74**, 4404 (1952).
74. Trent and Miller (1961).
83. K. K. Sun and R. T. Holman, *J. Amer. Oil Chem. Soc.* **45**, 810 (1968).
84. N. C. Rol, *Rec. Trav. Chim. Pays-Bas* **84**, 413 (1965).

85. W. H. McFadden, E. A. Day, and M. J. Diamond, *Anal. Chem.* **37**, 89 (1965).
86. B. J. Millard, *Org. Mass Spectrom.* **1**, 279 (1968).
87. J. H. Bowie, R. G. Cooks, G. E. Gream, and M. H. Laffer, *Aust. J. Chem.* **21**, 1247 (1968).
88. J. H. Bowie, D. H. Williams, S.-O. Lawesson, and G. Schroll, *J. Org. Chem.* **31**, 1384 (1966).

45. Ballantine and Pillinger (1968).
89. M. Vandewalle, N. Schamp, and M. Francque, *Org. Mass Spectrom.* **2**, 877 (1969).
90. D. A. Bak and K. Conrow, *J. Org. Chem.* **31**, 3608 (1966).
91. J. Cason and A. I. A. Khodair, *J. Org. Chem.* **32**, 575 (1967).

92. W. W. Christie and R. T. Holman, *Lipids* **1**, 176 (1966).

93. V. I. Kadentsev, B. M. Zolotarev, O. S. Chizov, Ch. Shachidayatov, L. A. Yanovskava, and V. F. Kucherov, *Org. Mass Spectrom.* **1**, 899 (1968).

SECTION B: CHO COMPOUNDS—CONT.

Name of class (series no.)	II	III	IV	V	MW range	Reference
Compounds with two O atoms—*cont.*						
0 Series —*cont.*						
Methyl and ethyl acrylates, crotonates	4	4		M	84–98	34
Unsaturated methyl esters	12	12		M	126–156	94
Propiolates, maleates, fumarates	30	11	N	C	84–126	95
1,3-Cyclohexanediones						96
Dimedone and 2-ethyldimedone	2	2		M	140–168	97
Doubly unstaurated carbonyl compounds	2	2	D	M	182	39
Esters of monoterpene alcohols	15	14		M	182–210	98
Monocyclic 2-pyrones	5	5	X	M	140–212	99
Isoflavonols	16	16		M	194–418	100
Flavan-3,4-diols	5	5		M	332–416	101
−2 Series						
Bicyclic γ-lactones	3	3		M	180–182	102
Alkenyl phenols, resorcinols, and methoxy compounds	9	12	D	M	320–402	52
Oxygenated furan derivatives	29	29	X	M	96–299	103
	13	13			96–180	50
Flavones	10	10		M		104
Chalcones, flavones, and isoflavones	13	13	X	M	208–222	105
−4 Series						
Aromatic esters	37	37		C	136–418	106
Aromatic esters of aliphatic acids	17	17		C	136–218	107
Aliphatic esters of aromatic acids	25	25		C	136–218	107
Benzoate type of methyl esters	7	7		C	136–178	108
Aliphatic esters of phenylalkanoic acids	7	7		M	150–192	93
Aromatic acids	13	13		C	122–178	106
	12	12		C	122–164	58
Aromatic carboxylic acids, benzoic, and terephthalic	5	5		O	122	109
					122	84
6-Phenyl-substituted alkanoic acids and esters	8	8	D	M	192–194	110
	4	4	D	M	206–208	110
Benzoic acid	5	5	D	M	122–124	111
Aromatic carbonyl compounds	33	8	D	M	150–152	112
(*ortho*, *peri* effects in methoxy-substituted)						113
Kawalactone (α-pyrone type)					108	114
α- and γ-pyrones				M	108	115

SECTION B: CHO COMPOUNDS—CONT.

Key to references

34. Bowles *et al.* (1969).
94. W. K. Rohwedder, A. F. Mabrouk, and E. Selke, *J. Phys. Chem.* **69**, 1711 (1965).
95. J. H. Bowie, D. H. Williams, P. Madsen, G. Schroll, and S.-O. Lawesson, *Tetrahedron* **23**, 305 (1967).
96. M. Vandewalle, N. Schamp, and H. De Wilde, *Bull. Soc. Chim. Belg.* **76**, 123 (1967).
97. T. Goto, A. Tatematsu, Y. Nakajima, and H. Tsuyama, *Tetrahedron* **21**, 757 (1965).

39. Thomas *et al.* (1969).
98. E. von Sydow, *Acta Chem. Scand.* **19**, 2083 (1965).
99. H. Nakata, Y. Hirata, and A. Tatematsu, *Tetrahedron* **21**, 123 (1965).
100. A. Pelter, P. Stainton, A. P. Johnson, and M. Barber, *J. Heterocycl. Chem.* **2**, 262 (1965).
101. S. E. Drewes, *J. Chem. Soc. C*, 1140 (1968).

102. P. H. Chen, W. F. Kuhn, F. Will, III, and R. M. Ikeda, *Org. Mass Spectrom.* **3**, 199 (1970).
52. Occolowitz (1964).
103. R. Grigg, M. V. Sargent, D. H. Williams, J. A. Knight, *Tetrahedron* **21**, 3441 (1965).
50. Heyns *et al.* (1966).
104. R. I. Reed and J. M. Wilson, *J. Chem. Soc.* 5949 (1963).

105. Y. Itagaki, T. Kurokawa, S. Sasaki, C. Chang, and F. Chen, *Bull. Chem. Soc. Jap.* **39**, 538 (1966).
106. F. W. McLafferty and R. S. Gohlke, *Anal. Chem.* **31**, 2076 (1959).
107. E. M. Emery, *Anal. Chem.* **32**, 1495 (1960).

108. T. Aczel and H. E. Lumpkin, *Anal. Chem.* **34**, 33 (1962).

93. Kadentsev *et al.* (1968).
106. McLafferty and Gohlke (1959).
58. Aczel and Lumpkin (1961).

109. J. H. Beynon, B. E. Job, and A. E. Williams, *Z. Naturforsch. A* **20**, 883 (1965).

110. S. Meyerson and L. C. Leitch, *J. Amer. Chem. Soc.* **88**, 56 (1966).

111. S. Meyerson and J. L. Corbin, *J. Amer. Chem. Soc.* **87**, 3045 (1965).
112. J. H. Bowie and P. Y. White, *J. Chem. Soc. B*, 89 (1969).

113. J. H. Bowie, P. Y. White, and P. J. Hoffmann, *Tetrahedron* **25**, 1529 (1969).
114. M. Pailer, G. Schaden, R. Hänsel, *Monatsh. Chem.* **96**, 1842 (1965).
115. C. S. Barnes and J. L. Occolowitz, *Aust. J. Chem.* **17**, 975 (1964).

SECTION B: CHO COMPOUNDS—CONT.

Name of class (series no.)	II	III	IV	V	MW range	Reference
Compounds with two O atoms—*cont.*						
−4 Series —*cont.*						
Benzoquinones and derivatives	20	15		M	108–240	116
1,2-Quinones	7	20	D	M	122–134	117
Doubly unsaturated carbonyl compounds						39
−6 Series						
Unsaturated aromatic esters	3	3		M	190–232	93
2,2′-Bifurans	5	5		M	162	118
Sesquiterpenoid lactones	4	4	D	M	232–236	119
Custunolide						120
Ethylene ketals	29	5	D	M	316–416	121
−8 Series						
Aliphatic 1,2-glycols	19	19		M	118–204	122
Acetals (symmetrical)	21	21		C	118–230	123
(Asymmetrical)	8	8		C	104–202	123
7-Substituted norbornenes	7	7	N	M	202–365	65
Benzocycloalkenol acetates	5	5	D	M	174–216	66
Chromones (naturally occurring)	6	6		M	146	124
Cinnamyl compounds	5	5		M	132–202	125
Dimethylchromenes, etc.	28			C	118–314	115
Alkyl peroxides	7	7		M	62–146	126
Triterpenes	8	8		O	426–456	127
Triterpenoid (friedelan-y-one-y-al)	3	3		O	426–442	128
−10 Series						
Acetals (unsaturated and aromatic)	9	9		M	158–242	123
Oxiranes (epoxides)	3	3	X	E	88–136	35
Alkyl hydroperoxides	16	16		M	88–130	129
1,2-Naphthoquinones	11	11	X	M	158–358	130
Substituted derivatives	22	22	D	M	158–241	131
1,4-Naphthoquinones	5	5	D	M	172–240	132
Substituted derivatives (−14)	5	5			436–458	132
1-Arylanthra-9,10-quinones and derivatives	21	21	NX	M	284–362	133
−12 Series						
Benzyl benzoates	4	4		C	240–268	108
Diphenylmethoxy esters	8	8		M	212	29
Metabolite acids (3,4,4-trimethyl-5-oxo-trans-2-hexenoic acid and derivative)	2	2		M	170	134

SECTION B: CHO COMPOUNDS—CONT.

Key to references

116. J. H. Bowie, D. W. Cameron, R. G. F. Giles, and D. H. Williams, *J. Chem. Soc. B*, 335 (1966).
117. S. Ukai, K. Hirose, A. Tatematsu, and T. Goto, *Tetrahedron Lett*. 4999 (1967).

39. Thomas *et al.* (1969).

93. Kadentsev *et al.* (1968).
118. R. Grigg, J. A. Knight, and M. V. Sargent, *J. Chem. Soc. C*, 976 (1966).
119. R. N. Sathe, G. H. Kulkarni, G. R. Kelkar, and K. G. Das *Org. Mass Spectrom.* **2**, 935 (1969).
120. D. G. B. Boocock and E. S. Waight, *Chem. Commun.* 90 (1966).
121. Z. Pelah, D. H. Williams, H. Budzikiewicz, and C. Djerassi, *J. Amer. Chem. Soc.* **86**, 3722 (1964).
122. A. M. Duffield, C. Djerassi, J. Kossanyi, J. P. Morizur, B. Furth, and J. Wiemann, *Org. Mass Spectrom.* **1**, 777 (1968).
123. W. H. McFadden, J. Wasserman, J. Corse, R. E. Lundin, and R. Teranishi, *Anal. Chem.* **36**, 1031 (1964).
65. Das *et al.* (1970).
66. Boocock and Waight (1968).
124. M. M. Badawi, M. B. E. Fayez, T. A. Bryce, and R. I. Reed, *Chem. Ind. (London)*, 498 (1966).
125. E. F. H. Brittain, J. P. Kelly, and W. L. Mead, *Org. Mass Spectrom.* **2**, 325 (1969).
115. Barnes and Occolowitz (1964).
126. R. T. M. Fraser, N. C. Paul, and L. Phillips, *J. Chem. Soc. B*, 1278 (1970).
127. J. P. Kutney, G. Eigendorf, and I. H. Rogers, *Tetrahedron* **25**, 3753 (1969).

128. J. S. Shannon, C. G. MacDonald, and J. L. Courtney, *Tetrahedron Lett.*, 173 (1963).

123. McFadden *et al.* (1964).
35. Wada and Kiser (1962).
129. A. R. Burgess, R. D. G. Lane, and D. K. Sen Sharma, *J. Chem. Soc. B*, 341 (1969).
130. R. W. A. Oliver and R. M. Rashman, *J. Chem. Soc. B*, 1141 (1968).
131. J. H. Bowie, D. W. Cameron, and D. H. Williams, *J. Amer. Chem. Soc.* **87**, 5094 (1965).
132. S. J. Di Mari, J. H. Supple, and H. Rapoport, *J. Amer. Chem. Soc.* **88**, 1226 (1966).

133. D. R. Buckle and E. S. Waight, *Org. Mass Spectrom.* **1**, 273 (1968).

108. Aczel and Lumpkin (1962).
29. Ballentine and Pillinger (1968).

134. R. Srinivasan and K. L. Rinehart, Jr., *J. Org. Chem.* **33**, 351 (1968).

SECTION B: CHO COMPOUNDS—CONT.

Name of class (series no.)	II	III	IV	V	MW range	Reference
Compounds with two O atoms—*cont.*						
−16 Series (−2)						
Anthraquinones and derivatives	21	21	XN	M	180–310	135
Compounds with three O atoms						
Ozonides (−2, −4, −8)	8	8		M	122–304	136
β-Keto esters (−10)	16	16		M	130–368	137
	18	18		M	116–368	138
Derivatives (+2)	6	6		M	98–368	138
Dihydroxy-, hydroxymethoxy-, and esters of benzophenones	15	15		M	214–258	45
Triterpenoid dehydration products	10	10			408–468	139
Furanocoumarins (−10)					186	115
Furanochromones (−10)	6	6		M	186	124
Cyclic carbonates (−10)	4	4			88–240	140
Diglycolic anhydride (+2)	1	1		M	114	141
Tetronic acid and derivatives						
Alkyl (+2)	6	6	D	M	100–252	142
Aryl	7	7		M		142
Oxy	3	3		M		142
Alkyl-hydroxy	8	8	D	M	100–190	143
Alkoxy	11	11	D	M	142–298	143
Furoate esters (0)	31	31	N	M	126–218	103
Furancarboxylic acids (0)	6	6		M	112	62
Derivatives	13	13				62
Salicylic acids (−2)	2	2		M	180–194	144
Derivatives	13	13		M	183–323	144
Methoxybenzoic acid (*o*- and *p*-) (−2)	2	2		M	152	58
Methoxy and *O*-hydroxy						145
Benzyl alcohol (−2)						
Methoxyphenyl acetates	10	10		M	124–254	146
Pyrethroids, alcohols (−2)	2	2		D	166–180	147
Polycyclic ketones (−2)	27	27		M	208–314	135
Hop components, phloroglucinol derivatives (−2, variable)	15	15		M	194–416	148
Chromones—alkoxy and related compounds (−6)	4	4		M	176+	124
Cyclohexanetriols (−8)	8	8		M	132–146	149
Alkoxyquinones—*peri* effects	17	7	D	M	202–296	150
Hydroxynaphthaquinones (−8)	16	16		M	188–260	151

SECTION B: CHO COMPOUNDS—CONT.

Key to references

135. J. H. Beynon and A. E. Williams, *Appl. Spectrosc.* **14**, 156 (1960).

136. J. Castonguay, M. Bertrand, J. Carles, S. Fliszar, and Y. Rousseau, *Can. J. Chem.* **47**, 919 (1969).
137. J. H. Bowie, S.-O. Lawesson, G. Schroll, and D. H. Williams, *J. Amer. Chem. Soc.* **87**, 5742 (1965).
138. R. I. Reed and V. V. Takhistov, *Tetrahedron* **23**, 2807 (1967).

45. Ballentine and Pillinger (1968).
139. M. H. A. Elgamal, M. B. E. Fayez, and T. R. Kemp, *Org. Mass Spectrom.* **2**, 175 (1969).
115. Barnes and Occolowitz (1964).
124. Badawi *et al.* (1966).
140. P. Brown and C. Djerassi, *Tetrahedron* **24**, 2949 (1968).
141. J. D. S. Goulden and B. J. Millard, *Org. Mass Spectrom.* **2**, 893 (1969).

142. L. J. Haynes, A. G. Loudon, A. Kirkien-Konasiewicz, and A. Maccoll, *Org. Mass Spectrom.* **1**, 743 (1968).

143. J. A. Ballantine, R. G. Fenwick, and V. Ferrito, *Org. Mass Spectrom.* **1**, 761 (1968).

103. Grigg *et al.* (1965).
62. Reed and Reid (1963).
62. Reed and Reid (1963).
144. W. M. Scott, M. E. Wacks, C. D. Eskelon, J. Towne, and C. Cazee, *Org. Mass Spectrom.* **1**, 847 (1968).

58. Aczel and Lumpkin (1961).
145. J. S. Shannon, *Aust. J. Chem.* **15**, 265 (1962).

146. C. B. Thomas, *J. Chem. Soc. B*, 430 (1970).
147. T. A. King and H. M. Paisley, *J. Chem. Soc. C*, 870 (1969).
135. Beynon and Williams (1960).

148. S. J. Shaw and P. V. R. Shannon, *Org. Mass Spectrom.* **3**, 941 (1970).

124. Badawi *et al.* (1966).
149. A. Buchs, E. Charollais, and Th. Posternak, *Helv. Chim. Acta* **51**, 695 (1968).
150. J. H. Bowie, P. J. Hoffmann, and P. Y. White, *Tetrahedron* **26**, 1163 (1970).
151. T. A. Elwood, K. H. Dudley, J. M. Tesarek, P. F. Rogerson, and M. M. Bursey, *Org. Mass Spectrom.* **3**, 841 (1970).

SECTION B: CHO COMPOUNDS—CONT.

Name of class (series no.)	II	III	IV	V	MW range	Reference
Compounds with four O atoms						
Peroxides (cyclic)	7	7		M	148–396	152
	2	2				153
Maleic and fumaric acids (−10)	3	3	D	M	116	154
1,4-Diketo-2,5-dimethyldioxan (cyclic esters of aliphatic α-hydroxy acids) (+2)	7	7		M	116–200	141
Acetylene dicarboxylates	8	8		M	140–226	95
Derivatives (+2)	15	15	N	M	84–266	95
Dimethoxyphenyl acetates (0)				M	196–254	146
Dicarboxylic aromatic acids, phthalic and methylterephthalic acids (−2)	7	7		M	166–180	58
Norbornane derivatives (−4)	14	14	D	M	136–262	38
6,7-Dimethoxycoumarin (−4)	1	1	D	M	206	155
Trihydroxy-, alkoxybenzophenones (−8)	8	8		M	230–272	45
Dibasic acids—malonic acids				M	104	62
Methyl esters of dibasic acids (−8)	8	8		M	272–482	156
Methoxy migrations in *n*-dimethyl esters (−8)	10	10	D	M	132–186	157
Compounds with five or more O atoms						
Gaillardin-A cytotoxic sesquiterpene lactone (−2)	1	1		M	306	158
Dunnione and hydrodonnione diacetate (−10, −8)	2	2		M	242–328	159
Acetyl derivatives of 3,4-dideoxyaldopentane (−4)	3	3		M	234	160
Lond-chain fatty acids and esters (−8)	5	5	D	M	286–806	83
Triglycerides of straight-chain fatty acids (also see lipids, Section H)	2	2		C	806–890	161
Squaric acid and croconic acids (+2, −10) (monocyclic polycarbonyl compounds)	7	7		M	114–312	162
Commic acids (var.)	7	7		M	396–504	163
Rotenoids, chromanochromanones ($C_{22}H_{22}O_6$) (−2)	10	10		M	382	104
2′-Hydroxyflavonoids (var.)	7	7		M	210–330	164

SECTION B: CHO COMPOUNDS—CONT.

Key to references

152. M. Bertrand, S. Fliszar, and Y. Rousseau, *J. Org. Chem.* **33**, 1931 (1968).
153. T. Ledaal, *Tetrahedron Lett.*, 3661 (1969).
154. J. L. Holmes, F. Benoit, and N. S. Isaacs, *Org. Mass Spectrom.* **2**, 591 (1969).

141. Goulden and Millard (1969).
95. Bowie *et al.* (1967).

146. Thomas (1970).

58. Aczel and Lumpkin (1961).
38. Dimmel and Wolinsky (1967).
155. R. H. Shapiro and C. Djerassi, *J. Org. Chem.* **30**, 955 (1965).

45. Ballantine and Pillinger (1968).
62. Reed and Reid (1963).
156. R. Ryhage and E. Stenhagen, *Ark. Kemi* **23**, 167 (1964).

157. I. Howe and D. H. Williams, *J. Chem. Soc.* 202 (1968).

158. S. M. Kupchan, J. M. Cassady, J. E. Kelsey, H. K. Schnoes, D. H. Smith, and A. L. Burlingame, *J. Amer. Chem. Soc.* **88**, 5292 (1966).

159. D. R. Buckle and E. S. Waight, *Org. Mass Spectrom.* **2**, 367 (1969).

160. M. Venugopalan and C. B. Anderson, *Chem. Ind.* (*London*) **46**, 370 (1964).

83. Sun and Holman (1968).

161. M. Barber, T. O. Merren, and W. Kelly, *Tetrahedron Lett.*, 1063 (1964).

162. S. Skujins, J. Delderfield, and G. A. Webb, *Tetrahedron* **24**, 4805 (1968).
163. A. F. Thomas and B. Willhalm, *Tetrahedron Lett.* 3177 (1964).

104. Reed and Wilson (1963).
164. A. Pelter and P. Stainton, *J. Chem. Soc. C*, 1933 (1967).

SECTION C: CHN AND/OR CHNO COMPOUNDS

Name of class (series no.)	II	III	IV	V	MW range	Reference
Compounds with one N atom						
Aliphatic amines (+3)	66	66		C	31–311	1
	2	2	D	M		2
	16	16		C		3
Trimethylamine (+3)	1	1		M	59	4
2-Alkyl-*N*-methylpyrrolidines (+1)	1	8			99	5
Aliphatic nitriles (−1)	17	17		C	41–265	6
		9			167–293	7
n-Alkyl cyanides (−1)						8
Alkyl cyanides (−1)	16	16		M		9
Isocyanides (−1)	8	8		C	41–83	10
Nitriles (−1)	5	5		M		11
Isohexylcyanide (−1)	4	4	D	M	111	12
Enamines (−3)	7	3		M	125	13
Pyrroles (−3)	55	23		C		14
Acetanilides (−5)	9	9		M	149–181	15
Monocyclic derivatives of pyrrole (−5)		53		M	135	16
Tetrahydroquinolines (−7)	4	11	D	M	133–137	17
		23			133–279	18
Azetidines (−7)		17			147–293	18
Methylindoles (−9)	30	7		M	131	19
Alkylindoles (−9)	11	11		M	103–231	20
Aromatic isocyanides (−9)	19	19		M	103–231	21
1*H*-Pyrrolo[2,3-*b*]pyridines (−9)		23		M	117	22
Alkylquinolines (−11)	19	19	D	M	143–177	23
Labeled methylquinolines (−11)	8	8	D	M	143–146	24
Labeled dimethylquinolines (−11)	9	9	D	M	157–160	25
Benzo[*a*]- and benzo[*c*]acridines (−23)					229	26
Compounds with two N atoms						
1-Carbamoyl-2-pyrazolines (0)	8	8	D	M	168–277	27
Pyrazoles (−2)		9		M	68–192	28
Amidines (−2)						29
Imidazole (−2)	1	1	D		68	30
1-Methylimidazole (−2)			D		82	30
Pyrimidines (−4)	6	10	D	M		31
Diphenyl pyrazoles (−4)	4		D	M	220	32

SECTION C: CHN AND/OR CHNO COMPOUNDS—CONT.

Key to references

1. R. S. Gohlke and F. W. McLafferty, *Anal. Chem.* **34**, 1281 (1962).
2. C. Djerassi and C. Fenselau, *J. Amer. Chem. Soc.* **87**, 5752 (1965).
3. C. A. Brown, A. M. Duffield, and C. Djerassi, *Org. Mass Spectrom.* **2**, 625 (1969).
4. G. Hvistendahl and K. Undheim, *Org. Mass Spectrom.* **3**, 821 (1970).
5. S. Osman, C. J. Dooley, and T. Foglia, *Org. Mass Spectrom.* **2**, 977 (1969).
6. F. W. McLafferty, *Anal. Chem.* **34**, 26 (1962).
7. M. C. Hamming, *Arch. Mass Spectral Data* **1**, 196–213 (1970).
8. W. Carpenter, Y. M. Sheikh, A. M. Duffield, and C. Djerassi, *Org. Mass Spectrom.* **1**, 3 (1968).
9. W. Heerma and G. Dijkstra, *Org. Mass Spectrom.* **3**, 379 (1970).
10. R. G. Gillis and J. L. Occolowitz, *J. Org. Chem.* **28**, 2924 (1963).
11. N. C. Rol, *Rec. Trav. Chim. Pays-Bas* **84**, 413 (1965).
12. R. Beugelmans, D. H. Williams, H. Budzikiewicz, C. Djerassi, *J. Amer. Chem. Soc.* **86**, 1386 (1964).
13. H. J. Jakobsen, S.-O. Lawesson, J. T. B. Marshall, G. Schroll, and D. H. Williams, *J. Chem. Soc. B*, 940 (1966).
14. A. H. Jackson, G. W. Kenner, H. Budzikiewicz, C. Djerassi, and J. M. Wilson, *Tetrahedron* **23**, 603 (1967).
15. R. H. Shapiro and K. B. Tomer, *Org. Mass Spectrom.* **2**, 579 (1969).

16. H. Budzikiewicz, C. Djerassi, A. H. Jackson, G. W. Kenner, D. J. Newman, and J. M. Wilson, *J. Chem. Soc.*, 1949 (1964).
17. P. M. Draper and D. B. MacLean, *Can. J. Chem.* **46**, 1499 (1968).
18. M. B. Jackson, T. M. Spotswood, and J. H. Bowie, *Org. Mass Spectrom.* **1**, 857 (1968).

19. J. C. Powers, *J. Org. Chem.* **33**, 2044 (1968).
20. J. H. Beynon and A. E. Williams, *Appl. Spectrosc.* **13**, 101 (1959).
21. B. Zeeh, *Org. Mass Spectrom.* **1**, 315 (1968).
22. R. Herbert and D. G. Wibberley, *J. Chem. Soc. B*, 459 (1970).
23. P. M. Draper and D. B. MacLean, *Can. J. Chem.* **46**, 1487 (1968).
24. P. M. Draper and D. B. MacLean, *Can. J. Chem.* **48**, 747 (1970).
25. P. M. Draper and D. B. MacLean, *Can. J. Chem.* **48**, 738 (1970).

26. N. P. Buu-Hoi, C. Orley, and M. Mangane, *J. Heterocycl. Chem.* **2**, 236 (1965).

27. S. W. Tam, *Org. Mass Spectrom.* **2**, 729 (1969).
28. T. Nishiwaki, *J. Chem. Soc. B*, 885 (1967).
29. A. K. Bose, I. Kugajevsky, P. T. Funke, and K. G. Das, *Tetrahedron Lett.* 3065 (1965).
30. R. Hodges and M. R. Grimmett, *Aust. J. Chem.* **21**, 1085 (1968).

31. T. Nishiwaki, *Tetrahedron* **22**, 3117 (1966).
32. B. K. Simons, R. K. M. R. Kallury, and J. H. Bowie, *Org. Mass Spectrom.* **2**, 739 (1969).

SECTION C: CHN AND/OR CHNO COMPOUNDS—CONT.

Name of class (series no.)	II	III	IV	V	MW range	Reference
Compounds with two N atoms—*cont.*						
Arylamidines (−6)	6	10	D	M	148–224	29
N,N-Dimethyl-*N′*-phenyl-formamidines and derivatives	17	17	D	M	148–216	33
Benzimidazoles (−8)	22	9	D	M	118–330	34
2-Alkylbenzimidazoles (−8)	10		D	M	118–174	35
2*H*-Cyclopenta[*d*]pyridazines (−8)	5	5	D	M	118–194	36
Azaindenes (−8), di-, tri-	9	3	D	M	118–133	37
Pyrazolo[1,5-*a*]pyridines (−8)	6	6		M	118–298	38
Cinnolines (−10)	4	4	D	M	130–172	39
Azomethines (+2)						40
Uracils (−14)	14	14		M	140–270	41
Azobenzenes (−14)	38	19	Br	M	182–254	42
Phenylhydrazones (−14)	5				196–272	43
β-Carbolines (−24)	15			C	186–244	44
Compounds with three N atoms						
2-Aminopyrimidines (+3)	10	6			185–241	45
Guanidines (+3)	10				59–149	46
asym-Triazines	7	7	D	M	131–279	46a
Compounds with four N atoms						
Pteridine, methylpteridines (−8)	4		D	M	132–160	47
Methyltetrazoles (−14)	16		D		70–152	48
Compounds with one N and one O atom						
Aliphatic amides (+3)	35	35		M	45–339	49
Dimethylaminoacetone (+3)	2		D		101	50
2-Dimethylaminocyclohexanone (+1)	2		D		141	50
Piperidine nitroxides (+3)	4				157–215	51
6-Azasteroids (+1)	4		D		281–283	52
Isocyanates (+1)	15				43–295	53
N-Alkyllactams (+1)	2		D	M	127–141	54
Nitrones (+1)	12			M	225	55
Alkylamine-alkene-ones (−1)	10				99–155	56
Aromatic amides (−1)	6		D		167–311	57
Alkyl oxazoles (−1)	5				69–233	58
Alkyl isoxazoles (−1)	16				97–255	59

SECTION C: CHN AND/OR CHNO COMPOUNDS—CONT.

Key to references

29. Bose *et al.* (1965).

33. H.-Pr. Grützmacher and H. Kushel, *Org. Mass Spectrom.* **3**, 605 (1970).
34. S.-O. Lawesson, G. Schroll, J. H. Bowie, and R. G. Cooks, *Tetrahedron* **24**, 1875 (1968).
35. T. Nishiwaki, *J. Chem. Soc. C*, 428 (1968).
36. D. M. Forkey, *Org. Mass Spectrom.* **2**, 309 (1969).
37. W. W. Paudler, J. E. Kuder, and L. S. Helmick, *J. Org. Chem.* **33**, 1379 (1968).
38. K. T. Potts and U. P. Singh, *Org. Mass Spectrom.* **3**, 433 (1970).
39. M. H. Palmer, E. R. R. Russell, and W. A. Wolstenholme, *Org. Mass Spectrom.* **2**, 1265 (1969).
40. M. Fischer and C. Djerassi, *Chem. Ber.* **99**, 1541 (1966).
41. R. W. Reiser, *Org. Mass Spectrom.* **2**, 467 (1969).
42. J. H. Bowie, G. E. Lewis, and R. G. Cooks, *J. Chem. Soc. B*, 621 (1967).
43. W. D. Crow, J. L. Occolowitz, and R. K. Solly, *Aust. J. Chem.* **21**, 761 (1968).
44. R. T. Coutts, R. A. Locock, and G. W. A. Slywka, *Org. Mass Spectrom.* **3**, 879 (1970).

45. T. Nishiwaki, *Tetrahedron* **23**, 1153 (1967).
46. J. H. Beynon, J. A. Hopkinson, and A. E. Williams, *Org. Mass Spectrom.* **1**, 169 (1968).
46a. T. Sasaki, K. Minamoto, M. Nishikawa, and T. Shima, *Tetrahedron* **25**, 1021 (1969).

47. T. Goto, A. Tatematsu, and S. Matsuura, *J. Org. Chem.* **30**, 1844 (1965).
48. D. M. Forkey and W. R. Carpenter, *Org. Mass Spectrom.* **2**, 433 (1969).

49. J. A. Gilpin, *Anal. Chem.* **31**, 935 (1959).
50. F. R. Stermitz and K. D. McMurtrey, *J. Org. Chem.* **33**, 1140 (1968).

51. A. Morrison and A. P. Davies, *Org. Mass Spectrom.* **3**, 353 (1970).
52. U. K. Pandit, W. N. Speckamp, and H. O. Huisman, *Tetrahedron* **23**, 1767 (1965).
53. J. M. Ruth and R. J. Philippe, *Anal. Chem.* **38**, 720 (1966).
54. A. M. Duffield, H. Budzikiewicz, and C. Djerassi, *J. Amer. Chem. Soc.* **87**, 2913 (1965).
55. B. S. Larsen, G. Schroll, S.-O. Lawesson, J. H. Bowie, and R. G. Cooks, *Chem. Ind. (London)*, **10**, 321 (1968).
56. M. Vandewalle, N. Schamp, and M. Francque, *Org. Mass Spectrom.* **2**, 877 (1969).
57. K. G. Das, P. T. Funke, and A. K. Bose, *J. Amer. Chem. Soc.* **86**, 3729 (1964).
58. J. H. Bowie, P. F. Donaghue, H. J. Rodda, R. G. Cooks, and D. H. Williams, *Org. Mass Spectrom.* **1**, 13 (1968).
59. M. Ohashi, H. Kamachi, H. Kakisawa, A. Tatematsu, H. Yoshizumi, H. Kano, and H. Nakata, *Org. Mass Spectrom.* **2**, 195 (1969).

SECTION C: CHN AND/OR CHNO COMPOUNDS—CONT.

Name of class (series no.)	II	III	IV	V	MW range	Reference
Compounds with one N and one O atom—*cont.*						
1-Alkyl-2-phenyl-3-aroylazetidines (−1)	9				293–395	60
Diphenyl isoxazoles (−3)	7				221–300	32
Acetanilides (−5)	9			M	149–181	15
Acetanilide	1					61
Benzamide (−5)	1				121–135	61
Aromatic amides (−7)	1		X	M	203	62
2-Azetidinones (−7)	8	5			127–237	18
Oxygenated quinolines (−9)	15	15	D	M	159+	63
Quinoline *N*-oxides (−9)	9				145+	64
Isoquinoline *N*-oxides (−9)	4				145+	64
Aryl oxazoles (−9)	11		D		145–297	58
Benzyl phenyl ketoxime	7	7	D	M	211+	65
Compounds with one N and two O atoms						
N-Alkyl succinimides (+1)	2	2	D		141–155	54
Cyanoacetates (+1)	8		D	M	99–169	66
Cyclic nitrones (−1)	1	1			181	67
Benzindole acids (−1)	2	2			165	68
1-Phenyl-1-nitroethane (−3)	3				151	69
N-Substituted cyclohexene–1,2-dicarboximides (−3)	9	7	D	M	151–207	70
o-Nitrotoluene (−3)	2		$D^{13}C$		137	71
o-Aminobenzoic acid and derivatives					137+	72
Ethyl *N*-phenylcarbamate (−3)	5				165+	73
δ-Hydroxyamides (−5), and	6	6			233–331	74
(−9, −11)					215–299	74
Quinolinehydroxamic acids (−7)	21				161–295	75
Imides (−7, −11) R—C(=O)—N—R—C(=O)—R	18	7	D		101–315	76
Aromatic imides (−7)	3			M	147–223	77
Aromatic imides (−7, −6, −5, −3)	17		X	M	97–251	78
Aromatic cyanoesters (−7)	10			M	189+	79
Unsaturated esters (−11)	10			M	195+	79
1-Nitropropane (−9)	4		D		89–92	80

SECTION C: CHN AND/OR CHNO COMPOUNDS—CONT.

Key to references

60. J.-L. Imbach, E. Doomes, N. H. Cromwell, H. E. Baumgarten, and R. G. Parker, *J. Org. Chem.* **32**, 3123 (1967).
32. Simons *et al.* (1969).
15. Shapiro and Tomer (1969).
61. J. L. Cotter, *J. Chem. Soc.* 5477 (1964).

62. R. A. W. Johnstone and D. W. Payling, *Chem. Commun.*, 826 (1967).
18. Jackson *et al.* (1968).
63. D. M. Clugston and D. B. MacLean, *Can. J. Chem.* **44**, 781 (1966).
64. O. Buchardt, A. M. Duffield, and R. H. Shapiro, *Tetrahedron* **24**, 3139 (1968).

58. Bowie *et al.* (1968).
65. B. K. Simons, B. Nussey, and J. H. Bowie, *Org. Mass Spectrom.* **3**, 925 (1970).

54. Duffield *et al.* (1965).
66. J. H. Bowie, R. Grigg, S.-O. Lawesson, P. Madsen, G Schroll, and D. H. Williams, *J. Amer. Chem. Soc.* **88**, 1699 (1966).
67. R. F. C. Brown, W. D. Crow, L. Subrahmanyan, and C. S. Barnes, *Aust. J. Chem.* **20**, 2485 (1967).
68. U. K. Pandit, H. J. Hofman, and H. O. Huisman, *Tetrahedron* **20**, 1679 (1964).
69. N. M. M. Nibbering and Th. J. de Boer, *Org. Mass Spectrom.* **2**, 157 (1969).

70. E. D. Mitchell and G. R. Waller, *Org. Mass Spectrom.* **3**, 519 (1970).
71. J. H. Beynon, R. A. Saunders, A. Topham, and A. E. Williams, *J. Chem. Soc.* 6403 (1965).

72. R. M. Teeter, *Anal. Chem.* **38**, 1736 (1966).
73. C. P. Lewis, *Anal. Chem.* **36**, 176 (1964).
74. E. M. Levi, C. L. Mao, and C. R. Hauser, *Can. J. Chem.* **47**, 3671 (1969).

75. R. T. Coutts and K. W. Hindmarsh, *Org. Mass Spectrom.* **2**, 681 (1969).

76. C. Nolde and S.-O. Lawesson, *Tetrahedron* **24**, 1051 (1968).

77. J. L. Cotter and R. A. Dine-Hart, *Chem. Commun.* 809 (1966).
78. T. W. Bentley and R. A. W. Johnstone, *J. Chem. Soc. C*, 2354 (1968).
79. V. I. Kadentsev, B. M. Zolotarev, O. S. Chizov, C. Shachidayatov, L. A. Yanovskava, and V. F. Kucherov, *Org. Mass Spectrom.* **1**, 899 (1968).
80. N. M. M. Nibbering, Th. J. de Boer, and H. J. Hofman, *Rec. Trav. Chim. Pays-Bas* **84**, 481 (1965).

SECTION C: CHN AND/OR CHNO COMPOUNDS—CONT.

Name of class (series no.)	II	III	IV	V	MW range	Reference
Compounds with one N and two O atoms—*cont.*						
Nitronaphthalenes (−9)	8				173+	81
Nitrite esters (−9)	31				91–203	82
Related compounds	7		X		131+	83
Ethyl *N*-ethylcarbamate (−9)	1				117	73
N-Substituted ethyl carbamates (−9)	25				89+	84
Maleimides (−9)	6				173+	85
Isomaleimides (−9)	3				173+	85
Methyl esters of aliphatic α-amino acids (−9)	6				103–145	86
Ethyl esters of α-amino acids (−9)	6				117–159	86
Benzacridines and benzocarbazoles with an α-pyrone ring (−21)	2				273+	87
Compounds with one N and three or more O atoms						
Aliphatic polynitro compounds	8				103–165	88
(−3, −9, −6, 0 series)					134–238	88
Dihydrobenzindole acids (−9)	5	5			211–243	68
Aromatic nitro compounds (−9), nitronaphthalene	7				173+	89
Nitrate esters (−7)				C		83
Pyridoxol and isomers (−13)	6	6		M		90
Compounds with two N atoms and one O atom						
Furoxans (0, −8)	7	7	X	M	98–230	91
Azabicyclolactams (0)	5	5		M	154–182	92
Indazolones (−6)	13	13		M	148–220	93
Phenazine *N*-oxide (−8) and aromatic *N*-oxides	6	6			146–262	94
Ureas (−10, −18)	18	18		M	60–248	95
Heterocyclic carbonamides (−12, −18)	20	14		M	198–354	96
Dibenzylnitrosamines (−12)	2			M	254+	97
Aromatic azoxy compounds (−12)	11	11		M	198–298	98
Diaziridinones (−12, −14)	4	4		M	170–322	99
5-Pyrazolones (−12)	14	14	D	M	98	100
Diphenyl oxadiazoles (−16)	2	2		M	222	101

SECTION C: CHN AND/OR CHNO COMPOUNDS—CONT.

Key to references

81. J. H. Beynon, B. E. Job, and A. E. Williams, *Z. Naturforsch.* **21**, 210 (1966).
82. R. T. M. Fraser and N. C. Paul, *J. Chem. Soc. B*, 659 (1968).
83. R. T. M. Fraser and N. C. Paul, *J. Chem. Soc. B*, 1407 (1968).
73. Lewis (1964).

84. C. P. Lewis, *Anal. Chem.* **36**, 1582 (1964).
85. W. J. Feast, J. Put, F. C. de Schryver, and F. C. Compernolle, *Org. Mass Spectrom.* **3**, 507 (1970).

86. C. O. Andersson, R. Ryhage, S. Ställberg-Stenhagen, and E. Stenhagen, *Ark. Kemi* **19**, 405 (1962).

87. N. P. Buu-Hoi, M. Mangane, P. Jacquignon, *J. Chem. Soc. C*, 50 (1966).

88. F. E. Saalfield, J. T. Larkins, L. Kaplan, *Org. Mass Spectrom.* **2**, 213 (1969).

68. Pandit *et al.* (1964).

89. J. Harley-Mason, T. P. Toube, and D. H. Williams, *J. Chem. Soc. B*, 396 (1966).
83. Fraser and Paul (1968).
90. D. C. DeJongh, S. C. Perricone, and W. Korytnyk, *J. Amer. Chem. Soc.* **88**, 1233 (1966).

91. H. E. Ungnade and E. D. Laughran, *J. Heterocycl. Chem.* **1**, 61 (1964).
92. A. M. Duffield, C. Djerassi, L. Wise, and L. A. Paquette, *J. Org. Chem.* **31**, 1599 (1966).
93. R. B. Johns and J. M. Desmarchelier, *Org. Mass Spectrom.* **2**, 37 (1969).

94. J. H. Bowie, R. G. Cooks, N. C. Jamieson, and G. E. Lewis, *Aust. J. Chem.* **20**, 2545 (1967).
95. M. A. Baldwin, A. Kirkien-Konasiewicz, A. G. Loudon, A. Maccoll, and D. Smith, *J. Chem. Soc. B*, 34 (1968).

96. W. Schafer and P. Neubert, *Tetrahedron* **25**, 315 (1969).
97. T. Axenrod and G. W. A. Milne, *Chem. Commun.* **2**, 67 (1968).
98. J. H. Bowie, R. G. Cooks, and G. E. Lewis, *Aust. J. Chem.* **20**, 1601 (1969).
99. I. Lengyel, F. D. Greene, and J. F. Pazos, *Org. Mass Spectrom.* **3**, 623 (1970).
100. J. M. Desmarchelier and R. B. Johns, *Org. Mass Spectrom.* **2**, 697 (1969).
101. J. L. Cotter, *J. Chem. Soc.* 5491 (1964).

SECTION C: CHN AND/OR CHNO COMPOUNDS—CONT.

Name of class (series no.)	II	III	IV	V	MW range	Reference
Compounds with two N and two O atoms						
3,6-Dialkyl-2,5-dioxopiperazines (0)						102
Sydnones (−4, −6)	11	11			162–220	103
Aromatic cyclic diazoketones (−8, −10, −12)	11		S	M		104
Methoxycarbonylhydrazones (−10, −16)	7			M	144–222	105
Dialkyl sydnones (−12) also bis compounds	31		S	M	100–170	106
Phthalonitrile (−12) and phthalocyanines	7		X	C	128–901	107
Compounds with two N and three or more O atoms						
Nitroethylnitrates (−6)	3	3	X	M	120+	83
Bismaleimides	9	9		C	220	85
Bisisomaleimides (−18)	3	3		C	248	85
Diazotetracyclotetraones	3	3		C	234	85
Compounds with three N and one or more O atoms						
Pyridopyrimidin-4(3*H*)-ones (−6)	14	14		M	148–205	108
Pyridopyrimidine-2,4(1*H*,3*H*)-diones (−4)	12	12		M	164–220	108
Nitrophenylhydrazines (−1)	4	4	D	M	153	109
2,4-Dinitrophenylhydrazines (+2)	4	4	D	M	198	109
Nitrophenylhydrazones (−5)	2	2	D	M	247–261	109
2,4-Dinitrophenylhydrazones	2	2	D	M		109
Benzotriazinones (−7)	5	5		M	147–241	110
Pyrazole-4-oxyloximes (−9, −15)	5	5		M	187–263	111
Nitrophenylhydrazones (−9) of various aliphatic aldehydes and ketones	17	17	D	M	188–285	112
Nitrophenylhydrazones				D		113
asym-Triazine nitroxides	10	10		M	147–325	46a
Compounds with four N atoms and one O atom						
Hydroxypteridines (−6, −4)	5	5	D	M	148–164	47

Key to references

102. N. S. Wulfson, V. A. Puchkov, Yu. V. Denisov, B. V. Rozynov, V. N. Bochkarev, M. M. Shemyakin, Yu. A. Ovchinnikov, and V. K. Antonov, *Khim. Geterotsikl. Soedin.* 614 (1966).
103. J. H. Bowie, R. A. Eade, and J. C. Earl, *Aust. J. Chem.* **21**, 1665 (1968).

104. D. C. DeJongh, R. Y. Van Fossen, L. R. Dusold, and M. P. Cava, *Org. Mass Spectrom.* **3**, 31 (1970).

105. J. K. MacLeod, D. Becher, and C. Djerassi, *J. Org. Chem.* **31**, 4050 (1966).

106. R. S. Goudie, P. N. Preston, and M. H. Palmer, *Org. Mass Spectrom.* **2**, 953 (1969).

107. H. C. Hill and R. I. Reed, *Tetrahedron* **20**, 1359 (1964).

83. Fraser and Paul (1968).
85. Feast *et al.* (1970).
85. Feast *et al.* (1970).
85. Feast *et al.* (1970).

108. I. R. Gelling, W. J. Irwin, and D. G. Wibberley, *J. Chem. Soc. B*, 513 (1969).

109. F. Benoit and J. L. Holmes, *Can. J. Chem.* **47**, 3611 (1969).

110. J. C. Tou, L. A. Shadoff, and R. H. Rigterink, *Org. Mass Spectrom.* **2**, 355 (1969).
111. I. L. Finar and B. J. Millard, *J. Chem. Soc. C*, 2497 (1969).

112. J. Seibl, *Org. Mass Spectrom.* **3**, 417 (1970).
113. C. Djerassi and S. D. Sample, *Nature* **208**, 1314 (1965).
46a. Sasaki *et al.* (1969).

47. Goto *et al.* (1965).

Section C: CHN and/or CHNO Compounds—cont.

Name of class (series no.)	II	III	IV	V	MW range	Reference
Compounds with six N and three O atoms						
Benzotrifuroxan (−20)	1	1			204	114
Compounds with six N and six O atoms						
Secondary aliphatic nitramines (explosive) RDX, HMX, etc.	3	3			222+	115

Section D: CHS Compounds with O and N if Present

Name of class (series no.)	II	III	IV	V	MW range	Reference
Structure–fragmentation of organosulfur compounds (11 subclasses)	116	110		G	62–298	1
	186	186		G	48–282	2
Elemental sulfur	2	2			224–320	3
Compounds with one S atom						
Alkanethiols (−8)	29	29		C	48–202	4
	34	34		C	48–202	1
						2
Alkylthiaalkanes (−8)	31	31		C	62–230	4
	24	24		C	62–146	1
						2
D-Label study of rearrangement	3	3	D	M	132–174	5
Cycloalkylthiols (−10)	7	7		C	102–130	2
Cycloalkylthiaalkanes (−10)	20	20		C	116–186	6
Alkylthiacycloalkanes (−10)	13	13		C	74–116	1
						2
Ethylene sulfide	1	1		E	60	7
Propylene sulfide	1	1		E	74	8
Thiacyclobutane and thiacyclopentane	2	2		E	74–88	9
Dicycloalkylthiaalkanes (−12), Dicyclopentylthiamethanes	2	2		C	170–184	6
2-Thiadecalins, 2-thiahexahydroindans	4	4		C	142–156	1
						2
Bicyclothiaalkanes	4	4		C	114–156	2
						10

SECTION C: CHN AND/OR CHNO COMPOUNDS—CONT.

Key to references

114. A. S. Bailey, C. J. W. Gutch, J. M. Peach, and W. A. Waters, *J. Chem. Soc. B*, 681 (1969).

115. S. Bolusu, T. Axenrod, and G. W. Milne, *Org. Mass Spectrom.* **3**, 13 (1970).

SECTION D: CHS COMPOUNDS WITH O AND N IF PRESENT

Key to references

1. G. L. Cook and N. G. Foster, *Proc. Amer. Petrol. Inst.* **41**, III, 199 (1961).
2. G. L. Cook and G. U. Dinneen, *U.S. Bur. Mines Rep. Invest.* No. 6698 (1965).
3. U. I. Zahorszky, *Angew. Chem.* **7**, 633 (1968).

4. E. J. Levy and W. A. Stahl, *Anal. Chem.* **33**, 707 (1961).
1. Cook and Foster (1961).
2. Cook and Dinneen (1965).
4. Levy and Stahl (1961).
1. Cook and Foster (1961).
2. Cook and Dinneen (1965).
5. S. Sample and C. Djerassi, *J. Amer. Chem. Soc.* **88**, 1937 (1966).
2. Cook and Dinneen (1965).
6. J. E. Dooley, R. F. Kendall, and D. E. Hirsch, *U.S. Bur. Mines Rep. Invest.* No. 7351 (1970).
1. Cook and Foster (1961).
2. Cook and Dinneen (1965).
7. E. J. Gallegos and R. W. Kiser, *J. Phys. Chem.* **65**, 1177 (1961).
8. B. G. Hobrock and R. W. Kiser, *J. Phys. Chem.* **66**, 1551 (1962).
9. E. J. Gallegos and R. W. Kiser, *J. Phys. Chem.* **66**, 136 (1962).
6. Dooley *et al.* (1970).

1. Cook and Foster (1961).
2. Cook and Dinneen (1965).
2. Cook and Dinneen (1965).
10. API-44 Nos. 1790–1792.

SECTION D: CHS COMPOUNDS WITH O AND N IF PRESENT—CONT.

Name of class (series no.)	II	III	IV	V	MW range	Reference
Compounds with one S atom—*cont.*						
Tricycloalkylthiaalkanes (−14)						
Thiaadamantane	1	1		O	154	11
Alkylthiophenes (−14)	56	56		C	84–224	1, 2
	27	27		D	84–154	12
Mono- and dialkyl substitution	23	23		C	112–224	13
	14	14		M	154–280	14
t-Butyls and di-*t*-butyls	4	4		M	140–196	15
D-Label study of *t*-butyls	4	4	D	M	149–205	16
^{13}C-Label study of *n*-alkyl	1	1	^{13}C	M	168–169	17
D-Label study of *n*-butyl	4	4	D	M	140–142	18
D-Label study of ring H scrambling	6	6	D	M	84–87	19
D-Label study	1	1	D	M	84–85	20
Dialkylthiophenes				E		21
4-Thiapyrones	1	1		M	112	22
Thiophenols, benzenethiols (−16)	1	1		C	110	2
	1	1	D	M	111	23
	14	14	D	M	110–180	24
Arylthiaalkanes (−16)	3	3		C	124–138	1, 2
	13	13		M	124–218	25
Dithienyls (−16)	1	1		C	166	26
Diphenylthiophenes (−16)	3	3		C	236	26
Indanthiols (−18)						
Thiaindans (−18)	2	2		O	136	1
Thiatetralins (−18)	2	2		O	150	1
Octahydrodibenzothiophenes (−18)	1	1		O	192	1
Phenyl vinyl sulfides (−18)	12	12		M	136–288	27
Thiaindenes, benzo[*b*]thiophenes (−20) and derivatives	11	11		C	134–190	2
	11	11	D	M	134–224	28
Hexahydrodibenzothiophene (−20)	1			C	190	1
Naphthalenethiol (−22)	1	1		C	160	2
Alkylthianaphthalenes (1- and 2-)	15	15		C	202–300	29
Phenylthiophenes (−22), label study	5	5	D	M	160–165	30

SECTION D: CHS COMPOUNDS WITH O AND N IF PRESENT—CONT.

Key to references

11. API-44 No. 940.
1. Cook and Foster (1961).
2. Cook and Dinneen (1965).
12. I. W. Kinney, Jr., and G. L. Cook, *Anal. Chem.* **24**, 1391 (1952).
13. N. G. Foster, D. E. Hirsch, R. F. Kendall, and B. H. Eccleston, *U.S. Bur. Mines Rep. Invest.* No. 6433 (1964).
14. N. G. Foster, D. E. Hirsch, R. F. Kendall, and B. H. Eccleston, *U.S. Bur. Mines Rep. Invest.* No. 6671 (1965).
15. N. G. Foster, *U.S. Bur. Mines Rep. Invest.* No. 6741 (1966).
16. R. G. Fowler, N. G. Foster, and R. W. Higgins, Paper No. 144, Seventeenth Annual Conference on Mass Spectrometry, ASTM Committee E-14, Dallas, Texas, 1969.
17. N. G. Foster and R. W. Higgins, *Org. Mass Spectrom.* **1**, 191 (1968).
18. N. G. Foster and R. W. Higgins, Paper No. 97, Fourteenth Annual Conference on Mass Spectrometry, ASTM Committee E-14, Dallas, Texas, 1966.
19. S. Meyerson and E. K. Fields, *Org. Mass Spectrom* **2**, 241 (1969).
20. D. H. Williams, R. G. Cooks, J. Ronayne, and S. W. Tam, *Tetrahedron* **24**, 1777 (1968).
21. V. I. Khvostenko and A. S. Sultanov, *Akad. Nauk Bashkirsk. Filia* **6**, 230 (1964).
22. J. Bonham, E. McLeister, and P. Beak, *J. Org. Chem.* **32**, 639 (1967).
2. Cook and Dinneen (1965).
23. D. G. Earnshaw, G. L. Cook, and G. U. Dinneen, *J. Phys. Chem.* **68**, 296 (1964).
24. S.-O. Lawesson, J. O. Madsen, G. Schroll, J. H. Bowie, and D. H. Williams, *Acta Chem. Scand.* **20**, 2325 (1966).
1. Cook and Foster (1961).
2. Cook and Dinneen (1965).
25. J. H. Bowie, S.-O. Lawesson, J. O. Madsen, G. Schroll, and D. H. Williams, *J. Chem. Soc. B*, 951 (1966).
26. J. H. Bowie, R. G. Cooks, S.-O. Lawesson, and C. Nolde, *J. Chem. Soc. B*, 616 (1967).

1. Cook and Foster (1961).

27. W. D. Weringa, *Tetrahedron Lett.*, 273 (1969).

2. Cook and Dinneen (1965).
28. Q. N. Porter, *Aust. J. Chem.* **20**, 103 (1967).
1. Cook and Foster (1961).
2. Cook and Dinneen (1965).
29. N. G. Foster, R. W. Higgins, and M. C. Hamming, Paper No. G9, Eighteenth Annual Conference on Mass Spectrometry, ASMS with ASTM Committee E-14, San Francisco, California, 1970.
30. S. Meyerson and E. K. Fields, *Org. Mass Spectrom.* **1**, 263 (1968).

SECTION D: CHS COMPOUNDS WITH O AND N IF PRESENT—CONT.

Name of class (series no.)	II	III	IV	V	MW range	Reference
Compounds with one S atom—*cont.*						
1-(Phenyl)-1-(2-thienyl)methane	1	1		M	174	31
Tetrahydrodibenzothiophenes (−22)	2	2		C	188–202	2 32
Tetrahydro-9-thiafluorenes (−22)	2	2		C	188–202	2 32
Diphenylthiamethane (−24)	1	1		O	186	33
2-Methyl-9-thiafluorene (−26)	1	1		C	198	2
3- and 4-Methyldibenzothiophenes	2	2			198	34
Compounds with two S atoms						
Alkyldithiaalkanes (RSSR) (−4)	11	11		C	94–178	1, 2
	13	13		C	94–278	35
R″SR′SR (−6)	17	17		M	108	36
Diphenyldithiaalkanes						
Phenyl–CH_2SSCH_2–phenyls	2	2		M	218–246	35
Benzodithiophenes (−6)	3	3			190	37
1-(2-Thienyl)-1-thiaalkanes (−10)	11	11		M	158–228	38
4,5,6,7-Tetrahydro-1-(2-benzo[*b*]-thienyl)-1-thiapentane	1	1		M	226	39
1-(2-Benzo[*b*]thienyl)-1-thiaalkanes	5	5		M	222–278	40
1-(3-Benzo[*b*]thienyl)-1-thiaalkanes	5	5		M	222–278	41 40 42
Compounds with three S atoms						
s-Trithianes		9		O	138–408	43
Compounds with one S and one O atom						
Alkyl sulfoxides (−4) R(S=O)—R	8	8	D	M	78–230	44
Thioesters (−8) (thioacetates to thiooctanoates)	46	46		C	90–230	45
Alkyltetrahydropyranylthioethers (−8)	4	4		C	146–174	46

SECTION D: CHS COMPOUNDS WITH O AND N IF PRESENT—CONT.

Key to references

31. J. P. Liao, N. G. Foster, and R. W. Higgins, presented at S.W. ACS sectional meeting, Dec. 5, 1968; API-44 No. 2165.
2. Cook and Dinneen (1965).
32. API-44 Nos. 1375, 1307.

33. API-44 No. 637.

34. API-44 Nos. 1306, 1344.

1. Cook and Foster (1961).
2. Cook and Dinneen (1965).
35. J. H. Bowie, S.-O. Lawesson, J. O. Madsen, C. Nolde, G. Schroll, and D. H. Williams, *J. Chem. Soc. B*, 946 (1966).
36. T. H. Shuttleworth, *Appl. Spectrosc.* **18**, 78 (1964).

35. Bowie *et al.* (1966).
37. API-44 Nos. 1343, 1496, 1497.
38. N. G. Foster, D. W-K. Shiu, and R. W. Higgins, Paper No. 71, Sixteenth Annual Conference on Mass Spectrometry, ASTM Committee E-14, Pittsburgh, Pennsylvania, 1968; API-44, Nos. 2123–2133.

39. J. P. Liao, M.S. thesis, Texas Woman's University, Denton, Texas, 1969; API-44 No. 2170.
40. J. P. Liao, N. G. Foster, and R. W. Higgins, Paper No. 162, Seventeenth Annual Conference on Mass Spectrometry, ASTM Committee E-14, Dallas, Texas, 1969.

41. API-44 Nos. 2166, 2168, 2171, 2173, 2175.
40. Liao *et al.* (1969).
42. API-44 Nos. 2167, 2169, 2172, 2174, 2176.

43. API-44 Nos. 2134–2142.

44. J. H. Bowie, D. H. Williams, S.-O. Lawesson, J. O. Madsen, C. Nolde, and G. Schroll, *Tetrahedron*, 3515 (1966).

45. W. H. McFadden, R. M. Seifert, and J. Wasserman, *Anal. Chem.* **37**, 560 (1965).

46. S. J. Isser, A. M. Duffield, and C. Djerassi, *J. Org. Chem.* **33**, 2266 (1968).

SECTION D: CHS COMPOUNDS WITH O AND N IF PRESENT—CONT.

Name of class (series no.)	II	III	IV	V	MW range	Reference
Compounds with one S and one O atom—*cont.*						
Methoxy- and ethoxythioacetates	2	2		M	90–104	47
4-Thiacyclohexanone (−14)	1	1		M	116	48
1-(2-Thienyl)alkylalkanones	2	2	D	M	154–156	49
Thioesters, derivatives (−14)	17	22		M	98–212	50
2-Aryl-1,3-dithianes (N, O derivatives) (−14)	11	14	D	M	196–241	51
Thiobenzoates, methyl and ethyl (−16) and derivatives	13	13	D	M	152–211	47
Phenylacetylenic sulfoxides (−18)	1	1		C	164	52
Cyclic aromatic sulfoxides (−18) (benzothiophene derivatives)	2	2		M	150	53
2-Aryl-1,3-dithiolanes (−22)	3	3	D	M	314–315	51
Thiophene derivatives, various oxygenated and halo compounds	30	22		M	112–400	26
S-Methyl xanthates (1 O, 2 S)	10	10	D	M	478	54
Compounds with one S and two O atoms						
Cyclic aromatic sulfones (−16)	2	2		M	166	53
Phenylacetylenic sulfones (−16)	1	1		C	180	53
Alkyl sulfones (−18)	7	7		M	94–216	44
Diaryl sulfones (−20)	24	24		M	218–302	55
Thioglycolic acids (−20)	11	11		M	120–148	56
Thioglycolic esters (−20)	6	6		M	134–162	56
4-Thiacyclohexanone 4,4-dioxide	2	2		M	132–198	48
Dibenzothiophene 5,5-dioxide (−22)	1	1		M	216	57
Thionessal dioxides (tetraphenyl-thiophene dioxide) (−30)	1	1		M	400	58
Hydroxythiophenes and thiolactones	9	9	D	M	158–248	59
(Derivatives with 3–5 O)	8	8				59
Compounds with S and three or more O atoms						
Dialkyl sulfites (−16)	7	7		M	110–194	60
Alkyl methanesulfonates (−16)	7	7		M	124–186	61
Cycloalkyl sulfites (−18)	7	7		M	108–184	62
Sulfonic acids and sulfonates						63

SECTION D: CHS COMPOUNDS WITH O AND N IF PRESENT—CONT.

Key to references

47. A. Ohno, T. Koizumi, Y. Ohnishi, and G. Tsuchihashi, *Org. Mass Spectrom.* **3**, 261 (1970).
48. A. A. Kutz and S. J. Weininger, *J. Org. Chem.* **33**, 4070 (1968).
49. N. G. Foster and R. W. Higgins, *Org. Mass Spectrom.* **2**, 1005 (1969).
50. J. H. Bowie, R. G. Cooks, P. Jakobsen, S.-O. Lawesson, and G. Schroll, *Aust. J. Chem.* **20**, 689 (1967).

51. J. H. Bowie and P. Y. White, *Org. Mass Spectrom.* **2**, 611 (1969).

47. Ohno *et al.* (1970).
52. T. H. Kinstle, W. R. Oliver, and L. A. Ochrymowycz, *Org. Mass Spectrom.* **3**, 241 (1970).

53. J. Heiss, K.-P. Zeller, and B. Zeeh, *Tetrahedron* **24**, 3255 (1968).
51. Bowie and White (1969).

26. Bowie *et al.* (1967).
54. W. Briggs and C. Djerassi, *J. Org. Chem.* **33**, 1612 (1968).

53. Heiss *et al.* (1968).
53. Heiss *et al.* (1968).
44. Bowie *et al.* (1966).
55. S. Meyerson, H. Drews, and E. K. Fields, *Anal. Chem.* **36**, 1294 (1964).
56. J. H. Bowie, J. O. Madsen, S.-O. Lawesson, and R. G. Cooks, *Org. Mass Spectrom.* **2**, 413 (1969).
48. Kutz and Weininger (1968).

57. E. K. Fields and S. Meyerson, *J. Amer. Chem. Soc.* **88**, 2836 (1966).

58. M. M. Bursey, T. A. Elwood, and P. F. Rogerson, *Tetrahedron* **25**, 605 (1969).

59. R. Grigg, H. J. Jakobsen, S.-O. Lawesson, M. V. Sargent, G. Schroll, and D. H. Williams, *J. Chem. Soc. B*, 331 (1966).

60. A. A. Gamble, J. R. Gilbert, J. G. Tillett, R. E. Coombs, and A. J. Wilkinson, *J. Chem. Soc. B*, 655 (1969).
61. W. E. Truce and L. W. Christensen, *J. Org. Chem.* **33**, 2261 (1968).
62. P. Brown and C. Djerassi, *Tetrahedron* **24**, 2949 (1968).
63. A. Heywood, A. Mathias, and A. E. Williams, *Anal. Chem.* **42**, 1272 (1970).

SECTION D: CHS COMPOUNDS WITH O AND N IF PRESENT—CONT.

Name of class (series no.)	II	III	IV	V	MW range	References
Derivatives used with sugars						
Dithioacetals	12	12		M	222–328	64
Dithioketal peracetates	1	1		M	496	65
Diethyldithioacetal peracetates	6	6		M	424–496	65
Ethylenedithioacetal peracetates	4	4		M	394–466	66
Compounds with S and N atoms						
Isothiocyanates	40	40		G	73–283	67
Allyl, C_3 to 4-pentenyl (+1)	6	6			99–127	67
Aryl (−5)	8	8			135–177	67
Alkylthioisothiocyanates (−7)	8	8			119–217	67
Alkyl (normal) (−11)	11	11			73–283	67
Isoalkyl (−11)	7	7			101–185	67
Alkylisothiocyanates (−11)	8	8		M	73–119	68
Isothiazoles (and O, Br, derivatives) (+1)	15	15		M	85–195	69
Thiazoles (+1)	14	17		M	85–207	70
Benzothiazoles (−5) and derivatives	19	19		M	135–208	71
3-Phenylisothiazoles (−7)	27	27	X	M	161–249	72
Phenothiazines—derivatives (−9)	26	26	X	M	199–446	73
	14	14		M	229–376	74
Thiosemicarbazone derivatives of aldehydes and ketones (−9)	10	10		M	145–215	75
1-Thiocarbamoyl-2-pyrazolines (−5)	3	3	D	M	275–277	76
Thiourea derivatives (−8, −10)	7	7		M	104–172	77
Compounds with S, N, and O atoms						
Alkylsulfonylthioureas (0)	4	4		M	182–238	78
Arylsulfonylthioureas (−8)	5	5		M	230–292	78
1-Substituted 3-phenyl-2-thioureas	12	12	D	M	152–270	79
1,1-Disubstituted 3-phenyl-2-thioureas	11	11		M	180–332	79
Alkylsulfonylhydrazones (−4)	6	6		M	226–300	80
Arylsulfonylhydrazones (+2)						80
2-Amino-Δ^2-thiazolines and related compounds	5	5		M		81
Arylsulfinylamines (−1)	16	16		M	153–200	82
Thiotheophilline (−13)	4	1		M	197	83

SECTION D: CHS COMPOUNDS WITH O AND N IF PRESENT—CONT.

Key to references

64. D. C. DeJongh, *J. Org. Chem.* **30**, 1563 (1965).
65. D. C. DeJongh, *J. Amer. Chem. Soc.* **86**, 3149 (1964).
66. D. C. DeJongh, *J. Amer. Chem. Soc.* **86**, 4027 (1964).
67. A. Kjaer, M. Ohashi, J. M. Wilson, and C. Djerassi, *Acta Chem. Scand.* **17**, 2143 (1963).
68. K. A. Jensen, A. Holm, C. Wentrup, and J. Moller, *Acta Chem. Scand.* **20**, 2107 (1966).
69. B. J. Millard, *J. Chem. Soc. C*, 1231 (1969).
70. G. M. Clarke, R. Grigg, and D. H. Williams, *J. Chem. Soc. B*, 339 (1966).
71. B. J. Millard, *Org. Mass Spectrom.* **1**, 285 (1968).
72. T. Naito, *Tetrahedron* **24**, 6237 (1968).
73. J. N. T. Gilbert and B. J. Millard, *Org. Mass Spectrom.* **2**, 17 (1969).
74. A. M. Duffield, J. C. Craig, L. R. Kray, *Tetrahedron* **24**, 4267 (1968).
75. Y. M. Sheikh, A. M. Duffield, and C. Djerassi, *Org. Mass Spectrom.* **1**, 633 (1968).
76. S. W. Tam, *Org. Mass Spectrom.* **2**, 729 (1969).
77. T. Kinoshita and C. Tamura, *Tetrahedron Lett.*, 4963 (1969).
78. A. M. Duffield, C. Djerassi, R. Neidlein, and E. Heukelbach, *Org. Mass Spectrom.* **2**, 641 (1969).
79. R. H. Shapiro, J. W. Serum, and A. M. Duffield, *J. Org. Chem.* **33**, 243 (1968).
80. A. Bhati, R. A. W. Johnstone, and B. J. Millard, *J. Chem. Soc. C*, 358 (1966).
81. D. L. Klayman and G. W. A. Milne, *J. Org. Chem.* **31**, 2349 (1966).
82. J. H. Bowie, F. C. V. Larsson, G. Schroll, S.-O. Lawesson, and R. G. Cooks, *Tetrahedron* **23**, 3743 (1967).
83. M. Chaigneau, G. Valdener, and J. Seyden-Penne, *C.R. Acad. Sci.* **260**, 3965 (1965).

SECTION E: CHX COMPOUNDS WITH O, N, AND S IF PRESENT

Name of class (series no.)	II	III	IV	V	MW range	Reference
Mixed halogenated compounds	8	8		G		1
	22	22		G		2
				G		3
	2	2	D	G		4
Aliphatic halogenated compounds	103	103		G	34–380	5
Fluoroalkanes		7	F		34–118	
Chloroalkanes		27	Cl		50–288	
Bromoalkanes		21	Br		94–332	
Iodoalkanes		27	I		142–380	
Trihalomethanes (F, Cl)	4	4		E		6
Halocyclopropanes (F, Cl, Br)	7	7		M		7
Aromatic halogenated compounds	88	88		G	96–254	8
Fluoroalkanes		8	F			
Chloroalkanes		43	Cl			
Bromoalkanes		27	Br			
Iodoalkanes		10	I			
Fluorocarbons						
Hexafluoroethane	3	3		E	70–138	9
1,1,1-Trifluoroethane						9
Trifluoromethane						9
Fluorinated alkanes	10	10		M		10
Perfluorocycloalkanes	3	3		E	200–300	11
Perfluoroalkylbenzenes	10	10		M	228–328	12
Trifluoromethylgeminal systems						13
Aromatic fluorocarbons				E		14
Fluorocarbons—positive and negative ions						15
Disubstituted fluorobenzenes				E		16
Aliphatic ketones of form $CF_3—C(=O)—R$	4	4		M	112–250	17
Fluorine-containing dimethyl esters	4	4		M	418–568	18
Polyfluorinated aromatic and heterocyclic compounds	5	5		M	196–362	19
Chloro, bromo, and iodo compounds						
Norbornenyl and nortricyclyl chlorides						20
exo- and *endo*-2-Norbornyl bromides						21
1,1-Dichlorocyclopropanes and derivatives	21	21		M		22

SECTION E: CHX COMPOUNDS WITH O, N, AND S IF PRESENT—CONT.

Key to references

1. F. L. Mohler, E. G. Bloom, J. H. Lengel, and C. E. Wise, *J. Amer. Chem. Soc.* **71**, 337 (1949).
2. F. L. Mohler, V. H. Dibeler, and R. M. Reese, *J. Res. Nat. Bur. Stand.* **44**, 343 (1952).
3. J. Collin, *Bull. Roy. Sci. Liege* **25**, 426 (1956).
4. W. H. McFadden and M. Lounsbury, *Can. J. Chem.* **40**, 1965 (1962).
5. F. W. McLafferty, *Anal. Chem.* **34**, 2 (1962).
6. B. G. Hobrock and R. W. Kiser, *J. Phys. Chem.* **68**, 575 (1964).
7. M. L. Deem and J. Roboz, *Org. Mass Spectrom.* **3**, 155 (1970).
8. F. W. McLafferty, *Anal. Chem.* **34**, 16 (1962).
9. K. A. G. MacNeil and J. C. J. Thynne, *Int. J. Mass Spectrom. Ion Phys.* **2**, 1 (1969).
10. E. R. McCarthy, *Org. Mass Spectrom.* **1**, 81 (1968).
11. P. Natalis, *Bull. Roy. Sci. Liege* **29**, 94 (1960).
12. D. M. Gale, *Tetrahedron* **24**, 1811 (1968).
13. R. G. Kostyanovskii and V. P. Nechiporenko, *Teor. Eksp. Khim.* **2**, 558 (1966).
14. J. L. Cotter, *J. Chem. Soc.*, 1520 (1965).
15. M. M. Bibby and G. Carter, *Trans. Faraday Soc.* **59**, 2455 (1963).
16. I. Howe and D. H. Williams, *J. Amer. Chem. Soc.* **91**, 7137 (1969).
17. W. Carpenter, A. M. Duffield, and C. Djerassi, *Org. Mass Spectrom.* **2**, 317 (1969).
18. J. L. Cotter, *Org. Mass Spectrom.* **1**, 913 (1968).
19. M. I. Gorfinkel, V. S. Kobrin, and V. A. Koptyug, *Zh. Obshch. Khim.* **38**, 1815 (1969).
20. W. C. Steele, B. H. Jennings, G. L. Botyos, and G. O. Dudek, *J. Org. Chem.* **30**, 2886 (1965).
21. D. C. DeJongh, and S. R. Shrader, *J. Amer. Chem. Soc.* **88**, 3881 (1966).
22. D. S. Weinberg, C. Stafford, and M. W. Scoggins, *Org. Mass Spectrom.* **2**, 567 (1969).

SECTION E: CHX COMPOUNDS WITH O, N, AND S IF PRESENT—CONT.

Name of class (series no.)	II	III	IV	V	MW range	Reference
Chloro, bromo, and iodo compounds —*cont.*						
Polychlorinated bridged-ring compounds	25	25		M	278–715	23
Aryl halo compounds						24, 25
Triaryl haloethylenes (Cl, Br)	17	17		C	256–334	26
*Iso*hexyl bromide	6	6	D	M	164–172	27
Iodocyclopentane	1	1		M	196	28
Halogenated hetero atom compounds						
Fluorinated aliphatic nitriles	3	5		C	70–140	29
Fluorinated 1,2,4-oxadiazoles	4	4		EM	314–440	30
Fluorinated 1,3,4-oxadiazoles	3	3		EM	314–514	31
Chloro-unsaturated esters	5	5		M	248–282	32
Halogenated alcohols						33
Aromatic anhydrides	4	4		M		34
Halogenated steroid derivatives of methyl haloacetone ketals	12	12		C		35
Sulfur esters of perfluoropinacol						36
Chloro–aminochloro- and ethylaminochloro-*s*-triazines	4	4		C		37
Perfluoroester derivatives						
Fluoroalcohol esters	15	15		M	186–794	38
α,α,ω-Trihydro-						
N-Alkyltrifluoroacetamines	3	3		M	141–197	39
N-Aryltrifluoroacetamines	10	10		M	189–234	40
Nitro, chloro, and oxygenated						
Alditol trifluoroacetate	10	10		M	267–758	41
Derivatives of C_3–C_6 sugars						41
N-Trifluoroacetylamino acid methyl esters	2	2	D	M		42

SECTION E: CHX COMPOUNDS WITH O, N, AND S IF PRESENT—CONT.

Key to references

23. R. Binks, K. Mackenzie, and D. L. Williams-Smith, *J. Chem. Soc. C*, 1528 (1969).
24. L. D'Or, J. Momigny, and A. M. Wirtz-Cordier, *Bull. Cl. Sci. Acad. Roy. Belg.* **47**, 811 (1961).
25. V. H. Dibeler, R. M. Reese, and F. L. Mohler, *J. Chem. Phys.* **26**, 304 (1957).
26. J. Moller and C. Pedersen, *Acta Chem. Scand.* **22**, 706 (1968).
27. D. H. Williams, C. Beard, H. Budzikiewicz, and C. Djerassi, *J. Amer. Chem. Soc.* **86**, 877 (1964).
28. A. J. Lorquet, *Bull. Cl. Sci. Acad. Roy. Belg.* **49**, 1225 (1964).

29. J. B. Flannery and G. J. Janz, *J. Chem. Eng. Data* **10**, 387 (1965).
30. J. L. Cotter, *J. Chem. Soc. B*, 1271 (1967).
31. J. L. Cotter, *J. Chem. Soc. B*, 6842 (1965).
32. D. H. Williams, R. G. Cooks, J. H. Bowie, P. Madsen, G. Schroll, and S.-O. Lawesson, *Tetrahedron* **23**, 3173 (1967).
33. E. R. McCarthy, *J. Org. Chem.* **31**, 2042 (1966).
34. M. P. Cava, M. J. Mitchell, D. C. DeJongh, and R. Y. Van Fossen, *Tetrahedron* **22**, 2947 (1966).

35. G. A. Sarfaty and H. M. Fales, *Anal. Chem.* **42**, 288 (1970).
36. M. Allan, A. F. Janzen, and C. J. Willis, *Chem. Commun.*, 55 (1968).

37. J. A. Ross and B. G. Tweedy, *Org. Mass Spectrom.* **3**, 219 (1970).

38. R. M. Teeter, *Anal. Chem.* **39**, 1742 (1967).

39. M. J. Saxby, *Org. Mass Spectrom.* **2**, 33 (1969).
40. M. J. Saxby, *Org. Mass Spectrom.* **2**, 835 (1969).

41. O. S. Chizhov, B. A. Dmitriev, B. M. Zolotarev, A. Ya. Chernyak, and N. K. Kochetkov, *Org. Mass Spectrom.* **2**, 947 (1968).

42. M. S. Manhas, R. S. Hsieh, and A. K. Bose, *J. Chem. Soc. C*, 116 (1970).

SECTION F: CHM COMPOUNDS WITH O, N, S, AND HALOGEN ATOMS IF PRESENT

Name of class	II	III	IV	V	MW range	Reference
Volatile hydrides						
Silylphosphine	4	1		E	64	1
Silylgermane	5	1		E	108	2
Alkyl metallic compounds						
Organoaluminum compounds	5	5		M	58–114	3
Trimethylaluminum	1	1		E	72	4
Trimethylantimony	1	1		E		
Trimethylzinc	1	1		E		
Tetramethylsilicon	1	1		E	88	5
Tetramethyltin	1	1		E		
Tetramethyllead	1	1		E		
Tetramethylgermanium	1	1		E		6
Trimethylsilane	1	1		E		
Dimethylmercury	1	1		E		
Tetraorganogermanes, stannanes, and leads	6	6		C		7
Dimethylthallium derivatives	7	7	X	C	249–656	8
Vinyl triphenyl derivatives of silicon, germanium, tin						9
π-Bonded complexes						
Allylic						
π-Allylic rhodium	2	2		C		10
π-Allylic palladium	2	2				
Cyclopentadienyl						
Cobalt complexes	6	6				11
Chromium (binuclear)	6	5		D	358–522	12
						13
Ferrocenes				M		14
Ferrocenes, fragmentation				C		15
Polyfluoroferrocenes	8	8		D		16
Alkylferrocenes and biferrocenyls	17	17	D	MC	200–446	17
Tricarbonyl π complexes						
Cymantrenes (Mn)	18	18		C	232–428	18
1,3-Dieniron complexes	10	10		C	208–302	19
Cobalt, manganese, and vanadium complexes	3	3		E		20
Halides	20	20		C		21
Organometallic, Fe, Mo	14	14		D	220–390	22
Halide derivatives, W, Mo, Fe, and Pd	6	6		D	364–636	23
Organonitrogen derivatives of Fe, Mo, W, and Mn	13	13		D	249–490	24

SECTION F: CHM COMPOUNDS WITH O, N, S, AND HALOGEN ATOMS IF PRESENT—CONT.

Key to references

1. F. E. Saalfield and H. J. Svec, *Inorg. Chem.* **3**, 1442 (1964).
2. F. E. Saalfield and H. J. Svec, *J. Phys. Chem.* **70**, 1753 (1966).
3. D. B. Chambers, G. E. Coats, F. Glocking, and M. Weston, *J. Chem. Soc. C*, 1618 (1969).
4. R. E. Winters and R. W. Kiser, *J. Organometal. Chem.* **10**, 7 (1967).
5. B. G. Hobrock and R. W. Kiser, *J. Phys. Chem.* **65**, 2186 (1961).
6. B. G. Hobrock and R. W. Kiser, *J. Phys. Chem.* **66**, 155 (1962).
7. D. B. Chambers, F. Glocking, J. R. C. Light, M. Weston, *Chem. Commun.*, 282 (1966).
8. A. G. Lee, *Int. J. Mass Spectrom. Ion Phys.* **3**, 239 (1969).
9. P. N. Preston, P. J. Rice, and N. A. Weir, *Int. J. Mass Spectrom. Ion Phys.* **1**, 303 (1968).
10. M. S. Lupin and M. Cais, *J. Chem. Soc. B*, 3095 (1968).
11. K. Yasafuku, and H. Yamazaki, *Org. Mass Spectrom.* **3**, 23 (1970).
12. F. J. Preston and R. I. Reed, *Org. Mass Spectrom.* **1**, 71 (1968).
13. F. J. Preston and R. I. Reed, *Chem. Commun.*, 51 (1966).
14. R. I. Reed and F. M. Tabrizi, *Appl. Spectrosc.* **17**, 124 (1963).
15. A. Mandelbaum and M. Cais, *Tetrahedron Lett.* 3847 (1964).
16. M. I. Bruce, *Org. Mass Spectrom.* **2**, 997 (1969).
17. I. J. Spilners and J. G. Larson, *Org. Mass Spectrom.* **3**, 915 (1970).
18. M. Cais, M. S. Lupin, N. Maoz, J. Sharvit, *J. Chem. Soc. A*, 3086 (1968).
19. W. G. Dauben and M. E. Lorber, *Org. Mass Spectrom.* **3**, 211 (1970).
20. R. E. Winters and R. W. Kiser, *J. Organometal. Chem.* **4**, 190 (1965).
21. M. I. Bruce, *Int. J. Mass Spectrom. Ion Phys.* **1**, 141 (1968).
22. R. B. King, *J. Amer. Chem. Soc.* **90**, 1417 (1968).
23. R. B. King, *Org. Mass Spectrom.* **2**, 401 (1969).
24. R. B. King, *Org. Mass Spectrom.* **2**, 387 (1969).

SECTION F: CHM COMPOUNDS WITH O, N, S, AND HALOGEN ATOMS IF PRESENT—CONT.

Name of class	II	III	IV	V	MW range	Reference
Tricarbonyl π complexes —*cont.*						
Cyclopentadienone complexes (Fe and Ru)	4	4		D		25
Cyclooctatetraene complexes (Fe and Ru)	6	6		D		26
Metal carbonyl complexes						
Nickel tetra-, and iron pentacarbonyls	2	2		E		27
Chromium, molybdenum, and tungsten hexacarbonyls	3	3		E		28
Nickel, iron, chromyl, molybdenum, and tungsten carbonyls	5	5		E		29
Binuclear manganese, rhenium and mixed decacarbonyls	3	3		E		30
Metal fluorocarbon complexes (Mn, Re, Fe, Ru)	10	10		D		31
Polyfluorophenyl derivatives						
Pentafluorophenyl derivatives (Si, Ge, Sn, and Pb)	4	4		D		32
Polyfluorobenzenes, transition-metal derivatives	14	9		D		33
Perfluorophenyl derivatives	3	3		D	444–492	34
Polyfluoro aromatic derivatives				D	270–520	35
Metal chelates and other complexes						
Monothio-β-diketones and their Ni(II) chelates	1	1		D		36
MS study of *cis*-1,2-ethylene-dithiolate–metal complexes						37
2-Carbamoyldimedone and *N*-phenyl-2-carbamoyldimedone complexes of Cu(II), Zn(II), and Fe(III)	6	6		M		38
Metal acetylacetonates						39
Fluorine-substituted acetyl-acetonates (Fe, Cu, Zn)	6	6		E		40
Dibenzoylmethane complexes	14	14		M		41
Miscellaneous compounds						
Copper phthalocyanines	6	6	Cl	M	575–901	42
For porphyrins and related compounds, see Section H						

SECTION F: CHM COMPOUNDS WITH O, N, S, AND HALOGEN ATOMS IF PRESENT—CONT.

Key to references

25. M. I. Bruce, *Int. J. Mass Spectrom. Ion Phys.* **1**, 335 (1968).

26. M. I. Bruce, *Int. J. Mass Spectrom. Ion Phys.* **2**, 349 (1969).

27. R. E. Winters and R. W. Kiser, *Inorg. Chem.* **3**, 699 (1964).
28. R. E. Winters and R. W. Kiser, *Inorg. Chem.* **4**, 157 (1965).

29. G. A. Junk and H. J. Svec, *Z. Naturforsch. B* **23**, 1 (1968).

30. H. J. Svec and G. A. Junk, *J. Amer. Chem. Soc.* **89**, 2836 (1967).

31. M. I. Bruce, *Org. Mass Spectrom.* **2**, 63 (1969).

32. J. M. Miller, *Can. J. Chem.* **47**, 1613 (1969).

33. M. I. Bruce and M. A. Thomas, *Org. Mass Spectrom.* **1**, 835 (1968).
34. S. C. Cohen and A. G. Massey, *J. Organometal. Chem.* **10**, 471 (1967).
35. M. I. Bruce, *J. Organometal. Chem.* **10**, 495 (1967).

36. S. H. H. Chaston, S. E. Livingstone, T. H. Lockyer, V. A. Pickles, and J. S. Shannon, *Aust. J. Chem.* **18**, 673 (1965).

37. S. M. Bloom and G. O. Dudek, *Inorg. Nucl. Chem. Lett.* **2**, 183 (1966).

38. E. P. Dudek and M. Barber, *Inorg. Chem.* **5**, 375 (1966).
39. C. G. MacDonald and J. S. Shannon, *Aust. J. Chem.* **19**, 1545 (1966).

40. C. Reichert, F. M. Bancroft, and J. B. Westmore, *Can. J. Chem.* **48**, 1362 (1970).
41. M. J. Lacey, C. G. MacDonald, and J. S. Shannon, *Org. Mass Spectrom.* **1**, 115 (1968).

42. H. C. Hill and R. I. Reed, *Tetrahedron* **20**, 1359 (1964).

SECTION G: CHB, P, Si, Se, Te, ETC. (ESSENTIALLY THE NONMETALS)

Name of class	II	III	IV	V	MW range	Reference
Boron compounds						
Alkyl borates	3	3		C	146–230	1
Boron hydrides	3	3	D	M		2
Boroxines	4	4		M	252–312	3
Cyclotetrazenoboranes	5	5		M	99–176	4
	4	4		M	97–222	5
Heteroaromatic boron compounds (borazerenes)	12	12		M	143–372	6
Isotopically labeled tetraborames	4	4		D	50–60	7
1,3,2-Oxazaborolidines	9	9		M	161–251	8
Pyrolysis of boranes						9
Trimethyl borate						10
A six-boron carborane						11
Phosphorus compounds						
Phosphorus	4	4		D	P_1–P_4	12
Phosphorus oxides	8	8		D	47–284	12
Trimeric chlorobromophosphonitriles	7	7		D	345–609	13
Alkyl hydrogen phosphites						14
Triphenylphosphine	4	4	D	M	262–277	15
Triphenylphosphine oxide	4	4	D	M	278–291	15
Triphenylphosphine sulfide	1	1		M	294	15
Dialkyl alkylphosphonates	11	11		M		16
Alkyl-substituted phosphines	4	4		E	76–118	17
Diaryl phosphinates						18
Trialkyl phosphates						19
						20
Esters of phosphorus and phosphonic acids	13	13		M	126–310	20
Alkyl phosphate esters	3	3		E	124–182	21
Fluoroalkylphosphorus compounds	11	11			102–258	22
Tetra(trifluoromethylphosphine) sulfide	4	4		M	476–508	23
	1	1		M	432	23
β-Ketoalkylidenephosphoranes	6	6		M		24
Organophosphorus esters	6	6		M	110–190	25
Organophosphorus pesticides (see Section H)						

SECTION G: CHB, P, Si, Se, Te, ETC. (ESSENTIALLY THE NONMETALS)—CONT.

Key to references

1. P. J. Fallon, P. Kelly, and J. C. Lockhart, *Int. J. Mass Spectrom. Ion Phys.* **1**, 133 (1968).
2. T. P. Fehlner and W. S. Koski, *J. Amer. Chem. Soc.* **86**, 581 (1964).
3. C. J. W. Brooks, D. J. Harvey, and B. S. Middleditch, *Org. Mass Spectrom.* **3**, 231 (1970).
4. E. F. H. Brittain, J. B. Leach, and J. H. Morris, *Org. Mass Spectrom.* **1**, 459 (1968).
5. E. F. H. Brittain, J. B. Leach, and J. H. Morris, *J. Chem. Soc. A*, 340 (1968).

6. R. C. Dougherty, *Tetrahedron* **24**, 6755 (1968).
7. A. D. Norman, R. Schaeffer, A. B. Baylis, G. A. Pressley, Jr., and F. E. Stafford, *J. Amer. Chem. Soc.* **88**, 2151 (1966).
8. C. J. W. Brooks, B. S. Middleditch, and G. M. Anthony, *Org. Mass Spectrom.* **2**, 1023 (1969).
9. A. B. Baylis, G. A. Pressley, Jr., M. E. Gordon, and F. E. Stafford, *J. Amer. Chem. Soc.* **88**, 929 (1966).
10. Y. Wada and R. W. Kiser, *J. Phys. Chem.* **68**, 1588 (1964).
11. R. E. Williams and G. F. James, *J. Amer. Chem. Soc.* **87**, 3513 (1965).

12. L. W. Daasch, J. N. Wever, M. A. Ebner, and G. Sparrow, *Int. J. Mass Spectrom. Ion Phys.* **2**, 503 (1969).

13. G. E. Coxon, T. F. Palmer, and D. B. Sowerby, *J. Chem. Soc. A*, 358 (1969).
14. H. R. Harless, *Anal. Chem.* **33**, 1387 (1961).
15. D. H. Williams, R. S. Ward, and R. G. Cooks, *J. Amer. Chem. Soc.* **90**, 966 (1968).

16. T. Nishiwaki, *Tetrahedron* **22**, 1383 (1966).
17. Y. Wada, and R. W. Kiser, *J. Phys. Chem.* **68**, 2290 (1964).
18. P. Haake, M. J. Frearson, and C. E. Diebert, *J. Org. Chem.* **34**, 788 (1969).
19. F. W. McLafferty, *Anal. Chem.* **28**, 306 (1956).
20. J. Occolowitz and G. White, *Anal. Chem.* **35**, 1179 (1963).

21. D. A. Bafus, E. J. Gallegos, and R. W. Kiser, *J. Phys. Chem.* **70**, 2614 (1966).
22. R. G. Cavell and R. C. Dobbie, *Inorg. Chem.* **7**, 101 (1968).

23. R. G. Cavell and R. C. Dobbie, *Inorg. Chem.* **7**, 690 (1968).

24. R. T. Aplin, A. R. Hands, and A. J. H. Mercer, *Org. Mass Spectrom.* **2**, 1017 (1969).
25. J. E. Pritchard, *Org. Mass Spectrom.* **3**, 163 (1970).

SECTION G: CHB, P, Si, Se, Te, ETC. (ESSENTIALLY THE NONMETALS)—CONT.

Name of class	II	III	IV	V	MW range	Reference
Miscellaneous compounds						
Tetraethyl distibine, diethyl diselenide, diethyl ditelluride, and the corresponding mono-derivatives						26
Benzoselenopheno[2,3-*b*]-benzoselenophenes						27
2-Amino-Δ^2-selenazoline	1	1		M		28
(Adamsite)10-chloro-5,10-dihydrophenarsazine and 10-methyl-5,10-dihydrophenarsazine	2	2		M		29
Phenoxarsine derivatives	9	9		M	278–602	30
Silicon compounds						
Silicon tetrahalides	2	2	X	M	344–536	31
Silicon derivatives of hydrocarbons						32
Silanes, mono- and di-	2	2		E		33
Methylsilanes	9	9		C	32–88	34
Silicon–methylene compounds						35
Linear and cyclic silicon–methylene compounds						36
Saturated and unsaturated organosilicon compounds						37
Pyrolysis of methylchlorosilanes and compounds from–						38
Trimethylsilyl (TMS) derivatives						
Of alditols	34	34		G	104–230	39
	6	6		C	308–526	40
Of ethers of aliphatic glycols	14	14		M	161–290	41
Of ethers	11	11			160–256	42
Of amines	1	1		M	179	42
Of sulfides	2	2			196–210	42
Of ω-phenoxyalkanoic acids	11	11		M		43
Of tris(TMS) phosphate	2	2	D	M		44
Of various ethers		32				45
Inositol TMS ethers and acetate esters	8	8				46
TMS sugar phosphates, GC–MS			D			47
Steroids	5	5	D	M	348–500	48
Tetrasilaadamantanes, tetrachloro derivative	3	3	D	M	336	49

SECTION G: CHB, P, Si, Se, Te, ETC. (ESSENTIALLY THE NONMETALS)—CONT.

Key to references

26. G. M. Bogolyubov, N. N. Grishin, and A. A. Petrov, *J. Gen. Chem. USSR* **39**, 2190 (1969).
27. D. Elmaleh, S. Patai, and Z. Rappoport, *J. Chem. Soc. C*, 939 (1970).
28. D. L. Klayman and G. W. A. Milne, *J. Org. Chem.* **31**, 2349 (1966).
29. N. P. Buu-Hoi, M. Mangane, and P. Jacquignon, *J. Heterocycl. Chem.* **3**, 149 (1966).
30. J. C. Tou and C. S. Wang, *Org. Mass Spectrom.* **3**, 287 (1970).
31. H. J. Svec and G. R. Sparrow, *J. Chem. Soc. A*, 1162 (1970).
32. R. A. Khmel'nitskii, A. A. Polyakova, and A. A. Petrov, *Inst. Organ. Khim. Akad. Nauk Frunze*, 236 (1962).
33. P. Potzinger and F. W. Lampe, *J. Phys. Chem.* **73**, 3912 (1969).
34. R. P. Van der Kelen, O. Volders, H. Van Onckelen, and Z. Eeckhaut, *Z. Anorg. Allg. Chem.* **338**, 106 (1965).
35. G. Fritz, H. Buhl, J. Grobe, F. Aulinger, and W. Reerink, *Z. Anorg. Allg. Chem.* **312**, 201 (1961).
36. F. Aulinger, *Colloq. Spectrosc. Int.*, *8th*, 267 (1959).
37. F. Aulinger and W. Reerink, *Colloq. Spectrosc. Int.*, *9th*, 24 (1961).
38. F. Aulinger and W. Reerink, *Z. Anal. Chem.* **197**, 24 (1963).
39. A. G. Sharkey, Jr., R. A. Friedel, and S. H. Langer, *Anal. Chem.* **29**, 770 (1957).
40. G. Petersson, *Tetrahedron* **25**, 4437 (1969).
41. J. Diekman, J. B. Thomson, and C. Djerassi, *J. Org. Chem.* **33**, 2271 (1968).
42. J. Diekman, J. B. Thomson, and C. Djerassi, *J. Org. Chem.* **32**, 3904 (1967).
43. J. Diekman, J. B. Thomson, and C. Djerassi, *J. Org. Chem.* **34**, 3147 (1969).
44. M. Zinbo and W. R. Sherman, *J. Amer. Chem. Soc.* **92**, 2105 (1970).
45. J. A. McCloskey, R. A. J. M. Leemans, and P. O. Prochaska, *Arch. Mass Spectral Data* **1**, 16–113 (1970).
46. W. R. Sherman, N. C. Eilers, and S. L. Goodwin, *Org. Mass Spectrom.* **3**, 829 (1970).
47. M. Zinbo and W. R. Sherman, *Tetrahedron* **25**, 2811 (1969).
48. J. Diekman and C. Djerassi, *J. Org. Chem.* **32**, 1005 (1967).
49. R. S. Gohlke and R. J. Robinson, *Org. Mass Spectrom.* **3**, 967 (1970).

SECTION H: SUBSTANCES OF BIOLOGICAL INTEREST

Name of class	II	III	IV	V	MW range	Reference[a]
Carbohydrates						
Structure analysis						1, 2
Pentoses and hexoses	14		D	M	390–393	3
Permethylated *N*-Acetylamino sugars						4
Characterization of amino sugars						5
Monosaccharide derivatives and amino sugars						6
Methyl ethers of disaccharides						7
Polysaccharides						8
Methylated methyl glycosides						9
Hexosamines						10
Sugars from antibiotics						11
O-Isopropylidene derivatives						12, 13
Dithioacetals of amino sugars						5
Methyl haloacetone ketals as derivatives in GC–MS						14
Lipids and steroids						
Application in lipid chemistry						15
Triglycerides of straight-chain fatty acids	2				806–890	16
Cyclopropane fatty acid esters	5				296–374	17
1,2-Monomyristic acids and their TMS derivatives	6				302–496	18
4,4-Dimethyl-5α-androstan-3-one and related ketones	1			M		19
Reactions of a typical 3-keto steroid, 5α-androstan-3-one	1			M		20
Enolization studies on 7-keto-5α-androstanes						21
Location of the double bond in steroid systems						22

[a] For many additional references of work in areas of interest to the biochemist and biologist see G. R. Waller (ed.), "Biochemical Applications of Mass Spectrometry," Wiley, New York, 1972.

SECTION H: SUBSTANCES OF BIOLOGICAL INTEREST—CONT.

Key to references

1. K. Heyns, *Staerke* **18**, 261 (1966).
2. K. Heyns, H.-P. Grützmacher, H. Scharmann, and D. Muller, *Fortschr. Chem. Forsch.* **5**, 448 (1966).
3. K. Biemann, D. C. DeJongh, and H. K. Schnoes, *J. Amer. Chem. Soc.* **85**, 1763 (1963).

4. K. Heyns, *Tetrahedron* **21**, 3151 (1965).
5. D. C. DeJongh and S. Hanessian, *J. Amer. Chem. Soc.* **87**, 3744 (1965).

6. K. Heyns and H. Scharmann, *Ann. Chem.* **667**, 183 (1963).
7. O. S. Chizhov, L. A. Polyakova, and N. K. Kochetkov, *Dokl. Akad. Nauk SSSR* **158**, 685 (1964).
8. H. Bjorndal, C. G. Hellerqvist, B. Lindberg, and S. Svensson, *Angew. Chem. Int. Ed.* **9**, 610 (1970).
9. N. K. Kochetkov and O. S. Chizhov, *Tetrahedron* **21**, 2029 (1965).
10. N. K. Kochetkov, O. S. Chizhov, and B. M. Zolotarev, *Carbohydrate Res.* **2**, 89 (1966).
11. K. L. Rinehart, Jr., R. F. Schimbor, T. N. Kinstle, *Antimicrob. Ag. Chemother.* 119 (1965).
12. D. C. DeJongh and K. Biemann, *J. Amer. Chem. Soc.* **86**, 67 (1964).
13. J. A. McCloskey and M. McClelland, *J. Amer. Chem. Soc.* **87**, 5090 (1965).
5. DeJongh and Hanessian (1965).

14. G. A. Sarfaty and H. M. Fales, *Anal. Chem.* **42**, 288 (1970).

15. R. T. O'Connor, *J. Amer. Oil Chem. Soc.* **41**, Suppl. 4, 14, 20, 22, 25, 55 (1964).

16. M. Barber, T. O. Merren, and W. Kelly, *Tetrahedron Lett.*, 1063 (1964).
17. W. W. Christie and R. T. Holman, *Lipids* **1**, 176 (1966).

18. C. B. Johnson and R. T. Holman, *Lipids* **1**, 371 (1966).

19. R. H. Shapiro and C. Djerassi, *Tetrahedron* **20**, 1987 (1964).

20. R. H. Shapiro, D. H. Williams, H. Budzikiewicz, C. Djerassi, *J. Amer. Chem. Soc.* **86**, 2837 (1964).

21. R. Beugelmans, R. H. Shapiro, L. J. Durham, D. H. Williams, H. Budzikiewicz, and C. Djerassi, *J. Amer. Chem. Soc.* **86**, 2832 (1964).

22. N. S. Wulfson, V. I. Zaretskii, V. G. Zaikin, G. M. Segal, I. V. Torgov, and T. P. Fredkina, *Tetrahedron Lett.*, 3015 (1964).

SECTION H: SUBSTANCES OF BIOLOGICAL INTEREST—CONT.

Name of class	II	III	IV	V	MW range	Reference
Lipids and steroids—*cont.*						
Monohydroxyprogesterones	8			M	330	23
Pregnanes and pregnenes	30	30		C	298–404	24
Keto steroids	9	20	D	M	302–306	25
	16	16		M	274–386	26
Alcohols of progesterone series	8	8			302–346	27
Progesterone steroids	21			C	314–328	28
Hydroxy steroids	14					28
Proteins and amino acids						
Amino acids	24			C	103–209	29
α-Amino acids						30
Twenty-two free amino acids						31
Determination of amino acid sequences—oligopeptides						32
Phenylthiohydantoins of alanine, tyrosine, and proline						33
Peptides containing monoamino dicarboxylic acid residues						34
Methyl esters of 2,4-dinitrophenyl-amino acids						35
Amino acid sequence determination						36
Esters of *N*-acetylamino acids						37
	5	5		M		38
Peptides	25	25		G	131–221	39
Amino acids and peptides	11	7		M	203–421	40
Esters of amino acids	6			M	159–189	41
N-Adamantol peptides	21			M	350–597	42
Dipeptides	22		D	C		43
N-Acetyl peptides of simple monoaminocarboxylic acids						44
Cyclopeptides						45, 46

SECTION H: SUBSTANCES OF BIOLOGICAL INTEREST—CONT.

Key to references

23. M. F. Grostic and K. L. Rinehart, Jr., *J. Org. Chem.* **33**, 1740 (1968).
24. L. Peterson, *Anal. Chem.* **34**, 1781 (1962).
25. L. Tokes, R. T. La Londe, and C. Djerassi, *J. Org. Chem.* **32**, 1020 (1967).
26. C. Beard, J. M. Wilson, H. Budzikiewicz, and C. Djerassi, *J. Amer. Chem. Soc.* **86**, 269 (1964).
27. V. I. Zaretskii, N. S. Wulfson, V. G. Zaikin, M. Kogan, N. E. Voishvillo, and I. V. Torgov, *Tetrahedron* **22**, 1399 (1966).
28. V. I. Zaretskii, N. S. Wulfson, and V. G. Zaikin, *Tetrahedron* **23**, 3667, 3683 (1967).

29. K. Biemann, J. Seibl, and F. Gapp, *J. Amer. Chem. Soc.* **83**, 3795 (1961).
30. G. A. Junk and H. J. Svec, *J. Amer. Chem. Soc.* **85**, 839 (1963).
31. N. Martin, *NASA Accession* No. N66-14246, Rep. No. NASA-CR-68768 (1965).

32. M. Barber, W. A. Wolstenholme, U. Guinand, G. Michel, B. C. Das, and E. Lederer, *Tetrahedron Lett.*, 1331 (1965).

33. V. M. Stepanov, N. S. Wulfson, V. A. Puchkov, and A. M. Zyakun, *Zh. Obshch. Khim.* **34**, 3771 (1964).

34. M. M. Shemyakin, Yu. A. Ovchinnikov, A. A. Kiryushkin, A. I. Miroshnikov, and B. V. Rozynov, *J. Gen. Chem. USSR* **40**, 408 (1970).

35. T. J. Penders, H. Copier, W. Heerma, G. Dijkstra, and J. F. Arens, *Rec. Trav. Chim. Pays-Bas* **85**, 216 (1966).

36. J. P. Kamerling, W. Heerma, T. J. Penders, and J. F. G. Vliegenthart, *Org. Mass Spectrom.* **1**, 345 (1968).
37. J. H. Beynon, G. R. Lester, R. A. Saunders, and A. E. Williams, *Trans. Faraday Soc.* **57**, 1259 (1961).
38. C. O. Andersson, R. Ryhage, and E. Stenhagen, *Ark. Kemi* **19**, 417 (1962).
39. E. Lederer, *Pure Appl. Chem.* **N**, 489 (1969).
40. R. T. Aplin and J. H. Jones, *J. Chem. Soc. C*, 1770 (1968).
41. K. Biemann, J. Seibl, and F. Gapp, *Biochem. Biophys. Res. Commun.* **1**, 307 (1959).
42. I. Lengyel, R. A. Salomone, and K. Biemann, *Org. Mass Spectrom.* **3**, 789 (1970).
43. H. J. Svec and G. A. Junk, *J. Amer. Chem. Soc.* **86**, 2278 (1964).

44. K. Heyns and H.-P. Grützmacher, *Ann. Chem.* **669**, 189 (1963).
45. Yu. V. Denisov, V. A. Puchkov, N. S. Wulfson, T. E. Agadzhanyan, V. K. Antonov, and M. M. Shemyakin, *Zh. Obshch. Khim.* **38**, 770 (1968).
46. B. V. Rozynov, V. M. Burikov, V. V. Shilin, and A. A. Kiryushkin, *J. Gen. Chem. USSR* **38**, 2002 (1969).

SECTION H: SUBSTANCES OF BIOLOGICAL INTEREST—CONT.

Name of class	II	III	IV	V	MW range	Reference
Nucleoside derivatives						
Analogs of adenosine	7			G	284–474	47, 48
Drugs						
Penicillins						49
Penicillin derivatives						50
Antiepileptic agents	12				141–252	51
Sulfonamide						52
Barbiturates						53
Barbituric acid derivatives						54
Hashish compounds	6				310–318	55
Hashish, XIII						56
Analysis of opium alkaloids						57
Assay of codeine phosphate						58
Morphine alkaloids	18				146–204	59
Fragmentation of morphinan						60
Fragmentation in *cis* and *trans* B:C ring-fused morphine derivatives						61
Alkaloid–peptide compounds						62
Protopine alkaloids						63
Furoquinoline alkaloids	6				199–273	64
Tetrahydroprotoberberine alkaloids	10				323–371	65
Bisbenzyltetrahydroisoquinoline alkaloids	7				594–650	66
Bisbenzylisoquinoline alkaloids	16	9			606–636	67
Structure of amaryllisine						68
Muramine	1				385	69
Tropane alkaloids	10				139–289	70
Thalmine and related alkaloids	8					71
Piperidine alkaloids						72
Lycurenine alkaloids	15	11			299–347	73

Key to references

47. J. J. Dolhun and J. L. Wiebers, *Org. Mass Spectrom.* **3**, 669 (1970).
48. S. J. Shaw, D. M. Desiderio, Jr., K. Tsuboyama, and J. A. McCloskey, *J. Amer. Chem. Soc.* **92**, 2510 (1970).

49. V. N. Bochkarev, N. S. Ovchinnikova, N. S. Wulfson, E. M. Kleiner, and A. S. Khokhlov, *Dokl. Chem.* **172**, 142 (1967).
50. S. Kukolja, R. D. G. Cooper, and R. G. Morin, *Tetrahedron Lett.*, 3381 (1969).
51. R. A. Locock and R. T. Coutts, *Org. Mass Spectrom.* **3**, 735 (1970).
52. G. Spiteller and R. Kaschnitz, *Monatsh. Chem.* **94**, 964 (1963).
53. H. M. Fales, G. W. A. Milne, T. Axenrod, *Anal. Chem.* **42**, 1432 (1970).
54. H.-P. Grützmacher and W. Arnold, *Tetrahedron Lett.*, 1365 (1966).
55. H. Budzikiewicz, R. T. Aplin, D. A. Lightner, C. Djerassi, R. Mechoulam, and Y. Gaoni, *Tetrahedron* **21**, 1881 (1965).
56. U. Claussen and F. Korte, *Tetrahedron Lett.*, 2067 (1967).
57. A. Tatematsu and T. Goto, *Yakugaku Zasshi* **85**, 152 (1965).
58. A. Tatematsu and T. Goto, *Yakugaku Zasshi* **85**, 786 (1965).
59. H. Audier, M. Fetison, D. Ginsburg, A. Mandelbaum, and T. Rull, *Tetrahedron Lett.*, 13 (1965).
60. H. Nakata, Y. Hirata, A. Tatematsu, H. Tada, and Y. Sawa, *Tetrahedron Lett.*, 829 (1965).

61. A. Mandelbaum and D. Ginsburg, *Tetrahedron Lett.*, 2479 (1965).
62. K. T. Poroshin, Y. V. Denisov, V. K. Burichenko, and S. B. Davidyants, *Dokl. Akad. Nauk Tadzh USSR* **9**, 23 (1966).
63. L. Dolejš, V. Hanuš, J. Slavik, *Collect. Czech. Chem. Commun.* **29**, 2479 (1964).
64. D. M. Clugston and D. B. MacLean, *Can. J. Chem.* **43**, 2516 (1965).

65. C.-Y. Chen and D. B. MacLean, *Can. J. Chem.* **46**, 2501 (1968).

66. D. C. DeJongh, S. R. Shrader, and M. P. Cava, *J. Amer. Chem. Soc.* **88**, 1052 (1966).
67. M. Tomita, T. Kikuchi, K. Fujitani, A. Kato, H. Furukawa, Y. Aoyagi, M. Kitano, and T. Ibuka, *Tetrahedron Lett.*, 857 (1966).
68. A. L. Burlingame, H. M. Fales, R. J. Highet, *J. Amer. Chem. Soc.* **86**, 4976 (1964).
69. A. D. Cross, L. Dolejš, V. Hanuš, M. Maturova, and F. Santavy, *Collect. Czech. Chem. Commun.* **30**, 1335 (1965).
70. E. C. Blossey, H. Budzikiewicz, M. Ohashi, G. Fodor, and C. Djerassi, *Tetrahedron* **20**, 585 (1964).
71. J. Baldas, Q. N. Porter, I. R. C. Bick, G. K. Douglas, M. R. Falco, J. X. de Vries, and S. Y. Yunusov, *Tetrahedron* **24**, 6315 (1968).
72. M. Spiteller-Friedmann and G. Spiteller, *Monatsh. Chem.* **96**, 104 (1965).
73. H. K. Schnoes, D. H. Smith, A. L. Burlingame, P. W. Jeffs, and W. Dopke, *Tetrahedron* **24**, 2825 (1968).

SECTION H: SUBSTANCES OF BIOLOGICAL INTEREST—CONT.

Name of class	II	III	IV	V	MW range	Reference
Drugs —*cont.*						
Ajmaline and related alkaloids	1				326	74
Pesticides						
Key fragments in mass spectra of pesticides						75
Carbamate pesticides	14				179–222	76
Organophosphorus pesticides	23				215–456	77
Chlorinated pesticidal compounds	9					78
Chlorinated aromatic pesticide compounds	12				266–368	79
Pesticide-residue analysis by GC–MS						80
Porphyrins and related compounds				G		81
Porphyrins and chlorins	27					82
Porphyrins and chlorins, 7,8-dihydro-	13				422–492	83
ms porphyrins	3				614–750	84
Oxidation products of porphyrins	6	4		M		85
Comparison of porphyrins in shale oil, oil shale, and petroleum				G	366–522	86
Miscellaneous compounds						
Gibberellins	8	3			314–362	87
Kynurenines (wing pigment of butterflies)	6	6			191–235	88
Guanidines (vitamin B_6)	6					89
Naphthoquinones (vitamin K_{20})						90
Ubiquinones and ubiquinols						91
Ubiquinones	2		D	M	794–796	92
Pyridoxol and *O*-isopropylidene derivatives			D	M		93
Natural products						
Constituents of hop oil						94
Volatiles from oranges						95

SECTION H: SUBSTANCES OF BIOLOGICAL INTEREST—CONT.

Key to references

74. K. Biemann, P. Bommer, A. L. Burlingame, and W. J. McMurray, *J. Amer. Chem. Soc.* **86**, 4624 (1964).

75. J. Joerg, R. Houriet, and G. Spiteller, *Monatsh. Chem.* **97**, 1064 (1966).
76. J. N. Damico and W. R. Benson, *J. Ass. Offic. Agr. Chem.* **48**, 344 (1965).
77. J. N. Damico, *J. Ass. Offic. Anal. Chem.* **49**, 1027 (1966).
78. J. N. Damico, R. P. Barron, and J. M. Ruth, *Org. Mass Spectrom.* **1**, 331 (1968).

79. J. A. Sphon and J. N. Damico, *Org. Mass Spectrom.* **3**, 51 (1970).

80. F. J. Biros and A. C. Walker, *J. Agr. Food. Chem.* **18**, 425 (1970).

81. G. R. Lester, *in* "Mass Spectrometry" (R. I. Reed, ed.), pp. 153–181, Academic Press, New York, 1965.
82. D. R. Hoffman, *J. Org. Chem.* **30**, 3512 (1965).

83. A. H. Jackson, G. W. Kenner, K. M. Smith, R. T. Aplin, H. Budzikiewicz, and C. Djerassi, *Tetrahedron* **21**, 2913 (1965).
84. A. D. Adler, J. H. Green, and M. Mautner, *Org. Mass Spectrom.* **3**, 955 (1970).
85. D. B. Boylan, *Org. Mass Spectrom.* **3**, 339 (1970).

86. J. R. Morandi and H. B. Jensen, *J. Chem. Eng. Data* **11**, 81 (1966).

87. N. S. Wulfson, V. I. Zaretskii, I. B. Papernaja, E. P. Serebryakov, and V. F. Kucherov, *Tetrahedron Lett.*, 4209 (1965).

88. K. S. Brown, Jr. and D. Becher, *Tetrahedron* **23**, 1721 (1967).
89. J. H. Beynon, J. A. Hopkinson, and A. E. Williams, *Org. Mass Spectrom.* **1**, 169 (1968).
90. S. J. Di Mari, J. H. Supple, H. Rapoport, *J. Amer. Chem. Soc.* **88**, 1226 (1966).
91. R. F. Muraca, J. S. Whittick, G. D. Daves, Jr., P. Friis, and K. Folkers, *J. Amer. Chem. Soc.* **89**, 1505 (1967).
92. T. M. Farley, G. D. Daves, Jr., and K. Folkers, *J. Org. Chem.* **33**, 905 (1968).
93. D. C. DeJongh, S. C. Perricone, W. Korytnyk, *J. Amer. Chem. Soc.* **88**, 1233 (1966).

94. R. G. Buttery, W. H. McFadden, R. Teranishi, M. P. Kealy, and T. R. Mon, *Nature* **200**, 435 (1963).
95. T. H. Schultz, R. Teranishi, W. H. McFadden, P. W. Kilpatrick, and J. Corse, *J. Food Sci.* **29**, 790 (1964).

SECTION H: SUBSTANCES OF BIOLOGICAL INTEREST—CONT.

Name of class	II	III	IV	V	MW range	Reference
Natural products						
Fragmentation of some lignans						96
Heterocycles	6	6			386–404	97
Lignans of the 1-phenyl-1,2,3,4-tetrahydronaphthalene series	7				340–414	98
Ellagitannins (4-gallic acid derivatives)	4				170–302	99
Gossypol hexamethyl ether (cotton seed pigment)	1				950	100
Analysis of bituminous coal by ozonolysis, GC–MS						101
Regulators of cell divisions in plant tissues (zeatin)						102
Aromatic carboxylic acids from the Colorado Green River formation						103
Analysis of tobacco ash						104

SECTION H: SUBSTANCES OF BIOLOGICAL INTEREST—CONT.

Key to references

96. A. M. Duffield, *Heterocycl. Chem.* **4**, 16 (1967).
97. A. Pelter, *J. Chem. Soc. C*, 1376 (1967).
98. A. Pelter, *J. Chem. Soc. C*, 74 (1968).
99. P. F. Nelson and Q. N. Porter, *Holzforschung* **21**, 104 (1967).
100. M. G. Abou-Donia, J. W. Dieckert, C. M. Lyman, *J. Agr. Food Chem.* **18**, 534 (1970).
101. Sr. M. C. Bitz and B. Nagy, *Anal. Chem.* **39**, 1310 (1967).
102. D. S. Letham, J. S. Shannon, I. R. C. McDonald, *Tetrahedron* **23**, 479 (1967).
103. P. Haug, H. K. Schnoes, and A. L. Burlingame, *Geochim. Cosmochim. Acta* **32**, 358 (1968).
104. R. M. Jones, W. F. Kuhn, and C. Varsel, *Anal. Chem.* **40**, 10 (1968).

APPENDIX **II**

INTERPRETATION MAPS

The discussion covering the use of these maps is presented in Chapter 7.

-11	-10	-9	-8	-7	-6	-5	-4	-3	-2	-1	+0	+1	+2
-	-	-	-	-	-	-	10	11	12	13	14	15	16
17	18	19	20	21	22	23	24	25	26	27	28	29	30
31	32	33	34	35	36	37	38	39	40	41	42	43	44
45	46	47	48	49	50	51	52	53	54	55	56	57	58
59	60	61	62	63	64	65	66	67	68	69	70	71	72
73	74	75	76	77	78	79	80	81	82	83	84	85	86
87	88	89	90	91	92	93	94	95	96	97	98	99	100
101	102	103	104	105	106	107	108	109	110	111	112	113	114
115	116	117	118	119	120	121	122	123	124	125	126	127	128
129	130	131	132	133	134	135	136	137	138	139	140	141	142
143	144	145	146	147	148	149	150	151	152	153	154	155	156
157	158	159	160	161	162	163	164	165	166	167	168	169	170
171	172	173	174	175	176	177	178	179	180	181	182	183	184
185	186	187	188	189	190	191	192	193	194	195	196	197	198
199	200	201	202	203	204	205	206	207	208	209	210	211	212
213	214	215	216	217	218	219	220	221	222	223	224	225	226
227	228	229	230	231	232	233	234	235	236	237	238	239	240
241	242	243	244	245	246	247	248	249	250	251	252	253	254
255	256	257	258	259	260	261	262	263	264	265	266	267	268
269	270	271	272	273	274	275	276	277	278	279	280	281	282
283	284	285	286	287	288	289	290	291	292	293	294	295	296
297	298	299	300	301	302	303	304	305	306	307	308	309	310
311	312	313	314	315	316	317	318	319	320	321	322	323	324
325	326	327	328	329	330	331	332	333	334	335	336	337	338
339	340	341	342	343	344	345	346	347	348	349	350	351	352
353	354	355	356	357	358	359	360	361	362	363	364	365	366
↓	↓	↓	↓	↓	↓	↓	↓	↓	↓	↓	↓	↓	↓
Map Nos.													
1, 6, 11	1, 6	2, 7	2, 7, 11	2, 7, 12	2, 7	3, 8, 12	3, 8	3	3, 9	4, 9, 15	4, 9	5, 10, 15	5, 10, 15
16, 19, 26	11, 19, 27	11, 16, 28	16, 20, 29	16, 20, 30	12, 21, 31	17, 21, 32	13, 22, 33	8, 22, 34	14, 23, 35	17, 23, 36	15, 24, 37	18, 24, 38	18, 25, 39

MAP A

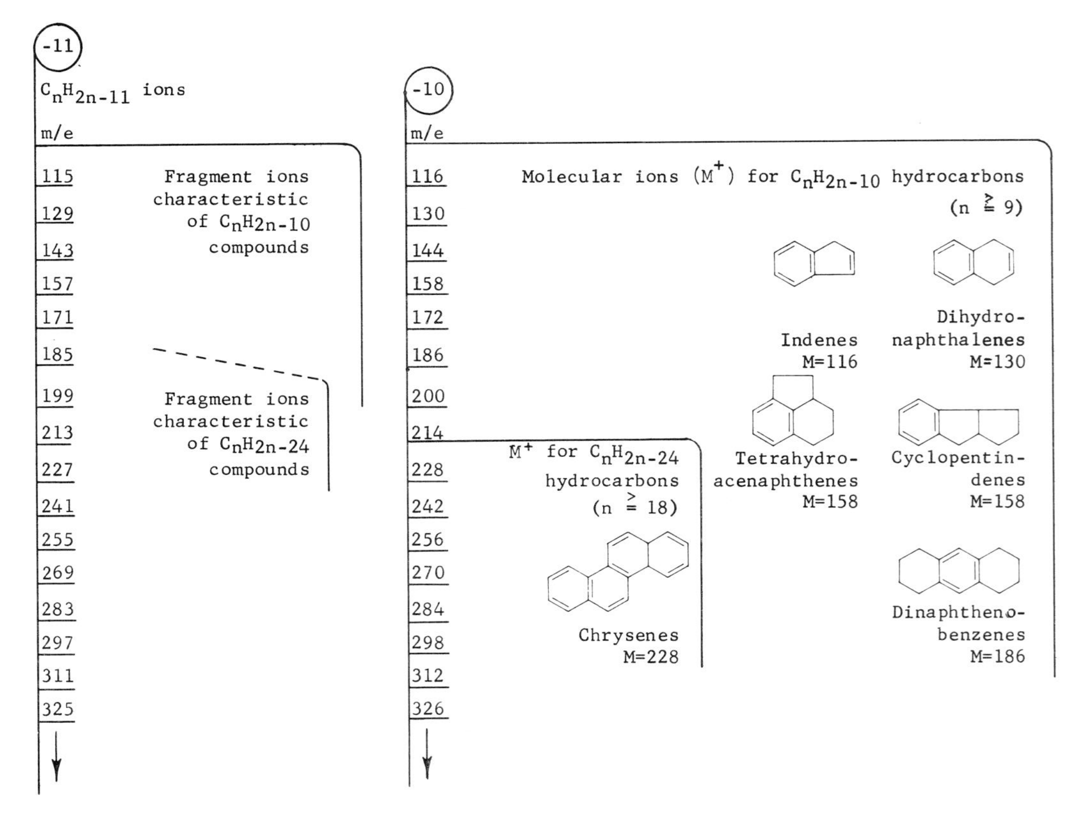
-11
CnH2n-11 ions
m/e
115
129
143
157
171
185
199
213
227
241
255
269
283
297
311
325
Fragment ions characteristic of CnH2n-10 compounds
Fragment ions characteristic of CnH2n-24 compounds
-10
m/e
116
130
144
158
172
186
200
214
228
242
256
270
284
298
312
326
Molecular ions (M+) for CnH2n-10 hydrocarbons (n ≥ 9)
Indenes M=116
Dihydro-naphthalenes M=130
Tetrahydro-acenaphthenes M=158
Cyclopentin-denes M=158
Dinaphtheno-benzenes M=186
M+ for CnH2n-24 hydrocarbons (n ≥ 18)
Chrysenes M=228

MAP 1

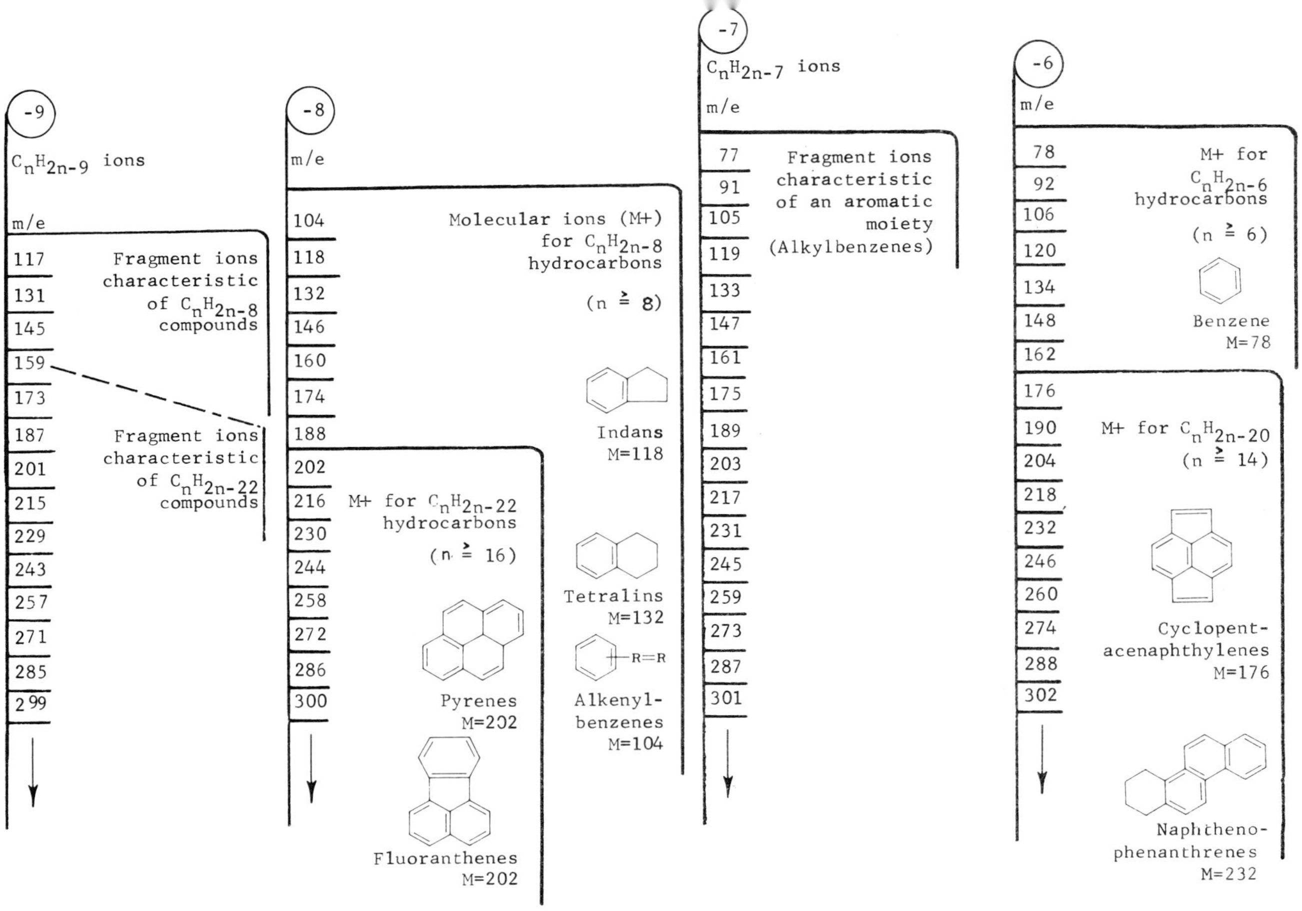

-9: C_nH_{2n-9} ions

m/e	
117	Fragment ions characteristic of C_nH_{2n-8} compounds
131	
145	
159	
173	
187	Fragment ions characteristic of C_nH_{2n-22} compounds
201	
215	
229	
243	
257	
271	
285	
299	

-8

m/e	
104	Molecular ions (M+) for C_nH_{2n-8} hydrocarbons ($n \geq 8$): Indans M=118; Tetralins M=132; Alkenylbenzenes M=104 (R=R)
118	
132	
146	
160	
174	
188	
202	M+ for C_nH_{2n-22} hydrocarbons ($n \geq 16$): Pyrenes M=202; Fluoranthenes M=202
216	
230	
244	
258	
272	
286	
300	

-7: C_nH_{2n-7} ions

m/e	
77	Fragment ions characteristic of an aromatic moiety (Alkylbenzenes)
91	
105	
119	
133	
147	
161	
175	
189	
203	
217	
231	
245	
259	
273	
287	
301	

-6

m/e	
78	M+ for C_nH_{2n-6} hydrocarbons ($n \geq 6$): Benzene M=78
92	
106	
120	
134	
148	
162	
176	M+ for C_nH_{2n-20} ($n \geq 14$): Cyclopentacenaphthylenes M=176; Naphthenophenanthrenes M=232
190	
204	
218	
232	
246	
260	
274	
288	
302	

MAP 2

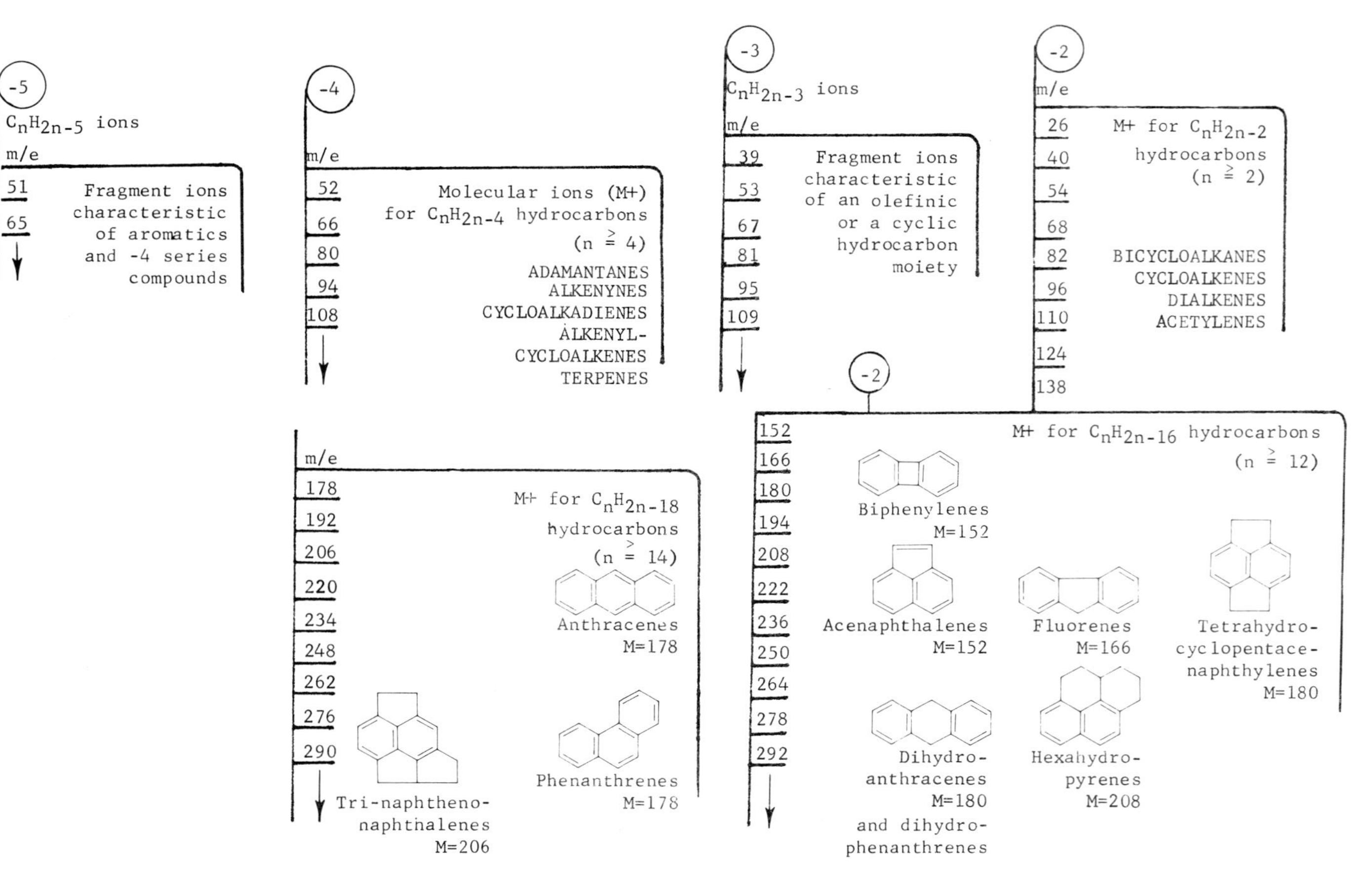
-5
C_nH_{2n-5} ions
m/e
51
65
Fragment ions characteristic of aromatics and -4 series compounds
-4
m/e
52
66
80
94
108
Molecular ions (M+) for C_nH_{2n-4} hydrocarbons ($n \geq 4$)
ADAMANTANES
ALKENYNES
CYCLOALKADIENES
ALKENYL-CYCLOALKENES
TERPENES
-3
C_nH_{2n-3} ions
m/e
39
53
67
81
95
109
Fragment ions characteristic of an olefinic or a cyclic hydrocarbon moiety
-2
m/e
26
40
54
68
82
96
110
124
138
M+ for C_nH_{2n-2} hydrocarbons ($n \geq 2$)
BICYCLOALKANES
CYCLOALKENES
DIALKENES
ACETYLENES
-2
152
166
180
194
208
222
236
250
264
278
292
M+ for C_nH_{2n-16} hydrocarbons ($n \geq 12$)
Biphenylenes M=152
Acenaphthalenes M=152
Fluorenes M=166
Tetrahydro-cyclopentace-naphthylenes M=180
Dihydro-anthracenes M=180 and dihydro-phenanthrenes
Hexahydro-pyrenes M=208
m/e
178
192
206
220
234
248
262
276
290
M+ for C_nH_{2n-18} hydrocarbons ($n \geq 14$)
Anthracenes M=178
Phenanthrenes M=178
Tri-naphtheno-naphthalenes M=206

MAP 3

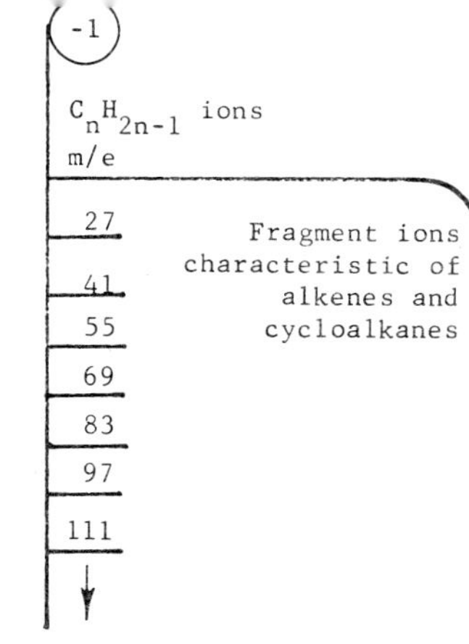
-1
C_nH_{2n-1} ions
m/e
27
41
55
69
83
97
111
Fragment ions characteristic of alkenes and cycloalkanes

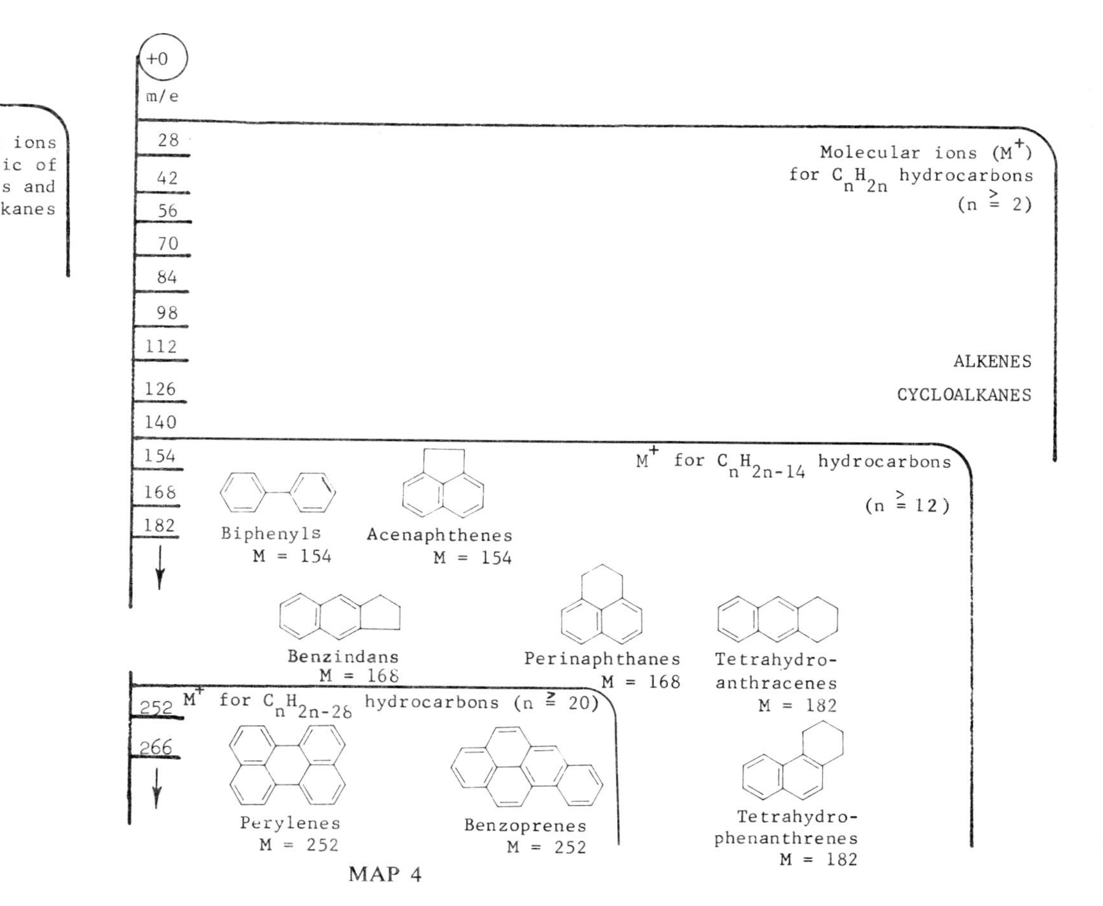
+0
m/e
28
42
56
70
84
98
112
126
140
154
168
182
Molecular ions (M^+) for C_nH_{2n} hydrocarbons ($n \geq 2$)
ALKENES
CYCLOALKANES
M^+ for C_nH_{2n-14} hydrocarbons ($n \geq 12$)
Biphenyls M = 154
Acenaphthenes M = 154
Benzindans M = 168
Perinaphthanes M = 168
Tetrahydro-anthracenes M = 182
Tetrahydro-phenanthrenes M = 182
M^+ for C_nH_{2n-28} hydrocarbons ($n \geq 20$)
252
266
Perylenes M = 252
Benzoprenes M = 252

MAP 4

+1

C_nH_{2n+1} ions

m/e	
15	C_nH_{2n+1} fragment ions characteristic of a saturated hydrocarbon moiety
29	
43	
57	
71	
85	
99	
113	
127	
141	C_nH_{2n-13} fragment ions characteristic of naphthalenes
155	
169	
183	
↓	

+2

m/e	
16	Molecular ions (M+) for C_nH_{2n+2} hydrocarbons ($n \geq 1$) ALKANES
30	
44	
58	
72	
86	
100	
114	
128	M+ for C_nH_{2n-12} hydrocarbons ($n \geq 10$) — Naphthalenes M=128; Trinaphthene-benzenes M=240
142	
156	
170	
184	
↓	
254	M+ for C_nH_{2n-26} hydrocarbons ($n \geq 20$) DINAPHTHYL-ALKANES AND CHOLANTHRENES
268	
↓	

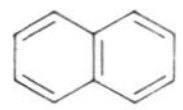

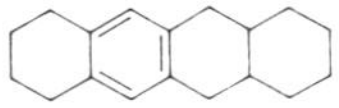

MAP 5

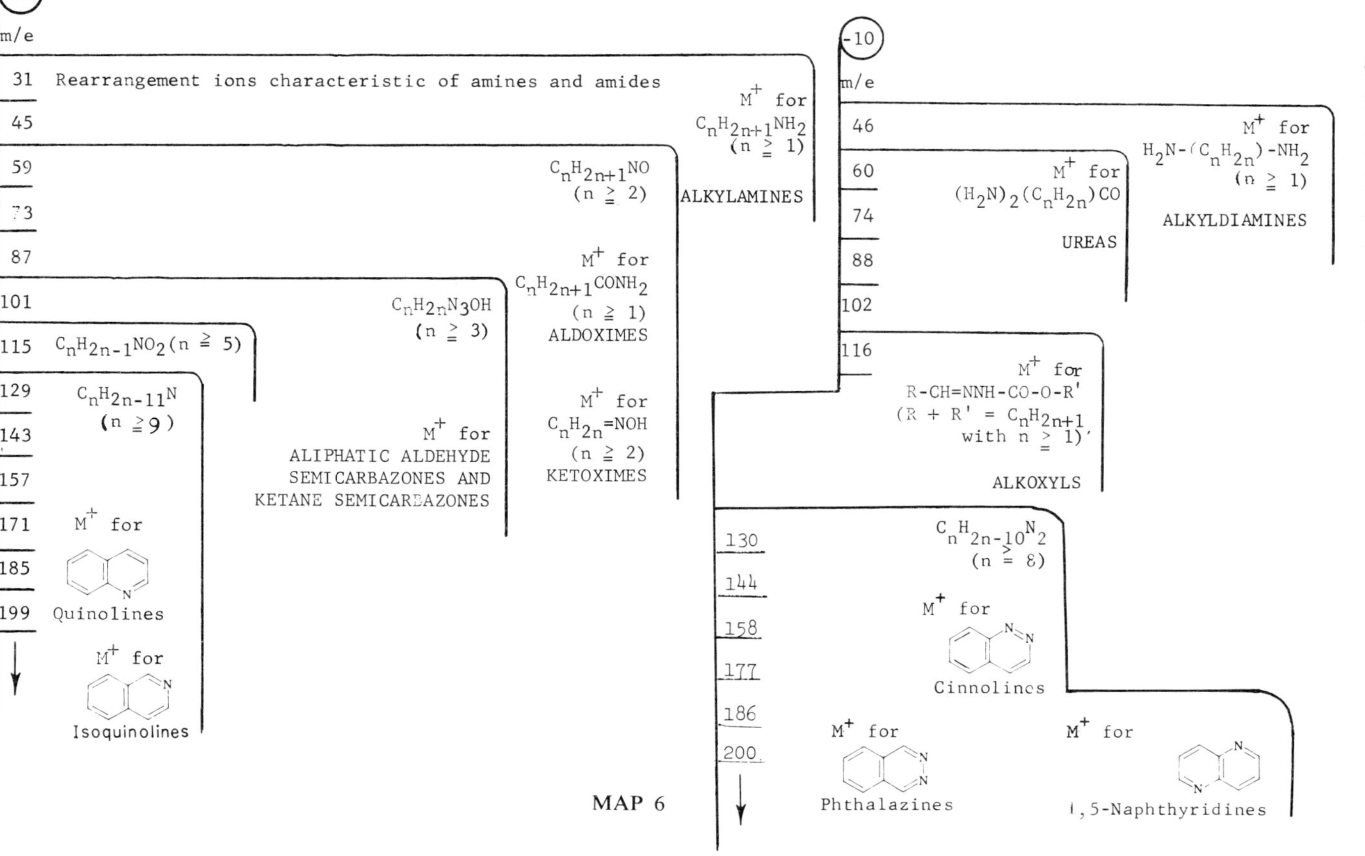
-11
m/e
31
45
59
73
87
101
115
129
143
157
171
185
199
Rearrangement ions characteristic of amines and amides
M$^+$ for $C_nH_{2n+1}NH_2$ ($n \geq 1$)
ALKYLAMINES
$C_nH_{2n+1}NO$ ($n \geq 2$)
M$^+$ for $C_nH_{2n+1}CONH_2$ ($n \geq 1$)
ALDOXIMES
M$^+$ for $C_nH_{2n}=NOH$ ($n \geq 2$)
KETOXIMES
$C_nH_{2n}N_3OH$ ($n \geq 3$)
M$^+$ for
ALIPHATIC ALDEHYDE
SEMICARBAZONES AND
KETANE SEMICARBAZONES
$C_nH_{2n-1}NO_2$ ($n \geq 5$)
$C_nH_{2n-11}N$ ($n \geq 9$)
M$^+$ for
Quinolines
M$^+$ for
Isoquinolines
-10
m/e
46
60
74
88
102
116
130
144
158
177
186
200
M$^+$ for $H_2N-(C_nH_{2n})-NH_2$ ($n \geq 1$)
ALKYLDIAMINES
M$^+$ for $(H_2N)_2(C_nH_{2n})CO$
UREAS
M$^+$ for R-CH=NNH-CO-O-R'
(R + R' = C_nH_{2n+1} with $n \geq 1$)
ALKOXYLS
$C_nH_{2n-10}N_2$ ($n \geq 8$)
M$^+$ for
Cinnolines
M$^+$ for
Phthalazines
M$^+$ for
1,5-Naphthyridines

MAP 6

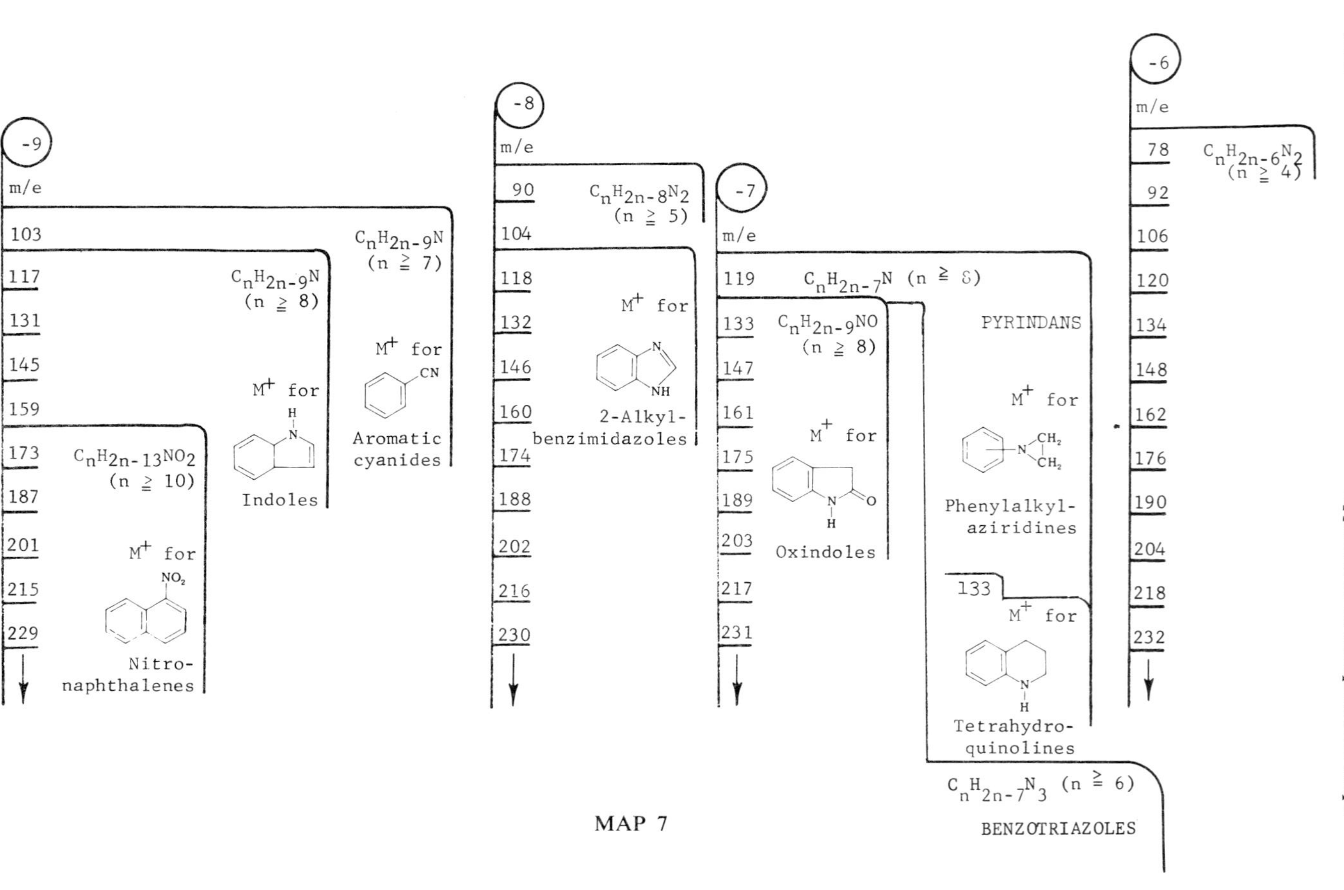

-9
m/e
103 117 131 145 159 173 187 201 215 229
$C_nH_{2n-9}N$ (n ≥ 7)
M^+ for
Aromatic cyanides
$C_nH_{2n-9}N$ (n ≥ 8)
M^+ for
Indoles
$C_nH_{2n-13}NO_2$ (n ≥ 10)
M^+ for
Nitro-naphthalenes
-8
m/e
90 104 118 132 146 160 174 188 202 216 230
$C_nH_{2n-8}N_2$ (n ≥ 5)
M^+ for
2-Alkyl-benzimidazoles
-7
m/e
119 133 147 161 175 189 203 217 231
$C_nH_{2n-7}N$ (n ≥ 8)
$C_nH_{2n-9}NO$ (n ≥ 8)
M^+ for
Oxindoles
PYRINDANS
M^+ for
Phenylalkyl-aziridines
133
M^+ for
Tetrahydro-quinolines
$C_nH_{2n-7}N_3$ (n ≥ 6)
BENZOTRIAZOLES
-6
m/e
78 92 106 120 134 148 162 176 190 204 218 232
$C_nH_{2n-6}N_2$ (n ≥ 4)

MAP 7

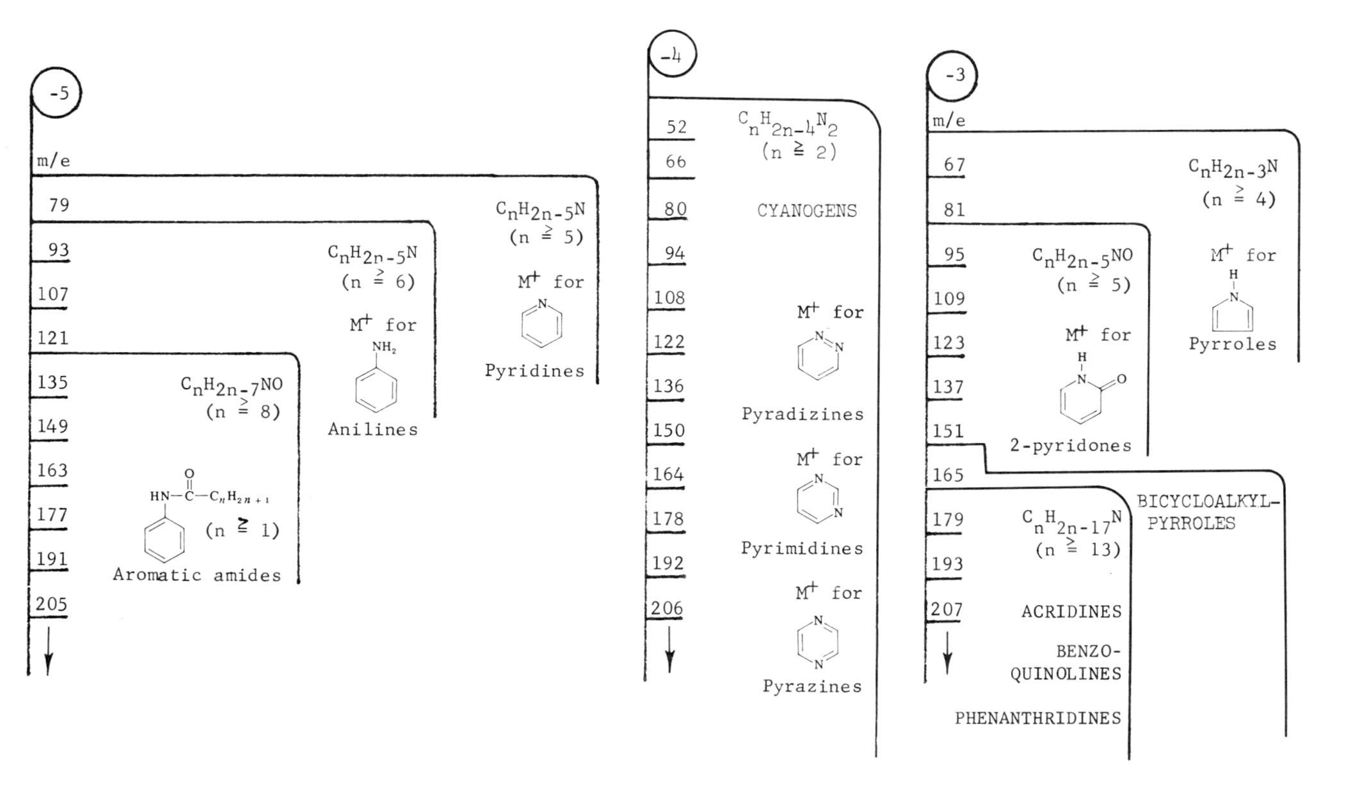

-5
m/e
79
93
107
121
135
149
163
177
191
205
$C_nH_{2n-5}N$
(n ≥ 5)
M+ for
Pyridines
$C_nH_{2n-5}N$
(n ≥ 6)
M+ for
Anilines
$C_nH_{2n-7}NO$
(n ≥ 8)
HN-C-C_nH_{2n+1}
(n ≥ 1)
Aromatic amides
-4
52
66
80
94
108
122
136
150
164
178
192
206
$C_nH_{2n-4}N_2$
(n ≥ 2)
CYANOGENS
M+ for
Pyradizines
M+ for
Pyrimidines
M+ for
Pyrazines
-3
m/e
67
81
95
109
123
137
151
165
179
193
207
$C_nH_{2n-3}N$
(n ≥ 4)
M+ for
Pyrroles
$C_nH_{2n-5}NO$
(n ≥ 5)
M+ for
2-pyridones
BICYCLOALKYL-
PYRROLES
$C_nH_{2n-17}N$
(n ≥ 13)
ACRIDINES
BENZO-
QUINOLINES
PHENANTHRIDINES

MAP 8

-2 m/e	
40	$C_nH_{2n}CN$ Fragment ions characteristic of nitriles and cyanides
54	
68	$C_nH_{2n-2}N_2$ $(n \geq 3)$ M+ for Pyrazoles
82	
96	
110	
124	
138	
152	
166	
180	$C_nH_{2n-16}N_2$ $(n \geq 12)$ M+ for 4,7-Phenanthrolines; M+ for Benzo(c)cinnolines
194	
208	
222	
236	

-1 m/e	
41	$C_nH_{2n-1}N$ $(n \geq 2)$ M+ for $C_nH_{2n+1}CN$ compounds $(n \geq 1)$ ALKYL CYANIDES NITRILES
55	
69	$C_nH_{2n-3}NO$ $(n \geq 3)$ M+ for Oxazoles
83	
97	
111	
125	
139	
153	
167	$C_nH_{2n-15}N$ $(n \geq 12)$ M+ for Carbazoles
181	
195	
209	
223	
237	

+0 m/e	
42	Fragment ions characteristic of $C_nH_{2n+1}N$ and $C_nH_{2n-1}NO$ compounds
56	
70	
84	
98	
112	
126	
140	
154	
168	
182	
196	
210	
224	$C_nH_{2n-8}N_4O_4$ $(n \geq 8)$ M+ for $C_nH_{2n+1}-CH=N-NH-C_6H_3(NO_2)_2$ Dinitrophenylhydrazones
238	

MAP 9

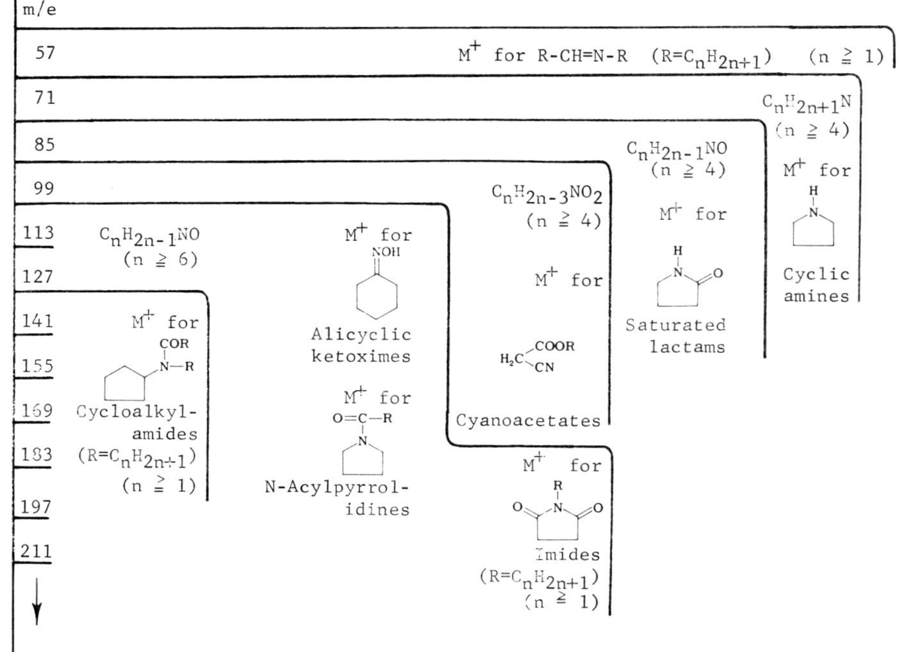
+1
m/e
57 71 85 99 113 127 141 155 169 183 197 211
M^+ for R-CH=N-R (R=C_nH_{2n+1}) (n ≧ 1)
$C_nH_{2n+1}N$ (n ≧ 4)
M^+ for
Cyclic amines
$C_nH_{2n-1}NO$ (n ≧ 4)
M^+ for
Saturated lactams
$C_nH_{2n-3}NO_2$ (n ≧ 4)
M^+ for
Cyanoacetates
M^+ for
Imides (R=C_nH_{2n+1}) (n ≧ 1)
M^+ for
Alicyclic ketoximes
M^+ for
N-Acylpyrrolidines
$C_nH_{2n-1}NO$ (n ≧ 6)
M^+ for
Cycloalkylamides (R=C_nH_{2n+1}) (n ≧ 1)

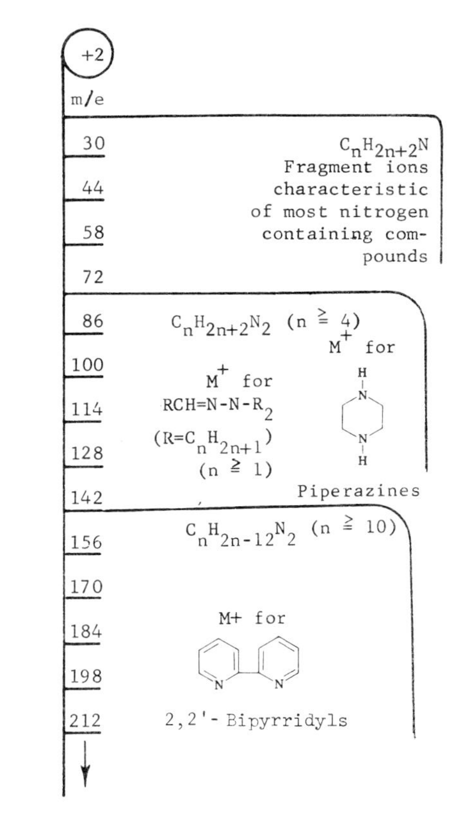
+2
m/e
30 44 58 72 86 100 114 128 142 156 170 184 198 212
$C_nH_{2n+2}N$
Fragment ions characteristic of most nitrogen containing compounds
$C_nH_{2n+2}N_2$ (n ≧ 4)
M^+ for RCH=N-N-R_2 (R=C_nH_{2n+1}) (n ≧ 1)
M^+ for
Piperazines
$C_nH_{2n-12}N_2$ (n ≧ 10)
M+ for
2,2'- Bipyrridyls

MAP 10

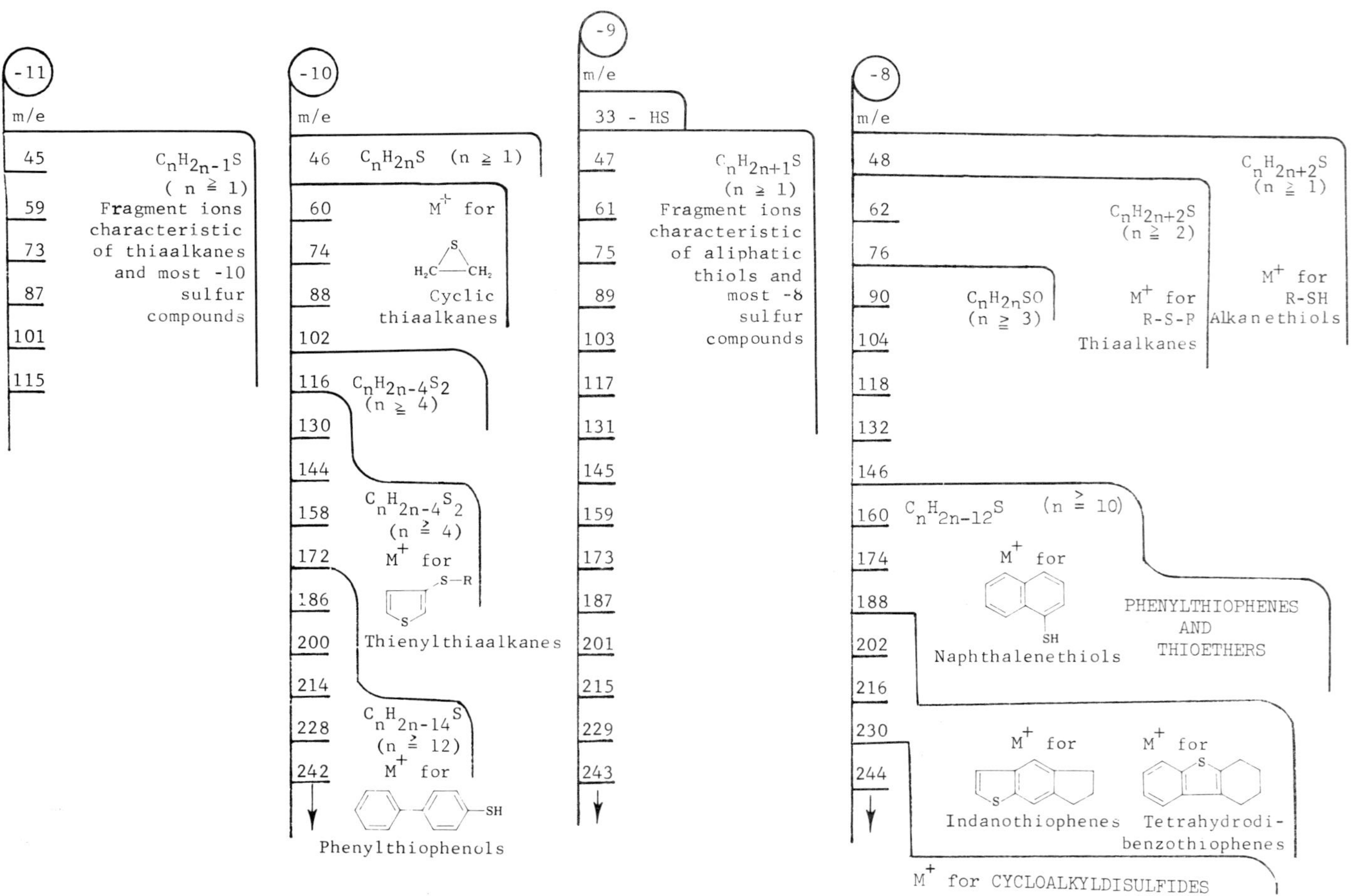

-11
m/e
45
59
73
87
101
115
$C_nH_{2n-1}S$ (n ≧ 1)
Fragment ions characteristic of thiaalkanes and most -10 sulfur compounds
-10
m/e
46 $C_nH_{2n}S$ (n ≧ 1)
60
74
88
102
116
130
144
158
172
186
200
214
228
242
M^{+} for
S
H_2C
CH_2
Cyclic thiaalkanes
$C_nH_{2n-4}S_2$ (n ≧ 4)
$C_nH_{2n-4}S_2$ (n ≧ 4)
M^{+} for
S—R
S
Thienylthiaalkanes
$C_nH_{2n-14}S$ (n ≧ 12)
M^{+} for
SH
Phenylthiophenols
-9
m/e
33 - HS
47
61
75
89
103
117
131
145
159
173
187
201
215
229
243
$C_nH_{2n+1}S$ (n ≧ 1)
Fragment ions characteristic of aliphatic thiols and most -8 sulfur compounds
-8
m/e
48
62
76
90
104
118
132
146
160
174
188
202
216
230
244
$C_nH_{2n+2}S$ (n ≧ 1)
M^{+} for R-SH
Alkanethiols
$C_nH_{2n+2}S$ (n ≧ 2)
M^{+} for R-S-P
Thiaalkanes
$C_nH_{2n}SO$ (n ≧ 3)
$C_nH_{2n-12}S$ (n ≧ 10)
M^{+} for
SH
Naphthalenethiols
PHENYLTHIOPHENES AND THIOETHERS
M^{+} for
S
Indanothiophenes
M^{+} for
S
Tetrahydrodibenzothiophenes
M^{+} for CYCLOALKYLDISULFIDES

MAP 11

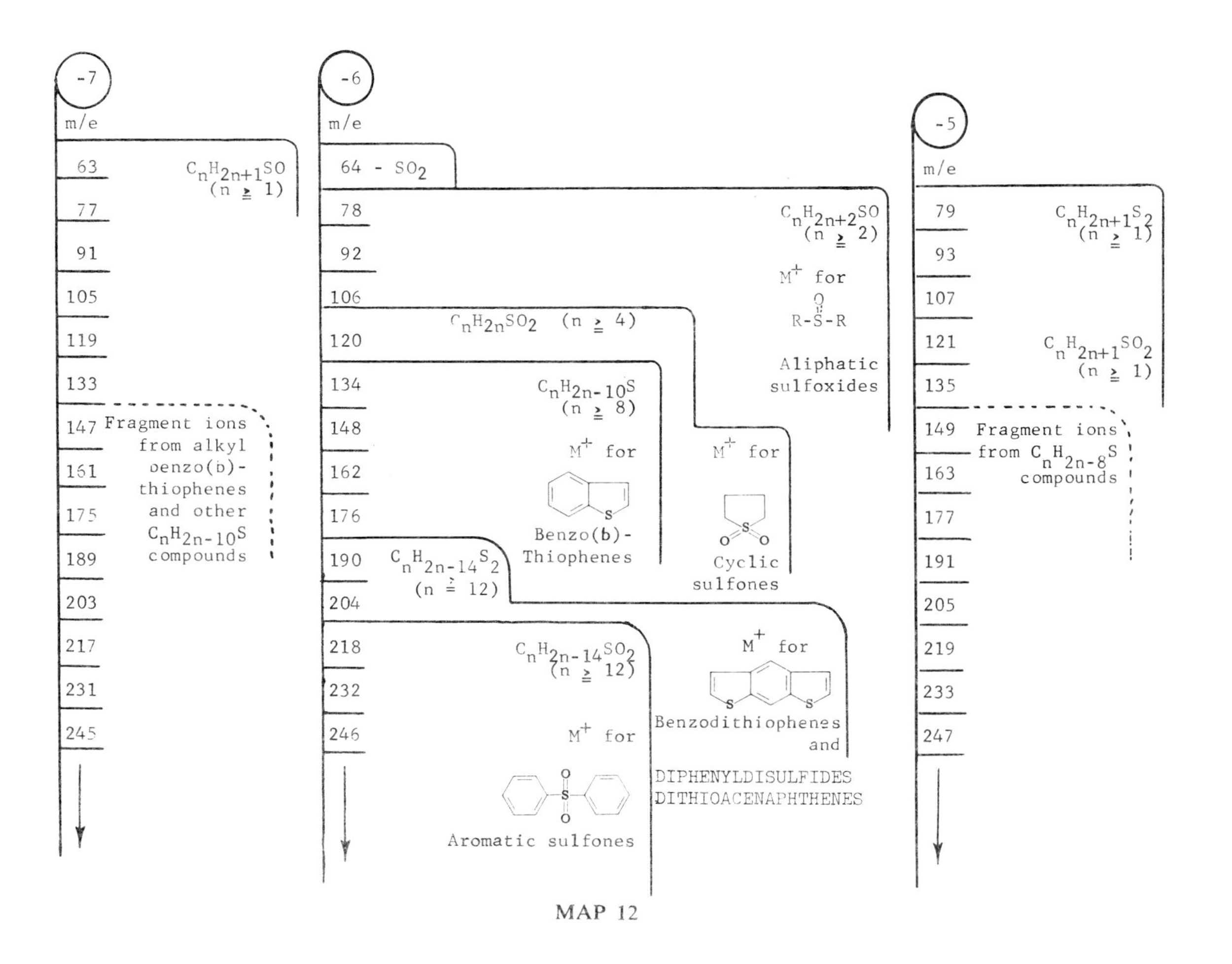
-7
m/e
63
77
91
105
119
133
147
161
175
189
203
217
231
245
$C_nH_{2n+1}SO$ (n ≥ 1)
Fragment ions from alkyl benzo(b)-thiophenes and other $C_nH_{2n-10}S$ compounds
-6
m/e
64 - SO_2
78
92
106
120
134
148
162
176
190
204
218
232
246
$C_nH_{2n+2}SO$ (n ≥ 2)
M+ for R-S(O)-R
Aliphatic sulfoxides
$C_nH_{2n}SO_2$ (n ≥ 4)
M+ for
Cyclic sulfones
$C_nH_{2n-10}S$ (n ≥ 8)
M+ for
Benzo(b)-Thiophenes
$C_nH_{2n-14}S_2$ (n ≥ 12)
M+ for
Benzodithiophenes and
DIPHENYLDISULFIDES
DITHIOACENAPHTHENES
$C_nH_{2n-14}SO_2$ (n ≥ 12)
M+ for
Aromatic sulfones
-5
m/e
79
93
107
121
135
149
163
177
191
205
219
233
247
$C_nH_{2n+1}S_2$ (n ≥ 1)
$C_nH_{2n+1}SO_2$ (n ≥ 1)
Fragment ions from $C_nH_{2n-8}S$ compounds

MAP 12

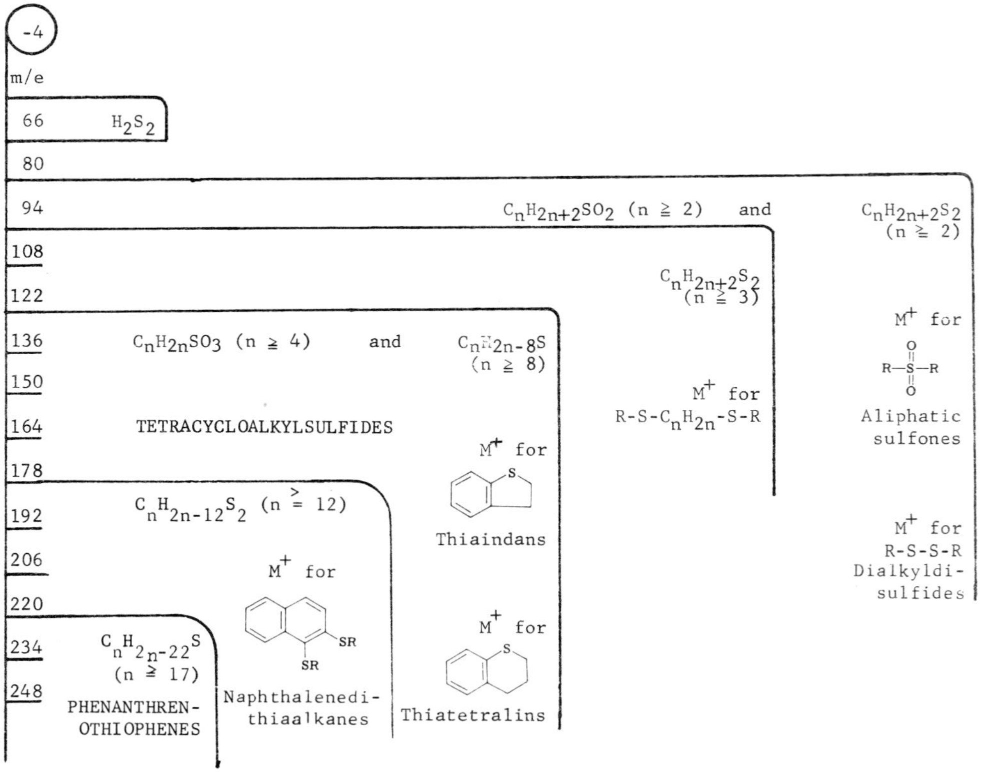

(-4)

m/e: 66, 80, 94, 108, 122, 136, 150, 164, 178, 192, 206, 220, 234, 248

66: H_2S_2

94: $C_nH_{2n+2}SO_2$ ($n \geq 2$) and $C_nH_{2n+2}S_2$ ($n \geq 2$)

M^+ for R–S(=O)(=O)–R

Aliphatic sulfones

M^+ for R-S-S-R

Dialkyldi-sulfides

108: $C_nH_{2n+2}S_2$ ($n \geq 3$)

M^+ for R-S-C_nH_{2n}-S-R

136: $C_nH_{2n}SO_3$ ($n \geq 4$) and $C_nH_{2n-8}S$ ($n \geq 8$)

TETRACYCLOALKYLSULFIDES

M^+ for

Thiaindans

M^+ for

Thiatetralins

192: $C_nH_{2n-12}S_2$ ($n \geq 12$)

M^+ for

SR

SR

Naphthalenedi-thiaalkanes

234: $C_nH_{2n-22}S$ ($n \geq 17$)

PHENANTHREN-OTHIOPHENES

MAP 13

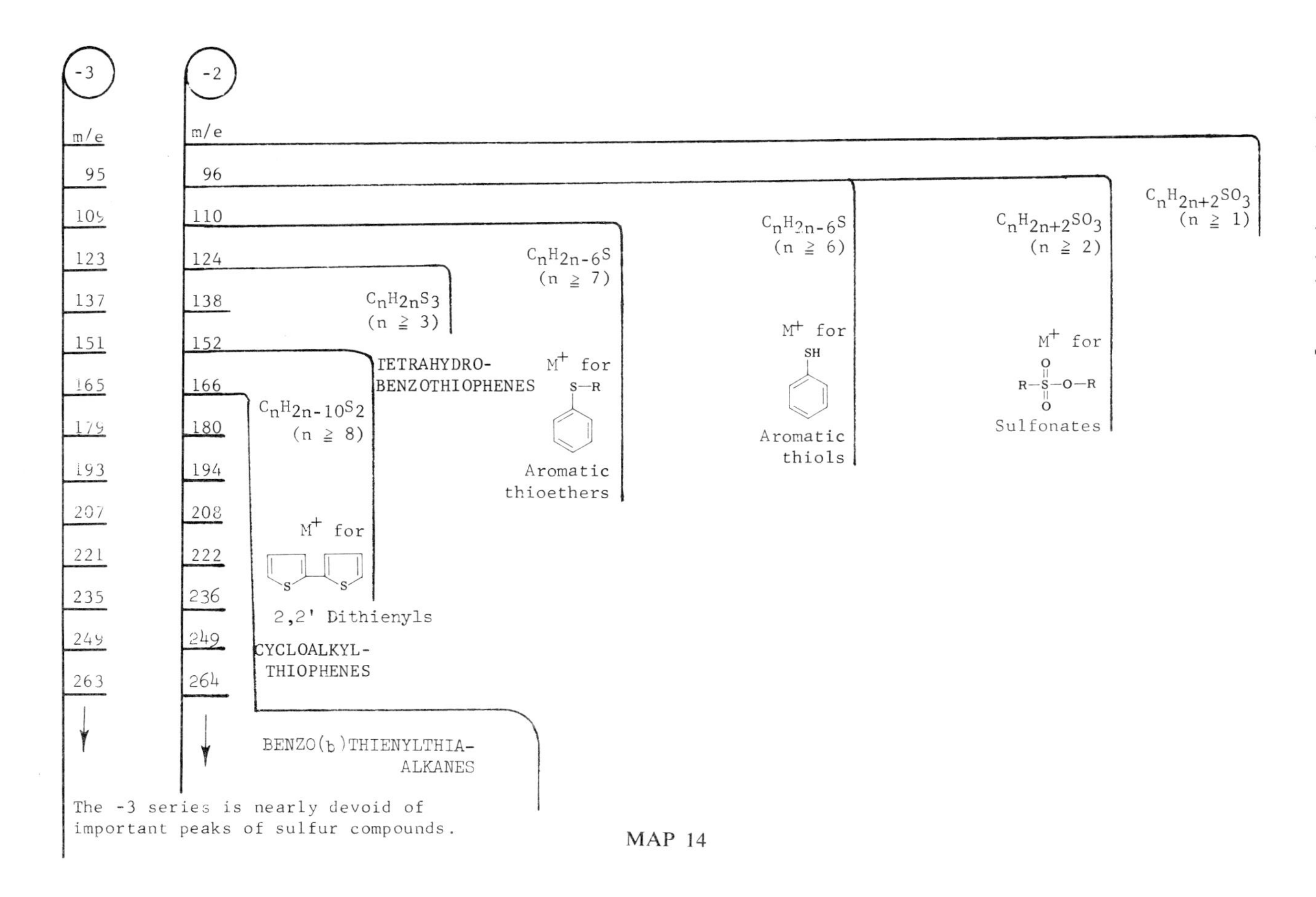

-3
m/e
95
109
123
137
151
165
179
193
207
221
235
249
263
-2
m/e
96
110
124
138
152
166
180
194
208
222
236
249
264
$C_nH_{2n+2}SO_3$ $(n \geq 1)$
$C_nH_{2n+2}SO_3$ $(n \geq 2)$
M^+ for
Sulfonates
$C_nH_{2n-6}S$ $(n \geq 6)$
M^+ for
Aromatic thiols
$C_nH_{2n-6}S$ $(n \geq 7)$
M^+ for
Aromatic thioethers
$C_nH_{2n}S_3$ $(n \geq 3)$
TETRAHYDRO-BENZOTHIOPHENES
$C_nH_{2n-10}S_2$ $(n \geq 8)$
M^+ for
2,2' Dithienyls
CYCLOALKYL-THIOPHENES
BENZO(b)THIENYLTHIA-ALKANES
The -3 series is nearly devoid of important peaks of sulfur compounds.

MAP 14

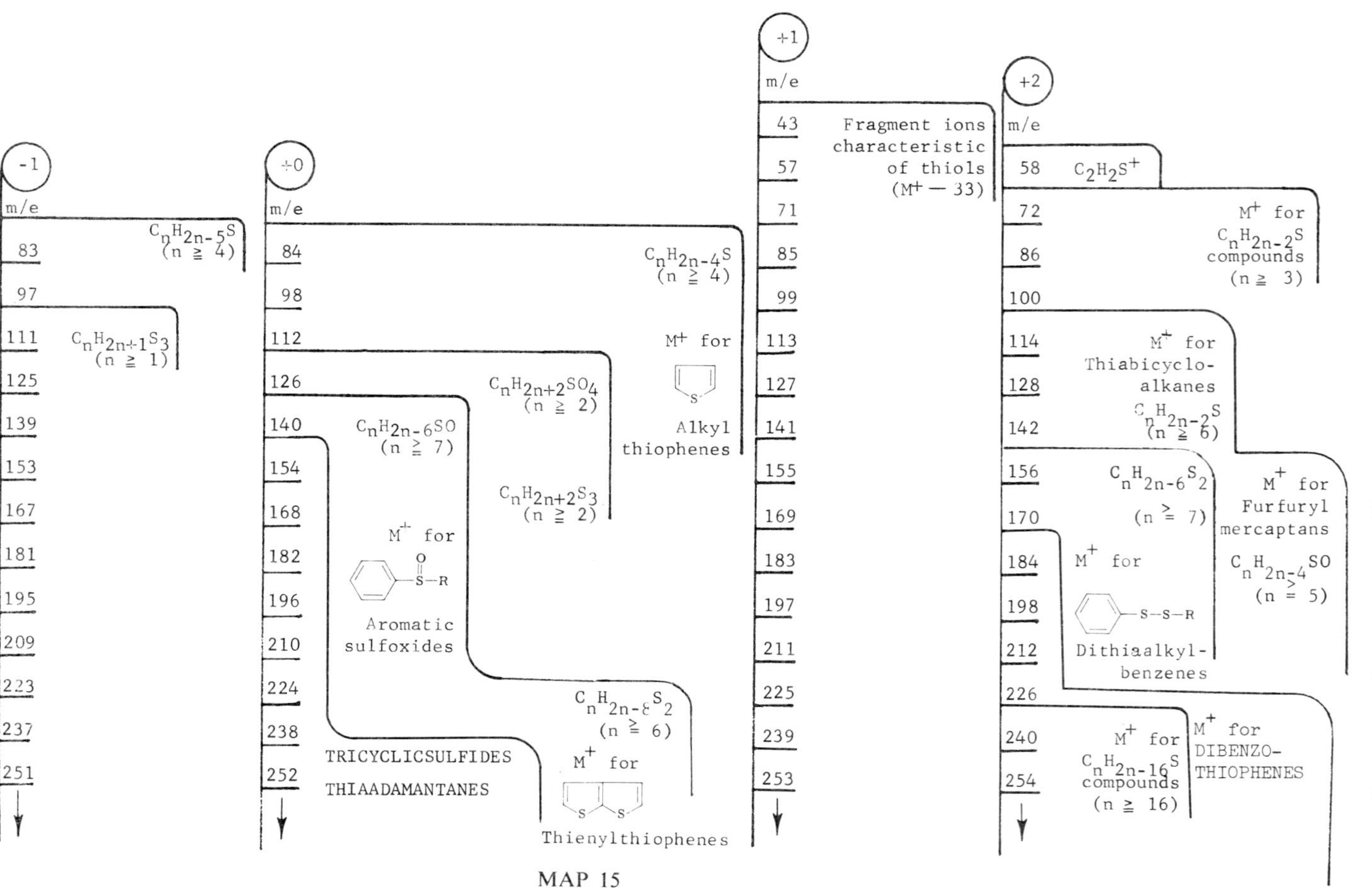
-1
m/e
83
97
111
125
139
153
167
181
195
209
223
237
251
CnH2n-5S (n ≥ 4)
CnH2n+1S3 (n ≥ 1)
±0
m/e
84
98
112
126
140
154
168
182
196
210
224
238
252
CnH2n-4S (n ≥ 4)
M+ for
S
Alkyl thiophenes
CnH2n+2SO4 (n ≥ 2)
CnH2n+2S3 (n ≥ 2)
CnH2n-6SO (n ≥ 7)
M+ for
O
S—R
Aromatic sulfoxides
CnH2n-8S2 (n ≥ 6)
M+ for
S
S
Thienylthiophenes
TRICYCLICSULFIDES
THIAADAMANTANES
+1
m/e
43
57
71
85
99
113
127
141
155
169
183
197
211
225
239
253
Fragment ions characteristic of thiols (M+ — 33)
+2
m/e
58
72
86
100
114
128
142
156
170
184
198
212
226
240
254
C2H2S+
M+ for CnH2n-2S compounds (n ≥ 3)
M+ for Thiabicyclo-alkanes CnH2n-2S (n ≥ 6)
M+ for Furfuryl mercaptans
CnH2n-4SO (n ≥ 5)
CnH2n-6S2 (n ≥ 7)
M+ for
S—S—R
Dithiaalkyl-benzenes
M+ for DIBENZO-THIOPHENES
M+ for CnH2n-16S compounds (n ≥ 16)

MAP 15

-11

m/e	
59	CHNS
73	$C_nH_{2n+1}SCN$ ($n \geq 1$)
87	
101	
115	
129	M^+ for R-SCN Thiocyanates
143	and
157	
171	M^+ for R-NCS Isothio-cyanates
185	
199	
213	
227	
241	
255	↓

-9

m/e	
75	$C_nH_{2n+1}CNHSH$ ($n \geq 1$)
89	
103	
117	
131	
145	M^+ for 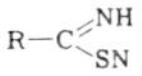
159	$R-C(=NH)SN$
173	Thioacetamide
187	
201	
215	
229	
243	
257	↓

-8

m/e	
90	$C_nH_{2n+1}NHCSNH_2$ ($n \geq 1$)
104	
118	
132	
146	M^+ for R-NH-CS-NH_2 Thioureas
160	
174	
188	
202	
216	
230	
244	
258	↓

-7

m/e	
161	$C_nH_{2n-11}NS$ ($n \geq 9$)
175	
189	
203	M^+ for 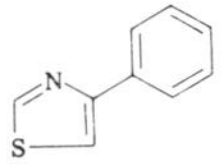
217	[phenylthiazole structure]
231	
245	Phenyl-thiazoles
259	↓

MAP 16

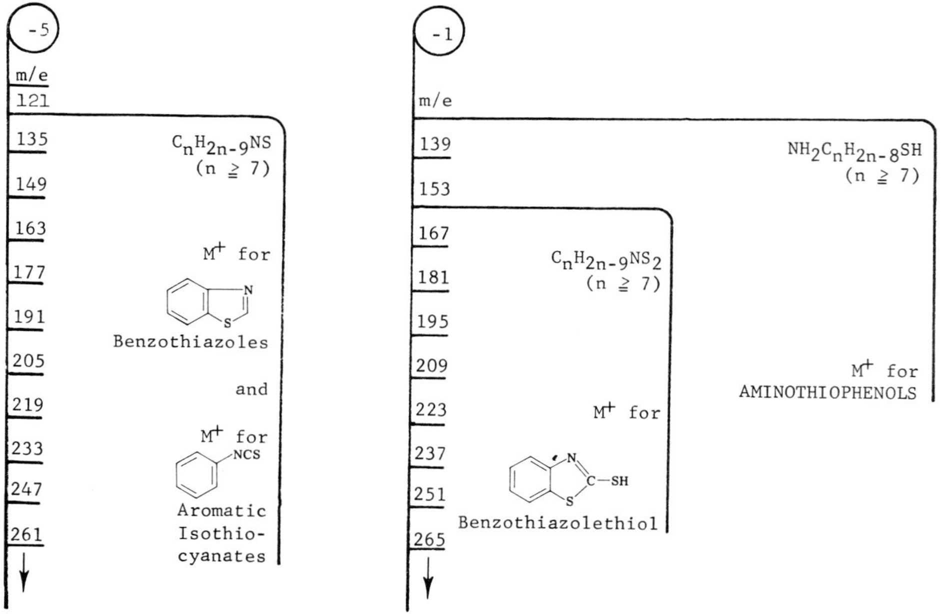

-5

m/e

121

135 — $C_nH_{2n-9}NS$ (n ≧ 7)

149

163

177 — M^+ for Benzothiazoles and M^+ for Aromatic Isothiocyanates (C₆H₅NCS)

191

205

219

233

247

261

↓

-1

m/e

139 — $NH_2C_nH_{2n-8}SH$ (n ≧ 7)

153

167 — $C_nH_{2n-9}NS_2$ (n ≧ 7)

181

195

209

223

237

251

265

↓

M^+ for Benzothiazolethiol (at 167)

M^+ for AMINOTHIOPHENOLS (at 139)

MAP 17

+1

m/e	
85	$C_nH_{2n-3}NS$ ($n \geq 3$)
99	
113	
127	
141	M^+ for Thiazoles
155	
169	
183	
197	
211	$C_nH_{2n-17}NS$ ($n \geq 13$)
225	
239	M^+ for Phenylbenzothiazoles
253	
267	

+2

m/e	
72	$C_nH_{2n-2}NS$ ($n \geq 2$)
86	
100	$C_nH_{2n-2}N_2S$ ($n \geq 3$)
114	
128	
142	M^+ for 2-Aminothiazole
156	
170	
184	
198	
212	
226	
240	
254	
268	

MAP 18

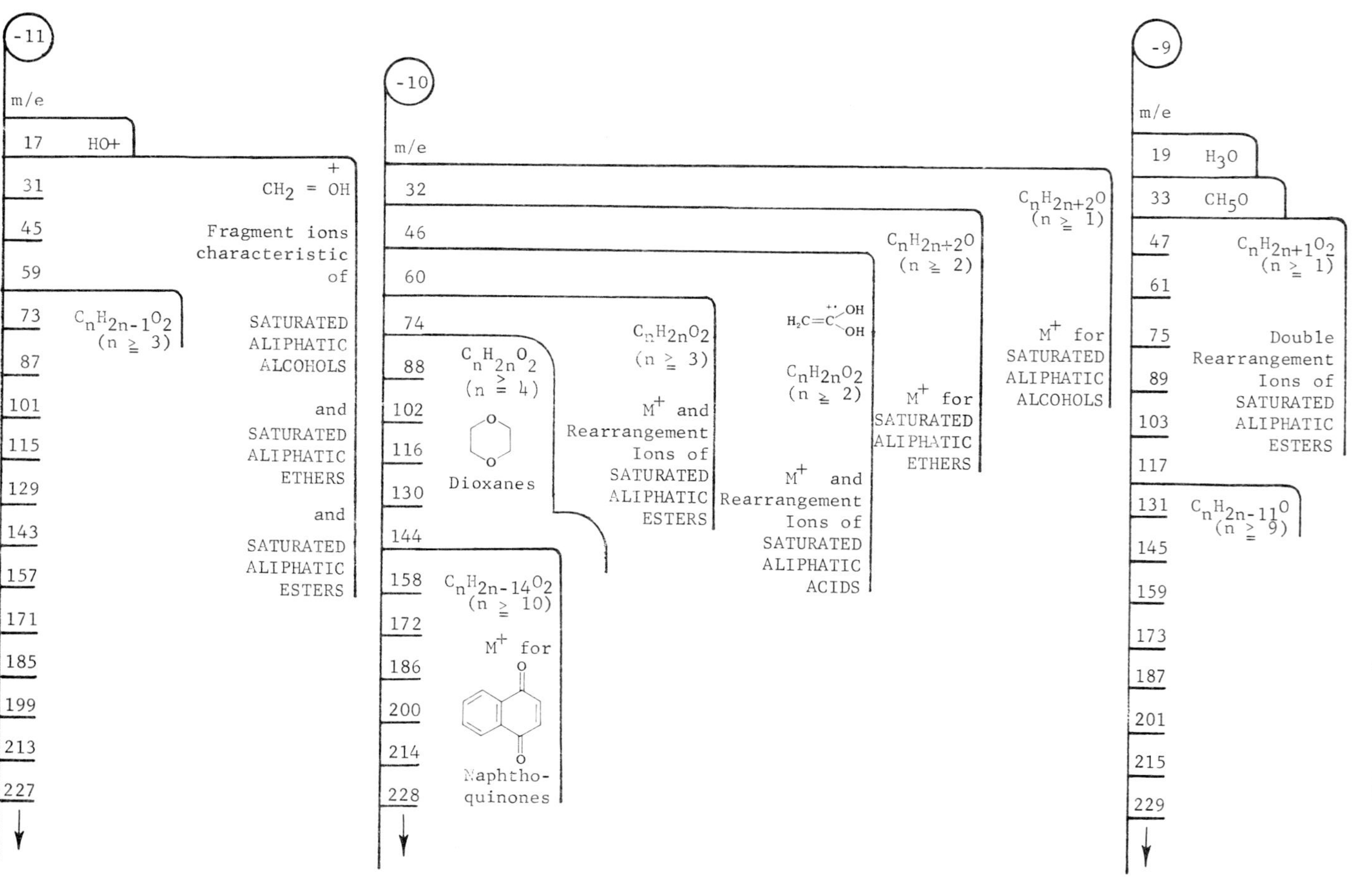
-11
m/e
17
HO+
31
CH2 = OH +
Fragment ions characteristic of
SATURATED ALIPHATIC ALCOHOLS
and
SATURATED ALIPHATIC ETHERS
and
SATURATED ALIPHATIC ESTERS
45
59
73
CnH2n-1O2 (n ≥ 3)
87
101
115
129
143
157
171
185
199
213
227
-10
m/e
32
CnH2n+2O (n ≥ 1)
M+ for SATURATED ALIPHATIC ALCOHOLS
46
CnH2n+2O (n ≥ 2)
M+ for SATURATED ALIPHATIC ETHERS
60
H2C=C(OH)OH
CnH2nO2 (n ≥ 2)
M+ and Rearrangement Ions of SATURATED ALIPHATIC ACIDS
74
CnH2nO2 (n ≥ 3)
M+ and Rearrangement Ions of SATURATED ALIPHATIC ESTERS
88
CnH2nO2 (n ≥ 4)
Dioxanes
102
116
130
144
158
CnH2n-14O2 (n ≥ 10)
M+ for
Naphtho-quinones
172
186
200
214
228
-9
m/e
19
H3O
33
CH5O
47
CnH2n+1O2 (n ≥ 1)
Double Rearrangement Ions of SATURATED ALIPHATIC ESTERS
61
75
89
103
117
131
CnH2n-11O (n ≥ 9)
145
159
173
187
201
215
229

MAP 19

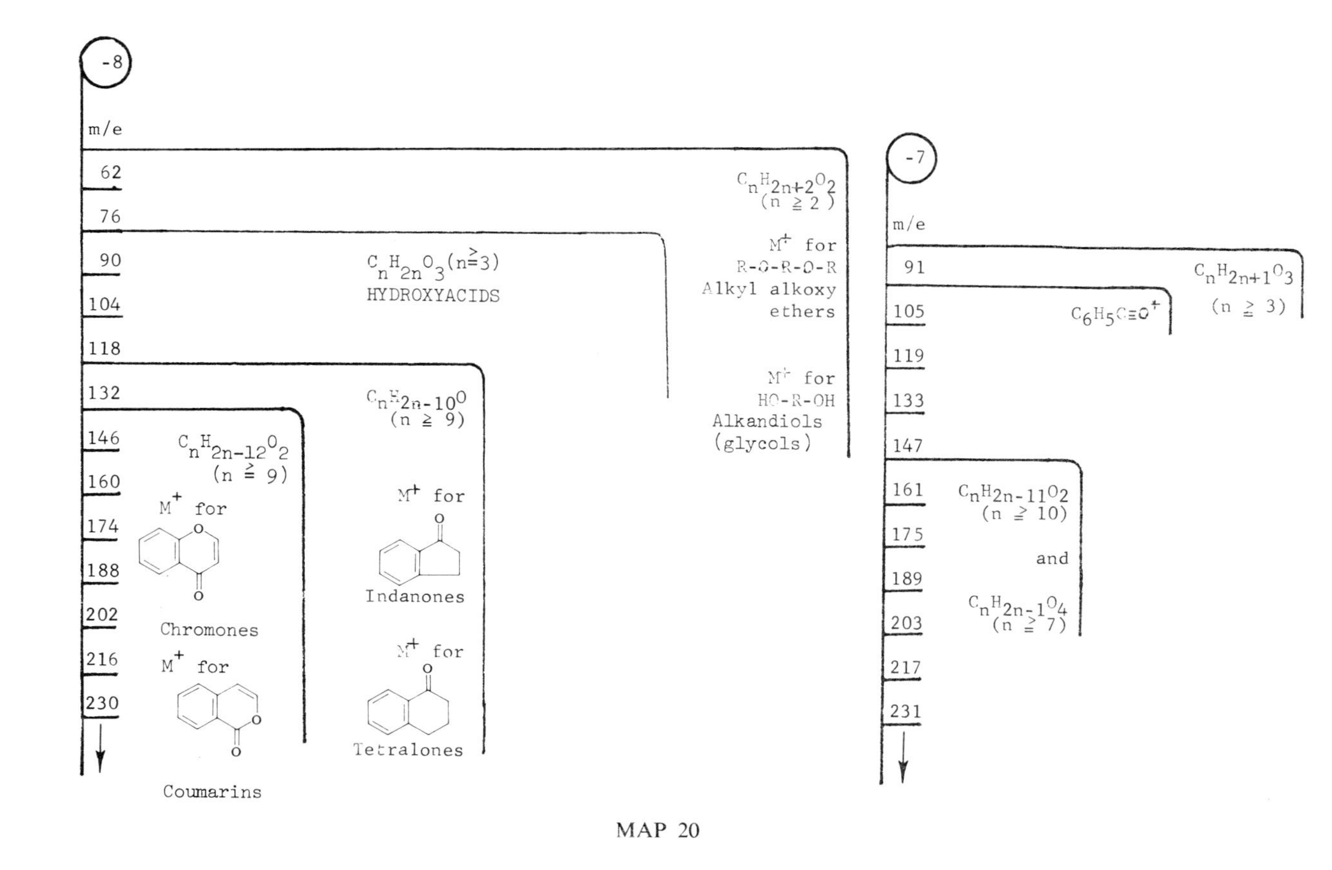
-8
m/e
62
76
90
104
118
132
146
160
174
188
202
216
230
C_nH_2n+2O_2 (n ≥ 2)
M+ for R-O-R-O-R
Alkyl alkoxy ethers
M+ for HO-R-OH
Alkandiols (glycols)
C_nH_2nO_3 (n ≥ 3)
HYDROXYACIDS
C_nH_2n-10O (n ≥ 9)
M+ for
Indanones
M+ for
Tetralones
C_nH_2n-12O_2 (n ≥ 9)
M+ for
Chromones
M+ for
Coumarins
-7
m/e
91
105
119
133
147
161
175
189
203
217
231
C_nH_2n+1O_3 (n ≥ 3)
C_6H_5C≡O+
C_nH_2n-11O_2 (n ≥ 10)
and
C_nH_2n-1O_4 (n ≥ 7)

MAP 20

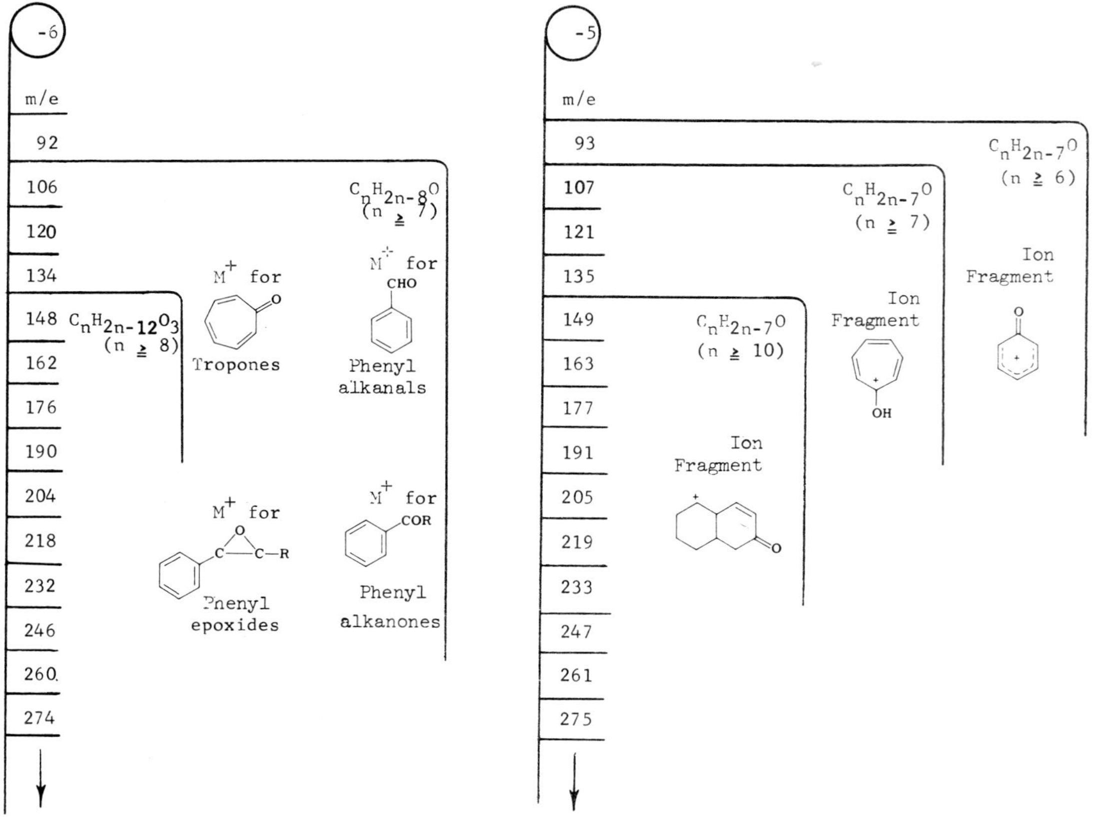

(-6)

m/e

92

106 — $C_nH_{2n-8}O$ ($n \geq 7$)

120

134

148 — $C_nH_{2n-12}O_3$ ($n \geq 8$)

162

176

190

204

218

232

246

260

274

M^+ for Tropones

$M^{+\cdot}$ for CHO — Phenyl alkanals

M^+ for — Phenyl epoxides

M^+ for COR — Phenyl alkanones

(-5)

m/e

93 — $C_nH_{2n-7}O$ ($n \geq 6$)

107 — $C_nH_{2n-7}O$ ($n \geq 7$)

121

135

149 — $C_nH_{2n-7}O$ ($n \geq 10$)

163

177

191

205

219

233

247

261

275

Ion Fragment (93)

Ion Fragment (107)

Ion Fragment (149)

MAP 21

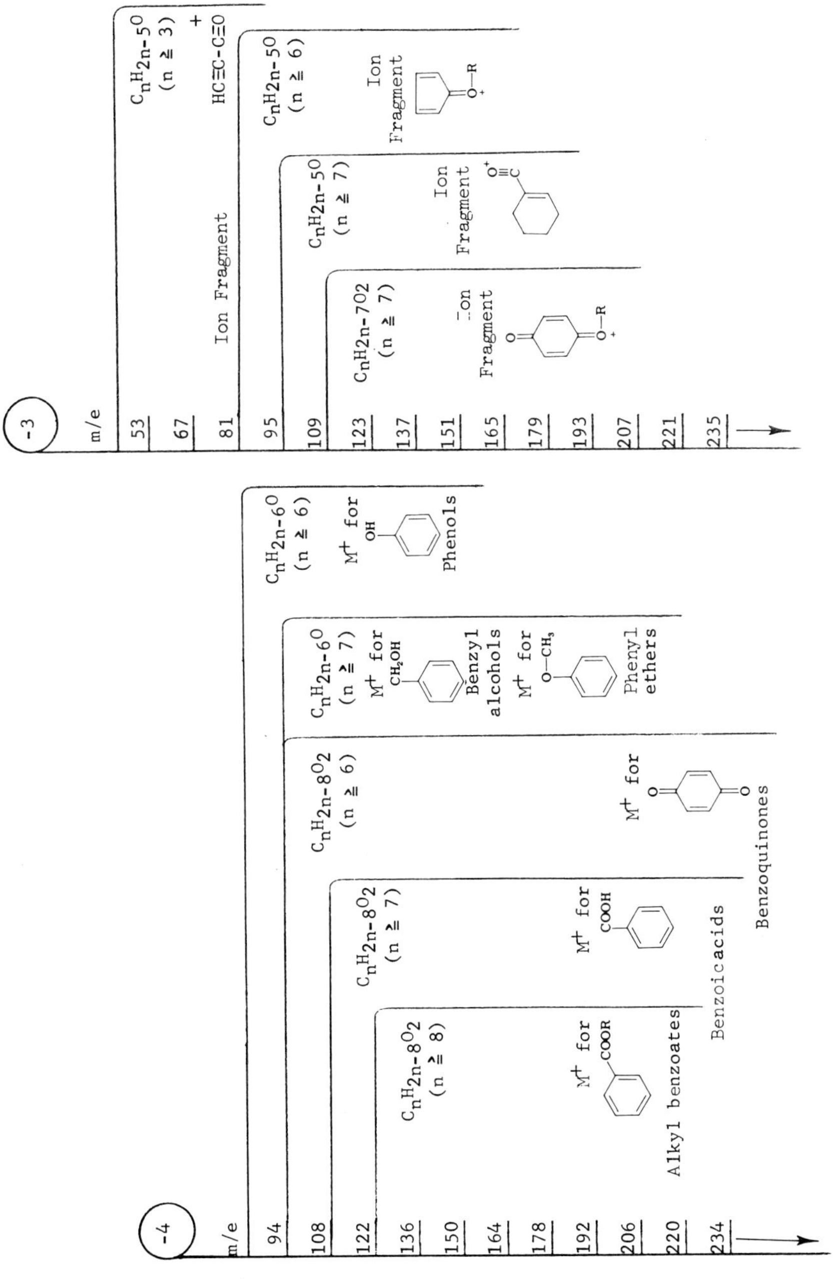

(-3)

m/e: 53, 67, 81, 95, 109, 123, 137, 151, 165, 179, 193, 207, 221, 235

- $C_nH_{2n-5}O$ (n ≧ 3), from m/e 53: Ion Fragment $HC{\equiv}C{-}C{\equiv}O^+$
- $C_nH_{2n-5}O$ (n ≧ 6), from m/e 95: Ion Fragment (cyclopentadienone, =O⁺–R)
- $C_nH_{2n-5}O$ (n ≧ 7), from m/e 109: Ion Fragment (cyclohexenyl–C≡O⁺)
- $C_nH_{2n-7}O_2$ (n ≧ 7), from m/e 123: Ion Fragment (quinone, =O⁺–R)

(-4)

m/e: 94, 108, 122, 136, 150, 164, 178, 192, 206, 220, 234

- $C_nH_{2n-6}O$ (n ≧ 6), from m/e 94: M^+ for (C₆H₅–OH) Phenols
- $C_nH_{2n-6}O$ (n ≧ 7), from m/e 108: M^+ for (C₆H₅–CH_2OH) Benzyl alcohols; M^+ for (C₆H₅–O–CH_3) Phenyl ethers
- $C_nH_{2n-8}O_2$ (n ≧ 6), from m/e 108: M^+ for (O=C₆H₄=O) Benzoquinones
- $C_nH_{2n-8}O_2$ (n ≧ 7), from m/e 122: M^+ for (C₆H₅–COOH) Benzoic acids
- $C_nH_{2n-8}O_2$ (n ≧ 8), from m/e 136: M^+ for (C₆H₅–COOR) Alkyl benzoates

MAP 22

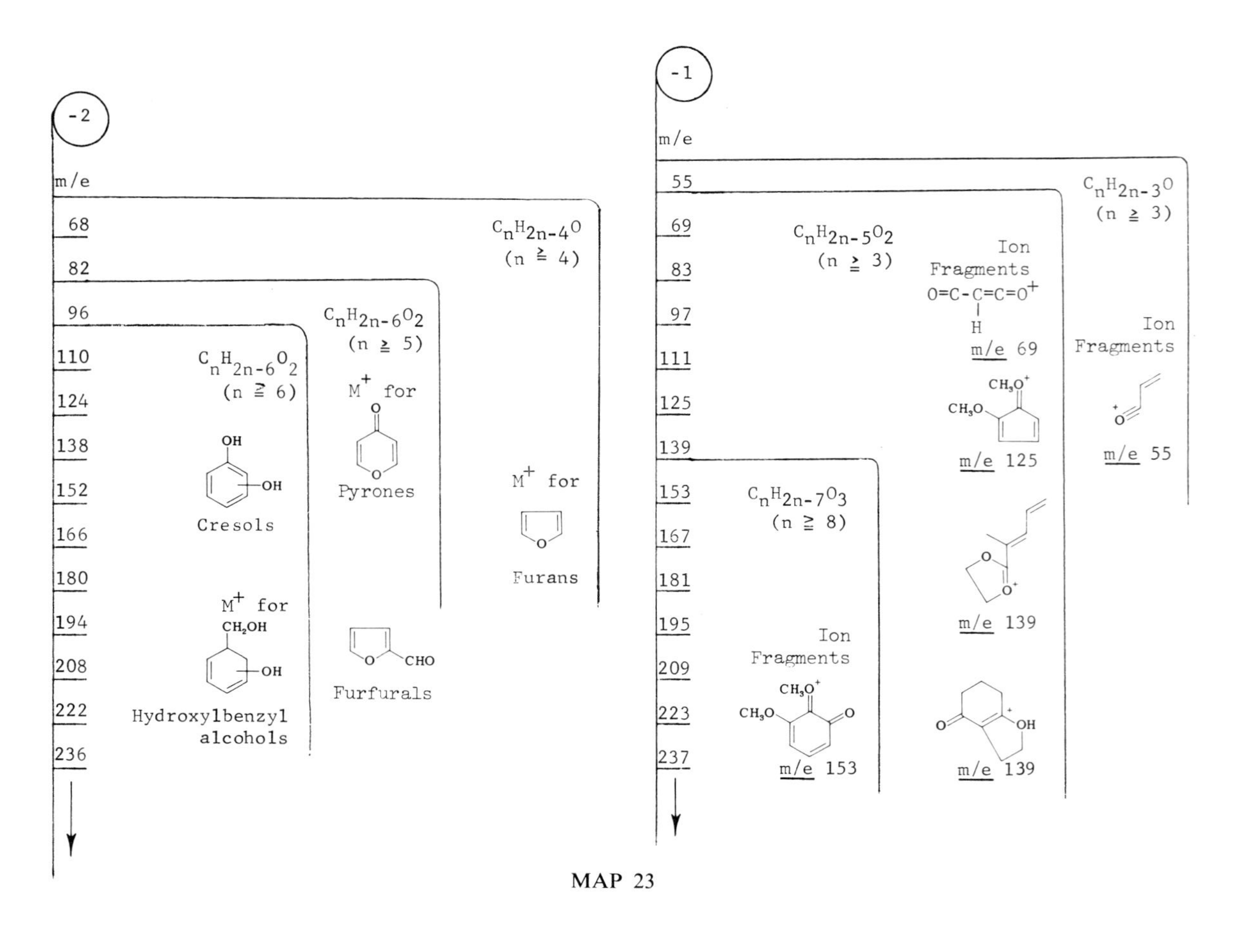
-2
m/e
68
82
96
110
124
138
152
166
180
194
208
222
236
$C_nH_{2n-4}O$
(n ≧ 4)
$C_nH_{2n-6}O_2$
(n ≥ 5)
$C_nH_{2n-6}O_2$
(n ≧ 6)
OH
OH
Cresols
M^+ for
CH_2OH
OH
Hydroxylbenzyl alcohols
M^+ for
Pyrones
CHO
Furfurals
M^+ for
Furans
-1
m/e
55
69
83
97
111
125
139
153
167
181
195
209
223
237
$C_nH_{2n-3}O$
(n ≥ 3)
Ion Fragments
m/e 55
$C_nH_{2n-5}O_2$
(n ≥ 3)
Ion Fragments
$O=C-C=C=O^+$
H
m/e 69
CH_3O^+
CH_3O
m/e 125
m/e 139
OH
m/e 139
$C_nH_{2n-7}O_3$
(n ≧ 8)
Ion Fragments
CH_3O^+
CH_3O
m/e 153

MAP 23

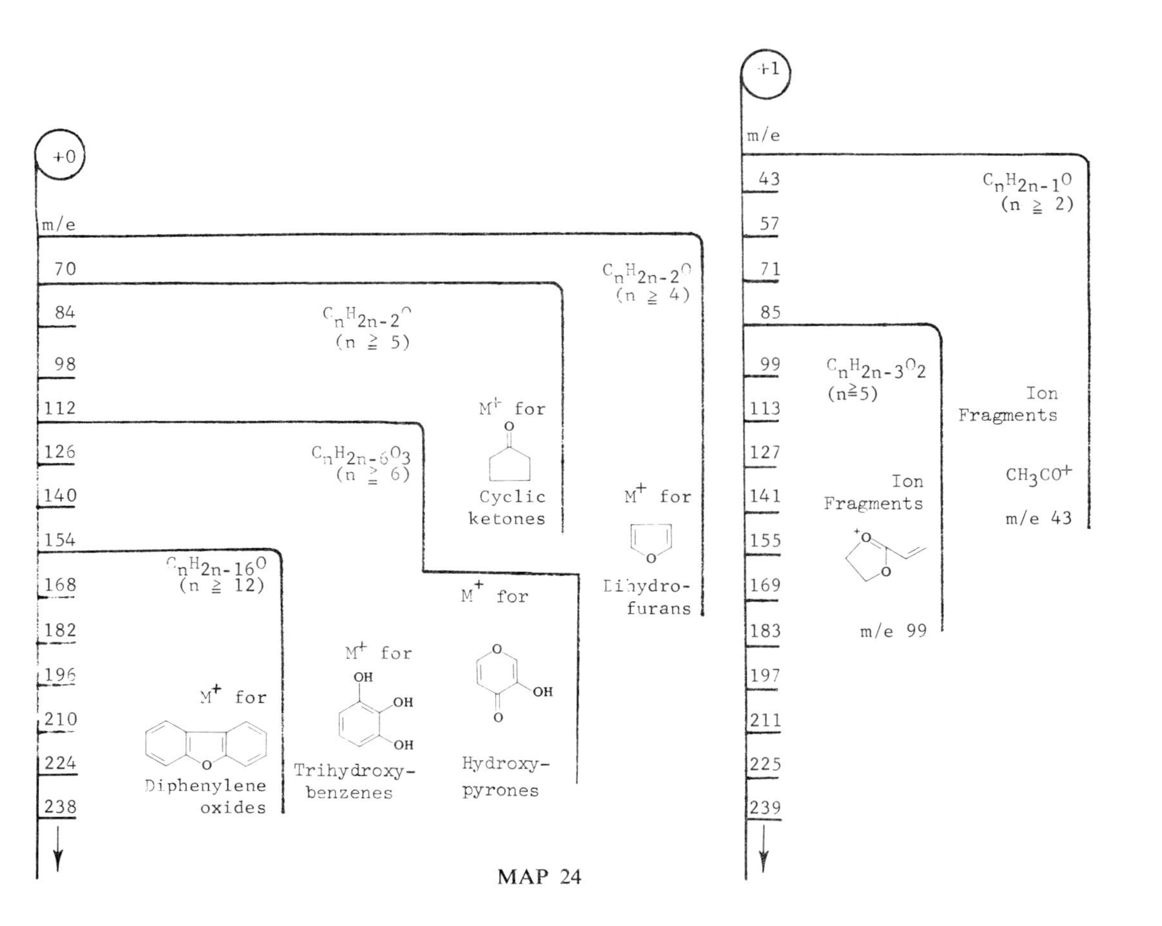
+0
m/e
70
84
98
112
126
140
154
168
182
196
210
224
238
CnH2n-2O
(n ≥ 4)
M+ for
Dihydro-
furans
CnH2n-2O
(n ≥ 5)
M+ for
Cyclic
ketones
CnH2n-6O3
(n ≥ 6)
M+ for
Trihydroxy-
benzenes
M+ for
Hydroxy-
pyrones
CnH2n-16O
(n ≥ 12)
M+ for
Diphenylene
oxides
+1
m/e
43
57
71
85
99
113
127
141
155
169
183
197
211
225
239
CnH2n-1O
(n ≥ 2)
Ion
Fragments
CH3CO+
m/e 43
CnH2n-3O2
(n≥5)
Ion
Fragments
m/e 99

MAP 24

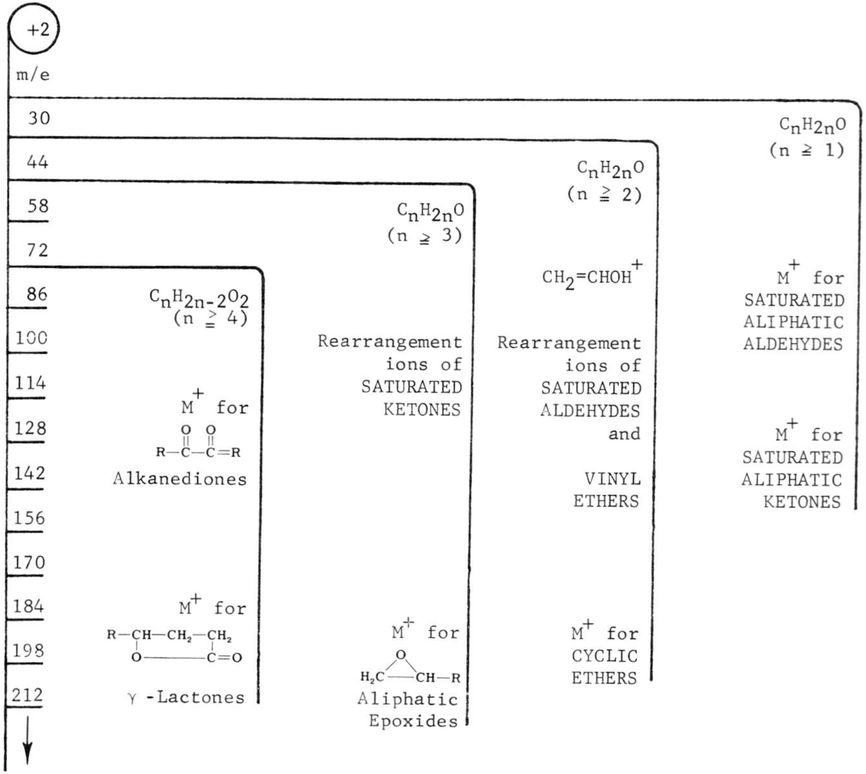

+2

m/e

30 — $C_nH_{2n}O$ (n ≧ 1)
- M^+ for SATURATED ALIPHATIC ALDEHYDES
- M^+ for SATURATED ALIPHATIC KETONES

44 — $C_nH_{2n}O$ (n ≧ 2)
- $CH_2=CHOH^+$
- Rearrangement ions of SATURATED ALDEHYDES and VINYL ETHERS
- M^+ for CYCLIC ETHERS

58 — $C_nH_{2n}O$ (n ≧ 3)
- Rearrangement ions of SATURATED KETONES
- M^+ for H_2C—O—CH—R (three-membered ring) Aliphatic Epoxides

72

86 — $C_nH_{2n-2}O_2$ (n ≧ 4)
- M^+ for R—C(=O)—C(=O)—R Alkanediones
- M^+ for R—CH—CH₂—CH₂—C(=O)—O— (ring) γ-Lactones

100

114

128

142

156

170

184

198

212

MAP 25

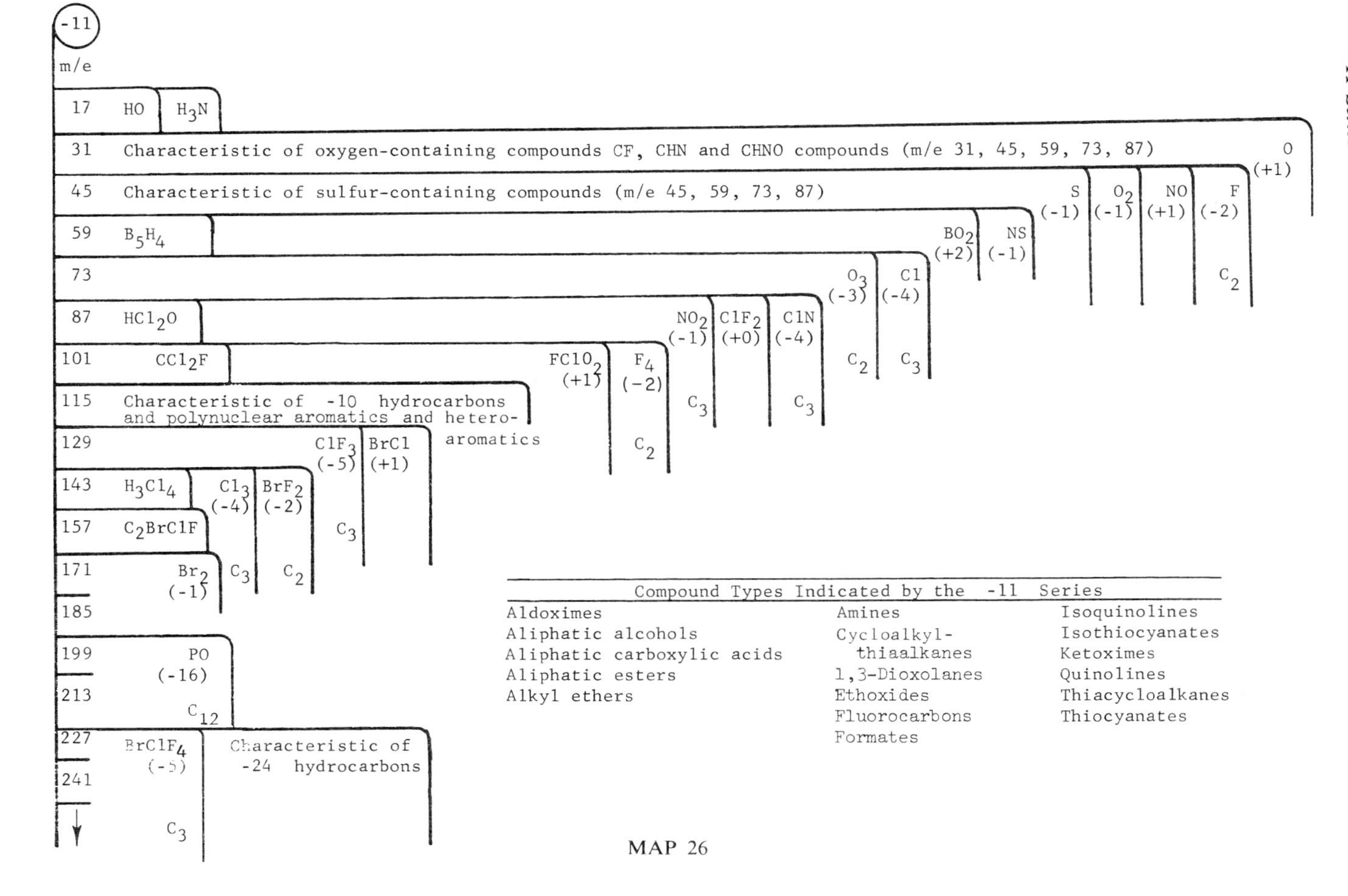
-11
m/e
17 HO H3N
31 Characteristic of oxygen-containing compounds CF, CHN and CHNO compounds (m/e 31, 45, 59, 73, 87)
O (+1)
45 Characteristic of sulfur-containing compounds (m/e 45, 59, 73, 87)
S (-1) O2 (-1) NO (+1) F (-2)
C2
59 B5H4
BO2 (+2) NS (-1)
73
O3 (-3) Cl (-4)
C2 C3
87 HCl2O
NO2 (-1) ClF2 (+0) ClN (-4)
C3 C3
101 CCl2F
FClO2 (+1) F4 (-2)
C2
115 Characteristic of -10 hydrocarbons and polynuclear aromatics and hetero-aromatics
129
ClF3 (-5) BrCl (+1)
C3
143 H3Cl4
Cl3 (-4) BrF2 (-2)
157 C2BrClF
171 Br2 (-1)
C3 C2
185
199 PO (-16)
213
C12
227 BrClF4 (-5)
Characteristic of -24 hydrocarbons
241
C3
Compound Types Indicated by the -11 Series
Aldoximes
Aliphatic alcohols
Aliphatic carboxylic acids
Aliphatic esters
Alkyl ethers
Amines
Cycloalkyl-thiaalkanes
1,3-Dioxolanes
Ethoxides
Fluorocarbons
Formates
Isoquinolines
Isothiocyanates
Ketoximes
Quinolines
Thiacycloalkanes
Thiocyanates

MAP 26

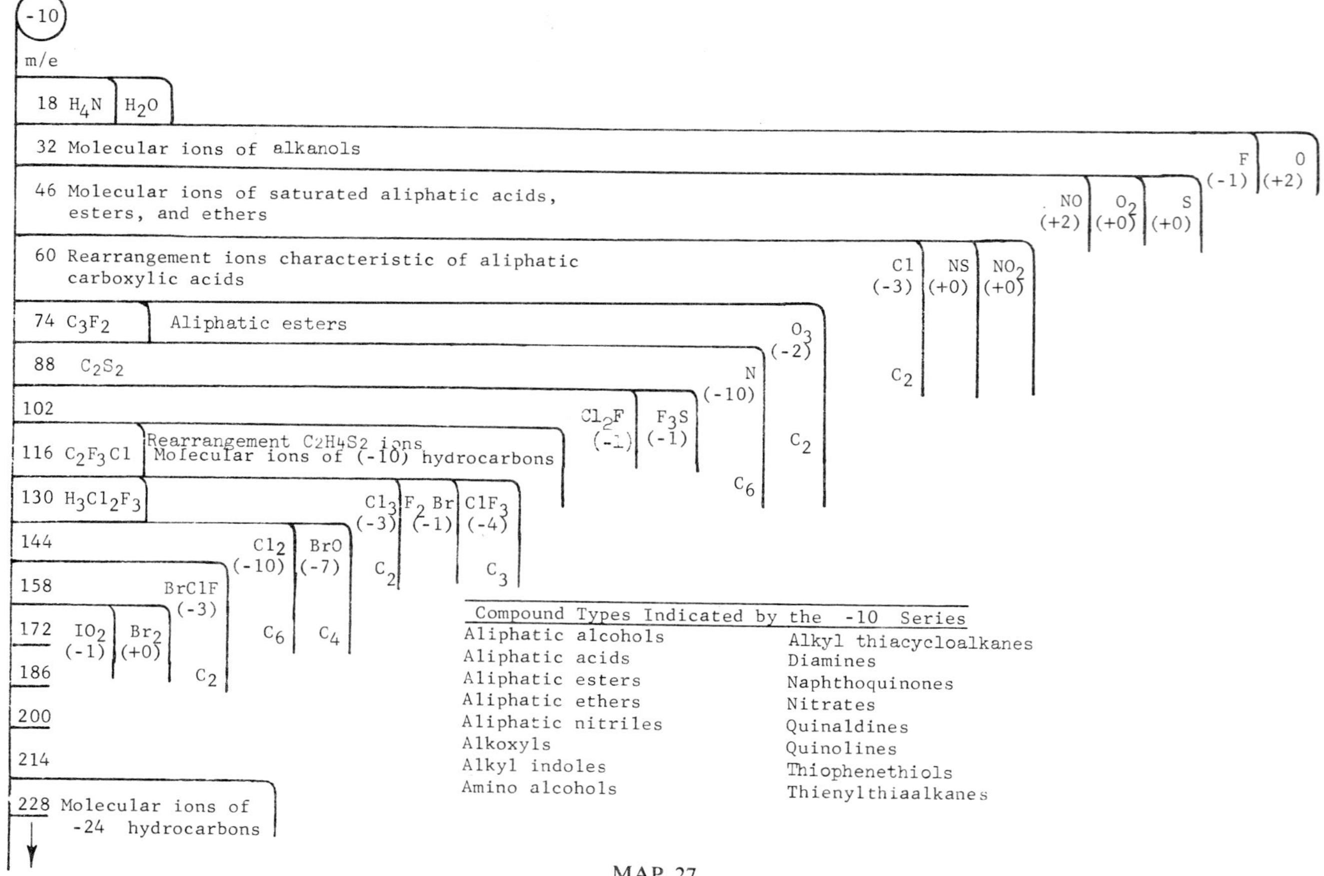
-10
m/e
18 H4N
H2O
32 Molecular ions of alkanols
F (-1)
O (+2)
46 Molecular ions of saturated aliphatic acids, esters, and ethers
NO (+2)
O2 (+0)
S (+0)
60 Rearrangement ions characteristic of aliphatic carboxylic acids
Cl (-3)
NS (+0)
NO2 (+0)
C2
74 C3F2
Aliphatic esters
O3 (-2)
C2
88 C2S2
N (-10)
C6
102
Cl2F (-1)
F3S (-1)
116 C2F3Cl
Rearrangement C2H4S2 ions
Molecular ions of (-10) hydrocarbons
130 H3Cl2F3
Cl3 (-3)
F2Br (-1)
ClF3 (-4)
C2
C3
144
Cl2 (-10)
BrO (-7)
C6
C4
158
BrClF (-3)
C2
172
IO2 (-1)
Br2 (+0)
186
200
214
228 Molecular ions of -24 hydrocarbons
Compound Types Indicated by the -10 Series
Aliphatic alcohols
Aliphatic acids
Aliphatic esters
Aliphatic ethers
Aliphatic nitriles
Alkoxyls
Alkyl indoles
Amino alcohols
Alkyl thiacycloalkanes
Diamines
Naphthoquinones
Nitrates
Quinaldines
Quinolines
Thiophenethiols
Thienylthiaalkanes

MAP 27

-9

m/e

19 F H_3O

33 SH

47 H_3B_4 CFO Double rearrangement ions characteristic of esters

CH_2-SH Fragment ions characteristic of alkanethiols and thioalkanes

O_2 (+1) S (+1) F (+0)

61 O_3 (-1) NO_2 (+1) Cl (-2) C_2

75 N (-9) F_2 (-5) ClN (-2) C_5 C_3 C_2

89 O (-11) C_6

103 N (-9) ClF_2O (+0) C_7

117 Characteristic of -8 hydrocarbons ClF_3 (-3) C_2

131 C_3F_5 BrF_2 (+0) Cl_3 (-2) C_2

145 $C_2F_3S_2$

159 HBr_2 BrClF (-2) C_2

173 NO_2 (-13) C_{10}

187

201

215 SNO (-15) C_{12}

229

Compound Types Indicated by the -9 Series	
Acetals	Fluorocarbons
Aliphatic esters	Glycols
Alkyl indoles	Nitronaphthalenes
Aromatic cyanides	Thioacetamide
Cinnamates	Thioalkanes
Dinitrotoluenes	Alkanethiols
Diols	

MAP 28

-8

m/e

34 H_2S ... F (+1)

48 SO | H_4B_4 | Molecular ions of alkanethiols ... S (+2) | O_2 (+2) | Cl (-1)

62 C_2F_2 | Molecular ions of thioalkanes ... NO_2 (+2) | O_3 (+0)

76 C_2NF_2 | CS_2 ... C10 (-3) C_2 | N (-8) C_5 | NO_3 (+0)

90 ... N_2 (-8) C_5 | O (-10) C_6

104 ... Br (-3) C_2

118 Molecular ions of -8 hydrocarbons | ClF_3 (-2) C_2

132 $C_2Cl_2F_2$ | C_2SF_4

146

160 Molecular ions of Naphthalenethiols and Naphthalene—thiaalkanes | BrF -8 C_5 | S (-12) C_{10}

174

188

202 Molecular ions of -22 Hydrocarbons

216

230

244

Compound Types Indicated by the -8 Series	
Aliphatic nitrates	Coumarones
Alkanethiols	Ethylene glycols
Alkyl alkoxy ethers	Fluorides
2-Alkylbenzimidazoles	Phenylethyl esters
Coumarins	Phenylthiophenes
	Tetralones
	Tetrahydrodiabenzothiophenes
	Thioalkanes

MAP 29

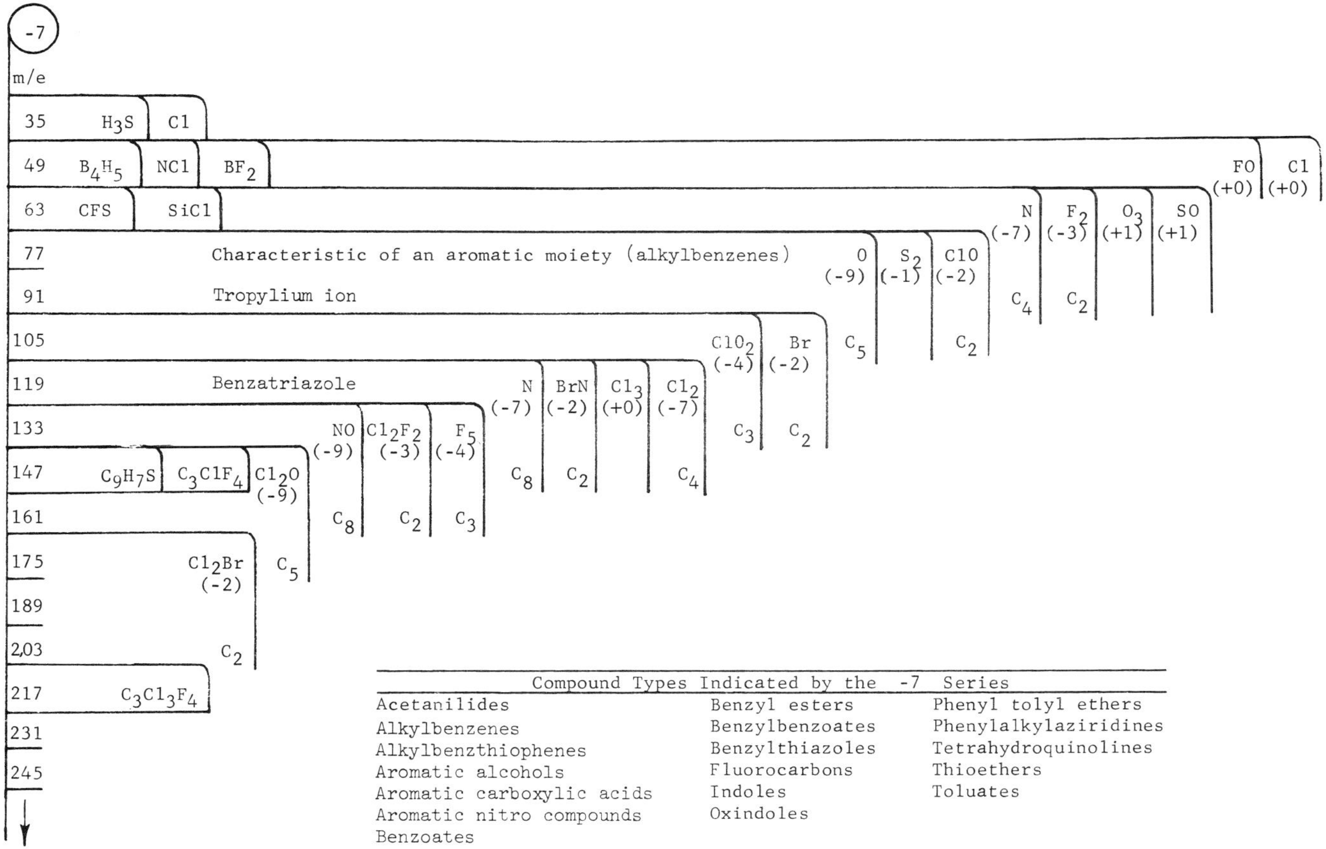

-7

m/e

35 H_3S | Cl

49 B_4H_5 | NCl | BF_2 | FO (+0) | Cl (+0)

63 CFS | SiCl | N (-7) C_4 | F_2 (-3) C_2 | O_3 (+1) | SO (+1)

77 Characteristic of an aromatic moiety (alkylbenzenes) | O (-9) C_5 | S_2 (-1) | ClO (-2) C_2

91 Tropylium ion

105 | ClO_2 (-4) C_3 | Br (-2) C_2

119 Benzatriazole | N (-7) C_8 | BrN (-2) C_2 | Cl_3 (+0) | Cl_2 (-7) C_4

133 | NO (-9) C_8 | Cl_2F_2 (-3) C_2 | F_5 (-4) C_3

147 C_9H_7S | C_3ClF_4 | Cl_2O (-9) C_5

161

175 Cl_2Br (-2) C_2

189

203

217 $C_3Cl_3F_4$

231

245

Compound Types Indicated by the -7 Series		
Acetanilides	Benzyl esters	Phenyl tolyl ethers
Alkylbenzenes	Benzylbenzoates	Phenylalkylaziridines
Alkylbenzthiophenes	Benzylthiazoles	Tetrahydroquinolines
Aromatic alcohols	Fluorocarbons	Thioethers
Aromatic carboxylic acids	Indoles	Toluates
Aromatic nitro compounds	Oxindoles	
Benzoates		
Benzodithiophenes		

MAP 30

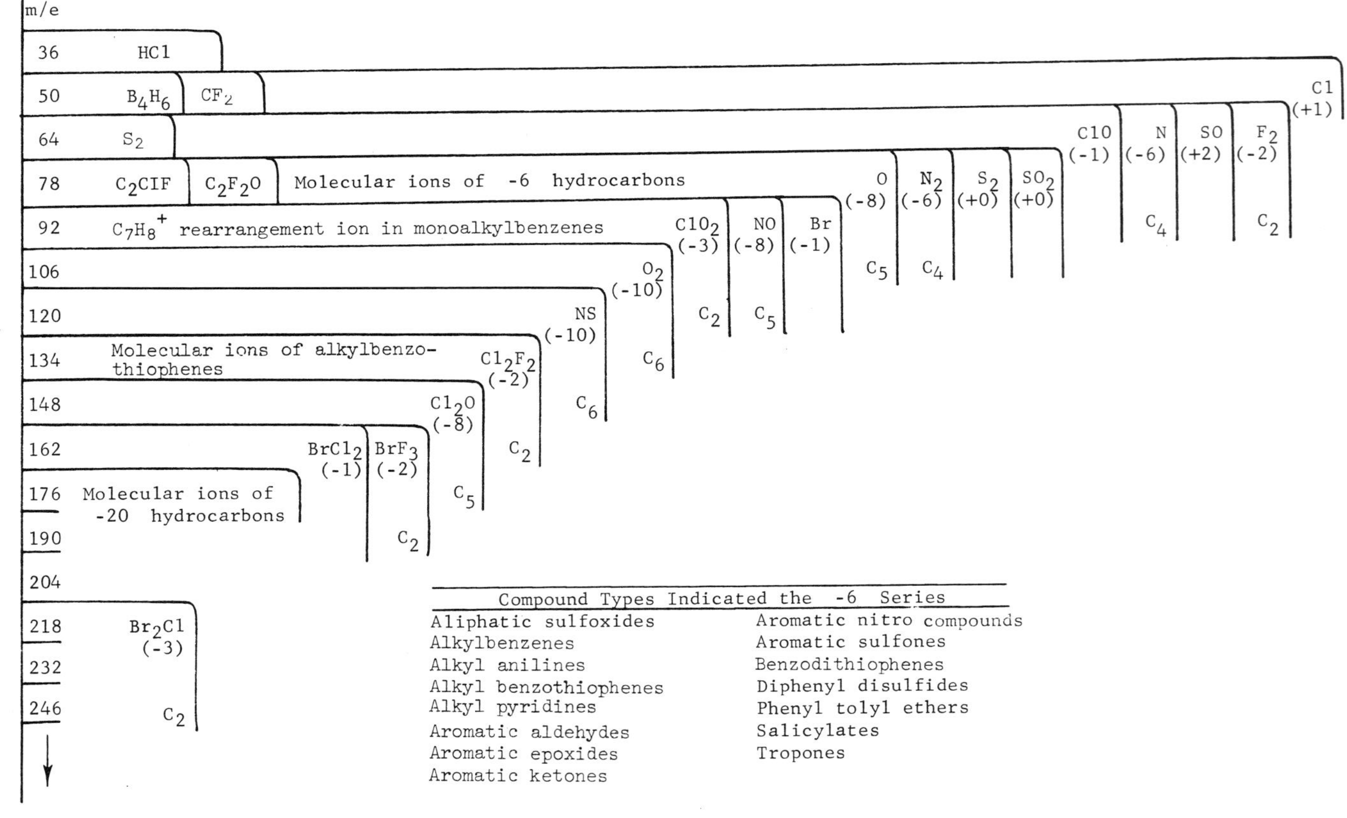
-6
m/e
36
50
64
78
92
106
120
134
148
162
176
190
204
218
232
246
HCl
B4H6
CF2
S2
C2ClF
C2F2O
Molecular ions of -6 hydrocarbons
C7H8+ rearrangement ion in monoalkylbenzenes
Molecular ions of alkylbenzo-thiophenes
Molecular ions of -20 hydrocarbons
Br2Cl (-3)
BrCl2 (-1)
BrF3 (-2)
Cl2O (-8)
Cl2F2 (-2)
NS (-10)
O2 (-10)
ClO2 (-3)
NO (-8)
Br (-1)
O (-8)
N2 (-6)
S2 (+0)
SO2 (+0)
ClO (-1)
N (-6)
SO (+2)
F2 (-2)
Cl (+1)
C2
C4
C5
C6
Compound Types Indicated the -6 Series
Aliphatic sulfoxides
Alkylbenzenes
Alkyl anilines
Alkyl benzothiophenes
Alkyl pyridines
Aromatic aldehydes
Aromatic epoxides
Aromatic ketones
Aromatic nitro compounds
Aromatic sulfones
Benzodithiophenes
Diphenyl disulfides
Phenyl tolyl ethers
Salicylates
Tropones

MAP 31

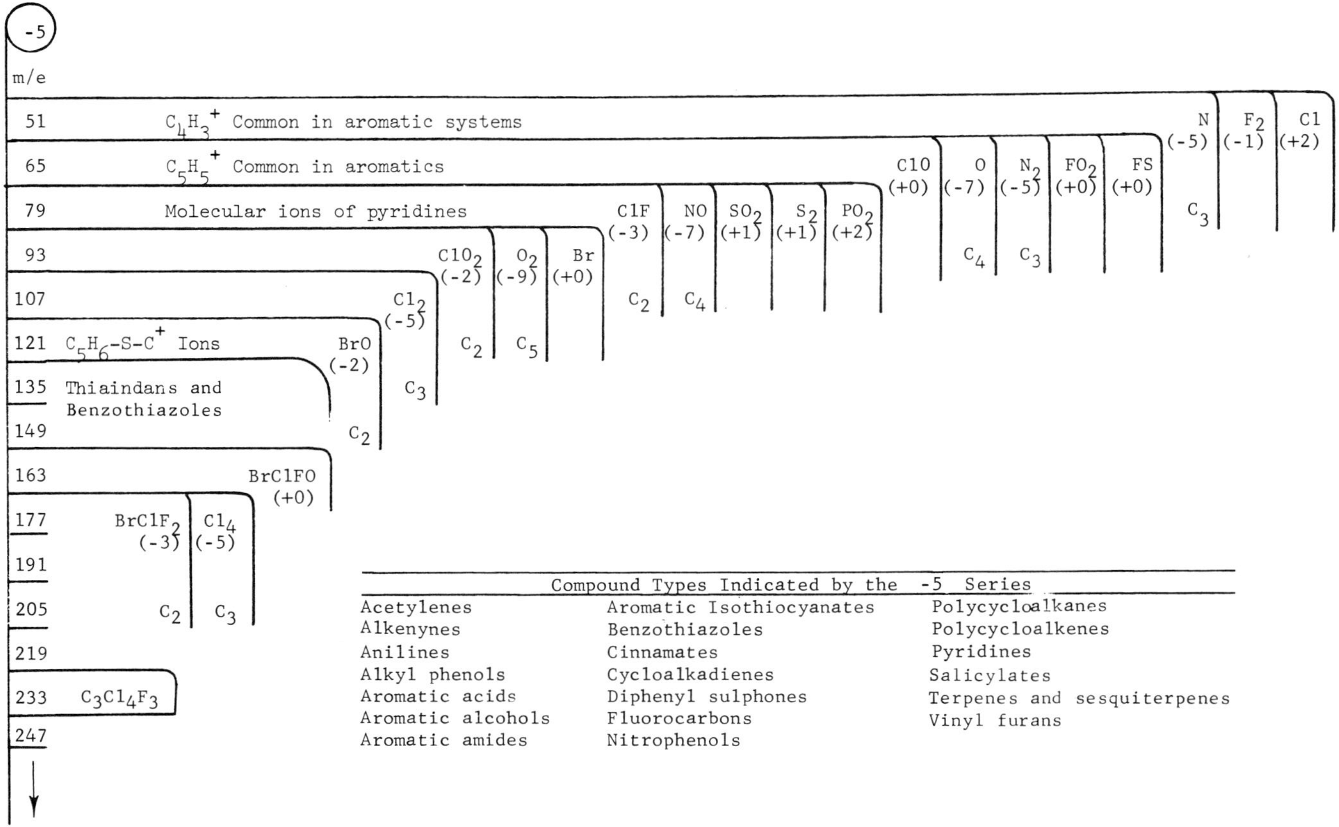

-5

m/e

51 $C_4H_3^+$ Common in aromatic systems — N (-5) C_3; F_2 (-1); Cl (+2)

65 $C_5H_5^+$ Common in aromatics — ClO (+0); O (-7) C_4; N_2 (-5) C_3; FO_2 (+0); FS (+0)

79 Molecular ions of pyridines — ClF (-3) C_2; NO (-7) C_4; SO_2 (+1); S_2 (+1); PO_2 (+2)

93 — ClO_2 (-2) C_2; O_2 (-9) C_5; Br (+0)

107 — Cl_2 (-5) C_3

121 C_5H_6-S-C^+ Ions — BrO (-2) C_2

135 Thiaindans and Benzothiazoles

149

163 — BrClFO (+0)

177 — $BrClF_2$ (-3) C_2; Cl_4 (-5) C_3

191

205

219

233 $C_3Cl_4F_3$

247

Compound Types Indicated by the -5 Series		
Acetylenes	Aromatic Isothiocyanates	Polycycloalkanes
Alkenynes	Benzothiazoles	Polycycloalkenes
Anilines	Cinnamates	Pyridines
Alkyl phenols	Cycloalkadienes	Salicylates
Aromatic acids	Diphenyl sulphones	Terpenes and sesquiterpenes
Aromatic alcohols	Fluorocarbons	Vinyl furans
Aromatic amides	Nitrophenols	

MAP 32

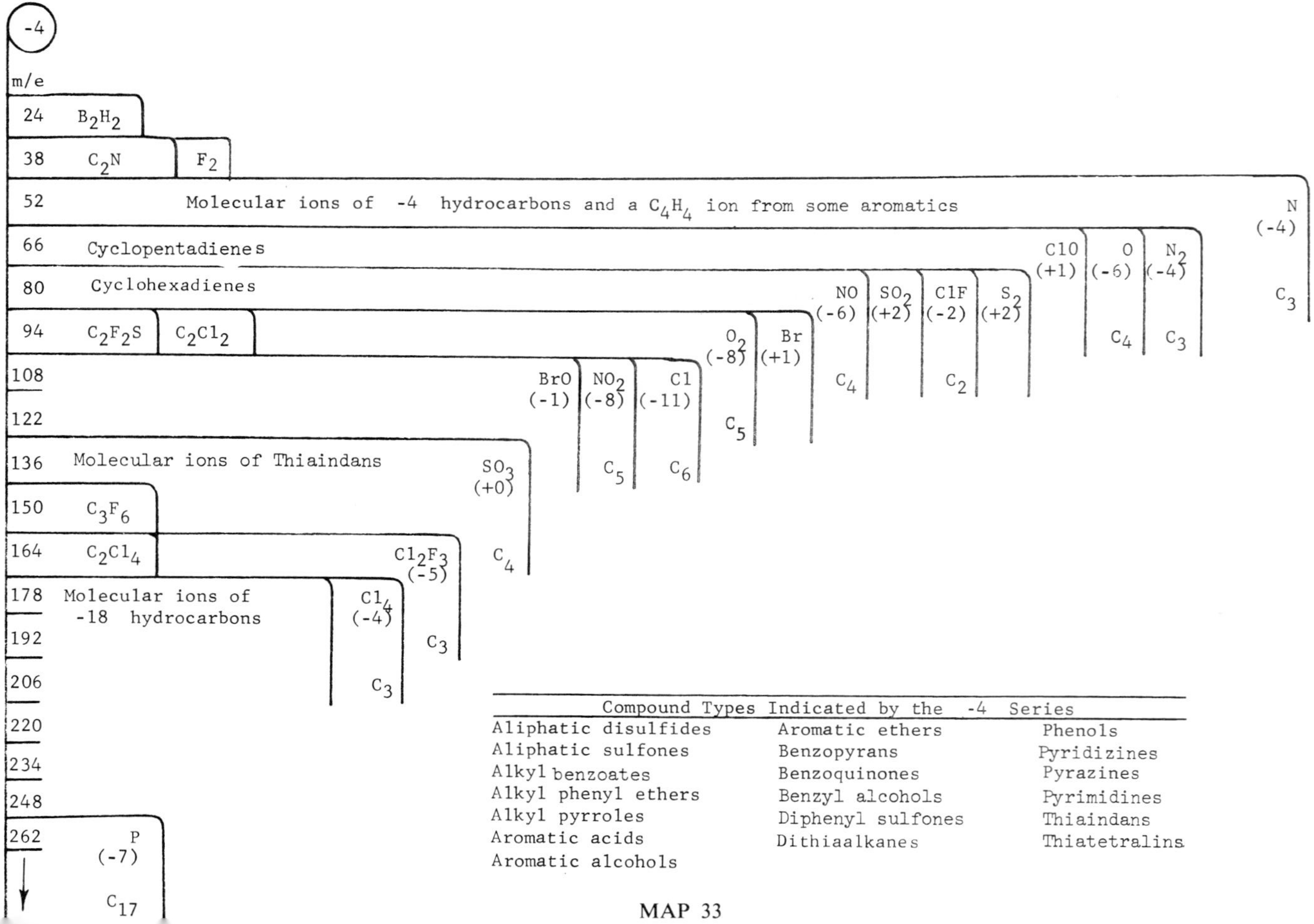

(-4)

m/e

24 B_2H_2

38 C_2N — F_2

52 Molecular ions of -4 hydrocarbons and a C_4H_4 ion from some aromatics — N (-4) C_3

66 Cyclopentadienes — C10 (+1) — O (-6) C_4 — N_2 (-4) C_3

80 Cyclohexadienes — NO (-6) C_4 — SO_2 (+2) — ClF (-2) C_2 — S_2 (+2)

94 C_2F_2S — C_2Cl_2 — O_2 (-8) C_5 — Br (+1)

108 BrO (-1) — NO_2 (-8) C_5 — Cl (-11) C_6

122

136 Molecular ions of Thiaindans — SO_3 (+0) C_4

150 C_3F_6

164 C_2Cl_4 — Cl_2F_3 (-5) C_3

178 Molecular ions of -18 hydrocarbons — Cl_4 (-4) C_3

192

206

220

234

248

262 P (-7) C_{17}

Compound Types Indicated by the -4 Series		
Aliphatic disulfides	Aromatic ethers	Phenols
Aliphatic sulfones	Benzopyrans	Pyridizines
Alkyl benzoates	Benzoquinones	Pyrazines
Alkyl phenyl ethers	Benzyl alcohols	Pyrimidines
Alkyl pyrroles	Diphenyl sulfones	Thiaindans
Aromatic acids	Dithiaalkanes	Thiatetralins
Aromatic alcohols		

MAP 33

-3

m/e

25 B_2H_3

39 C_3H_3 Common in aromatic compounds - a ring fragment ion

53 C_4H_5 Common to aromatics and unsaturated hydrocarbons

67 Characteristic of a cyclic hydrocarbon moiety

81 C_6H_9

95

109

123

137

151

165

179 $C_3Cl_3F_2$

193

207

221

235

N (-3) C_2

O (-5) C_3

F_2O (-1)

NO (-5) C_3

ClF (-1)

FO_2 (+2)

ClO (+2)

FS (+2)

ClO_2 (+0)

O_2 (-7) C_4

F_2O_2 (-3) C_2

F_3 (-4) C_3

Br (+2)

Cl_2 (-3) C_2

BrO (+0)

SO_3N (+1)

O_3 (-9) C_5

ClF_2 (-6) C_4

Cl_4 (-3) C_2

Cl_2F_3 (-4) C_3

Compound Types Indicated by the -3 Series	
Decalins	Sesquiterpenones
Furans	2-pyridones
Monoterpenones	Pyrroles

MAP 34

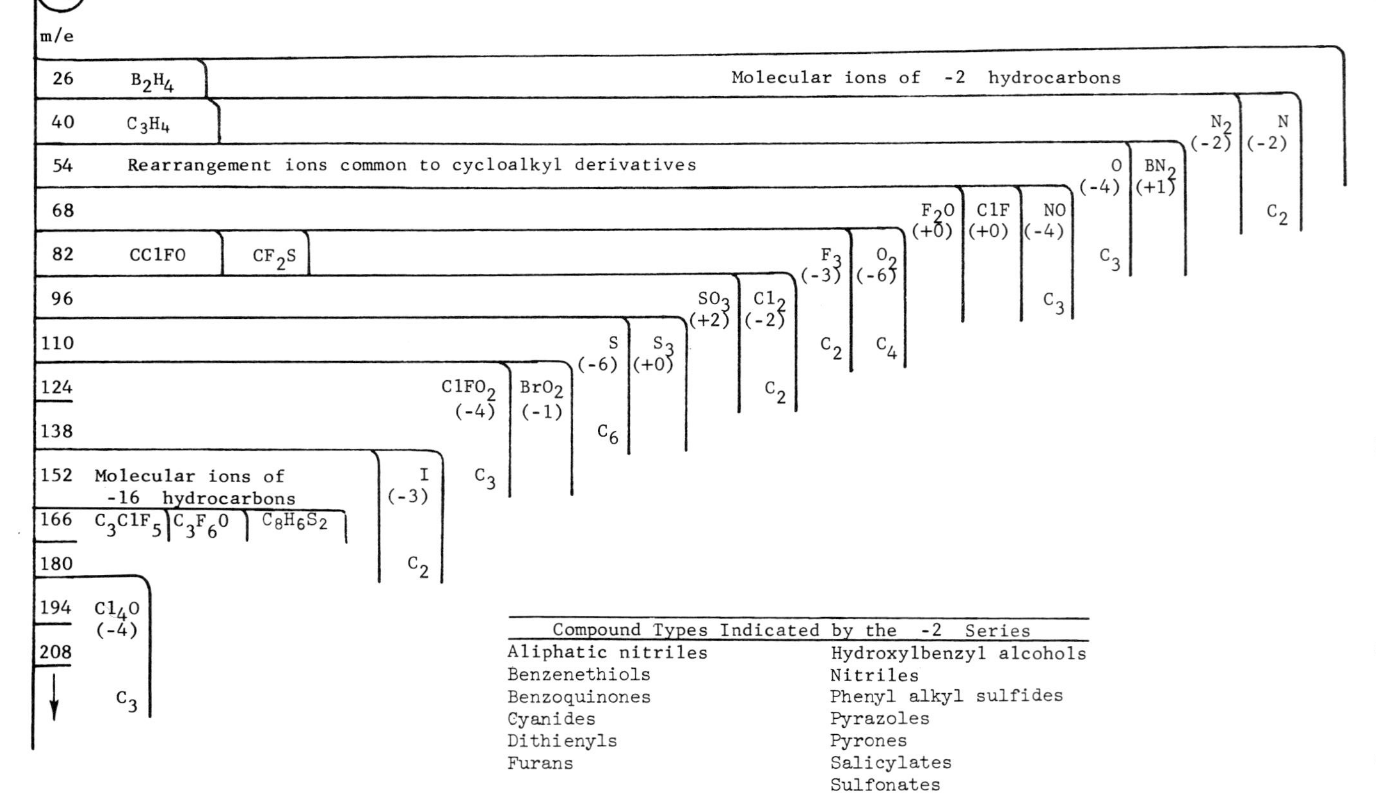
-2
m/e
26
B_2H_4
Molecular ions of -2 hydrocarbons
40
C_3H_4
N_2 (-2)
N (-2)
54
Rearrangement ions common to cycloalkyl derivatives
O (-4)
BN_2 (+1)
68
F_2O (+0)
ClF (+0)
NO (-4)
C_2
82
CClFO
CF_2S
F_3 (-3)
O_2 (-6)
C_3
96
SO_3 (+2)
Cl_2 (-2)
C_3
110
S (-6)
S_3 (+0)
C_2
C_4
124
$ClFO_2$ (-4)
BrO_2 (-1)
C_2
138
C_6
152
Molecular ions of -16 hydrocarbons
I (-3)
C_3
166
C_3ClF_5
C_3F_6O
$C_8H_6S_2$
180
C_2
194
Cl_4O (-4)
208
C_3
Compound Types Indicated by the -2 Series
Aliphatic nitriles
Benzenethiols
Benzoquinones
Cyanides
Dithienyls
Furans
Hydroxylbenzyl alcohols
Nitriles
Phenyl alkyl sulfides
Pyrazoles
Pyrones
Salicylates
Sulfonates

MAP 35

(−1)

m/e	Ion / branch labels
27	B_2H_5 — Characteristic of alkylcyclohydrocarbons or an alkenyl moiety — N (−1); B (+2)
41	C_3H_5 — N_2 (−1); O (−3) C_2
55	NO (−3) C_2
69	CF_3 — ClF (+1); F_2O (+1); F (−6) C_4; O_2 (−5) C_3; S (−5) C_3
83	SO_2F — ClO_2 (+2); Cl_2 (−1); F_3 (−2) C_2
97	C_2F_3O; C_2ClF_2; C_6H_5S — Cl (−8) C_5
111	Cl_2O (−3) C_2; BrF (−1); S_3 (+1)
125	ClO (−10) C_6
139	BrCl (−3) C_2
153	
167	$C_2Cl_3F_2$ — N (−15) C_{12}
181	
195	
209	

Compound Types Indicated by the −1 Series	
Aliphatic nitriles	Carbazoles
Alkenes	Cycloalkanes
Aminothiophenols	

MAP 36

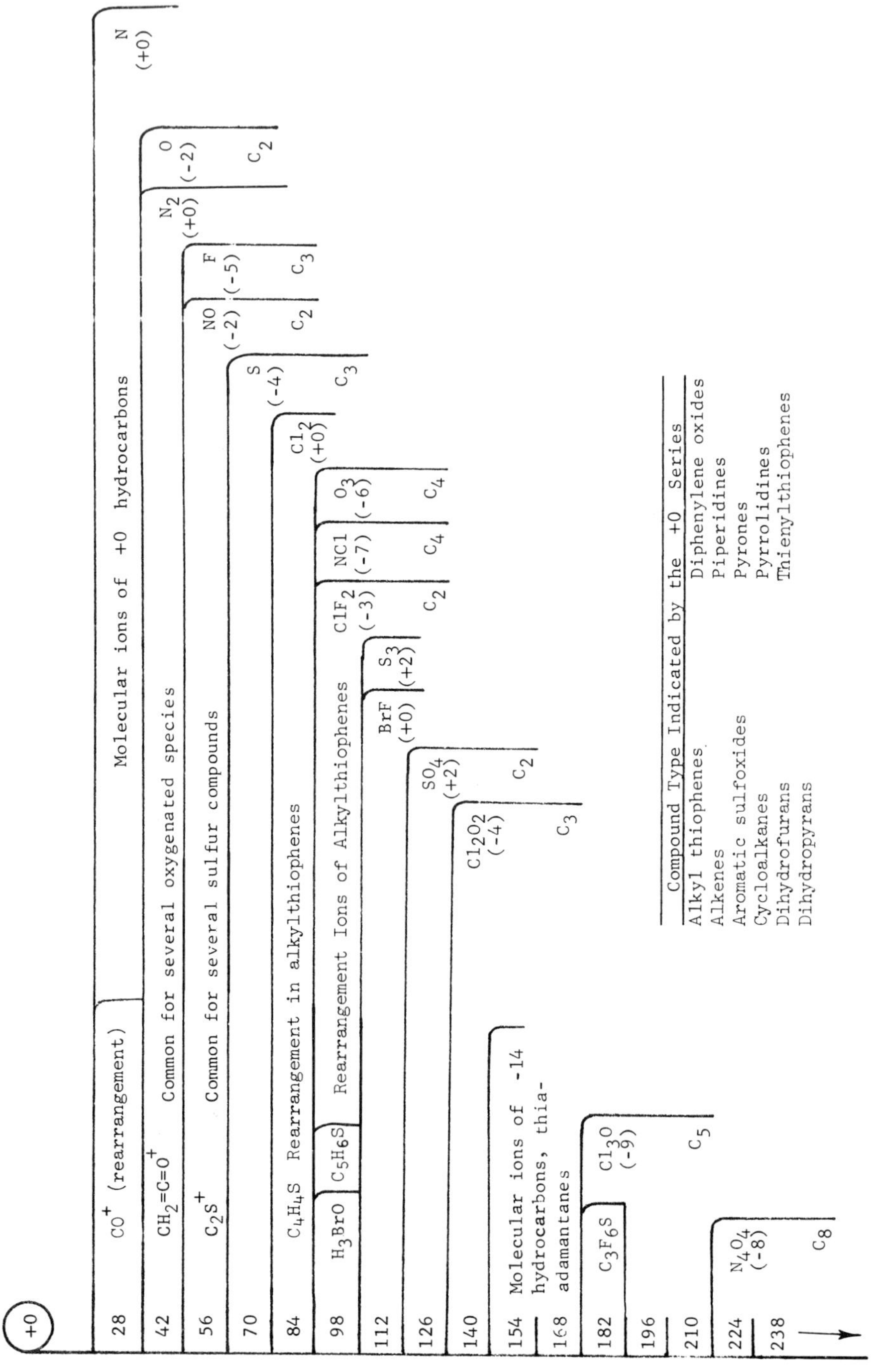

+0

28 CO^+ (rearrangement) — Molecular ions of +0 hydrocarbons — N (+0)

42 $CH_2{=}C{=}O^+$ Common for several oxygenated species — N_2 (+0) — O (-2) C_2

56 C_2S^+ Common for several sulfur compounds — NO (-2) C_2 — F (-5) C_3

70 — S (-4) C_3

84 C_4H_4S Rearrangement in alkylthiophenes — Cl_2 (+0)

98 H_3BrO — C_5H_6S — Rearrangement Ions of Alkylthiophenes — ClF_2 (-3) C_2 — NCl (-7) C_4 — O_3 (-6) C_4

112 — BrF (+0) — S_3 (+2)

126 — SO_4 (+2) C_2

140 — Cl_2O_2 (-4) C_3

154 Molecular ions of -14 hydrocarbons, thia-adamantanes

168

182 C_3F_6S — Cl_3O (-9) C_5

196

210

224 N_4O_4 (-8) C_8

238

Compound Type Indicated by the +0 Series	
Alkyl thiophenes	Diphenylene oxides
Alkenes	Piperidines
Aromatic sulfoxides	Pyrones
Cycloalkanes	Pyrrolidines
Dihydrofurans	Thienylthiophenes
Dihydropyrans	

MAP 37

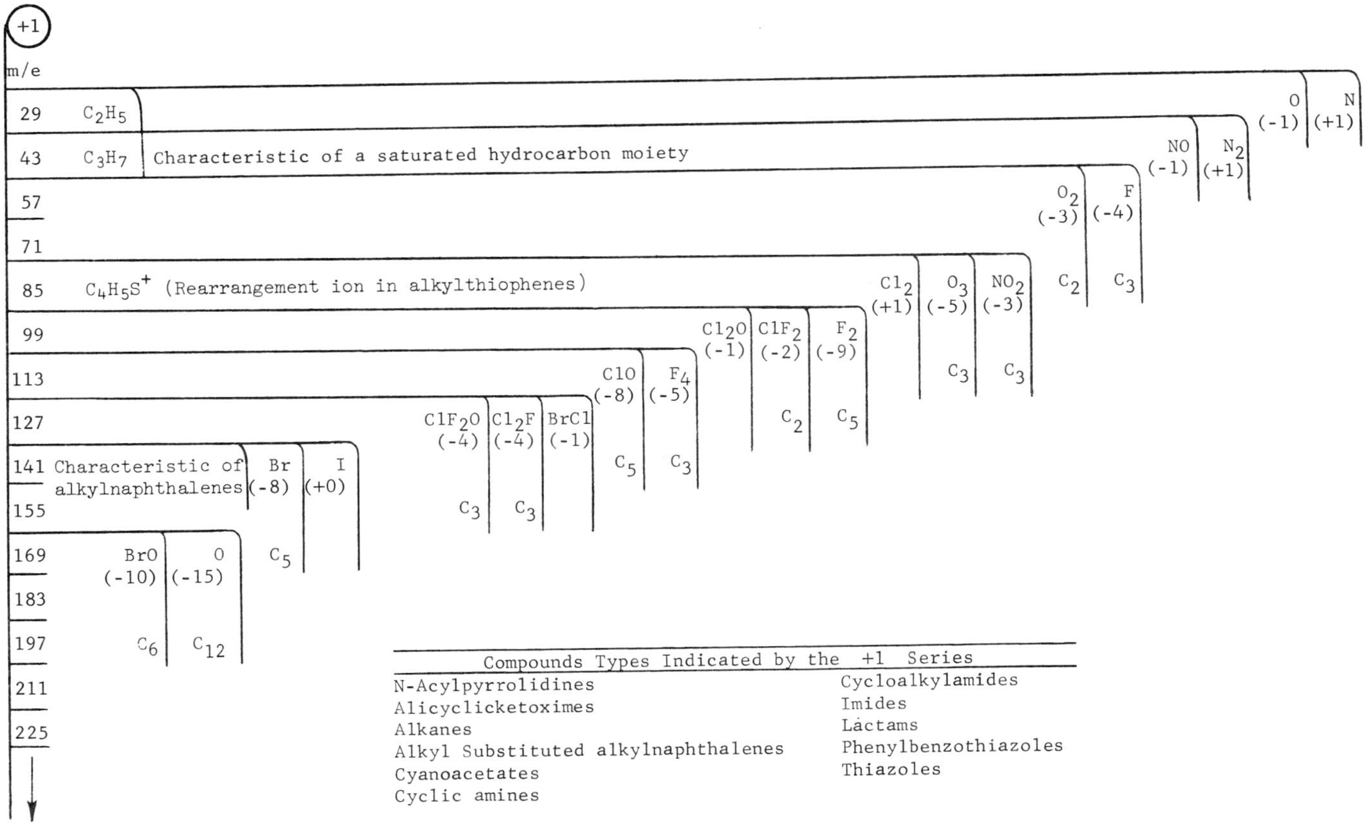

+1

m/e	
29	C_2H_5
43	C_3H_7 Characteristic of a saturated hydrocarbon moiety
57	
71	
85	$C_4H_5S^+$ (Rearrangement ion in alkylthiophenes)
99	
113	
127	
141	Characteristic of alkylnaphthalenes
155	
169	
183	
197	
211	
225	

Branches from 29: O (-1); N (+1)

Branches from 43: NO (-1); N_2 (+1)

Branches from 57: O_2 (-3) C_2; F (-4) C_3

Branches from 85: Cl_2 (+1); O_3 (-5) C_3; NO_2 (-3) C_3

Branches from 99: Cl_2O (-1); ClF_2 (-2) C_2; F_2 (-9) C_5

Branches from 113: ClO (-8) C_5; F_4 (-5) C_3

Branches from 127: ClF_2O (-4) C_3; Cl_2F (-4) C_3; BrCl (-1)

Branches from 141: Br (-8) C_5; I (+0)

Branches from 169: BrO (-10) C_6; O (-15) C_{12}

Compounds Types Indicated by the +1 Series	
N-Acylpyrrolidines	Cycloalkylamides
Alicyclicketoximes	Imides
Alkanes	Lactams
Alkyl Substituted alkylnaphthalenes	Phenylbenzothiazoles
Cyanoacetates	Thiazoles
Cyclic amines	

MAP 38

(+2)

m/e

16 Molecular ions of (+2) hydrocarbons

30 H_2Si | $CH_2=NH_2$ | Rearrangement ions characteristic of alkyl amines | N (+2) | O (+0)

44 CO_2 | CS | NO (+0) | N_2 (+2) | F (-3) C_2

58 Rearrangement ions characteristic of alkyl-ketones (CH_3COCH_3) | O_2 (-2) C_2

72 S (-2) C_3 | Cl (-5) C_3 | NS (-2) C_2

86 ClF_2 (-1)

100 C_2F_4 | ClO (-7) C_4

114 Cl_2F (-3) C_2

128 Molecular ions of (-12) hydrocarbons | BrCl (+0)

142 BrF_2 (-3) C_2 | I (+1)

156

170 O (-14) C_{12}

184 Molecular ion of Dibenzo-thiophenes | Br (-7) C_8

198 $C_3Cl_3F_3$

212

226

240

Compound Types Indicated by the +2 Series		
Aliphatic aldehydes	Amides	Piperazines
Aliphatic amines	2-Aminothiazole	Thiabicycloalkanes
Aliphatic epoxides	Cyclic ethers	Thiocyanates
Aliphatic ketones	Lactones	Vinyl alkyl ethers
Alkanediones	Malonic acids	Vinyl ethers
Alkyl indoles	Quinolines	

MAP 39

APPENDIX III

CURRENT CATALOGS AND COMPILATIONS OF MASS SPECTRAL DATA

ASTM (1963). "Index of Mass Spectral Data," STP-356, 244 pp. American Society for Testing and Materials, Philadelphia, Pennsylvania.

ASTM (1969). "Index of Mass Spectral Data," AMD 11, 632 pp. American Society for Testing and Materials, 1916 Race Street, Philadelphia, Pennsylvania. Includes the Göteborg University mass spectral data collection. (Contact R. M. Sherwood.)

"Catalog of Selected Mass Spectral Data" (1947 to date), American Petroleum Institute Research Project 44 and Thermodynamics Research Center (formerly MCA Research Project), Texas A & M University, College Station, Texas. Dr. Bruno Zwolinski, Director, functions in collecting evaluated or selected data.

Cornu, A., and Massot, R. (1966). "Compilation of Mass Spectral Data," Heyden, London. First and second supplements issued.

"Eight Peak Index of Mass Spectra" (1970), British Information Services, 845 Third Ave., New York. A compilation of essential mass spectral data indexed from over 17,000 mass spectra contributed by a large number of mass spectrometrists from all over the world.

Gohlke, R. S. (1963). "Uncertified Mass Spectral Data." Data originated at Dow Chemical Co., Midland, Michigan. Collection is part of the uncertified mass spectral data of ASTM Committee E-14 on Mass Spectrometry.

Organic Mass Spectrometry, Chemical Compound Index. Compounds for which spectral data are given in this journal are indexed by molecular weights with molecular formula, compound name, and initial page of paper having the spectral data in tabular or bar graph form.

Ridley, R. G., and Quayle, A., Mass Spectrometry Data Centre of the Atomic Weapons Research Establishment (AWRE) Aldermaston, England (about 8200 spectra on magnetic tape). The organization functions in cooperation with ASTM as a worldwide central collection for mass spectra.

Stenhagen, E., Abrahamsson, S., and McLafferty, F. W. (1969). "Atlas of Mass Spectral Data," Vols. I, II, and III, 2266 pp. Wiley (Interscience), New York. Contains about 6000 different spectra taken from ASTM Committee E-14, the American Petroleum Institute, and other sources. Atlas also available on magnetic tape.

Stenhagen, E., Abrahamsson, S., and McLafferty, F. W., eds. (1969). *Archives of Mass Spectral Data*, Vol. I, 784 pp. This journal provides newly acquired spectra on a quarterly basis.

Tatematsu, A., and Tsuchiya, T. (1968). "Structure Indexed Literature of Organic Mass Spectra—1966," 275 pp. Academic Press of Japan, Japan.

Tatematsu, A., and Tsuchiya, T. (1969). "Structure Indexed Literature of Organic Mass Spectra—1967," 496 pp. Academic Press of Japan, Japan.

Zwolinski, B. J. (1968). "Comprehensive Index API 44–TRC Selected Data of Thermodynamics and Spectroscopy," 507 pp. Thermodynamics Research Center, Texas A & M University, College Station, Texas.

APPENDIX **IV**

TABLES OF ISOTOPIC ABUNDANCES

Part A. An Abundance Ratio Table of Some Elements Commonly Encountered in Mass Spectra of Organic Compounds

Part B. Intensity Contribution Tables of Natural Isotopes

APPENDIX IV, Part A

Element[a]	Natural abundance (%)[b]	Mass[b]
^{12}C	98.8930	12.0000000
^{13}C	1.1070	13.0033554
^{1}H	99.9850	1.0078246
^{2}H (D)	0.0149	2.0141021
^{16}O	99.7590	15.9949141
^{17}O	0.0374	16.9991322
^{18}O	0.2039	17.9991616
^{14}N	99.6337	14.0030732
^{15}N	0.3663	15.0001088
^{31}P	Monoisotopic	30.9737640
^{32}S	95.0000	31.9720727
^{33}S	0.750	32.9714635
^{34}S	4.215	33.9678628
^{36}S	0.017	35.967089
^{28}Si	92.2100	27.9769290
^{29}Si	4.71	28.976491

APPENDIX IV, *Part A—cont.*

Element[a]	Natural abundance (%)[b]	Mass[b]
^{30}Si	3.12	29.973761
^{10}B	19.6	10.012940
^{11}B	80.3900	11.0093070
^{19}F	Monoisotopic	18.9984022
^{35}Cl	75.5290	34.9688531
^{37}Cl	24.4710	36.9659034
^{79}Br	50.5370	78.9183900
^{81}Br	49.4630	80.9164200
^{127}I	Monoisotopic	126.904660

[a] For additional elements, the reader is referred to J. H. Beynon, R. A. Saunders, and A. E. Williams, "The Mass Spectra of Organic Molecules," pp. 476–477. Amer. Elsevier Publishing Company, New York, 1968.

[b] E. G. Owens and A. M. Sherman, "Technical Report No. 265." Lincoln Laboratory, MIT, Cambridge, Massachusetts, 1962, and/or J. A. Konig, J. H. E. Mattauch, and A. H. Wapstra, *Nucl. Phys.* **31**, 18 (1962).

APPENDIX IV, Part B

The use of these tables is illustrated with an example. Consider a peak at m/e 76 of an ion having the elemental composition of C_3H_8S. The three carbons are 36, the eight hydrogens are 8, and the one sulfur is 32, or a total of 76 mass units (m) for the C_3H_8S ion. To find the intensity contributions of the stable isotopes at $m + 1$, $m + 2$, etc., for this ion assign to each of these three isotopes, $^{12}C_3{}^1H_8{}^{23}S$, an intensity of 100.0. Start from the most abundant isotope (100.0) to find the total contribution of intensities to the isotopic peaks. In this case, it can be determined from the table that the $^{12}C_3$ contributes 3.4 units (related to the 100.0 units), the H_8 contributes 0.1, and ^{32}S contributes 0.8 or a total of 4.3 units for the $m + 1$ isotope at m/e 77. The second isotopic peak having no significant contribution from $^{12}C_3$ or $^1H_8$ is due primarily to ^{34}S, which gives a 4.4 contribution for the intensity of the peak at m/e 78. The numbers in the column designated m are an aid in checking the mass values.

It is intended that the user of these tables will find them convenient in calculating the contributions in terms of the m peak height and then ignore anything at the threshold level of detection, perhaps 0.2–1.0 arbitrary division.

Some of the elements commonly encountered in mass spectrometry are monoisotopic and are thus not included. Among these are ^{19}F, ^{31}P, ^{127}I, as well as some metals such as ^{197}Au, ^{75}As, ^{59}Co, ^{55}Mn, ^{27}Al, ^{23}Na, ^{133}Cs, ^{209}Bi, ^{45}Sc, and ^4Be and a few of the "rare-earth" elements.

APPENDIX IV, Part B[a]

Carbon	m (100.0)	% at $m+1$	% at $m+2$	% at $m+3$	Carbon	m (100.0)	% at $m+1$	% at $m+2$	% at $m+3$
C	12	1.12			C_{17}	204	19.03	1.70	0.10
C_2	24	2.24	0.01		C_{18}	216	20.15	1.92	0.12
C_3	36	3.36	0.04		C_{19}	228	21.27	2.14	0.14
C_4	48	4.48	0.07		C_{20}	240	22.39	2.39	0.16
C_5	60	5.60	0.12		C_{21}	252	23.51	2.63	0.19
C_6	72	5.72	0.19		C_{22}	264	24.62	2.90	0.22
C_7	84	7.84	0.26		C_{23}	276	25.74	3.16	0.25
C_8	96	8.95	0.35	0.01	C_{24}	288	26.86	3.46	0.29
C_9	108	10.07	0.45	0.01	C_{25}	300	27.98	3.76	0.32
C_{10}	120	11.19	0.56	0.01	C_{26}	312	29.10	4.07	0.36
C_{11}	132	12.31	0.69	0.02	C_{27}	324	30.22	4.40	0.41
C_{12}	144	13.43	0.82	0.03	C_{28}	336	31.34	4.74	0.46
C_{13}	156	14.55	0.98	0.03	C_{29}	348	32.46	5.08	0.51
C_{14}	168	15.67	1.14	0.04	C_{30}	360	33.58	5.45	0.57
C_{15}	180	16.79	1.31	0.05	C_{31}	372	34.70	5.85	0.63
C_{16}	192	17.90	1.51	0.08	C_{32}	384	35.82	6.24	0.69

[a] Based upon abundance ratios from the compilation of E. G. Owens and A. M. Sherman, "Technical Report No. 265," Lincoln Laboratory, MIT, Cambridge, Massachusetts, 1962, and/or J. H. Beynon and A. E. Williams, "Mass and Abundance Tables for Use in Mass Spectrometry," 570 pp., Amer. Elsevier, New York, 1963. In these tables, m is the nominal mass, in atomic mass units, of species containing the most abundant isotope.

APPENDIX IV, Part B[a]*—cont.*

Hydrogen	m (100.0)	% at $m+1$	Hydrogen	m (100.0)	% at $m+1$	Hydrogen	m (100.0)	% at $m+1$
H	1	0.02	H_{12}	12	0.18	H_{23}	23	0.35
H_2	2	0.03	H_{13}	13	0.20	H_{24}	24	0.36
H_3	3	0.05	H_{14}	14	0.21	H_{25}	25	0.38
H_4	4	0.06	H_{15}	15	0.23	H_{26}	26	0.39
H_5	5	0.08	H_{16}	16	0.24	H_{27}	27	0.40
H_6	6	0.09	H_{17}	17	0.26	H_{28}	28	0.42
H_7	7	0.11	H_{18}	18	0.27	H_{29}	29	0.43
H_8	8	0.12	H_{19}	19	0.29	H_{30}	30	0.45
H_9	9	0.14	H_{20}	20	0.30	H_{31}	31	0.46
H_{10}	10	0.15	H_{21}	21	0.32	H_{32}	32	0.47
H_{11}	11	0.17	H_{22}	22	0.33			

Oxygen	m (100.0)	% at $m+1$	% at $m+2$	Nitrogen	m (100.0)	% at $m+1$	% at $m+2$
O	16	0.04	0.20	N	14	0.37	
O_2	32	0.07	0.41	N_2	28	0.74	
O_3	48	0.11	0.61	N_3	42	1.11	
O_4	64	0.15	0.82	N_4	56	1.47	0.01
O_5	80	0.19	1.01	N_5	70	1.83	0.01
O_6	96	0.20	1.22	N_6	84	2.21	0.02
O_7	112	0.26	1.43	N_7	98	2.58	0.03

Sulfur	m (100.0)	% at $m+1$	% at $m+2$	% at $m+3$	% at $m+4$
S	32	0.80	4.44		0.01
S_2	64	1.60	8.89	0.07	0.22
S_3	96	2.40	13.34	0.21	0.64
S_4	128	3.20	17.81	0.43	1.24
S_5	160	4.00	22.27	0.71	1.98

Silicon	m (100.0)	% at $m+1$	% at $m+2$	% at $m+3$	% at $m+4$
Si	28	5.10	3.35		
Si_2	56	10.20	6.96	0.34	0.12
Si_3	84	15.29	10.83	1.03	0.36
Si_4	112	20.39	14.97	2.10	0.77

Boron	m (100.0)	% at $m-4$	% at $m-3$	% at $m-2$	% at $m-1$
B	11				24.39
B_2	22			5.96	48.79
B_3	33		1.44	17.84	73.16
B_4	44	0.36	5.79	35.70	97.58

Combinations with chlorine and bromine	m (100.0)	% at $m+2$	% at $m+4$	% at $m+6$	% at $m+8$
Cl	35	32.40			
CCl_2	82	64.80	10.50		
CCl_3	117	97.19	31.49	3.41	
Br	79	97.86			
C_2Br_2	182	195.66	95.72		
C_2Br_3	261	293.38	287.19	93.54	
Br_4	316	391.70	574.98	375.14	91.87

APPENDIX **V**

COMPUTER PROGRAM

The application of this computer program is discussed in Chapter 8, Section III, B.

```
C
C.....PROGRAM  INVER
C.....THIS PROGRAM INVERTS A MATRIX WHICH IS SET UP
C.....FROM MASS SPECTROMETRIC DATA OF COMPOUNDS
C.....WHICH COMPOSE A SAMPLE.
C.....THE SOLUTIONS OF THE MATRIX YIELD THE
C.....UN-NORMALIZED COMPONENTS IN THE SAMPLE.
C.....THESE SOLUTIONS ARE THEN NORMALIZED TO 100 PERCENT.
C
      DIMENSION A(20,41), B(40)
      REAL*8  A
C
C.....READ SAMPLE IDENTIFICATION
C
      READ(5,1) B
    1 FORMAT(20A4)
C
C.....READ NUMBER OF ROWS OF MATRIX
C
      READ(5,2) N
    2 FORMAT(I2)
      NN=2*N+1
      NJ=N+1
C
C.....READ MATRIX COEFFICIENTS INTO ARRAY
C
      READ(5,3) ((A(I,J),J=1,NJ),I=1,N)
    3 FORMAT(6F10.3)
C
C.....PRINT SAMPLE IDENTIFICATION
C
      WRITE(6,4) B
    4 FORMAT(1H1,2(20A4/1X))
C
C.....PRINT MATRIX FOR ANALYSIS
C
      WRITE(6,5)
    5 FORMAT(1H0,19HMATRIX FOR ANALYSIS/)
      DO 6 I=1,N
    6 WRITE(6,7) (A(I,J),J=1,NJ)
    7 FORMAT(10F10.3)
C
C.....BUILD IDENTITY MATRIX
C
      N2=N+2
      DO 10 II=1,N
      DO 10 MM=N2,NN
      J1=N+II+1
      IF(MM-J1)8,9,8
    8 A(II,MM)=0.0
      GO TO 10
    9 A(II,MM)=1.0
   10 CONTINUE
C
C.....INVERT AUGMENTED MATRIX
C
      I=1
   11 K=1
      J=I+1
      DO 12 LL=J,NN
      IF (A(I,I).NE.0.) GO TO 12
      GO TO 30
   12 A(I,LL)=A(I,LL)/A(I,I)
   13 IF(K-I)14,16,14
   14 DO 15 KK=J,NN
   15 A(K,KK)=A(K,KK)-A(I,KK)*A(K,I)
   16 K=K+1
      IF(K-N)13,13,17
   17 I=I+1
      IF(I-N)11,11,18
```

```
C
C.....PRINT INVERSE OF MATRIX
C
   18 WRITE(6,19)
   19 FORMAT(1H0,15HINVERTED MATRIX/)
      DO 20 I5=1,N
   20 WRITE(6,21) (A(I5,I6),I6=N2,NN)
   21 FORMAT(10F10.4/)
C
C.....PRINT SOLUTIONS TO MATRIX
C
      WRITE(6,22)
   22 FORMAT(1H0,21HUN-NORMALIZED FACTORS/)
      DO 23 I4=1,N
   23 WRITE(6,24) A(I4,NJ)
   24 FORMAT(F10.2/)

C
C.....NORMALIZE SOLUTIONS TO 100 PERCENT
C
      SUM=0.0
      DO 25 I7=1,N
   25 SUM=SUM+A(I7,NJ)
      DO 26 I8=1,N
   26 A(I8,NJ)=(A(I8,NJ)/SUM)*100.
C
C.....PRINT NORMALIZED COMPONENTS
C
      WRITE(6,27)
   27 FORMAT(1H0,11X,18HNORMALIZED RESULTS/4X,
     110HCOMPONENTS,13X,7HPERCENT/)
      DO 28 I9=1,N
   28 WRITE(6,29) A(I9,NJ)
   29 FORMAT(18X,F10.1/)
      GO TO 32
   30 WRITE (6,31)
   31 FORMAT (1H1,36HDIAGONAL DATA POINT COMPUTES TO ZERO)
   32 STOP
      END
/*
```

AUTHOR INDEX

Numbers in italics refer to the pages on which the complete references are listed.

C

E

F

G

H

I

J

K

L

R

S

T

U

V

Y

SUBJECT INDEX

A

N